D1194162

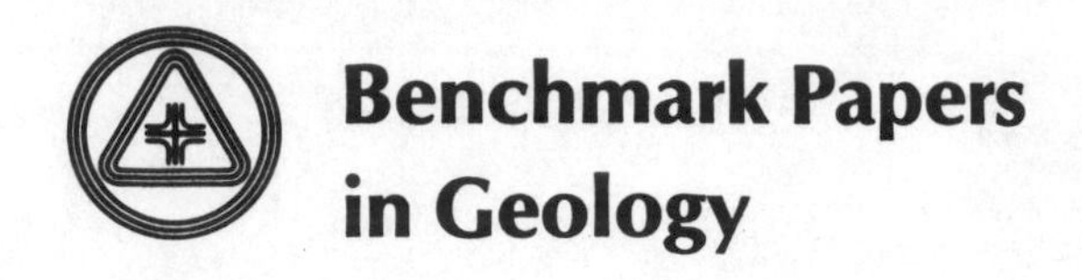

Series Editor: Rhodes W. Fairbridge
Columbia University

A selection from the published volumes in this series

Volume

A complete listing of volumes published in this series begins on p. 445.

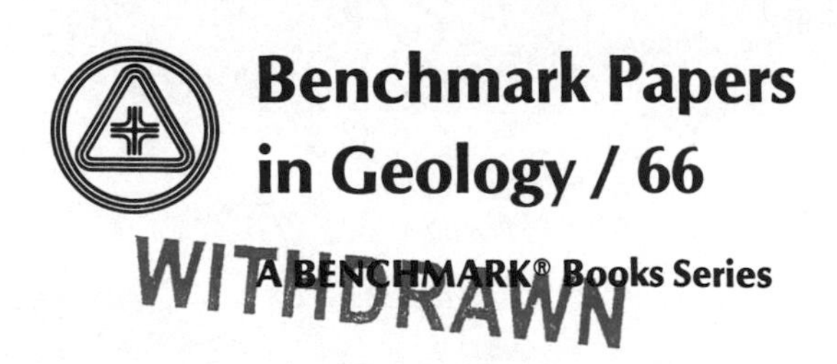

OPHIOLITIC AND RELATED MÉLANGES

Edited by
G. J. H. McCALL
Fellow, Liverpool University
Department of Geology

Hutchinson Ross Publishing Company
Stroudsburg, Pennsylvania

Benchmark Papers in Geology, Volume 66
Library of Congress Catalog Card Number: 81-13490
ISBN: 0-87933-421-5

85 84 83 1 2 3 4 5
Manufactured in the United States of America.

LIBRARY OF CONGRESS CATALOGING IN PUBLICATION DATA
Main entry under title:
Ophiolitic and related mélanges.
(Benchmark papers in geology; 66)
Includes bibliographical references and index.
1. Melanges (Petrology) 2. Ophiolites. I. McCall, G. J. H. (Gerald Joseph Home), 1920- II. Series.
QE471.15.M4406 552 81-13490
ISBN 0-87933-421-5 AACR2

Distributed worldwide by Van Nostrand Reinhold Company Inc., 135 W. 50th St., New York, NY 10020.

CONTENTS

SERIES EDITOR'S FOREWORD

The philosophy behind the Benchmark Papers in Geology is one of collection, sifting, and rediffusion. Scientific literature today is so vast, so dispersed, and, in the case of old papers, so inaccessible for readers not in the immediate neighborhood of major libraries that much valuable information has been ignored by default. It has become just so difficult, or so time consuming, to search out the key papers in any basic area of research that one can hardly blame a busy person for skimping on some of his or her "homework."

This series of volumes has been devised, therefore, as a practical solution to this critical problem. The geologist, perhaps even more than any other scientist, often suffers from twin difficulties—isolation from central library resources and immensely diffused sources of material. New colleges and industrial libraries simply cannot afford to purchase complete runs of all the world's earth science literature. Specialists simply cannot locate reprints or copies of all their principal reference materials. So it is that we are now making a concerted effort to gather into single volumes the critical materials needed to reconstruct the background of any and every major topic of our discipline.

We are interpreting "geology" in its broadest sense: the fundamental science of the planet Earth, its materials, its history, and its dynamics. Because of training in "earthy" materials, we also take in astrogeology, the corresponding aspect of the planetary sciences. Besides the classical core disciplines such as mineralogy, petrology, structure, geomorphology, paleontology, and stratigraphy, we embrace the newer fields of geophysics and geochemistry, applied also to oceanography, geochronology, and paleoecology. We recognize the work of the mining geologists, the petroleum geologists, the hydrologists, and the engineering and environmental geologists. Each specialist needs a working library. We are endeavoring to make the task of compiling such a library a little easier.

Each volume in the series contains an introduction prepared by a specialist (the volume editor)—a "state of the art" opening or a summary of the object and content of the volume. The articles, usually some twenty to fifty reproduced either in their entirety or in significant extracts, are selected in an attempt to cover the field, from the key papers of the last century to fairly recent work. Where the original works are in foreign languages, we have endeavored to locate or commission translations. Geologists, because of their global subject, are often acutely aware of the oneness of our world. The selections cannot therefore be restricted to any one country, and whenever possible an attempt is made to scan the world literature.

To each article, or group of kindred articles, some sort of "highlight commentary" is usually supplied by the volume editor. This commentary should serve to bring that article into historical perspective and to emphasize its particular role in the growth of the field. References, or citations, wherever possible, will be reproduced in their entirety—for by this means the observant reader can assess the background material available to that particular author, or, if desired, he or she too can double check the earlier sources.

A "benchmark," in surveyor's terminology, is an established point on the ground that is recorded on our maps. It is usually anything that is a vantage point, from a modest hill to a mountain peak. From the historical viewpoint, these benchmarks are the bricks of our scientific edifice.

RHODES W. FAIRBRIDGE

PREFACE

Geology has undergone a revolution during the last decade with the advent of the concept of global tectonics. It is becoming increasingly clear that lateral movements of enormous plates, genesis of new crust at mid-ocean ridges, and subduction of that new crust to the mantle beneath the leading edge of an overriding plate are mechanisms that cannot be extrapolated back into the earliest Precambrian (or Archean); quite different geotectonic mechanisms appear to have been operative then and immense oceans probably did not exist. Nevertheless, much of the modern geologist's preoccupation is with the search for fossil plate boundaries in Phanerozoic systems. Ophiolites have been equated with such boundaries, though it is apparent that ophiolites can be developed in a wide variety of geotectonic zones at or near convergent plate margins. It is also possible that they can develop quite apart from such margins. The ophiolite problem is certainly more complex than was at first supposed.

A peculiar type of tectonic deformation, that characteristic of *mélanges,* is widespread in ancient convergent plate margins. Mélanges are a common feature of these ancient sutures and are not normally found elsewhere. They occur in a wide variety of zones: trench, fore-arc, arc, marginal basin, peripheral rifted, and possibly in oceanic fracture zones. The nongenetic term "mélange" denotes any chaotic body of rock that constitutes a mappable unit and contains exotic blocks. There is a wide spectrum of mélange types ranging from the truly ophiolitic serpentinite and colored mélanges, through argillite-, turbidite-, clay-, or slate-matrix mélanges with little ophiolite content, to sediment-matrix mélanges with no ophiolite content at all. They include small-scale olistostromes (debris flows) produced by synsedimentational and surficial gravity slides on steep sea floor slopes. There are also descriptions of more extensive, regional-scale olistostrome mélanges, which are thicker and contain immense rafts and blocks. These mélanges are related to major geotectonic zones—such as advancing allochthonous sheets and their disintegration at the toe or sole with the shedding of fragments onto steep, unstable sea-floor slopes—and may also be related to the immense sea-floor slides recognized by oceanologists. Mélanges can also be purely tectonic, related to compressive nappe or schuppen regimes and transcurrent or other faulting. Many tectonic mélanges may be composed of offscraped slabs that, at shallow crustal depths, resisted passage down the subduction zone. Others appear related to processes of protrusion of disseminated rafts and blocks up fault zones in regimes of extreme compression. Mélanges are commonly polykinematic and may derive from the superposition of different processes of fragmentation and mixing.

This volume is an attempt to present a selection of descriptions and interpretations of mélanges from the worldwide literature. Included are 30 papers covering both ancient and recent rock systems by authorities from America, Britain, Canada, France, Italy, Japan, New Zealand, Russia, Switzerland, and Yugoslavia. There is no concensus of opinion among the authorities represented but there is a wealth of excellent observation and argument. Many other important papers are mentioned in the editorial commentaries.

The mélange problem remains open to more detailed study, both in the field and by modern laboratory methods, and particularly to the more specialized, expert structural studies of small areas and material components. This volume has been restricted to mélanges preserved in the terrestrial rock record, but it is appreciated that there is a need for a related study of the oceanological aspect of the same problem—of the immense, present-day submarine slides and fracture zones. Whether oceanological studies are sufficiently advanced at the present time for such a study to be meaningful remains questionable. Correlation between on-land geological studies and offshore geological studies is something that is, as yet, only weakly represented in the literature.

G. J. H. McCALL

CONTENTS BY AUTHOR

OPHIOLITIC AND RELATED MÉLANGES

INTRODUCTION

OPHIOLITES

This volume is not concerned primarily with the broader problems of *ophiolites* but with the specific problem of widespread *mélange* developments commonly found to be closely associated with ophiolites in zones of ancient ocean-continent plate convergence. The ophiolites themselves are summarized by Coleman (1977). They were first mentioned by Brongniart (1827), the term being derived from the Greek root *ophi-* (serpentine). Later, ophiolites were the subject of the conceptual thinking of Steinmann (1927), who envisaged a genetic assemblage of submarine volcanics, including serpentinites with deep-water sediments—a feature of the early stage of geosynclinal development. Benson (1926), Bowen (1927), Hess (1938, 1955) and Wyllie (1967) produced additional new and fundamental ideas about the ophiolites. All of these ideas were reevaluated with the advent of the "new global tectonics." For example, the Alpine *green rocks* were reinterpreted as fragments of oceanic lithosphere and mantle thrust up over the continental margins at consuming plate boundaries (Gass, 1967; Coleman, 1971; Dewey and Bird, 1971; Davies, 1971; Moores and Vine, 1971; Church, 1972). In an attempt to link the new thinking with the older, mainly European-originated concepts, the term *ophiolite* was redefined at a Penrose Conference (Anon., 1972, p. 25). To paraphrase, "ophiolite" refers to a distinct assemblage of mainly mafic and ultramafic rocks, which, from bottom to top, consists of (1) an ultramafic complex: dunite, harzburgite and lherzolite, more or less serpentinised and commonly displaying tectonite fabric; (2) a gabbroic complex, ordinarily with cumulate textures, that may contain cumulus peridotites, pyroxenites, and troctolites and is commonly less deformed than the first subdivision; (3) a complex of mafic sheeted dikes; (4) a mafic volcanic complex, commonly of pillow lavas with an association of pelagic sediments (chert, siltstone, shale, and limestone, including radiolarite). Chromitite may occur in the dunite and the gabbro may be associated with subordinate developments of sodic and felsic intrusive rocks (trondhjemite, tonalite, and so on). (Intermediate and acid members are now known to be not uncommon in the sheeted dike and volcanic suites and are of the same generally sodic character.) This definition is of an idealized model and most ophiolite sequences are disrupted by faulting, with whole sections missing. Dismembered sequences are more prevalent than complete sequences. They commonly display only the effects of static hydrothermal metamorphism to zeolite or greenschist facies assemblages, but may have a regional metamorphic overprint.

Since the Penrose definition was advanced (Anon., 1972), studies have revealed that ophiolites are not restricted to fossil mid-ocean ridge assemblages and associated rocks of the abyssal plain but that they also incorporate rocks of arc and marginal basin provenance. The classic ophiolite sequence is being recognized in marine sedimentary and eruptive sequences from peripheral rift-like depressions of the continental hinterland to the main plate suture. It has become evident that ophiolites represent the results of a spectrum of oceanic activities—eruptivity and, mainly, pelagic sedimentation—throughout the main spreading centers of the ancient oceans, the arcs and marginal basins, and the extreme peripheral and only partially developed ancillary spreading zones of the continent itself.

The Penrose definition made notably little mention of the associated mélanges. Coleman and Irwin (1974, p. 925) remarked that, in general, ophiolitic mélanges are structures within the suture zones parallel to the trend of the zones, comprising ophiolite mixed with sedimentary and metamorphic rocks. Such mélanges commonly mark boundaries of suture zones. This statement appears to oversimplify what emerges as a highly complex problem, even though one subsidiary to the main ophiolite problem. However, the complexity of the mélange problem was acknowledged by these authors as indicated by the following passage:

> The early Paleozoic ophiolites . . . . have undergone repeated deformation and metamorphism, which often transforms the layered fragments into tectonic mixtures or mélanges. . . . Transformation of the metamorphic peridotites (usually dunite-harzburgite) into serpentinite early in the tectonic evolution of ophiolites produces a material whose response to tectonic movements is somewhat analogous to that of salt. Sheared serpentinites are less dense than the rocks surrounding them and commonly provide a plastic medium in which fragments of ophiolites, and sedimentary and metamorphic country rocks are immersed. . . . . Once a mélange with a predominantly serpentine matrix is developed, it is vulnerable to any future tectonic event. . . . [responding] to tectonism by movement, complicating its inner structure, and incorporating blocks of country rock through which it passes. (Coleman and Irwin, 1974, p. 926)

The concept stated by Coleman and Irwin is that mélanges are disrupted ophiolite sequences with serpentinite matrixes. However, an extensive literature search indicates that there is a continuum from serpentinite mélanges (truly ophiolitic) at one end of the spectrum, through colored mélanges (also ophiolitic) and mélanges containing insignificant ophiolitic material, to those that contain none at all. All of these varieties of mélange relate to the major problem of mélange genesis within the framework of plate convergence.

It also became apparent from the literature search that restricting the term *mélange* to those assemblages of *olistostromic* (gravity slide, sedimentary) or *tectonic* origins is unrealistic. There are authorities (for example, Ballance and Spörli, Paper 28) who would eliminate the term *mélange* altogether, but Coleman (1977, p. 146) clearly includes both types of assemblage in the *ophiolitic* mélange:

> Considerable confusion surrounds the significance of mélange and its relationship to the emplacement of ophiolite slabs. . . . Where gravity sliding is important, the presence of mélange is considered to be part of a chaotic mixture of blocks produced by the effects of gliding down slope (olistostrome). In other situations the associated mélange has been interpreted as a remnant of an ancient subduction zone operative during the emplacement of the ophiolite. . . .

Preferable to the inclination of Ballance and Spörli is the adoption of a *nongenetic* usage of the term *mélange* as applied to any chaotic rock assemblage, with additional characterization by means of a prefix denoting inferred origin (as favored by Berkland et al. in Paper 4 and Hibbard and Williams in Paper 13). However, this nongenetic usage should be adopted, bearing in mind that there may well be a degree of convergence, because it is possible that immense-scale submarine gravity slides involve not only the immediate sea floor zone but also the underlying zone of deep burial where incipient metamorphism is operative. The problem of mélange genesis within the framework of plate convergence is complicated by the fact that there are both small-scale, localized, mainly interstratified olistostrome mélange (*debris flows*) and immense-scale submarine collapses of regional extent. The latter phenomenon apparently precedes large-scale advancing thrust or schuppen sheets and is intimately related to tectonic mélanges that occur at the sole of such sheets. Tectonic mélange may be in primary relationship to thrust or schuppen sheets or in secondary relationship to olistostromes. They also may be related to transcurrent and other fault dislocations; however, most tectonic mélanges are developed on a regional scale (though some fault-related mélanges are of localized development).

The whole continuum of mélanges is to be found at or close to convergent plate margins, and these margins may be *subductive* or *obductive* in type. Some mélanges have been formed in deep marine basins occupying rifted depressions marginal to the main zone of convergence (e.g., some of the inland colored mélange developments in Iran). Both tectonic and olistostrome mélanges occur at convergent plate margins, as do mélanges that appear to be combinations of the chaotic effects of both tectonic and gravity slide activities (though the problem of possible simulation of gravity slide activities by tectonic deformations of partial consolidated rocks throws some doubt on many examples of multiple, poly-type deformation).

MÉLANGES: HISTORICAL

In a study of Anglesey, Greenly (1919) coined the term *mélange* to denote a chaotic assemblage of rocks. He added the characterizing term *autoclastic,* believing the Gwna mélange to be of tectonic origin, thus giving no genetic implication to the term *mélange* itself. He also distinguished a polyphase (polykinematic) mélange. Bailey and McCallien (Paper 18, 1953) revived the term mélange for the Ankara mélange, also deemed tectonic, and Shackleton (1951), appreciating the nongenetic nature of the term, applied it to a chaotic mudflow (lahar) assemblage in Rusinga Island.

Four papers have been selected that represent the modern base literature of the mélange question. Hsü (1968, Papers 2 and 3) believed that contemporaneous submarine gravity sliding had been overemphasized in interpretations of the Franciscan and Alpine mélanges, and he related them to oceanward migration of the Benioff zone at convergent plate margins, believing them to be essentially tectonic though they could incorporate olistostromes of more local development more or less fortuitously. He separated mélanges from olistostromes, which he defined as stratigraphic units grading up through boulder beds from turbidites; mélanges were of greater extent and block size. Deformation in mélanges took place beneath an overburden, not at or close to the sea floor. He believed that the Argille Scagliose of Italy could represent sea floor deformations because of evidence of low confining pressures. Blocks in mélanges were essentially boudins rounded by extensional shear fracturing. *Exotic blocks* were an essential feature of mélanges; otherwise, the assemblage was no more than a *broken formation*. Mélanges were products of two essential processes, *fragmentation* and *mixing,* and many mélanges were *polykinematic*.

Berkland et al. (Paper 4) supplied excellent definitions for *tectonic* and *exotic* blocks in mélanges, which they regarded as mappable bodies of rock, characterized by inclusion of fragments and blocks of all sizes, both *exotic* and *native,* embedded in a fragmental and generally sheared *matrix* of more tractable material. Unlike Hsü, they favored a nongenetic usage plus qualifiers such as "tectonic-" or "olistostrome-" to denote the inferred origin. They defined *knockers* as erosionally resistant monolithic blocks.

Gansser (Paper 1) distinguished *ophiolitic* mélanges from purely sedimentary mélanges of olistostrome or tectonic origin. Such mélanges, composed of ophiolitic and other components, were characterized by serpentinite or sedimentary (commonly flyschoid) matrixes and were essentially mega-olistostromes modified by tectonic processes; they were never of purely tectonic origin. This viewpoint is the converse of that of Hsü, who considered mélanges to be essentially tectonic and olistostrome incorporation to be incidental and fortuitous. Gansser related ophiolitic mélanges to the consuming end of oceanic plates and restricted them to ophiolite belts outlining plate boundaries. He accepted secondary modifications, of both tectonic or gravity slide origin (recycling). He considered that final emplacement of a mélange could conceivably be bodily (i.e., by diapiric processes). He separated his mega-olistostromes from normal olistostromes—quite small, localized debris flow manifestations within normal stratigraphic sequences.

A REVIEW OF THE MÉLANGE PROBLEM

Reviewing the mélange problem in the light of the literature available at the end of 1980, neither the restrictive definition of Hsü nor that of Gansser appears acceptable. Only Berkland et al.'s nongenetic usage is acceptable (a usage for which an elegant plea has recently been advanced by Hibbard and Williams, Paper 13). The term *ophiolitic mélange* is useful, but only as a subsidiary term, as a qualifier. The mélange problem really concerns a much wider spectrum of chaotization than is covered by the ophiolite-dominated range of mélanges. Mélanges may in some cases be no more than complexly dismembered and mixed up ophiolite sequences, but these are a minority.

The supposition that subduction-related mélanges sample the sea floor, crust, and upper mantle of mid-ocean ridge and abyssal plain (pelagic sediments, radiolarites) of the downgoing plate, and flysch (turbidites) of the overriding plate, is probably an oversimplification. This concept (Moores, 1969; Bird and Dewey, 1970; Condie, 1976, p. 103) has been questioned by Blake and Jones (Paper 24). Though anomalous microfossil ages in pelagic sediments in some mélanges do indicate a possible wide spectrum of provenance [e.g., the colored mélange of the Makran, Iran (McCall, Paper 21) and the Kodiak mélange of Alaska (Connelly, Paper 26)], other mélanges appear to derive from quite local rock sources. Some appear to represent arc or marginal basin assemblages, and by no means all pillow basalts in mélanges are tholeiitic (abyssal) or characteristic of mid-ocean ridges; calc-alkaline and alkaline suites are also recorded. Pelagic sediments in mélanges may derive from fore-arc zone, arc, or marginal basin sedimentation, as well as abyssal plain sedimentation. Shelf limestones (Papers 15 and 20), flysch turbidites (Papers 2, 3, 21, 24, and 25), and anomalous metamorphic rocks (Paper 19) are common components of mélanges and appear at least in part to derive from carbonate and flysch fore-arc zones, the continental crust itself, and the shelf veneer of the overriding continental plate. Some mélanges appear to have been formed from a secondary sedimentary cover over a basement composed of an earlier accretion of offscraped mélange slabs [this has been suggested for some Franciscan mélanges (Paper 24) and also for the sedimentary mélange of the Makran, Iran (Paper 21)].

Ophiolitic mélanges were at one time considered to be trench deposits, but it is now apparent that the geotectonic instability that results in mélange formation, whether by tectonic or superficial gravity slide processes, or both combined, affects a much wider zone, covering rock masses of trench, fore-arc, arc, and marginal basins. It is suggested by Saleeby (Paper 23) that mélanges can also form along deep ocean fracture zones. Mélanges may well originate from tectonic or gravity slide processes affecting sedimentary-volcanic assemblages of deep rifted marine basins within the peripheral areas of the overriding continental plate. Mercier and Vergely (Paper 17) relate mélanges to major trancscurrent dislocations.

Mélanges feature not only in classic Alpine nappe regimes, but are also widespread in schuppen regimes characterized by steep, arcuate reverse faults [e.g., Makran, Iran, (Paper 21)]. Though many mélange occurrences can be related to allochthonous sheets, some appear to relate to steep faults, transcurrent, reverse, and possibly even normal faults. Protrusion up such steep faults to form a mélange is recognized in both the Makran (Paper 21) and the Oman (Paper 22).

Many definitions of mélange require a matrix as an essential component, but whereas some mélanges display matrixes of flyschoid material [Franciscan mélanges, (Papers 2, 3, 4, 24, and 25)], clay or argillite (Papers 5–10), or slate [older Caledonian mélanges (Papers 11–14)], the matrix component is less well defined in other mélanges and in others again totally absent. In the serpentinite mélanges, the matrix comprises a jumble of serpentinite fragments of all sizes, much more abundant than, and thus enclosing all other block lithologies, but not strictly a matrix. The colored mélanges of Iran and some of the Oman (Papers 21 and 22) are of *block-to-block* character, with no matrix, only sheared surfaces separating the blocks. Thus, a fine sedimentary matrix is a common, but by no means essential, feature of mélanges.

There seems to be no doubt that normal olistostromes, small in scale and commonly interstratified within sedimentary sequences, are a reality (Papers 1, 5, and 6), and the evidence suggests that the larger regional-scale manifestations of the submarine, synsedimentational gravity slide process—consisting of immense blocks and rafts developed within bodies of great areal extent and thickness—are also a reality (Gansser, Paper 1). The concept of precursor-type olistostromes (Elter and Trevisan, Paper 6) related to advancing thrust sheets seems adequately to account for these regional rather than local olistostromes. Indeed, a number of authorities included in this volume favor this model (Leonov, Paper 7; Audley-Charles, Paper 8; Robertson, Paper 10; Graciansky, Paper 19; Lippard et al., Paper 22; Knipper, Papers 29 and 30). The disintegration of the sole or toe of the sheet is succeeded in many such models by travel over soft sediments with gravity collapse of the intermingled sediments and fragments, as the body moves into the deeper parts of the basin.

Complications of superposition of tectonic deformation are widespread in the unstable zones of plate convergence in which mélanges occur and can confuse interpretation. A major problem is also posed by the possibility of simulation of soft rock deformations related to superficial gravity slide processes by effects of tectonic deformation of partially consolidated, still-water-charged, but buried sediments (Ballance and Spörli, Paper 28; Cowan, 1980). The converse may also be true.

Hsü's binary definition of fragmentation and mixing is realistic, but the mixing may take varied forms. Graciansky (Paper 19) described detachment of fragments from the sole of a thrust sheet and mingling in the overridden sediment, but an upward protrusion of exotics, up reverse faults, has been recognized in the sedimentary mélange of the Makran (Paper 21), and both processes may be equally important in mélange formation. Such protrusions can occur quite free from any lubricating serpentinite, though this is a common factor. Igneous, metamorphic, and sedimentary lithologies of a wide range of character and mechanical behavior can be so protruded. The process of protrusion is as yet little understood, but protrusion of very large rock masses is a feature of these very unstable suture zones, essentially taking the form of a squeezing up through the crustal rocks along faults and close to them, like pips.

Leonov (Paper 7) accepts that there may be actual transitions between tectonic and olistostrome mélanges; the same is suggested by Ballance and Spörli (Paper 28). One can easily visualize simultaneous superficial and deep (under burial pressure) effects in very large submarine slides, as well as tectonic and superficial processes so closely following one another that the distinction becomes blurred. Many mélanges, by virtue of their polyphase character, must defy simple genetic classification or characterization.

Some mélanges display a remarkable homoclinal structure and consistency of facing (pillows, etc.)—that is, they appear like thick sedimentary-volcanic sequences, disrupted and intermingled to varying degrees. They could be thick dismembered sequences with minor exotics, isoclinal repetitions with one limb cut out, or stacks of fault slices, all facing one way (imbrics). This is an aspect of mélanges that calls for much more detailed study. Such developments have been described from Iran, New Zealand, and Alaska (Papers 21 and 28 and Moore and Wheeler, 1978).

In such major crustal zones of instability there seems to be a special type of mélange tectonics, quite distinct from classic tectonic patterns. Fragmentation plus chaotic mingling are the essential characteristics of this mélange tectonics; processes involved include tectonic fragmentation, superficial gravity slide processes, and protrusion related to fault zones. In some regions such as the circum-Pacific zone, tectonic deformation seems to be the ruling process, olistostromes appear to be subsidiary. In other regions, the superficial gravity slide processes may well be dominant. The exact sequence of events recognized shows a wide variation, but there are two models: (1) tectonic deformation (break up of the allochthon) succeeded by superficial gravity sliding and further fragmentation and intermingling and (2) superficial gravity slide deformation on the steep fore-arc slope succeeded by later tectonic deformation, with further fragmentation and intermingling.

CONCLUSION

There can be no doubt that both purely tectonic and purely olistostromal mélanges exist, but the interpretation of the more complex, regional, polyphase mélanges remains confused. A need exists for much more specialized structural study of small areas in great detail in the case of such mélanges. Each mélange occurrence must be treated on its own merits; the literature on the subject is bedevilled by unwise extrapolations of perfectly good interpretations of one occurrence to cover mélanges generally. The lesson of this enquiry is that there are mélanges and mélanges (to paraphrase the editor's distinguished mentor, H. H. Read), and their preeminent character is their diversity.

Mélanges must be mapped as rock bodies in the field; the internal zonation of the body studied; the individual blocks studied; and native (cognate) and exotic blocks distinguished. The metamorphism, internal structure (including the facing in pillows and sediments) must be studied, and the tectonic deformation must be studied and analyzed in great detail. In particular, evidence of soft rock (surficial) submarine deformation and tectonic (under burial load) deformation must be weighed. All suitable material must be subjected to paleontological examination (especially radiolaria and foraminifera). Only when this is all complete can a valid assessment of the mode of origin be attempted; even then the characterization may not be unequivocal. There is probably much scope for trace geochemical and isotopic studies to discriminate the provenance and radiometric age of the components of mélanges, but such techniques have yet to be widely applied. The polyphase character of many mélanges may well resist even such techniques, and mélanges that incorporate fragments of an older mélange in a younger one will obviously be intractable.

Bodies chaotized simply by virtue of internal stress accommodation differentials and devoid of exotic components should not, in the editor's view, be termed *mélange*. The terms *broken* or *dislocated* formation favored by Hsü (Papers 2 and 3) and McCall (Paper 21) respectively should be applied. Some authorities, however, do appear to regard these as mélanges—for example, some of the occurrences described by Cowan (Paper 25) from the Franciscan appear to be virtually devoid of exotics, and Moore and Wheeler (1978) have also described exotic-free mélanges from Alaska. There is, perhaps, a need for a redefining conference aimed at mélanges, as in the case of ophiolites.

It has been suggested that mélanges comprise material actually chaotized within subduction zones deep in the crust. Many mélanges have manifestly not suffered appreciable subduction as they are quite unmetamorphosed or only of sub-phyllite grade of incipient metamorphism. Even ophiolitic mélanges such as the colored mélange of Iran (Paper 21), though probably an assemblage of offscraped slabs, display negligible metamorphism and cannot have suffered appreciable subduction. The mechanism of offscraping is as yet little understood—that is, it may take the form of slab production or disseminated protrusions—but most offscraping appears to be a near surface process, not a deep crust process. Whether any mélanges stem from chaotization deep in subduction zones is an open question. At least one Russian authority (see editorial comments, Papers 29 and 30) favors such a process. However, though high grade metamorphic components such as eclogite do figure among mélange blocks, especially in serpentinite mélanges, there seems to be little or no real evidence that actual deep crustal chaotization forms mélanges deep within subduction zones.

REFERENCES

Anon., 1972, Penrose Field Conference on Ophiolites, *Geotimes* **17:**24-25.

Bailey, E. B., and W. J. McCallien, 1953, Serpentine Lavas, the Ankara-Mélange and the Anatolian Thrust, *Royal Soc. Edinburgh Trans.* **62:**403-442.

Benson, W. N., 1962, The Tectonic Conditions Accompanying the Intrusion of Basic and Ultrabasic Igneous Rocks, *Natl. Acad. Sci. (USA) Mem.* **1:**1-90.

Bird, J. M., and J. F. Dewey, 1970, Lithosphere Plate Continental Margin Tectonics and the Evolution of the Appalachian Orogen, *Geol. Soc. America Bull.* **81:**1031-1060.

Bowen, N. L., 1927, The Origin of Ultrabasic and Related Rocks, *Am. Jour. Sci.* **14:**89-108.

Brongniart, A., 1827, *Classification et charactères mineralogiques des roches homogènes et hétérogenes,* F. G. Levrault, Paris.

Church, W. R., 1972, Ophiolite: Its Definition, Origin as Ocean Crust, and Mode of Emplacement in Orogenic Belts, with Special Reference to the Appalachians, *Canada Dept. Energy, Mines and Resources Publ.* **42:**71-85.

Coleman, R. G., 1971, Plate Tectonic Emplacement of Upper Mantle Peridotites Along Continental Edges, *Jour. Geophys. Research* **76:**1212-1222.

Coleman, R. G., 1977, *Ophiolites.* Springer-Verlag, Berlin, Heidelberg, New York, 229p.

Coleman, R. G., and W. P. Irwin, 1974. Ophiolites and Ancient Continental Margins, in *The Geology of Continental Margins,* C. A. Burk and C. L. Drake, eds., Springer-Verlag, Berlin, Heidelberg, New York, pp. 921-931.

Condie, K. L., 1976. *Plate Tectonics and Crustal Evolution,* Pergamon, New York, 288p.

Cowan, D. S., 1980, Deformation of Partly Dewatered and Consolidated Franciscan Sediments, Piedras Blancas and San Simeon, California, in *Trench and Fore-Arc Sedimentation and Tectonics in Modern and Ancient Subduction Zones, Abstract Volume,* Geological Society of London and British Sedimentological Research Group, pp. 21-22.

Davies, H. L., 1971, Peridotite-Gabbro-Basalt Complex in Eastern Papua. An Overthrust Plate of Oceanic Mantle and Crust, *Australia Bur. Mineral Resources Geology and Geophysics Bull. 128,* 48p.

Dewey, J. F., and J. M. Bird, 1971, Origin and Emplacement of the Ophiolite Suite: Appalachian Ophiolites in Newfoundland, *Jour. Geophys. Research* **76:**3179-3206.

Gass, L. G., 1967, The Ultrabasic Volcanic Assemblages of the Troodos Massif, Cyprus, in *Ultramafic and Related Rocks,* P. J. Wyllie, ed. Wiley, New York, pp. 39-42.

Greenly, E., 1919, The Geology of Anglesey, *Great Britain Geol. Survey Mem. 1,* 980p.

Hess, H. H., 1938, A Primary Peridotite Magma, *Am. Jour. Sci.* **35:**321-344.

Hess, H. H., 1955, Serpentinites, Orogeny and Epeirogeny, *Geol. Soc. America Spec. Paper 62,* pp. 391–407.

Hsü, K. J., 1968. Franciscan Mélanges and Their Bearing on the Franciscan-Knoxville Paradox, *Geol. Soc. America Bull.* **79:**1063-1074.

Moore, J. C., and R. L. Wheeler, 1978, Structural Fabric of a Mélange, Kodiak Island, Alaska, *Am. Jour. Sci.* **278:**739–765.

Moores, E. M., 1969, Petrology and Structure of the Vourinos Ophiolitic Complex, Northern Greece, *Geol. Soc. America Spec. Paper 118,* pp. 1–74.

Moores, E. M., and F. J. Vine, 1971, Troodos Massif, Cyprus, and Other Ophiolites as Oceanic Crust: Evaluation and Implication, *Royal Soc. London Philos. Trans.* **268A**443–466.

Shackleton, R. M., 1951, The Kavirondo Rift Valley, *Geol. Soc. London Quart. Jour.* **106:**345–392.

Steinmann, G., 1927, Die ophiolithischen Zonen in dem mediterranen Kettengebirge, *Internat. Geol. Congress, 14th, Madrid, Proc.* **2:**638–667.

Wyllie, P. J., 1967, *Ultramafic and Related Rocks,* Wiley, New York, 464p.

Part I

FOUNDATIONS

Editor's Comments on Papers 1 Through 4

Gansser (Paper 1) introduced the term *ophiolitic mélange* in place of his earlier "coloured mélange" (Gansser, 1955), though this term is deeply entrenched as a local term in Iran and survives as such there. He distinguished the ophiolitic mélange from the purely sedimentary mélange, of either olistostrome or tectonic type. Focusing the attention of geologists to this major problem, which relates to convergent plate margins in the context of the new global tectonics, he presented a historical and global review, identifying the roots of this problem in the observations at the end of the last century of Griesbach, Diener, and Middlemiss, in the Kiogar region of the central Himalaya—where a "wild jumble of flysch-type sediments together with foreign looking blocks from a few meters to several thousand meters in size, intricately mixed with basic and ultrabasic rocks" was recorded (Paper 1, p. 479). However, the term *mélange* was not actually coined until later (Greenly, 1919). The Kiogar area, a major subduction zone, was further described by Heim and Gansser (1939).

Particularly important in this review is the restatement of Kundig's (1959) outline of the general characteristics of ophiolitic mélanges:

1. The general variety and mixture of plutonic, hypabyssal, and volcanic rocks;
2. The occasional association, unexplained so far, of metamorphic (glaucophane) rocks and nonmetamorphic rocks;

3. An extremely complex, often chaotic imbrication structure, frequently accompanied by overthrust sheets and nappes;
4. The apparent scarcity or absence of magmatic channels and feeders to supply the effusions and intrusions;
5. The scarcity or absence of thermal contacts with the embedding sediments, most contacts being apparently tectonic;
6. The difficulty of age determination for the *mis-en-place* of the igneous rocks, owing to the chaotic relation with, and scarcity of fossils in, the accompanying sediments.

Gansser considered that such mélanges should not border on oceanic ridge but the consuming end of the plate. This led him to question certain interpretations of Cyprus, and he quoted Turner (1973): "the last word on Cyprus has not been said." New strong evidence suggests that the Troodos ophiolites relate to a marginal basin or island arc situation and that they are essentially a calc-alkaline suite (endorsing Gansser's doubts).

Gansser considered that the recognition by Hsü (1968) that mélanges should be treated as mappable units eliminated confusion due to practices of mapping only the individual blocks, but it must be stressed that study of individual block content in intrinsic petrographic, sedimentological, paleontological, and metamorphic aspects remains just as important as statistical approaches, study of deformations in blocks, their contacts and matrix (if any), and studies of block matrix relationships.

Gansser defined ophiolitic mélanges under eleven headings:

1. Mixture of ophiolitic rocks with nonophiolitic rocks (often of unknown provenance, e.g., exotic blocks);
2. Matrix ophiolitic (frequently tectonically sheared serpentinite) or sedimentary (commonly flyschoid);
3. No contact metamorphic relations between components (but some metasomatism);
4. Base invariably a tectonic contact; top often a transgression into pelagic or flyschoid sediments;
5. Mélange results from sedimentary (olistostromal) and tectonic mixture (he believes that a mélange cannot stem from purely tectonic processes);
6. Recycling, including olistostromal slumping, tectonization, and further incorporation of oceanic crust during emplacement, a part of the mixing being related to this episode;
7. Restriction to ophiolitic belts, which outline major plate boundaries;
8. Final emplacement by diapiric protrusion as a mélange body a

possibility, with disruption of original connections to members of the ophiolitic suite;

9. May outline plate boundaries like ophiolites when the normal ophiolite suite is missing;
10. Commonly overthrust by ultrabasic sheets in zones of strongly allochthonous character;
11. Distinct from normal olistostromes that are interbedded within a normal sequence (Paper 6).

Controversial aspects of these definitions are the statement that mélanges cannot be entirely tectonic in origin; many authorities would vehemently disagree. Gansser also fails to include the block-to-block mélange with no matrix in his definition, and this is a reality (Paper 21). His sixth definition is also questionable as being too vague. Though it may be generally true, more definition is needed to explain in exactly what geotectonic and paleogeographic situation the further slumping, tectonization, and incorporation of oceanic crust occur. The latter, in particular, must be defined in terms of subduction and obduction and of the downgoing and overriding plate. Gansser related ophiolitic mélanges to obduction, a specific connotation that would be argued by most presentday authorities, who would relate mélanges primarily to subduction, though they also occur in obduction situations.

Gansser emphasized the relationship between mélanges and major thrusting, and in this he supports the similar equation suggested by Russian authorities (Papers 29 and 30).

Hsü (Paper 2) has summed up his views on mélanges, developed in several papers based on the Franciscan mélanges of California (Hsü, 1968, 1971; Hsü and Ohrbom, 1969). He believes that interpretations in terms of contemporaneous gravity slide processes have been overemphasized; that mélanges relate to convergent plate margins, to the oceanward migration of the Benioff zone (they are absent, or relatively scarce, in the Andes sector where landward migration has probably occurred). He regards mélanges as mappable tectonic units bounded by shear surfaces and distinguishes them from olistostromes, usually separated from overlying and underlying units by depositional contacts, and thus stratigraphic units. Olistostromes, he believes, should show gradation from boulder beds to graded turbidites. There is, he believes, a contrast in the scale of development—that is, mélanges may be composed of much larger blocks (even up to several miles long, individually) and may be developed on a regional scale, whereas olistostromes must, by virtue of their very nature, be more localized in development, and there are limits to the size of block that can move downslope under submarine conditions, under gravity. Mélanges are, he believes, deformed under an overburden so that pelitic materials flow plastically and brittler materials fracture. Such an overburden is not possible in an olistostrome, except as a secondary effect after burial. Blocks in mélange are bounded by shear fractures, though tectonic processes can lead to some rounding; sedimentary blocks in olistostromes were rounded prior to transport. Argillaceous (or less commonly sandy) matrixes are a feature of olistostromes.

It is obvious that there is an almost diametrically opposite approach by Gansser and Hsü. Gansser regards an (ophiolitic) mélange as essentially of olistostrome origin though modified by tectonic processes; he rejects the possibility of a purely tectonic mélange. Hsü defines a mélange as tectonic; separates olistostromes from mélanges, and allows that mélanges may contain obscure tectonically modified olistostromes, just as they may contain other sediments.

Hsü admitted the problem that sheared olistostromes come to resemble tectonic mélanges and thus proposed the term *wildflysch* where a genetic distinction cannot be made. Hsü's solution to the dilemma of genetically uncertain chaotic bodies by invoking the term *wildflysch* is unsatisfactory. For example, it conflicts with the definition of wildflysch given by Dzulynski and Walton (1965; p. 189), and such ambiguous bodies may well have no flysch component at all. It seems that both authorities, despite their excellent observation that has immeasurably advanced the study of chaotic bodies of this type, have erred in seeking a genetic definition at all. The original definition of mélange by Greenly (1919) was not genetic: mélanges have been described that are of neither of the two origins preferred by these authors, and the literature is already so confused, with the term *mélange* being applied indiscriminately to tectonic mélanges and olistostromes, that the only reasonable usage is application of *mélange* to both tectonic and olistostrome mélanges and to those of mixed origin, with further characterization dependant on the nature of blocks and matrix (if any) structure and origin. Hsü, in the addendum to this paper, did accept such arguments for a nongenetic use of the term *mélange* (sensu lato) for all pervasively sheared chaotic deposits with exotic blocks. A further point is that Hsü based his arguments on the Franciscan mélange that has a matrix. Like Gansser, he does not appear to have envisaged the possibility of a matrix-free block-to-block mélange (Paper 21), and so to some extent both these discussions are incomplete.

Hsü regards the presence of exotic blocks as essential to mélange, and chaotic bodies devoid of these are referred to by him as broken formations, but Cowan (Paper 25) applies the term to bodies from the Franciscan apparently more or less devoid of exotics. Moore and Wheeler (1978) proposed that mélanges need not contain exotic blocks, but earlier, Moore (1973) described products entirely of strain-contrast chaotization, devoid of exotic blocks, as broken formations, following Hsü. McCall (Paper 21) has referred to bodies devoid of exotics, or with minor exotics, as dislocated flysch, not mélange, "dislocated flysch" being synonymous with broken formation.

In an earlier paper, Hsü (1968) emphasized that mélanges are the product of two essential processes, fragmentation and mixing. He also introduced the concept of a polykinematic mélange, one that involves two stages (i.e., an older mélange suffers later reworking, fragmentation, and mixing), and cited one of the Anglesey mélanges of Greenly (1919). The sedimentary mélange of the Makran, Iran (McCall, Paper 21), could be considered to be, in a sense, polykinematic.

In another paper (Hsü and Ohrbom, 1969), the important observation is made that boudins in the Franciscan mélange are rounded by extensional shear fracturing; such a pattern is consistent with deformation under considerable burial overload. In the Argille Scagliose of the Apennines, some classic boudinage with tension fractures normal to bedding is observed, and this probably indicates a very low confining pressure (such as would be consistent with an olistostrome model).

Hsü and Ohrbom (1969) quote Davoudzadeh (1969) as stating that the famous Coloured Mélange of central Iran has a serpentinite matrix. This mélange (McCall, Paper 21) is of immense extent of development, and the term probably embraces numerous ophiolitic mélange developments of diverse character. Certainly, in the Makran, no such continuous matrix is present and it is likely that such matrixes are the exception rather than the rule in the Iranian colored mélanges.

Hsü (Paper 3) presented a model for the relationship of the Franciscan mélanges and subduction, considering that the mélanges incorporate ophiolites emplaced at mid-ocean ridge sites, radiolarites deposited on an abyssal plain, and flysch turbidites poured off the overriding continental plate into the deep sea tranch. At the time at which Hsü was writing, ophiolites were being confidently and unquestionably equated with oceanic ridges, following Moores (1969) and Bird and Dewey (1970). Nowadays it is realized that ophiolites can be formed in other sites of spreading, including island-arc and marginal basin sites close to the subducting margin of the oceanic plate and widely separated from the mid-ocean spreading center. Also, radiolarites and related pelagic sediments could well be deposited in the arc-trench gap as well as the abyssal plain. The concept of mélanges as mixtures of oceanic rocks providing a complete spectrum from ridge to arc-trench gap, well stated by Condie (1976, p. 183), though elegant and attractive, may be an oversimplification—that is ophiolitic mélanges may be far more complex and highly variable in the provenance of their components and the locus of chaotization and deformation.

Berkland et al. (Paper 4) asked the question, "What is the Franciscan?" This paper is more than of purely local significance because it bears on the whole question of mélanges. In it they redefined tectonic and exotic blocks, and knockers (erosionally emphasized tectonic and exotic blocks). They redefined mélange as "a mappable body of rock characterised by the inclusions of fragments and blocks of all sizes, both exotic and native, embedded in a fragmented and generally sheared matrix of more tractable material." The words "of all sizes" are vague—"of variable size and commonly very large" would be preferable. These investigators regard a matrix as essential, thus excluding the possibility of a block-to-block mélange. Their definition of broken formations does not agree with the usage favored by Cowan (Paper 25) or by Moore and Wheeler (1978), who believe that mélanges need not contain exotics. These authors do, despite some limitations in their definitions, favor a nongenetic usage for the term *mélange*—the only really satisfactory usage (Paper 13).

Berkland (1972), in another paper, related the Franciscan mélange to subduction, invoking mixing within a trench (presumably due to gravity slide processes down the fore-arc/trench slope followed by tectonic mixing within the subduction zone). Hsü and Ohrbom (1969) suggested a similar two-phase effect, original olistostrome material being incorporated in the dominant tectonic mixing process. However, though some evidence exists from the Franciscan and the Apennines of such two-stage processes, it remains doubtful if this is essential to tectonic mélanges and many mélanges may well be purely tectonic, with no initial olistostrome development.

REFERENCES

Berkland, J. O., 1972, Paleogene "Frozen" Subduction Zone in the Coast Ranges, California, *Internat. Geol. Congress, 24th, Proc.* **3:**99–105.

Bird, J. M., and J. F. Dewey, 1970, Lithosphere Plate Continental Margin Tectonics and the Evolution of the Appalachian Orogen, *Geol. Soc. America Bull.* **81:**1031–1060.

Condie, K. L., 1976, *Plate Tectonics and Crustal Evolution,* Pergamon Press, New York, 288p.

Davoudzadeh, M., 1969, Geologie und Petrographie des Gebeites nördlich von Nain, Zentral-Iran, *Zürich (ETH) Geol. Inst. Mitt., new ser., No. 98,* 91p.

Dzulynski, S., and E. K. Walton, 1965, *Sedimentary Features of Flysch and Greywackes,* Developments in Sedimentology, vol. 7, Elsevier, Amsterdam, London, New York, 274p.

Gansser, A., 1955, New Aspects of the Geology of Central Iran, *World Petroleum Congress, 4th, Rome, Proc. Section 1,* pp. 279-300.

Greenly, E., 1919, The Geology of Anglesey, *Great Britain Geol. Survey Mem. 1,* 980p.

Heim, A., and A. Gansser, 1939, *Central Himalaya, Geological Observations of the Swiss Expedition, 1939,* Schweizer. Naturf. Gesell. Denskschr. 731p. [New Edition: Hindustan Publishing Corp. (India), Delhi, 1974.]

Hsü, K. J., 1968, Franciscan Mélanges and Their Bearing on the Franciscan-Knoxville Paradox, *Geol. Soc. America Bull.* **79:**1063-1074.

Hsü, K. J., 1971, Franciscan mélanges as a Model for Eugeosynclinal Sedimentation and Underthrusting Tectonics, *Jour. Geophys. Research* **76:**1162-1170.

Hsü, K. J., and R. Ohrbom, 1969, Mélanges of San Francisco Peninsula: Geologic Reinterpretation of Type Franciscan. *Am. Assoc. Petroleum Geologists Bull.* **53:**1348-1367.

Kundig, E., 1959, Eugeosynclines as Potential Oil Habitats, *World Petroleum Congress, 5th, New York, Proc. Section 1,* pp. 461-479.

Moore, J. C., 1973, Complex Deformation of Cretaceous Trench Deposits, Southwestern Alaska, *Geol. Soc. America Bull.* **84:**2005-2020.

Moore, J. C., and R. L. Wheeler, 1978, Structural Fabric of a Mélange, Kodiak Island, Alaska, *Am. Jour. Sci.,* **278:**739-765.

Moores, E. M., 1969, *Petrology and Structure of the Vourinos Ophiolitic Complex, Northern Greece,* Geol. Soc. America Spec. Paper 118, 74p.

Turner, W. M., 1973, The Cyprian Gravity Nappe and the Autochthonous Basement of Cyprus, in *Gravity and Tectonics,* Wiley, New York, pp. 287-301.

1

Reprinted from *Eclogae Geol. Helvetiae* **67**:479-507 (1974)

The Ophiolitic Mélange, a World-wide Problem on Tethyan Examples

By AUGUSTO GANSSER[1])

ABSTRACT

The ophiolitic mélange is defined as a olistostromal and tectonic mixture of ophiolitic material and sediments of oceanic origin with exotic blocks, reflecting areas which have subsequently disappeared. Being intimately related to ophiolitic belts they indicate important suturelines connected to plate boundaries. Ophiolitic mélanges occur mostly where continental type plates are subducted with obduction of oceanic material. These facts are discussed on examples from the Middle East and the Himalayas, where the mélanges are particularly well exposed.

1. General aspect

81 years ago, a famous trio of geologists, C. L. Griesbach, D. Diener and C. S. Middlemiss, sponsored by the Geological Survey of India, visited the Kiogar area at the border between Kumaon and Tibet in the Central Himalaya. In this fascinating, remote and barren area, lying between 5000 and 6000 m above sea level, they were puzzled by a most unexpected geology. Above competent folds in the well known Spiti sequence they observed a wild jumble of flysch-type sediments, together with foreign looking blocks from a few meters to several thousand meters in size, intricately mixed with basic and ultrabasic rocks. After a preliminary investigation of this area and a vivid discussion with his colleagues, well known for their great Himalayan experience, Diener came to the following conclusion (DIENER 1895, p. 604):

> «Fünf Momente sind für die Klippen von Chitichun und am Balchdhura bezeichnend: 1. Die von der Hauptregion des Himalaya abweichende Schichtfolge; 2. die bogenförmige, diagonal auf das Streichen der Himalaya-Falten verlaufende Streichrichtung; 3. ihr Auftreten innerhalb eines muldenförmigen, mit Flysch und Spiti Shales erfüllten Gebietes; 4. ihre innige Verbindung mit Eruptivgesteinen; 5. das Fehlen jedweder Art von Strandbildungen in ihrer Umgebung.»

He could not agree to a direct comparison with the Klippen of the Swiss Alps and Carpatians, but accepted in principle Bertrand's and Schardt's ideas as «lambeaux de recouvrement». His main difficulty was the interpretation of the presence of large

---

[1]) Geological Institute, Swiss Federal Institute of Technology and University of Zürich (Switzerland).

amount of basic rocks. Based on the available information he could not solve the problem as he stated at the end of his work (p. 606):

«Eine Lösung des tektonischen Problems der Klippen von Chitichun geben zu wollen, halte ich in Anbetracht der ungenügenden Kenntniss, die wir heute noch von der Ausdehnung jenes Phänomens besitzen, und bei dem Mangel einigermassen brauchbarer Nachrichten über den geologischen Bau der angrenzenden Theile von Tibet für verfrüht. Es mag genügen, hier auf jene Eigenthümlichkeiten hingewiesen zu haben, welche die tibetanische Klippenregion von allen bisher in Europa bekannten Klippenzügen unterscheiden und dieselbe zu einem der interessantesten und merkwürdigsten Theile des Himalaya stempeln, der wie kaum ein anderer noch auf lange Zeit hinaus ein dankbares Feld für weitere Forschungen abgeben wird.»

Though this discovery was of outstanding interest, eight years elapsed before new action had been taken in connection with this fascinating problem. Inspite of the scepticism of Dr. William King, then Director of the Geological Survey of India, A. von Krafft was ordered by the Survey to investigate this remote area during the summer season of 1900. During six weeks of intensive field work in the wider Kiogar area von Krafft gathered a wealth of most valuable information including rare fossil collections from various blocks, which, after a careful investigation by Diener, supported the foreign aspect of these faunas, unknown in this part of the Himalaya. Excellent panoramic sketches and photographs clearly outlined the contrast between lower and upper structural levels.

While DIENER (1895) compared the exotic Kiogar blocks to the Klippen of the Alps and the Carpathians, which at the time were newly discovered and thus of great general interest, VON KRAFFT (1902) saw no connection with Klippen and proposed as primary origin hughe volcanic explosions in Tibet through which the sediments were mixed with the basic volcanics. He admitted, however, that these arguments were not fully convincing on the still scanty available facts. His stimulating discussions of pro and contra of his ideas bring out many of the problems which crop up during the present discussions of mélange rocks. On page 169 he wrote:

"Those which assume structural causes would imply that the exotic blocks were brought into existence by the disturbances which lead to the upheaval of the Himalayas. This, however, is not the case. We have seen that the volcanics, which include the blocks, were folded together with the flysch. From this we can only conclude that the volcanics and consequently also the exotic blocks were preexistent to those disturbances, and any causal connection between the latter and the blocks is therefore impossible.

Nor is it at all conceivable that folding or faulting should have brought about those features, which were described in the above chapters. None of the theories alluded to would explain the presence of the volcanics and their intimate connection with the exotic blocks."

It is most unfortunate to science that von Krafft was not able to continue his fascinating studies. He died 1901 in Calcutta. Though von Krafft has not coined the word "mélange", his careful descriptions and discussions of a possible origin of the exotic rocks put him among the very first pioneers who stressed the importance of the mélange problem.

The first to mention ophiolitic rocks in southern Tibet north of the Kiogar region discussed above was Captain STRACHEY in 1851 (p. 308):

"A great outburst, in which are found hypersthene and bronzite, besides syenitic and ordinary greenstones, and various varieties of porphyry, occurs in the vicinity of the lakes which are found at the eastern extremity of the plateau. The greenstone is known to extend considerably to the west, and forms, at an elevation of about 17,600 feet, the summit of Balch, one of the Himalayan Passes into Tibet which I have crossed."

He stressed the structural difference between the granites of the Transhimalaya and the ophiolitic rocks (p. 309):

"The Granites appear to constitute lines of elevation, not of rupture; but there seems to be no specific action produced by them on the dip of the strata, which they appear to leave generally unchanged.

The Greenstones, on the other hand, usually follow lines of dislocation of the strata; being sometimes apparently contemporaneous, and at others intruded through rocks already consolidated."

SUESS, in his famous *Das Antlitz der Erde* (1901, p. 352 ff.), recognized the importance of the extraordinary findings of the Himalayan geologists and devoted some chapters to this problem. He realized the significance of thrust movements and of Eocene volcanics along the Indus:

«Vielleicht war hier im Eocän eine mit Vulcanen besetzte Disjunctiv-Linie vorhanden» (1901, p. 351).

He connected this line with the exotic area of Kiogar, and he stated (1909, p. 648):

«Diese Erfahrungen lehren, dass die Verfrachtung der tibetanischen Schollen und das Hervortreten der simischen Gesteine zur gleichen Zeit eingetreten sind»

and on the same page:

«Balchadura [the Kiogar area] stellt den Rand einer Bewegungsfläche erster Ordnung dar, an welcher sedimentäre Serien von abweichender Facies sich überlagern.»

It is significant, that this particular zone was later recognized as one of the most important subduction zones of modern geology.

The term "mélange" (autoclastic mélange) was introduced by GREENLY in 1919 when he described a complicated tectonic mixture in the Precambrian Mona complex of Anglesey island in Wales. From his careful descriptions it is evident that his autoclastic mélange has strictly tectonic implications with no relations to ophiolitic type rocks. The tectonic aspect of a mélange is still widely propagated (HSÜ 1968), but, as we will see, the question is much more complex and still most problematic.

The "mélange" has been revived in the Middle East, where excellent outcrops of this widespread phenomenon had eventually called the geologist's attention. BLUMENTHAL, when mapping in Central Anatolia, described a "Mésozoïque à facies tectonique brouillé" (1948). This same facies was mentioned by BAILEY & MCCALLIEN as Ankara Mélange (1950, 1953). They used the term mélange "out of respect for descriptions that Greenly and Matley have furnished of a similar tectonic mélange in Anglesey...". The detailed account of BAILEY & MCCALLIEN stressed the relation of exotic blocks to basic and ultrabasic rocks. They recognized Steinmann's "Trinity" after having discovered pillow lavas and stressed the world-wide occurrence of ophiolitic rocks, failed, however, to include the mélange formation into this general picture. Although not mentioning mélange in particular, KÜNDIG (1959) gave the following characteristics of these ophiolite occurrences (p. 469):

a) The great variety and mixture of hypabyssal, plutonic and volcanic rocks.
b) The occasional association – unexplained so far – of metamorphic (Glaucophane rocks) and non-metamorphic rocks.

c) An extremely complex, often chaotic imbricated structure frequently accompanied by overthrust sheets and nappes.
d) The apparent scarcity or absence of magmatic channels and feeders to supply the effusions and intrusions.
e) The scarcity or absence of thermal contacts with the embedding sediments; most contacts are apparently tectonic.
f) The difficulty of an age determination of the *mise en place* of the igneous rocks, owing to the chaotic relation with, and scarcity of fossils in, the accompanying sediments.

While discussing the ophiolites of Celebes the same author located ophiolites as follows (KÜNDIG 1956 *b*, p. 229):

"The consequence is that, with the appearance of ophiolites in an orogenic belt, i.e. with the transition from the shelf- or mio-geosynclinal- into the eu-geosynclinal zone, facies and thickness of sediments change abruptly.

As the transition zone coincides with the old trench-slope, which developed in a subsequent diastrophic phase into a thrust-line of the first order, it frequently occurs that the trough-contents, namely the eu-geosynclinal facies, with its masses of ophiolites, are thrust over the facies of the shelf edge. Such cases are known from Turkey, Kurdistan, Oman and in East-Celebes.

The trench-slope is, particularly during the ophiolitic extrusions at depth, an area of catastrophic events. It is not only the site of turbidity currents, but during the buckling down, vast tracts of the shelf edge mantle may glide down and accumulate as huge exotic masses (e.g. Mélange of BAILEY 1953) in a foreign environment."

When substituting miogeosyncline and eugeosyncline by continental and oceanic plates respectively, KÜNDIG's statements fit modern ophiolitic concepts remarkably well.

That the world-wide occurrence of the mélange rocks together with ophiolites is not sufficiently recognized is underlined by the fact that the recent Penrose Conference, dealing with ophiolites, did not include mélange in its terminology (anon. 1972). This unfortunate situation calls for reaction, and it is the purpose of this article to stress the world-wide importance of the mélange formations and their intimate relation to the ophiolite suite and thus their direct involvement in modern geotectonic theories. It is for this reason that the name *ophiolitic mélange* is proposed in order to distinguish it from purely sedimentary mélange of the olistostrome type or purely tectonic mélange in the original meaning of GREENLY.

My personal interest in the "mélange" arose in 1936 when I visited the Central Himalayan border regions, already described by VON KRAFFT, and followed these enigmatic outcrops into the "root zone" of the Himalaya in southern Tibet. They were described as an "Exotic Bloc zone" (HEIM & GANSSER 1939) and later, based on the excellent occurrences of the Middle East, as "Colored Mélange" (GANSSER 1955).

In order to stress the global importance of ophiolites with their related mélanges, the Commission of Structural Geology of the International Union of Geological Sciences chose this subject as its main working theme. During a special Ophiolite Symposium, held in USSR in 1973 and sponsored by the Academy of Science of the USSR as well as the Commission on Structural Geology, the mélange problem was the main topic and could be discussed on the basis of excellent outcrops in the western Tien-Shan and Kyzyl-Kum area (Paleozoic ophiolites) as well as in the little Caucasus (Mesozoic ophiolites). In both areas the relation of ophiolitic mélange to major tectonics is well exposed, and this topic received an excellent review in a recent

Russian publication by the members of the Academy of Sciences of the USSR, Moscow (1973), entitled *Developmental stages of folded belts and the problem of ophiolites* (see also Geological Newsletter 1973, Vol. 3, p. 211). Their acceptance of major thrusting in relation to mélange formations as a world-wide occurrence is new and most stimulating.

2. Definition of ophiolitic mélange

From the great amount of data published recently on ophiolites and their relations to global tectonics, the importance of the ophiolitic mélange is increasingly evident, and also the confusion regarding its definition. Hsü (1968) discussed the rock unit "mélange" in great detail and has the merit of stressing the structural importance of this complex formation. The recognition of the mélange as a mappable unit has helped to clear up many a confused situation, where before much care was taken to unravel a hopeless tectonic picture by carefully mapping single exotic blocks.

Hsü's definition does not, however, stress the intimate relation to ophiolitic belts. The mélange, as originally defined by Greenly (1919), could be a tectonic feature only or also a purely sedimentary olistostrome complex. I would like to use the term *ophiolitic mélange* as the more widespread occurrence without, however, substituting the general term mélange by ophiolitic mélange. For this term I offer the following definition, which is based on many occurrences known to me personally or described in the literature. This definition is open to discussion, and I would welcome constructive criticism:

Factual and genetic characteristics of an ophiolitic mélange:

1. We recognize a mixture of rocks of the ophiolitic suite together with non-ophiolitic rocks often of unknown origin (exotic blocks).
2. The matrix of the mélange could be ophiolitic (frequently tectonically sheared serpentines) or sedimentary, the latter mostly of a flyschoid facies.
3. No contact metamorphism exists between components of a mélange, but metasomatic reactions can be locally important.
4. The base of a mélange is always a tectonic contact; the top is often marked by transgression of pelagic calcareous or flyschoid sediments.
5. The ophiolitic mélange results from a sedimentary (olistostromal) *and* tectonic mixture. (I would like to stress that such a mélange cannot have only a tectonic origin. This could not explain the intimate mixture of exotic blocks of very varied composition. See also Graciansky, in press.)
6. During emplacement recycling occurs, including olistostromal slumping, tectonization and further incorporation of oceanic crust. A part of the mixing is related to this episode.
7. The ophiolitic mélange is restricted to ophiolitic belts and outlines zones of major structural significance (plate boundaries).
8. The final emplacement of a mélange body can be by diapiric protrusion and then the original connections to members of the ophiolitic suite become completely disrupted.

9. It thus can occur also when the classical ophiolitic suite is incomplete or totally missing and may outline geotectonic belts in the same way as an ophiolite zone would do.
10. In zones with strong allochthonous character the mélange horizon is often overthrust by ultrabasic sheets (Himalaya, Oman, Zagros, Taurus, Papua ?).
11. The ophiolitic mélange must be distinguished from normal olistostromes (ELTER & TREVISAN 1973), the latter being interbedded into a normal sedimentary sequence. These differences are for instant clearly expressed in southeastern Turkey (Fig. 3), though in strongly tectonized regions this distinction is admittedly difficult.

The broad characteristics and the regional setting of some ophiolitic mélange formations from Asian examples are discussed in the following:

3. General discussion of ophiolitic mélange formations

a) The Oman mélanges

The ophiolitic belt of Oman forms the eastern border of the Arabian shield. From here, mélange formations have been known for over 45 years through the excellent work of LEES (1928). His ideas about the allochthonous position of the Hawasina beds (mélange) and the separate ultrabasic nappe (Semails) have been confirmed by recent investigations (REINHARDT 1969, ALLEMANN & PETERS 1972, GLENNIE et al. 1973), while some authors follow a more autochthonous interpretation with uplifts and related gravity slides (MORTON 1959, TSCHOPP 1967, WILSON 1969). Based on these new though still incomplete investigations, we suggest an underthrusting (subduction) of the Arabian (African) plate, resulting in overthrusting (obduction) of the Oman oceanic slab on of the Arabian shield (Fig. 1). A parautochthonous sedimentary shield cover is "overthrust" by the Hawasina unit, a mélange nappe which was overridden by the relatively undisturbed Semail unit, where the classical oceanic sequences from ultrabasics to pillow lavas is locally preserved below the upper Maestrichtian transgression. The locally steeply dipping metamorphic slivers observed in northern Oman between Hawasina and Semail thrust sheets are difficult to place (ALLEMANN & PETERS 1972).

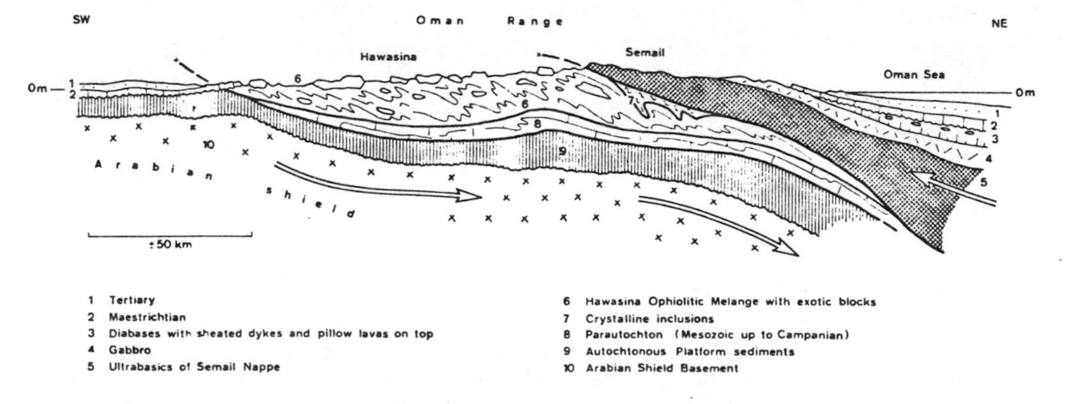

Fig. 1. Schematic section through Oman Mountains (compiled from various sources).

The youngest beds involved in the emplacement of the ophiolite nappes are Campanian to (?) lower Maestrichtian. In the Hawasina mélange matrix a planktonic Cenomanian to (?) lower Turonian fauna indicates the youngest age (ALLEMANN & PETERS 1972), while the exotics range from Ordovician to middle Cretaceous with a predominance of hughe Permian and Triassic blocks.

There is little doubt that the emplacement of the Oman ophiolitic nappes, including the mélange sheet, falls between lower Turonian and middle Maestrichtian times. Unlike other orogenic belts, the later movements in Oman were only of a small scale. Whereas the foreland features of Oman are relatively clear, the backland covered by the Oman sea is still highly enigmatic. An oceanic crust seems to be covered by a pile of 4000 m of mostly clastic sediments of Neogene age, the top of which lies over 3000 m below sea level (HINZ & CLOSS 1969). Similarly to the Beluchistan coast north of the Oman sea they mask the important pre-Maestrichtian structural history.

While Oman was little affected by post-Cretaceous movements, the history following the reactions of the northern Arabian (African) plate with its Eurasian counterpart is highly involed. This is witnessed by the mountain fronts bordering the shield from Turkey over Iran into Afghanistan–Pakistan, a zone exposing some of the world's most striking ophiolitic mélange occurrences (Fig. 2).

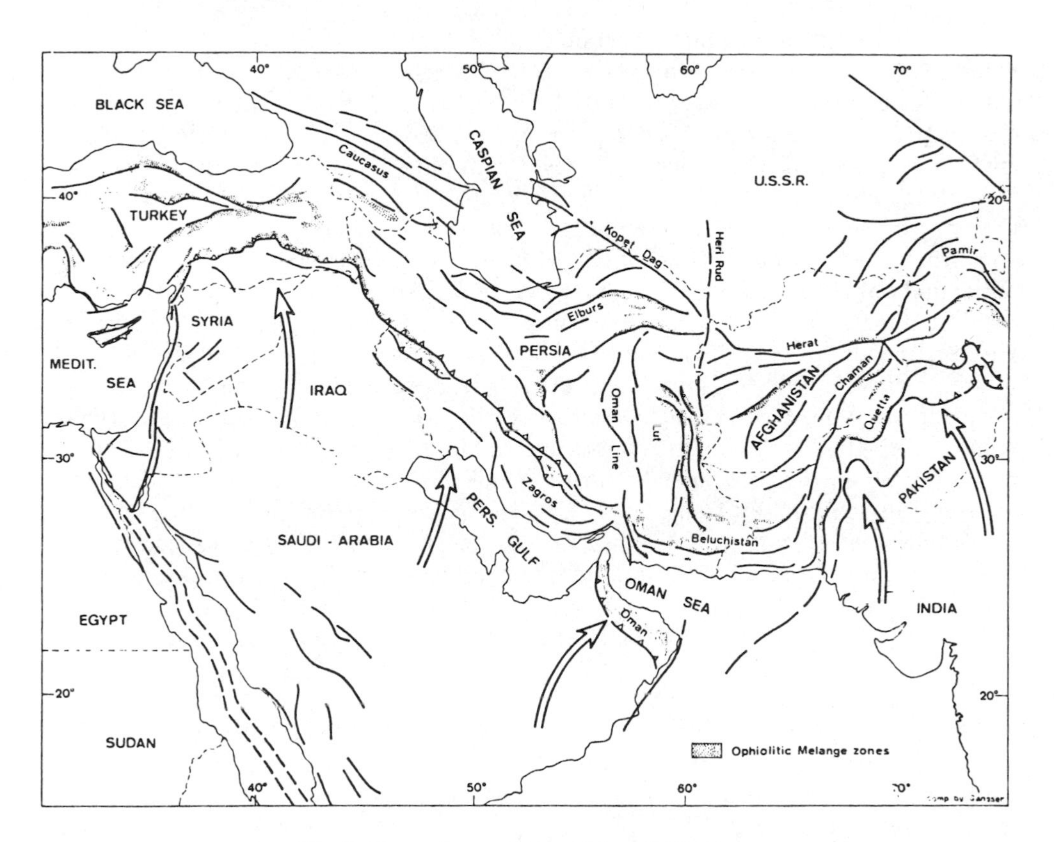

Fig. 2. The main structures of the Middle East.

b) The Anatolian mélanges including Cyprus

Turkey straddles the northern continuation of the western border of the Arabian shield, which contrasts strongly with Oman. This western edge coincides with the Levantine rift system, where Arabia has apparently moved in post-Cretaceous time relative to Africa over 150 km to the north (QUENNELL 1958, FREUND et al. 1968, 1970). Westwards from this structural north-south line conditions are strikingly different: The Mediterranean covers the actual African front and displays an oceanic ridge running in a west-east direction into Cyprus, which represents a part of this ridge rotated by 90°. The Kyrenia range in north Cyprus, with its large olistoliths, runs east-west and is thrust to the south forming a southernmost branch of the Taurids (DUCLOS 1964). This range occurs suspiciously near to our expanding oceanic ridge. Again quite unlike oceanic ridge conditions, south of the classical and well known Trodos mountains (MOORES & VINE 1971, VINE & MOORES 1972) and separated by them by a steep fault zone we find the Mammonia area (Paphos) with a classical ophiolitic mélange, which is transgressed by Maestrichtian marls. This area was studied in detail by LAPIERRE, who observed a pre-ophiolitic volcanism related to Triassic masses and separated this zone from the Trodos ophiolites (ROCCI & LAPIERRE 1969, LAPIERRE & PARROT 1972). Based on personal observations I prefer to regard the Trias with its volcanism as exotic blocks belonging to a Cretaceous Mammonia mélange. Only, and here comes the difficulty, a mélange should not border an oceanic ridge but should be expected at the consuming end of a plate. The last word on the eastern Mediterranean has certainly not yet been said (TURNER 1973).

To the east of the Levant lineament the active border of the northwestern Arabian shield is well exposed. In southeast Turkey, north of the Mardin area, the northwards drift of the shield produced an ophiolitic belt with classical mélange formations. The normal ophiolitic suite is here totally disrupted and mélanges predominate in the overthrust sheets, while erosion of the southwards advancing thrust produced olistostromal ophiolitic sedimentation over the stable plate. Here we observe the juxtaposition of mélange nappes in tectonic contact with secondary ophiolitic olistostromes (Fig. 3). In outcrops and particularly in well sections one sees the intercalation of ophiolitic material into turbidity sediments. These most interesting conditions, known already to KÜNDIG (1956), were described by RIGO DE RIGHI & CORTESINI (1964), without distinguishing ophiolitic nappes and ophiolitic olistostromes. The base of the mélange is again thrusted and the thrust plane has subsequently been folded. The mélange top is normally transgressed by Eocene, but in the higher ranges it was covered by the crystalline Bitlis thrust. A coherent ultrabasic thrust sheet is not developed and was probably sheared off by the higher crystalline thrust. Much of this material is now found in the olistostromes of the foreland (*Map Erzurum* with explanatory notes, 1963, Mad. Tetkik Arama Enst.).

This frontal ophiolite belt is not related to the famous "Ankara Mélange" of BAILEY & MCCALLIEN (1950, 1953). The latter occurs in a complicated, often steeply bordered, but mostly flat lying sheet and may have originated from the North-Anatolian fault zone. The presence of widespread and well exposed mélange zones along this important tectonic disturbance has been overlooked by most previous writers whose interest centered on the recent earthquake problems and possible strike slip movements. No convincing proof exists that along this line horizontal shifts of

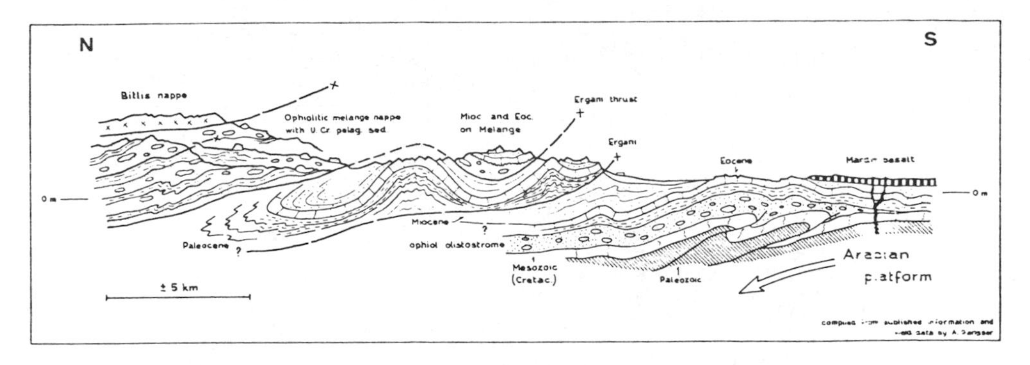

Fig. 3. Ophiolitic mélange nappe and ophiolitic olistostrome in southeast Turkey.

several 100 km accumulated (PAVONI 1961, AMBRASEYS & TCHALENKO 1969, NOWROOZI 1971, 1972). The presence of ophiolitic mélange suggests a rather large north–south compression which was active in pre-Eocene times along microcontinental margins, with obduction of ophiolitic material, of which the Ankara mélange may be a witness. In the more eastern sections of the North-Anatolian "fault zones" larger masses of serpentinized ultrabasics together with the mélange zones do occur (Fig. 4).

c) The Iranian and Afghan mélanges

They occur, similar to those of Anatolia, in two tectonically separate areas. While in Turkey the northern continuation of the western Arabian edge interferes as the Levant lineament, in Iran the eastern edge of the shield, the Oman range, interferes with its northern continuation, the still disputed *Oman line*. First propagated by FURON (1941) and later stressed by GANSSER (1955), estimates of its importance have fluctuated (STÖCKLIN 1968). Now, in connection with various plate boundaries, it has again aroused considerable interest (CRAWFORD 1972). This division is most striking if we compare the Zagros fold belt on the west of the Oman line to the Beluchistan ranges east of it. It is marked by a north–south striking fault zone or steep thrust zone with a considerable dextral displacement (GANSSER 1955; FALCON 1967, 1969; STÖCKLIN 1968).

The main north movement of the Arabian shield has been taken up by the *Zagros thrust* (TAKIN 1972). The underthrusting shield and a young (Pliocene) collision have obliterated a larger part of the previously present mélange rocks which were probably related to the subduction of Africa *and* Arabia previous to the opening of the Red sea (LAUGHTON 1966, 1970). At least two phases of post-Miocene movements can be recognized, one mainly thrusting and the other, still younger, with strike-slip elements (BRAUD & RICOU 1971) and vertical uplift.

Locally, ophiolitic mélange formations are still preserved along the western (Kermanshah region, see BRAUD 1970) and the eastern part of the Zagros thrust, the Neyriz area. In this latter area RICOU (1968, 1971) has revived the tectonic window hypothesis of GRAY (1949) by postulating 3 nappes which are piled on the Zagros foreland, all in front of the main Zagros thrust. He distinguishes above the upper Cretaceous foreland sediments a sedimentary nappe (Pichakun) ranging fromTrias to

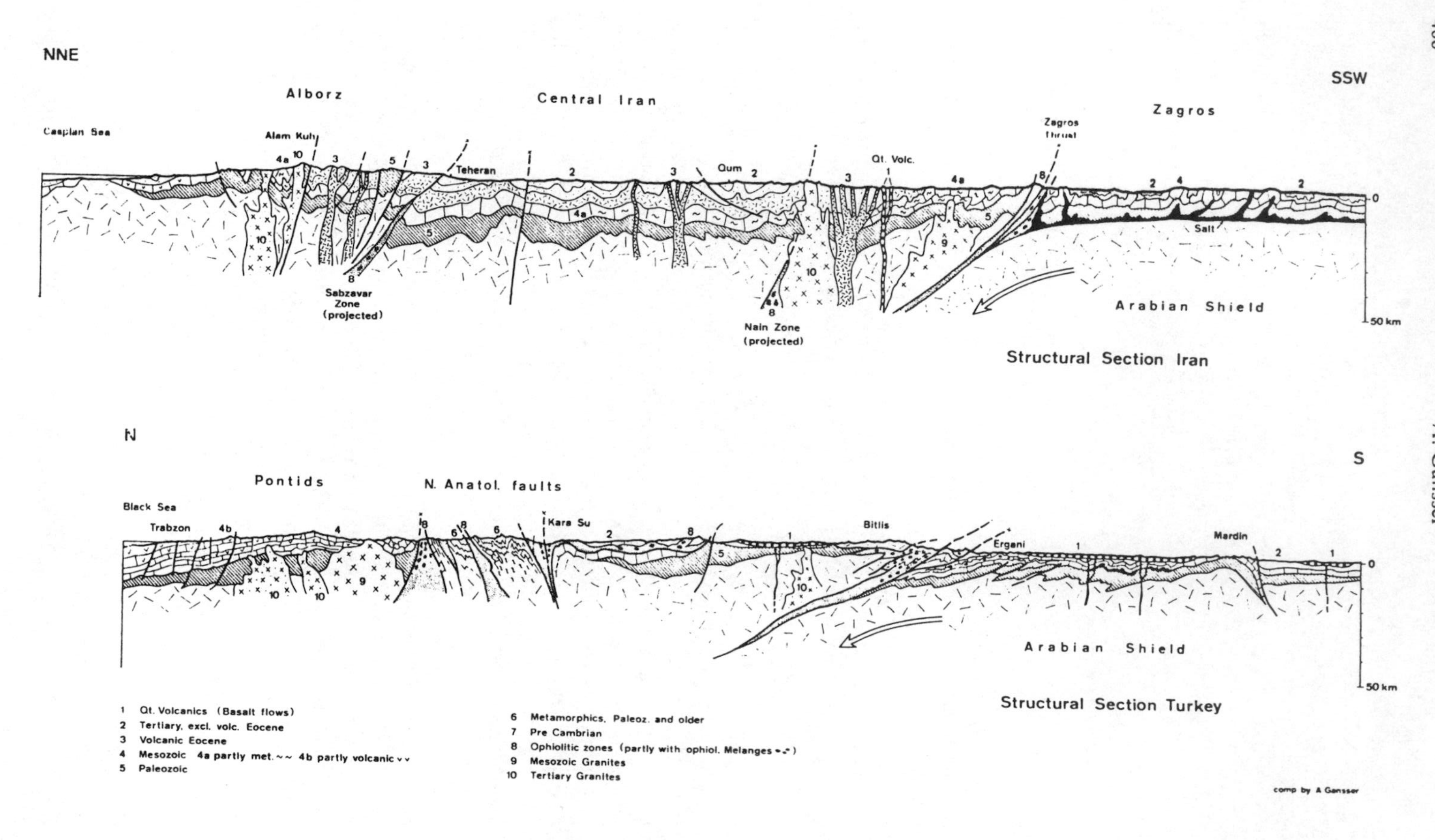

Fig. 4. Structural sections: Iran and Turkey.

middle Cretaceous, a mélange nappe with exotic blocks of metamorphic composition (marbles, amphibolites and biotite schists), Permian and Triassic limestones, radiolarites and Cretaceous pelagic limestones as well as basic lavas and serpentines, and on top a several km thick thrust-sheet of peridotites with gabbros. All nappes are transgressed by rudist limestones of Campano-Maestrichtian age, followed by Paleogene and Neogene sediments, themselves overridden by the young Zagros thrust (Fig. 5, Fig. 6). The various tectonic phases are well displayed. Here again, one may realize the effect of the Cretaceous drift of the African plus Arabian plate (pre-Maestrichtian Neyriz nappes) and the subsequent Arabian event in post-Miocene time (opening of the Red sea) which is responsible for the younger Zagros thrust. In spite of its pronounced structural complication, the Neyriz area displays a surprising similarity to Oman (a fact also stressed by RICOU 1971). Between the Cretaceous thrust sheets at Kermanshah and Neyriz, younger movements along the Central part of the Zagros thrust caused overriding of the older nappes. Similar conditions are known from the thrust-zone in Irak–Kurdistan between the Turkish and the Iranian ranges. Here amphibolite bodies together with serpentines suggest an incipient metamorphism, which also affected the upper Cretaceous flysch (GANSSER 1959). Mélange zones are not exposed and are again probably overridden or sheared off during the younger movements.

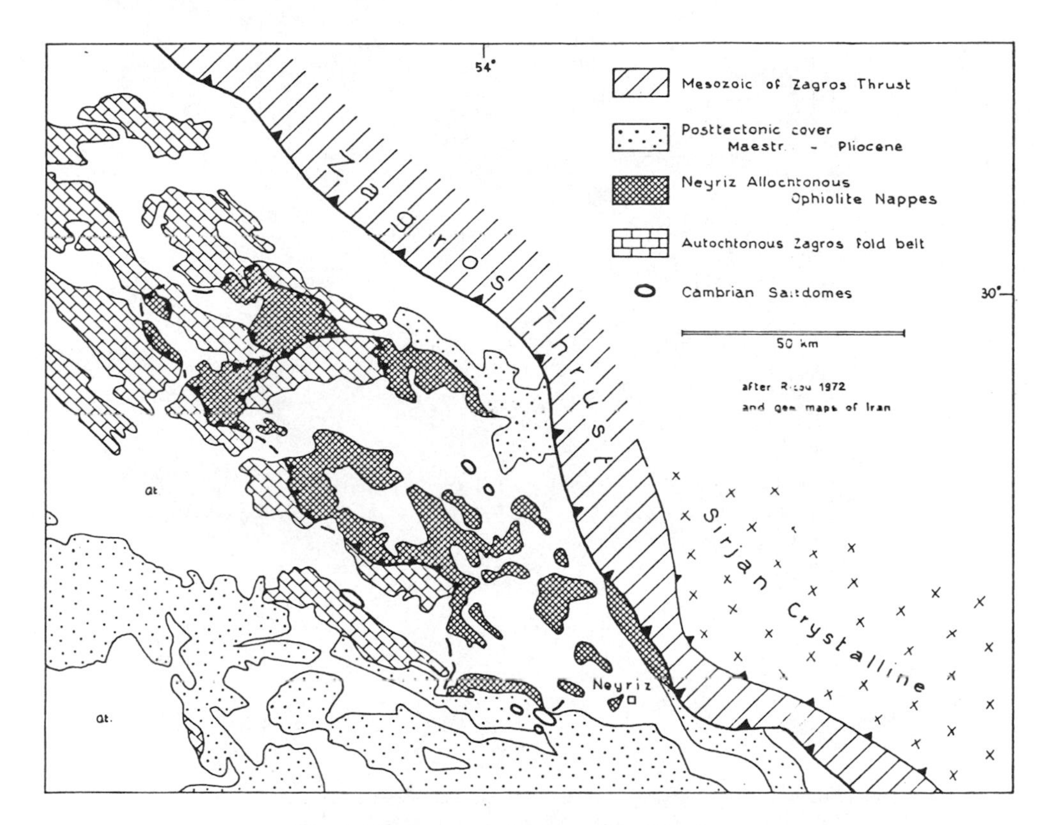

Fig. 5. Geological sketchmap of the Neyriz area.

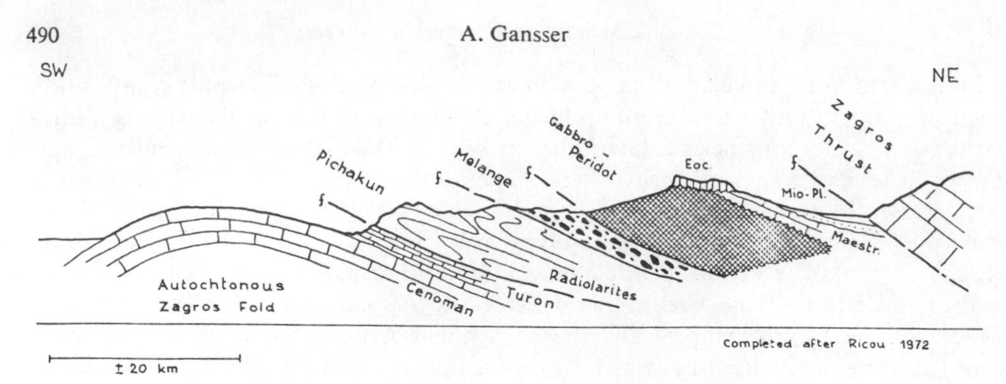

Fig. 6. Schematic section through the Neyriz nappes.

While the Zagros thrust has assimilated most of the north drift of the Arabian shield, in the wide and most complex area of Central Iran certain ophiolite mélange belts are surprisingly well preserved (Fig. 2). We find the best investigated area in the very center of Iran, in Nain (DAVOUDZADEH 1969). Here mélange belts are present

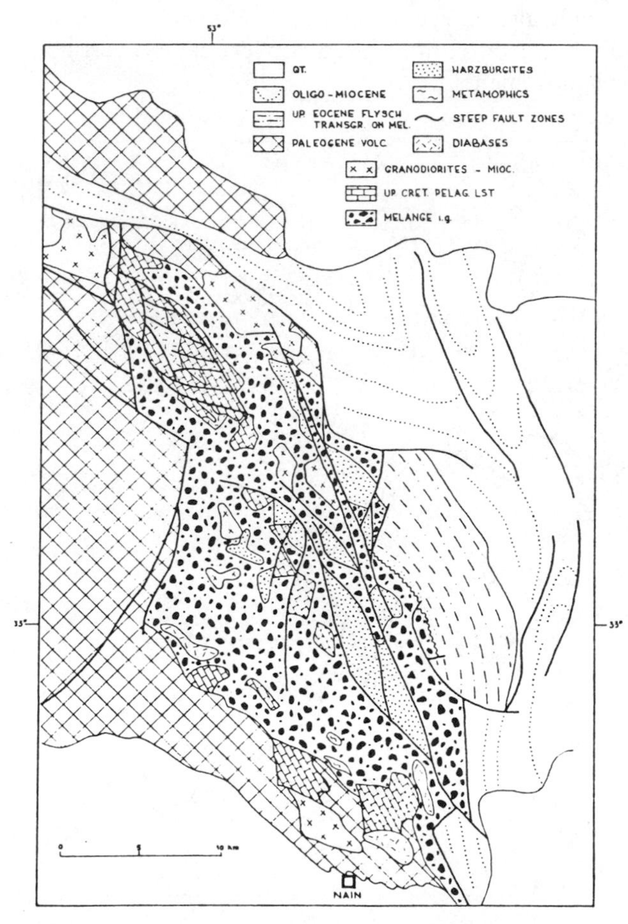

Fig. 7. The ophiolitic mélange of the Nain area, Central Iran (generalized after DAVOUZADEH 1969).

besides large homogeneous bodies of ultrabasics (Fig. 7). All contacts are steeply thrusted, but not overthrusted as they are in Neyriz, Oman or the Himalaya. Along a sharp south-east trending fault zone, reactivated in recent time (faulted alluvial fans), the Nain zone trends towards the mélange area of Neyriz already south of the Zagros thrust zone.

East of the Oman line the dominant geological feature is the *Lut block* (GANSSER 1969, STÖCKLIN et al. 1972), a classical microcontinent surrounded on its north, east and south side by more or less tectonized ophiolitic mélange zones. The northern area is at present being investigated by the University of Mashad, and mélange zones, mapped in detail, have provided a wealth of data (Davoudzadeh, personal information). Pink Globotruncana limestones and large patches of radiolarites are widely distributed, the latter apparently concentrated in the southern part of two belts (just north of the Doruneh fault zone). Rodingites in all possible stages of development are very frequent in the ultrabasic bodies (mostly harzburgites) while diabases are rather rare and top some of the higher mountains.

On the eastern edge of the Lut block ophiolitic mélanges form narrow bands which are limited by strike slip faults (FREUND 1970, GANSSER 1971, STÖCKLIN et al. 1972). They enter the northeastern part of the block as steep slices and may be associated with the large earthquakes within the northern Lut block (GANSSER 1969, AMBRASEYS & TCHALENKO 1969, NOWROOZI 1972). In spite of the strong tectonization within those slices, large masses of pillow lavas are locally preserved and show surprisingly little deformation.

The mélange zones increase towards the southeast of the Lut block within upper Cretaceous flysch. In the middle of one of those larger mélange areas erupted the still semi-active Taftan volcano (GANSSER 1971). This volcano, together with the Bazman volcanics, belongs to an east–west aligned volcanic belt separating the main Lut block from the Djaz Murian basin in the south. Along the southern border of this stable mass occur some of the largest ophiolitic belts in Iran which extend into the Makran region north of the Gulf of Oman. Unfortunately these remote areas are little known, and my information is mainly based on personal reconnaissance surveys (Fig. 8). Somewhat similar to the Nain region, though on a larger scale, massive basic (gabbro and diabase) and ultrabasic (serpentinized harzburgite) bodies are sharply separated from extensive mélange zones. The contacts are mostly steep,

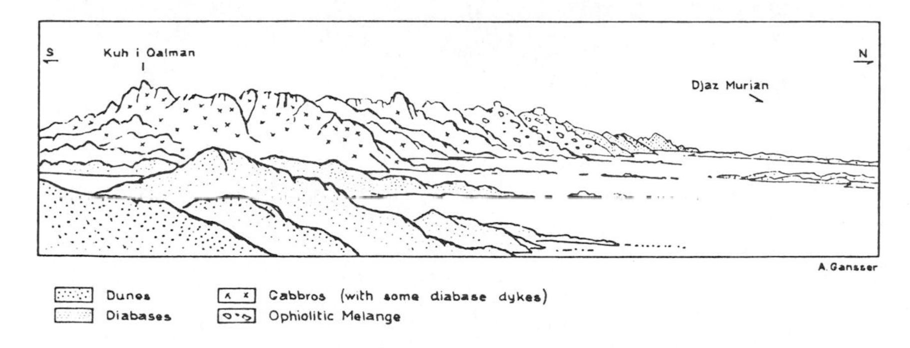

Fig. 8. The little known ophiolite belt south of Djaz Murian basin (view to west).

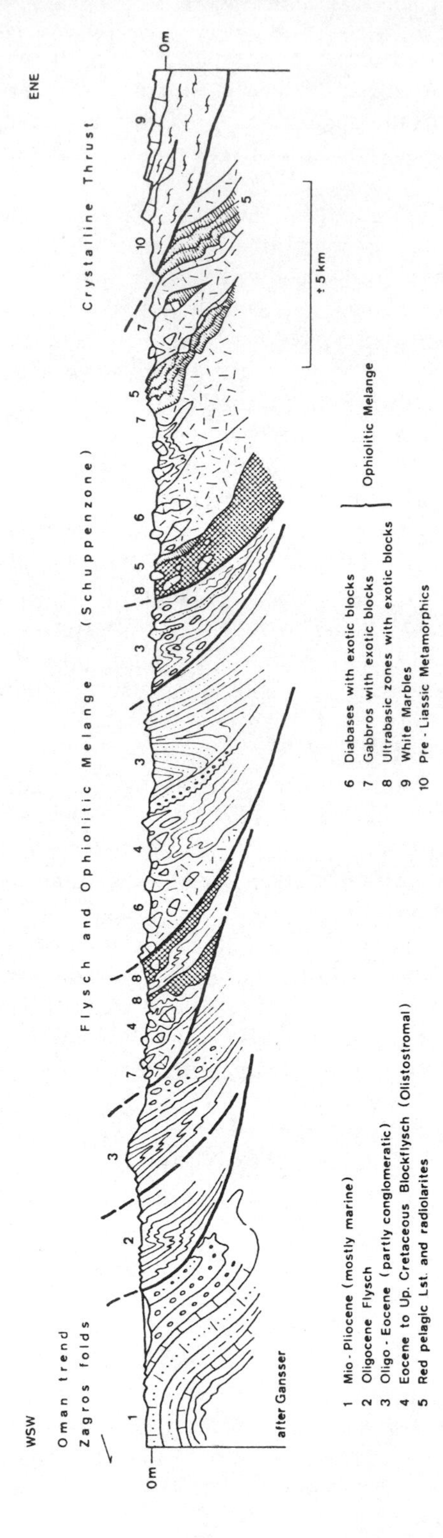

Fig. 9. Flysch/mélange section: north Minab, southeast Iran.

except for the western part, where one oberves westwards thrusting towards the eastern Zagros (Oman line) (Fig. 9).

This section is of considerable interest, since it displays a crystalline thrust sheet in the east which forms the highest unit. All tectonic contacts are steep at the surface but seem to flatten out at depth – a pattern of regional importance in this area. The crystalline thrust sheet consists of diaphthoretic chlorite-sericite schists and chloritic hornblende gneisses with intercalated small bands of gray marbles. They are topped by conspicuously white, sugary marbles, which are widespread as enclosures within the underlying mélange formations. In the Sargaz area the crystalline rocks are transgressed by flyschoid sediments which contain Liassic ammonites (unpublished report by Harrison 1936). In analogy to the Bitlis thrust sheet (Fig. 3) in southeastern Turkey, the pre-Liassic carbonates could also be Permian and not Precambrian as generally accepted. The lithological and even structural similarities are striking.

Within the mélange thrust sheets we find masses of diallage peridotites, serpentines with more or less developed ophicarbonates, diallage gabbros crisscrossed by gabbro-pegmatite dykes, various diabases and rarely pillow lavas. Apart from the ophiolitic components occur some reddish amygdaloid porphyries of a probable pre-ophiolitic age and sediments such as radiolarites, pink pelagic limestones, siliceous limestones together with white marbles, chlorite schists, epidote quartzites and hornblende-chlorite-schists of the type reminiscent of the crystalline thrust sheet. As youngest component occur *Globotruncana*-bearing pinkish argillaceous limestones.

All these blocks are again incorporated as olistostromal zones within the Eocene and lower Oligocene flysch. This fact is typical in the whole Beluchistan area and reminds again the conditions described in southeastern Turkey, only with more structural complications. In such areas the clear distinction between ophiolitic mélange formations and ophiolitic olistostromes is often difficult, and only a careful analysis of the youngest ages of matrix and blocks and of the structural relations can solve this problem. Most of these flysch sections show a highly compressed isoclinal folding. A younger east/west component with horizontal displacements is added to the structural complications. The flysch is transgressed unconformably by strongly deltaic clastic sediments which initiate the several kilometer thick Mio-Pliocene Makran group. The latter extends southeastwards to the Makran coast and beyond, and covers with its gentle folds a structurally highly complicated subsurface.

Since the Makran region is situated east of the Oman line, the effect of the underthrusting Arabian shield diminishes. In the gap between the Arabian and the Indian plate occur microcontinents such as the Lut block and the Sistan-Helmand blocks of central Afghanistan. A differential movement between these small but coherent masses seems evident and is well displayed around the Lut block with its well outlined fault zones. Along its eastern border they strongly affect the ophiolitic zones by shearing larger ultrabasic masses from the more mobile ophiolitic mélange formations. We again find similar movements in the major fault zones limiting central Afghanistan such as the Herat fault in the north and the Chaman fault in the southeast. Both these fault zones merge towards the east-northeast into the Kabul area and continue along the Panj Shir valley into the Hindukush. Between this triangular fault system all other structures are more and more compressed towards northeast, a narrowing effect, related to the stronger northwards drift of the northwestern Indian plate, as compared to the Arabian shield. The differential movement between the Arabian and Indian plates is mostly taken up by the large Chaman fault which separates two surprisingly different areas. While west of the fault zone the basement is covered by a complete Paleozoic and Mesozoic conformal section with some granitic intrusions, the eastern side displays large sedimentary gaps, a marked ophiolitic belt and a flysch facies of the uppermost Cretaceous and the Paleogene (DE LAPPARENT 1972). It is along this ophiolitic belt, south and southeast of Kabul, that well developed ophiolitic mélange zones are exposed, in sharp contact with the ultrabasic masses of the Logan Valley and along the Kabul river. They are transgressed by the basal conglomerates of the Eocene flysch, which locally contain ophiolitic olistostromes. These mélange formations along the Chaman fault and its secondary branches, well displayed in the Altimur region (MENNESSIER 1972), stress the importance of this structural trend and underline the striking facies changes related to it (Fig. 10).

The Quarternary volcanism, widespread in the west, disappears northeastwards, and its last occurrence was only recently discovered in the Dasht i Nawar region of central Afghanistan by BORDET (1972). No Quarternary or Recent volcanism is known within the orogenic belt around the northwards drifting Indian plate.

d) The Himalayan mélanges

The Chaman fault trend as well as the more eastern Quetta line, which separates the Indian platform sediments from the ophiolitic Beluchistan trend (GANSSER 1966), can be followed through the western Himalayan syntaxis into the Indus zone of the northern Himalayas, where the relation of ophiolitic mélange nappes to one of the best preserved plate boundaries is excellently exposed. Unfortunately this area is still very little known. We have now returned to our starting point, where 82 years ago the famous trio of geologists of the Indian Geological Survey were staring at a then not yet recognized ophiolitic mélange front.

Along the Indus zone, north of the Himalaya main range, we know so far three areas where the ophiolitic belt, which outlines the important suture line (plate boundary), has been studied. From west to east we distinguish the wider Ladak zone with the Dras volcanic belt (DE TERRA 1935, WADIA 1937), the ophiolites of the Rupshu area (BERTHELSEN 1951, 1953) and the Kiogar Kailas area (VON KRAFFT 1902, HEIM & GANSSER 1939, GANSSER 1964). It is in the latter region where we find, apart from the steep "root zone", a large extension of ophiolitic mélange nappes with a visible thrusting (or gliding) of over 80 km (Fig. 11). As far as we know this remote but highly interesting area has not been reinvestigated lately – at least no newly published information is available. However, reappraisals of previous findings in the light of modern global tectonics, paleomagnetic informations and comparisons with the well outlined ophiolitic mélange zones of the Middle East stress the intimate relations of the Himalayan mélanges to the zone of collision and subsequent obduction along the Indian and Eurasian (Tibetan) plate boundary (GANSSER 1966, 1973, DEWEY & BIRD 1970, CRAWFORD 1972, POWELL & CONAGHAN 1973, ATHAVALE 1973, CHANG CHEN-FA & ZENG SHI-LANG 1973). The divergence of opinion in the modern publications on this subject is understandable, since only a very few of these authors have seen the facts in the field.

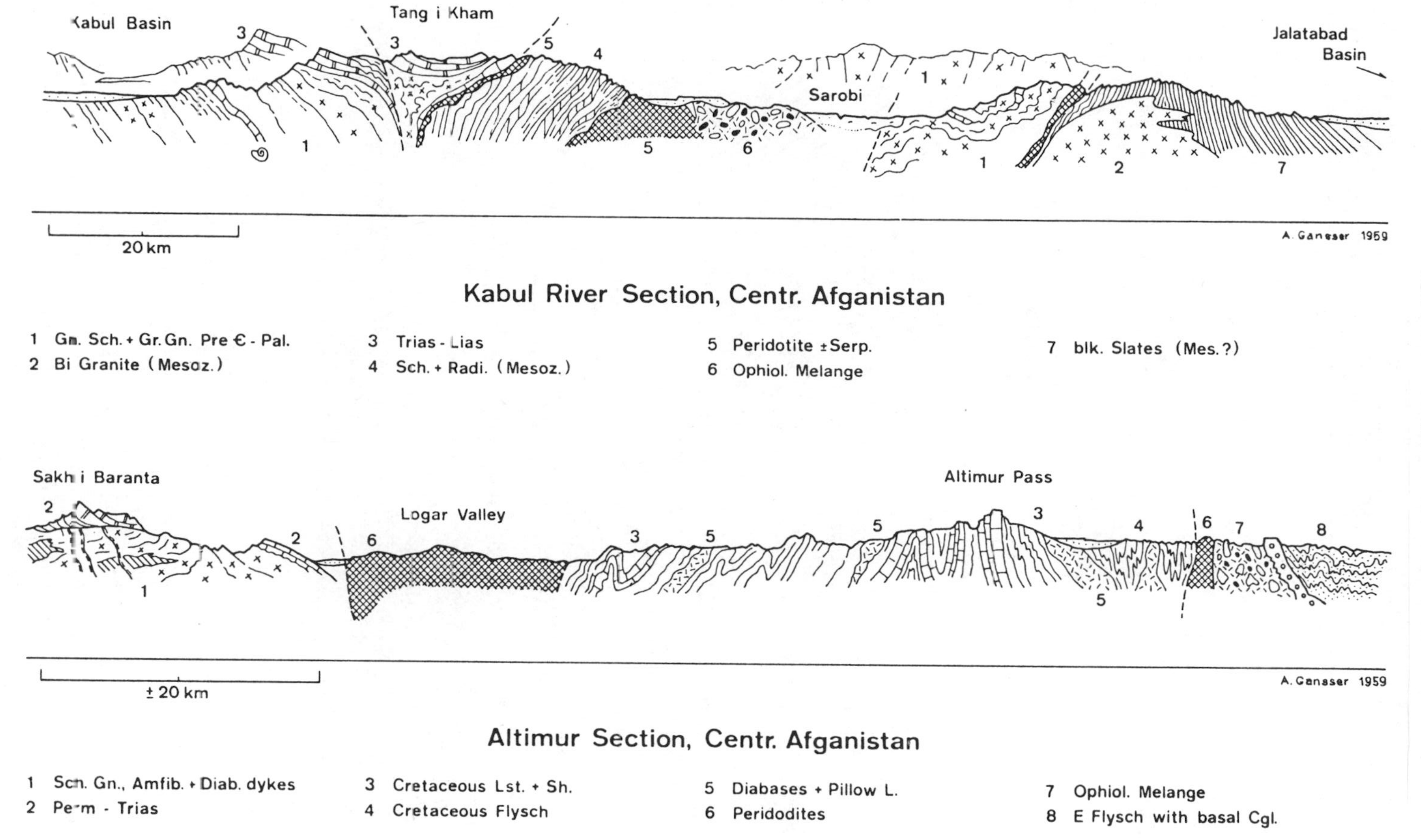

Fig. 10. Kabul River section and Altimur section (Central Afghanistan).

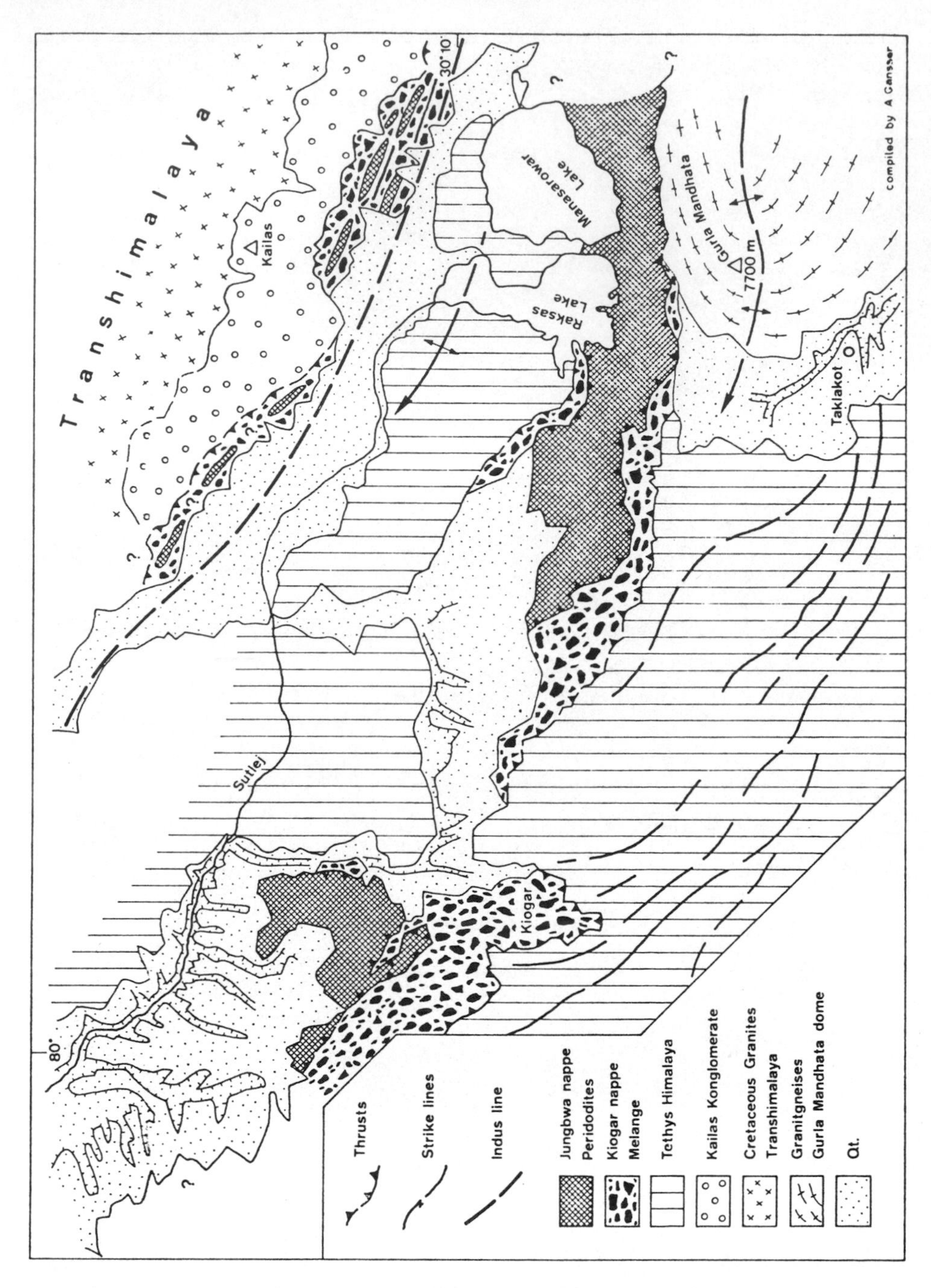

Fig. 11. The ophiolitic nappes in the Northern Central Himalaya (scale appr. 1 : 1 200 000).

Between the Indian/Tibetan border and the Transhimalaya south of the holy Kailas mountain we can recognize three major structural units: 1. *The Tethys Himalayan fold belt*, which in respect to the overlying ophiolitic nappes forms a neo-autochthonous base. 2. An *ophiolitic mélange nappe*, which, following VON KRAFFT (1902), will be locally called the Kiogar nappe. 3. An *ultrabasic nappe*, which forms the highest structural element of the Himalaya, previously referred to as the Jungbwa unit, more recently as the Jungbwa nappe (GANSSER 1964). Each unit is limited by major thrusts, which again can be complicated by subsequent imbrications (Fig. 12).

The sediments of the *Tethys Himalaya* display a Jura type folding and include the middle Cretaceous Jumal sandstones. Following above the pelagic and ammonite-rich Spiti shales, the Jumal sandstones indicate the beginning of the first orogenetic Himalayan phase. The disturbance of the long conformable history of Tethys sedimentation, spanning the time from Eocambrian to the middle Cretaceous, is reflected by the incoming of detrital sediments which lead upwards into a flyschoid facies. These sediments were folded and subsequently sharply overthrust by the Kiogar ophiolitic mélange nappe, which cuts the already existing folds and postdates the Tethys folding phase.

The Kiogar ophiolitic mélange nappe exposes one of the best mélange sequences on our globe, well described on its southern margin by VON KRAFFT already in 1902 (Fig. 13). We also find some of the largest single units known, such as the Triassic Kiogar limestone covering as a single "block" an area of 20 km^2 with a thickness of up to half a km. For a detailed account of this area I refer to HEIM & GANSSER (1939, with a new edition 1974). Of special interest is the fact that the numerous sedimentary blocks display a facies unknown otherwise in normal outcrops in the Himalaya, and they witness a sedimentary basin of oceanic dimensions which has completely disappeared during the collision between the Indian and Tibetan plates. This is particularly true for the Permian, Triassic, Liassic and upper Jurassic blocks, while radiolarites and flysch blocks are not unlike some of the uppermost Cretaceous of the normal Tethyan sediments. Flysch-type sediments as well as highly sheared basic and ultrabasic rocks form the matrix of the mélange.

The most frequent sedimentary inclusions are Permian crinoidal limestones, which contain brachiopods, and various types of Triassic rocks. Amongst the latter huge algal limestone blocks and slabs contrast with pelagic limestones of Hallstätter type. VON KRAFFT discovered a Carnian limestone block from which 40 species of ammonites have been described (DIENER 1912). Curiously enough some of these Triassic blocks are embedded in red and greenish, more or less amygdaloid porphyrites, some even with free quartz, which are strictly related to the Trias and display sometimes intrusive contacts. They can also occur as single blocks within the flysch. It seems most likely that these volcanics are only slightly younger than the Triassic limestones and that they have no relation to the ophiolitic suite. We observe here a certain similarity with the porphyritic rocks related to Triassic limestones in the Mammonia region of southern Cyprus (see p. 486). Unknown in the "normal" Himalayan sediments are blocks of Liassic limestone containing ammonites and resembling the Adneth facies of Austria. Again of a most peculiar and otherwise unknown facies are oolitic limestones imbricated with a Kiogar block containing Calpionellas. This rather unexpected association is of particular interest, apart from the fact that Calpionellas seem to be unknown in the eastern Tethys. The Kiogar Calpionellas may actually represent the easternmost occurrence within the Tethys belt (personal information from Michel Durand-Delga).

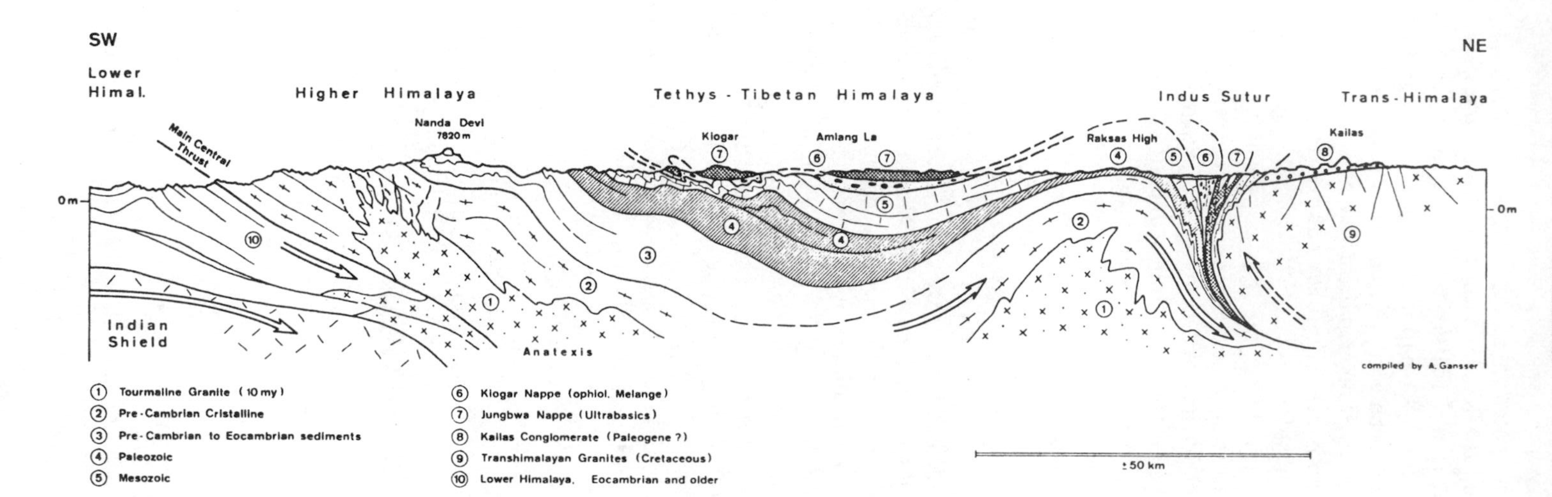

Fig. 12. Structural section: Northern Central Himalaya.

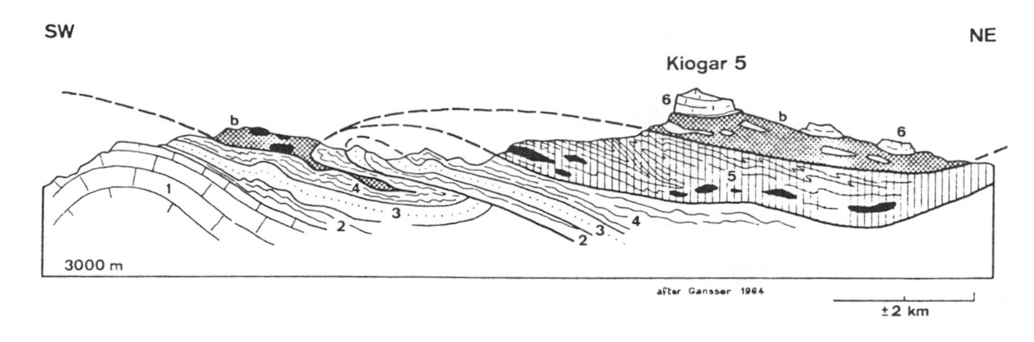

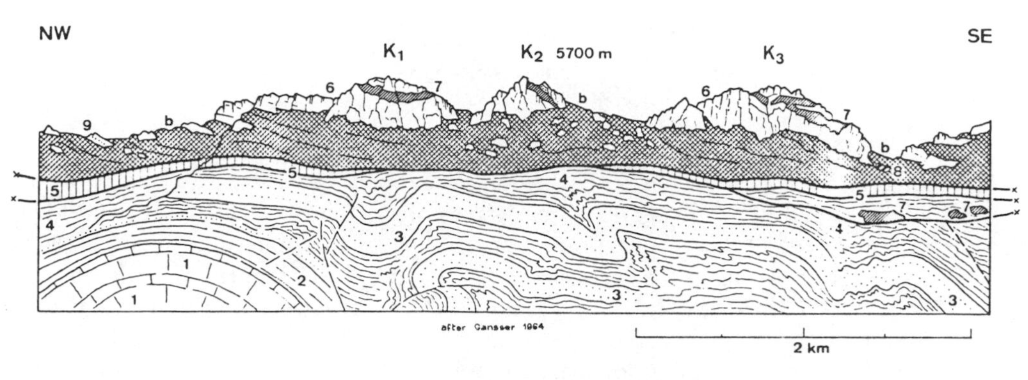

1 Kioto limestones (U. Triassic)
2 Spiti shales
3 Giumal sandstones
4 silty red and green shales covered by dark grey fucoid shales (Flysch)
5 siliceous sandstones with radiolarians (U. Cretac)
6 white Kiogar limestones (Trias ?)
7 oolitic limestones and calc shists with *Calpionella* (Up. Jur.)
8 Radiolarites
9 Tuffites
b basic and ultrabasic rocks with various exotic blocks from Permian to Cretaceous age

Fig. 13. The ophiolitic Kiogar thrustmass with exotic blocks, Northern Central Himalaya.

Of the younger sediments, radiolarites, siliceous shales and limestones and glauconitic sandstones have been noted. Apart from sediments the mélange exposes various types of basic and ultrabasic rocks within the flysch or replacing the latter as matrix but again appearing in some areas as blocks. Peridotites, serpentines, often with borders of ophicarbonate and diabase, are a frequent association. To complicate this geological jumble even more, we observe within some flysch horizons of the Amlang La region in southern Tibet layered intrusions of fine to medium grained augite monzonites with pigeonitic augite, alkalifeldspar, andesines and biotite. These curious rocks show thermal contacts with flysch limestones and form a post-flysch intrusion, probably of post-Cretaceous age. It may be younger and not related to the ophiolitic association.

The Kiogar ophiolitic mélange nappe is again overthrust by the highest structural unit of the whole Himalaya, the ultrabasic *Jungbwa nappe* (Fig. 14). There seems little

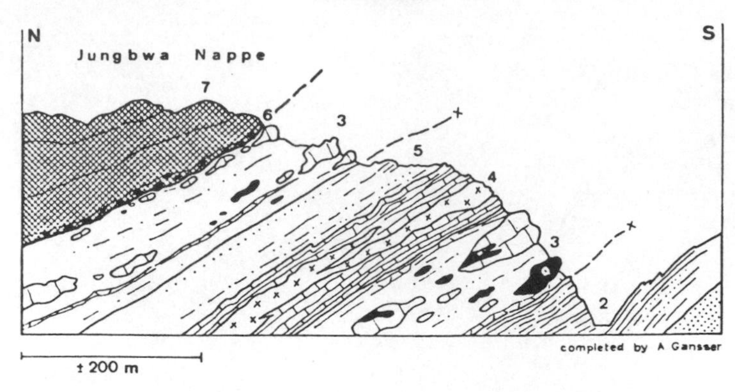

Fig. 14. Schematic profile: Southend of the ultrabasic Jungbwa nappe, Northern Central Himalaya.

doubt that here we witness the obduction and subsequent southward thrusting of a large mantle slab, squeezed out from the collision zone along the Indus line between the Indian and Tibetan plate. The visible extension of this sheet is about 3500 km^2 with a minimum thickness of 500 m, the upper part of this horizontal sheet having been eroded. Surprising is the widespread occurrence of harzburgites with only a local serpentinization in the form of antigorite. There is hardly any doubt that the composition of this basic sheet indicates an original high temperature and rather high pressure zone (under mantle conditions) and a much later "mise en place" in various not yet well recognized phases. The olivines (forsterite) are coarse-grained, with conspicuous kink bands, often parallel to the extinction. Less frequent are enstatite, approaching hypersthene and showing a strong lamellar segregation of monoclinic augites. The latter can also occur as single grains. Fe-Mg spinell as well as Cr spinell are locally enriched. In the Rupchu area to the northwest, BERTHELSEN (1953) reported from the same belt larger masses of chromite.

The base of the Jungbwa nappe consists of a highly sheared zone of ophicarbonates which even contain larger fragments of original peridotite. Metasomatic alterations against the flysch–mélange zone have partly obscured the original contacts.

Following the Kiogar and Jungbwa nappes northwards, one reaches the actual *Indus line*, where the same units can be recognized in a vertical position, but where everything has been highly squeezed and boudinaged, and where the ultrabasics occur as lenticular, serpentinized masses within mélange formations and slightly metamorphosed flysch-type deposits. Towards the Cretaceous biotite-hornblende granites of the Transhimalaya, this imbricated "root zone" is thrust northwards, partly on top of the Tertiary Kailas conglomerates which normally transgress the granites and are undisturbed apart from a local backfolding at the thrust contact (Fig. 12).

The relation of the Indus zone or line to the collision suture between the Indian and Tibetan plates is rather obvious, and the ophiolitic zone with more or less developed mélange horizons can be followed intermittently from the western Himalayan syntaxes over 900 km to the area under discussion in the northern Central Himalayas. One of the biggest open questions is the continuation of this major tectonic element to the east. That it continues with the same trend is indicated by the east–west striking topographical features, underlined by the suspiciously straight Tsangpo river system. A few observations corroborate this assumption. HAYDEN (1907) reported from the area south of Shigatse (southern Tibet) numerous basic intrusions in the form of diabases, noritic gabbros and serpentines within metamorphic Jurassic (Cretaceous) calcareous schists. He followed these basic rocks to the large Yamdrok Tso (lake) in the east. WIENERT (1947), in a little known publication, refers to his magnetic measurements during the botanical Schaefer expedition between Sikkim and Lhasa in 1938/39. Most surprising are the sudden large magnetic anomalies along the Tsangpo river ($+2600\ \gamma$, while just south of it, at Tsetang, the measurements drop to $-870\ \gamma$). Recently an ultrabasic belt along and paralleling the Tsangpo river was mentioned by Chinese geologists (CHANG CHEN-FA & ZENG SHI-LANG 1973). This information further supports an eastern extension of the ophiolitic belt of the northern Himalaya, the structural aspect of which is well exposed in the ERTS pictures covering this area. It may not take long before ophiolitic mélange formations are recognized also in this remote area, further corroborating the important suture zone related to the collision of the Indian and Eurasian plates.

4. Conclusions

The intimate relations of the ophiolitic mélange to the ophiolitic belts of world-wide distribution show us that most problems adherent to the ophiolites exist for the mélange formations. I stress the recognition of ophiolite belts as originally representing oceanic crust and mantel, a fact of prime importance when interpreting ophiolites in highly tectonized mountain ranges with nappe character (Alps, Himalaya) (DEWEY & BIRD 1970, 1971; MOORES 1970, 1973; GILLULY 1971; ROST 1971; VINE & MOORES 1972; ABBATE et al. 1972; GANSSER 1973; Academy of Science, Moscow 1973, I–III). The consumption of oceanic lithosphere in subduction zones along major mountain ranges is by now well known. While subduction is generally accepted, obduction (AVIAS 1967, COLEMAN 1971, MOORES 1973) brings us the same difficulty as the formation of ultrabasic nappes. Evidently most of the ophiolitic belts are now visible, because they are obducted and not subducted. This statement is particularly applicable to mélange formations, and we may go so far to state that an ophiolitic mélange may be explained through a mechanism related to obduction. A Cordilleran type orogenetic belt with the classical subduction of oceanic crust below a continental margin hardly produces mélange formations. They as well as the ophiolites are missing along 5000 km of Andean subduction (GANSSER 1972). Mélanges are weakly expressed along the northern Andes, but again developed under peculiar conditions along the central west coast of North America (HSÜ 1968, 1971). Substantially different conditions occur at the subduction of continental plates. Obduction of oceanic crust is here the rule, developed into conspicuous thrust-masses of ophiolitic rocks onto slabs of continental crust (COLEMAN 1971). Under these conditions mélange formations are particularly wide spread, as we have seen in many Asian ophiolitic belts, since our best examples were from the Middle East ranges and the Himalaya, and recently ophiolitic mélange formations have been recognized in the Paleozoic Tien-Shan and the Mesozoic of the Lesser Caucasus (USSR Academy of Science's publication 1973). Further obducted ophiolites, where mélange formations may have been obliterated by thrusting, are reported from Papua (DAVIES 1968, DAVIES & SMITH 1971), though the mode of occurrence has recently been disputed (ROD 1974).

The subduction mechanism of continental plates (shield masses) and the obduction of ozeanic crust belonging to basins obliterated during a collision phase is still highly disputed. Flipping of a subduction zone is hardly the answer. The acceptance of the blue schist facies as valid indicator of underthrusting is widely generalized, but to exhume the blue schists from great depth is still a major tectonical problem (COLEMAN 1972). Blue schists mostly occur in mélange formations as exotic blocks, but they are rather atypical for such formations. It is not the place to enter into these controverses, but we may stress the fact that obducted mélange formations are mostly related to underthrust continental slabs which do not sink to greater depth but flatten out at approximately 40–60 km. Earthquake centers seem to confirm this in many places (COLEMAN 1971; NOWROOZI 1971, 1972; MOLNAR et al. 1973). This is particularly well documented in the subduction of the Indian plate and its thrust below the Tibetan mass. Earthquakes are widely spaced below the Tibetan plate at depths not exceeding 60 km. Below the Himalaya the calculated focal mechanism is consistant with thrusting along a plane dipping gently to the north (FITCH 1970).

The subduction of the Indian plate, now reflected in the Indus suture line, produced one of the best ophiolitic mélange formations known on our globe. We can only hope that this challenging area will receive in the future the intense geological investigation it deserves, after the first steps in this direction had been taken by VON KRAFFT 80 years ago.

Acknowledgments

I thank Alan G. Milnes for valuable suggestions and corrections, Sylvia Spalinger for the preparation of the manuscript and Albert Uhr for completing the drawings.

REFERENCES

ABBATE, E., BORTOLOTTI, V., & PASSERINI, P. (1972): *Studies on mafic and ultramafic rocks. 2: Paleogeographic and tectonic considerations on the ultramafic belts in the Mediterranean area.* Boll. Soc. geol. ital. *91*, 239–282.

– (1973): *Major structural events related to ophiolites of the Tethys belt.* International Symposium on Ophiolites in the Earth's Crust, Moscow 1973. Centro Studi Geologia Appennino in rapporto alle Geosinclinali Mediterranee. Cons. naz. Ric. *31*, 1–43 (preprint).

ABDEL-GAWAD, M. (1971): *Wrench movements in the Baluchistan Arc and relation to Himalayan–Indian Ocean tectonics.* Bull. geol. Soc. Amer. *82*, 1235–1250.

Academia Sinica, Laboratory of Isotopic Geology, Kweiyang Institute of Geochemistry (1973): *Isotopic dating of the metamorphic rocks from the Mt. Jolmo Lumgma Area, China.* Scientia sinica *16*/3, 385–395.

Academy of Sciences of the USSR, Geological Institute (1973): *Developmental stages of Folded Belts and the problem of Ophiolites* (Part I–III). Moscow.

ALLEMANN, F., & PETERS, T. (1972): *The Ophiolite–Radiolarite Belt of the North-Oman Mountains.* Eclogae geol. Helv. *65*/3, 657–697.

AMBRASEYS, N. N. (1970): *Some characteristic features of the Anatolian fault zone.* Tectonophysics *9*, 143–165.

AMBRASEYS, N., & TCHALENKO, J. (1969): *The Dasht-e-Bajaz, Iran, earthquake of August 1968.* Bull. seismol. Soc. Amer. *59*, 1751–1792.

– (1972): *Seismotectonic Aspects of the Gediz, Turkey, Earthquake of March 1970.* Geophys. J. r. astron. Soc. *30*, 229–252.

ATHAVALE, R. N. (1973): *Inferences from recent Indian palaeomagnetic results about the northern margin of the Indian Plate and the tectonic evolution of the Himalayas.* Nat. geophys. Res. Inst. Hyderabad, India (preprint).

ATHAVALE, R. N., HANSRAJ, A., & VERMA, R. K. (1972): *Palaeomagnetism and age of Bhander and Rewa sandstones from India.* Geophys. J. r. astron. Soc. *28*, 499–509.

AVIAS, J. V. (1967): *Overthrust structure of the main ultrabasic New Caledonian massives.* Tectonophysics *4*, 531–541.

BAILEY, E. B. (1944): *Mountains that have travelled over volcanoes.* Nature (London) *154*, 752–756.

BAILEY, E. B., & MCCALLIEN, W. J. (1950): *The Ankara Mélange and the Anatolian Thrust.* Nature (London) *166*, 938–943, and Min. Res. Explor. Inst. Turkey *40*, 17–22.

– (1953): *Serpentine lavas, the Ankara Mélange and the Anatolian Thrust.* Trans. r. Soc. Edinburgh *62/2*, 403–442.

BERTHELSEN, A. (1951): *A geological section through the Himalayas. A preliminary report.* Medd. dansk geol. Foren. *12*, 102–104.

– (1953): *On the Geology of the Rupshu District N.W. Himalaya.* Medd. dansk geol. Foren. *12/3*, 350–414.

BIQ CHINGCHANG (1971): *Comparison of Mélange tectonics in Taiwan and in some other mountain belts.* Petroleum Geol. Taiwan *9*, 79–106.

BIRD, J. M., & DEWEY, J. F. (1970): *Lithospheric Plate-Continental margin tectonics and the evolution of Appalachian orogeny.* Bull. geol. Soc. Amer. *81*, 1031–1060.

BLUMENTHAL, M. (1948): *Un aperçu de la géologie des chaînes nordanatoliennes entre l'Ova de Bolu et le Kizilermak inferieur.* Publ. Inst. Etud. Rech. min. Turquie (B) *13*.

BOGDANOFF, A. A. (1967): *Colloque sur la Tectonique de la Zone Alpine de l'Iran et de la Turquie.* Congr. géol. int., Comm. Carte géol. Monde, Moscow, p. 1–20.

BORDET, P. (1972): *Le volcanism récent du Dacht-e Nawar.* Rev. Géogr. phys. Géol. dyn. *14/4*, 427–431.

BRAUD, J. (1970): *Géologie générale. – Les formations du Zagros dans la région de Kermanshah (Iran) et leurs rapports structuraux.* C.R. Acad. Sci. Paris (D) *271*, 1241–1244.

BRAUD, J., & BRUNN, J. H. (1972): *Présence de quartz-kératophyres et de brèches éruptives dans les roches vertes des unités almopiennes occidentales du Vermion (Grèce).* C.R. Soc. géol. France *10*, 29–30.

BRAUD, J., & RICOU, L.-E. (1971): *Tectonique. – L'accident du Zagros ou Main Thrust, un charriage et un coulissement.* C.R. Acad. Sci. Paris (D) *272*, 203–206.

BRUNN, J. H. (1959): *La dorsale médio-atlantique et les épanchements ophiolithiques.* C.R. Soc. géol. France *8*, 234–235.

– (1973): *Contribution à la discussion sur la mise en place – magmatique ou tectonique – des ophiolites.* Symposium sur les ophiolites dans la croute terrestre. Moscow, May–June 1973 (preprint).

CHANG CHENG-FA & ZENG SHI-LANG (1973): *Tectonic features of the Mount Jolmo Lungma region in southern Tibet, China.* Scientia geol. sinica *1*, 1–12.

COLEMAN, R. G. (1963): *Serpentinites, rodingites and tectonic inclusions in Alpine-type mountain chains.* Spec. Pap. geol. Soc. Amer. *73*.

– (1971): *Plate tectonic emplacement of upper mantle peridotites along continental edges.* J. geophys. Res. *76/5*, 1212–1222.

– (1972): *Blueschist metamorphism and plate tectonics.* 24th int. geol. Congr., Sect. 2: Petrology, p. 19–26.

– (1973): *Ophiolites in the earth's crust: A symposium, field excursions, and cultural exchange in the USSR.* Spec. Rep. Geology, geol. Soc. Amer., p. 51–54.

CRAWFORD, A. R. (1972): *Iran, continental drift and plate tectonics.* 24th int. geol. Congr., Sect. 3, p. 106–112.

– (in press): *The Indus suture line, the Himalaya, Tibet and Gondwanaland.*

DASARATHI, N. (1967): *Tectonic-setting of Indus ophiolites.* Upper Mantle Symposium, New Delhi 1964, p. 33–43 (Berlingske Bogtrykkeri, Copenhagen).

Davies, H. L. (1968): *Papuan ultramafic belt.* 23rd int. geol. Congr. Prague *1*, 209–220.

Davies, H. L., & Smith, I. E. (1971): *Geology of Eastern Papua.* Bull. geol. Soc. Amer. *82*, 3299–3312.

Davoudzadeh, M. (1969): *Geologie und Petrographie des Gebietes nördlich von Nain, Zentral-Iran.* Mitt. geol. Inst. ETH u. Univ. Zürich.

Dewey, J. F., & Bird, J. M. (1970): *Mountain belts and the new Global tectonics.* J. geophys. Res. *75*/14, 2625–2647.

– (1971): *Origin and emplacement of the Ophiolite suite: Appalachian ophiolites in Newfoundland.* J. geophys. Res. *76*/14, 3179–3206.

Diener, C. (1895): *Ergebnisse einer geologischen Expedition in den Central-Himalaya von Johar, Hundes, und Painkhanda (4. April 1895).* Denkschr. kais. Akad. Wiss. Wien, math-natw. Classe *62*, 533–607.

– (1912): *Trias of the Himalaya.* Mem. geol. Surv. India *36*/3, 1–159.

Dietz, R. S., & Holden, J. C. (1970): *Reconstruction of Pangaea: Breakup and dispersion of Continents, Permian to present.* J. geophys. Res. *75*, 4939–4956.

Douglas, J. A. (1950): *The Carboniferous and Permian faunas of south Iran and Iranian Baluchistan.* Mem. geol. Surv. India, Palaeontologia indica [n.s.] *22*, 7.

Duclos, C. (1964): *Notes of the geology of the Kyrenia range.* Annu. Rep. geol. Surv. Cyprus *1963/64*, 57–67.

Elter, P., & Trevisan, L. (1973): *Olistostromes in the tectonic evolution of the Northern Apennines.* In: *Gravity and Tectonics* (John Wiley & Sons, New York), p. 175–188.

Falcon, N. L. (1967): *The geology of the north-east margin of the Arabian basement shield.* Adv. Sci. (London) *24*/119, 31–42.

– (1969): *Problems of the relationship between surface structure and deep displacements illustrated by the Zagros Range. Time and Place in orogeny.* Spec. Publ. geol. Soc. London *3*, 9–22.

Fitch, T. J. (1970): *Earthquake mechanisms in the Himalayan, Burmese and Andaman regions and continental tectonics in Central Asia.* J. geophys. Res. *75*, 2699–2709.

Freund, R. (1970): *Rotation of strike slip faults in Sistan, Southeast Iran.* J. Geol. *78*/2, 188–200.

Freund, R., Garfunkel, Z., Zak, I., Goldberg, M., Weissbrod, T., & Derin, B. (1970): *The shear along the Dead Sea rift.* Phil. Trans. r. Soc. London (A) *267*, 107–130.

Freund, R., Zak, I., & Garfunkel, Z. (1968): *Age and rate of the sinistral movement along the Dead Sea rift.* Nature (London) *220*/5164, 253–255.

Furon, R. (1941): *Géologie du Plateau Iranien.* Mém. Mus. Hist. nat. Paris *7*/2,

Gansser, A. (1955): *New aspects of the geology in Central Iran.* Proc. 4th World Petroleum Congr. Rome, Sect. I/A/5, p. 279–300.

– (1959): *Ausseralpine Ophiolithprobleme.* Eclogae geol. Helv. *52*/2, 659–680.

– (1964): *Geology of the Himalayas* (Interscience Publ. John Wiley & Sons Ltd. London.)

– (1966): *The Indian Ocean and the Himalayas. A geological interpretation.* Eclogae geol. Helv. *59*/2, 831–848.

– (1969): *The large earthquakes of Iran and their geological frame.* Eclogae geol. Helv. *62*/2, 443–466.

– (1971): *The Taftan volcano (SE Iran).* Eclogae geol. Helv. *64*/2, 319–334.

– (1973a): *Facts and theories on the Andes.* J. geol. Soc. London *129*/2, 93–131.

– (1973b): *Orogene Entwicklung in den Anden, im Himalaya und den Alpen, ein Vergleich.* Eclogae geol. Helv. *66*/1, 23–40.

– (in press): *The Himalayan Tethys.*

Gilluly, J. (1971): *Plate tectonics and magmatic evolution.* Bull. geol. Soc. Amer. *82*/9, 2383–2396.

Glennie, K. W., Bœuf, M. G. A., Hughes Clarke, M. W., Moody-Stuart, M., Pilaar, W. F. H., & Reinhardt, B. M. (1973): *Late Cretaceous Nappes in Oman Mountains and their geologic evolution.* Bull. amer. Assoc. Petroleum Geol. *57*/1, 5–27.

Glennie, K. W., & Hughes Clarke, M. W. (1973): *Late Cretaceous Nappes in Oman Mountains and their geologic evolution: Reply.* Bull. amer. Assoc. Petroleum Geol. *57*/11, 2287–2290.

Graciansky, P. C. (in press): *Le problème des coloured mélanges à propos de formations chaotiques associées aux ophiolites de Lycie occidentale (Turquie).*

Gray, K. W. (1949): *A tectonic window in south-western Iran.* Quart. J. geol. Soc. London *105*(II)/418, 189–223.

Greenly, E. (1919): *The geology of Anglesey.* Mem. geol. Surv. G.B.

GRIESBACH, C. L. (1891): *Geology of the Central Himalayas.* Mem. geol. Surv. India *23.*

HAMILTON, W. (1969): *Mesozoic California and the underflow of Pacific mantle.* Bull. geol. Soc. Amer. *80,* 2409.

HAYDEN, H. H. (1907): *The Geology of the Provences of Tsang and Ü in Central Tibet.* Mem. geol. Surv. India *36*/2, 122–202.

HEIM, A., & GANSSER, A. (1939): *Central Himalaya. Geological Observations of the Swiss Expedition 1936.* Denkschr. schweiz. natf. Ges. *73*/1 [new edition: Hindustan Publishing Corp. (India), Delhi 1974].

HINZ, K., & CLOSS, H. (1969): *Auswertung und Ergebnisse der seismischen Messungen und ihre geologische Deutung. Ergebnisse seismischer Untersuchungen im nördlichen Arabischen Meer, Teil III.* Meteor Forschungsr. C/2, 9–28.

HSÜ, K. J. (1968): *Principles of mélanges and their bearing on the Franciscan-Knoxville paradox.* Bull. geol. Soc. Amer. *79,* 1063–1074.

– (1971): *Franciscan mélanges as a model for eugeosynclinal sedimentation and underthrusting tectonics.* J. geophys. Res. *76*/5, 1162–1170.

– (1972): *The concept of the geosyncline, yesterday and today.* Trans. Leicester lit. phil. Soc. *66,* 26–48.

Italconsult (1959): *Preliminary report geo-mining survey. Plan organisation of Iran.* Ist. graf. tiberino (Roma).

JACKSON, E. D., & THAYER, T. P. (1972): *Some criteria for distinguishing between stratiform, concentric and Alpine Peridotite-Gabbro complexes.* 24th int. geol. Congr., Sect. 2, p. 289–296.

KAY, M. (1972): *Dunnage melange and Lower Paleozoic deformation in northeastern Newfoundland.* 24th int. geol. Congr., Sect. 3, p. 122–133.

– (1973): *Tectonic Evolution of Newfoundland.* In: *Gravity and Tectonics* (Ed. K. A. DE JONG & R. SCHOLTEN; John Wiley Interscience), p. 313–326.

KRAFFT, A. VON (1902): *Notes on the "Exotic Blocks" of Malla Johar in the Bhot Mahals of Kumaon.* Mem. geol. Surv. India *32*/3, 127–185.

KÜNDIG, E. (1956a): *The position in time and space of the ophiolites with relation to orogenic metamorphism.* Geol. en Mijnb. [n.s.] *18*/4, 106–114.

– (1956b): *Geology and ophiolite problems of East-Celebes* (Gedenkboek H. A. Brower). Verh. k. nederl. geol. mijnbouwkd. Genoot. *16,* 210–235.

– (1959): *Eugeosynclines as potential oil habitats.* 5th World Petroleum Congr. New York, Sect. 1, p. 461–479.

LAPIERRE, H., & PARROT, J.-F. (1972): *Identité géologique des régions de Paphos (Chypre) et du Baër-Bassit (Syrie).* C. R. Acad. Sci. Paris (D) *274,* 1999–2002.

LAPPARENT, A. F. DE (1972a): *Esquisse géologique de l'Afghanistan.* Rev. Géogr. phys. Géol. dyn. *14*/4, 327–344.

– (1972b): *L'Afghanistan et la dérive du Continent indien.* Rev. Géogr. phys. Géol. dyn. *14*/4, 449–455.

LAUGHTON, A. S. (1966): *The birth of an Ocean.* New Scientist, p. 218–220.

LAUGHTON, A. S., WHITMARSH, R. B., & JONES, M. T. (1970): *The evolution of the Gulf of Aden.* Phil. Trans. r. Soc. London (A) *267,* 227–266.

LEES, G. M. (1928): *The geology and tectonics of Oman and of parts of southeastern Arabia.* Quart. J. geol. Soc. London *84*/336, 585–670.

LEMOINE, M. (1973): *Les Ophiolites des Alpes et la Tectonique des Plaques.* Symposium "Ophiolites in the Earth's Crust", Moscow 1973 (preprint).

LOCKWOOD, J. P. (1971): *Sedimentary and gravity-slide emplacement of Serpentinite.* Bull. geol. Soc. Amer. *82,* 919–936.

– (1972): *Possible Mechanisms for the Emplacement of Alpine-Type Serpentinite.* Mem. geol. Soc. Amer. *132,* 273–287.

Maden Tetkik ve Arama Enstitüsü Yayinlarindan (1963): *Explanatory text of the Geological Map of Turkey 1:500000: Erzurum.* Ankara.

MAXWELL, J. C. (1969): *"Alpine" mafic and ultramafic rocks. – The ophiolite suite: a contribution to the discussion of the paper "The origin of ultramafic and ultrabasic rocks" by P. J. Wyllie.* Tectonophysics *7*/5–6, 489–494.

MENNESSIER, G. (1970): *Découverte du Turonien et de l'Albo-Aptien au sein de la série schisteuse à ophiolites de la partie centrale de la chaîne d'Altimour, Afghanistan Oriental.* C.R. Acad. Sci. (Paris) *270*, 1427.
– (1972): *Géologie de la chaîne d'Altimour.* Rev. Géogr. phys. Géol. dyn. *14*/4, 345–355.
MIASHIRO, A. (1966): *Some aspects of peridotite and serpentinite in orogenic belts.* Jap. J. Geol. Geogr. *37*/1.
MOLNAR, P., FITCH, T. J., & WU, F. T. (1973): *Fault plane solutions of shallow earthquakes and contemporary tectonics in Asia.* Earth and planet. Sci. Lett. *19*, 101–112.
MOORES, E. M. (1970): *Ultramafics and orogeny, with models for the U.S. Cordillera and the Tethys.* Nature (London) *228*, 837–842.
– (1973a): *Geotectonic significance of ultramafic rocks.* Earth-Sci. Rev. *9*, 241–258.
– (1973b): *Plate tectonic significance of Alpine peridotite types.* Earth-Sci. Rev. *9*/2, 963–975.
MOORES, E. M., & VINE, F. J. (1971): *The Troodos massif, Cyprus and other ophiolites as oceanic crust: evaluation and implications.* Phil. Trans. r. Soc. London (A) 268, 443–466.
MORTON, D. M. (1959): *The geology of Oman.* 5th World Petroleum Congr., Sect. 1, *14*, 1–14.
MU AN-TZE, WEN SHIH-HSUAN, WANG YI-KANG, CHANG PING-KAO & YIN CHI-HSIANG (1973): *Stratigraphy of the Mount Jolmo Lungma region in Southern Tibet, China.* Scientia sinica *16*/1, 96–111.
NOWROOZI, A. A. (1971): *Seismo-tectonics of the Persian Plateau, Eastern Turkey, Caucasus, and Hindu-Kush regions.* Bull. seismol. Soc. Amer. *61*/2, 317–341.
– (1972): *Focal mechanism of earthquakes in Persia, Turkey, West Pakistan, and Afghanistan and Plate tectonics of the Middle East.* Bull. seismol. Soc. Amer. *62*/3, 823–850.
ODELL, N. E. (1967): *Geology of the Himalaya(s). Essay Review.* Geol. Mag. *104*/1, 86–91.
PASSERINI, P. (1965): *Rapporti fra le ofioliti e le formazioni sedimentarie fra Piacenza e il Mare Tirreno.* Boll. Soc. geol. ital. *84*/5.
PAVONI, N. (1961): *Die Nordanatolische Horizontalverschiebung.* Geol. Rdsch. *51*, 122–139.
Penrose Field Conference (1972): *Ophiolites.* Geotimes *17*/12, 24–25.
PETERS, TJ. (1969): *Rocks of the alpine ophiolitic suite.* Tectonophysics, spec. Issue *7*, 507–509.
POWELL, C. McA., CONAGHAN, P. J. (1973a): *Polyphase deformation in phanerozoic rocks of the Central Himalayan Gneiss, Northwest India.* J. Geol. *81*/2, 127–143.
– (1973b): *Plate tectonics and the Himalayas.* Earth and planet. Sci. Lett. *20*, 1–12.
PUSHCHAROVSKY, YU. M. (1972): *Tectonics of continental margins of the Pacific Ocean.* 24th int. geol. Congr., Sect. 8, 53–57.
QUENNELL, A. M. (1958): *The structural and geomorphic evolution of the Dead Sea rift.* Quart. J. geol. Soc. London *114*, 1–24.
QURESHY, M. N. (1969): *Thickening of a basalt layer as a possible cause for the uplift of the Himalayas – a suggestion based on gravity data.* Tectonophysics 7/2, 137–157.
REINHARDT, B. M. (1969): *On the genesis and emplacement of Ophiolites in the Oman Mountains geosyncline.* Schweiz. mineral. petrogr. Mitt. *49*/1, 1–30.
RICOU, L.-E. (1968): *Une coupe à travers les séries à radiolarites des monts Pichakun (Zagros, Iran).* Bull. Soc. géol. France (7) *10*, 478–485.
– (1971): *Le croissant ophiolitique péri-arabe, une ceinture de nappes mises en place au Crétacé supérieur.* Rev. Géogr. phys. Géol. dyn. (2) *13*/4, 327–350.
RIGO DE RIGHI, M., & CORTESINI, A. (1964): *Gravity tectonics in foothills structure belt of southeast Turkey.* Bull. amer. Assoc. Petroleum Geol. *48*/12, 1911–1937.
ROCCI, G., & LAPIERRE, H. (1969): *Etude comparative des diverses manifestations du volcanisme préorogénique au sud de Chypre.* Schweiz. mineral. petrogr. Mitt. *49*/1, 31–46.
ROD, E. (1974): *Geology of Eastern Papua: Discussion.* Bull. geol. Soc. Amer. *85*, 653–658.
ROST, F. (1971): *Die alpinotypen Ultramafitite und ihre Bedeutung für den Tiefgang der alpinen Orogenese.* Verh. geol. Bundesanst. (Wien) *2*, 266–286.
SCHROEDER, J. W. (1944): *Essai sur la structure de l'Iran.* Eclogae geol. Helv. *37*/1, 37–81.
STEINMANN, G. (1905): *Geologische Beobachtungen in den Alpen.* Ber. natf. Ges. Freiburg *16*.
– (1927): *Die ophiolitischen Zonen in den mediterranen Kettengebirgen.* 14e Congr. Géol., sect. Espagne, p. 637–668.
STÖCKLIN, J. (1968): *Structural history and tectonics of Iran: a review.* Bull. amer. Assoc. Petroleum Geol. *52*/7, 1229–1258.

STÖCKLIN, J., EFTEKHAR-NEZHAD, J., & HUSHMAND-ZADEH, A. (1972): *Central Lut reconnaissance, East Iran.* Rep. geol. Surv. Iran *22*, 1–62.

STRACHEY, R. (1851): *On the Geology of Part of the Himalaya Mountains and Tibet.* Quart. J. geol. Soc. London *7*/1, 292–330.

SUESS, E. (1901): *Das Antlitz der Erde* (Bd. 3/I). Tempsky, Prag–Wien; Freytag, Leipzig.

– (1909): *Das Antlitz der Erde* (Bd. 3/II). Tempsky, Wien; Freytag, Leipzig.

TAKIN, M. (1972): *Iranian geology and continental drift in the Middle East.* Nature (London) *235*/5334, 147–150.

TERRA, H. DE (1935): *Geological studies in the North-west Himalaya between the Kashmir and Indus valleys.* Mem. Connecticut Acad. Arts Sci. *8*, 18–76.

TSCHOPP, R. H. (1967): *The General Geology of Oman.* Proc. 7th World Petroleum Congr. Mexico *2*, 231–241.

TURNER, W. M. (1973): *The Cyprian Gravity Nappe and the Autochthonous Basement of Cyprus.* In: *Gravity and Tectonics* (John Wiley & Sons, New York), p. 287–301.

VINE, F. J., & MOORES, E. M. (1972): *A model for the gross structure, petrology, and magnetic properties of Oceanic crust.* Mem. geol. Soc. Amer. *132*, 195–205.

VUAGNAT, M. (1964): *Remarques sur la trilogie serpentinites gabbros-diabases dans le bassin de la Méditerranée occidentale.* Geol. Rdsch. *53*/1, 336–357.

WADIA, D. N. (1937): *The Cretaceous volcanic series of Astor-Deosai, Kashmir, and its intrusions.* Rec. geol. Surv. India *72*/2, 151–161.

WELLMAN, H. W. (1966): *Active wrench faults of Iran, Afghanistan and Pakistan.* Geol. Rdsch. *55*/3, 716–735.

WIENERT, K. (1947): *Preliminary report on the magnetic results of a journey to Sikkim and Southern Tibet.* J. geophys. Res. *52*/4, 505–521.

WILLIAMS, H., & SMYTH, W. R. (1973): *Metamorphic aureoles beneath ophiolite suites and Alpine peridotites: Tectonic implications with West Newfoundland Examples.* Amer. J. Sci. *273*, 594–621.

WILSON, H. H. (1969): *Late Cretaceous Eugeosynclinal Sedimentation, Gravity Tectonics, and Ophiolite Emplacement in Oman Mountains, Southeast Arabia.* Bull. amer. Assoc. Petroleum Geol. *53*/3, 626–671.

WYLLIE, P. J. (1969): *The origin of Ultramafic and Ultrabasic rocks.* Tectonophysics *7*/5–6, 437–455.

2

Reprinted by permission of the publisher from pages 321-333 of *Modern and Ancient Geosynclinal Sedimentation,* R. H. Dott, Jr. and R. H. Shaver, eds., Soc. Econ. Paleontologists and Mineralogists Spec. Pub. No. 19, 1974, 388p.

MELANGES AND THEIR DISTINCTION FROM OLISTOSTROMES

K. J. HSÜ
Geologisches Institut, ETH Zürich, Switzerland

INTRODUCTION

Sedimentary formations are commonly characterized by laterally persistent beds. This observation was made by Nicolas Steno in the 17th Century and was formulated as one of the three fundamental laws of sedimentary sequences. The law states (Gilluly and others, 1968, p. 92):

> "A water-laid stratum, at the time it is formed, must continue laterally in all directions until it thins out as a result of non-deposition, or until it abuts against the edge of the original basin of deposition."

In some deformed terranes, one is impressed by the fact that few rock bodies can be traced to any large lateral extent. Rocks of different types lie jumbled in close juxtaposition. A classical example is the wildflysch in the Swiss Alps. The wildflysch consists of blocks of flysch sandstones and other rocks in a shaly matrix. Kaufmann (1886) assigned to descriptive prefix "wild" to those rock bodies, because of the wild, or undisciplined nature of the bedding in the wildflysch in contrast to the so-called "quiet" (*ruhig*) stratification of the type Flysch, which is characterized by a regular alternation of very evenly bedded graywackes and shales.

The origin of the wildflysch has been a subject of unending controversy during the last hundred years. Even today, my colleagues and I might stand on the same outcrop and not agree on its genesis. Several reviews have been published (e.g., Cadisch, 1953, p. 182–186; Badoux, 1967). The dispute has been focused on the mode of emplacement of the blocks in the wildflysch. One school considered the blocks to be sedimentary, being mixed in a muddy matrix during submarine slumping, sliding, or during some other sedimentary events (Schardt, 1898; Lugeon, 1916). Another school pointed out that the wildflysch also included tectonic blocks that were picked up and modified during some tectonic event, such as folding, thrusting, or pervasive shearing (e.g., Adrian, 1915; Häfner, 1924). As a matter of fact, what has been called wildflysch includes both sedimentary olistostromes and tectonic mélanges. Furthermore, pervasively sheared olistostromes might appear identical to a mélange, and the two might be indistinguishable in the field. If a wildflysch deposit is to be considered as a sedimentary formation, the law of original continuity applies. On the other hand, if a body of wildflysch is a tectonic unit, the principles of mélanges must be invoked (see Hsü, 1968). The two alternative interpretations for a given wildflysch unit could thus lead to completely different geological conclusions. Thus, it is important to attempt to distinguish them. This fact did not escape Marshall Kay (1970) when he pondered if his Newfoundland wildflysch should be considered as a slump breccia (e.g., Patrick, 1956) or a tectonic mélange (e.g., Dewey, 1969). On the occasion of the symposium honoring Kay, I was invited by the convener to outline some characteristic features of mélanges, to discuss their significance, and to present some criteria for differentiating mélanges from olistostromes.

NATURE OF MÉLANGES

Mélanges are bodies of deformed rocks characterized by the inclusion of tectonically mixed fragments or blocks, which may range up to several miles long, in a pervasively sheared matrix (Greenly, 1919; Hsü, 1968).[1] Each mélange includes both exotic and native blocks and a matrix. Native blocks are disrupted brittle layers, which were once interbedded with the ductilely deformed matrix. Exotic blocks are tectonic inclusions detached from some rock-stratigraphic units foreign to the main body of the mélange. A body of pervasively sheared strata that contains no exotic elements may be called a broken formation, because such a body, regardless of its broken state, functions as a rock-stratigraphic unit.

To make a mélange involves two processes: fragmentation and mixing of broken fragments in a ductilely deformed matrix. A broken formation resulting from fragmentation, but no mixing of exotic elements, may be considered as a preparatory stage in the genesis of a mélange.

The broken formations intercalated within the Franciscan mélanges of the Santa Lucia Range, California, provide some of the best illustrations of progressive fragmentation. Those formations are interbedded graywackes and shales. Graywackes appear to be the more brittle and tend to fracture, whereas the shales have been deformed by flowage. Fragmentation commonly occurred along shear fractures, which formed under compression subparallel to the bedding planes. Thus, they may be recognized as miniature thrust faults.

In contrast, fragmentation resulting from extension parallel to the bedding surface resulted in shear fractures inclined at a high angle (60° or more) to the bedding plane (fig. 1). In the field, one might recognize examples representing progressive stages of separation: starting from shear joints, to miniature-grabens, to lozenge-shaped boudins (name given by Rast, 1956), and to completely separated, isolated, and rotated fragments. Typical boudins are bounded by extensional fractures, normal to the bedding. Yet the boudins in the Franciscan mélanges are almost exclusively bounded by extensional shear fractures. I did observe, however, typical boudins in the *argille scagliose* of the Apennines. The occurrence of extensional fracture is indicative of brittle, or very brittle behavior, which in turn attests to the very low effective confining pressure during the time of fragmentation.

Broken formations commonly show evidence of having been subjected to alternate tension and compression parallel to bedding surfaces. Lenticular masses bounded by compressional shears are often internally traversed by extensional shear joints (fig. 2).

Fragmentation *per se* is not sufficient proof of brittle deformation, as brittleness is a relative measure of deformation behavior. A material is regarded as very brittle, brittle, moderately brittle, moderately ductile, or ductile, if the total strain before the fracturing is less than 1, 1 to 5, 2 to 8, 5 to 10, and more than 10 percent respectively (Handin, 1966, p. 226). Not only brittle rocks fracture, for rocks undergoing ductile deformation may eventually fracture after the total strain has exceeded a certain undefined limit. The occurrence of phacoids and boudins is a measure of the relative ductility between the fragmented layers and the host matrix.

The fragments of a mélange are autoclasts. Autoclasts bounded by compressional shear surfaces have commonly been modified to produce phacoids (Greenly, 1919; Cadisch, 1953, p. 185). Autoclasts produced by stretching were originally boudins. Many, if not most of the autoclasts no longer betray their tectonic heritage. Ductile stretching of phacoids produced tails. Rotation of lozenges led to tectonic rounding. Ultimately, the autoclasts may assume such ir-

[1] Editors' note: Some workers urge use of "mélange" as a nongenetic term for any chaotic deposit; they would invoke genetic modifiers, for example, "tectonic" or "sedimentary" mélange. In such usage, "mélange" is synonymous with Hsü's definition of "wildflysch." (See also papers in this volume by Blake and Jones, Wood, and Kay.)

FIG. 1.—Lozenge-shaped boudin in a broken formation, Franciscan in age.

FIG. 2.—Boudins traversed by extensional shear fractures and bounded by compressional shears, indicative of alternate extension and compression, Franciscan in age.

FIG. 3.—Partially rounded exotics in mélange, Franciscan in age.

regular shapes of rounded outlines that they may become indistinguishable from sedimentary boulders in an olistostrome (fig. 3).

The matrix of the mélange is characteristically a material that could undergo very large permanent deformation without fracturing. The shales were apparently more ductile than graywackes or limestones under the conditions of mélange genesis. The matrix of mélanges is most commonly a pelitic rock subjected to different degrees of regional metamorphism. The matrix of the Franciscan mélanges in the Santa Lucia Range is a shale or a slightly metamorphosed phyllite (Hsü, 1969); The Franciscan mélange matrix in the northern Coast Ranges has been locally altered to lawsonite schist or to metamorphic rocks of higher grade. Locally it contains a high proportion of sheared serpentinite. Rarely some other rock may form the mélange matrix. The famous Colored Mélange of central Iran (e.g., Gansser, 1955) has a serpentinite matrix (Davoudzadeh, 1969). In metamorphic terranes, tectonic fragments of calcsilicate rocks or of dolomite are found in marble matrix. In rare instances, the matrix of a mélange is gypsum (Leine and Egeler, 1962).

TECTONIC SIGNIFICANCE OF MÉLANGES

The fragmented nature of a mélange testifies to the fact that such a body is incapable of transmitting stress. This consideration played a critical role in the interpretation of the *argille scagliose* of the Apennines. Obviously, those scaly shales are too weak to have been moved as a sheet by a push on one side (Page, 1963, p. 669). One school of thought considers the *argille scagliose* as an olistostrome, or a "gigantic submarine gravity slide" (Maxwell, 1959b, p. 2711; written communication, 1972). Yet, the *alberesi* blocks in some of the *argille scagliose* are "mechanically derived," as Page (1963, p. 664) demonstrated. My own field observations in the Apennines convinced me that a large part of this allochthonous complex is a tectonic mélange, although olistostrome beds are present as allochthonous slabs.

If a mélange has been displaced as a body of jumbled blocks, an interpretation of tectonic gravity sliding seems to be the only possible mechanism. Page (1963) reached this conclusion for the *argille scagliose*. I once also invoked this hypothesis to account for the genesis of the Franciscan mélanges (e.g., Hsü, 1967a; 1968; Hsü and Ohrbom, 1969). On the other hand, if we regard a mélange as a composite schuppen zone at the toe of a thrust plate, we no longer have to appeal to gravity as the prime driving force. Instead, allochthonous materials could have been transported for long distances as an integral part of a lithospheric plate. The plate was broken as it was descending along a subduction zone. We might picture a freight train carrying piles of lumber whistling its way into a tunnel with a low ceiling. The part of the lumber piled too high would be sheared off. The broken and splintered planks would accumulate as a mélange in front of the tunnel. Such a mélange would be formed *in situ,* although the lumber would have come a long way from its original supply depot.

In a paper analyzing thrusting mechanics, I have demonstrated that the toe effect of thrusting under compression is to produce a mélange having densely spaced shear planes if the thrust plate is displaced along a zone of flowage (Hsü, 1970). This mechanical analysis is in agreement with the postulate of plate-tectonics theory that interprets mélanges as products of pervasive shearing along consuming plate margins (e.g., Hamilton, 1969; Ernst, 1970; Dewey and Bird, 1970; Dickinson, 1970; Hsü, 1971a).

The Franciscan mélanges of the California Coast Ranges were very probably produced as a late Mesozoic subduction zone migrated oceanwards (e.g., Hamilton, 1969; Hsü, 1971a). The presence of a mélange could thus indicate the prior existence of a convergent plate margin. Its absence from some circum-Pacific mountains, such as the central Andes, is noteworthy. A probable explanation lies in the fact that the subduction zone there may have migrated landward (e.g., James, 1971); older mélanges went down with the descending plate into the mantle. I have also pointed out that the Alpine Wildflysch is volumetrically insignificant compared to the Franciscan mélanges. In the Alps the cover thrusts constitute coherent nappes, and the deformed sediments between thrust planes have largely retained their stratal continuity. The subordinate role of mélange deformation might be considered as evidence that plate subduction was far less important as a mechanism in the Alps than in California (Hsü, 1972).

In contrast to the largely underthrusting movement of the oceanic plates under the North American plate, the plate movement between Africa and Europe, which was responsible for the genesis of the Alps and the Mediterranean, had a large lateral component (e.g. Smith, 1971; Hsü, 1971b; Pitman and Talwani, 1972).

OLISTOSTROMES

The name "olistostrome" was introduced by Flores in 1955 during the discussion of a paper presented by E. Beneo on the oil geology of Sicily. His (Flores') definition is quoted as follows (p. 122):

> "By olistostromes, we define those sedimentary deposits occurring within normal geological sequences that are sufficiently continuous to be mappable, and that are characterized by lithologically or petrographically heterogenous material, more or less intimately admixed, that were accumulated as a semifluid body. They show no true bedding, except for possible large inclusions of previously bedded material. In any olistostrome we distinguish a 'binder' or 'matrix' represented by prevalently pelitic, heterogenous material containing dispersed bodies of harder rocks. The latter may range in size from pebble to boulder and up to several cubic km."

The name "olistolith" was applied to the masses included as individual elements within the binder. These masses had been previously referred to as "exotics" or "erratics."

There is no doubt that Flores intended the term to designate sedimentary deposits. In 1959, Flores referred to olistostromes as chaotic accumulations, "vertically delimited by underlying and overlying series of normal marine sediments, . . . that contain fossils in situ of identifiable age and environment" (p. 261). He added, "the evidence collected indicates that the olistostrome emplacement must have occurred in an *aqueous medium,* as shown by the associated turbidity current and mudflow depositional phenomena observed" (p. 261).

The Tertiary strata of Sicily include both normally deposited sedimentary sequences and chaotic rocks. Beneo (1951) considered the whole of Sicily, with the exceptions of two small areas, as being one overthrust mass, resulting from gravity sliding. We see that Flores and later Marchetti agreed to the gravity-sliding origin of the chaotic deposits. However, they emphasized that the chaotic rocks constitute olistostromes, which are formations, or "regular and mappable units of the sedimentary series" (Marchetti, 1957, p. 220). Instead of one large mélange envisioned by Beneo, the other authors found in Sicily a normal sedimentary series including a few chaotic intercalations.

From the sketches and descriptions provided by Flores and Marchetti, we might attempt to delineate the difference between a conceptual olistostrome and a conceptual mélange (table 1).

Olistostromes do exist in the Apennines, as I was shown during the 1964 American Geological Institute field trip to Italy. Those deposits were illustrated by some superb photographs by Abbate and others (1970). Identical or similar deposits have been called pebbly mudstones, tilloids, fluxoturbidites, or simply slides or slump beds. Badoux, (1967, p. 403) suggested that the olistostromes contain exotic blocks, whereas the submarine slides have a matrix that encloses blocks of the same age. On the other hand, the Italian school prefers to call both olistostromes

TABLE 1.—CONCEPTUAL DIFFERENCE BETWEEN MELANGES AND OLISTOSTROMES

Item	Mélange	Olistostrome
Is underlain by	a formation not pervasively sheared	a formation slightly older than the olistostrome
Is overlain by	a formation not pervasively sheared	a formation slightly younger than the olistostrome
At time of chaotic emplacement	tectonic emplacement is under overburden of hundreds or thousands of meters	Sedimentary emplacement of upper surface of olistostrome was then subaerial or submarine
Site of chaotic emplacement	is a major shear zone, in some cases the consuming plate margin	is a major submarine topographic depression as a suitable receptacle for a large olistostrome
Duration of chaotic emplacement of one unit	is duration of shearing movement, which may have been short or long	is geologically very short, such as for a sedimentary layer

(e.g., Elter and Raggi, 1965).

Abbate and others (1970) modified the definition of Flores. However, they agreed on the essential principle that olistostromes are sedimentary bodies. Those authors "usually regarded as olistostromes chaotic units up to 100–200 m thick' (p. 524), and very large chaotic structures were excluded. The term was applied to designate several different kinds of coarse sedimentary deposits. Typically, they have subrounded cobbles and pebbles embedded in a marly or shaly matrix (fig. 4). These are similar to pebbly mudstones (Crowell, 1957). Others have isolated or scattered angular fragments in an argillite. These are comparable to tilloids (cf. Lindsay, 1966) or boulder-bearing shales (cf. Cline 1970). Some are simply calcareous breccias containing angular clasts and little binding material. Still others consist of subangular or subrounded cobbles and boulders cemented by muddy sands (fig. 5).

Görler and Reutter (1968, figs. 1 and 2) proposed to restrict the term to those coarse deposits having an argillaceous matrix and suggested viscous flow as the mechanism of transport to distinguish olistostrome from other coarse sedimentary deposits. The authors also recognized that olistostromes of the Apennines are commonly composite deposits, each consisting of many individual olistostrome beds. Thus, a unit might be more than 1 km thick, but each olistostrome bed only a few meters thick. The attempt to characterize olistostromes by their postulated transport mechanism is not satisfactory, especially in view of the fact that Dott (1963) had emphasized the plastic behavior of olistostrome transport (by subqueous sliding) in contrast to the viscous behavior of turbidity flows.

Despite the somewhat varied usages and interpretations, "olistostrome" is defined as a sedimentary deposit. Such chaotic masses move downslope under gravity, are emplaced as a sedimentary stratum of finite dimensions, and are intercalated between other, more normal sedimentary units. The different types of olistostromes illustrated by Abbate and others and by Görler and Reutter (1968) are indisputedly sedimentary deposits. Their distinction from a mélange should constitute no problem. Yet the *argille scagliose* includes not only olistostromes; also present in this chaotic complex are large allochthonous slabs as much as hundreds of cubic kilometers in volume (see Maxwell, 1959b; Hsü 1967b). Those large allochthonous sheets were commonly not considered olistoliths, and "the movement of the sheets . . . is generally attributed to a tectonic process" (Abbate and others, 1970, p. 526). Görler and Reutter (1968) also attempted to discriminate a large olistolith from a gliding thrust sheet on the basis of the thickness of the slab, those thicker than 1 km being called thrust sheets.

Is a tectonic process to be distinguished from a sedimentary process solely on the basis of a geometric scale? Is there a fundamental genetic difference? If so, what are the criteria for distinction? Those are the questions I explore in the following sections.

NEED FOR DISTINCTION

It may not seem particularly necessary to make a great distinction between olistostromes and mélanges inasmuch as both have been referred to as the product of gravitysliding. Abbate and others (1970, p. 524) commented on this problem:

"In the common geological practice, tectonics and

FIG. 4.—Olistostrome in *argille scagliose* resembling pebbly mudstone (after Abbate and others, 1970).

FIG. 5.—Olistostrome in *argille scagliose* having sandy matrix.

sedimentation are considered two classes of distinct phenomena. When slides are involved, however, no sharp boundary can be drawn between the two: a small olistostrome is unquestionably a sedimentary feature and the emplacement of a large, even chaotic nappe is commonly considered a tectonic process. Yet there exists a continuous transition between these extreme cases."

They chose to distinguish the two primarily on the basis of size. However, this criterion is arbitrary. Some geologists have made no distinctions at all and referred to tectonic mélanges as olistostromes (e.g., Rigo di Rhigi and Cortesini, 1964). For paleogeographical reconstructions, mélanges have to be distinguished from olistostromes. Consider the zone of carbonate blocks under the Vermont high Taconic sequence as an example. If those blocks are olistoliths, we could interpret the high taconic material as a normally superposed sequence above the autochton, a hypothesis favored by one school of geologists (e.g., McFayden, 1956). If, however, those blocks are exotics in a mélange, the high Taconic sequence must be allochthonous (Zen and Ratcliff, 1966).

The wildflysch problem provides another illustration. Wildflysch rocks were originally thought to be sedimentary breccias or olistostromes (e.g., Kaufmann, 1886). The discovery of the exotic nature of some blocks argued against a normal stratigraphic succession (Schardt, 1898; Beck, 1908). Micropaleontological research proved that the wildflysch of central Switzerland is thrust under older flysch formations (e.g., Boussac, 1912; Leupold, 1942). Meanwhile, the origin of the exotics continued to be a favorite subject of dispute (see Trümpy, 1960, p. 888–890). One school of thought traced their origin to an advancing nappe front (Schardt, 1898). The opposition invoked cordillera in the Ultrahelvetic realm itself (e.g., Lugeon, 1916; Tercier, 1928; Leupold, 1942; Hsü, 1960). The question has been particularly puzzling concerning the source of a pelagic limestone, known locally as the Leimernkalk, that occurs as phacoids in a wildflysch deposit. The phacoids range from Turonian to Paleocene in age and are in a matrix that yielded late Eocene (Priabonian) fossils. The classical approach assumed the existence of a late Eocene wildflysch trench fringed by a coastal range of Leimernkalk terranes. Because the Leimern phacoids resemble the *couches rouges* of the Klippen Nappe, some Alpine geologists postulated that the front of the Klippen had advanced to the southern margin of the wildflysch trench to supply the phacoids (e.g., Gigon, 1952). Another school adopted the concept of an Ultrahelvetic cordillera and considered the resemblance of the *Leimern* and *couches rouges* coincidental. Their key argument was that the Klippen could not have come so far north during the late Eocene to dump exotic blocks into the wildflysch trench (e.g., Herb, 1966). My own field observations led me to the belief that the Leimern phacoids have been tectonically mixed with the Helvetic *Globigerina* Mergel to produce the wildflysch. If the wildflysch is a mélange, not an olistostrome, then we need not postulate the existence of an Eocene wildflysch trench in central Switzerland as the depositional site for the Leimern phacoids. Instead, this mélange was formed from tectonic mixing of rocks at the base of the Klippen Nappe with the Priabonian *Globigerina* Mergel at the top of the Helvetic sequence. The time of the mixing, or the arrival of the Klippen into the Helvetic realm, could be dated as post-Eocene, thus avoiding the necessity of assuming an overhasty itinerary for the Klippen travel.

It is not my intention to settle the controversial wildflysch problem in this article. I have only mentioned the various alternative interpretations to emphasize the need to distinguish a mélange from an olistostrome and to counter the arguments of some of my colleagues that such an effort is merely a semantic exercise.

CRITERIA FOR DISTINCTION

General statment.—Typical mélanges and typical olistostromes are so different that a glance at figures 1 through 5 would suffice to make a distinction. Yet, large and prevasively sheared olistostromes are all but indistinguishable from mélanges. I cannot offer any infallible rules but shall discuss several different lines of approach to the problem.

Shape of clasts.—Broken formations and mélanges can be distinguished from olistostromes by the shape of fragments. Broken strata exhibit progressive fragmentation through boudinage. Some mélange autoclasts are also bounded by fracture surfaces, especially by compressional and extensional shears. They may be modified by pervasive shearing and converted to phacoids having delicate tails trailing into the matrix. The occurrence of very rounded autoclasts in a mélange would suggest derivation from disintegration of an olistostrome (fig. 6).

Nature of clasts.—An overthrust, like a bulldozer, may pick up exotic blocks from an underlying tectonic element. The Taconic wildflysch blocks, for example, have been derived from the underlying autochthon (Zen and Ratcliff, 1966). The Alpine wildflysch may include blocks from both the underlying and the overlying tectonic

FIG. 6.—A pervasively sheared olistostrome in Franciscan rocks. Subangular fragment in middle is glaucophane schist.

elements (see Cadisch, 1953). In contrast, olistoliths are derived from an advancing thrust or gravity-slide mass, but not from a buried autochthon.

Blocks in a narrow mélange zone are commonly recognizable as having been detached from adjacent tectonic elements. Blocks that cannot be identified as such, like those of the Habkern Granite in the Alpine Wildflysch, are most probably olistoliths introduced by gravity deposition. This particular interpretation can be reinforced by the fact that the same granite furnished considerable debris to turbidite sandstones associated with the wildflysch (see Hsü, 1960).

Blocks in a broad mélange zone, such as the Franciscan, may also trace their origins to so-called "phantom-stratigraphical units," which had been broken and mixed to yield the mélange (see Hsü and Ohrbom, 1969). Exotic blocks completely foreign to other components in the mélange may have been derived from a disintegrated olistostrome. For example, only a few scattered blocks of glaucophane schist have been found in the Franciscan mélange of the Santa Lucia Range, California (fig. 6). They might have been introduced as blocks in Cretaceous olistostromes, which have since been broken up, leaving the originally sedimentary clasts of high-pressure metamorphics in an unmetamorphosed mélange matrix.

Blocks yielding fossils younger than the matrix fauna are in all probability mélange exotics. Not recognizing this simple rule has led to confusion, giving rise to the misnomer "Einsiedler Flysch." In Sihltal, near Einsiedeln, Switzerland, lenses of nummulitic limestones are apparently intercalated in a shaly formation, and the Einsiedler complex was mapped by Kaufmann (1886) as part of his Wildflysch. The nummulitids are typically early Tertiary image. Yet, Late Cretaceous faunas, including an ammonite specimen, were eventually found in the shales (Rollier, 1912; Heim, 1923). Making the implicit assumption that the blocks were sedimentary and that they could not carry a fauna younger than the matrix, Rollier (1923) and Heim (1923) questioned the stratigraphical value of the nummulitic faunas and spoke of Late Cretaceous *Nummulites*. Eventually, the geology was worked out and the Einsiedler Flysch was recognized as a tectonic mixture in a schuppen zone, where Tertiary limestone lenses were shoved into an underlying Upper Cretaceous shale.

Fabric of matrix.—The matrix of mélanges is pervasively sheared. The ductilely deformed materials have flowed into the irregular spaces between blocks that are arranged chaotically or in a subparallel fashion. In places the pelitic ma-

Fig. 7.—Olistostrome in *argille scagliose,* a serpentinite breccia containing ophiolite blocks in matrix of ophiolite debris (after Abbate and others, 1970).

terials were intruded into the cracks or fractures of disintegrating blocks. The orientation of shear surfaces tends to be parallel to edges of the blocks. Mélanges containing irregular blocks show random orientation of local shear surfaces (see Hsü and Ohrbom, 1969, fig. 4). Those containing phacoids commonly acquired a preferred orientation parallel to regional tectonic trend.

The matrix of olistostromes is not necessarily pervasively sheared (e.g. fig. 4). The shaly matrix may acquire a fissility under compaction, or it may acquire a cleavage during tectonic deformation (Abbate and others, 1970).

Nature of matrix.—Both mélanges and olistostromes typically have a pelitic matrix. However, some olistostromes have a sandy matrix. Abbate and others (1970) referred to serpentinite breccias as olistostromes (fig. 7). I have found similar breccias in the Franciscan rocks (Hsü, 1969). The breccia beds, each a few tens of feet thick, are composed of blocks of ophiolite in a sandy matrix of similar composition. Those olistostromes are interbedded with coarse-grained sandstones, and they are easily distinguished from serpentinite-bearing mélanges, which have pelitic or antigoritic matrix.

Nature of contacts with adjacent units.—Mélanges are tectonic units. Their contacts with adjacent units are almost invariably shear surfaces. In certain regions like the Pacific Coast Ranges, the base of a mélange is nowhere to be seen, and the thickness of the mélange piles may reach several miles. The upper contact is, as a rule, definable but may seem locally anomalous to geologists schooled in overthrust tectonics. One of the cardinal rules in mapping overthrust terranes states that the highest tectonic element in an allochthonous pile has been transported the farthest (see Bailey, 1935). It was thus most puzzling that the highest tectonic element in western California is the Great Valley sequence, which is autochthonous with respect to the North American continent. A parallel situation exists in the Northern Apennines. The Mount Antola unit is the highest element of the Ligurian Nappe, which is the highest nappe of the Northern Apennines. Yet the Mount Antola is autochthonous with respect to the Po Valley block (Abbate and Sagri, 1970, p. 319). This paradox is now resolved if we assume the underthrusting of mélanges, for the highest tectonic element in an underthrust pile is autochthonous or parautochthonous with respect to the upper or the overthrust plate.

Another paradox is the observation that the upper contact of a mélange with an upper autochthonous element seems to be a shear surface at one place and a depositional contact at another (Hsü, 1968, p. 1070). The long controversy over the Franciscan-Knoxville problem could be traced to this observation. In the northern Coast Ranges, slabs of the Great Valley sequence were detached and mixed with the Franciscan rocks. The slabs are separated from the mélange by shear surfaces, which have been collectively described as an overthrust (e.g., Irwin, 1964; Bailey and others, 1964; Blake and others, 1967). However, on the west side of the Great Valley the contact between the Great Valley sequence and its underlying ophiolite, which has been traditionally mapped as Franciscan, is depositional (Hsü, 1969; Bailey and others, 1970; Page, 1972). For this reason, some current workers propose to include the ophiolite slab in the Great Valley sequence (e.g., Blake and Jones, this volume).

Olistostromes are stratigraphic units intercalated in a normal sedimentary sequence. The contact with overlying and underlying units should be depositional and conformable. Flores (1959) noted a discordant lower contact of the Sicilan *argille scagliose,* apparently overlapping rocks ranging from Aquitanian to Tortonian in age. He rationalized the observation by assuming uncommon erosive power for the postulated submarine slide.

Regional synthesis.—A distinction between a mélange and an olistostrome is not always possible on the basis of observable criteria. A final judgment depends upon consideration of all available geological data. The current consensus interpreting the Franciscan as a mélange is based not only upon field observations, but also upon regional synthesis. Such an interpretation clarified a long-standing paradox of Coast Range geology. The olistostrome hypothesis took root in the Apennines because it accounted for many facets of the regional geology there. Nevertheless, the hypothesis also proved inadequate to solve some puzzles. Would a mélange hypothesis lead to a better understanding there?

SUMMARY

Mélanges are among the most important bodies of rocks in the mountain ranges of the world. Yet I have found no reference to mélanges, in any of the current textbooks on structural geology. The concept of tectonic mixing has been much neglected by our profession, considering that the term was coined by Greenly in 1919. Prior to 1965, only a few European geologists resorted to the use of this expression (e.g., Bailey and McCallien, 1950; 1963; Gansser, 1955). The term became popular after parts of the Franciscan were identified (e.g., Hsü, 1967a) and were referred to as such (e.g., Ernst, 1965; Hsü, 1968; Hamilton, 1969; Page, 1970; Swe and Dickinson, 1970). The idea of mélange genesis through underthrusting found a cozy niche in the scheme of plate-tectonic theory (e.g., Crowell, 1968; Bailey and Blake, 1969; Page, 1969; Hamilton, 1969; Ernst, 1970; Dewey and Bird, 1970; Dickinson, 1970; Dott, 1971; Hsü, 1971a). Today, one rarely picks up a geological journal without finding frequent recurrence of "mélange." Meanwhile, the concept of olistostrome survives to describe the deposition of coarse, chaotic sedimentary units. Both mélanges and olistostromes are common in regions of underthrusting tectonics and so are often indistinguishable in zones of pervasive shearing. This article defines their distinction. Where distinction is impossible, some nongenetic, collective term such as "wildflysch" is suggested.

ADDENDUM

During an international excursion attended by more than a hundred colleagues in early June 1973 to examine the ophiolitic mélanges in Tienshan and Caucasus, it became apparent that an unsheared olistostrome could be distinguished easily from a mélange. On the other hand, it is not always clear in the field if a mélange is a sheared olistostrome or if the fragmentation and mixing have been entirely tectonic. The use of the term wildflysch as I suggested in the text of my paper has been objected to by many colleagues because the alpine Wildflysch, in contrast to most mélanges, generally does not include ophiolite blocks. I now tend to agree to a suggestion made by the editors in the footnote of my text that the term mélange *sensu lato* be used as a descriptive term to designate all pervasively sheared chaotic deposits with exotic blocks. Sedimentary mélanges are sheared olistostromes, which include blocks fragmented and mixed by sedimentary processes that are now embedded in a pervasively sheared matrix.

REFERENCES

ABBATE, E., AND SAGRI, M., 1970, Development of the Northern Apennines Geosyncline—the eugeosynclinal sequences: Sed. Geology, v. 4, p. 251–340.

——, BORTOLOTTI, V., AND PASSERINI, P., 1970, Development of the Northern Apennines Geosyncline—olistostromes and olistoliths: *ibid.*, v. 4, p. 521–557.

ADRIAN, H., 1915, Geologische Untersuchung der beiden Seiten des Kandertals im Berner Oberland: Eclog. Geol. Helvetiae, v. 13, p. 238–351.
BADOUX, H., 1967, De quelques phénomènes sédimentaires et gravifiques liés aux orogenèse: *ibid.*, v. 60, p. 399–406..
BAILEY, E B., 1935, Tectonic essays, mainly alpine: Oxford, England, Oxford University, 200 p.
———, 1963, Liguria nappe: Northern Apennines: Royal Soc. Edingburgh, Trans., v. 65, p. 315–333.
———, AND MCCALLIEN, W. J., 1950, The Ankara mélange and the Anatolian thrust: Nature, v. 166, p. 938–940.
BAILEY, E. H., AND BLAKE, M. C., 1969, Tectonic development of western California during the late Mesozoic: Geotektonika, v. 3, p. 17–30, v. 4, p. 24–34.
———, ———, AND JONES, D. L., 1970, On-land Mesozoic oceanic crust in California Coast Ranges: U.S. Geol. Survey Prof. Paper 700-C, p. 70–81.
———, IRWIN, W. P., AND ———, 1964, Franciscan and related rocks and their significance in the geology of western California: California Div. Mines Geol. Bull. 183, 177 p.
BECK, P., 1908, Vorläufige Mitteilung über Klippen and exotische Blöcke östlich des Thunersees: Naturf. Gesell. Bern, Mitt., fur 1908.
BENEO, E., 1951, Sull' identià tettonica esistente fra la Sicilia e il Rif: Boll. Service Geol. d'Italia, v. 72, Note 1, Roma.
BLAKE, M. C., IRWIN, W. P., AND COLEMAN, R. G., 1967, Upside-down metamorphic zonation, blueschist facies, along a regional thrust in California and Oregon: U.S. Geol. Survey Prof. Paper 575-C, p. 1–9.
BOUSSAC, J., 1912, Études stratigraphiques sur le Nummulitique alpin: Carte Géol. France, Mém., 662 p.
CADISCH, J., 1953, Geologie der Schweizer Alpen: Basel, Switzerland, Wepf et Cie., 480 p.
CLINE, L. M., 1970, Sedimentary features of late Paleozoic flysch, Ouachita Mountains, Oklahoma, *in* LAJOIE, J. (ed.), Flysch sedimentology in North America, Geol. Assoc. Canada, Special Paper 7, p. 85–102.
CROWELL, J. C., 1957, Origin of pebbly mudstones: Geol. Soc. America Bull., v. 67, p. 993–1010.
———, 1968, Movement histories of faults in the transverse ranges and speculations on the tectonic history of California, *in* DICKINSON, W. R., AND GRANTZ, A. (eds.), Proceedings of conference on geologic problems of San Andreas Fault System, 11: Stanford, California, Stanford Univ. Pub. Geol. Sci., 374 p.
DAVOUDZADEH, M., Geologie und Petrographie des Gebietes nördlich von Näin, Zentral-Iran: Geol. Inst. ETH, Mitt., new ser., v. 98, 91 p.
DEWEY, J. F., 1969, Evolution of the Appalchian/Caledonian Orogen: Nature, v. 222, p. 124–129.
———, AND BIRD, J.,1970, Mountain belts and the new global tectonics: Jour. Geophys. Research, v. 75, p. 2625–2647.
DICKINSON, W. R., 1970, Second Penrose Conference: The new global tectonics: Geotimes, v. 15, no. 4, p. 18–22.
DOTT, R. H., 1963, Dynamics of subaqueous gravity depositional processes: Am. Assoc. Petroleum Geologists Bull., v. 47, p. 104–128.
———, 1971, Geology of the southwestern Oregon coast west of the 124th Meridian: Oregon Geol. and Min. Industries Bull. 69, 63 p.
ELTER, P., AND RAGGI, G., 1965, Contributo alla conoscenza dell'Appennino ligure: 1. Osservazioni sul problema degli olistostromi: Soc. Geol. Italia, Boll., v. 84, p. 303–322.
ERNST, W. G., 1965, Mineral paragenesis in Franciscan metamorphic rocks, Panoche Pass, California: Geol. Soc. America Bull., v. 76, p. 879–914.
———, 1970, Tectonic contact between the Franciscan melange and the Great Valley sequence—crustal expression of a late Mesozoic Benioff zone: Jour. Geophys. Research, v. 75, p. 887–901.
FLORES, G., 1955, Discussion, *in* BENEO, E., Les resultats des etudes pour la recherche petrolifere en Sicilie (Italie): 4th World Petroleum Cong., Rome, Proc. sect. 1, p. 121–122.
———, 1959, Evidence of slump phenomena (olistostromes) in areas of hydrocarbons exploration in Sicily: 5th World Petroleum Cong., New York, Proc. sect. 1, p. 259–275.
GANSSER, A., 1955, New aspects of the geology in central Iran: 4th World Petroleum Cong., Rome, Proc. sect. 1, p. 279–300.
GIGON, W., 1952, Geologie des Habkerntales und des Quellgebiets der Grossen Emme: Naturf. Gesell. Basel, Verh., v. 63, p. 49–136.
GILLULY, J., WATERS, A. C., AND WOODFORD, A. O., 1968, Principles of geology, 2nd ed.: San Francisco, Freeman and Co., 534 p.
GÖRLER, K., AND REUTTER, K. J., 1968, Entstehung und Merkmale der Olistostrome: Geol. Rundschau, v. 57, p. 484–514.
GREENLY, E., 1919, The geology of Angelsey: Great Britain Geol. Survey Mem., 980 p.
HÄFNER, W., 1924, Geologie des südöstlichen Rätikon (zwischen Klosters und St. Antönien): Geol. Karte Schweiz, Beitr., new ser., v. 54, p. 1–33.
HAMILTON, W., 1969, Mesozoic California and the underflow of Pacific mantle: Geol. Soc. America Bull., v. 80, p. 2409–2430.
HARDIN, J., 1966, Strength and ductility, *in* CLARK, S. P., JR. (ed.), Handbook of physical constants: Geol. Soc. America Mem. 97, p. 223–289.
HEIM, A., 1923, Der Alpenrand zwischen Appenzell und Rheintal (Fähnern-Gruppe) und das Problem der Kreide-Nummuliten: Geol. Karte Schweiz, Beitr., new ser., v. 53, p. 1–51.
HERB, R., 1966, Geologie von Amden mit besonerer Berücksichtingung der Flyschbildung: *ibid.*, 144, p. 1–130.
HSÜ, K. J., 1960, Paleocurrent structures and paleogeography of the Ultrahelvetic Flysch basins: Geol. Soc. America Bull., v. 71, p. 577–610.
———, 1967a, Mesozoic geology of California Coast Ranges—a new working hypothesis, *in* SCHAER, J. P. (ed.), Étages tectoniques: Neuchâtel, Switzerland, a la Baconnière, p. 279–296.

———, 1967b, Origin of large overturned slabs of Apennines, Italy: Am. Assoc. Petroleum Geologists Bull., v. 51, p. 65–72.
———, 1968, Principles of mélange and their bearing on the Franciscan-Knoxville paradox: Geol. Soc. America Bull., v. 79, p. 1063–1074.
———, 1969, Preliminary report and geologic guide to Franciscan mélanges of the Morro Bay-San Simeon area, California: California Div. Mines and Geology Special Pub. 35, 46 p.
———, 1970, Cohesive strength, toe effect, and the mechanics of imbricated thrusts: 2nd Internat. Rock Mechanics Conf., Proc. Paper 3-36, 4 p.
———, 1971a, Franciscan mélanges as a model for eugeosynclinal sedimentation and underthrusting tectonics: Jour. Geophys. Research, v. 76, p. 1162–1170.
———, 1971b, Origin of the Alps and western Mediterranean: Nature, v. 233, p. 44–48.
———, 1972, Alpine flysch in a Mediterranean setting: 24th Internat. Geol. Cong., Montreal, Proc. sect. 6, p. 67–74.
———, and Ohrbom, R., 1969, Mélanges of San Francisco Peninsula: geologic reinterpretation of type Franciscan: Am. Assoc. Petroleum Geologists Bull., v. 53, p. 1348–1367.
Irwin, W. P., 1964, Late Mesozoic orogenies in the ultramafic belts of northwestern California and southwestern Oregon: U.S. Geol. Survey Prof. Paper 501C, p. 1–9.
James, D. E., 1971, Plate tectonic model for the evolution of the central Andes: Geol. Soc. America Bull., v. 82, p. 3325–3346.
Kaufman, F. J., 1886, Emmen- und Schlierengegend nebst Kantone Schwyz und Zug und Bürgenstocks bei Stanz: Geol. Karte Schweiz, Beitr., new ser., v. 24.
Kay, M., 1970, Flysch and bouldery mudstone in northeastern Newfoundland, *in* Lafoie, J. (ed.), Flysch sedimentology in North America: Geol. Assoc. Canada, Special Paper 7, p. 155–164.
Leine, L., and Egeler, C. G., 1962, Preliminary note on the origin of the so-called "konglomeratische Mergel" and associated "Rauhwackes," in the region of Menas de Seròn, Sierra de los Filabres (SE Spain): Geologie en Mijnb., v. 41, p. 305–314.
Leupold, W., 1942, Neue Beobachtungen zur Gliederung der Flyschbildung der Alpen zwischen Reuss and Rhein: Eclog. Geol. Helvetiae, v. 35, p. 247–291.
Lindsay, J. F., 1966, Carboniferous subaqueous mass-movement in the Manning-Macleay Basin, Kempsey, New South Wales: Jour. Sed. Petrology, v. 36, p. 719–732.
Lugeon, M., 1916, Sur l'origin des blocs exotiques du Flysch préalpin: Eclog. Geol. Helvetiae, v. 14, p. 217–221.
McFadyen, J. A., Jr., 1956, The geology of the Bennington area, Vermont: Vermont Geol. Survey Bull. 7, 72 p.
Marchetti, M. P., 1957, The occurrence of slide and flowage materials (olistostromes) in the Tertiary series of Italy: 20th Internat. Geol. Cong., Mexico City, 1956, sec. 5, v. 1, p. 209–225.
Maxwell, J. C., 1959, Turbidite, tectonics and gravity transport, northern Apennine Mountains, Italy: Am. Assoc. Petroleum Geologists Bull., v. 43, p. 2701–2719.
Page, B. M., 1963, Gravity tectonics near Passo della Cisa, northern Apennines, Italy: Geol. Soc. America Bull., v. 74, p. 655–672.
———, 1969, Relation between ocean floor spreading and structure of the Santa Lucia Range, California: *ibid.*, Abs. with Programs, pt. 3 (Cordilleran Sec.), p. 51–52.
———, 1970, Sur-Nacimiento Fault zone of California: continental margin tectonics: *ibid.*, Bull., v. 81, p. 667–690.
———, 1972, Oceanic crust and mantle fragment in subduction complex near San Luis Obispo, California: *ibid.*, v. 83, p. 957–971.
Patrick, T. O. H., 1956, Comfort Cove, Newfoundland (geologic map with marginal notes): Geol. Survey Canada Paper 55-31.
Pitman, W. C., and Talwani, M., 1972, Sea-floor spreading in the North Atlantic: Geol. Soc. America Bull., v. 83, p. 619–646.
Rast, N., 1956, The origin of significance of boudinage: Geol. Mag., v. 93, p. 401–408.
Rigo di Rhigi, M., and Cortesini, A., 1964, Gravity tectonics in foothills structure belt of southeast Turkey: Am. Assoc. Petroleum Geologists Bull., v. 48, p. 1911–1937.
Rollier, L., 1912, Ueber obercretazischen Pyritmergel (Wang und Seewener-Mergel) der Schwyzeralpen. Eclos. Geol. Helvetiae, v. 22, p. 178–180.
———, 1923, Supracrétacique et Nummulitique dans les Alpes suisses orientales: Karte Geol. Schweiz, new ser. 53, p. 53–85.
Schardt, H., 1898, Les régions exotiques du versant Nord des Alpes Suisse (Préalpes du Chablais et du Stockhorn et les Klippes): Soc. vaudoise Sci. nat. Bull., v. 34, p. 113–219.
Smith, A. G., 1971, Alpine deformation and the oceanic areas of the Tethys, Mediterranean and Atlantic: Geol. Soc. America Bull., v. 82, p. 2039–2070.
Suppe, J., 1969, Times of metamorphism in the Franciscan terrain of the northern Coast Ranges, California: *ibid.*, Bull., v. 80, p. 134–142.
Swe, W., and Dickinson, W. R., 1970, Sedimentation and thrusting of late Mesozoic rocks in the Coast Ranges, California: *ibid.*, v. 81, p. 165–187.
Tercier, J., 1928, Géologie de la Berra: Karte Geol. Schweiz, new ser. 60, 111 p.
Trümpy, R., 1960, Paleotectonic evolution of the central and western Alps: Geol. Soc. America Bull., v. 71, p. 843–908.
Zen, E. A., and Ratcliff, N. M., 1966, A possible breccia in southwestern Massachusetts and adjoining areas and its bearing on the existence of the Taconic allochton: U.S. Geol. Survey Prof. Paper 550-D, p. 39–46.

3

Reprinted from *Jour. Geophys. Research* **76:**1162–1170 (1971)

Franciscan Mélanges as a Model for Eugeosynclinal Sedimentation and Underthrusting Tectonics

K. Jinghwa Hsü

Laboratory for Experimental Geology, Geological Institute
Swiss Federal Institute of Technology, Zurich

The traditional view representing eugeosynclines as elongate troughs having active volcanism is challenged. Eugeosynclinal rocks are commonly tectonic mélanges, representing mixtures of rocks derived from more than one realm of deposition. The Franciscan mélange includes ophiolites emplaced at a mid-ocean ridge, radiolarites deposited on an abyssal plain, and flysch turbidites poured into a deep-sea trench. The Franciscan mélange was pervasively sheared when the North American plate rode over a Cretaceous Neo-Franciscan plate. The plate junction is not a single overthrust surface. The densely spaced shear surfaces within the mélange signify a westward migration of the plate junction as the Franciscan was being underthrust. The Franciscan glaucophane schists, metamorphosed under very high pressure 130–150 m.y. ago, record a pre-Cretaceous Franciscan history. Geological evidence suggests that those once deeply buried rocks were squeezed out during a Late Jurassic alpine type of deformation (Nevadan orogeny), when the North American continent collided with a microcontinent (Salinia). The Cretaceous underthrusting along a consuming plate margin resulted in mélange deformation and in lawsonite-schist metamorphism of the neo-Franciscan rocks, which are coeval with the Great Valley sequence.

In several segments of the Tethyan and circum-Pacific mountain systems, large areas are underlain by bodies of sheared and broken strata. The stratal continuity has been disrupted by innumerable dislocation surfaces, and rocks of diverse origins have been mixed, in the form of chaotically arranged blocks in a pelitic matrix. The term 'mélange' has been used to designate such bodies [*Greenly*, 1919; *Hsü*, 1968). The group of graywackes, radiolarian cherts, and ophiolites in the California Coast Ranges constitute such a mélange. These rocks, commonly referred to as the Franciscan group, have been considered eugeosynclinal, and their mélange structure has been cited as evidence of deformation within an ancient Benioff zone. However, these generalizations only provide semantic havens for unanswered questions: What is a eugeosyncline? Where do we find an actualistic example? How are mélanges formed within a Benioff zone? How did the mantle rocks manage to find their way to the earth's surface? This paper attempts to clarify the answers to these questions.

Eugeosyncline and Sea-Floor Spreading

The Mesozoic rocks of the California Coast Ranges include (1) the Coast Range batholiths and associated metamorphic rocks, (2) the Franciscan group, and (3) the Great Valley sequence (Figure 1). The first represents the remobilized sialic basement of an ancient microcontinent, Salinia [*Hsü*, 1967]. The latter two include flysch turbidites that range in age from Late Jurassic to Late Cretaceous: Yet the two have been distinguished by the facts that the Franciscan is characterized by a mélange structure [*Hsü*, 1968], and that this mélange includes ophiolites and radiolarian cherts, in addition to flysch turbidites.

It has been commonly supposed that there were two late Mesozoic troughs of deposition: a Franciscan eugeosyncline on the west and a Great Valley miogeosynclinal on the east [e.g., *Bailey and Blake*, 1969] (see Figure 2*a*). This reconstruction is an illustration of the classical concept of 'eu-miogeosyncline couples' [*Aubouin*, 1965]. The miogeosynclinal prism has been compared to thick clastic wedges on continental margins. But what is a 'eugeosyncline'?

Eugeosynclinal belts were once 'believed to

have been similar to modern island arcs and associated troughs' [*Kay*, 1951, p. 72]. Yet the andesitic volcanics of island-arc systems are genetically distinct from the eugeosynclinal ophiolites. In fact, the island-arc setting provides an actualistic model for flysch paleogeography [*Hsü and Schlanger*, 1971], and the alpine flysch is not eugeosynclinal.

The term 'eugeosyncline,' proposed by *Stille* [1941] and defined as a 'surface that has subsided deeply in a belt having active volcanism' [*Kay*, 1951, p. 4], assumed that such a surface once existed. What has been overlooked is the fact that eugeosynclinal rocks commonly constitute mélanges: examples, in addition to the Franciscan, are the mélange at Angelsey [*Greenly*, 1919], the Zermatt-Saas and Arosa Schuppen zones in the Alps [*Bearth*, 1967; *Steinmann*, 1906], the Ligurian mélange of the Apennines [*Bailey and McCallien*, 1963], the Ankara mélange of Turkey [*Bailey and McCallien*, 1950], the colored mélange of Iran [*Davoudzadeh*, 1959], and the Exotic Blocks zone of the Himalayas [*Gansser*, 1964]. Such

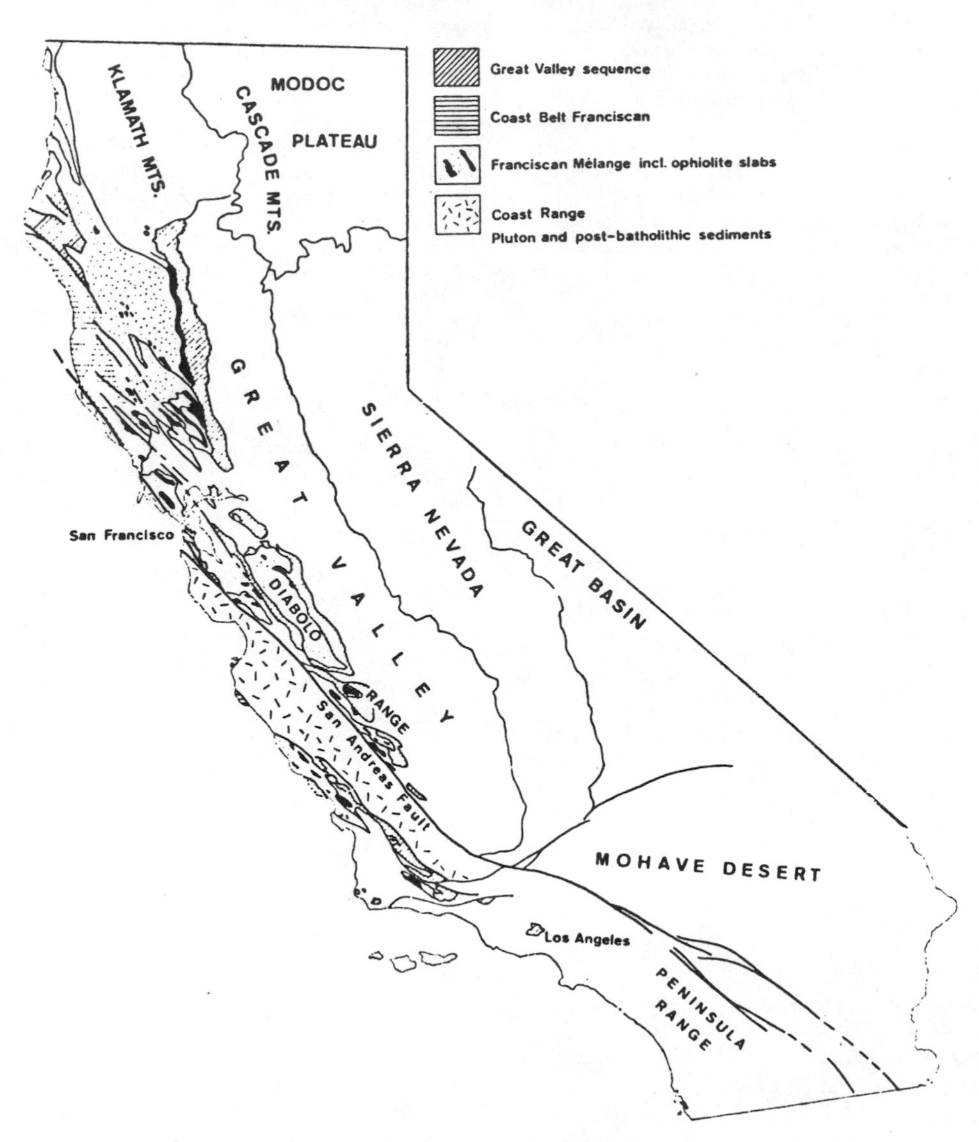

Fig. 1. Tectonic provinces of California.

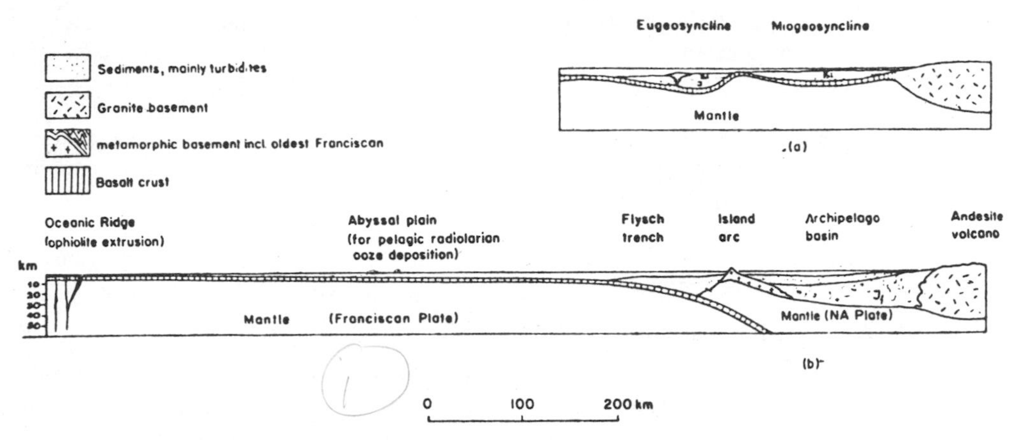

Fig. 2. Reconstruction of the Franciscan eugeosyncline: (*a*) as the outer member of a eu-miogeosynclinal couple, traditional model [after *Bailey and Blake*, 1969]; (*b*) as a segment of the Pacific Ocean, plate tectonics model.

mélanges represent mixtures of rocks derived from more than one realm of deposition that were hundreds or thousands of kilometers apart.

The Franciscan mélanges include mainly three major rock types: the ophiolites, the biogenic pelagic sediments, and the deep-water clastic sediments:

The ophiolite consists of (in descending order of their original superposition) basalt tuff, pillow breccia, pillow basalt, diabase, fine-grained gabbro, and several kinds of ultramafic rocks. The mafic rocks, though locally altered to spilites, have been derived from normal tholeiitic magmas [*Bailey et al.*, 1964, p. 51]. The actualistic model for the generation of such ophiolites is provided by the volcanism of the mid-oceanic ridges, where tholeiitic magmas are extruded [*Moores*, 1969; *Bird and Dewey*, 1970]. The ultramafic rocks, which include pyroxenites and serpentinized peridotites, have been traversed by many dikes of diabase (now rodingite) [*Hsü*, 1969]; such a complex may well constitute the so-called mantle-crust mixed zone, which exists under the present mid-ocean ridges [e.g., *Talwani et al.*, 1965].

The Franciscan biogenic sediments are mainly radiolarian cherts. Locally, but rarely, pelagic limestones are present. The cherts, being recrystallized radiolarian oozes, must have been deposited on an abyssal plain below the calcite compensation depth, whereas calcareous planktons, accumulated on local rises or seamounts, formed the limestones.

The Franciscan deep-water clastics are typically flysch, deposited in a trench with an island-arc setting. The sources for the Franciscan flysch include not only products of contemporary andesitic volcanism, but also debris of older Franciscan rocks then exposed in newly uplifted chains of islands. The Upper Cretaceous Franciscan flysch contain much K-feldspar [*Hsü and Ohrbom*, 1969], when the newly unroofed discordant plutons of the Coast Range batholith began to contribute debris to the Franciscan trough.

The paleogeographic reconstruction I propose is shown by Figure 2*b*. Instead of the traditional concept of an eu-miogeosynclinal couple, the Franciscan rocks were shown to have been deposited in three different oceanic provinces: the mid-ocean ridge, where ophiolites were being extruded, may have been at a distance of some 1000 km from the flysch trench at plate junction while radiolarian oozes were being deposited on a wide abyssal plain.

Petrographical evidence indicates that the Cretaceous Franciscan flysch was derived from an island chain where Franciscan rocks were exposed [*Hsü*, 1969]. These islands, a part of an island-arc system, separated the Franciscan flysch from the flysch turbidites of the Great Valley sequence, which were deposited in a deep-sea basin behind the arc. Farther to the east were andesitic volcanoes, which were rooted in discordant plutons.

Franciscan paleogeography, thus interpreted,

did not consist of a narrow elongate trough, which received sediments and submarine volcanics. Rather, the various 'eugeosynclinal' deposits were laid down on a segment of the Northwest Pacific, which can be called the Mesozoic Neo-Franciscan plate. Since the plate was plunged under the North American continent, the only vestiges of this once extensive plate are the blocks of the various Franciscan rocks now intimately mixed in the mélange.

MÉLANGE DEFORMATION RELATED TO BENIOFF ZONE

The Great Valley sequence was deposited on a pre-Tithonian (latest Jurassic) Franciscan basement [*Hsü*, 1967]. On the west side of the Sacramento Valley and in the Santa Lucia Range, this Franciscan basement is an ophiolitic complex [*Hsü*, 1969; *Bailey et al.*, 1970]. In the Diablo Range, the Great Valley sequence has been superposed upon metamorphosed Franciscan rocks; locally (e.g., at Pacheco Pass) unmetamorphosed sediments of the former lie directly above Franciscan jadeitic graywackes [*McKee*, 1962]. A fault commonly intervenes where the Great Valley sequence and the Franciscan rocks are in contact [*Page*, 1966], and this fault has been interpreted as a major overthrust [e.g., *Bailey et al.*, 1964; *Dickinson*, 1966; *Hamilton*, 1969; *Ernst*, 1970]. However, the major crustal displacement did not take place along this, or any other, single surface of discontinuity, but along innumerable shear surfaces within the Franciscan mélange [*Hsü*, 1968]. In fact, the mélange is the result of such severe penetrative strain.

The order of magnitude of the shearing strain can be estimated on the basis of a kinematic consideration. Magnetic stratigraphy suggests a half-spreading rate of about 4 cm/yr for the North Pacific in Late Cretaceous [*Heirtzler et al.*, 1968]. At this rate, the crustal displacement during the last 25 m.y. of Cretaceous deformation would amount to 1000 km. If such displacement did not take place along any single surface but was distributed as penetrative shear over a 10-km-thick Franciscan package, the resulting shearing strain would be 10^2. Such large magnitude of shearing strain is indicated by the tectonic style of the mélange: Brittle layers of the Franciscan graywackes have been broken along extensional shear fractures, and the lozenge-shape boudines so formed have been rotated 90 degrees, or completely overturned, through the ductile deformation of the shaly matrix. A 90-degree rotation of rigid boudines in a ductile matrix could only have resulted when shearing strains reach the orders of 10 or 10^2 [*Ramsay*, 1967, p. 223].

The Franciscan mélange has been interpreted as a fossilized Benioff zone [e.g., *Hamilton*, 1969; *Ernst*, 1970]. According to this hypothesis, the Great Valley sequence was the leading edge of the Mesozoic North American plate, and its thrust fault contact with the Franciscan was the plate junction. Yet the thrust contact was nearly horizontal and extended nearly 100 km across the Northern Coast Ranges before this surface was later folded [*Bailey et al.*, 1964; *Bailey and Blake*, 1969]. Such a horizontal 'great overthrust' could hardly have been the upper boundary of an inclined Benioff zone along which the oceanic crust and mantle rode relatively eastward beneath the continent and its fringing sedimentary shelf. The common postulate to equate the mélange to the inclined 'Benioff seismic zone of tectonic intermixing of crustal and mantle rocks,' as shown by Figure 3*a*, is oversimplified. If such a postulate were correct, the 'great overthrust' would have had to be tilted 20 degrees or more from an inclined to a nearly horizontal position after the mélange formation, implying thus that the Great Valley rocks were once buried to 50-km depth. This is obviously incorrect. An alternative model is schematically shown by Figure 3*b*: A mélange belt some 100 km wide or more resulted from a migration of the plate junction. As the Franciscan flysch trough occupied successively more westerly positions, the older mélanges were accreted to the North American plate and were uplifted as islands to provide detritus for younger Franciscan flysch. In this model, the Benioff zone was not a single shear zone at a fixed plate junction, but a parallel set of shear surfaces, marking the successive traces of a migrating plate junction.

MANTLE ROCKS IN MÉLANGES

The exotic blocks in the Franciscan mélange that may have been derived from the mantle or from the deepest parts of the crust are: (1) ultramafic rocks of the ophiolite suite, (2)

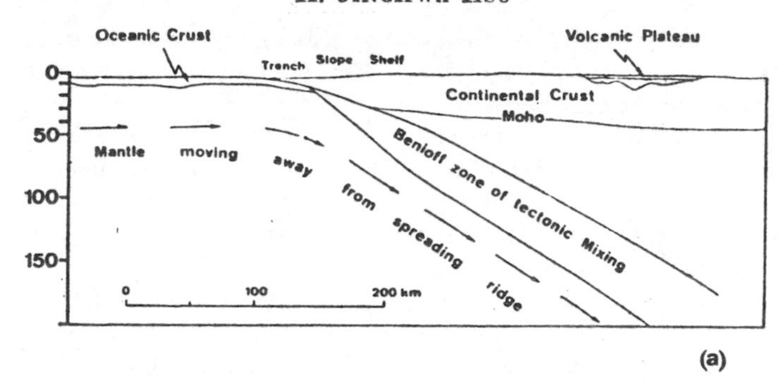

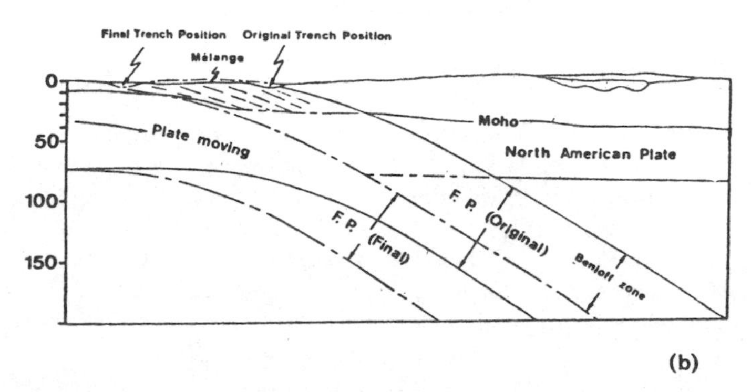

Fig. 3. Origin of Franciscan mélanges: (*a*) as an inclined zone of tectonic mixing at a fixed plate junction [after *Hamilton*, 1969], (*b*) as sub-parallel traces of a migrating plate junction (F.P., Franciscan plate).

eclogites, and (3) crustal rocks metamorphosed under very high pressures.

The occurrence of ultramafic ophiolites, which may represent a part of the upper mantle, presents no problem; the postulated model of Franciscan deformation accounts for the mixing of blocks detached from a shallow oceanic mantle with crustal rocks. The mixing of eclogites would also be expected, if they were present in the mantle as segregations [e.g., *Ringwood*, 1962]. The puzzle is represented by the presence of jadeitic and aragonitic metasedimentary and meta-volcanic rocks. Experimental evidence indicates that these rocks should have been metamorphosed at depths of 20 or 30 km or more [e.g., *Ernst*, 1965; *Brace et al.*, 1970]. That those rocks could have been brought down to such depths by underthrusting is not surprising. Yet, how did they find their way back to the surface and come into contact with unmetamorphosed Great Valley rocks?

The Coast Ranges have been uplifted since the mélanges were formed, but the magnitude of the vertical displacement probably did not exceed 10 km. In any case, vertical uplift could not explain the fact that jadeitic graywackes directly underlie unmetamorphosed Great Valley sequence in some parts of the Diablo Range [e.g., *McKee*, 1962]. *Gilluly* [1969] used the expression 'squeezed upward' to explain the presence of the high-pressure metamorphics in the mélange. Yet the movement along the Benioff zone is downward. Also, if the jadeitic rocks of the Pacheco Pass have been squeezed up some 30 km, it is surprising indeed that the adjacent Great Valley rocks show no evidence of being intensively sheared.

The 'squeezed upward' type of deformation is more typical of the Alpine or the Himalaya type of orogeny, involving collision of continents, whereby parts of the deeply buried ophiolitic sequences were squeezed out, while the thin ocean plate between two thicker continental plates was being consumed. Examples are the Pennine zone of the Alps, or the Exotic Blocks zone of the Himalayas [*Mitchell and Reading*, 1969]. Is it possible that the Franciscan jadeitic rocks were squeezed out during an earlier event before the Cretaceous underthrusting along the Benioff zone? Several lines of evidence suggest that this might well be the case:

Radiometric dates indicate that there have been two or more episodes of Franciscan metamorphism [*Coleman*, 1967; *Suppe*, 1969; *Ernst*, 1970]: a Jurassic episode, between 130 and 150 m.y., for the *unfossiliferous* glaucophane schists in the Diablo Range and its northern extension, and a Cretaceous episode, between 110 and 130 m.y., for the sparsely fossiliferous lawsonite schists in the Northern Coast Ranges. A younger 75- to 80-m.y. event, involving albitization of older glaucophane schists, has also been reported [*Keith and Coleman*, 1968]. The evidence thus suggests that the metamorphism, resulting in the formation of the very high-pressure jadeite-bearing rocks, took place before the deposition of the earliest Great Valley sequence in Tithonian (latest Jurassic). The contact between the unmetamorphosed Knoxville and the jadeitic-bearing Franciscan in the Diablo Range could thus be interpreted as a major unconformity modified by subsequent faulting or minor thrusting, rather than a great overthrust involving 100 km or more displacement [*Hsü*, 1967].

According to this interpretation [*Hsü*, 1967], the metamorphism of the jadeite-bearing glaucophane schists took place during the Late Jurassic. This event, which may be considered a part of the Nevadan orogeny, was an alpine type of deformation, involving the collision of the North American continent and a microcontinent Salinia. The schists, and other deeply buried rocks such as serpentinites and eclogites, were squeezed out and thrust over the Salinia. The deformed and metamorphosed older Franciscan rocks were partially eroded before the first (Tithonian) Great Valley and neo-Franciscan sediments were laid down. During the Cretaceous, blocks and slabs of glaucophane schists and mantle rocks were emplaced into the neo-Franciscan trench as sedimentary breccias (olistostromes), which were subsequently mixed with other Franciscan rocks to form tectonic mélanges.

MESOZOIC FRANCISCAN HISTORY

I have presented detailed geological arguments in a series of papers to prove that the emplacement of the ophiolite and the deposition of flysch west of the North American continent started long before Cretaceous, and that the Franciscan had a pre-Tithonian history [*Hsü*, 1967, 1968, 1969; *Hsü and Ohrbom*, 1969]. The Mesozoic history, interpreted in terms of new global tectonics, is shown schematically by Figure 4:

1. *Early Mesozoic sphenoclastic rifting (part 1 of Figure 4).* During the early Mesozoic, the accreting plate margin was between the main North American continent and a peninsula (or microcontinent) Salinia. (This type of continental margin has its actualistic counterpart in the Gulf of California, which is being opened by a spreading ridge). Pre-Tithonian ophiolites and radiolarites were laid down in such a Paleo-Franciscan gulf or strait.

2. *Jurassic plate-consumption (part 2 of Figure 4).* During the Jurassic, the Paleo-Franciscan plate plunged under the North American plate. (The Paleo-Franciscan plate, located between a spreading ridge axis and the North American continent, can be compared to the actualistic Juan de Rica plate, except the latter does not have a microcontinent to the west.) Pre-Tithonian Franciscan graywackes were deposited in a trench at the consuming plate margin. Jurassic andesitic volcanoes, underlain by discordant plutons, were present in the Sierra Nevada region.

3. *Late Jurassic Nevadan orogeny (part 3 of Figure 4).* During the Late Jurassic, the North American continent collided with the peninsula or microcontinent Salinia: the Paleo-Franciscan plate was eliminated, and the older Franciscan rocks (San Francisco unit, *Hsü and Ohrbom* [1969]) were deformed. Deeply buried metamorphic and mantle rocks were squeezed out and partly thrust over the Salinia during this orogenic episode.

4. *Cretaceous plate-consumption (part 4 of Figure 4).* Beginning from latest Jurassic (Tithonian) and continuing until Late Cretaceous, and probably until early Tertiary, the western

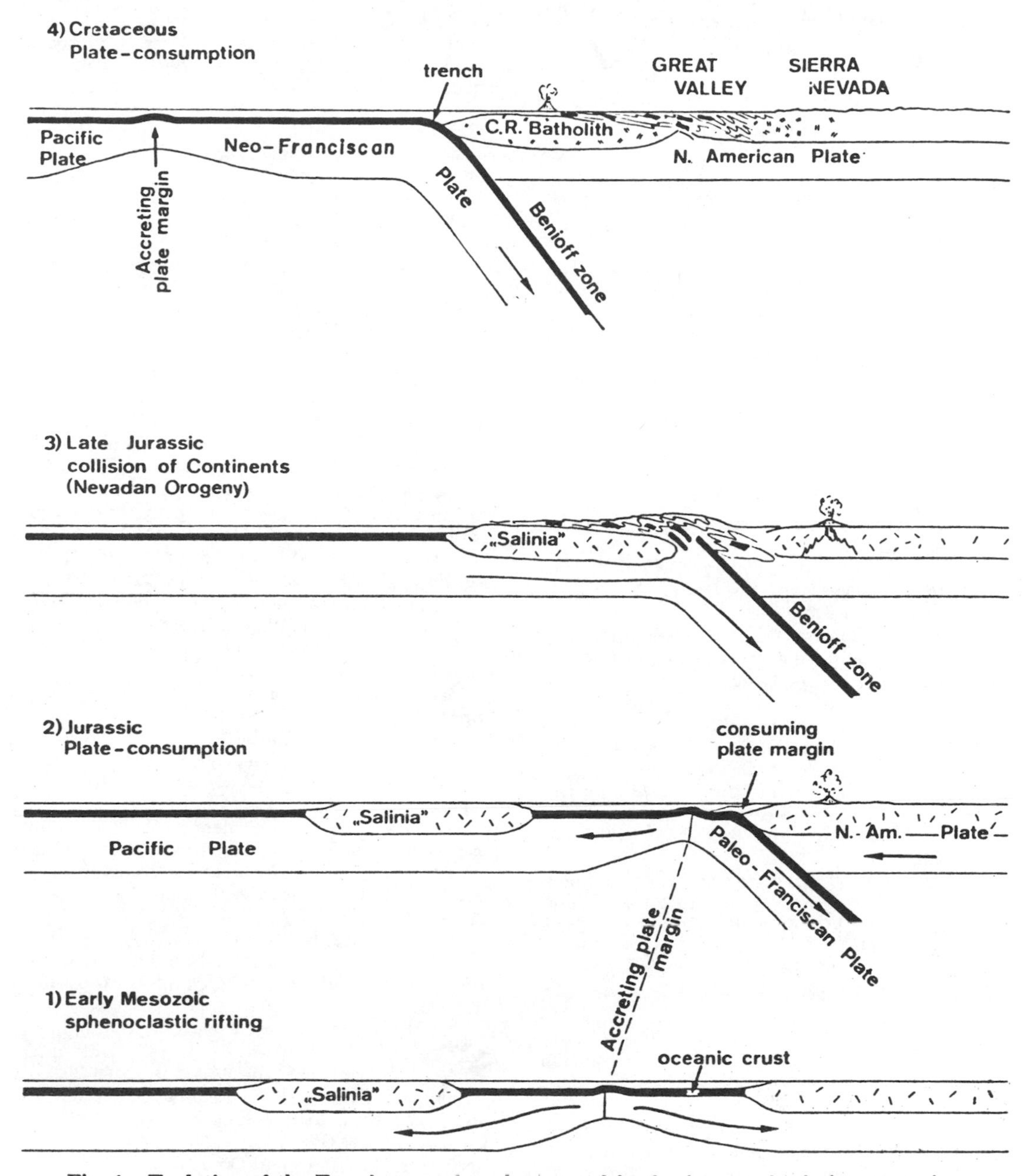

Fig. 4. Evolution of the Franciscan geology interpreted in the framework of plate tectonics.

North American margin had a trench. A Neo-Franciscan plate was present between a spreading ridge and the North American plate. Ophiolites were extruded at the ridge, radiolarites accumulated on an abyssal plain, and flysch graywackes were poured into the trench. The neo-Franciscan rocks (San Mateo unit, *Hsü and Ohrbom* [1969]) are coeval to the Great Valley sequence. Mixing of these rocks under the trench, and along the Benioff zone, produced the mélanges. Meanwhile, Cretaceous andesitic volcanoes were present in Salinia, as this microcontinent was converted by magmatic activities into the Coast Range batholith.

The postulated tectonic history is thus considerably more complicated than the current view illustrated by part 4 of Figure 4 (Cretaceous plate consumption). This postulate is necessary, however, to account for the pre-Tithonian Franciscan history, which has been commonly ignored.

Acknowledgment. This research was supported by the National Science Foundation, NSF grant GP 4634. The author has benefited from discussions with John Crowell, Eldridge Moores, John Dewey, Jerry Winterer, Bill Dickinson, Bob Coleman, and many others during a field trip to the Franciscan terrain before the Penrose Conference on New Global Tectonics, Monterey, 1969. Needless to say, the author alone is responsible for any unorthodox interpretations or inaccuracies that may exist in this paper.

REFERENCES

Aubouin, Jean, *Geosynclines*, 335 pp., Elsevier, Amsterdam, 1965.

Bailey, Sir E. B., and W. J. McCallien, The Ankara mélange and the Anatolian thrust, *Nature, 166*, 938–940, 1950.

Bailey, Sir E. B., and W. J. McCallien, Liguria nappe: Northern Apennines, *Roy. Soc. Edinburgh Trans., 65*, 315–333. 1963.

Bailey, E. H., and M. C. Blake, Tectonic development of western California during the late Mesozoic, *Geotektonika*, pt. 3, 17–30, pt. 4, 24–34, 1969.

Bailey, E. H., M. C. Blake, and D. L. Jones, On-land Mesozoic oceanic crust in California Coast Ranges, *U.S. Geol. Surv. Prof. Pap. 700-C*, 70–81, 1970.

Bailey, E. H., W. P. Irwin, and D. L. Jones, Franciscan and related rocks and their significance in the geology of western California, *Calif. Div. Mines Geol. Bull., 183*, 177 pp., 1964.

Bearth, P., Die Ophiolithe der Zone von Zermatt-Saas Fee, *Beitr. geol. Karte Schweiz, n.f. 132*, 130 pp., 1967.

Bird, J. M., and J. F. Dewey, Lithosphere plate-continental margin tectonics and the evolution of the Appalachian orogen, *Bull. Geol. Soc. Amer., 81*, 1031–1060, 1970.

Brace, W. F., W. G. Ernst, and R. W. Kallberg, An experimental study of tectonic overpressure in Franciscan rocks, *Bull. Geol. Soc. Amer., 81*, 1325–1338, 1970.

Coleman, R. G., Glaucophane schists from California and New Caledonia, *Tectonophysics, 4*, 479–498, 1967.

Davoudzadeh, M., *Geologie und Petrographie des Gebietes nördlich von Nain, Zentral-Iran*, Diss. ETH Zürich, Zimmermann, Uster, 91 pp., 1969.

Dickinson, W. R., Table Mountain serpentinite extrusion in California Coast Ranges, *Bull. Geol. Soc. Amer., 77*, 451–472, 1966.

Ernst, W. G., Mineral parageneses in Franciscan metamorphic rocks, Panoche Pass, California, *Bull. Geol. Soc. Amer., 76*, 879–914, 1965.

Ernst, W. G., Tectonic contact between the Franciscan Mélange and the Great Vally Sequence: Crustal expression of a Late Mesozoic Benioff zone, *J. Geophys. Res., 75*, 886–901, 1970.

Gansser, A., *Geology of the Himalayas*, Wiley, New York, 289 pp., 1964.

Gilluly, J., Oceanic sediment volumes and continental drift, *Science, 166*, 992–994, 1969.

Greenly, E., The Geology of Angelsey, *Great Britain Geol. Surv. Mem.*, 980 pp., 1919.

Hamilton, W., 1969, Mesozoic California and the underflow of Pacific mantle, *Bull. Geol. Soc. Amer., 80*, 2409–2430, 1969.

Heirtzler, J. R., G. O. Dickson, E. M. Herron, W. C. Pitman, III, and X. Le Pichon, Marine magnetic anomalies, geomagnetic field reversals, and motion of the ocean floor and continents, *J. Geophys. Res., 73*, 2119–2136, 1968.

Hsü, K. J., Mesozoic geology of the California Coast Ranges: A new working hypothesis, in *Etages tectoniques*, edited by J. P. Schaer, pp. 216–296, Baconnière, Neuchâtel, Suisse, 1967.

Hsü, K. J., Principles of mélange and their bearing on the Franciscan-Knoxville paradox, *Bull. Geol. Soc. Amer., 79*, 1063–1074, 1968.

Hsü, K. J., Preliminary report and geologic guide to Franciscan mélanges of the Morro Bay–San Simeon area, California, *Calif. Div. Mines, Geol. Spec. Publ. 35*, 46 pp., 1969.

Hsü, K. J., and R. Ohrbom, Mélanges of San Francisco Peninsula: Geologic reinterpretation of type Franciscan, *Bull. Amer. Assoc. Petrol. Geol., 53*, 1348–1367, 1969.

Hsü, K. J., and S. O. Schlanger, Ultrahelvetic Flysch sedimentation and deformation related to plate-tectonics, *Bull. Geol. Soc. Amer.*, 1971.

Kay, M., North American geosynclines, *Geol. Soc. Amer. Mem., 48*, 143 pp., 1951.

Keith, T. C., and R. G. Coleman, Albite-pyroxene-glaucophane schist from Valley Ford, California, *U.S. Geol. Surv. Prof. Paper 600-C*, 13–17, 1968.

McKee, B., Widespread occurrence of jadeite, lawsonite, and glaucophane in central California, *Amer. J. Sci., 260*, 596–610, 1962.

Mitchell, A. H., and H. G. Reading, Continental margins, geosynclines, and ocean floor spreading, *J. Geol., 77*, 629–646, 1969.

Moores, E. M., Petrology and structure of the Vourinos ophiolitic complex of northern Greece, *Geol. Soc. Amer. Spec. Paper 118*, 74 pp., 1969.

Page, B. M., Geology of the Coast Ranges of California, *Calif. Div. Mines Geol. Bull., 190*, 255–276, 1966.

Ramsay, J. R., *Folding and Fracturing of Rocks*, McGraw-Hill, New York, 568 pp., 1967.

Ringwood, A. G., A model for the upper mantle, *J. Geophys. Res., 67*, 857–867, 4473–4478, 1962.

Steinmann, G., Geologische Beobachtungen in den Alpen: die schardtsche Ueberfaltungs-theorie und die geologische Bedeutung der Tiefseeabsätze und der ophiolithischen Massengesteine, *Ber. Natf. Ges. Freiburg i.B., 16,* 1–49, 1906.

Stille, H., *Einführung in den Bau Amerikas,* Bornträger, Berlin, 717 pp., 1941.

Suppe, J., Times of metamorphism in the Franciscan terrain of the Northern Coast Ranges, California, *Bull. Geol. Soc. Amer., 80,* 135–142, 1969.

Talwani, M., X. Le Pichon, and M. Ewing, Crustal structure of the Mid-ocean Ridges, 2, Computed model from gravity and seismic refraction data, *J. Geophys. Res., 70,* 341–352, 1965.

(Received July 30, 1970.)

4

Reprinted from *Am. Assoc. Petroleum Geologists Bull.* **56**:2295-2302 (1972)

What is Franciscan?[1]

JAMES O. BERKLAND,[2] LOREN A. RAYMOND,[2] J. CURTIS KRAMER,[3]
ELDRIDGE M. MOORES,[3] and MICHAEL O'DAY[3]
Boone, North Carolina 28607, and Davis, California 95616

INTRODUCTION

Every geologist studying rocks called "Franciscan" in western California must contend with the fundamental question—what is Franciscan? This point was emphasized by Hsu and Ohrbom (1969, p. 1365). Transformations in concepts of the Franciscan are reflected by variations in nomenclature (*e.g.,* Series, Formation, Group, assemblage, mélange) since geologists began working with these rocks more than 100 years ago. Thus, the nature of the Franciscan has been closely bound to a secondary problem—what should it be called? Even today there is no generally accepted formal name, and the recent literature contains such terms as "the Franciscan," Franciscan suite, Franciscan sequence, Franciscan assemblage, and Franciscan terrane. Hsu and Ohrbom raised the question primarily with regard to the age of the Franciscan, whereas this paper seeks to elucidate the nature and the nomenclature of the Franciscan.

Almost by definition, Franciscan rocks consistently have been considered to be of Mesozoic age. During the first half of this century, most authors regarded Franciscan as an unfossiliferous Upper Jurassic stratigraphic unit. This was partly explained by Reed (1933, p. 93-94), "When a supposed Franciscan locality yields fossils, it is nearly always transferred forthwith to some other formation." Such fossil occurrences were particularly questioned when they dated the rocks as younger than Jurassic. Thus age was a criterion for distinguishing Franciscan rocks.

Irwin (1957) demonstrated that Franciscan rocks range at least from Late Jurassic to Late Cretaceous in age and are, therefore, temporally equivalent to much of the Great Valley sequence. Although this important work clearly documented that age alone could not be used to define Franciscan rocks, the dogma—that Franciscan rocks are strictly of late Mesozoic age—often surfaces in written or oral discussions. Recently discovered fossils (Kramer *et al.,* 1972; D. L. Jones, 1972, personal commun.) have established an early Cenozoic age for parts of the "coastal belt" of the Franciscan. Therefore, the question arises among some workers as to whether the "coastal belt" should be considered as Franciscan. To respond adequately to this question, *we must first determine what* Franciscan *is.*

Examination of this problem reveals that interpretation of the nature of the Franciscan and use of the many terms applied to Franciscan rocks are almost as varied as the rocks themselves. Effective communication between geologists working with Franciscan problems is critical to concise evaluation and synthesis of the diverse aspects of Franciscan geology, and requires that some agreement be reached on the meaning of terms used in descriptions and discussions. Toward that end, we present, herein, a glossary of Franciscan terms, hopefully to serve as a basis for more complete communication between investigators of Franciscan and Franciscan-like

[1] Manuscript received, March 15, 1972; accepted, May 20, 1972.

[2] Department of Geography-Geology, Appalachian State University.

[3] Department of Geology, University of California at Davis.

The writers express their appreciation to students and faculty of the University of California at Davis for many stimulating discussions on a wide range of subjects, including those covered in this paper. Special thanks are given to E. H. Bailey of the U. S. Geological Survey at Menlo Park, California, for his thoughtful and constructive criticisms of an earlier version of this paper. Many of his suggestions have been incorporated in the present article, but the responsibility for interpretations and the viewpoints given are our own. Part of the costs of field work and laboratory studies involved in background data for this paper was defrayed by NSF Grant GA 14298.

rocks. Included is a redefinition of the Franciscan that we feel is broad enough to be factual, yet explicit enough to be descriptively useful.

Discussion of Terminology

Tectonic and exotic blocks—Bailey *et al.* (1964, p. 93, 97) and Ernst *et al.* (1970, Ch. V) used the term "tectonic block" for a tectonically emplaced rock, from a few feet to more than several hundred feet in diameter, occurring in a lithologic association foreign to that in which the mass formed. A similar definition is given by Hsu (1968) for "exotic block." We believe that a distinction between the two types of blocks should be utilized. Although we concur with Hsu that exotic blocks are foreign to the main body of rock with which they are associated, we recognize that such blocks may be introduced into a terrane or rock unit by sedimentary, as well as tectonic processes. Thus, we redefine exotic blocks as *variably sized masses of rock occurring in a lithologic association foreign to that in which the mass formed.* Tectonic blocks are here considered more restricted in origin and are redefined as *variably sized masses of rock that have been transported with respect to adjacent rock masses through the operation of tectonic processes.* In lithology or metamorphic grade, tectonic blocks may be either compatible or incompatible with the host rocks. Clearly, not all tectonic blocks are exotic, and not all exotic blocks are tectonic.

Knocker—"Knocker" is a colloquial field term which denotes a *resistant, rounded monolith, ranging from a few feet to several hundred feet in diameter, that generally stands out prominently above the level of the surrounding terrane.* Such a mass usually consists of a hard, resistant lithologic type, such as schist, eclogite, greenstone, or chert. Bailey (written commun.) pointed out that some geologists consider as knockers only metamorphic rocks. We feel this is unduly restrictive and limits the usefulness of the field term. Ghent (1964, p. 103) used the term in a broad, field sense in describing "knockers of igneous rock." A practical advantage of such usage is seen where Franciscan terranes are viewed at a distance or on a photograph. Areas might be designated as having abundant knockers, but restricted use of the term would require relabeling of the nonmetamorphic blocks following field or laboratory examination. We find it much more convenient to use "knocker" in the general, nongenetic sense, for both tectonic and exotic blocks, where their true nature has not been determined.

Mélange—The term mélange (French, mixture) was introduced by Greenly (1919; see Hsu, 1968) to describe terranes characterized by lentils or phacoids of more durable rocks embedded in a sheared matrix of more tractable materials, the original sedimentary, igneous, or metamorphic junctions having been destroyed, apparently as a result of tectonic processes. As noted by Greenly (1919), a mélange may include rocks derived from two stratigraphic units separated by an unconformity. Hsu (1968, p. 1065) explicitly included tectonism as the formative agent and redefined the term as follows:

> . . . mélanges are thus defined as mappable bodies of deformed rocks characterized by the inclusion of tectonically mixed fragments or blocks, which may range up to several miles long, in a pervasively sheared, fine-grained, and commonly pelitic matrix. Each mélange includes both exotic and native blocks and a matrix.

In restricting the term *mélange* to *tectonically mixed* masses of rock, Hsu designated mixtures produced by sedimentary processes as olistostromes. However, as noted by Hsu (1968, p. 1066), ". . . distinction between an olistostrome and a mélange becomes difficult, if not impossible, if the olistostrome has undergone mélange deformation." Because we feel that basically chaotic rock units, as well as rock-stratigraphic units, should be divorced as much as possible from inferred geologic history, we suggest a redefinition of the term "mélange" as follows. *A mélange is a mappable body of rock characterized by the inclusion of fragments and blocks of all sizes, both exotic and native, embedded in a fragmented and generally sheared matrix of more tractable material.*

If the origin of the rock body is known, mélanges may be subdivided into tectonic mélanges, produced by tectonic processes, and olistostromes, produced by sedimentary processes. The advantages of the new definition are: (1) it insures that the term, "mélange," remains a useful field term by retaining Hsu's criterion that it be a mappable body of rock, (2) it allows chaotic units of unknown origin to be assigned a lithologic designation that is free of genetic implications, and (3) it removes the unnecessary—although characteristic—condition that the matrix be fine grained.

Broken formation—The definition of a broken formation is accepted here as presented by Hsu (1968, p. 1065-1066).

> *(A mappable) body of broken strata that contains no exotic elements . . . is called a broken formation,* because such a body, regardless of its broken state, functions as a rock-stratigraphic unit.

Broken formations may contain tectonic blocks, but not exotic blocks.

Basement terrane—Within a given region, basement terrane may be defined as the lowest mappable mass of rock, generally with complex structure, that structurally underlies the other major rock sequences of the area. This definition differs from that offered by the AGI (1962) for the term, "basement," in that we specify mappability but do not require an unconformable upper contact—a restriction that does not allow for *Abscherung zones, décollements,* or faults. Our designation of basement terrane is useful in the California Coast Ranges and we do not attempt to resolve existing widely different uses of the term basement (*e.g.*, any pre-Tertiary rock, granite or gneiss, igneous rock underlying sediments, or geophysical acoustic basements). The issue as to whether any of the Franciscan terranes constitute basement is resolved by noting that neither top nor base of the Franciscan has ever been recognized. If the base of a mappable unit is not known, then that unit is, for all practical purposes, the basement terrane. Our concept of Franciscan basement terrane parallels that of Suppe (1970, p. 3254).

"The Franciscan"—Lawson (1895) named the rocks of the San Francisco Peninsula, including the San Francisco Sandstone of Blake (1858), the Franciscan series. No type locality has been formally designated, but the San Francisco Peninsula is generally considered the type area (Bailey *et al.*, 1964; Hsu and Ohrbom, 1969). The rock types of the area include graywacke, siltstone, conglomerate, limestone, radiolarian chert, basic volcanic rocks, serpentinite, and glaucophane schist. Lawson noted similar lithologic associations over broad areas of the California Coast Ranges. During subsequent investigations he and other workers extended the name Franciscan to similar rocks, which together comprise the greater part of the basement terrane of the Coast Ranges.

Various workers between 1895 and 1964 considered the Franciscan to be a rock-stratigraphic unit and designated it as the Franciscan Series, Franciscan Formation, or Franciscan Group; however, the Franciscan does not fit the accepted definitions of these terms (ACSN, 1961). Bailey *et al.* (1964) signaled a new awareness of the nature of the Franciscan by suggesting that it represents an *assemblage* "representative of a particular sedimentary and tectonic environment."

Page (1966, p. 260) also emphasized the significant tectonic style of the Franciscan, noting that "the structure of the Franciscan is as characteristic as the petrology." Page further stated (1966, p. 260):

> Viewed in broad perspective, the entire Franciscan might be regarded as a gigantic tectonic zone at the fringe of the continent.

This description of the Franciscan, although interesting in terms of later developed plate-tectonic concepts, led to such undesirably generalized extensions of the name Franciscan as that proposed by Bogdanov (1969) for *all* ocean floor-trench sequences that border the Pacific Ocean.

In a series of articles, Hsu (1967, 1968, 1969, 1971; and Hsu and Ohrbom, 1969) emphasized shearing as the major deformational aspect of the Franciscan. Hsu (1967, p. 283) denied that mappable folds are present, asserting that, in the Franciscan, "nowhere was a fold ever mapped on the basis of tracing correlative stratigraphic horizons." Instead he indicated that the Franciscan terranes are characterized structurally by mélanges and broken formations. In contrast, however, other workers have shown that some areas of the Franciscan contain relatively coherent rock units that, locally, exhibit mappable folds (Coleman and Lee, 1963; Bailey and Everhardt, 1964; Blake, 1970; Ernst *et al.*, 1970; Raymond, 1970a,c). Such folds are isoclinal to moderately open structures having amplitudes of 1-2 km and wavelengths of 1-10 km or more (Ernst *et al.*, 1970; Raymond, 1970a,b; Ernst, 1971; Kerrick and Cotton, 1971). The relative ages of the shearing and folding are generally unknown, but in parts of the northeastern Diablo Range, isoclinal folding and shearing were followed by a later period of open folding. Thus, the Franciscan may contain normal rock-stratigraphic units—informally designated as formations (*e.g.*, Lawson, 1914; Ghent, 1964)—as well as broken formations and mélanges (Hsu, 1968, 1969). The various units often exhibit structures that reflect one or more periods of deformation (*e.g.*, Ghent, 1964; Blake, 1965).

As a result of our own studies, we concur with the idea that the Franciscan has both lithologic and structural significance. Therefore, it is appealing to define the Franciscan in terms of a conceptual, genetic model, such as the currently popular plate-tectonic model (Bailey and Blake, 1969; Hamilton, 1969; Ernst, 1970; Moores, 1970). We consider this concept to be highly probable, but neither it nor any other theory should be a part of the definition of a geologic unit. Thus we refrain from defining the Franciscan in terms such as "the deposits of a subducted oceanic plate." Instead we herein define the Franciscan Complex as *the folded, faulted, and stratally disrupted basement terrane of the California Coast Ranges—including extensions into Oregon and Mexico—which is composed of graywacke, shale, minor conglomerate, radiolarian chert and siliceous shale, minor limestone, volcanic rocks, mafic-ultramafic plutonic rocks and their zeolite-to-blueschist-facies metamorphic equivalents.*

In offering this definition we are rejecting earlier proposals, such as that presented by Hsu and Ohrbom (1969), who hoped to "provide a unique definition" of the Franciscan in suggesting:

> ...all rocks of the Coast Ranges which contributed tectonic inclusions or allochthonous slabs to the Franciscan mélanges during late Mesozoic deformation be included as part of the Franciscan, *sensu lato.*

According to this definition, the Franciscan includes metamorphic and plutonic rocks of the Salinian block, as well as the lower parts of the Great Valley sequence, as both terranes have been considered as provenances for fragments in Franciscan mélanges (*cf.* Hsu and Ohrbom, 1969; Page, 1970, 1972; Raymond and Christensen, 1971). Also, the definition of Hsu and Ohrbom (1969) specified that only those units contributing tectonic elements during late Mesozoic deformation be included in the Franciscan. However, the time of formation of most tectonic mélanges is unknown, and newly determined Tertiary ages for parts of the Franciscan Complex suggest that Tertiary time cannot be excluded as an era of mélange formation in the Franciscan. Thus, although the definition of Franciscan proposed by Hsu and Ohrbom succeeds in being unique, it fails to be useful because it relies on interpretations as to *when* mélanges formed, and it may include nearly all of the Mesozoic rocks in the California Coast Ranges.

Franciscan Designation

The Franciscan cannot be described as a normal rock-stratigraphic unit (*e.g.*, formation or group) for several reasons: (1) Franciscan terranes contain tectonic mélanges, as well as rock-stratigraphic units; (2) Franciscan terranes commonly are characterized by stratal disruption—a tectonic, rather than a lithologic criterion; and (3) Franciscan terranes consist of individual areas differing in both lithology and structure. Nevertheless, much of the literature continues to perpetuate the fallacious term "Franciscan Formation," even for rocks outside the type area. For this reason we emphasize the need for an appropriate and formalized designation for the Franciscan, which will supersede "Franciscan Formation" but will not preclude the informal use of terms such as Franciscan rocks, assemblage, or terrane.

The term "assemblage" used by Bailey *et al.* (1964) has no formal definition, but usage of the term has gained wide acceptance, together with the terms Franciscan rocks or, simply, Franciscan. We consider use of "assemblage" undesirable because it connotes a compatible collection, such as "fossil assemblage" or "mineral assemblage." Also it may be confused with "assemblage zone," a formally adopted biostratigraphic unit. Furthermore, "Franciscan assemblage" does not describe adequately the lithologic and structural complexity of the Franciscan and does not provide sufficient contrast in nomenclature to that used for the contemporaneous but regularly bedded rocks of the Great Valley sequence (note: Bailey *et al.*, 1964, p. 7, refer to "both the Franciscan and Great Valley assemblages...").

The term "series," employed by Lawson (1895), has been redefined as a time-stratigraphic term (ACSN, 1961), and should not be used to designate rock-stratigraphic units. The term "sequence," used previously to describe the Franciscan by Raymond (1970a, b) and by Cotton (1971), is also inappropriate because, in both the formal and the informal sense, sequence is a rock-stratigraphic term commonly used to describe unconformity-bounded masses of strata representing the lithologic record of a major tectonic cycle (Krumbein and Sloss, 1963, p. 34-35). Neither cyclic tectonism nor bounding unconformities have been demonstrated for the Franciscan.

The appellation "mélange" was used variously by Hsu to describe the Franciscan as a whole (Hsu, 1968, p. 1066) and to designate individual chaotic rock units that, together, comprise some Franciscan terranes (Hsu, 1969; Hsu and Ohrbom, 1969). Used in the former sense, the term Franciscan mélange implies that all coherent units within the Franciscan are engulfed in a pervasively sheared matrix—an unlikely possibility. Thus, if the entire Franciscan were designated as a mélange, the term would be interpretive rather than descriptive. We wish to restrict the amount of interpretive and genetic implications involved in naming the Franciscan; therefore, we restrict the term mélange to individual units within Franciscan terrane.

To remain within the confines of the Code of Stratigraphic Nomenclature (ACSN, 1961) and, at the same time, to use a term that adequately reflects the observable lithologic and tectonic characteristics of the Franciscan, we propose that the Franciscan be designated a *complex.* The Code of Stratigraphic Nomenclature (ACSN, 1961, Article 6j) states:

> If a mass of rock is composed of diverse types of any class or classes, or is characterized by highly complicated structure, the word "complex" may be used as a part of the formal name instead of a lithologic or rank term....

Page (1966), Hsu (1967), Crowell (1968), Pessagno (1971), and Berkland (1971) used the term "complex" to describe the Franciscan, but did not formally define it as such. Bailey and Blake (1969) also appeared to use the term Franciscan complex repeatedly, but this was an artifact of English-Russian-English translations and was not used in the original manuscript (E. H. Bailey, written commun.). Hsu (1968) reconsidered his usage of the term complex and abandoned it in regard to the Franciscan in the belief that complexes do not encompass chaotic units. We disagree and support the usage of "complex" for several reasons: (1) it adequately describes the Franciscan, as defined above; (2) genetic connotations are avoided in the original definition of complex; and (3) the term complex has been formalized by the Code of Stratigraphic Nomenclature.

Geographic Subdivisions

The Franciscan Complex has great lateral variation in lithology, mineralogy, and structure. The type area of the San Francisco Peninsula is not typical of the Franciscan Complex because of unusually thick and extensive limestones, as well as the poor development of regional metamorphism in that area. Also the formational stratigraphy conceived by Lawson (1914) during mapping of the San Francisco Peninsula has proved inaccurate on stratigraphic grounds (Bailey *et al.*, 1964), and because much of that type area consists of mélange (Hsu and Ohrbom, 1969). However, despite the general lack of continuity in rocks of the Franciscan Complex, it has been divided into several geographic zones, where rocks are similar in lithology, structure, or both (*e.g.*, Bailey and Irwin, 1959).

Coastal belt—A western zone of the Franciscan Complex in northern California was named informally by Bailey and Irwin (1959, p. 2799-2802) as the "Coastal belt." Bailey and Irwin originally excluded this zone from the Franciscan Complex; however, in all subsequent papers they, and most other workers, have considered the Coastal belt to be a relatively young geographic subdivision of the Franciscan. Bailey and Irwin (1959) described this zone as:

> ... the belt extend(ing) southeast along the coast from Cape Mendocino to Point Arena, a distance of 100 miles; from Point Arena nearly to the Russian River it is bounded on the west by the San Andreas fault and separated from the coast by a thin wedge of other rocks. How much farther south it extends is not known. The eastern boundary of the belt is generally a few miles west of U.S. Highway 101. Its contact with Franciscan rocks in most places seems to be a fault. The rocks of this belt can not yet be properly assigned to any established formation; they are referred to herein as rocks of the coastal belt.

Later, Bailey *et al.* (1964, p. 13) described the lithologic and structural character of the Coastal belt as follows:

> This belt contains medium-grained, massively bedded graywackes, with minor amounts of shale and conglomerate, all structurally deformed in such a fashion that they closely resemble the rocks typical of the Franciscan Formation (of the "type area"). However, the belt contains only a little greenstone, chert, serpentine, and blueschist-grade metamorphic rock.

Our preliminary studies of the Coastal belt substantiate these characteristics, and we would include also the abundant potassium feldspar (Bailey and Irwin, 1959), and the common zeolite-facies metamorphic mineral, laumontite (Bailey *et al.*, 1964). Most workers recognize these mineralogic criteria for the typical rocks of the Coastal belt. Lithologic and structural data for these rocks, as well as the interpretation that they constitute the basement terrane of the area, include the Coastal belt rocks within our definition of the Franciscan Complex. Additional studies may permit establishment of all or part of the Coastal belt as a rock-stratigraphic unit.

Eastern belt—An Eastern belt of the Franciscan Complex in the northern Coast Ranges was referred to as the "metamorphosed Franciscan (?) formation" by Bailey and Irwin (1959). It long has been recognized that rocks of this zone are generally metamorphosed (*e.g.*, Fairbanks, 1893), but according to Irwin (1960, p. 38), "At some places the foliation is so weakly developed that decisions as to whether the rocks should be mapped as part of the metamorphic assemblage were arbitrary." The experience of Berkland (unpub. data) in many parts of the eastern zone substantiates Irwin's earlier observations. However, many unfoliated rocks contain lawsonite, and such rocks definitely are metamorphosed. Berkland estimates that at least 75 percent of the Eastern-belt terrane consists of metaclastic rocks having metamorphic fabrics and/or containing the diagnostic blueschist facies mineral, lawsonite. Locally the rocks are lower grade containing pumpellyite, or higher grade containing jadeitic pyroxene (*cf.* Ghent, 1964; Bailey *et al.*, 1964; Blake, 1965).

Because of the structure and lithology of the rocks, coupled with the interpretation that these rocks form the basement terrane in the area, the "metamorphosed Franciscan (?) formation" unquestionably should be considered part of the Franciscan Complex. The metamorphosed Franciscan (?) formation cannot justifiably be given a rock-stratigraphic designation, any more than can the Franciscan Complex as a whole. Therefore, we suggest that the area occupied by the metamorphosed Franciscan (?) formation be redesignated as part of the Franciscan Complex, the Eastern belt—a term employed by Blake (1965, p. 12), but not adopted by other authors. The Eastern belt is defined as *a zone of the Franciscan Complex lying east of an approximate line drawn along the east edge of Clear Lake, northwest to Mt. Sanhedrin, and then north along the eastern edge of Round Valley to beyond the confluence of Hull's Creek with the North Fork of the Eel River. The Eastern belt is bounded on the east by the Great Valley sequence. The rocks are dominantly metaclastic, characteristically having a metamorphic fabric and/or containing high-pressure, low-temperature metamorphic minerals, especially lawsonite.*

Central belt—Bailey and Irwin (1959) designated all Franciscan rocks east of the Coastal belt as part of their Franciscan belt. Blake (1965, p. 12) and we believe that the Franciscan belt can be subdivided into an eastern and a central zone. Blake (1965) did not describe the rocks of the central zone or fix its limits. By definition the rocks of this zone lie between the Coastal belt and the Eastern belt.

Lithologic types in the central zone of the northern California Franciscan Complex cover the whole range of Franciscan rocks, including graywacke, chert, greenstone, gabbro, serpentine, rodingite, limestone, eclogite, as well as prehnite-pumpellyite and blueschist-facies metamorphic rocks. The characteristic rock is sheared shale, which forms the matrix of a mélange in which more resistant rocks are enclosed. The matrix generally is not metamorphosed beyond the prehnite-pumpellyite facies and does not contain potassium feldspar, except as rare veins of adularia. We define the Central belt as *a zone of the Franciscan Complex in the northern California Coast Ranges which lies (approximately) between State Highway 101 and a line drawn along the east edge of Clear Lake, northwest to Mt. Sanhedrin and then north along the eastern edge of Round Valley to beyond the confluence of Hull's Creek with the North Fork of the Eel River. The zone is characterized by mélange, with abundant high-grade blueschist blocks enclosed in a pumpellyitic sheared-shale matrix.* The Central belt is marked by almost ubiquitous landsliding.

Other areas of the Franciscan Complex—In addition to the Coastal, Central, and Eastern belts of the Franciscan Complex, many other areas have been considered to contain Franciscan rocks. All of the following regions appear to qualify as part of the Franciscan Complex as defined above: (1) certain islands and coastal areas of Baja California as far south as Magdalena Bay (Beal, 1948; Wisser, 1954; Cohen *et al.*, 1963; Kilmer, 1961); (2) Catalina Island and the Palos Verdes Hills of southern California (Woodford, 1924; Bailey, 1941); (3) Diablo Range (Leith, 1949; Maddock, 1964; Ernst, 1965, 1971; Raymond, 1970a); (4) Nacimiento block (Bailey *et al.*, 1964; Hsu, 1969; Page, 1969, 1970, 1972); (5) San Francisco Bay area (Lawson, 1895, 1914; Weaver, 1949; Gluskoter, 1962; Berkland, 1969a); (6) Humboldt-Del Norte Coast Ranges (Manning and Ogle, 1950; Ogle, 1953; Blake *et al.*, 1967; Bailey *et al.*, 1970); (7) southwestern Oregon (Taliaferro, 1942; Koch, 1966; Hotz, 1969). This list is far from exhaustive; rather it is a sampling of Franciscan literature that will enable the reader to obtain an understanding of the nature and distribution of the Franciscan Complex. Although we recognize the three geographic zones of rather distinctive Franciscan terranes in northern California, we see no particular value in attempting to extend these names to other geographic areas at this time.

Relation of Franciscan Complex to Great Valley Sequence

The Great Valley sequence was named informally by Bailey *et al.* (1964, p. 13) to encompass more than 40,000 ft of dominantly clastic, marine sandstone, shale, and conglomerate that border the western edge of the Sacramento-San Joaquin valley. Later work (Moiseyev, 1966, 1970; Bailey *et al.*, 1970; Raymond, 1970b; Page, 1972) demonstrated that in the Coast Ranges the basal part of the Great Valley sequence contains chert, volcaniclastic, volcanic, and mafic-ultramafic plutonic rocks, forming an ophiolite complex. The lower part of the Great Valley sequence is metamorphosed to the zeolite facies (Dickinson *et al.*, 1969), to the prehnite-pumpellyite facies (Moiseyev, 1966), and to the blueschist facies (McKee, 1966). Thus the Great Valley sequence and the Franciscan Complex are lithologically similar in containing clastic and metaclastic rocks, chert, volcanic rocks, and mafic-ultramafic rocks. As the Great Valley sequence ranges in age from Late Jurassic through Late Cretaceous (to Paleocene in the southern area), it is also coeval with the Franciscan Complex, excepting the Eocene part of the Coastal belt. The major contrast between Franciscan and Great Valley rocks is structural. Whereas the former are characterized by stratal disruption and multiple periods of deformation, the latter generally show only homoclinal tilting. An additional contrast is provided by the greater degree and extent of metamorphism in the Franciscan Complex than in the Great Valley sequence. The former contains common blueschist-facies metamorphic rocks, whereas in the latter the metamorphic grade rarely exceeds the prehnite-pumpellyite facies.

The Great Valley sequence is considered by most authors as structurally overlying the Franciscan Complex along a major fault that may represent the upper boundary of a fossil subduction zone (Bailey and Blake, 1969; Ernst, 1970; Berkland, 1972; Page, 1972). However, some workers still regard the contact as depositional and intergrading (Maxwell and Raney, 1970), much the same as was earlier believed by Taliaferro (1943, p. 194) and Enos (1963). In nearly every case, depositional contacts have been shown later to be either tectonic (Bailey *et al.*, 1964, p. 146), or between clastic and ophiolitic parts of the lower Great Valley sequence (Bailey and Blake, 1969; Moiseyev, 1970; Bailey *et al.*, 1970; Raymond, 1970b).

References Cited

American Commission on Stratigraphic Nomenclature, 1961, Code of stratigraphic nomenclature: Am. Assoc. Petroleum Geologists Bull., v. 45, no. 5, p. 645-665.

American Geological Institute, 1962, Dictionary of geological terms: Dolphin Reference Books ed.: Garden City, New York, Doubleday.

Bailey, E. H., 1941, Mineralogy, petrology, and geology of Santa Catalina Island, California: Ph.D. thesis, Stanford Univ.

——— and M. C. Blake, Jr., 1969, Late Mesozoic tectonic development of western California: Geotectonics, no. 3, p. 148-154, 225-230.

——— and D. L. Everhardt, 1964, Geology and quicksilver deposits of the New Almaden district, Santa Clara County, California: U.S. Geol. Survey Prof. Paper 360, 206 p.

——— and W. P. Irwin, 1959, K-feldspar content of Jurassic and Cretaceous graywackes of northern Coast Ranges and Sacramento Valley, California: Am. Assoc. Petroleum Geologists Bull., v. 43, no. 12, p. 2797-2809.

——— M. C. Blake, and D. L. Jones, 1970, On-land Mesozoic oceanic crust in California Coast Ranges: U.S. Geol. Survey Prof. Paper 700-C, p. C70-C81.

——— W. P. Irwin, and D. L. Jones, 1964, Franciscan and related rocks, and their significance in the geology of western California: California Division Mines and Geology Bull. 183, 177 p.

Beal, C. H., 1948, Reconnaissance of the geology and oil possibilities of Baja California, Mexico: Geol. Soc. America Mem. 31, 138 p.

Berkland, J. O., 1969a, Geology of the Novato quadrangle, Marin County, California: M.S. thesis, San Jose State College, 146 p.

——— 1969b, Late Mesozoic and Tertiary sequence near the proposed Garrett tunnel, Mendocino and Lake Counties, California (abs.): Geol. Soc. America Abs. with Programs for 1969, pt. 3, p. 4-5.

——— 1971, New occurrence of Cretaceous and Paleocene strata within Franciscan terrane of the northern Coast Ranges, California (abs.): Geol. Soc. America Abs. with Programs, v. 3, p. 81-82.

——— 1972 (in press), Paleogene "frozen" subduction zone in the Coast Ranges of northern California: 24th Internat. Geol. Cong., Montreal, Sec. 3, p. 99-105.

Bezore, S. P., 1969, The Mount Saint Helena ultramafic-mafic complex of the northern California Coast Ranges (abs.): Geol. Soc. America Abs. with Programs for 1969, pt. 3, p. 5-6.

Blake, M. C., Jr., 1965, Structure and petrology of low-grade metamorphic rocks, blueschist facies, Yolla Bolly area, northern California: Ph.D. thesis, Stanford Univ., 91 p.

——— 1970, Different facies in Franciscan rocks and their significance in the late Mesozoic history of western California (abs.): Geol. Soc. America Abs. with Programs, v. 2, no. 2, p. 73.

——— W. P. Irwin, and R. G. Coleman, 1967, Upside-down metamorphic zonation, blueschist facies, along a regional thrust in California and Oregon: U. S. Geol. Survey Prof. Paper 575-C, p. 1-9.

Blake, W. P., 1858, Report of a geological reconnaissance in California: U.S. War Dept., Exploration and Surveys for Railroad, Mississippi to Pacific Ocean Repts., v. 5, pt. 2, 370 p.

Bogdanov, N. A., 1969, Thalassogeosynclines of the circumpacific ring: Geotectonics, no. 3, p. 141-147.

Brice, J. C., 1953, Geology of Lower Lake quadrangle, California: California Div. Mines Bull. 166, 72 p.

Cohen, L. H., K. C. Condie, L. J. Kuest, G. S. Mackenzie, F. H. Meister, P. Pushkar, and A. M. Steuber, 1963, Geology of the San Benito Islands, Baja California, Mexico: Geol. Soc. America Bull., v. 74, no. 11, p. 1355-1369.

Coleman, R. G., and D. E. Lee, 1963, Glaucophane-bearing metamorphic rock types of the Cazadero area, California: Jour. Petrology, v. 4, no. 2, p. 260-301.

Cotton, W. R., 1971, Franciscan stratigraphy of the northwestern portion of the Diablo Range, central California (abs.): Geol. Soc. America Abs. with Programs, v. 3, no. 2, p. 103-104.

Crowell, J. C., 1968, The California Coast Ranges, *in* A coast to coast tectonic survey of the United States: Univ. Missouri, Rolla, UMR Jour., no. 1, p. 135-156.

Dickinson, W. R., R. W. Ojakangas, and R. J. Stewart, 1969, Burial metamorphism of the late Mesozoic Great Valley sequence, Cache Creek, California: Geol. Soc. America Bull., v. 80, no. 3, p. 519-526.

Enos, P., 1963, Jurassic age of Franciscan Formation south of Panoche Pass, California: Am. Assoc. Petroleum Geologists Bull., v. 47, no. 1, p. 158-163.

Ernst, W. G., 1965, Mineral paragenesis in Franciscan metamorphic rocks, Panoche Pass, California: Geol. Soc. America Bull., v. 76, no. 8, p. 879-914.

——— 1970, Tectonic contact between the Franciscan mélange and the Great Valley sequence; crustal expression of a late Mesozoic Benioff zone: Jour. Geophys. Research, v. 75, no. 5, p. 886-901.

——— 1971, Petrologic reconnaissance of Franciscan metagraywackes from the Diablo Range, central California Coast Ranges: Jour. Petrology, v. 12, no. 2, p. 413-437.

——— Y. Seki, H. Onuki, and M. C. Gilbert, 1970, Comparative study of low-grade metamorphism in the California Coast Ranges and the outer metamorphic belt of Japan: Geol. Soc. America Mem. 124, 276 p.

Fairbanks, H. W., 1893, Notes on the geology and mineralogy of portions of Tehama, Colusa, Lake, and Napa Counties: California Mining Bur. 11th Rept., p. 54-75.

Ghent, E. D., 1964, Petrology and structure of the Black Butte area, Hull Mountain and Anthony Peak quadrangles: Ph.D. thesis, Univ. California at Berkeley, 224 p.

Gluskoter, H. J., 1962, Geology of a portion of western Marin County, California: Ph.D. thesis, Univ. California at Berkeley, 184 p.

——— 1969, Geology of a portion of western Marin County, California: California Div. Mines and Geology Map Sheet 11.

Greenley, E., 1919, The geology of Angelsey: Great Britain Geol. Survey Mem., v. 1, 980 p.

Hamilton, W., 1969, Mesozoic California and the underflow of

Pacific Mantle: Geol. Soc. America Bull., v. 80, no. 12, p. 2409-2430.

Hotz, P. E., 1969, Relationships between the Dothan and Rogue Formations, southwestern Oregon: U.S. Geol. Survey Prof. Paper 650-D, p. 131-137.

Hsü, K. J., 1967, Mesozoic geology of the California Coast Ranges—a new working hypothesis, *in* Étages tectoniques: Neuchâtel Univ. Inst. Geol., Neuchâtel, Switzerland, La Baconniere, p. 279-296.

——— 1968, Principles of mélanges and their bearing on the Franciscan-Knoxville paradox: Geol. Soc. America Bull., v. 79, no. 8, p. 1063-1074.

——— 1969, Preliminary report and geologic guide to Franciscan mélanges of the Morro Bay-San Simeon area, California: California Div. Mines and Geology Spec. Pub. 35, 46 p.

——— 1971, Franciscan mélanges as a model for eugeosynclinal sedimentation and underthrusting tectonics: Jour. Geophys. Research, v. 76, no. 5, p. 1162-1170.

——— and R. Ohrbom, 1969, Mélanges of San Francisco Peninsula—geologic reinterpretation of type Franciscan: Am. Assoc. Petroleum Geologists Bull., v. 53, no. 7, p. 1348-1367.

Irwin, W. P., 1957, Franciscan Group in Coast Ranges and its equivalents in Sacramento Valley, California: Am. Assoc. Petroleum Geologists Bull., v. 41, no. 10, p. 2284-2297.

——— 1960, Geologic reconnaissance of the northern Coast Ranges and Klamath Mountains, California, with a summary of the mineral resources: California Div. Mines and Geology Bull. 179, 80 p.

Kerrick, D. M., and W. R. Cotton, 1971, Stability relations of jadeite pyroxene in Franciscan metagraywackes near San Jose, California: Am. Jour. Sci., v. 271, no. 4, p. 350-369.

Kilmer, F. H., 1961, Anomalous relationship between the Franciscan Formation and metamorphic rocks, northern Coast Ranges, California (abs.): Geol. Soc. America Spec. Paper 68, p. 210.

Koch, J. G., 1966, Late Mesozoic stratigraphy and tectonic history, Port Orford-Gold Beach area, southwestern Oregon Coast: Am. Assoc. Petroleum Geologists Bull., v. 50, no. 1, p. 25-71.

Kramer, J. C., W. R. Evitt, and M. O'Day, 1972, Tertiary coastal belt: Geol. Soc. America Bull., v. 83 (in press).

Krumbein, W. C., and L. L. Sloss, 1963, Stratigraphy and sedimentation, 2d ed.: San Francisco, W. H. Freeman, 660 p.

Lawson, A. C., 1895, Sketch of the geology of the San Francisco Peninsula (California): U.S. Geol. Survey 15th Ann. Rept., p. 399-476.

——— 1914, Description of the San Francisco district; Tamalpais, San Francisco, Concord, San Mateo, and Haywards quadrangles: U.S. Geol. Survey Geol. Atlas, Folio 193, 24 p.

Leith, C. J., 1949, Geology of the Quien Sabe quadrangle, California: California Div. Mines Bull. 147, p. 7-35.

Maddock, M. E., 1964, Geologic map and sections of the Mount Boardman quadrangle, Santa Clara and Stanislaus Counties, California: California Div. Mines and Geology Map Sheet 3.

Manning, G. A., and B. A. Ogle, 1950, Geology of the Blue Lake quadrangle, California: California Div. Mines Bull. 148, 36 p.

Maxwell, J. C., and J. Raney, 1970, Interfingering contact of Franciscan and Great Valley rocks (abs.): EOS (Am. Geophys. Union Trans.), v. 51, no. 11, p. 825.

McKee, B., 1966, Knoxville-Franciscan contact near Paskenta, western Sacramento Valley, California (abs.): Geol. Soc. America Spec. Paper 87, p. 215-216.

Moiseyev, A. N. (Moisseeff), 1966, Geology and geochemistry of the Wilbur Springs quicksilver district, Colusa and Lake Counties, California: Ph.D. thesis, Stanford Univ., 246 p.

——— 1970, Late serpentine movements in the California Coast Ranges: new evidence and its implications: Geol. Soc. America Bull., v. 81, no. 6, p. 1721-1731.

Moores, E., 1970, Ultramafics and orogeny, with models of the U.S. Cordillera and the Tethys: Nature, v. 228, no. 5274, p. 837-842.

Ogle, B. A., 1953, Geology of Eel River Valley area, Humboldt County, California: California Div. Mines Bull. 164, 128 p.

Page, B. M., 1966, Geology of the Coast Ranges of California, *in* Geology of northern California: California Div. Mines and Geology Bull. 190, p. 255-276.

——— 1969, Relation between ocean-floor spreading and structure of the Santa Lucia Range, California (abs.): Geol. Soc. America Abs. with Programs for 1969, pt. 3, p. 51-52.

——— 1970, Time of completion of underthrusting of Franciscan beneath Great Valley rocks west of Salinian block, California: Geol. Soc. America Bull., v. 81, no. 9, p. 2825-2833.

——— 1972, Oceanic crust and mantle fragment in subduction complex near San Luis Obispo, California: Geol. Soc. America Bull., v. 83, no. 3, p. 957-972.

Pessagno, E. A., Jr., 1971, Radiolarian zonation of the Upper Cretaceous portion of the Great Valley sequence (abs.): Geol. Soc. America Abs. with Programs, v. 3, no. 2, p. 177.

Raymond, L. A., 1970a, Cretaceous sedimentation and regional thrusting, northeastern Diablo Range, California: Geol. Soc. America Bull., v. 81, no. 7, p. 2123-2128.

——— 1970b, Del Puerto keratophyres and Lotta Creek tuffs; Great Valley not Franciscan rocks (abs.): Geol. Soc. America Abs. with Programs, v. 2, no. 2, p. 133.

——— 1970c, Relationships between blueschist facies metamorphism, folding, and faulting in Franciscan rocks, Seegers Ranch area, northeastern Diablo Range, California (abs.): Geol. Soc. America Abs. with Programs, v. 2, no. 2, p. 133-134.

——— 1972, Tesla-Ortigalita fault, Coast Range thrust fault, and Franciscan metamorphism, northeast Diablo Range, California: Geol. Soc. America Bull. (in press).

——— and W. P. Christensen, 1971, Petrographic reconnaissance of Franciscan rocks, Occidental-Guerneville area, Sonoma County, California, *in* Geologic guide to the northern Coast Ranges, Point Reyes region, California: Geol. Soc. Sacramento, Ann. Field Trip Guidebook, p. 38-46.

Reed, R. D., 1933, Geology of California: Am. Assoc. Petroleum Geologists, 355 p.

——— and J. S. Hollister, 1936, Structural evolution of southern California: Am. Assoc. Petroleum Geologists Bull., v. 20, no. 12, p. 1529-1704.

Silberling, N. M., and R. J. Roberts, 1963, Pre-Tertiary stratigraphy and structure of northwestern Nevada: Geol. Soc. America Spec. Paper 72, 58 p.

Suppe, J., 1969, Times of metamorphism in the Franciscan terrain of the northern Coast Ranges, California: Geol. Soc. America Bull., v. 80, no. 1, p. 135-142.

——— 1970, Offset of late Mesozoic basement terrains by the San Andreas fault system: Geol. Soc. America Bull., v. 81, no. 11, p. 3253-3258.

Taliaferro, N. L., 1942, Geologic history and correlation of the Jurassic of southwestern Oregon and California: Geol. Soc. America Bull., v. 53, no. 1, p. 71-112.

——— 1943, Franciscan-Knoxville problem: Am. Assoc. Petroleum Geologists Bull., v. 27, no. 2, p. 109-219.

Weaver, C. E., 1949, Geology of the Coast Ranges immediately north of the San Francisco Bay region, California: Geol. Soc. America Mem. 35, 242 p.

Wisser, E. H., 1954, Geology and ore deposits of Baja California, Mexico: Econ. Geology, v. 49, no. 1, p. 44-76.

Woodford, A. O., 1924, The Catalina metamorphic facies of the Franciscan series: California Univ. Dept. Geol. Sci. Bull., v. 15, p. 49-68.

Part II

OLISTOSTROME MÉLANGES

Editor's Comments on Papers 5, 6, and 7

5 **ABBATE, BORTOLOTTI, and PASSERINI**
Excerpts from *Olistostromes and Olistoliths*

6 **ELTER and TREVISAN**
Olistostromes in the Tectonic Evolution of the Northern Apennines

7 **LEONOV**
Olistostromes and Their Origin

The three papers in this part define olistostrome mélange occurrences and seek a genetic model. The terms *olistostrome* and *olistolith,* originally defined by Flores (1955), were slightly revised by Abbate, Bortolotti, and Passerini (Paper 5). For the chaotic deposits of the Apennines, the olistostrome/olistolith interpretations replaced earlier tectonic interpretations in which they were considered to be the crushed base of major allochthonous blankets and the binding material within the main sheets of such blankets (Merla, 1951). These authors favor a simple sedimentary gravity slide interpretation (though allowing that there are some tectonically produced chaotic bodies). Elter and Trevisan (Paper 6) accept olistostromes produced by pure submarine gravity slumping, as well as *precursory* olistostromes related to the toe of a thrust sheet advancing over soft sediments. Leonov (Paper 7) relegates the olistostromes produced by pure slumping to a minor role and considers that most olistostromes are related to either spalling from the toe of thrust sheets moving over soft sediments or derived from fragmentation at the base of such sheets.

The concept of sedimentary slides was derived from the discovery of chaotic terrains as well defined and repeated layers in sedimentary sequences. Abbate and his co-workers state that olistostromes are developed on larger scale than submarine slumps and differ from them in the presence of some extraformational (exotic) components. They accept a gradation between greater slumps and *intraformational* olistostromes and differentiate olistostromes from turbidites. These researchers differ from Flores in not requiring a rigid condition of intercalation within a sedimentary sequence, and obviously this is correct because cases must exist in which it is not so (e.g., where the olistostrome is at the top of a sequence with nothing above it or where it occurs within a body with faulted or tectonized contacts). Even so, the lack of any normal stratigraphic relationships over wide areas must still throw doubt on olistostrome interpretations, as must lack of evidence of formation at one restricted period of time during sedimentation (a case in point is the sedimentary mélange of the Makran, Iran, described in Paper 21). Abbate and his associates do not dispute the size limitation on olistostrome bodies relative to tectonic mélange bodies (see Hsü, Paper 2), though they point out that their acceptance of this limitation is for reasons of everyday practice rather than any theoretical constraint. A thickness limit to individual olistostromes of 100-200 m is suggested, along with a limit to the dimensions of contained blocks; there must be a limit to the size of blocks that can roll downslope under gravity. Many of the Cretaceous-Eocene olistostromes of the Apennines contain abundant ophiolitic material and are validly characterized as ophiolitic olistostrome mélanges in the nongenetic usage adopted in this volume.

The Oligocene and Miocene olistostromes are essentially argillaceous and calcareous and are essentially sedimentary olistostrome mélanges with only minor ophiolite exotic block content. All are characterized by a sedimentary matrix, and in the older mélanges this is mainly calcareous shale. Geologists rarely deny that sedimentary olistostromes exist in the Apennines, but they still argue about the relative roles of tectonic and sedimentary chaotization and the sequential chronology of such processes in the generation of these chaotic bodies (Hsü Paper 2; Elter and Trevisan, Paper 6; Leonov, Paper 7). Hsü and Ohrbom (1969) suggested that the chaotization is dominantly tectonic, produced in a composite schuppen zone at the toe of a thrust plate and that slabs of olistostromes are caught up in the allochthonous assemblage. They suggested that mélange terminology (i.e., with a tectonic implication) might be more suitable to the Apennine bodies. It is interesting to note that their model has a diametrically reversed sequence of events to the models—also related to advancing thrust sheets—proposed by Elter and Trevisan (Paper 6) and Leonov (Paper 7), models in which gravity sliding in a soft sedimentary medium follows on tectonic chaotization.*

*Fairbridge (1942) appears to have suggested the essential concept of olistostromes but referred to them as *sedimentary klippen*—his chronology was identical with that of Abbate et al.—"primary sedimentary features lodged within the flysch envelope by gravitational slides . . . then overriden by major thrusts" (Fairbridge, 1980).

Elter and Trevisan (Paper 6) presented models for the generation of olistostromes in the Apennines, very well illustrated by sectional diagrams. They related the exotic block material to incorporation of subjacent rocks in the sole of an olistostrome and believed that this could occur in a simple slump type of olistostrome. However, they proposed a precursor type of olistostrome, in which the material was largely derived from the advancing toe of an allochthonous nappe, moving over the sedimentary basin of deposition of the soft sediments that formed the matrix and was itself disrupted. Travel distances of up to 50 k were suggested for such olistostromes, and they could carry enormous blocks (mostly diabase or peridotite, on a scale of cubic kilometers), considered to be enormous clasts that slid over the muddy sea bottom. Considering the evidence for the tectonic protrusion of such immense blocks in schuppen regimes now available from Iran (McCall, Paper 21), the doubt must remain whether at least some of these immense blocks described by these authors are not better explained by a complication of tectonic protrusion.

Leonov (Paper 7) relates olistostromes primarily to the destruction of tectonic nappes and overthrusts and the sedimentary (gravity slide) dispersion of material derived mainly from the basal parts of the thrust sheets, moving over basins of soft sediments, into the deeper parts of the basin. He considers that the essential process of fragmentation is tectonic and that the slump features only reflect the mode of further transport of the material so produced. He notes that surfaces of fragments in olistostromes commonly exhibit slickensides and are commonly cemented by tectonic gouge (features surely indicating fragmentation under considerable burial pressure, not in a virtually unloaded sedimentary slump environment). Olistostromes, he believes, are commonly reworked by tectonic processes, but some are not; thus, these features have to be related to the primary genesis. He lists seven features, anomalous in terms of the simple slump mechanism:

1. Concentration along large fault zones (of overthrusting);
2. Incorporation of rocks from the allochthon and piling up in front of it, or overriding by it;
3. Definite time relationship to periods of overthrusting;
4. Very widespread development (over hundreds or thousands of kilometers) in one and the same time interval;
5. Common superposition by tectonic nappes and common reworking by tectonic processes;
6. Content of dynamically reworked material and flattened, smoothed over fragments;
7. Direct relationship between chaotic breccias in olistostromes and overthrust slabs and gradual transitions from one to another.

Leonov suggests that disintegration of tectonic sheets, especially the frontal parts of them, is in many cases a preferable alternative to fragmentation during slumping. He illustrates his arguments with summaries of several occurrences including those of the Lycian Taurus of Turkey (Graciansky, Paper 19) where transitions downward from thrust sheets into mélange are recorded.

Leonov's arguments are extremely important and pose further questions. The model seems to be equally applicable to a classic Alpine nappe and a schuppen system. Such a model allows the possibility that in some cases there might be no further complication by gravity sliding into the deeper part of the basin; then we would get a purely tectonic mélange, spatially related to the thrust dislocations at the soles of the sheets. Also, it must be expected that on some occasions the allochthon would move over consolidated, not soft, sediments or that the schuppen structure would be developed in such a medium, and here we would get fragmentation of consolidated rocks beneath the thrust (in fact, the sort of sedimentary mélange described from Iran in Paper 21).

REFERENCES

Fairbridge, R. W., 1942, Subaqueous Sliding and Slumped Blocks, D.Sc. thesis, University of Western Australia, 468p.

Fairbridge, R. W., 1980, The Concept of the Sedimentary Klippe, Internat. Geol. Congress, 26th, unpublished note.

Flores, G., 1955, 4th Discussion, *World Petroleum Congress, 4th, Rome, Proc. Section A2,* pp. 120–121.

Hsü, K. J., and R. Ohrbom, 1969, Mélanges of San Francisco Peninsula: Geologic Reinterpretation of Type Franciscan, *Am. Assoc. Petroleum Geologists Bull.* **53:**1348–1367.

Merla, G., 1951, Geologia dell'Appennino settentrionale, *Soc. Geol. Italiana Boll.* **70:**95–382.

5

Reprinted from pages 521-529, 533-535, 537, 539-540, 542-544, and 551-557 of
Sed. Geology **4**:521-557 (1970)

OLISTOSTROMES AND OLISTOLITHS

ERNESTO ABBATE, VALERIO BORTOLOTTI AND PIETRO PASSERINI
Istituto di Geologia, Università di Firenze, Florence (Italy)

(Received May 30, 1969)

SUMMARY

The Northern Apennines are a typical area of slide deposits. Sliding phenomena gave rise to various products ranging from gravity nappes to olistostromes and olistoliths. The latter differ from gravity nappes with regard to size and internal structure.

Current research in the Northern Apennines suggests using the terms "olistostrome" and "olistolith" with a somewhat different meaning from that originally proposed by FLORES (1955). The main differences concern the size limits and the relative position in the sedimentary sequences. Olistostromes occur in Jurassic to Pliocene deposits pertaining to eu-, mio-, late and postgeosynclinal sequences. They are particularly common in Upper Cretaceous to Miocene formations. The material of the Cretaceous and Eocene olistostromes generally comes from the ophiolites and the rocks overlying the ophiolitic suite. The olistostromes occur as thick layers or breccias or paraconglomerates, and, like the olistoliths, are intercalated in the eugeosynclinal flysch formations.

The Oligo–Miocene olistostromes are also made of material from the eugeosynclinal sequences, but they are interbedded in the miogeosynclinal flysch or in the late geosynclinal formations. They generally appear as argillaceous bodies with scattered rock fragments (mostly limestones). The genesis of olistostromes and olistoliths is strictly related to the migration of the flysch basins from west to east. Slumping was caused by the eastward orogenic wave: olistostromes were discharged from the uplifted areas and/or from the front of the advancing nappes.

INTRODUCTION

In the Northern Apennines chaotic complexes crop out extensively. They usually consist of an argillaceous matrix enclosing variously shaped rock components (limestones, sandstones, ophiolites). The clasts range in size from a few millimetres to several kilometres and represent more or less disrupted remnants of original stratified sequences.

The chaotic complexes were previously interpreted as the result of a tectonic process (see ABBATE et al., 1970). They were considered to be the crushed base of the major allochthonous blankets and also the binding material between the main sheets in the allochthon (MERLA, 1951).

Further research has shown that the chaotic terrains are also found as well-defined and repeated layers in sedimentary sequences (SIGNORINI, 1940; JACOBACCI et al., 1959a; ELTER and SCHWAB, 1959; ABBATE and BORTOLOTTI, 1961; etc.). These intercalations have been regarded as sedimentary and the mechanism of deposition has been referred to submarine slides. The chaotic intervals have been thus considered olistostromes (JACOBACCI, 1965). This does not imply that all chaotic terrains are olistostromes: tectonic grinding can be responsible for chaotic zones, especially when these zones are not interbedded in sedimentary sequences. Olistostrome is synonymous of submarine slide; an olistostrome with a muddy composition is synonymous of tilloid, or, more exactly, of submarine mudflow. In this case, which is the most common, the descriptive term is paraconglomerate. Olistostrome differs from submarine slump, as the term has been applied, for the greater scale of the movement involved and generally for the extraformational origin of the components. Furthermore, in the olistostromes the original sedimentary structures are lost; this does not happen in some intraformational slumps (soft sediment deformation, PETTIJOHN, 1957).

However, a continuous series of intermediate types can be found. The distinction disappears if intraformational endolistostromes (see below) and greater slumps are compared.

Some chaotic slide deposits, which strongly resemble olistostromes, were formerly called "chaos" or "megabreccia" (see DUNBAR and RODGERS, 1957). The correspondence of an olistostrome-rich formation with a wildflysch was pointed out by JACOBACCI et al. (1959).

In spite of the existence of other descriptive terms, we shall rather use the more general name "olistostrome" in order to include the wider range of gravity phenomena that occur in the Northern Apennines.

Definition of olistostrome

The name olistostrome was originally introduced by FLORES (1955) who defined it in the following way: "Olistostrome, from the Greek words *olistomai* (to slide) and *stroma* (accumulation). By analogy with biostrome (accumulation due to life), olistostrome indicates an accumulation due to sliding. By olistostromes we define those sedimentary deposits occurring within normal geologic sequences that are sufficiently continuous to be mappable, and that are characterized by lithologically or petrographically heterogeneous material, more or less intimately admixed, that were accumulated as a semifluid body. They show no true bedding, except for possible large inclusions of previously bedded material. In any olistostrome we distinguish a binder or matrix represented by prevalently pelitic, heterogeneous material containing dispersed bodies of harder rocks.

The latter may range in size from pebble to boulder and up to several cubic kilometres. There is no constant ratio between the total volume of the inclusion and that of the binding mass.

In some olistostromes one or more types of accumulation due to flowage can be recognized, ranging from the chaotic disposition of coarser elements which were bodily detached from their original position, down to graded bedding due to turbidity currents."

A discussion of the term olistostrome, including an extensive bibliography, is found in JACOBACCI (1965).

Current research in the Northern Apennines and in Turkey has led us to use the term olistostrome in a somewhat different sense than the original. The difference requires a discussion of FLORES's (1955) definition.

(*1*) In the original definition, an olistostrome is interposed between "sequences of normal deposits". Flores presumably considered "normal" any marine sediment differing from slide deposits and turbidites. It has to be noted that Flores regarded turbidites as olistostromes (see below). As to the enclosing sediments, in our opinion, they may be both normal deposits and turbidites, for the latter are not to be considered olistostromes (JACOBACCI, 1965).

(*2*) The original assumption that olistostromes must be "intercalated" in a sedimentary sequence seems unpractical and unnecessary: also at the top of a sedimentary sequence or between two sedimentary sequences, one of which is tectonically superimposed on the other, slide deposits occur, which are in all respects similar to the interbedded ones. We realize that in this case it is more difficult to distinguish between chaotic terrains of slide origin (olistostromes) and crushed units due to tectonic shearing. So far it has not been possible to identify features that distinguish clearly between the two modes of origin, although attempt was made by PASSERINI (1965).

In the Northern Apennines the basal part of the allochthonous cover on top of the miogeosynclinal sequences often consists of chaotic shales and limestones. These can be interpreted either as shear basal breccia or as an olistostrome that preceded the Allochthon. Although we are aware of these difficulties of distinction, we do not agree with the limiting geometric condition put by Flores; slide sediments are to be expected in any position in a sedimentary sequence, including the top. We do not see any convenience in separating the top slide deposits from those intercalated, and not to consider them as olistostromes.

(*3*) Flores pointed out that the olistostromes must be mappable bodies. In recent mapping practice, the term olistostrome has been applied also to small slide deposits, by far under the limit of representation on topographic maps of 1/25,000–1/100,000 scale (see Fig.13, 15). We incline toward this use, as it implies a genetic-sedimentological rather than a formal stratigraphical connotation.

(*4*) The use of the term olistostrome raises a problem, for it can be easily applied also to large-scale tectonic phenomena, such as gravity nappes, orogenic landslides (MIGLIORINI, 1933), etc. The analogy between olistostromes and large-scale tectonic slides was emphasized by FLORES (1959) and by MERLA (1964). "Olistostrome" has been used for large-scale chaotic structures by GANSSER (1959) and RIGO DE RIGHI and CORTESINI (1964). This matter requires a further discussion.

In the common geological practice, tectonics and sedimentation are considered two classes of distinct phenomena. When slides are involved, however, no sharp boundary can be drawn between the two: a small olistostrome is unquestionably a sedimentary feature and the emplacement of a large, even chaotic, nappe is commonly considered a tectonic process. Yet there exists a continuous transition between these extreme cases. We do not call very large-scale chaotic structures olistostromes. This statement is not supported by any theoretical reason, but only by the habit of everyday practice. We favour the term olistostrome for sedimentary bodies, although opportunity may arise to consider olistostromes some chaotic masses otherwise regarded as tectonic.

As to the size limits, we usually regard as olistostromes chaotic units up to 100–200 m thick, but we think the size criterion must be integrated with the analysis of the internal structure. When the bulk of the unit is composed of finely disrupted material, suggesting movement in a semifluid state, the upper size limit for olistostromes may rise considerably. On the contrary, when the structure is relatively undisturbed, or very large, though disrupted, masses are piled up, we favour a tectonic interpretation, even for units of smaller thickness.

(*5*) Flores stated that olistostromes are "characterized by lithologically or petrographically heterogeneous material". We shall include beyond "olistostrome" also monolithological deposits (cf., JACOBACCI, 1965), e.g., serpentinite or diabase slides of Cretaceous age (see Fig.11).

(*6*) The condition of a semifluid behaviour is generally satisfied and it is closely related to the presence of a pelitic matrix. Nevertheless, olistostromes entirely composed of hard rock clasts may occur (cf., JACOBACCI, 1965), in which the binder is scarce and mostly sandy. The lack of an abundant pelitic matrix does not suggest a semiliquid behaviour; these bodies might have moved in the manner of hard-rock slides. However, a number of breccia layers, previously considered olistostromes (PASSERINI, 1965), are bedded, a poor suggestion in the context of hard-rock sliding. They might be something like fluxoturbidites rather than olistostromes. In the absence of distinctive features, doubts may remain in several cases.

(*7*) According to Flores, the rock fragments dispersed in the matrix range in size from pebbles up to masses of several cubic kilometres. Actually a lot of smaller elements are present, but they are subordinate (see below).

As to the upper size limit, the presence of slabs of such volumes as cubic kilometres implies a slide with a size largely exceeding the limits given above for olistostromes.

(*8*) In our opinion turbidites cannot be considered olistostromes (cf., JACOBACCI, 1965) as proposed by FLORES (1955) and by BENEO (1956a). Slides and turbidity currents are generally distinct phenomena, although intermediate cases have been described (KUENEN, 1956).

In a few cases olistostromes may include at the very top a thin zone of graded turbidite sandstone, the components of which are lithologically similar to the clasts dispersed in the olistostrome. The turbidity current was evidently connected with the final stages of the slide (BORTOLOTTI, 1964b; JACOBACCI, 1965).

As to the mechanics of transport, according to FLORES (1959) and to JACOBACCI et al., (1959a), the emplacement of an olistostrome is, in some cases at least, almost instantaneous. JACOBACCI (1965) attributed the chaotic structure to rapid deposition. In the majority of the olistostromes of the Northern Apennines the rapidity of emplacement cannot be established. The chaotic structure could result either from quick sliding or creeping. In some cases the preservation of delicate sedimentary structures, like siltstone lamination in beds immediately underlying thick olistostromes, rather suggest a slow, possibly laminar flow.

ELTER and RAGGI (1965a) proposed a distinction between *endolistostromes* and *allolistostromes*. In the first type the material is derived from formations of the same sequence in which the olistostrome is enclosed (e.g., clasts of Tuscan miogeosynclinal rocks in olistostromes of the Tuscan sequence). The second type is characterized by components derived from a different depositional basin, such as olistostromes of Allochthon material in the miogeosynclinal sequences.

In the Northern Apennines, according to Elter and Raggi, endolistostromes are caused by early tectonic movements in a depositional basin, whereas the allolistostromes slid down the front of advancing allochthon masses into a depositional basin lying ahead.

The statement that allolistostromes are related to advancing Allochthon units is valid in a number of cases, such as those of the Oligocene and Miocene olistostromes, but there are exceptions.

In some instances, like the eugeosynclinal Cretaceous and Eocene ophiolitic olistostromes, it is not always possible to determine which type is present.

Definition of olistolith

FLORES (1955) wrote: "The name olistolith, from the Greek words *olistomai*– to slide, and *lithos*–rock, is applied to the masses included as individual elements within the binder. These masses were previously referred to as *exotics* or *erratics*." Later BENEO (1956a) suggested that the name clasts *(pezzame)* should be limited to the smaller elements, like pebbles, cobbles and small boulders. RIGO DE RIGHI (1956) distinguished macro-olistoliths, above 10^5 m^3, meso-olistoliths, of some thousand of cubic metres, and micro-olistoliths, those a few tens of cubic metres.

In the original definition the olistoliths must be "included . . . within the binder". MARCHETTI (1956) added that olistoliths may be also masses "floating on an olistostrome". JACOBACCI (1965) regarded as olistoliths also masses locally independent from an olistostrome, provided there are elements to suggest their emplacement by sliding. We have adopted the latter meaning.

As to the size limits of olistoliths, more details are needed. Usually we call olistoliths components bigger than boulders, i.e., more than 4 m (Wentworth scale, modified by DUNBAR and RODGERS, 1957). All smaller-sized elements are referred to as clasts. The upper limit is somewhat problematic, in the same way as for the olistostromes. In our mapping practice in the Northern Apennines it is usually convenient to consider any slid mass up to 200–300 m thick and 1–2 km wide to be an olistolith. Allochthonous sheets, some tens of kilometres wide, have not been considered olistoliths. For intermediate sizes we cannot give any definite rule. The problem is merely a matter of nomenclature.

Olistoliths, as defined in our practice, include products that are currently considered sedimentary and others tectonic. The sliding of a small olistolith is to be considered a sedimentary process; the movement of sheets that can be mapped at scales smaller than 1 : 100,000 is generally attributed to a tectonic process (see Fig.4). The sliding of blocks may be sedimentary or tectonic according to the size. Hence there is a transition between the sedimentary and the tectonic phenomenon and the setting of limits can only be conventional.

STRATIGRAPHIC DISTRIBUTION OF OLISTOSTROMES IN THE NORTHERN APENNINES

Olistostromes are a widespread type of deposit in the Northern Apennines, where they can even form thick stratigraphic units. They occur mainly in sequences of Cretaceous to Miocene age, although a few have been found in Jurassic and in Pliocene formations. Four principal phases of submarine sliding can be recognized.

(*1*) A few olistostromes are enclosed in Jurassic formations. They are of intraformational type, generally inside marly formations, and comprise clasts of possibly semiconsolidated rocks. This pre-orogenic stage of sliding has been recognized so far only in the miogeosynclinal area.

(*2*) Olistostromes were a frequent occurrence in the Cretaceous and the Eocene. The pre-flysch and flysch formations of the eugeosynclinal zone contain thick sequences of calcareous-shaly and ophiolitic olistostromes and numerous large ophiolitic olistoliths. Olistostromes and olistoliths are also found in the Cretaceous and Eocene rocks of the Tuscan sequence of the miogeosynclinal area; their scale is much smaller and the components are mostly carbonate rocks.

(*3*) Many argillaceous-calcareous olistostromes are interbedded in the Oligocene and Miocene arenaceous flysch of the Apennines (miogeosynclinal area).

(*4*) A few occasional olistostromes have been reported in the transgressive Pliocene deposits of southwestern Tuscany and, more frequently, in the southern Po plain.

Phases (*2*) and (*3*) are the most important and conspicuous.

Jurassic olistostromes

The first deposits that can be clearly referred to submarine sliding are found in the *Posidononia* marls (PASSERINI, 1964a; BOCCALETTI and BORTOLOTTI, 1965; BORTOLOTTI and PASSERINI, 1965) together with the first occurrence of graded beds (calcarenites). There are intraformational olistostromes with a marly or shaly-marly matrix and pebbles of marlstone or marly limestone, sometimes closely resembling the matrix (Fig.1). The pebble shapes suggest that at the time of

Fig.1. Olistostrome with marly matrix and calcareous clasts, *Posidonia* marls. Monte Cetona, Siena.

slumping the sediment was poorly consolidated. The olistostromes are only a few metres thick at most. In the same formation (Monte Cetona, Siena, PASSERINI 1964a) there are also breccias and intraformational conglomerates, which at places pass into graded calcarenites that have no matrix but a calcite cement; these rocks are not like typical olistostromes, but could have been formed by small slumps or by turbidity currents. The roundness of the clasts decreases with size and it is possible that both rounding and fragmentation are related to a slumping or creeping movement in a poorly consolidated material.

The occurrence of olistostromes and the thickness variation (0–50 m) of the

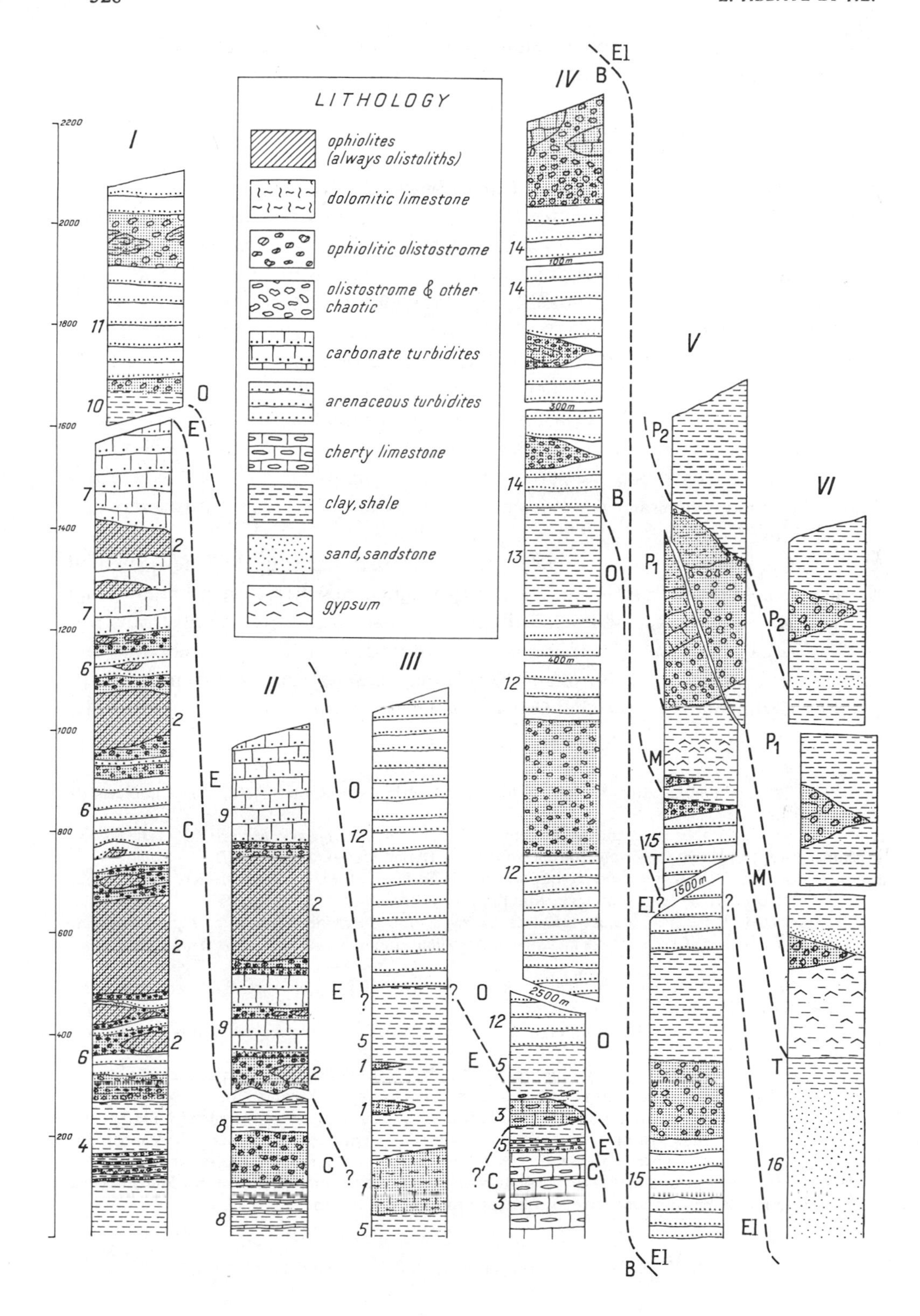
LITHOLOGY
ophiolites (always olistoliths)
dolomitic limestone
ophiolitic olistostrome
olistostrome & other chaotic
carbonate turbidites
arenaceous turbidites
cherty limestone
clay, shale
sand, sandstone
gypsum
I
II
III
IV
V
VI
2200
2000
1800
1600
1400
1200
1000
800
600
400
200
100m
300m
400m
2500m
1500m
O
E
C
B
El
P2
P1
M
T

Posidononia marls indicates an undulating sea floor, the material being removed from the higher places and deposited not far away.

Cretaceous to Eocene olistostromes

The olistostromes of the eugeosynclinal sequences. In the Cretaceous began in the present Tyrrhenian area the intercalation, in the eugeosynclinal sequences (now allochthonous), of a series of submarine slides (Fig.2–6; 9–19), essentially made of ophiolites and of rocks from the overlying sequence up to the Palombini shales (BORTOLOTTI, 1961, 1962a, 1962b, 1963, 1964a, 1964b; GIANNINI, 1962; PASSERINI, 1962, 1964b, 1965; PAGE, 1962; BORTOLOTTI and LAZZERI, 1964; ELTER and RAGGI, 1965a, 1965b; MICHELI, 1965; RAGGI, 1965; LAZZAROTTO and MAZZANTI, 1966; MAZZANTI, 1966a,b; CERRINA FERONI and MAZZANTI, 1966).

The thickness of these olistostromes is quite variable, from layers of breccia a few centimetres thick to a few tens of metres. It is not clear whether the very thick beds represent only one, or several successive slumps. Thicknesses over a few hundred metres generally result from the aggregation of several olistostromes and olistoliths.

Ophiolitic olistoliths (Fig.4) of very different sizes (up to a few hundred metres in thickness, and even 1–2 km in length), together with psammitic and psephitic ophiolitic turbidites (see Fig.15) are associated with the olistostromes.

(*a*) Components. The olistostromes are made of igneous (serpentinite, diabase, less commonly gabbro and very rarely granite) and sedimentary rocks (micritic limestones from the *Calpionella* limestones and the Palombini shales

---

Fig.2. Main olistostromes, olistoliths and Chaotic Allochthonous sheets in the Northern Apennines.

Section I. Eastern Liguria and Emilia Apennines; lower part: eastern Liguria, Trebbia Supergroup; upper part: Monte Barigazzo, Parma (After RADRIZZANI, 1964.)

Section II. Calvana Supergroup, Pieve S. Stefano, Arezzo. (After BORTOLOTTI, 1963.)

Section III. Tuscan sequence, Mt. Pisano. (After GIANNINI et al., 1967.)

Section IV. Tuscan sequence, Lima Valley-Monte Cervarola, Tuscan and Emilian Apennines. (Lower part after BOCCALETTI and SACRI, 1966; upper part after *Carta Geologica d'Italia*, Sheet 97, 2nd ed.)

Section V. Umbrian sequence; lower part: Sillaro Valley; upper part: Savio and Marecchia valleys. (After RUGGIERI, 1958.)

Section VI. Outer foothills of the Emilian Apennines, and adjacent plain. (After LUCCHETTI et al., 1962.)

1. Burano Formation; *2.* Ophiolites (olistoliths); *3.* Maiolica; *4.* Lavagna shales; *5.* "Scisti Policromi"; *6.* Casanova Complex; *7.* Gottero sandstones; *8.* Sillano Formation; *9.* Monte Morello Formation; *10.* Montepiano marls; *11.* Ranzano sandstones; *12.* Macigno; *13.* Pievepelago marls; *14.* Cervarola sandstones; *15.* Marnoso-arenacea Formation; *16.* Salsomaggiore sandstones.

C = Cretaceous; E = Eocene; O = Oligocene; B = Aquitanian and Burdigalian; El = Helvetian; T = Tortonian; M = Messinian; P_1 = Lower Pliocene; P_2 = Middle Pliocene; white: normal sedimentation; shaded: olistostromes, olistoliths and other allochthon rocks.

[*Editor's Note:* Figures 3 through 7 have been omitted.]

formations, radiolarian cherts from the Monte Alpe cherts, and sandstones and marls of unknown origin).

The predominant size of the components is from coarse pebbles to cobbles (i.e., a few centimetres to a few decimetres). Elements larger than cobbles may be of any size, up to large olistoliths. Smaller size classes are less represented. Thus in the olistostromes with abundant argillaceous matrix the grain size distribution is bimodal (Fig.6).

We generally attribute the finer mode to the sliding of loose pelitic sediments, whereas the coarser mode derives from the fragmentation of already lithified rocks. This is quite evident when both the matrix and the elements come from the same formation; for example, the shales and hard beds of the Palombini shales.

The clasts generally are angular (Fig.5) to subrounded (see Fig.9, 17B); limestones sometimes appear as fragments of beds (Fig.6). Components that are almost spherical, yet poorly rounded, are also frequent.

In the argillaceous or marly olistostromes the clasts tend to lie with the major axes in the plane of stratification. This is probably a compaction effect rather than an original texture, because no preferred orientation occurs in the matrix-poor olistostromes.

(*b*) Matrix. The matrix may be argillaceous (Fig.6, 13, 16) or, less commonly, psammitic (Fig.9, 15, 17a) or even marly. In the first case it consists of silty or

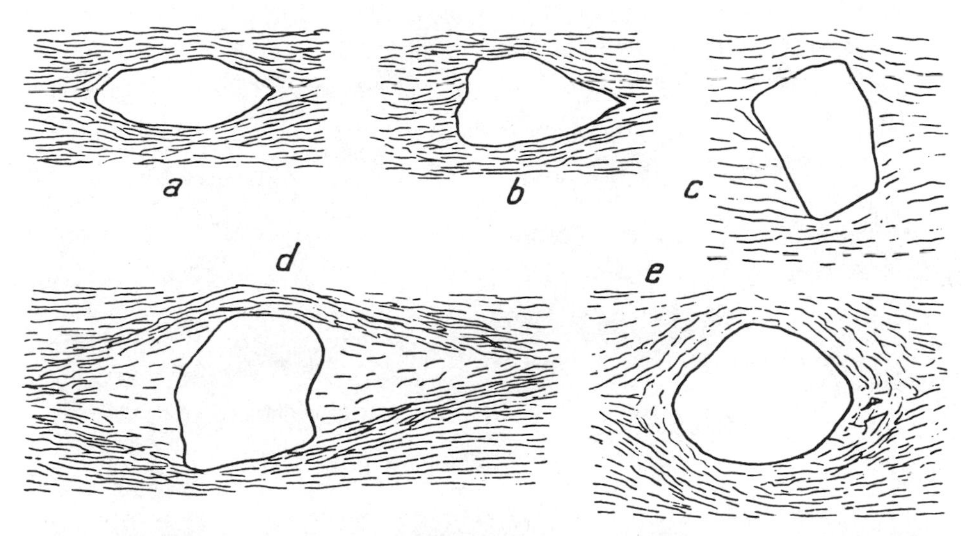

Fig.8. Cleavage in the matrix of the argillaceous olistostromes. The cleavage follows the surface of the clasts (*a*) is sharply interrupted by the clast surface (*b*, *c*). *a* and *c*: Forcella Pass, eastern Liguria, Lavagna shales. *b*: Roveta, Firenze, Macigno. These are the most common patterns of cleavage in argillaceous olistostromes and it is thought to be due to simple compaction. *d*: the cleavage becomes confused on the two sides of a clast: Casanova, Piacenza, Casanova Complex (Fig.7D). This pattern is referred to as shear cleavage. *e*: the cleavage envelopes the clasts: a strong suggestion of flow cleavage (Forcella Pass, eastern Liguria, Lavagna shales).

sandy shales with a cleavage parallel to the bedding of the enclosing sequence. It is worthwhile to examine the relations between cleavage and clast (Fig.7, 8). When a clast face is at right angle to the cleavage, the latter does not deviate, but when the face makes a small angle with the cleavage, the latter tends to bend around the clast margin. Cleavage is evidently related to loading. This is the most frequent structure recognizable in the olistostromes of the Northern Apennines.

Sometimes shear cleavage is also present (eastern Liguria, PASSERINI, 1965): the cleavage envelopes the clasts even in the case of a right angle between a clast face and the general cleavage plane. In other occurrences the cleavage becomes confused and faint and even whirled on the sides of the clasts, where the rigid body prevented the regular flow of the matrix under the shearing stress.

Tectonized olistostromes may be quite similar to tectonic breccias. They bear stretched pebbles elongated parallel to the lineation, and the cleavage planes may not be parallel to the bedding of the enclosing sequence.

A distinction between undisturbed olistostromes and tectonic breccias has been attempted (PASSERINI, 1965) on the basis of relations between cleavage and clasts, although no precise criterion has been found and conclusions need additional field evidence.

The psammitic matrix often is a coarse sand mainly composed of ophiolitic grains with abundant secondary calcite. The sandy matrix (Fig.9) and the bigger clasts generally have a similar composition, both are made of serpentinite or diabase.

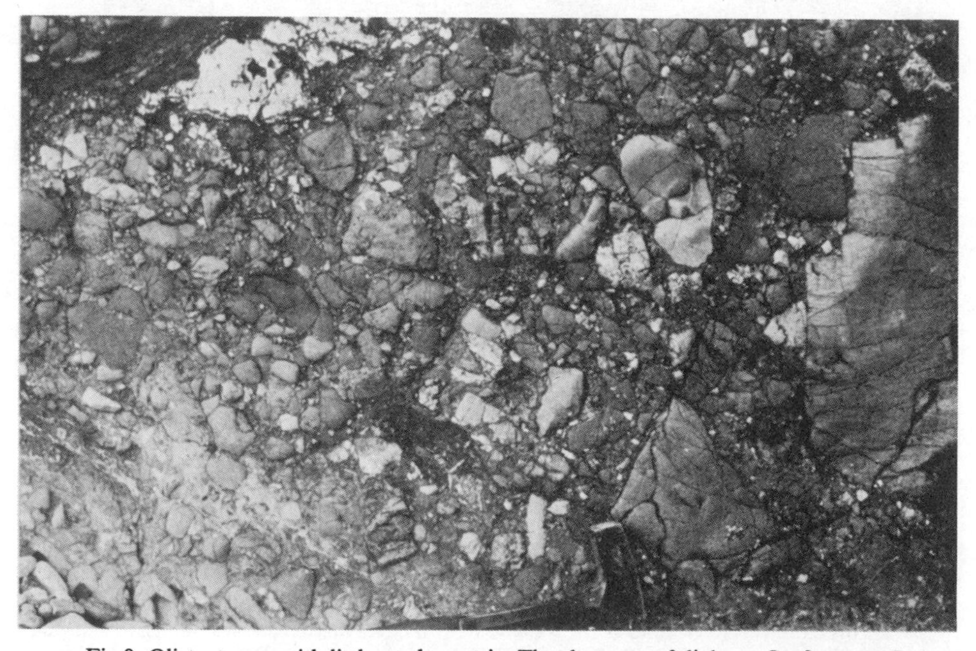

Fig.9. Olistostrome with little sandy matrix. The clasts are of diabase. Garfagnana, Lucca.

The olistoliths of the eugeosynclinal sequences. The olistoliths are almost exclusively made of ophiolitic rocks (Fig.4, 10, 11), locally with granite as a minor constituent. In a few exceptional cases the olistolith includes parts of the original sedimentary cover of the ophiolites, i.e., Monte Alpe Chert and *Calpionella* Limestone. The ophiolites in the olistoliths are mainly diabase and serpentinite; gabbro is less common. Serpentinite may be highly fractured, diabase much less so.

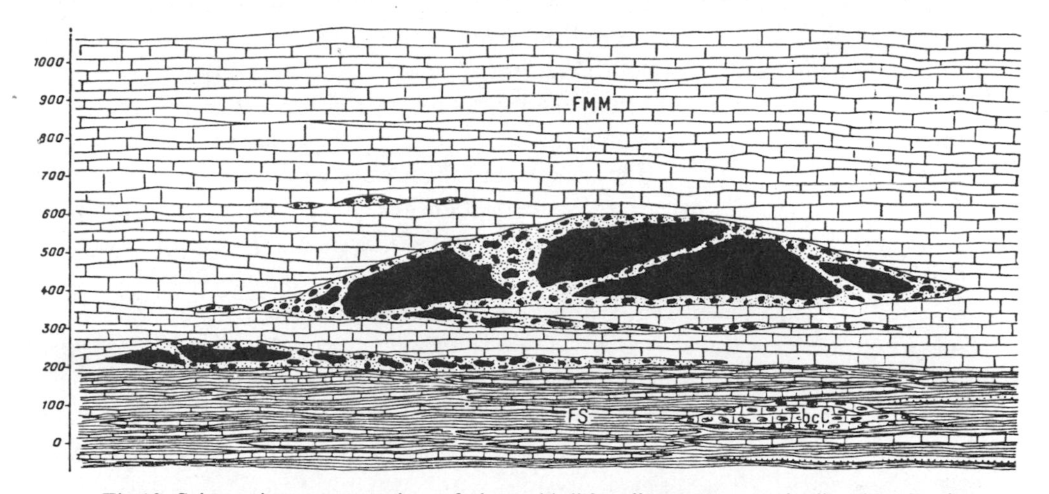

Fig.10. Schematic reconstruction of the ophiolitic olistostromes and olistoliths in the Calvana Supergroup, Viamaggio Pass, Arezzo. Limestone/diabase olistostromes (*bcC*) in the Sillano Formation (*FS*, Upper Cretaceous). At the transition between the Sillano Formation and the Monte Morello Formation (*FMM*, Eocene) small ophiolitic olistostromes (black dots) and olistoliths (black). In the body of the Monte Morello Formation a thick unit made of ophiolitic olistostromes and large serpentinite olistoliths.

The olistoliths generally are lense-shaped and lie parallel with the bedding of the enclosing formation (Fig.4, 10). Dimensions vary considerably and may reach 200–300 m in thickness and 1,000–2,000 m in width. As mentioned before, this is the upper size limit conventionally used for olistoliths. Masses exceeding these size limits should not be called olistoliths; in fact they are absent or are very rare in the eugeosynclinal olistostrome sequences. There are also fractured masses of larger size, a few hundred metres thick and a few kilometres long, but they possibly represent several olistoliths close together.

(*a*) Relations. Olistostromes and olistoliths are frequently associated (Fig.2, 4, 10, 11). The latter are generally lacking in the sequences that bear only small olistostromes. The ophiolitic clasts of the olistostromes show the same composition of the adjacent olistoliths (mainly diabase or serpentinite).

The relations of olistoliths and olistostromes with the enclosing sediments are rather variable. There is, however, a type sequence in which ophiolitic olistoliths or large olistostromes, interbedded in flysch formations, are directly underlain

[*Editor's Note:* Figures 11 and 12 have been omitted.]

and covered by ophiolitic turbidites and minor olistostromes (Fig.2, 4, 10). The latter increase in thickness and frequency toward the olistolith or the larger olistostrome, and grade above and below into the flysch through alternations. The sandstone or marly limestone beds of the flysch may also contain ophiolitic fragments (Fig.12), which are more abundant in the vicinity of the olistostrome.

Olistostromes and olistoliths generally form a minor part of the enclosing sequence. In the Casanova Complex (Upper Cretaceous) and the Lanciaia Formation (Paleocene-Lower Eocene) olistostromes, olistoliths and ophiolitic turbidites form the major part of the unit (Fig.2).

(*b*) Stratigraphic occurrence. Olistostromes or groups of olistostromes are interbedded (Fig.2, *I* and *II*): (*1*) in the Cretaceous sequence Palombini shales–Lavagna shales–Casanova Complex; (*2*) in the Upper Cretaceous Monte Antola Formation; (*3*) in the Upper Cretaceous to Eocene Calvana Supergroup; (*4*) in the Lanciaia Formation of Paleocene–Lower Eocene age; and (*5*) in the Canetolo Complex.

(*1*) Olistostromes begin to appear in the Palombini shales (Lower Cretaceous) or in the Lavagna shales (since Lower Cretaceous) (Fig.13, 14, 15) (PASSERINI, 1965). Most are calcareous–clayey (Fig.13) but some are ophiolitic. At the top of the Lavagna shales and especially in the overlying Casanova Complex the frequency and thickness of the olistostromes increase and there are olistoliths of all

Fig.13. Calcareous-clayey olistostrome in the Lavagna shales, Forcella Pass Genova.

[*Editor's Note:* Figures 14 and 15 have been omitted.]

sizes. The olistostromes of the Casanova Complex are mainly ophiolitic (Fig.15), but calcareous–clayey olistostromes are also abundant (PASSERINI, 1965). At many places olistostromes and olistoliths form most of the Complex, elsewhere they are interbedded with ophiolite-bearing turbidites.

(*2*) The olistostromes of the Antola Formation are sometimes the continuation of those of the Casanova Complex or of other basal units (Berceto, Parma: BORTOLOTTI, 1963). In the Antola Formation itself, olistostromes of all types and ophiolitic olistoliths may occur at several levels (Fig.16) (Castellina Marittima, Livorno: GIANNINI, 1962; Montaione, Firenze: BORTOLOTTI and LAZZERI, 1964; Sassa, Grosseto: BORTOLOTTI, 1964a; Val di Vara, La Spezia: ELTER and RAGGI, 1965a, b). A peculiar olistostrome made chiefly of granite clasts (Fig.17) was found near Passo della Cisa, Parma (PAGE, 1962, 1963; BORTOLOTTI, 1963).

(*3*) Thin ophiolitic olistostromes are also intercalated in the Calvana Supergroup (Upper Cretaceous–Eocene). The older ones occur in the Pietraforte at Impruneta, Firenze (BORTOLOTTI and LAZZERI, 1964) and near the top of the Sillano Formation (Talla and Sarna, Arezzo). At the base and in the lower part of the Monte Morello Formation (Paleocene to Lower–Middle Eocene) there are several groups of ophiolitic and calcareous-clayey olistostromes with large ophiolitic olistoliths (Pieve S. Stefano and Monte Rognosi, Arezzo: BORTOLOTTI, 1961, 1962a,b, 1963; Fig.5, 10, 19).

Fig.16. Argillaceous olistostrome in the Antola Formation. Montaione, Firenze.

Fig.17. A bed of the Antola Formation overlies an olistostrome with granite clasts and sandy matrix, Passo della Cisa, Parma.

(*4*) The Lanciaia Formation is made of small ophiolitic olistostromes (Fig.18) with a few small olistoliths and ophiolitic turbidites associated with marls and calcarenites. This formation lies over large ophiolitic masses referable to a gravity nappe on top of the Antola Formation (SIGNORINI et al., 1963; BORTOLOTTI, 1963, 1964a; LAZZAROTTO and MAZZANTI, 1966; MAZZANTI, 1966a,b; CERRINA FERONI and MAZZANTI, 1967).

(*5*) Very few small ophiolitic olistostromes and ophiolitic turbidites are also found in the Canetolo Complex.

In conclusion, the sliding of ophiolites in the eugeosynclinal sequence started in the Lower Cretaceous and lasted until the Eocene.

Interpretation of the ophiolitic olistostromes. In the preceding paragraphs it has been assumed that many ophiolitic breccias and ophiolitic masses are respectively olistostromes and olistoliths. Moreover, the age of the ophiolitic magmatism is considered to be pre-Neocomian (BORTOLOTTI and PASSERINI, 1970). Some writers, however, do not agree with these conclusions.

In particular, the ophiolitic breccias (or at least some of them) have been interpreted either as pyroclastic rocks (LOTTI, 1910; LABESSE, 1962; CORTEMIGLIA, 1963; etc.) or as tectonic breccias (PRINCIPI, 1924; ROVERETO, 1939). The intercalation of some of the ophiolitic masses in the Cretaceous to Eocene flysch units

[*Editor's Note:* Figure 18 has been omitted.]

Fig.19. Faulted olistostromes in the Monte Morello Formation, Viamaggio Pass Arezzo. Light coloured: calcareous olistostrome; dark coloured: mainly diabasic olistostrome.

has been related to: (*a*) post-Cretaceous or post-Eocene intrusions (ROVERETO, 1939); (*b*) Upper Cretaceous and Eocene extrusions (LOTTI, 1910; RICHTER, 1962; LABESSE, 1962; PAREA, 1965); and (*c*) to tectonic displacements. These theories require a brief discussion.

The interpretation of the ophiolitic masses in the Upper Cretaceous to Eocene sequences as lava flows must be excluded, whenever the ophiolites are still connected with remnants of their original sedimentary cover, and this is older than the enclosing formation. For instance, the Monte Morello Formation of the Monte Rognosi, Arezzo, includes ophiolitic masses overlain by patches of Malm–Lower Cretaceous Monte Alpe cherts and *Calpionella* limestones (BORTOLOTTI, 1962b).

As to the breccias that are associated with the ophiolitic masses, obviously a pyroclastic origin must be excluded for all those that are made entirely of sedimentary clasts. These breccias often grade into, or alternate with, the ophiolitic ones, and display similar features. The sedimentary components of the breccias belong mostly to the original sedimentary cover of the ophiolites from the Monte Alpe cherts to the Palombini shales.

There are other arguments against the extrusive nature of the ophiolite intercalations in the Upper Cretaceous–Eocene sequences. The serpentinites, particularly the lherzolitic types that are common in the Apennines, are almost generally believed to be derived from peridotites. Even admitting the possibility of peridotite flows, the fact that diabase and serpentinite bodies repeatedly alternate in the flysch formations would imply a magmatism producing, at the same time and in the same place, very different and sharply separated rocks. A strong evidence for a non-extrusive secondary emplacement are the rare gabbro masses (commonly coarse-grained) enclosed in the sedimentary sequences.

No definite evidence of contact metamorphism nor of magmatic feeders has been reported in the Upper Cretaceous–Eocene flysch units. Nevertheless, this is not a proof against magmatism, as these conditions are common in basic extrusives.

Some deposits associated with the ophiolites sequence do have obvious pyroclastic characters, but they are older (Tythonian–Neocomian) and their petrographic features and occurrence are quite different from those of the breccias, here considered olistostromes.

Volcanic breccias may be also present in the Upper Cretaceous–Eocene flysch formations, but they probably form olistoliths.

According to GÖRLER and REUTTER (1963) the hypothesis that breccias may be olistostromes would not exclude the extrusive emplacement of the related ophiolitic masses within the flysch sequences. The olistostromes would be caused by the lava flow itself. It is clear that the arguments offered above against a pyroclastic origin of the breccias do not exclude this view, but the objection against the interpretation of some ophiolites in the flysch sequence as lava flows, must be maintained, at least in the cases mentioned above.

A tectonic origin of the breccias here considered olistostromes, is excluded by their evident sedimentary features. A tectonic mechanism for the intercalations of ophiolitic masses is acceptable only in the sense of gravitational sliding. Other tectonic mechanisms, such as faulting, squeezing etc., do not explain the presence of sedimentary breccias and their position with respect to the ophiolitic masses.

The intrusive nature of the ophiolitic masses is equally disproved by the association with sedimentary ophiolitic breccias.

Olistostromes and olistoliths of the miogeosynclinal sequences. In the miogeosynclinal area sliding also began to be of some importance in the lower part of the Upper Cretaceous. Already in the Lower Cretaceous Maiolica Formation, coarse-graded breccias, possibly turbidites, were formed (Lima Valley, Lucca, BOCCALETTI and SAGRI, 1964b, 1966, 1967). Only some breccias found at Monte Cetona (Siena) at the contact between Diaspri and Scisti Policromi (JACOBACCI et al., 1959; JACOBACCI, 1962; 1963) could be interpreted as olistostromes. They are not graded and are poor in cement (PASSERINI, 1964a).

In the basal part of the "Scisti Policromi" (Cenomanian) near Mommio, Massa, GÜNTER and REUTTER (1966) found a few olistoliths made of small stratified masses of "Calcare Selcifero" and "Rosso Ammonitico", which are enclosed in thin bands of breccia (Fig.20).

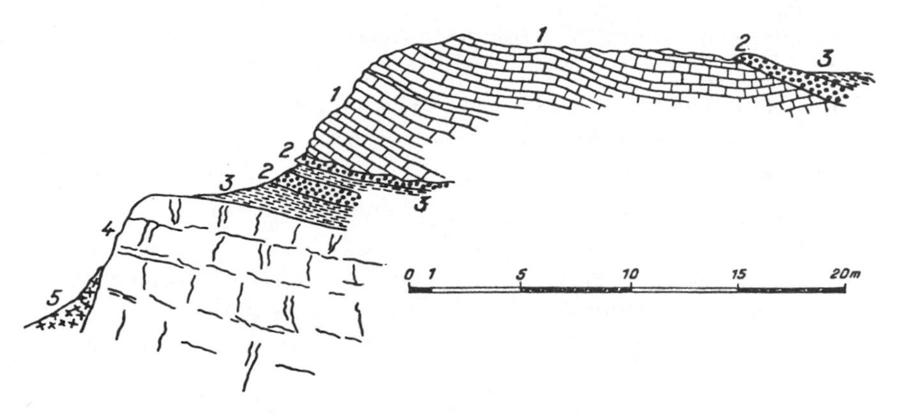

Fig.20. Olistolith of Rosso Ammonitico (*1*) and limestone clasts olistostromes (*2*) in the Scisti Policromi (*3*). *4*. Calcare Selcifero; *5*. talus. (After GÜNTHER and REUTTER, 1966.)

In the Lima Valley there are olistoliths 500 m long and 50 m thick of Maiolica Limestone (Fig.2, *IV*), in the basal part of the Scisti Policromi (BOCCALETTI and SAGRI, 1967); lenses of breccia mark the lower contact. In the upper part of the Scisti Policromi, olistoliths and olistostromes have been found in various places of Tuscany. The olistoliths (in the Alpi Apuane, at Monte Pisano, Montagnola Senese, and Campiglia Marittima, GIANNINI et al., 1967) consist of Calcare Cavernoso and dolomitic limestones of Late Triassic age; they are never associated with breccias (Fig.2, *III*).

The olistostromes of the Chianti Mountains, and near Castellazzara, Grosseto, are made of rounded clasts of Liassic to Late Cretaceous limestones, generally of neritic facies (CANUTI and PIRINI, 1964; CANUTI et al., 1965). The clasts are variably dispersed in a red marlstone matrix very similar in lithology and microfauna to the marlstones of the enclosing Scisti Policromi.

There are also conglomerates (generally 2–10 m thick) with an abundant pelitic matrix, which may or may not be graded; the graded ones have been interpreted as turbidity current deposits (Brucciano and Fegana Valley, Lucca, DALLAN, 1966; cf. SESTINI, 1964). Pebbly mudstones and pelitic conglomerates are probably evidence of various types of movement from dense turbidites to slides.

The source of the olistoliths is probably in uplifted parts of the Tuscan Series. In this case the olistoliths are endolistoliths. The clasts of the olistostromes may have come from shallow-water areas of neritic deposition, equivalents of which are found in the Southern Apennines and in the Venetian–Dinaric region (SESTINI, 1964; CANUTI et al., 1965; DALLAN, 1966).

[*Editor's Note:* Material referring to Oligocene to Miocene olistostromes has been omitted.]

OLISTOSTROME ENVIRONMENT AND GENESIS

Although olistostromes are not necessarily bound to any sedimentary environment, in the Northern Apennines most of them were formed in flysch troughs. Olistostromes and turbidites seem to be closely connected, as nearly all the turbidite sequences are associated to a more or less rich olistostrome suite. From a tectonic point of view the olistostromes suggest uplift of a flank of the basin. In a few cases the uplift generated both olistostromes and turbidites (e.g., Casanova Complex); more frequently the source areas were different. Filling of the basins was likely to be longitudinal for the turbidites and lateral for the olistostromes.

The fragmentation of the material of the olistostrome may be referred to several processes. The sliding movement itself may account for the intense distruption of the more or less consolidated rocks (Fig.25).

Tectonic crushing prior to sliding might have occurred at the front of the advancing nappes. Subaereal or wave erosion could also have supplied the detrital material.

Fragmentation due to sliding or to tectonic crushing seems likely when coarse angular clasts are dispersed in a clay matrix, whereas subaereal or wave erosion can explain the presence of a sandy matrix and the rounding of some clasts.

The frequent interposition of olistostromes in the eugeosynclinal Cretaceous or Eocene flysch is the record of the earliest general orogenic phase of the Alpine cycle. Following the hypothesis that the olistostrome-bearing flysch are allochthonous from the west, the flysch basins were located in the Ligurian and Tyrrhenian area. In the eugeosynclinal basin the ophiolites were partially uplifted in the Cretaceous and Lower Eocene. The uplift caused the sliding from the topographically higher parts into the flysch basins of large quantities of materials from the ophiolitic sequence. The slides produced either olistostromes with large olistoliths or small olistostromes and turbidites made of finer ophiolitic material (in part possibly derived from subaerial erosion). Flysch deposition generally continued after the emplacement of each olistostrome. The sliding is thought to be eastward, by analogy with better-documented later movements.

An Upper Cretaceous uplift in the Tyrrhenian area was formerly advocated by MIGLIORINI (1945), because of the occurrence of Cretaceous turbidites in the eugeosynclinal sequences. He called this orogenic phase "paleo-Apenninic". The name can be maintained, although it is now evident that many turbidity currents came from an Alpine area and not from an early Apenninic ridge. A paleo-Apenninic phase was supposed also by BORTOLOTTI (1962a, 1963) and PASSERINI (1962, 1965) in order to account for ophiolitic olistoliths and olistostromes. ELTER and RAGGI (1965a,b) assumed, for the Ligurian Apennines, the existence of only one ridge which they called "Ruga del Bracco".

The Oligocene and Miocene olistostromes are probably related to the east-

Fig.25. Two aspects of disrupted boulders of Palombino limestone in an olistostrome of the Lavagna shales. Note the peculiarly fragmented margins. Semiconsolidated conditions during emplacement are strongly suggested. The outcrop gives a good instance of a process by which the olistostrome clasts can be produced. Forcella Pass, Genoa.

ward and northeastward advancement of the allochthon (ELTER and SCHWAB, 1959; ABBATE and BORTOLOTTI, 1961). Sliding involved especially the more shaly units: limestone or sandstone of the Monte Morello Formation or of the Antola Formation are much less common than those from the Canetolo Complex, the Sillano Formation and the Palombini shales, which are much more argillaceous.

The Miocene and Pliocene olistostromes of the outer margin of the Apennines are related to the last orogenic pulses. They differ substantially from the older ones because they are not embedded in flysch formations and many of them clearly occurred in a shallow-water environment.

As far as the small olistostromes of southern Tuscany are concerned, they are not connected with any orogenic phases, but only with postgeosynclinal block tectonics.

OLISTOSTROMES IN THE MEDITERRANEAN AREA

Olistostromes and olistoliths are found in the Southern Apennines and in Sicily, in formations of about the same age range as those of the Northern Apennines (BENEO, 1956a, b, c, 1957; RIGO DE RIGHI, 1956; ARDIGO', 1957; FLORES, 1959; SCHMIDT DI FRIEDBERG et al., 1959; SCHMIDT DI FRIEDBERG, 1964–65; CATENACCI, 1965; JACOBACCI, 1965; etc.). Ophiolitic olistostromes and olistoliths have been reported in the Upper Cretaceous and Eocene of the Pontic Ranges and Toros Mountains, Turkey (RIGO DE RIGHI and CORTESINI, 1964; BOCCALETTI et al., 1966; BORTOLOTTI and SAGRI, 1968; SESTINI and CANUTI, 1968).

GANSSER (1959) also described extensive olistostromes from the Himalayas to Turkey; some of them are of much larger scale, and may rather be considered a complex of large olistoliths emplaced by gravitational chaotic nappes ("frane orogeniche" or "frane tettoniche"—orogenic landslides or tectonic landslides—MIGLIORINI, 1933; MERLA, 1951).

Outside the Mediterranean area, a submarine slide origin has been suggested for a number of occurrences of boulder beds. Most of these are associated with flysch formations and orogenic belts, like in the Mediterranean area. Recent lists of references are in HSU (1968), SESTINI (1968) and WILSON (1969).

REFERENCES

ABBATE, E. and BORTOLOTTI, V., 1961. Tentativo di interpretazione dei livelli di "Argille scagliose" intercalati nella parte alta del macigno lungo d'allineamento M. Prado-Chianti (Appennino settentrionale) mediante colate sottomarine. *Boll. Soc. Geol. Ital.*, 80(2): 335–342.

ABBATE, E., BORTOLOTTI, V., PASSERINI, P. and SAGRI, M., 1970. Development of the Northern Apennines geosyncline—Introduction to the geology of the Northern Apennines. *Sediment. Geol.*, 4: 207–249.

ARDIGO', G., 1957. Osservazioni geologiche sulle alte valli del Calore e dell'Ofanto (Appennino meridionale). *Boll. Serv. Geol. Italia*, 79: 67–100.

Baldacci, F., Brandi, G. P., Nardi, R., Squarci, P. and Taffi, L., 1967. Sulla giacitura dei calcari cavernosi e dei gessi di Sassalbo, del Passo del Cerreto e della Val di Secchia (Appennino Tosco-Emiliano). *Mem. Soc. Geol. Ital.*, 6: 199–211.

Beneo, E., 1956a. Accumuli terziari da risedimentazione (olisthostroma) nell'Appennino centrale e frane sottomarine. *Boll. Serv. Geol. Italia*, 78: 291–319.

Beneo, E., 1956b. The results of the studies on petroleum exploration in Sicily. *Boll. Serv. Geol. Italia*, 78: 27–50.

Beneo, E., 1956c. Il problema delle "Argille scagliose"–"flysch" in Italia e sua probabile risoluzione — Nuova nomenclatura. *Boll. Soc. Geol. Ital.*, 75(4) 53–68.

Beneo, E., 1957. Sull'olistostroma quaternario di Gela. *Boll. Serv. Geol. Italia*, 79: 5–16.

Boccaletti, M. and Bortolotti, V., 1965. Lacune della serie toscana, 1. Serie stratigrafiche giurassico-eoceniche nelle zone di Roggio e Trassilico sul versante orientale delle Alpi Apuane. *Boll. Soc. Geol. Ital.*, 84(5): 271–356.

Boccaletti, M. and Sagri, M., 1964a. Strutture caotiche dell'Appennino, I. Età, assetto e giacitura del complesso argilloso-calcareo affiorante nella parte occidentale del F° 129 "S. Fiora". *Boll. Soc. Geol. Ital.*, 83(4): 461–524.

Boccaletti, M. and Sagri, M., 1964b. Sulla presenza di una breccia alla base del complesso scaglia-nummulitico in Val di Lima (Provincia di Lucca). *Boll. Soc. Geol. Ital.*, 83(4): 339–352.

Boccaletti, M. and Sagri, M., 1966. Lacune della Serie Toscana, 2. Brecce e lacune al passaggio Maiolica—Gruppo degli Scisti Policromi in Val di Lima. *Mem. Soc. Geol. Ital.*, 5(1): 19–66.

Boccaletti, M. and Sagri, M., 1967. Olistostromi e olistoliti di Maiolica negli Scisti Policromi della Val di Lima (Provincia di Lucca). *Boll. Soc. Geol. Ital.*, 86(3) 525–536.

Boccaletti, M., Bortolotti, V. and Sagri, M., 1966. Richerche sulle ofioliti delle catene alpine, I. Osservazioni sull'Ankara Mélange nella zona di Ankara. *Boll. Soc. Geol. Ital.*, 85: 485–508.

Bortolotti, V., 1961. Sui rapporti ofioliti–alberese tra Pieve S. Stefano e Borgo Sansepolcro (Arezzo). *Boll. Soc. Geol. Ital.*, 80(3): 269–274.

Bortolotti, V., 1962a. Stratigrafia e tettonica dei terreni alloctoni (ofioliti e alberese) nei dintorni di Pieve S. Stefano (Arezzo). *Boll. Soc. Geol. Ital.*, 81(3): 257–306.

Bortolotti, V., 1962b. Sulla giacitura della serie ofiotilifera dei Monti Rognosi (Arezzo). *Boll. Soc. Geol. Ital.*, 81(3): 313–322.

Bortolotti, V., 1963. Sulla posizione delle rocce della serie ofiolitifera nell'Appennino settentrionale a SE del Taro e della Magra. *Boll. Soc. Geol. Ital.*, 82(2): 151–166.

Bortolotti, V., 1964a. Osservazioni preliminari sulla posizione delle rocce ofiolitiche nelle zone di Berceto (Parma), di Boccassuolo (Modena), dei Monti Livornesi e di Pomarance (Pisa). *Boll. Soc. Geol. Ital.*, 83(2): 259–264.

Bortolotti, V., 1964b. Geologia dell'alta Garfagnana tra Poggio, Dalli e Gramolazzo. *Boll. Soc. Geol. Ital.*, 83(4): 25–154.

Bortolotti, V. and Lazzeri, L., 1964. Sulla giacitura delle rocce della serie ofiolitifera nelle zone di Gambassi e dell'Impruneta (Firenze). *Boll. Serv. Geol. Italia*, 85: 11–22.

Bortolotti, V. and Passerini, P., 1963. Sulla presenza di depositi da frane sottomarine nella parte settentrionale della Corsica. *Boll. Soc. Geol. Ital.*, 82(2): 167–172.

Bortolotti, V. and Passerini, P., 1965. Segnalazione della presenza delle formazioni dal Calcare Selcifero ai Calcari Variegati ad Aptici alla base della Serie Toscana a Cintoia (Chianti settentrionale). *Boll. Soc. Geol. Ital.*, 84(6): 37–40.

Bortolotti, V. and Passerini, P., 1970. Development of the Northern Apennines geosyncline — Magmatic activity. *Sediment. Geol.*, 4: 599–624.

Bortolotti, V. and Sagri, M., 1968. Ricerche sulle ofioliti delle catene alpine. 4. Osservazioni sulla giacitura e l'età delle ofioliti tra Smirne ed Erzurum (Turchia). *Boll. Soc. Geol. Ital.*, 87(4): 661–666.

Braga, G., 1964. Stratigrafia e tettonica delle formazioni implicate nella struttura della valle del T. Spettine (Appennino di Piacenza). *Mem. Soc. Geol. Ital.*, 4. 93–112.

Canuti, P. and Pirini, C., 1964. Microfossili liassici in ciottoli negli "Scisti Policromi" dei monti del Chianti. *Paleontol. Ital.*, 59: 35–52.

CANUTI, P., FOCARDI, P. and SESTINI, G., 1965. Stratigrafia, correlazione e genesi degli Scisti Policromi dei Monti del Chianti (Toscana). *Boll. Soc. Geol. Ital.*, 84(6): 93–166.

CATENACCI, E., 1965. Sulla presenza di masse calcaree mesozoiche incluse nei sedimenti miocenici della media valle del T. Ausente (Lazio meridionale). *Boll. Soc. Geol. Ital.*, 83(3): 247–254.

CERRINA FERRONI, A. and MAZZANTI, R., 1966. Geologia della parte meridionale dei Monti Livornesi in Toscana. *Atti Soc. Toscana Sci. Nat. Pisa, Proc. Verbali, Mem.*, 73(2): 412–468.

CORTEMIGLIA, G. C., 1963. Esempio di ofiolitismo cretaceo-superiore presente nei calcari marnosi ad elimintoidi e fucoidi della Val d'Aveto (Foglio Rapallo). *Atti Ist. Geol. Univ. Genova*, 1(1): 321–346.

DALLAN, L., 1966. Le microfacies dei ciottoli del conglomerato presente nella "Scaglia Toscana" in alcuni affioramenti della Val di Serchio (prov. di Lucca). *Mem. Soc. Geol. Ital.*, 5: 387–424.

DUNBAR, C. O. and RODGERS, J., 1957. *Principles of Stratigraphy*. Wiley, New York, N.Y., 356 pp.

EBERHARDT, R., FERRARA, G. and TONGIORGI, E., 1962. Détermination de l'âge absolu des granites allochtones de l'Apennin septentrional. *Bull. Soc. Géol. France*, 4: 666–667.

ELTER, P. and RAGGI, G., 1965a. Contributo alla conoscenza dell'Appennino ligure: 1. Osservazioni preliminari sulla posizione delle ofioliti nella zona di Zignago (La Spezia); 2. Considerazioni sul problema degli olistostromi. *Boll. Soc. Geol. Ital.*, 84(3): 303–322.

ELTER, P. and RAGGI, G., 1965b. Contributo alla conoscenza dell'Appennino ligure: 3. Tentativo di interpretazione delle brecce ofiolitiche cretacee in relazione con movimenti orogenici nell'Appennino ligure. *Boll. Soc. Geol. Ital.*, 84(5): 1–12.

ELTER, P. and SCHWAB, K., 1959. Nota illustrativa della carta geologica all' 1/50.000 della regione Carro–Zeri–Pontremoli. *Boll. Soc. Geol. Ital.*, 78(2): 157–188.

FAZZINI, P., 1965. La geologia dell'alta val Dolo. *Boll. Soc. Geol. Ital.*, 84(6): 213–238.

FLORES, G., 1955. Discussion. *World Petrol. Congr., Proc., 4th, Rome, 1955*, A2: 120–121.

FLORES, G., 1959. Evidence of slump phenomena (olistostromes) in areas of hydrocarbons exploration in Sicily. *World Petrol Congr., Proc., 5th, N.Y., 1959*, 13: 259–275.

GANSSER, A., 1959. Ausseralpine Ophiolithprobleme. *Eclogae Geol. Helv.*, 52(2): 659–680.

GIANNINI, E., 1962. Geologia del bacino della Fine. *Boll. Soc. Geol. Ital.*. 81(2): 99–224.

GIANNINI, E., LAZZAROTTO, A. and NARDI, R., 1967. Ipotesi sulla giacitura di lembi di dolomie triassiche negli scisti sericitici varicolori della serie toscana metamorfica. *Boll. Soc. Geol. Ital.*, 86(1): 39–48.

GÖRLER, K. and REUTTER, K. J., 1963. Die stratigraphische Einordnung der Ophiolithe des Nordapennins. *Geol. Rundschau*, 53: 358–375.

GÜNTER, K. and REUTTER, K. J., 1966. Submarine Brekzienbildung als Folge cenomaner orogenetischer Tätigkeit in dem Toskaniden II bei Mommio (Provinz Massa-Carrara). *Neues Jahrb. Geol. Paläontol., Monatsh.*, 124(3): 241–253.

HSU, J. K., 1968. Principles of melange and their bearing on the Franciscan-Knoxville paradox. *Bull. Geol. Soc. Am.*, 79(8): 1063–1074.

JACOBACCI, A., 1962. La serie rovesciata del Monte Cetona. *Boll. Serv. Geol. Italia*, 83: 33–49.

JACOBACCI, A., 1963. Nuovi orientamenti nello studio del flisch appenninico. *Mem. Soc. Geol. Ital.*, 4(2): reprint 23 pp.

JACOBACCI, A., 1965. Frane sottomarine nelle formazioni geologiche. Interpretazione dei fenomeni olistostromici e degli olistoliti nell'Appennino e in Sicilia. *Boll. Serv. Geol. Italia*, 86: 65–85.

JACOBACCI, A., MARTELLI, G., MALFERRARI, N. and PERNO, U., 1959a. Osservazioni e considerazioni sulle formazioni terziarie prepontiche affioranti nel Foglio 129 "S. Fiora". *Boll. Serv. Geol. Italia*, 81: 181–197.

JACOBACCI, A., MARTELLI, G., MALFERRARI, N. and PERNO, U., 1959b. Gli olistostromi di età pliocenica nel foglio 129 "(S. Fiora)". *Boll. Serv. Geol. Italia*, 81: 407–436.

KUENEN, PH. H., 1956. The difference between sliding and turbidity flow. *Deep-Sea Res.*, 3: 134–139.

LABESSE, B., 1962. Sur les ophiolites et les brèches associées dans l'Apennin septentrional. *Bull. Soc. Géol. France*, 4: 867–870.

LAZZAROTTO, A. and MAZZANTI, R., 1966. Studio geologico e micropaleontologico di una sezione tra Castelnuovo in Val di Cècina e Monte Gabbri (in provincia di Pisa). *Atti Soc. Toscana Sci. Nat. Pisa, Proc. Verbali Mem.*, 73(2): 330–375.
LOTTI, B., 1910. Geologia della Toscana. *Mem. Descrit. Carta Geol. Ital.*, 13: 1–484.
LUCCHETTI, L., ALBERTELLI, L., MAZZEI, R., THIEME, R., BONGIORNI, D. and DONDI, L., 1962. Contributo alle conoscenze geologiche del Pedeappennino padano. *Boll. Soc. Geol. Ital.*, 81(4): 5–245.
MARCHETTI, M. P., 1956. The occurrence of slide and flowage materials (olistostromes) in the Tertiary series of Sicily. *Congr. Geol. Intern. Compt. Rend., 20e*, 1: 209–225.
MAXWELL, J. C., 1962. Structural geology of the Ottone area, Piacenza and Genova. *Mem. Soc. Geol. Ital.*, 4: 69–92.
MAZZANTI, R., 1966a. Geologia della zona di Monteverdi Marittimo – Canneto (in provincia di Pisa). *Atto Soc. Toscana Sci. Nat. Pisa, Prov. Verbali, Mem.*, 73(2) 469–490.
MAZZANTI, R., 1966b. Geologia della zona di Pomarance – Larderello (prov. di Pisa). *Mem. Soc. Geol. Ital.*, 5(2): 105–138.
MERLA, G., 1951. Geologia dell'Appennino settentrionale. *Boll. Soc. Geol. Ital.*, 70(1): 95–382.
MERLA, G., 1952. Ricerche tettoniche nell'Appennino settentrionale. *Intern. Geol. Congr., 19th, London, 1948, Rept.*, 13: 178–185.
MERLA, G., 1957. Essay on the geology of the Northern Apennines. *Meeting Gas Field W. Europe, 1957, Milan*, 2: 629–651.
MERLA, G., 1964. Centro di Studio per la Geologia dell'Appennino. I Sezione-Firenze. Attività svolta nel periodo 1951–1963. *La Ric. Sci., Suppl.*, 34(2): 107–126.
MERLA, G. and BORTOLOTTI, V., 1967. *Note Illustrative della Carta Geologica d'Italia, Foglio 113, Castelfiorentino.* Serv. Geol. Italia, Roma, 62 pp.
MERLA, G., BORTOLOTTI, V. and PASSERINI, P., 1967. *Note Illustrative della Carta Geologica d'Italia, Foglio 106, Firenze.* Serv. Geol. Italia, Roma, 61 pp.
MICHELI, P., 1965. Osservazioni sui rapporti fra ofioliti e alberese nella zona di Collalto (Siena). *Boll. Serv. Geol. Italia*, 86: 113–122.
MIGLIORINI, C. I., 1933. Considerazioni su di un particolare effetto dell'orogenesi. *Boll. Soc. Geol. Ital.*, 52: 293–304.
MIGLIORINI, C. I., 1945. Le fasi orogeniche nell'Appennino settentrionale. *Boll. Soc. Geol. Ital.*, 64: 46–48.
NARDI, R., 1964. Contributo alla geologia dell'Appennino Tosco-Emiliano, V. La geologia della valle dello Scoltenna tra Pievepelago e Montecreto (Appennino Modenese). *Boll. Soc. Geol. Ital.*, 83(4): 353–400.
NARDI, R. and TONGIORGI, M., 1962. Contributo alla geologia dell'Appennino Tosco–Emiliano, 1. Stratigrafia e tettonica dei dintorni di Pievepelago (Appennino modenese). *Boll. Soc. Geol. Ital.*, 81(3): 1–76.
PAGE, B. M., 1962. Geology south and east of Passo della Cisa, Northern Apennines. *Boll. Soc. Geol. Ital.*, 81(3): 147–194.
PAGE, B. M., 1963. Gravity tectonics near Passo della Cisa, Northern Apennines, Italy. *Geol. Soc. Am. Bull.*, 74(6): 655–672.
PAPANI, G., 1963. Su un olistostroma di "argille scagliose" intercalato nella serie Oligomiocenica del subappennino reggiano. *Boll. Soc. Geol. Ital.* 82(3): 195–202.
PAREA, G. C., 1965. Evoluzione della parte settentrionale della Geosinclinale appenninica dall'Albino all'Eocene superiore. *Atti Mem. Accad. Naz. Sci. Lettere Arti, Modena, Ser. 6*, 7: 3–97.
PASSERINI, P., 1962. Giacitura delle ofioliti tra il M. Aiona e Rovegno (Appennino ligure). *Boll. Soc. Geol. Ital.*, 81(3): 139–146.
PASSERINI, P., 1964a. Il Monte Cetona (provincia di Siena). *Boll. Soc. Geol. Ital.*, 83(4): 219–338.
PASSERINI, P., 1964b. Examination of the ophiolites, flysch and poligenic breccias East of Fontanigorda. In: *Guidebook – Italy 1964, Intern. Field. Inst., A.G.I.*, Washington: XI, 7–8.
PASSERINI, P., 1965. Rapporti fra le ofioliti e le formazioni sedimentarie fra Piacenza e il Mare Tirreno. *Boll. Soc. Geol. Ital.*, 84(5): 95–176.
PETTIJOHN, F. J., 1957. *Sedimentary Rocks.* Harper, New York, N.Y., 718 pp.
PRINCIPI, P., 1924. I terreni terziari dell'Alta Valle del Tevere. *Boll. Soc. Geol. Ital.*, 43(1): 64–80.

RADRIZZANI, S., 1964. Presenza di colate sottomarine nella placca oligocenica del M. Barigazzo. *Mem. Soc. Geol. Ital.*, 4(1): 273–280.

RAGGI, G., 1965. Contributo alla conoscenza dell'Appennino ligure, 4. Osservazioni sulla posizione delle ofioliti del massiccio del Monte Penna (Alta Val di Taro) e considerazioni sui complessi di base dei flysch del Monte Gottero e del Monte Caio. *Boll. Soc. Geol. Ital.*, 84(6): 15–28.

RICCI LUCCHI, F., 1965. Alcune strutture di risedimentazione nella Formazione marnoso-arenacea romagnola. *Giorn. Geol., Ann. Museo Geol. Bologna*, 33(1): 265–283.

RICCI LUCCHI, F., 1967. Trasporti gravitativi sinsedimentari nel tortoniano dell'Appennino Romagnolo (Valle del Savio). *Giorn. Geol., Ann. Museo Geol. Bologna*, 34(1): 1–30.

RICHTER, M., 1962. Das Alter der Serpentinite östlich von Florenz. *Neues Jahrb. Geol. Palaeontol., Monatsh.*, 130–142.

RIGO DE RIGHI, M., 1956. Olistostromi neogenici in Sicilia. *Boll. Serv. Geol. Italia*, 75(3): 185–215.

RIGO DE RIGHI, M. and CORTESINI, A., 1964. Gravity tectonics in foothills structure belt of Southeast Turkey. *Bull. Am. Assoc. Petrol. Geologists*, 48(12): 1911–1937.

ROVERETO, G., 1939. Liguria geologica. *Mem. Soc. Geol. Ital.*, 2: 1–743.

RUGGIERI, G., 1958. Gli esotici neogenici della colata gravitativa della Val Marecchia. *Atti Accad. Sci. Lettere Arti Palermo, Ser. 4*, 17: 1–169 (reprint).

SCHMIDT DI FRIEDBERG, P., 1964–65. Litostratigrafia petrolifera della Sicilia. *Riv. Mineraria Siciliana.*, 88–90: 198–217; 91–93: 50–71.

SCHMIDT DI FRIEDBERG, P., BARBIERI, F. and GIANNINI, G., 1959. La geologia del gruppo montuoso delle Madonie (Sicilia centro-settentrionale). *Boll. Serv. Geol. Italia*, 81: 73–140.

SESTINI, G., 1964. Paleocorrenti eoceniche nell'area tosco-umbra. *Boll. Soc. Geol. Ital.*, 83(1): 291–344.

SESTINI, G., 1968. Notes on the internal structure of the major Macigno Olistostrome (Oligocene, Modena and Tuscany Apennines). *Boll. Soc. Geol. Ital.*, 87(1): 51–64.

SESTINI, G. and CANUTI, P., 1968. Flysch facies in the Pontic Mountains of Turkey. *Boll. Soc. Geol. Ital.*, 87(2): 317–332.

SIGNORINI, R., 1940. Sulla tettonica dell'Appennino Romagnolo. *Atti R. Accad. Ital. Rend. Cl. Sci. Fis. Mat. Nat., Ser. 7*, 1: 370–383.

SIGNORINI, R., CENTAMORE, E. and CONATO, V., 1963. La formazione di Lanciana in Val di Cècina. *Boll. Serv. Geol. Italia*, 84: 83–100.

WILSON, H. H., 1969. Late Cretaceous and eugeosynclinal sedimentation, gravity tectonics and ophiolite emplacement in Oman Mountains, Southwest Arabia. *Bull. Am. Assoc. Petrol. Geologists*, 53(3): 626–672.

6

Reprinted from pages 175-188 of *Gravity and Tectonics*, K. A. De Jong and R. Scholten, eds., Wiley-Interscience, New York, 1973, by permission of John Wiley & Sons, Inc.

Olistostromes in the Tectonic Evolution of the Northern Apennines

P. ELTER
L. TREVISAN

Istituto di Geologia e Paleontologia
University of Pisa
Pisa, Italy

In this paper various aspects of diverse types of olistostromes in the Northern Apennines will be described, and the authors will attempt to interpret their structure and kinematics in the framework of the tectonic evolution of the Northern Apennines.

In 1955 Flores introduced the term "olistostrome" to indicate heterogeneous, more or less intimately mixed deposits that are interbedded with normal sedimentary rocks and interpreted them as the result of submarine sliding. The phenomenon for which this new term was created had been observed in Sicily in various manifestations and dimensions, but subsequently the term olistostrome was applied to comparable phenomena in the Apennines and other mountain chains.

Some authors have tried to give new descriptions and definitions of the terms olistostrome and olistolite, discussing the genesis and mode of occurrence of these phenomena (e.g., Abbate and others, 1970). We do not intend to enter into these discussions but prefer to illustrate some types of submarine slumping and their relation to the causative tectonic events.

In the Northern Apennines Elter and Schwab (1959) were the first to interpret as olistostromes some deposits of breccia intercalated between flysch beds. Previously, all deposits for which no explanation had been found, and which were simply called "chaotic" in view of their complex structure, were generally named *argille scagliose* (scaly clays or shales), because of the frequent occurrence of shales that break down in outcrop into splinters of several millimeters. Hence an old term with a purely lithological meaning was expanded to indicate allochthonous and commonly strongly tectonized units. Today the term "argille scagliose" should no longer be used in this generalized sense, because the stratigraphy and structure of the various shale-rich allochthonous complexes have been studied in sufficient detail in almost the entire Apennine chain.

Types of Slides in Relation to the Provenance of the Material

The gravitative phenomena we want to examine belong to a group which, in its simplest aspect, involves slumping. Other, more complex types deserve to be better known. Figure 1*A* shows a model of a slump. We include in this term intercalated deposits made up either of beds deformed by differential sliding or of mud flows with a brecciated character, providing that the material is derived from the same units with which they are interbedded. For this process to occur, a slight tilting of the sea floor during sedimentation suffices.

Figure 1*B* shows a model that differs from the preceding model in that the slide unit includes in addition materials derived from formations older than the sequence in which the slide occurs.

Figure 2 shows an actual example from Sicily. To obtain this type of structure a slight folding or some other kind of structural deformation appears to be required. The slide seems to occur in front of the fold. This kind of phenomenon was first named *argille brecciate* (brecciated shales) and later olistostromes.

In other instances the materials intercalated in a stratigraphic sequence with evidence of sliding movement belong at least partly to the sediments of a different depositional basin. This third type of slide is evidently related to the emplacement of nearby allochthonous masses (Fig. 1*C*), which follows the slide after a brief interval. For these slides we propose the term "precursory olistostromes."

The three types of slides are consequently distinguished from each other by the different provenances of the constituent materials. These different provenances may be attributed to different conditions of the "tectonic landscape" at the moment the slides occurred.

In the next section we shall examine several types of olistostromes in relation to the evolution in time and space of the tectonic landscape of the Northern Apennines. The distribution of olistostromes and of the major tectonic units in this region is shown in Fig. 3.

Evolution of the Tectonic Landscape of the Northern Apennines

Character of Sedimentary Basins

In the geosyncline of the Northern Apennines an internal domain (southwest of the present mountain chain) can be distinguished from an external domain by the considerable differences between their sediments (Fig. 4*A*).

The internal area, called the Ligurian domain, is characterized by the fact that the oldest sediments (radiolarites) belong to the Malm (Upper Jurassic). These pelagic sediments rest directly on a substratum of basic and ultrabasic igneous rocks: more or less serpentinized peridotites, gabbros, and discontinuous diabase flows. In view of this, one may imagine that during Malm time the sialic crust had been stretched and that an oceanic bottom appeared from below (Decandia and Elter, 1969). The radiolarites and the overlying *Calpionella* limestone, with their noteworthy variability in thickness and local absence, seem to indicate relatively rapid upward and downward movements of the basin floor. The oldest sediments often contain intercalations of detritic ophiolitic material. Until the start of the Cenomanian, sedimentation continues with shales interbedded with gray pelagic limestones (*palombini*). The subsequent flysch sedimentation extends until the end of the Paleocene in the innermost part of the Ligurian domain, and until the end of the Eocene toward the northeastern margin of the domain.

The external part of the northern Apennine

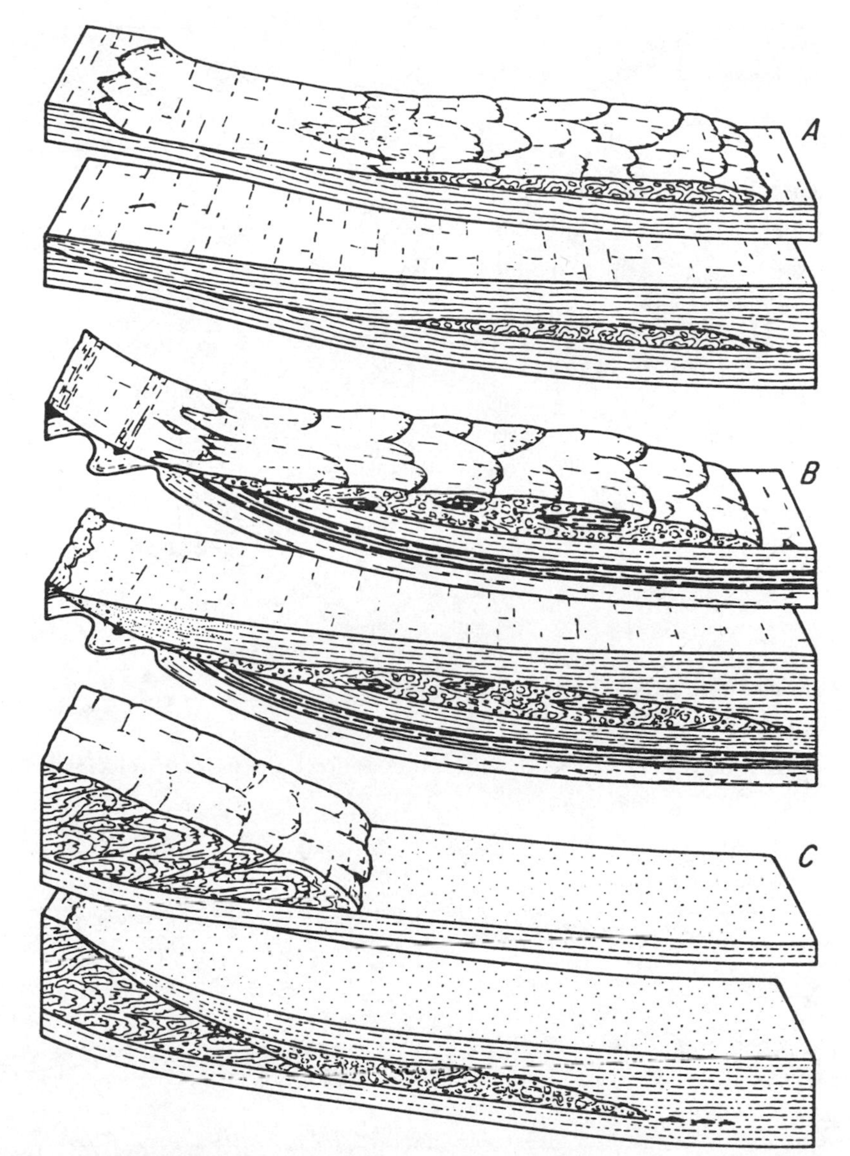

FIG. 1 Schematic representation of three different types of submarine slides. (*A*) Slumping (materials derived from same formation); (*B*) olistostromes (materials derived from other formations in the same sedimentary basin); (*C*) precursory olistostromes (materials derived from the front of an advancing allochthonous sheet).

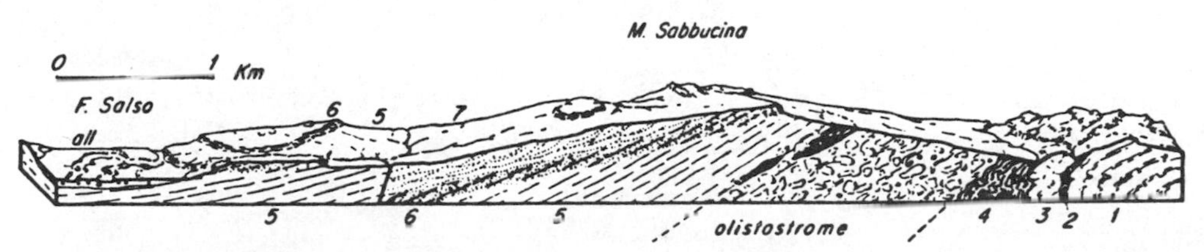

FIG. 2 Example of olistostromes near Caltanissetta (Sicily). (1) Shales (Tortonian: middle Miocene); (2) Tripoli; (3) gypsum (Messinian: upper Miocene); (4) white marls with foraminifera "trubi" (lower Pliocene); (5) blue shales of the Piacenza facies (Pliocene); (6, 7) yellow calcarenites and sands of the Asti facies (Pliocene). The olistostromes have the aspect of conglomerates with components of Tortonian shales and sequences of "trubi" strata.

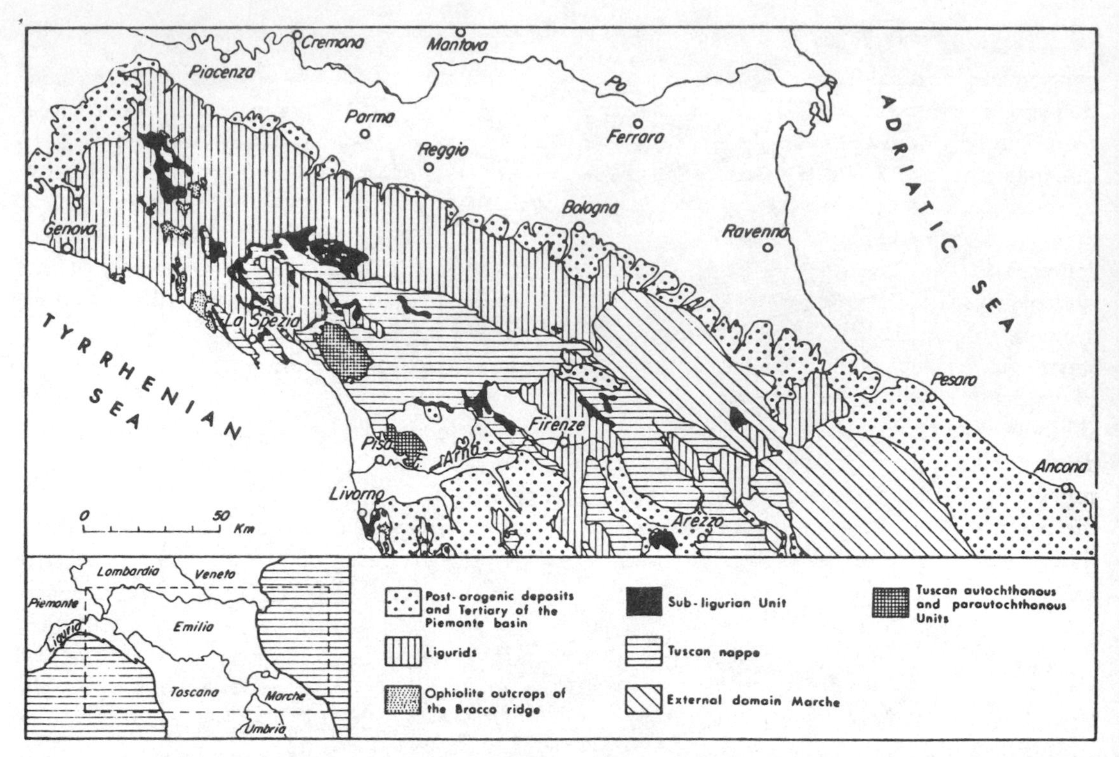

FIG. 3 Schematized map showing the distribution of the tectonic units of the Northern Apennines.

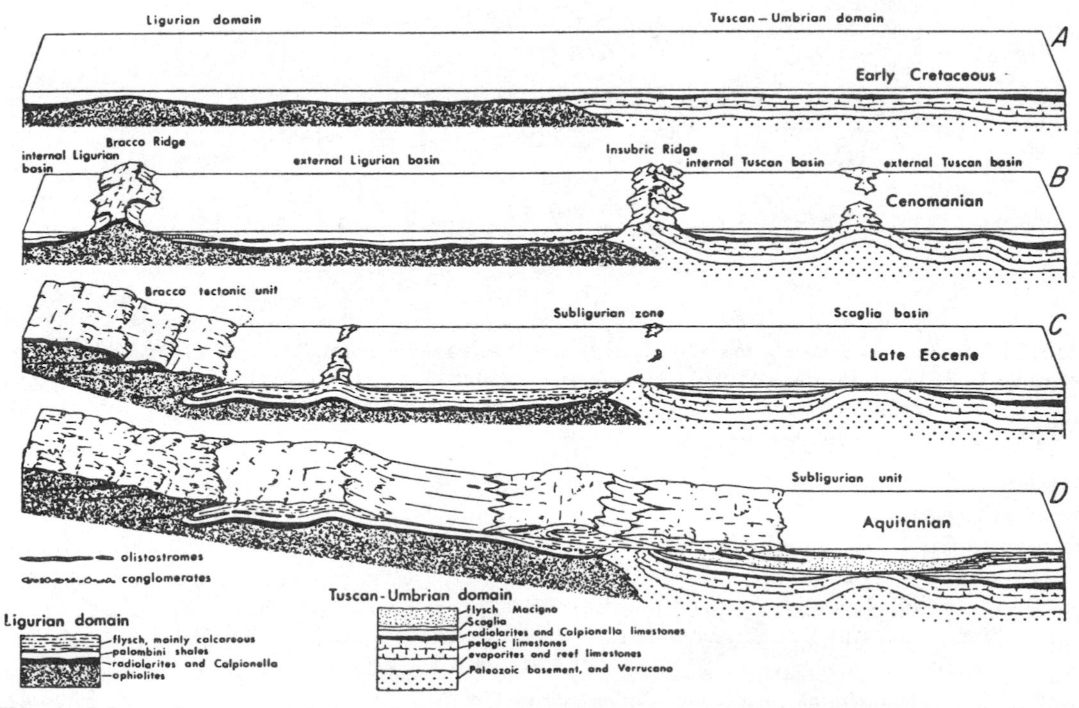

FIG. 4 Schematic representation of the tectonic evolution of the Northern Apennines from the Cretaceous to the Aquitanian.

geosyncline, called the Tuscan-Umbrian domain, is characterized by a sedimentary sequence in which the Middle Triassic rests discordantly on a metamorphosed continental-type basement involved in the Hercynian orogeny. The Upper Triassic is represented by evaporites and reef deposits, and the Jurassic by pelagic, calcareous, and siliceous rocks, locally interrupted by diastems. At the end of the Malm, formations of radiolarites and *Calpionella* limestone units similar to those of the Ligurian domain, were deposited in the Tuscan-Umbrian domain, suggesting that no barrier existed between the two areas at that time. Flysch sedimentation begins in the external domain in Oligocene time.

Cretaceous Tectonic Phase

The first tectogenic movements in the Northern Apennines manifest themselves toward the end of the Early Cretaceous (Fig. 4*B*). Several uplifts in the external domain are revealed by local discordances in the Cenomanian *scaglia rossa* (red scaly shale) and by an hiatus which corresponds to the Albian and Aptian stages or which may locally be even larger. Cenomanian conglomerates in the eastern part of the Ligurian domain, at the border between the internal and external domains, demonstrate that during the Cenomanian an emergent ridge (Insubric ridge of Elter and others, 1966) existed between the two areas. Crystalline pebbles in the conglomerate show that erosion had progressed to the metamorphic basement. The rise of this ridge is probably related to important tectonic movements which occurred in the Austroalpine region farther north.

A significant event in the Ligurian area, perhaps unrelated to the Insubric ridge, is the uplift of the Bracco ridge (Elter and Raggi, 1965), which consists of ophiolites and their sedimentary cover. This ridge divides the Ligurian domain into two parts, so that from then on there existed an internal and an external Ligurian basin. In both basins turbiditic flows came in both from raised portions of the Austroalpine domain and from the Hercynian massifs of Corsica and Sardinia. Thus began the flysch sedimentation of the Late Cretaceous.

The Bracco ridge was rapidly uplifted and attained its maximum development in the Cenomanian and Turonian. The elongate ridge with locally high and low parts was at some places above sea level, as shown by conglomerate lenses. In the two basins separated by the Bracco ridge sedimentation was different. The Val Lavagna shales were deposited in the internal basin; these marly-silty units tend to assume in their higher parts the aspects of flysch, with intercalations of sandstone beds. Away from the ridge a coarse sandy flysch, the Monte Gottero sandstone, is equivalent to the upper part of the Val Lavagna shales. In the external basin (NE of the ridge) flysch was deposited on shaly and sandy rocks of Cenomanian to Turonian age. This is the *Helminthoid* flysch of the Northern Apennines; Senonian to Paleocene in age, it is mainly calcareous to marly in composition and, because of common facies changes, is known under different formation names: Monte Caio, Monte Cassio, Serramazzoni, and others.

Olistostromes of the Cretaceous Phase

The uplift of the Bracco ridge is documented mainly by the appearance of detrital units in the basin sediments on the two sides of the ridge. These intercalated deposits have different aspects: in places they are turbidite sandstones with dominantly ophiolitic material, but more commonly they consist of layers or lenses of breccia with a pelitic matrix which are interpreted as olistostromes (Fig. 5). The breccias include clasts of ophiolitic rocks and Upper Jurassic to Cretaceous sediments that were originally on top of the ophiolites.

The locally abundant matrix is derived primarily from the source rock, but it appears in a few instances enriched with the sediments

FIG. 5 Aspect of an olistostrome of Cenomanian age composed of elements derived from ophiolites and their sedimentary cover, with abundant clay matrix.

over which sliding occurred. Breccias composed of ophiolites, *Calpionella* limestone, and radiolarites without a pelitic matrix are also known, and in some cases the matrix consists of sandstone or reworked and commonly chloritized hyaloclastites. The olistostromes with little clay matrix are localized in the vicinity of the ridge from which they originated, whereas those with a more abundant clay matrix may have moved over distances on the order of 50 km.

Associated with the latter type of olistostrome, but not necessarily enclosed in them, are large blocks, occasionally enormous in size (up to several cubic kilometers), and composed primarily of diabase and more or less serpentinized peridotite. In view of their relation to the surrounding sediments, these rocks are regarded as enormous clasts which slid over the muddy sea bottom.

The olistostromes are clearly more numerous in proximity to the ridge, and in many cases they substitute for the normal flysch sediments and the underlying rocks.

Olistostromes of this type are present on both sides of the Bracco ridge, but they are more frequent and extend farther out on the external side. This may indicate an asymmetry of the ridge, which is probably related to the future northeasterly vergence of the Apennines.

The olistostromes are generally intercalated in Cenomanian sediments, but in some areas as large as several square kilometers they rest directly on the ophiolites. Figure 6 provides an explanation of this phenomenon.

It seems certain that various generations

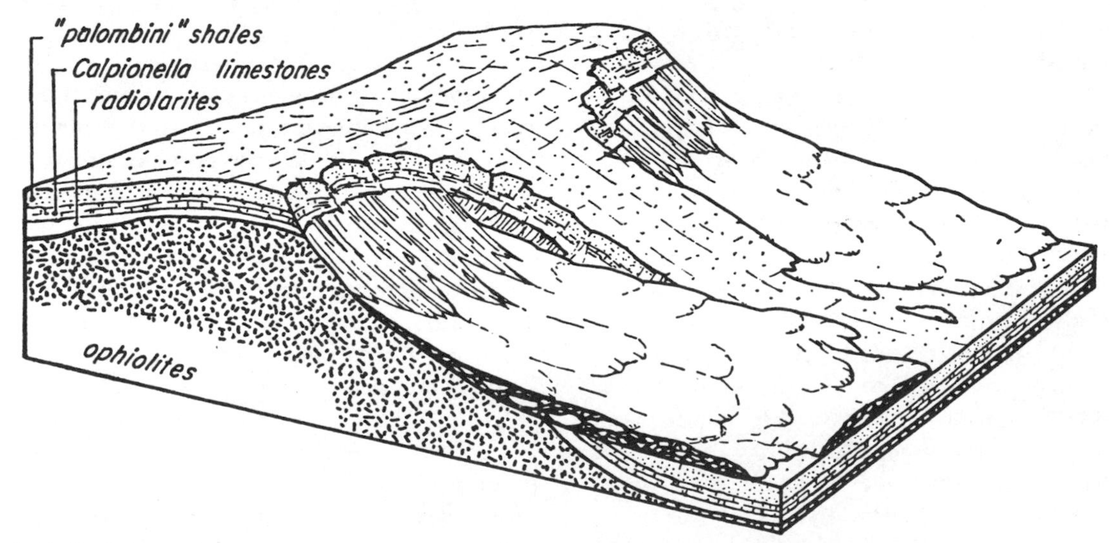

FIG. 6 Schematic representation of olistostromes on the NE flank of the Bracco ridge, showing the relationships to the substratum.

FIG. 7 Aspect of an olistostrome near the Bracco ridge. The arrows indicate a brecciated component derived from a preceding olistostrome.

of olistostromes occurred during the Cretaceous tectonic phase. It is, in fact, not uncommon to find fragments of older olistostromes within a younger one (Fig. 7). It can be concluded that the Bracco ridge has been the focus of increasing mobility during the Cenomanian, which may be considered as the precursor of the great Eocene paroxysmal phase in the Ligurian domain.

Eocene Tectonic Phase (Ligurian Phase)

Sedimentation in the Ligurian domain continued until the end of the Paleocene, predominantly in a flysch-type facies. During the Eocene an important phase of folding affected the internal Ligurian basin, the Bracco ridge, and part of the external basin (Fig. 4*C*). The upper part of the Bracco ridge moved toward the east in the form of a recumbent fold, as a result of which the sediments and the intercalated olistostromes that had been deposited on the exterior side of the ridge just prior to this movement were covered over an area of many square kilometers by the ophiolites that constitute the core of the fold. Other recumbent folds of a more internal origin, likewise with ophiolites in their cores, were piled on top of this fold. Part of this folded terrane was subsequently covered by sediments of late Eocene to middle Miocene (Helvetian) age. The most important formation of this unconformable sequence is the Oligocene Ranzano sandstone

Sedimentation was continuous in the extreme external part of the Ligurian basin, which did not participate in these structural movements. The dominantly calcareous flysch of this area grades laterally without apparent discontinuity into the Ranzano sandstone.

Olistostromes of the Eocene Phase

A part of the relief formed during the Ligurian phase was undoubtedly emergent and became the source area of the sandstones and associated conglomerates of the Ranzano formation. With respect to the Ligurian phase the Ranzano sandstone constitutes a molasse-type sediment. Since the marine domain had become more restricted, olistostromes are much less abundant. Nevertheless, several are intercalated in the Ranzano sandstone, and others occur sporadically in the external Eocene flysch. Some of these contain very large ophiolite blocks as well as portions of the basal complex, and they appear identical in composition to the Cenomanian ones. It is probable that they are "inherited" olistostromes, that is, due to reworking of the older ones.

Whatever the origin of the olistostromes of post-Ligurian age, they were successively formed in areas that are ever more external, and they are intercalated in ever younger sediments, thus indicating the relatively slow progress of the tectogenic movements toward the external side.

Aquitanian Tectonic Phase (Early Miocene)

The second or Ligurian tectonic phase was accompanied by large-scale overthrusting but remained confined to the internal part of the Ligurian domain. During the Cretaceous and the Eocene no important movements occurred in the Tuscan-Umbrian domain, apart from some vertical mobility as demonstrated by partial hiati in the formations of the *scaglia* (scaly shales). In the Oligocene a trough was formed in the external domain, which was filled by thick deposits of sandy flysch, the Macigno sandstone. The first overthrusting that involved the advancement of an allochthonous sheet of internal domain rocks over the external domain took place at the end of the Oligocene and the beginning of the Miocene (Fig. 4*D*).

The first tectonic unit to reach the Tuscan-Umbrian domain consists almost completely of Eocene rocks, composed subordinately of calcareous rocks and sandstones, but predominantly of a thick succession of shales with interbedded limestones. With some reservation the most external part of the Ligurian domain may be considered as the source area of this tectonic unit. Most probably its upper part advanced still farther, and it is now found in areas farther to the east as the calcareous flysch of "Alberese." The mechanism of this forward movement over the external domain is considered to be gravitational, if for no other reason than that a rock complex consisting mainly of shales cannot transmit a large pushing force from the rear.

This tectonic unit, previously called the shale-limestone unit, is here named the Subligurian unit. Its forward movement into the sedimentation trough of the Macigno was not instantaneous, but more or less gradual, as demonstrated by the presence of olistostromes in various horizons of the Macigno formation.

Olistostromes of the Aquitanian Phase

When the Subligurian unit advanced into the Macigno trough, small and large submarine slumps detached themselves from the frontal part, which advanced separately over many kilometers, and which may be observed today as brecciated deposits interbedded with the Macigno turbidites. All of the Eocene formations of the Subligurian unit can be found in the clastic components of the olistostromes, with a slight prevalence of limestones over sandstones. Angular fragments of shale are also encountered, indicating that diagenesis was already in progress prior to tectogenesis.

FIG. 8 Olistostromes of very small size included in Oligocene flysch (Macigno) of the Tuscan sedimentary basin. The calcareous components are slightly rounded.

FIG. 9 Same as Fig. 8.

The olistostromes have various aspects. The easiest recognizable is that of a lenticular mass composed of clastic elements predominantly in the range between 10 and 20 cm, and embedded in a disorganized manner within an abundant shale matrix. The corners of the clasts are slightly rounded, suggesting a rolling movement over a relatively long distance. The smaller olistostromes in particular have this aspect (Figs. 8 and 9).

The second type of olistostrome is composed of blocks most commonly between 0.2 and 1 m in size, with angular corners and to some degree fractured. The matrix consists of the underlying sediment with some traces of stratification. In some places, the blocks are separated from each other. It is thought that this type of olistostrome was transported over a short distance only, without rolling movement.

The olistostromes of the third type are characterized by very large blocks composed of many beds (Fig. 10). The lower boundary commonly appears tectonic in nature, and it is almost invariably marked by stretching of sedimentary beds as a result of traction. In many blocks the beds are overturned, which, however, does not necessarily indicate rolling movements because the possibility that they were derived from the overturned

FIG. 10 Aspect of an olistostrome with preserved sequences of strata.

flanks of recumbent folds cannot be excluded.

The various characteristics just described may be found together in a single olistostrome and, in some cases at least, may indicate the distance to the area of origin, whether nearby or far away.

Figure 11 shows a cross section between the Vara and Taro rivers in the Northern Apennines with two olistostromes of respectable dimensions. Mapping of the area made it possible to reconstruct the sequence of events (Fig. 12).

The truncation of the rear part of the two olistostromes is visible in outcrops a short distance south of the area represented in Fig. 11. Renewal of Macigno sedimentation after the second olistostrome was followed by the arrival of the allochthonous masses of the Subligurian unit. Figure 13, based on many field observations, shows schematically the relative positions of the olistostromes in the Macigno turbidite unit, which, including its highest members, exceeds 1500 m in thickness. It may be concluded that the advance of the Subligurian unit into the Macigno basin was gradual and extended over a relatively long period of time. These major olistostromes very likely correspond to renewals or accentuations of the forward movement of the allochthonous mass. As forerunners and indicators of overthrust movements they therefore have a special significance in the evolution of the tectonic landscape and deserve to be distinguished as "precursory" olistostromes.

The precursory olistostromes also provide testimony (and perhaps the only testimony) to the long time involved in the gravitative gliding of an allochthonous mass over an appreciable distance. Assuming that the Macigno trough had a width of 50 km and taking the length of the late Oligocene (i.e., the period of Macigno deposition) at 5 million years, the average velocity of the allochthonous mass would be in the order of 1 cm/yr.

The Tuscan Tectonic Phase (Tortonian: Middle Miocene)

A paroxysmal tectonic phase, characterized by large-scale overthrusting, occurred during the middle Miocene, particularly during the late part of this stage, the Tortonian. During this phase the tectonic units of the Northern Apennines reached their present positions (Fig. 3) and the large overthrust movements eventually ceased. The sediments of the internal Ligurian basin (sandstone and limestone flysch of Late Cretaceous age) were thrust across the ophiolite-bearing tectonic units derived from the Bracco ridge, which were stretched and laminated in the process. The tectonic pile of mutually overthrust Ligurian units moved as a whole toward the east and covered the external Tuscan-Umbrian domain. During this advance, overthrusting occurred in the external domain as well, resulting in structural repetition of

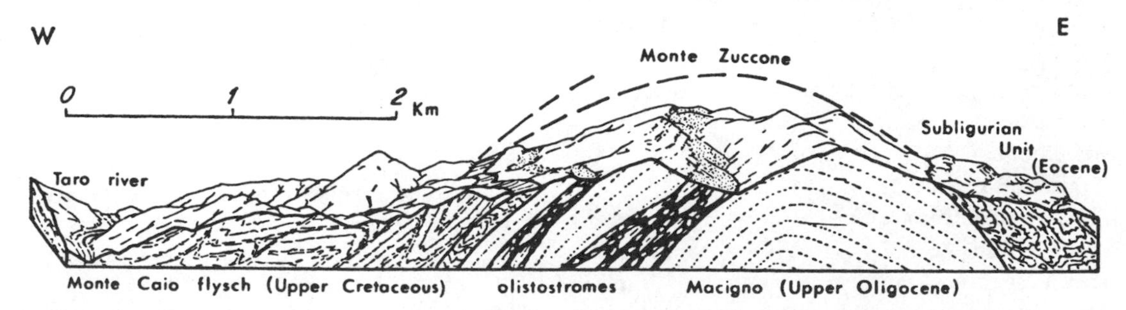

FIG. 11 Section showing two precursory olistostromes in the Oligocene flysch (Macigno) in relation to the tectonic window of the Taro south of Piacenza.

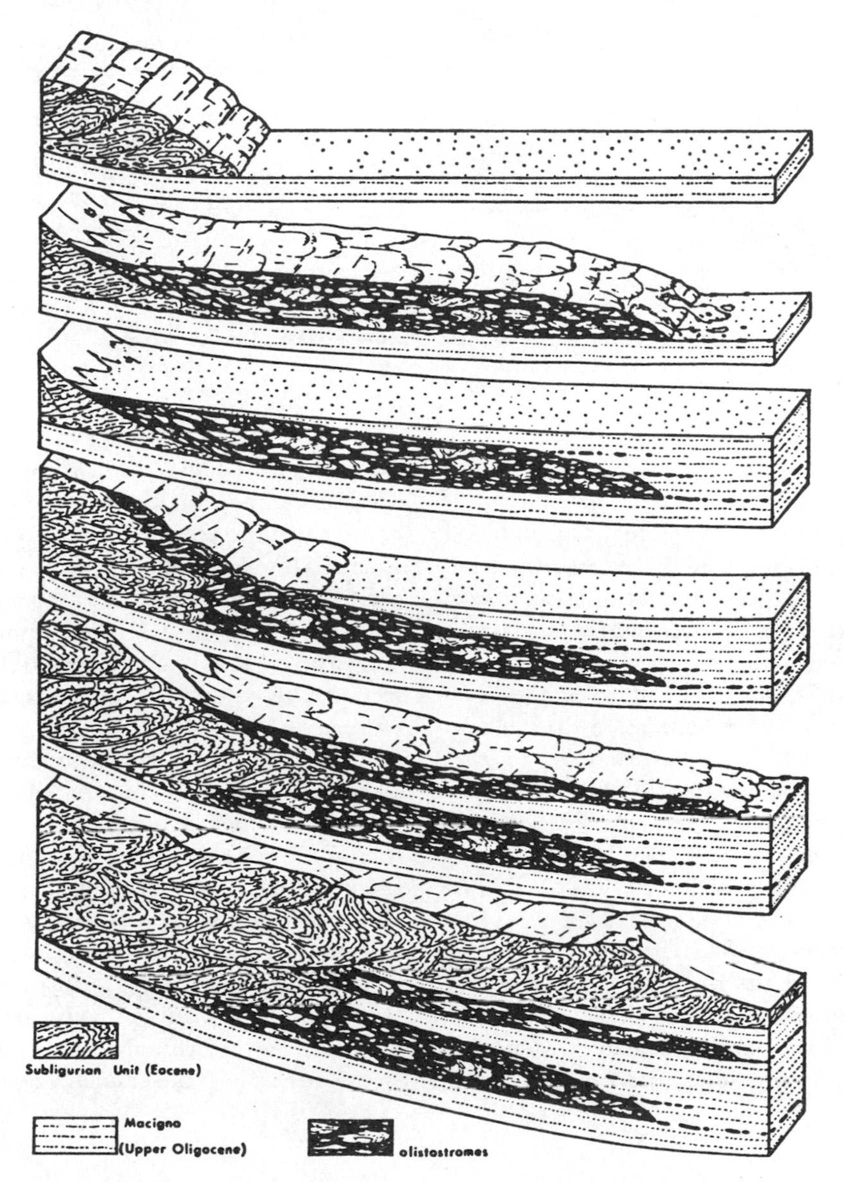

FIG. 12 The block diagrams show, from top to bottom, the reconstructed emplacement mechanism of the two olistostromes of Fig. 10. In the first two diagrams the front of the Subligurian Unit advances into the Macigno trough and gives rise to the first olistostrome. Subsequently a further advance of the allochthonous unit causes the detachment of a second olistostrome which includes reworked parts of the first. In the end, the ultimate advance of the Subligurian Unit interrupts completely the deposition of the Macigno flysch.

the Tuscan sequence over large areas (Fig. 14). The individuality of the Tuscan overthrust sheet is particularly manifest in the Apuane Alps, where the autochthonous terrane, exposed in a window, has the appearance of a metamorphic nucleus.

During this same period great horizontal translations occurred south of the Arno

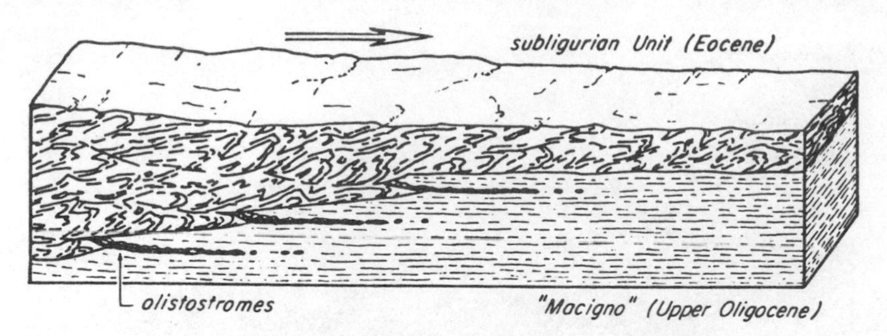

FIG. 13 Schematic representation of the relation between the gradual advance of the Subligurian Unit into the Macigno sedimentary basin and the distribution of the olistostromes.

River as well. Moreover, large areas of tectonic denudation occur here behind overthrust Tuscan units that were detached from their substratum and displaced eastward by sliding (Fig. 15). As a result, Tuscan rocks replaced the original cover of Miocene flysch in the Umbrian domain over a distance of several tens of kilometers ("cover substitution"). The tectonic denudation to the rear is demonstrated by extensive gaps in the Tuscan domain, where allochthonous Ligurian units rest directly on the Tuscan substratum of Upper Triassic evaporites, replacing the original Tuscan cover. These large gaps in the wake of advanced cover sheets are taken as proof that the movements were gravitative in origin.

Although olistostromes already reached into parts of the external domain during the advance of the Subligurian unit in the Aquitanian tectonic phase, it was not until the Tortonian tectonic phase that they arrived in the most external part of the Apennines, intercalating themselves in the autochthonous sediments of the Adriatic and Po basins. They had the character of large frontal slides, consisting of highly chaotic material derived from all Ligurian tectonic units, and mixed with scrambled slivers of the most recent autochthonous units of the external domain. In some cases they include extensive sheets of little-deformed sedimentary beds. They are large-dimensional gravitative forerunners of the front of the allochthonous Ligurian mass. Whether these phenomena qualify as olistostromes is open to discussion, for in this case it becomes difficult and arbitrary to define the boundary between olistostromes and gravitative overthrust sheets.

In any case, the allochthonous masses

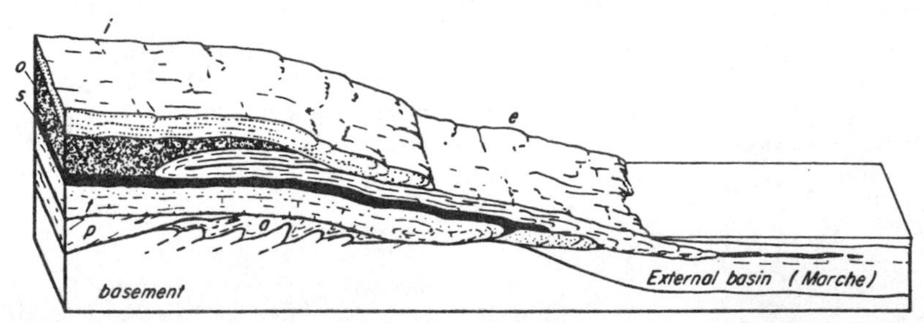

FIG. 14 Schematic representation of the relations between the various tectonic units at the end of the Tortonian phase. The olistostromes formed in the most external belt of the Apennines, corresponding to the Tuscan-Umbrian basin of sedimentation. (*i*) Internal Ligurian Unit; (*e*) External Ligurian Unit; (*o*) ophiolites; (*s*) Subligurian Unit; (*t*) Tuscan nappe; (*p*) Tuscan parautochthon (metamorphic); (*a*) Tuscan autochthon (metamorphic).

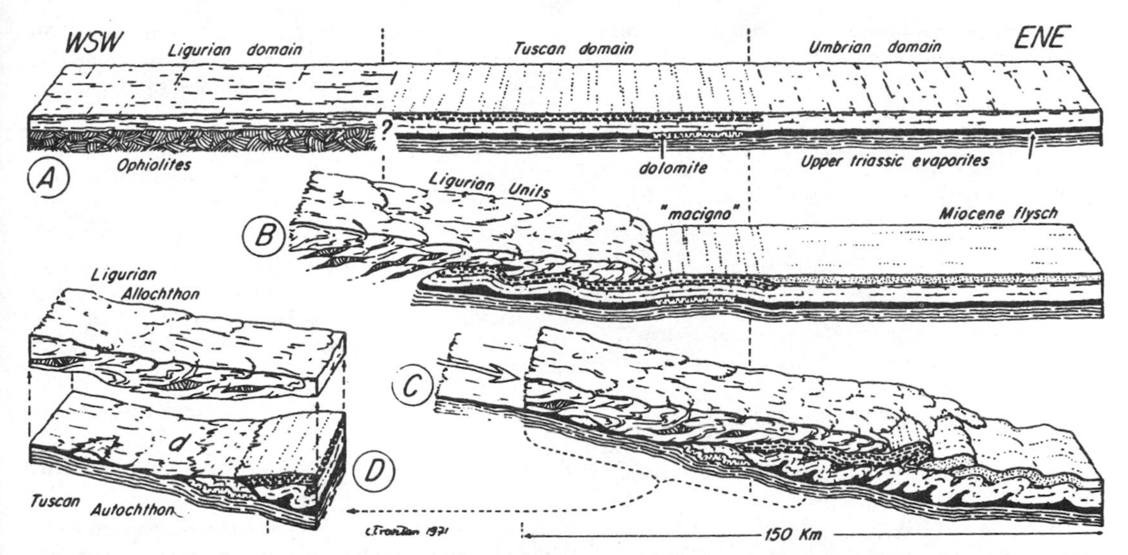

FIG. 15 Schematic representation of gravitative tectonics in a transversal section across the Apennines about 50 km south of Florence. (*A*) The three basins of deposition; (*B*) overthrusting of the Ligurian units onto the Tuscan domain; (*C*) gravitational sliding and folding along the Upper Triassic evaporites, whereby the Tuscan cover and the Ligurian units come to rest on the Umbrian area and, behind them, the cover of the substratum in the Tuscan domain is replaced by Ligurian units; (*D*) detail of *C*, showing the gap left by tectonic denudation (*d*) in the Tuscan domain.

arrived at the external Po-Adriatic margin of the Northern Apennines at the beginning of the Tortonian and during the Messinian (late Miocene) after a stepwise journey. These movements brought to a close the last episode of the gravitative horizontal movements in the Northern Apennines.

The Last Tectonic Pulsations and the Final Tensile Phase

After the Tortonian there were only compressive phenomena in the Marchesian part of the external domain. These take the form of folds which slowly die out in the direction of the Adriatic Sea. In Tuscany, on the other hand, extensional tectonism prevailed instead, characterized by normal faults with subsidence of graben in which accumulated neo-autochthonous sediments of late Miocene to Pliocene age, and uplift of horsts that carried Pliocene deposits more than 80 m above sea level. Late Miocene to Quaternary volcanic processes are related to these extensional movements. Meanwhile, the west coast of the Northern Apennines north of the mouth of the Arno subsided. These late movements of a clearly extensional and rigid style on the one hand, and of an attenuated compressive nature on the other, modeled the present-day topography of the Northern Apennines and are the cause of the asymmetry between the Tyrrhenian and Adriatic drainage patterns.

References

Abbate, E., Bortolotti, V., and Passerini, P., 1970, Introduction to the geology of the Northern Apennines: *Sed. Geol.*, v. 4, no. 3/4, p. 207–250, 521–558.

Baldacci, F., Elter, P., Giannini, E., Giglia, G., Lazzarotto, A., Nardi, R., and Tongiorgi, M., 1967, Nuove osservazioni sul problema della falda toscana e sulla interpretazione dei flysch arenacei tipo macigno dell'Appennino settentrionale: *Mem. Soc. Geol. Ital.*, v. 6, no. 6, p. 213–244.

Beneo, E., 1955, Les résultats des études pour la recherche petrolifère en Sicile: *Proc. 4th World Petrol. Congr., Roma*, p. 1–13.

———, 1956, Accumuli terziari da risedimentazione

(olistostroma) nell'Appennino centrale e frane sottomarine: Estensione tempospaziale del fenomeno: *Boll. Serv. Geol. d'Italia*, v. 78, no. 1–2, p. 291–319.

Decandia, F. A., and Elter, P., 1969, Riflessioni sul problema delle ofioliti nell'Appennino settentrionale (nota preliminare): *Atti Soc. Toscana Sc. Natur. (Pisa)*, serie A, v. 76, no. 1, p. 1–9.

Elter, G., Elter, P., Sturani, C., and Weidmann, M., 1966, Sur la prolongation du domaine ligure de l'Apennin dans le Monferrat et les Alpes et sur l'origine de la nappe de la Simme s. l. des préalpes romandes et chablaisiennes: *Arch. Sci. Genève*, v. 19, no. 3, p. 279–378.

Elter, P., and Raggi, G., 1965, Tentativo di interpretazione delle brecce ofiolitiche cretacee in relazione con movimenti orogenetici nell'Appennino Ligure: *Boll. Soc. Geol. Ital.*, v. 84, no. 5, p. 1–12.

Elter P., and Schwab, K., 1959, Nota illustrativa della carta geologica all'1:50.000 della regione Carro-Zeri-Pontremoli: *Boll. Soc. Geol. Ital.*, v. 78, no. 2, p. 157–187.

Flores, G., 1956, Lettera al presidente della Società geologica Italiana: *Boll. Soc. Geol. Ital.*, v. 75, no. 3, p. 221–222.

———, 1959, Evidence of slump phenomena (olistostromes) in areas of hydrocarbons exploration in Sicily: *Proc. 5th World Petrol. Congr., New York*, sec. I, p. 259–275.

Jacobacci, A., 1965, Frane sottomarine nelle formazioni geologiche: Interpretazione dei fenomeni olistostromici e degli olistoliti nell'Appennino e in Sicilia: *Boll. Serv. Geol. d'Italia*, v. 86, p. 65–85.

Merla, G., 1964, Centro di studio per la geologia dell'Appennino, I sez. Firenze: Attività svolte nel periodo 1951–1963: *La Ricerca scientifica*, suppl. v. 3, no. 3, ser. 2, Consiglio Nazionale delle Ricerche, Roma, p. 107–126.

Rigo de Righi, F., 1956, Olistostromi neogenici in Sicilia: *Boll. Soc. Geol. Ital.*, v. 75, no. 3, p. 185–215.

Tongiorgi, E., and Trevisan, L., 1953, Livret-guide de l'excursion AS (Sicile): *IV Congrès INQUA*, p. 1–36.

Trevisan, L., 1960–1963, La paléogéographie du Trias de l'Apennin septentrional et central et ses rapports avec la tectogénèse: *Livre a la mém. du Prof. Paul Fallot, Soc. Géol. Fr.*, v. 2, p. 217–225.

———, 1962, Considérations sur deux coupes à travers l'Apennin septentrional: *Bull. Soc. Géol. Fr.*, v. 4, p. 675–681.

7

Reprinted from *Geotectonics* **12:**333-342 (1978)

Olistostromes and Their Origin

M. G. LEONOV

Olistostromes originate in disintegration of tectonic nappes and overthrusts pushed into a sedimentation basin. They form when the basal parts of the sheets are broken and sliced up and tectonic breccias are produced in the body of the original allochthon and its frontal escarpment collapses. Olistostromes are tectonic-sedimentary formations. Their coarse clastic material is the product of tectonics, and the slump features merely reflect the manner of transportation. As such, olistostromes illustrate the unity and interrelationship of tectonic and sedimentation processes.

INTRODUCTION

The term, "olistostrome," introduced by Flores (1955), now is applied to spatially well-defined geologic bodies made up of a complex of deposits whose characteristic feature is the presence of large heaps of unsorted, chaotically thrown together material often associated with fine-grained areno-argillaceous sediments.

Although these special complexes of deposits have been known for almost 100 years, interest in their geology and distribution has significantly increased only recently when their genetic relationship with tectonic movements, often leading to the development of tectonic nappes, was discovered.

Much has been learned in the geology of olistostrome deposits. However, even now some points remain either obscure or controversial. One much-discussed problem is the origin of olistostromes. Without an understanding of this problem — namely the mechanism of origin of coarse clastic material, its redeposition, and burial — a number of general questions of regional and theoretical geology cannot be answered.

Without considering the complicated evolution of views on the origin of olistostromes, I only note that all concepts on this subject can be combined in four groups.

1. Olistostromes are purely sedimentary formations that originated in submarine, and, at times, subaerial, slumps down the slope of the sedimentation basin. Hence the term, olistostrome — layered-slump. The idea of a purely sedimentary origin for olistostromes has been and is entertained by many investigators, including Flores, the author of the term (Signorini, 1935; Soder, 1949; Merla, 1951; Flores, 1955; Beneo, 1958; Jacobacci, 1965; Abbate, Bortolotti, and Passerini, 1970; and others).

2. In addition to slumping, tectonics plays a definitive role in the development of olistostromes. However, the role of this factor is auxiliary; tectonics constitutes a sort of trigger for slumping processes (steepening of the slope of the floor of the sedimentation basin; earthquakes; growth of folds with olistostromes sliding down their steep slopes, etc.) (Migliorini, 1950; Marchetti, 1957; Gorler and Reutter, 1968; Richter, 1968; Cherenkov, 1973; and others).

3. The principal role in olistostrome development belongs to tectonic movements. Investigators holding this view (Kraus, 1951; Gigon, 1952; Belostotskiy, 1970; Porshnyakov, 1973; Leonov, 1975; Luk'yanov et al., 1975; Sokolov, 1975; and others) believe that olistostromes develop as the frontal parts of tectonic sheets disintegrate during their progress into the sedimentation basin. The slumping processes are considered merely as a factor in the subsequent displacement of the material.

4. Not all olistostromes have the same origin. Some of them are the products of slumping; others are controlled, to a considerable extent, by tectonics (as understood in the second view); still others originate in a disintegration of tectonic sheets. This view is held by many investigators; it is best presented by Elter and Trevisan (1973).

The reason for such a diversity of opinion will be clear from the following exposition. It is obvious nonetheless that all these hypotheses, particularly the "tectonic," fail to explain how huge masses of coarse-clastic material originated. The reason is that, with few exceptions (Leonov, 1975; Shcherba, 1975; Luk'yanov et al., 1975), no special study has been made of the manner of origin of the body of coarse-clastic rocks that compose the olistostrome.

Study of olistostromes in various regions of Alpine and Paleozoic folded systems, in conjunction with analysis of published references, makes it possible now to draw fairly substantiated conclusions on the subject. These conclusions are set forth in the present article. But before turning to a discussion of the origin of olistostromes, it is necessary to describe briefly their morphology, with emphasis on those structural details and relationships with the enclosing deposits that have bearing on their origin.

MORPHOLOGY OF THE OLISTOSTROMES

The morphological features of olistostromes have been repeatedly described. The most comprehensive reference is by Luk-yanov, Leonov, and Shcherba (1975), from which is largely borrowed this description, of olistostromes, which is necessary for understanding their genesis.

Olistostromes are usually included in normal sedimentary deposits; they are known from flysch, molasse, ophiolitic, and other formations with which they are connected by gradual transitions. At the sites of olistostrome development, sedimentary sequences become markedly heterogeneous, and exotic non-stratified coarse-clastic rocks appear in the groundmass material characterizing the flysch, molasse, or other type of deposit. The exotic inclusions occur as lenses, beds, and thick strata of conglomerates and breccias consisting, as a rule, of fragments of older rocks. The latter occur also in isolated olistoliths: within the surrounding deposits as fragments, chunks, lenses, and large layers.

The conglomerates and breccias occur in the enclosing rocks as lenses, stringers, and thick (up to 1 km) strata. Their contacts with the underlying and overlying formations are usually sharp, without a coarsening of clastic material in the underlying layer. Even where the underlying layer contains a sizable amount of fragments, its contact with the main body of breccias is sharp. The breccias often truncate the underlying beds, whose segments and fragments are partly incorporated into the breccias. The breccias appear to plow over the underlying deposits. Isolated chunks are pushed into the underlying sediments whose areno-argillaceous and carbonate material fills the space between them. The breccias often change laterally to normal sedimentary deposits; they and the conglomerates appreciably thin down and finally wedge out. However, thick breccia zones often continue over tens and hundreds of kilometers, whereas the conglomerates are distributed only locally.

There are two principal components of breccias: the ground mass (matrix) and the fragments. The groundmass consists either of silty to sandy argillates with variable additions of carbonate material, or else of a vivid red to raspberry calcareous argillaceous mass. Only when the breccia is made up wholly of limestone fragments is the groundmass largely carbonate but even so with some clastic material. As a rule, the breccias are non-stratified or else stratified very crudely and indistinctly. The non-stratified groundmass contains fragments, chunks, and slabs of various rocks, usually older than the enclosing groundmass. However, fragments of almost contemporaneous underlying deposits get into the breccias. Breccias made of a single type of rocks, usually limestones, are widespread. The fragments are poorly rounded, if at all. The breccia fragments, particularly the small ones, usually are sharp-angular; the larger ones often exhibit smoothed-over edges.

The fragments range in size over very wide limits, from a few centimeters to tens and hundreds of meters across. The breccia material is unsorted. The fragments, chunks, and pebbles are haphazardly distributed without any evident pattern. The fragment-cement ratio is inconsistent. Locally, the fragments are heaped together, and the chunks are fitted closely together; in other places, chunks are rare and are spaced a considerable distance apart, floating in the groundmass as it were, and, lending a pudding aspect to the breccia. The surfaces of most fragments exhibit slickensides, and the cement is often tectonic gouge or mylonitized rocks.

The isolated fragments, blocks, and slabs often occur outside the breccia horizons, directly within the enclosing rocks. They vary in size from a few centimeters to tens and hundreds of meters across and many kilometers long. The smaller fragments (up to several meters) usually are smooth, rounded to lenticular; the large ones are greatly elongated along the trend of the enclosing rocks, as slabs. The slabs thin down toward the margins and gradually wedge out. They often are associated with chaotic breccias made of their own material. The slabs are often strongly brecciated and broken up into separate minor slices.

A characteristic feature of olistostromes is their intense reworking by tectonic movements: strong crumpling, imbrication, lamination, and boudinage of the slabs and fragments. However, in many instances, the olistostromes are either unaffected by subsequent deformations or affected very insignificantly. But even then the olistostrome material shows a well-expressed evidence of tectonization: slickensided fragment surfaces, mylonitized groundmass, strong cataclasis, the leno-like squeezed-up form of chunks and fragments; and the presence of a calcite "jacket" around them. The large blocks and slabs are broken up into small slices, brecciated, and boudinaged.

THE ORIGIN OF OLISTOSTROMES

Analysis of the geometry of olistostrome bodies and their relationship with the enclosing formations shows that their development usually was affected by two types of sedimentation; first, normal sedimentation in the flysch, molasse, or other basins; second — the introduction into this "banal" (in the parlance of French geologists) sedimentation of huge masses of coarse-clastic material which is then deposited in lenses, stringers, and layers of chaotic block breccias. The coarse-clastic material does not come in gradually, over a long period of time, but rather is dumped all at once, rapidly, in recurrent batches, so that its contact with the enclosing rocks is always sharp, and some of the chunks appear to be pushed into the underlying, probably still unconsolidated sediments. At the same time, large blocks and slabs of older rocks were dumped into the basin, some of them large enough to be regarded as true tectonic nappes. That the blocks and slabs got into the deposits as they formed is corroborated by their presence not only as independent blocks but also as components of conglomerates and breccias, and especially by the presence of stratigraphic contacts between the older slabs, faunally dated, and the younger enclosing deposits. Finally, the slumping and internal deformation of the latter were just as episodic as the interduction of the coarse-clastic material. However, this episodicity is not invariable, although quite typical, especially in the case of flysch olistostromes.

Thus it is fairly obvious that slump phenomena play a decisive role in the development of olistostromes. Indeed, normal stratigraphic contacts of olistostromes with the under- and overlying deposits, a chaotic internal structure; plowing up of the underlying beds and their subsequent incorporation into the olistostrome;

the presence of submarine slump structures and of a redeposited shallow-water fauna; and the very presence of large piles of unsorted coarse material frequently amidst fairly deep-water sediments, unambiguously indicates — as believed by the majority of investigators — a submarine slump (or, less commonly, subaerial) character of olistostromes. In this variant, the origin of coarse-clastic masses is connected with the collapse of a relatively high escarpment, gravitationally unstable, and the subsequent creep of this material into the deeper reaches of the sedimentation basin.

However, certain features of structure and distribution of olistostromes cannot be explained by a simple slump mechanism. The principal of these are as follows:

1. Olistostromes tend to concentrate in large fault zones along which overthrust displacements are either proven or indicated by indirect evidence.
2. As a rule, the olistostromes and olistoliths conconsist of rocks from the allochthon. They are piled up in front of it or else are overridden by it.
3. The olistostromes are associated with definite time periods when large overthrust displacements of rock masses have been identified (Leonov, 1975, 1976).
4. Within one and the same time interval, the olistostromes are extremely widespread, extending with interruptions over many hundreds and thousands of kilometers.
5. The olistostrome complexes often are overlain by tectonic nappes and are themselves strongly reworked by tectonics.
6. The olistostromes contain dynamically reworked material: compressed, smoothed over, striated, flattened fragments.
7. The chaotic breccias of olistostromes are directly associated with the overthrust slabs and exhibit gradual transitions to them.

All these features force us to look beyond mere slumping to other factors to explain the development of olistostromes. At present, more and more investigators come to the conclusion that olistostromes originate in a disintegration of tectonic sheets. H. Schardt (1898) was first to point out a relation between "wild flysch" (olistostromes) and disintegration of overthrusts, connecting the central Switzerland windflysch with the breakup of the frontal part of the klippen nappe. He cited the presence close to klippen of fragments of rocks of klippen facies in the underlying wildflysch. In later years, this hypothesis of olistostrome origin has attracted increasing numbers of advocates.

It is not necessary to describe specific regions where the relationship between olistostromes and overthrusts has been observed, as there are many published references on this subject (bibliography in Leonov, 1975, 1976).

Indeed, the special conditions of sedimentation, the relationship between inclusions in the olistostromes and the overthrusts; the common strong tectonic reworking of the olistostromes, the presence of cosedimented tectonic sheets olistostrome-bearing sequences; coincidence in time of these sediments with strong tectonic movements leading to development of overthrusts; and the spatial relations between

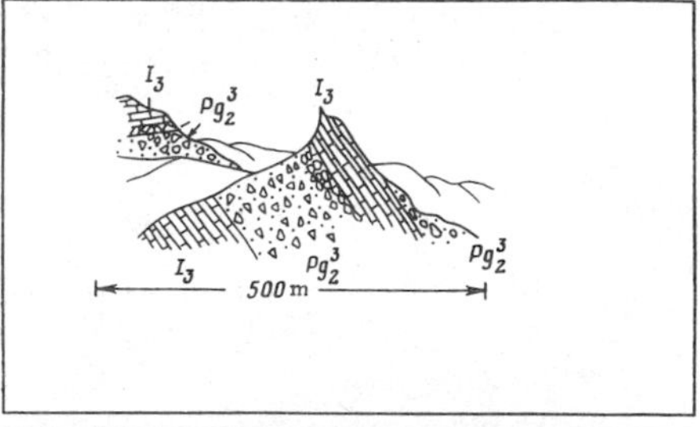

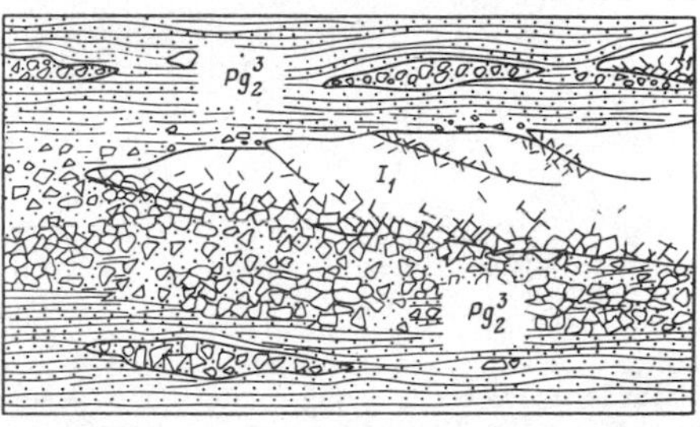

Fig. 1. Disintegration of basal parts of cosedimented overthrusts and slabs in their movement over the basin floor (Upper Eocene olistostromes on the south slope of the Greater Caucasus).

olistostromes and tectonic sheets — all these suggest that such deposits are associated with the disintegration of frontal parts of advancing overthrusts. Particularly illustrative in that respect are direct transitions from a tectonically broken up sheet, through tectonic breccias to typical olistostromes. These transitions are described below. Most of all, the distribution of olistostromes of one and the same age over distances measuring hundreds and thousands of kilometers hardly can be explained by simple slumping. There must be some reason for slumping over such a wide extent. This reason may be found in strong tectonic movements known at the time of olistostrome development.

Having established a connection between development of the fragmental olistostrome material and disintegration of the overthrust, however, we now consider the specific mechanism of this disintegration, i.e. precisely the point usually neglected by investigators.

1. The olistostrome sequences often contain slabs of older rocks, more competent than the enclosing groundmass. These slabs may be tens of kilometers long, in which case they usually are regarded as cosedimented overthrusts. Such slabs (olistoplaques, gliding overthrusts) have been described from olistostrome deposits in many localities, such as slabs of Laimerie limestones in the Upper Eocene wildflysch of central Switzerland (Gigon, 1952), Permian limestones in Neogene molasse of the Darvaz Range (Shcherba, 1975), ophiolite slabs in Upper Cretaceous

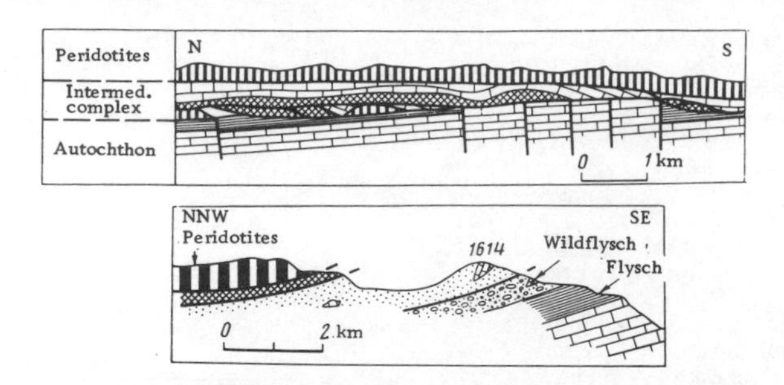

Fig. 2. Deformations at the base of the overthrust sheet. Lycian Taurus (Graciansky, 1973). Explanation in text.

olistostromes of the Lesser Caucasus (Sokolov, 1975), and others.

Other illustrative examples are Upper Jurassic and Upper Liassic slabs the author has studied in Upper Eocene flysch on the south slope of the Greater Caucasus (Fig. 1). These slabs occur directly within the enclosing flysch. They are made up of stratified Upper Liassic sandstones and shales and are greatly elongated along the stratification. One of them is about 20-25 km long; another, on the order of 4-5 km. Their maximum thickness is, correspondingly, 700-800 and 100-150 m. These slabs are cosedimented overthrusts that got into the basin during Upper Eocene sedimentation (Leonov, 1975). Smaller slabs (50-200 m) have been observed. At their margins they gradually thin down and wedge out. As a rule, their lower contacts are complicated by disruptions, fragmentation, and brecciation of their component rocks. The upper contacts with the overlying breccias and conglomerates usually are normal, with evidence of erosion. At the same time, the overlying and underlying flysch horizons constitute a single sequence. Brecciation and even cataclasis are present within the slabs. The continuity of the slabs often is disturbed by numerous faults that break up the slabs into assorted slices. The fragmentation of the slabs becomes more pronounced toward the edges and downward, and, the fragmented rocks pass imperceptibly into tectonic breccia that gradually takes on the aspect of sedimentary breccia. Here the fragments are haphazardly arranged, and; a cementing material appears. The sedimentary breccias make up a train around the slabs and continue within the enclosing flysch, usually as stringers conformable with stratification; but in some cases truncate the underlying layers sharply incorporate their material. Often they become separated from the slab by a process of slumping. The slab-bordering breccias consist of the component rocks of the slab. The fragments are not at all rounded but sharp-angular, often "splinters" of the shales and sandstones. They are cemented with a fine-grained areno-argillaceous aggregate of the same composition as the slabs and breccias.

The most typical features of these breccias are their monomict character; the very close association with the rocks of the slabs (often without a definite boundary between the brecciated slab, the tectonic breccias, and the sedimentary breccias, i.e. the olistostrome proper); the brecciated character of the fragments themselves; the presence of slickensides on their raw faces; and locally a mylonitization of the groundmass.

We see in this instance that the origin of the coarse clastic material is tectonic, the result of fragmentation and brecciation of the slab during its progress over the basin floor. Subsequently, the breccias are partly spatially associated with the slabs, partly separated from them and transferred by simple slumping away from the site of their origin, into deeper reaches of the sedimentation basin. Although they acquire the morphological character of a slump, at the same time they also maintain the evidence of their tectonic origin: slickensides and brecciation.

Thus, tectonic disintegration of the basal parts of large slabs (cosedimented overthrusts), in their progress over the basin floor, is one of the mechanisms of developing coarse clastic material for the olistostrome. The size of these slabs — thrust sheets varies within a very broad range and attains tens of kilometers along the long axis.

Disintegration of the basal part also occurs in larger overthrusts. Those described from the Lycian Taurus, Turkey, contain olistostrome sequences at the base (Graciansky, 1968, 1972, 1973). Here the succession of rock complexes tectonically overriding one another is illustrated in Fig. 2. The autochthonous basal formations begin with a triassic carbonate series overlain by flysch. The flysch section is augmented by Upper Cretaceous olistostromes.

The wildflysch contains among other rocks, blocks of basalts and radiolarian cherts in an aerno-argillaceous matrix. These are followed by mélange; the transition from the wildflysch corresponds to a change in the aspect and composition of the enclosing groundmass: areno-argillaceous near the base, it becomes substantially tuffaceous upward. According to Graciansky, the transition between wildflysch and mélange is "progressive," and normal, over a distance

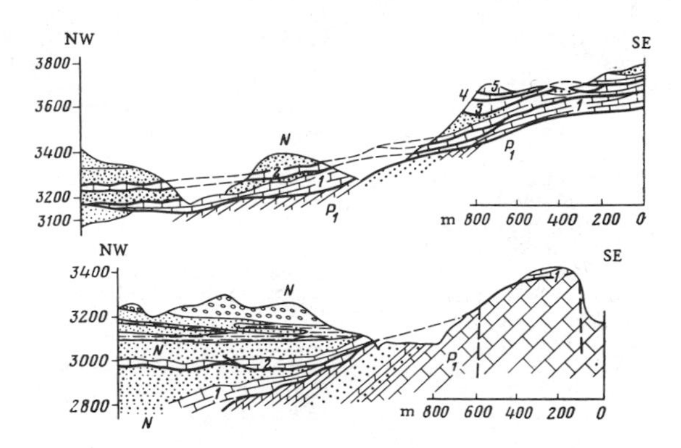

Fig. 3. Olistoliths and olistostromes in Neogene deposits of the Darvaz Range (Shcherba, 1975). Explanations in text.

of 5 m. The melange is tectonically overlain by a nappe of peridotites, locally resting directly on wildflysch, in other places on mélange, in others on an "intermediate" complex of separate, small to large slices of various rocks, including peridotites of the overthrust. Although Graciansky's description of these relationships is rather vague, it shows nonetheless that the lower surface of the allochthon is affected by spalling and fragmentation to produce a melange that changes downward to an olistostrome complex. In other words, we have here a picture similar to that we observed in disintegration of the smaller slabs and overthrusts.

Disintegration and brecciation of the base of the overthrust have been observed in the Rumanian Carpathians (Binaz River valley) where a tectonic nappe of Jurassic to Cretaceous rocks rests upon an olistostrome of fragments of the overthrust. Nevertheless, to draw a sharp boundary between the two is often impossible.

2. An excellent illustration of another mode of development of coarse clastic breccias and olistoliths is in the Neogene deposits of the Darvaz Range (Shcherba, 1975; Luk'yanov et al., 1975) (Fig. 3). Here, thick-bedded unsorted breccias occur at and above the contact between Neogene molasse and Paleozoic limestones. either adjoining the Paleozoic limestones or unconformable upon them. The Paleozoic rocks make up a steep escarpment, made of several low-angle imbricate slices separated by tectonic breccias. A zone of strong brecciation accounts for almost half the section. Brecciation is expressed in two ways: as calcite and pelitic veins that cut the rocks and are streamlined around the undisplaced fragments of a single rock mass; and as limestone breccias with the fragments rotated and displaced relative to one another. The brecciated limestones and the calcareous breccias at the boundaries of the slabs grade into each other, indicating the tectonic character of brecciation. Also visible here are gradual transitions from tectonic breccias separating the slices of the Paleozoic basement and unsorted sedimentary breccias — the olistostromes. The imbricate slices, initiated in the Paleozoic rocks, pass directly — like the breccias — into the olistostrome sequence where they occur as giant olistoliths with areas of tens of square kilometers.

The direct transition from tectonic breccias to sedimentary, and from tectonic sheets or nappe to olistoliths, unambiguously indicates a tectonic origin for the olistostrome material in this area. To quote Sherba (1975, p. 107), "A considerable part of the olistostromes ... is initiated not at the surface of the uplifts but within the layers composing it, in zones of tectonic fragmentation, as such, it is endogenous." And according to Luk'yanov (1975, p. 47), "This example shows that the olistostromes develop ... in strongly brecciated rocks and tectonic breccias squeezed out from under the overthrust sheet, carrying large slabs and lenses of less disturbed rocks."

Thus, the morphological features of ancient olistostromes lead to the conclusion that some of the olistostromes (one of their types, in any event) have a tectonic origin. However, many aspects of olistostrome development in the geological past are not sufficiently clear, owing to subsequent tectonic reworking, disappearance of some of the transitions, etc. With this in mind, we now consider several examples of development of present olistostromes.

The present disintegration of frontal parts of large overthrusts, leading to a development of olistostrome deposits (or their present analogs), is illustrated in a thrust of flysch on the south slope of the Greater Caucasus over the Georgian Block and the Gagry-Dzhav zone. The overthrust character of this fault, over 200 km long in Georgia, has long since been quite definitely established (Gamkrelidze, 1964; Milanovskiy and Khain, 1963; Leonov, 1975; and others). This fault has been active from the Late Sarmatian to the present (Milanovskiy, 1968). Its present activity is morphologically expressed in an escarpment, locally smoothed over but mostly quite conspicuous in relief, particularly in the basins of the Lekhura, Medzhuda, and Malaya Liakhva rivers, where slumps and rock chutes can be observed along the fault line. Generally, an overthrust of flysch on formations of various ages is established by geological mapping. In many localities, large and small talus, mainly of limestones from the flysch complex, can be observed at the foot of the fault escarpment. The smaller fragments (5-10 to 40-50 cm) are cemented with travertine and often turned into monolithic breccia. The fragments are not rounded, sharp-angular, often with a calcite jacket, unsorted, and unstratified. Their carbonate tuff cement contains imprints of oak, nut, and other contemporaneous plant leaves. The fragments, particularly the larger ones, exhibit slickensides, especially well expressed on the calcite jacket. These breccias could have originated in two ways.

At first glance, some of them appear to be mere slumps. However, weathering and gravity are not the only factors controlling the disintegration of the front of the overthrust. The principal factor is tectonic fragmentation observable within the flysch sequence where it is expressed in the presence of slickensides in the surface of blocks and chunks. Judging from field observations, the fragmentation proceeds as follows. As a result of regional tectonic movements, the flysch beds are fractured into blocks, best expressed in the carbonate rocks. The fractures are filled with calcite or travertine; these envelop each block, breaking up the monolithic continuity of rocks but stopping short of completely disintegrating it. As the result of the general movement, the blocks become progressively more separated and acquire a certain degree of freedom of individual movement. They begin to rotate relative to one another. Their original stratification even, if intact at the time of the initial fracturing, now is disturbed and finally disappears. The fragments, originally bound together by carbonate material (calcite, travertine), now rotate fairly freely relative to one another, and the originally stratified rock is turned into breccia. This breccia, however, is not yet torn off the parent rocks but remains at the site of its origin or is but slightly displaced. A continued tectonic brecciation and movement within these zones is indicated by the presence of slickensides on the secondary calcite jackets. Such breccias are not accompanied by primary mylonitization and do not contain, as a rule, tectonic gouge, provided they do not fall a second time into a fracture.

Finally, the intensity of brecciation becomes so high that the monolithic character of the breccia is broken up on the outcrop by weathering and gravity, and the fragments slump down the slope where they are again cemented either by travertine or by arenoargillaceous-carbonate material, depending on local conditions. However, the bulk of breccia disappears: its components are rapidly destroyed by subaerial processes, and the material is completely disintegrated and carried away by rivers whose channels mainly are parallel to the trend of the fracture.

The second means of breccia development is similar to that of the Darvaz olistostromes described above. The front part of the overthrust, consisting of various flysch units, is broken up into numerous large and small tectonic slices and blocks. Tectonic breccia, centimeters to meters thick, is developed at boundaries of slices within the fracture zone; these boundaries may be low-angle fractures or cross-cutting normal, reverse, and strike-slip faults. The breccias consist of fragments of the rocks in the wall of the fracture. A zone of more intense fracturing appears during the initial stage of brecciation, and the fractures fill with calcite. The fragments become separated; the gaps between them are filled either by secondary calcite or by ground-up carbonate-clay substance. Where the fracture zone is exposed, the breccias spill out and develop piles of seree. In contrast to common seree, not associated with a fault zone, the fragments often are recemented by travertine. The cement may be so strong that these breccias are readily mistaken for older ones, such as Meso-Cenozoic, as can be observed in many areas of the Caucasus.

As a rule, these breccias differ from the first type in the uniformity of the material and in the presence of tectonic gauge and mylonites.

It must be mentioned, however, that the two types of breccia grade into each other and that their material may be mixed. They are almost contemporaneous and interrelated.

The example cited illustrates once more that some of the olistostrome sequences are of a tectonic origin, and that clastic material develops within the massive monolith rather than merely in a disintegration of the frontal escarpment. However, a prerequisite for development of clastic material is an internal tectonic reworking of rocks, expressed in the development of tectonic slices and tectonic layering within the frontal part of the overthrust.

A similar picture of development of Quaternary breccias, the morphological analogs of ancient olistostromes, can be observed in the Tanymas overthrust zone, which separates the North- and South Pamir structures. The author visited this zone at the kind invitation of A.V. Luk'yanov in whose company a number of traverses were completed. In the Kyzyl-Tukoy ravine (a left tributary of the Tanymas), the Tanymas overthrust zone is a belt of assorted breccias, 1-2 km wide (Fig. 4). On the south, it is in fault contact with Triassic deposits of the Central Pamirs and is overlain on a high-angle thrust by schists of the Northern Pamirs. The breccias consist of fragments of various rocks: gray limestones, tuffs, a sugary quartz aggregate, and mafic volcanics. The fragments are predominantly sharp-angular, although some are rounded, with smoothed-over edges; they measure from 0.5 cm to 0.5 m and more. Some blocks are 10-100 m across. The fragments occur without any visible pattern, either snug to one another or separated by groundmass. Their surfaces are slickensided to rough. The larger chunks and blocks are brecciated, lenticular, broken up into slices, and floating in the groundmass of finer-grained breccias. Going up the slope, toward the upper fault, the number and size of blocks slightly increase, and so does the intensity of rocks disintegration.

A large block of mafic rocks exhibits a stronger fragmentation near the margins, also tectonic brecciation and a gradual transition to trains of monomict breccias that wedge into polymict. As the large slabs

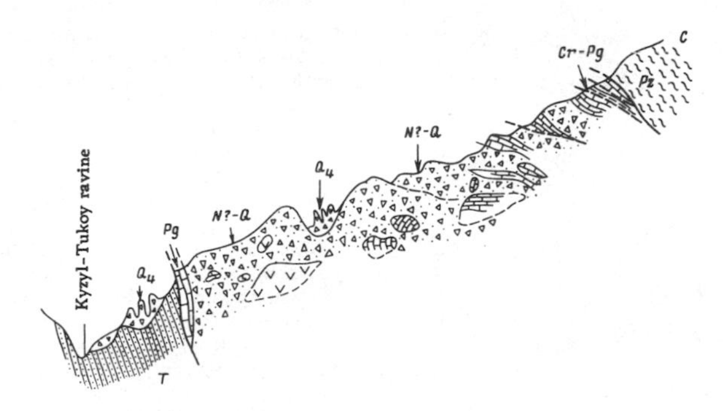

Fig. 4. Breccias within the Tanymas overthrust zone (Pamirs). Explanation in text.

and blocks disintegrate, their parts are displaced relative to one another, and material of the surrounding breccias gets into the fractures. The development and filling of the fractures are accompanied by brecciation and crushing of rocks within the blocks, sharp-angular fragments of which are incorporated into the breccias.

The filling of the breccias is represented by gravelstone and coarse-grained sandstone of the same material as the fragments. The cement is quartz-limonite to carbonate-limonite; it is well-consolidated. The breccia material is unsorted.

The breccia groundmass (both fragments and cement) shows evidence of its tectonic origin: cataclasis, crushing, mylonitization. The very fact of their occurrence within the zone of fracture substantiates their tectonic character. However this zone also contains talus breccias, practically indistinguishable in constitution and composition from those described above. It is not always possible to draw a sharp line between breccias of the fracture zone and those of the slumps, particularly in the upper part of the slope where the primary breccias are weathered at the surface — a process facilitated by their original tectonic inhomogeneity — and begin to slide down the slope. Such instances exhibit a gradual transition from unstratified breccias of the first type to cryptolayered slump breccias. The layering in the latter always dips down the slope, and the breccias are displaced by gravity to the lower parts of the slope and to its foot, where they build up independent bodies unconformable on the tectonic breccias and other rocks.

This is another demonstration that "purely sedimentary" breccias developed by slumping and rock falls down a slope are the product of preceding tectonic processes in the overthrust zone, leading to brecciation, fragmentation, and mylonitization of rocks introduced into this zone.

The frontal parts of the overthrusts disintegrate in a somewhat different way when the dynamic stress is relaxed in the gliding front, as illustrated by recent overthrusts in the mountain border of Fergana

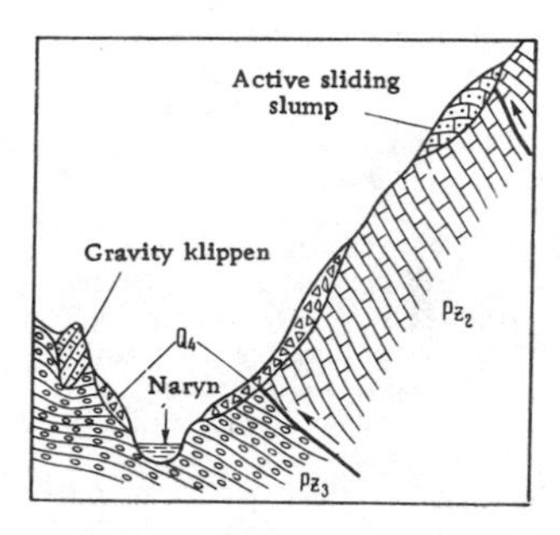

Fig. 5. Disintegration of the frontal part of an overthrust (after G. N. Pshenin, 1973). Explanation in text.

(Pshenin, 1973). According to that author, the gliding front is overstressed as the overthrust escarpment grows higher (Fig. 5). That leads to development of gravity slopes, side fractures, and sagging — all taking advantage, as a rule, of the tectonic fracturing. The horizontal movements steepen up the front of the overthrust, thereby decreasing its dynamic stability. Large blocks of the main monolith break off on gliding planes and are displaced toward the foot of the frontal slope. Broken up by faulting, these blocks are transformed, wholly or in part, to talus. The large tabular to lenticular blocks and slabs (gliding slumps) disintegrate in their downward movement, producing smaller blocks, 15,000-25,000 m^3 (the original volumes of shearing slumps usually exceed a million cubic meters), and screes of blocks and rubble. Some of the blocks reach the foot of the slope where they "either cut into the unconsolidated piedmont rocks as exotic blocks (gravity klippen) or come to rest on rocks without a significant mechanical disturbance" (Pshenin, 1973, pp. 51-52).

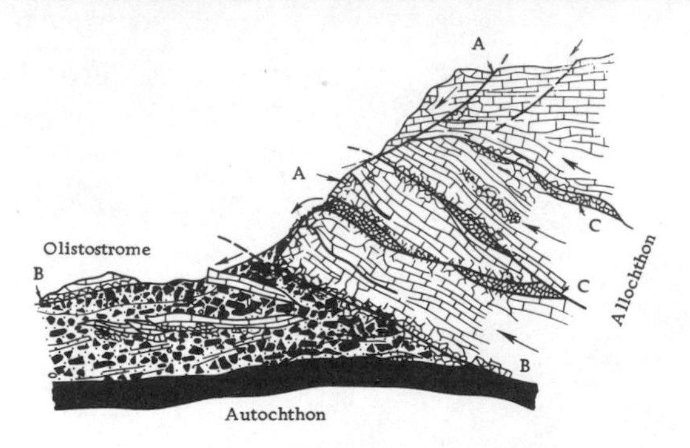

Fig. 6. Development of olistostrome breccias by disintegration of an overthrust. Explanation is in text.

Although Pshenin himself does not associate this process with olistostrome development, it is clear that he deals here precisely with this group of phenomena. Consequently, we have here olistostrome development in processes ostensibly purely gravitational. However, as emphasized by that author, and as is implicit in his exposition, the development of gravity-induced disintegration, and its intensity, are associated with activity of the overthrust itself. In other words, here, too, we have a close relationship between olistostrome development and tectonic movements, namely a disintegration of the frontal escarpment of the overthrust.

The examples cited show that present development of chaotic block accumulations is associated with interior tectonic reworking of the parent massif, the body of a "live" overthrust. The volume of material involved indicates the intensity of the thrust during the period of olistostrome development. In their structure (their geometry, distribution, and the relationship of fragments among themselves and with the main body), the present olistostromes are close to, and even identical with, the ancient ones. At the same time, the present olistostromes clearly exhibit many of the same features that have led us, in the study of ancient olistostromes, to assure their origin in the disintegration of tectonic sheets.

CONCLUSIONS

The material presented demonstrates fairly convincingly that olistostromes (at least a considerable part of them) are developed by the disintegration of overthrusts. The coarse-clastic material originates in various ways (Fig. 6).

1. Collapse of the frontal escarpment of the overthrust, owing to tectonic disintegration of the allochthon with participation of submarine or subaerial weathering (Fig. 6A).

2. Tectonic fragmentation, spalling, and brecciation of the basal part of the overthrust during its movement over the floor of the sedimentation basin and at boundaries of different sheets (Fig. 6-B).

3. Tectonic spalling, brecciation, and formation of tectonic breccias in the body of the parent allochthon, with a subsequent spilling out or tectonic squeezing-out of breccias and slabs to the surface or to the sea floor (Fig. 6C).

These three ways of developing clastic material are closely interrelated and are controlled by activity of the tectonic sheet or overthrust. In the majority of instances, coarse-clastic material is developed by a combination of the three, perhaps with one of them predominant, depending on the allochthon movement, so that it is difficult to impossible to distinguish their respective products. Subsequently, the clastic material may remain at the foot of the disintegrating overthrust massif, preserving the features of its tectonic origin, or it may be displaced by gravity (slumping) into the basin where sediments of a different type are laid down. During slumping and burial, the coarse-clastic sequences acquire new features typical of slumps and are turned into true olistostromes (slump layers). However, as pointed out before, their internal structure maintains features of their early tectonic origin.

Finally, olistostromes often are again overridden by the thrusts by whose disintegration they have originated. In this operation, they are tectonically reworked for a second time, often turning into a melange. Thus, genetic stages of olistostrome development, from the time of initiation to a final structure, are as follows:

Tectonic Process	Slumping	Tectonic Process
Subdivision of parent massif into slices; tectonic fragmentation and brecciation; development of coarse-clastic material	Displacement down the slope of sedimentation basin; additional mixing, introduction of basin sediments	Thrusting of massif over the oliostromes; their secondary tectonic reworking and transformation to tectonic melange

It is clear, then, that slumping features of the olistostromes do not reflect the circumstances of their origin but merely the manner of transportation for their material. The origin of their coarse-clastic material is tectonic. Their tectono-sedimentary character is practicularly well substantiated by a gradual transition, in the absence of a definite boundary, from tectonically originated chaotic breccias to their sedimentary analogs. The geological significance of olistostromes is that their internal structure reflects their tectono-sedimentary character. In other words, olistostromes embody the unity and interdependence of tectonic and sedimentation processes, more precisely the endogenous and the exogenous — because the slumping processes, strictly speaking, evidently cannot be thought of as sedimentary.

In conclusion, we would like to bring up two additional points:

1. There are olistostromes not directly associated with the disintegration of tectonic sheets: these are rather uncommon formations of two types. The first comprises the so-called endo-olistostromes, typically slump deposits originating in slopes of sedimentation basins as the result of seismic shocks or else as the result of accumulated excessive volumes of sediments which begin to creep downward under their own weight. They are characterized by the absence of

evidence of a tectonic origin for the material, as described above; and by the fact that they consist of the fragments of rocks developed in the same basin and often not quite consolidated.

The second type includes simple slump breccias that originated at morphologically expressed escarpments, the products of denudation, or else along large faults of the normal type. These deposits, like endo-olistostromes, often show no evidence of a tectonic origin; their volumes are insignificant because of a rapid flattening of the escarpment. Also, analysis of material bearing on olistostromes shows that, in the very few instances where their origin is ascribed to vertical movements, vertical faulting either is not substantiated at all except by the presence of the coarse-clastic materials or else leads itself to a dual interpretation (see for example the description of Upper Eocene olistostromes in Bulgaria; Leonov, 1975).

2. The term, "olistostrome," reflects only the manner of transportation of the material (slumping) and its burial, rather than the origin of these chaotic deposits. There is no need to substitute another term for it, because it is firmly established among geologists everywhere, but it should be kept in mind that, genetically, olistostromes are tectono-slumps rather than simply slumps. The main difference between them and sedimentary coarse-clastic rocks (river and shelf conglomerates, fanglomerates, etc.) associated with exogenous processes lies in the tectonic character of clastic material.

My sincere gratitude to my collaborators at the Geological Institute, USSR Academy of Sciences, A.V. Luk'yanov and I.G. Shcherba, with whom I repeatedly discussed the problems of olistostrome development.

REFERENCES

BELOSTOTSKIY, I.I. Zones of mélange and chaotic structures. In: Ocherki strukturnoy geologii slozhno dislotsirovannykh tolshch (Outlines of the geology of complexly deformed sequences). Nedra, 1970.

GAMKRELIDZE, P.D. Deep faults and the tectonics of Georgia. In: Gimalayskiy i Al'piyskiy orogenez (The Himalayan and Alpine orogenies). Nedra, 1964.

LEONOV, M.G. Disintegration of the frontal part of overthrusts, Dokl. Akad. nauk SSSR, 193, No. 3, 1970.

LEONOV, M.G. Wildflysch in the Alpine region. Tr. Geol. Inst. Akad. nauk SSSR, No. 199, 1975.

LEONOV, M.G. Tectonic conditions of the time of olistostrome development. Geotektonika, No. 3, 1976.

LUK'YANOV, A.V., M.G. LEONOV and I.G. SHCHERBA. Olistostrome development and the problem of psuedotillites. Litol. i polezn. iskop., No. 4, 1975.

MILANOVSKIY, YE.YE. Noveyshaya tektonika Kavkaza (Recent tectonics of the Caucasus). Izd. Mosk. univ., 1968.

MILANOVSKIY, YE.YE. and V.YE. KHAIN. Geologicheskoye stroyeniye Kavkaza (Geology of the Caucasus). Izd. Mosk. univ., 1963.

PORSHNYAKOV, G.S. Tectonic position of "mixed fauna" limestones in the Middle Carboniferous of Asia. In: Voprosy regional'noy geologii (Problems of regional geology). Izd. Leningr. univ., 1968.

PSHENIN, G.I. Development of relief of frontal parts of a recent overthrust in the mountain border of Fergana, Geomorfologiya, No. 2, 1973.

SOKOLOV, S.G. Olistostromovyye tolshchi i pozdnemelovyye tektonicheskiye pokrovy ofiolitovykh zon Malogo Kavkaza (Olistostrome sequences and Late Cretaceous tectonic sheets in ophiolite zones of the Lesser Caucasus). Avtoref. kandid. dissert, Geol. inst., Akad. nauk SSSR, 1975.

CHERENKOV, I.N. Verkhnepaleozoyskaya flishevaya formatsiya Gissaro-Alaya (Upper Paleozoic flysch development of the Gissar-Alai). Domish, Dushanbe, 1973.

SHCHERBAM, I.G. Neogene olistostromes in the Darvaz Ridge, Geotektonika, No. 5, 1975.

ABBATE, E., V. BORTOLOTTI and P. PASSERINI. Olistostromes and olistoliths. Sedimentary geology, 4, No. 3/4, 1970.

BEÑEO, E. Sull'olistostroma guanternarion di Gela (Sicilia meridionale). Boll. serv. geol. Italia, LXXIX, fasc. 1-2, 1958.

ELTER, P. and L. TREVISAN. Olistostromes in the tectonic evolution of the Northern Apennines. "Gravity tectonics", Wiley Interscience, 1973.

FLORES, G. Discussion. IV World Petrol. Congr. Rome, 1955.

GIGON, W. Geologie des Habkerntales und des Quellgebietes der Grossen Emme. Verhandl. naturforsch. Ges. Basel, 63, No. 1, 1952.

GORLER, K. and K.-J. REUTTER. Entstehung und Merkmale der Olistostrome. Geol. Rundschau, 57, H. 2, 1968.

GRACIANSKY, P.CH. Stratigraphie des unites superposees dans le Taurus licien et place dans l'arc dinarotaurique. Bull. miner. rech. and explorat. inst. Turkey, No. 71, 1968.

GRACIANSKY, P.-CH., M. LEMOIN, J. SIGAL and J.-P. THIEULOY. Sur L'existence de lentilles calcaires d'age barremient et dedoulien interstratifiees dans les marnes gargasiennes du sinclinae de Barrem (Alpes de Haute Provence), C. r. Acad. Sci., Vol. 247. ser. D, 1972.

GRACIANSKY, P.-CH. Le probleme des couloure melanges a propos de formations chaotiques associees aux ophiolites de Licie occidentale (Turquie). Rev. geogr. phisigue et geol. dyn., Vol. 15, fasc. 5, 1973.

JACOBACCI, A. Frane sotomarine nelle formazioni geologiche. Interpretazione dei Fenomeni olistostromici e degli olistolite nell-Appenino e in Sicilia. Boll. serv. geol. Italia, 86, No. 3, 1965.

KRAUS, E. Die Baugeschichte der Alpen, teil II. Berlin, Akad. Verlag, 1951.

MARCHETTI, M. The occurrence of slide and flowage material (olistostromes) in the Tertiery series of Sicily. XXX sess. Congr. Geol. Internat. Mexico, 1957.

MERLA, G. Geologia dell-Apenino settentrionale. Boll. Soc. geol. Italia, 70, No. 5, 1951.

MIGLIORINI, C. Considerazioni su di un particolare effecto della arenarie del macingo. Soc. Toscana Sci. Natur. Alti, mem. 57-A, 1950.

RICHTER, M. Bemerkungen sur Geologie Nord- und West-Siziliens. Neues Jahrb. Geol. Paleontol., Monatsh. No. 2, 1968.
SCHARDT, H. Die exotischen Gebiete, Klippen und Blocke am Nordzand der Schweizeralpen. Ecolog. helv., Vol. 5, 1898.
SIGNORINI, R. Linee tectoniche transversali dell'-Apennino settentrionale. Rned. Acc. Naz. Lincei, No. 6, 1935.
SODER, P. Geologische Untersuchung der Schrattenfluh und des sudlich anschlissenden Teiles der Habkern-Milde (Kt.) Luzern. Ecolog. geol. helv, 42, No. 4, 1949.

Geological Institute
USSR Academy of Sciences

Received February 8, 1977

Editor's Comments on Papers 8, 9, and 10

8 **AUDLEY-CHARLES**
A Miocene Gravity Slide Deposit from Eastern Timor

9 **NAGLE**
Chaotic Sedimentation in North-Central Dominican Republic

10 **ROBERTSON**
The Moni Mélange, Cyprus: An Olistostrome Formed at a Destructive Plate Margin

In this group are three papers that describe, in detail, deposits that are very closely related to the type examples from the Appenines that provide the basis for the previous group of papers.

The Miocene gravity slide deposit described by Audley-Charles (Paper 8) closely resembles the Argille Scagliose of the Appenines and occurs in a zone of plate convergence of extreme complexity. Again, there is a sedimentary matrix of soft, scaly, and variegated clay. There are structures in this matrix consistent with compaction after emplacement of the blocks, also slickensiding. The uniformity of the clay matrix indicates very rapid formation; the youngest contained microfossils are Upper Miocene, the presumed age of the deposit, but it contains older microfossils (Eocene, Cretaceous, Permian) and must be considered to incorporate older clays. Bedding is absent in this matrix. There is an unconformity at the base of this deposit, and it is overlain by an Upper Miocene to Pliocene sedimentary formation. It rests on all older formations.

Exotics are mainly Permian, with some pre-Permian representation, and range up to Lower Miocene. They are mostly of rocks known in East Timor, and the largest are 0.5 km long. Most are angular or subangular, but a few large blocks are rounded. Distribution is chaotic, orientation is random, and sorting is lacking. This deposit is undoubtedly an olistostrome mélange. Blocks include igneous rocks, but ophiolitic material, if present, is certainly only a minor exotic component. This mélange would seem to represent an excellent example of an olistostrome/olistolith mélange that has suffered no significant subsequent tectonic modification. It seems to be a simple sedimentary slumping product, but there is probably a genetic relationship to the overthrusting from the north at this time, the deposit slightly postdating such thrusting. Audley-Charles suggests two possible mechanisms, one related broadly to uplift of the volcanic inner arc and downwarp off the outer arc with culminating overthrusting, the second related directly to actual push of the clay deposit in front of the thrust. In the second model, the possibility of an initial push by the advancing thrust front and a subsequent gravitational slide downslope is suggested.

Nagle (Paper 9) describes a similar deposit from the Dominican Republic. Again this is clearly a gravity slide deposit, with a gray, structureless clay matrix. The matrix is dated, on the basis of pelagic foraminifera, as Paleocene or early Eocene, and the exotic blocks, which range up to 1.5 km long, include pre-Paleocene to middle Eocene rocks (i.e., both older than and apparently younger than the matrix). Serpentinite, gabbro, pillow volcanics, andesite, and tuffs suggest that there is a considerable ophiolitic exotic content, though this mélange would be classified as an olistostrome mélange, a sedimentary mélange with incidental ophiolitic and other exotic blocks. The older exotics are believed to have been torn loose from the submarine slope beneath the gravity slide, and the younger ones, from overlying beds. The suspicion arises, however, considering the spectrum of older foraminiferal ages obtained from the matrix of the East Timor mélange (Paper 8), whether the matrix may not be slightly younger than supposed, incorporate older clays, and not have yielded any planktonic foraminifera of the youngest possible age. This mélange is in a convergent plate margin situation, close to the Puerto Rico trench, a zone of continuing tectonic instability. Nagle suggests the possibility that this mélange includes rocks from the oceanic plate scraped up at a plate boundary.

Space limitations prevent the inclusion of descriptions of mélanges quite similar to those described by Audley-Charles (Paper 8) and Nagle (Paper 9). Such occurrences are described by Marchetti (1957) from Sicily, and Kear and Waterhouse (1967) from New Zealand. The New Zealand mélange has a sedimentary matrix of variegated mudstone, of Paleocene to Eocene age, and has been compared with the Argille Scagliose. This olistostrome mélange is almost entirely sedimentary in content, but rootless serpentinite bodies again testify to some ophiolitic content, though obviously minor. Again the situation is a convergent plate boundary, and such a situation is applicable to yet another occurrence of similar chaotic deposits comparable to the Miocene Argille Scagliose described by Brunnschweiler (1966) from

Burma (Arakan). These block clays display primary slumping features as well as pronounced tectonic deformation. Ophiolitic blocks of late Cretaceous age are a feature of these deposits, which are olistostrome mélanges of sedimentary origin, considerably tectonically deformed, and carrying some ophiolitic exotics.

Robertson (Paper 10) described an occurence of olistostrome mélange from Cyprus. The maximum thickness of this mélange is 200 m, and the blocks within it may be up to 1.5 km diameter. The matrix is of noncalcareous clays and siltstones and forms a para-autochthonous assemblage that is equivalent to the Kannaviou Formation above the Troodos ophiolites. It is dated on the basis of contained radiolaria as Campanian, but indirect evidence exists of extension up into the Maastrichtian. The environment of sedimentation was pelagic, in relatively deep water. The olistoliths are mainly derived from Mammonia Complex near Paphos and include sedimentary rocks of such known provenance and also rocks not otherwise known on Cyprus. The representation is from the Mammonia Complex and the Antalya nappe of Turkey. The serpentinite sheets are believed to be tectonically derived, not olistoliths. Whereas the olistoliths are supposed to come from a Troodos-type oceanic crust, they are believed to come from the continental margin, which had suffered pretectonic disruption before gravity sliding down the paleoslope into unconsolidated clays. Robertson thus invokes a combined olistostrome and tectonic process. Ophiolitic material is restricted to the tectonically derived serpentinite sheets. The mélange was apparently formed during quite short-lived subduction at a convergent plate margin during the Maastrichtian. Pelagic marls and chalks that succeeded the mélange show no deformation, and tectonic deformation clearly did not continue. Robertson believes that this mélange is of quite different character to those associated with prolonged, steady-state subduction (e.g., those of Java).

Mention is made in Paper 10 of the nearby Kathikas mélange (Swarbrick and Naylor, 1980). This is quite a small-scale occurrence in terms of size of debris fragments and extent; it is really a debris flow. It shows well preserved internal structure, reflecting its gravity sliding origin. When one considers the extreme range of scale of occurrence embraced by the term *mélange* (the extremes would be this mélange and the Coloured Mélange of Iran described in Paper 21, with blocks several kilometers long and no matrix), one is conscious of how eclectic the term *mélange* has become.

REFERENCES

Brunnschweiler, R. O., 1966, On the Geology of the Indo-Burman Ranges, *Geol. Soc. Australia Jour.* **13:**137–194.

Kear, D., and B. C. Waterhouse, 1967, Onerahi Chaos-Breccia of Northland, *New Zealand Jour. Geology and Geophysics* **10:**629–646; and **20:**205–209.

Marchetti, M. P., 1957, The Occurrence of Shale and Flowage Materials (Olistostromes) in the Tertiary Series of Sicily, *Internat. Geol. Congress, 20th, Proc.* **5:**209–225.

Swarbrick, R., and M. A. Naylor, 1980, The Kathikas Mélange, S. W. Cyprus: Late Cretaceous Submarine Debris Flows, *Sedimentology,* **27:**63–78.

8

Reprinted from *Geol. Mag.* **102:**267–276 (1965)

A Miocene Gravity Slide Deposit from Eastern Timor

By M. G. Audley-Charles

(PLATES VIII–IX)

Abstract

A formation composed of unbedded scaly bentonitic clay containing abundant scattered exotic blocks, boulders, pebbles, and smaller fragments is described from eastern Timor. It is proposed to call this formation the Bobonaro Scaly Clay. This formation is compared closely with the Argille scagliosa of the northern Apennines. Its origin is ascribed to submarine sliding of an unstable clay mass from the area north of Timor under the influence of gravity, closely associated with the emplacement of large overthrust slices of strata that marked a climax of the mid-Tertiary orogeny in Timor. The discovery of this formation has helped to bring about a radical revision of the published views of the stratigraphy and structure of Timor.

I. Introduction

The Bobonaro Scaly Clay is more widespread and outcrops over a larger area than any other formation in eastern Timor (Portuguese Timor); its area of outcrop exceeding 4,000 sq. km. There are only two large areas in the eastern half of Timor where it is not exposed: the eastern tip of the island where autochthonous Permian and Mesozoic rocks are covered by terraces of Pleistocene reef limestones, and in the north-west where there are large overthrust slices of Permian strata (see the distribution map in Text-fig. 1).

The type locality for this formation is beside the River Lomea, east of the village of Bobonaro (125° 20′ E. 9° 02′ S.). Here steep cliffs provide an almost continuous outcrop for about 8 km.

II. Description

Lithologically the Bobonaro Scaly Clay has two principal constituents: (*a*) a scaly clay matrix, and (*b*) a wide variety of unsorted, angular and subangular exotic blocks derived from other formations.

(*a*) *The Scaly Clay Matrix.*—This is remarkably uniform in character, being always soft, scaly, and variegated. The clay is generally dark reddish brown, but a green (particularly a dark olive green) variety is quite common. Black, grey, yellow, and bright red clays also occur. Contorted divisions of these contrasting clays indicate plasticity and flowing movement. Lines of flowage and slickensiding are particularly common around the exotic blocks; since these features surround the blocks they can probably be attributed to compaction subsequent to emplacement. It is most unlikely that they were caused by blocks falling into soft mud.

The composition of this clay matrix is fairly complex. Montmorillonite is the predominant clay mineral, forming up to 35 per cent of the whole,

[*Editor's Note:* Plates VIII and IX have been omitted.]

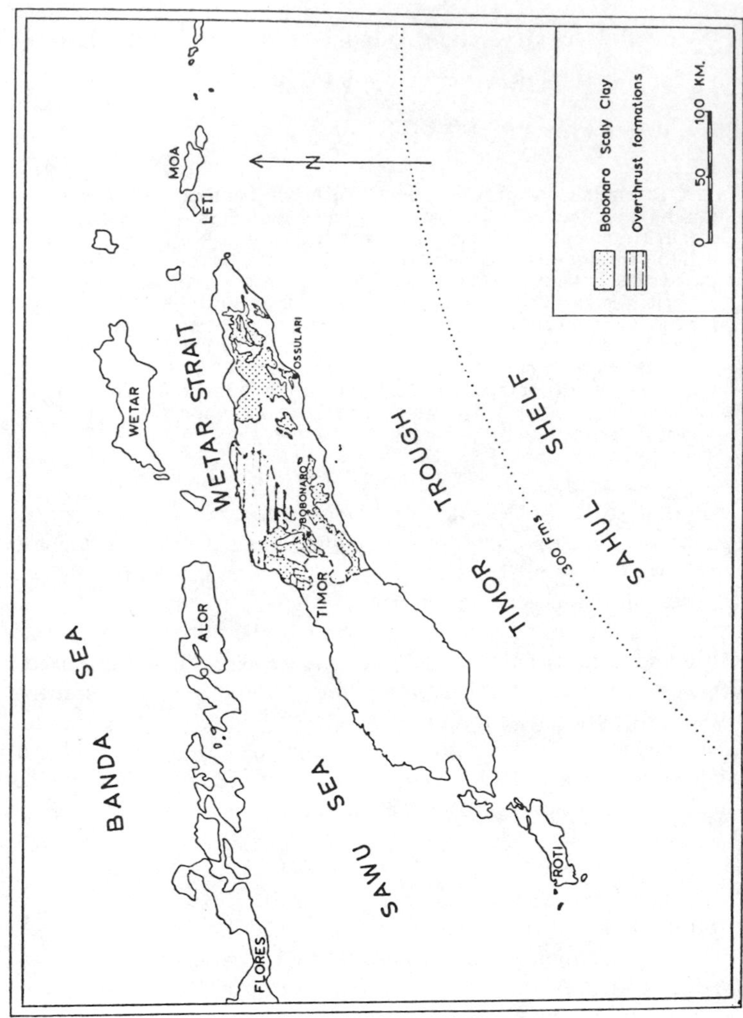

TEXT-FIG. 1.—General map of the Timor region showing the distribution of outcrops of the Bobonaro Scaly Clay and the sheets of Permian strata that were thrust into their present position during mid-Miocene times.

with lesser amounts of sepiolite and another mineral that is probably palygorskite. The quartz content is noteworthy, being about 10 per cent, while calcite is a variable constituent.

(*b*) *The Exotic Material.*—The exotic blocks and smaller fragments vary in age from pre-Permian to Lower Miocene, the most common material being Permian. More that 90 per cent of the exotic material seems to have been derived from the same rock formations as outcrop in eastern Timor. Among those rocks that cannot be matched is a cephalopod limestone of Lower Triassic age. Similar rocks have, however, been reported from western Timor (Umbgrove, 1938). The size of the exotic material varies from blocks more than 0·5 km. long down to silt grade particles (see Plates VIII and IX). Most of the exotic material is angular or subangular, but a few huge blocks show a remarkable degree of rounding (see Plate IX). The distribution of exotic material within the clay matrix is chaotic and has random orientation. The only pattern that can be discerned in the distribution is that locally exotic material of a particular formation is much more abundant than elsewhere. For example, in the type locality there are two adjacent cliffs ; in one cliff the exotic material is predominantly Upper Triassic limestones and shales, while the other cliff is characterized by red, crinoidal Permian limestones.

Where the exotic material is relatively easily eroded, for example, blocks of soft shales, the scaly clay matrix frequently takes on the colour of these shales due to the presence in the clay matrix of fine grade fragments of them. Where hard limestones or lavas are present they usually have no apparent effect on the colour of the clay matrix.

(*c*) *Texture and Structure.*—The quantity of exotic material mixed with the scaly clay matrix is very variable. Many areas are known where the Bobonaro Scaly Clay is exposed as cliffs without exotic material of any grade being found. Elsewhere the amount of exotic material forms a considerable proportion of the whole.

The erosion of this formation under tropical conditions, during the rapid elevation of the island since the beginning of the Pliocene, is very severe. It gives rise to a characteristic topography of deep gullies, landslips, and rugged, nobbly hillsides where the much more resistant blocks of hard rock protrude above the soft clay matrix.

The high percentage of montmorillonite (bentonite) in the scaly clay gives it a waxy appearance when fresh. After it has weathered it develops what Grim (1953) described as the " characteristic 'jigsaw puzzle' set of fractures " due to shrinking on drying (Plate VIII).

There is a total absence of recognizable bedding in this formation. In those exposures where there is little or no exotic material, some scaly fissility of the clay suggests a flat lying disposition. This is, however, probably a secondary feature due to the compaction of the soft clay.

The problem of deciding whether a large block of hard rock that protrudes through the mantle of scaly clay is an exotic block or a buried topographic peak (inlier) is frequently encountered. The evidence is always equivocal wherever the base of the hard rock is not discernible.

Since this formation is unstratified, and no marker horizons have been recognized, an accurate determination of its thickness is not possible. The base of the formation is a highly irregular unconformity. Locally the formation can be seen to thin out rapidly, while the greatest thickness that can be deduced is the section in the well Ossulari No. 1, which encountered 1,906 metres of this formation. It seems probable that the thickness of the formation beneath the south coastal plains may locally exceed 3,000 metres, and that in general the formation increases in thickness from north to south across the island.

III. Stratigraphic Position

The base of the formation is an irregular unconformity that locally is probably very rugged. The Bobonaro Scaly Clay forms a blanket that rests somewhere on every older formation. It is overlain by the Viqueque Formation which is Upper Miocene and Pliocene in age, the base of which probably belongs to the Tg Indo-Pacific stage (van der Vlerk and Umbgrove, 1927), and elsewhere by various Pleistocene beds.

Microfossils varying in age from Permian to Upper Miocene are found in the scaly clay part of this formation. The individual exotic blocks have been found to contain both macrofossils and microfossils whose ages vary from Permian to Lower Miocene. Since the evidence is clear that the exotic material arrived in its present position at the same time as the clay matrix two dates are pertinent: the age of the clay at the time of its formation, and the age of its emplacement in its present position, since it is clearly an allochthonous formation.

The exceptional uniformity of the clay matrix, and its particular nature (discussed below) indicate that it probably formed very rapidly. The fact that it contains Foraminifera ranging in age from Permian to Upper Miocene means that its age can only be stated in terms of its youngest fossils, namely Upper Miocene. Since the youngest rocks on which this formation rests (unconformably) are Lower Miocene of Te stage, and the oldest rocks that overlie (unconformably) the Bobonaro Scaly Clay belong to the Viqueque Formation of the Upper Miocene (Tg stage), then the Bobonaro Scaly Clay must clearly have been emplaced in its present position between late Te and early Tg stages. There is no evidence to suggest that this formation was anywhere emplaced outside this time interval. However, it is not certain whether the whole formation achieved its present position as a single event, or by a succession of small emplacements. Part of the clay matrix (perhaps a large part) must be Upper Miocene in age since the indigenous microfauna is Upper Miocene. The presence of derived microfossils of Eocene, Cretaceous, and Permian ages suggests that the clay matrix consists in part of clay derived from pre-Miocene rocks.

The following Foraminifera, that indicate an Upper Miocene age, have been determined from the clay matrix of this formation by D. J. Belford, of the Bureau of Mineral Resources, Canberra, Australia (in unpublished reports dated 1960 and 1961): *Globigerinoides quadrilobatus quadrilobatus* (d'Orbigny); *G. quadrilobatus immaturus* Le Roy; *G. quadrilobatus trilobus* (Reuss); *G. quadrilobatus irregularus* Le Roy; *G. ruber* (d'Orbigny); *Globigerina bulloides* (d'Orbigny); *G. subcretacea* Lomnicki; *Globorotalia cultrata* (d'Orbigny); *G. scitula* (Brady); *G. truncatulinoides* (d'Orbigny); *Pulleniatina obliquilocata* (Parker and Jones); *Sphaeroidinella dehiscens* (Parker and Jones); *Bolivinita quadrilatera* (Schwager); *Pullenia bulloides* (d'Orbigny); *Nonion pompilioides* (Fichtel and Mull); *Cassidulina laevigata* (d'Orbigny); *Ceratobulinina pacifica* Cushman and Harris; *Uvigerina* sp.; *Stilostomella lepidula* (Schwager); *Bulimina aculeata* d'Orbigny; *Astrononion* sp.; *Hyalinea balthica* (Schroeter); *Sphaeroidina bulloides* (d'Orbigny); *Bulimina inflata* Seguenza; *Bolivina robusta* (Brady); *Laticarinina pauperata* (Parker and Jones); *Hoglundina elegans* (d'Orbigny).

IV. Origin

The marked unconformity at the base of the Bobonaro Scaly Clay, together with the absence of any bedding planes and the prevalence of contorted flow structures, suggest that this formation was emplaced in its present position by sliding or slumping, and is therefore an allochthonous sedimentary formation. Other evidence that supports this view includes the variety of scattered, unsorted, exotic material that could not possibly have arrived in its present position by any known sedimentary mechanism other than sliding or slumping. The scaly nature of the clay matrix suggests stress and flow. The whole formation is very similar to the Argille scagliosa of the northern Apennines of Italy as described by Merla (1951). The nature and distribution of the exotic material suggests comparison with the Wildflysch of the Swiss Alps. There is, however, a major difference between the Wildflysch and the Bobonaro Scaly Clay, in that associated with the Wildflysch are beds of greywacke and shale commonly having graded bedding, convolute bedding, loadcasts, and current bedding (Crowell, 1957); such rocks are never found bedded with the Bobonaro Scaly Clay, nor apparently with the Argille scagliosa.

Kuenen (1956) pointed out that deposits resulting from slide movements are usually characterized by a fair proportion of lutite, and that they may contain huge blocks. He stated that the original structure varies from slightly deformed to entirely mixed, and that there is a piling up at the lower end of the slide. There is no grading and no segregation of larger particles in slide deposits. By contrast, he found that deposits that result from turbidity currents are characterized by " more or less graded granular bed(s), poorly sorted at each level. Coarsest grains do not exceed 10 cm. Upper part (of the deposit) may be laminated and current ripple laminated. One current produces one single bed ". Clearly the Bobonaro Scaly Clay is a typical slide deposit and shows none of the features of a turbidity current deposit.

There is general agreement that the Swiss Wildflysch, the northern Apennine Argille scagliosa, and similar deposits described by Crowell (1957) as " pebbly mudstones " have been generated by unstable masses of sediment moving down a submarine slope under gravity. From various parts of the world deposits originating in this manner have also been described by Lugeon (1916), Bailey *et al.* (1928), Merla (1951), Dorreen (1951), Kugler (1953), Renz *et al.* (1955). and Kuenen (1956).

Because of the high proportion of montmorillonite (bentonite) in the Bobonaro Scaly Clay it would have had the ability to flow very easily, as bentonitic clays are, according to Grim (1953), generally highly colloidal and plastic. Kuenen (1956) has shown that a submarine slope of 1 or 2 degrees is sufficient " under certain favourable conditions " for a slide to be set off and kept going. Many workers have considered that tectonic activity and earthquake shocks could provide a trigger mechanism to initiate the sliding of the unstable sediment down the slope.

In eastern Timor during the time interval between late Te and early Tg stages (when the Bobonaro Scaly Clay was emplaced) the great overthrust sheets of principally Permian strata achieved their present position (see Text-fig. 1) by being overthrust from the north; the same direction from which (as will be shown below) the Bobonaro Scaly Clay was derived. In the field the Bobonaro Scaly Clay is always found overlying the thrust sheets and never below them. Obviously the movement of the thrust sheets and the emplacement of the Bobonaro Scaly Clay are closely related. The overthrust movements formed the climax of the mid-Tertiary orogeny in Timor. The emplacement of the scaly clay probably occurred after the emplacement of the thrust sheets because they have not been found superimposed on the clay. There seem to be two possible mechanisms by which the Bobonaro Scaly Clay could have been emplaced. One is as a gravity slide down a submarine slope, brought about by the orogenesis that elevated the volcanic inner arc and downwarped the outer arc where folding of the pre-Upper Miocene culminated in the overthrusting described above. The other mechanism involves the thrust front that moved from north to south, pushing the bentonitic clay before it. It is possible that the submarine sliding of the clay mass was initiated by the thrust front to a position where a slope was created down which the clay could slide freely.

V. Environment

There are four environmental problems associated with this formation in Timor: (1) The location of the basin in which the clay matrix of the Bobonaro Scaly Clay was originally deposited; (2) The determination of the source area of the exotic material; (3) The nature of the submarine slope down which the gravity slides moved to their present position; (4) The determination of the bathymetry and submarine topography of the area now occupied by the land mass of eastern Timor at the time of the emplacement of the Bobonaro Scaly Clay. Each of these problems is discussed below.

(1) The presence of montmorillonite as the most important constituent indicates that a large part of the clay was probably derived from the submarine weathering of volcanic ash. The volcanic inner arc of islands is to-day about 50 km. north of the north coast of Timor (see Text-fig. 1). This volcanic arc opposite Timor (islands of Alor and Wetar) produced effusive igneous rocks during the Lower Miocene. This volcanic activity ceased temporarily in mid-Miocene times when this part of the volcanic arc was uplifted (van Bemmelen, 1949).

About 130 km. south of Timor is the northern edge of the Australian continental shelf (Sahul Shelf). Between the Sahul Shelf and Timor is a trough about 3,000 metres deep. Although the age of this trough is still uncertain, there is no evidence at all to suggest that the clay slide was derived from this direction, so that we can be sure that the clay was originally deposited in the region between the inner and outer arcs, that is in the region of the Wetar Strait (see Text-fig. 1). This accords with the conclusion reached by van Bemmelen (1949) who considered that the exotic blocks (without any scaly clay matrix), described from the island of Leti (see Text-fig. 1) by Molengraaff (1915), were derived by submarine gravity sliding of a pebbly mudstone from the area north of Leti. It is also highly significant that the youngest dated exotic block on Leti is (as in Timor) Lower Miocene. Both Molengraaff and van Bemmelen were unaware of the existence of the Bobonaro Scaly Clay in eastern Timor.

(2) The exotic material of the Bobonaro Scaly Clay (and the exotic blocks from Leti) consist mainly of four types of rock: Permian limestones, Upper Triassic limestones and shales, Cretaceous pelagic sediments, and a wide variety of lavas and tuffs. Much of the exotic material can be matched with formations that outcrop in eastern Timor. Together the Permian limestones and the volcanic rocks probably form about 70 per cent of all the exposed exotic material.

The mass of unstable clay sediment, once it began to move down the submarine slope, could be expected to remove any obstruction in its path. Since the sliding of the clay either accompanied the overthrusting of the Permian strata or followed immediately after, it is to be expected that the sea floor was littered with debris from the thrust front itself (Permian), and from the rocks over which the thrust sheets moved. This debris would be rapidly incorporated in a sliding clay mass. Kuenen (1950) arguing the case for submarine canyons having been cut by turbidity currents, admits the possibility of sliding clay also effecting such erosion. Ericson *et al.* (1952) appear to hold very similar views. Thus apart from picking up the loose debris resulting from thrusting, it seems very likely that some of the exotic material was incorporated into the sliding mass by its erosive action on the sea floor in that region that now separates the two arcs, and to a lesser extent from the Timor area.

(3) The submarine slope down which the gravity slide moved must have been between the inner and outer arcs (and closely associated with the thrust front), increasing in depth towards the outer arc. The indigenous microfauna of the scaly clay suggests that the clay was deposited in a basin environment in which the depth of water was not extreme.

During Lower Miocene times, before the emplacement of the Bobonaro Scaly Clay, the region now occupied by much of the central land area of eastern Timor formed a platform on which Bahaman-type limestones were deposited. On the site of the present south coast, where pelagic limestones were deposited, conditions were those of a deeper basin. There was therefore in existence at that time a steep submarine slope from the present central longitudinal axis of eastern Timor down to the present south coast.

Between the Bahaman-type platform that occupied much of eastern Timor and the volcanic inner arc (i.e. in the region now occupied by the Wetar Strait), the sea must have been deeper than it was over the Timor " platform ", probably considerably deeper. In order for gravitational forces to move the unstable clay mass southward from the Wetar Strait a submarine slope, increasing in depth southward, must have been created during the Tf time interval. Thus the region of Timor must have subsided relative to the area to the north (now occupied by the Wetar Strait), from which the clay and much of the exotic material were derived. This movement involved folding of the pre-Tf rocks and the overthrusting of large sheets of Permian strata in Timor, while the evidence from the islands of Alor and Wetar (van Bemmelen, 1949) indicates that the inner arc was elevated in mid-Miocene times. That the Timor region was downwarped by the Upper Miocene (Tg stage) is confirmed by the deeper water facies of the lower part of the Viqueque Formation (Tg), which contrasts with the very shallow Bahaman-type limestones of Lower Miocene times (Te stage).

VI. History and Synonyms

Earlier workers, Grunau (1953) and Gageonnet and Lemoine (1958), grouped together, under the title of the " Bibiliu Series ", the following sedimentary rocks: Cretaceous radiolarites and marls, Eocene limestones, and a block clay (herein defined as the Bobonaro Scaly Clay).

Gageonnet and Lemoine (1958), in their geological sketch map of Portuguese Timor (R.F. 1 : 500,000), mapped a number of very large areas as Mesozoic " Complexe Triasico-Jurassique " that are undoubtedly occupied by the Bobonaro Scaly Clay containing much Mesozoic exotic material.

I. B. Freytag, in an unpublished report in 1959, was the first to recognize the unity of this formation that he called the Tuca River Block Clay. The term " block clay " is being replaced by the term " scaly clay " because while exotic blocks are not always present, the clay is always scaly.

VII. Some Consequences of this Stratigraphic Revision

Apart from the stratigraphic importance of the proper division of the " members " of the Bibiliu Series of Grunau (1953) (a term it is suggested is now dropped), the recognition of the Bobonaro Scaly Clay has great relevance to the tectonic problems of Timor. As I. B. Freytag first pointed out in his unpublished report, many " klippen " identified by earlier workers (who usually called them " Fatus ") are not klippen but large exotic blocks of various ages embedded in a matrix of Miocene clay, and these blocks belong stratigraphically to the Bobonaro Scaly Clay not to any nappe. In the light of this discovery, the remapping of all of eastern Timor by the present author has led to a complete revision of the tectonic and stratigraphic history; it is hoped to publish these results shortly.

It seems highly probable that the Bobonaro Scaly Clay will be found in western Timor (formerly Dutch now Indonesian Timor), and that its discovery there will lead to a similar revision of the present view of its stratigraphy and structure.

VIII. Acknowledgments

The author is grateful to the Director of the Bureau of Mineral Resources, Canberra, Australia, for permission to include the palaeontological determinations made by D. J. Belford. The Fullers' Earth Union Ltd., Redhill, Surrey, kindly analysed the clay samples. Discussions with Dr. W. F. Schneeberger, Dr. D. V. Ager, Mr. D. J. Carter, and especially Professor W. D. Gill helped clarify some of the problems.

The author thanks Dr. D. V. Ager for helpful criticism of the manuscript. Gratitude is expressed to the Board of Directors of Timor Oil Ltd., Sydney, Australia, for permission to publish this material. The support of D.S.I.R. Research Studentship is gratefully acknowledged. The author wishes to acknowledge his debt to the unpublished (1959) report of Mr. I. B. Freytag.

Department of Geology,
Imperial College,
London, W. 7.

IX. REFERENCES

Bailey, E. B., L. W. Collet, and R. M. Field, 1928. Paleozoic submarine landslips near Quebec City. *J. Geol.*, **36**, 577–614.

Crowell, J. C., 1957. Origin of pebbly mudstones. *Bull. geol. Soc. Amer.*, **68**, 993–1010.

Dorreen, J. M., 1951. Rubble bedding and graded bedding in Talara formation of Northwestern Peru. *Bull. Amer. Ass. Petrol. Geol.*, **35**, 1829–1849.

Ericson, D. B., M. Ewing, and B. C. Heezen, 1952. Turbidity currents and sediments in the North Atlantic. *Bull. Amer. Ass. Petrol. Geol.*, **36**, 489–511.

Gageonnet, R., and M. Lemoine, 1958. Contribution á la connaissance de la géologie de la province portugaise de Timor. *Junta de Investigacões do Ultramar. Estud. Ens. Doc.*, *48*.

Grim, R. E., 1953. *Clay Mineralogy*. New York.

Grunau, H. R., 1953. Geologie von Portugiesisch Osttimor. Eine kurze Uebersicht. *Ecl. geol. Helv.*, **46**, 29–37.

Kuenen, Ph. H., 1950. *Marine Geology*. New York.

—— 1956. The difference between sliding and turbidity flow. *Deep-Sea Res.*, **3**, 2, 134–9.

Kugler, H. G., 1953. Jurassic to recent sedimentary environments in Trinidad. *Bull. Ass. Suisse des Géol. et Ing. du Pétrole*, **20**, 59, 27–60.

Lugeon, M., 1916. Sur l'origine des blocks exotiques du Flysch préalpin. *Ecl. geol. Helv.*, **14**, 217–221.

Merla, G., 1951. Geologia dell'Appennino Settentrionale. *Boll. Soc. geol. Ital.*, **70**, 95–382.

Molengraaff, G. A. F., 1915. Geografische en geologische beschrijving van het eiland Letti. *Jaarb. Mijnb. Ned. Ind.*, **43**, 1–87.

Renz, O., R. Lakeman, and E. van der Meulen, 1955. Submarine sliding in western Venezuela. *Bull. Amer. Ass. Petrol. Geol.*, **39**, 2053.

Umbgrove, J. H. F., 1938. Geological History of the East Indies. *Bull. Amer. Ass. Petrol. Geol.*, **22**, 1–70.

van Bemmelen, R. W., 1949. *The Geology of Indonesia*. The Hague.

van der Vlerk, I. M., and J. H. F. Umbgrove, 1927. Tertiaire gidsforaminiferen van Nederlandsch Oost-Indië. *Wet. Meded. Dienst. Mijnb. Ned.—Oost-Indië*, **6**.

9

Reprinted from pages 415–428 of *Geol. Soc. America Mem. 132,* 1972

Chaotic Sedimentation in North-Central Dominican Republic

FREDERICK NAGLE
Department of Geology and Rosenstiel School of Marine and Atmospheric Science
University of Miami, Miami, Florida 33149

ABSTRACT

Chaotic, allochthonous, submarine gravity slide deposits (olistostromes) have been reported with increasing frequency during the past 15 yrs in stratigraphic columns throughout the world. One such unit, estimated to be several hundred meters thick, is distributed over 90 sq km in the Puerto Plata area of north-central Dominican Republic.

The gray, clay-sized, structureless matrix of this unit behaves as a quick-clay under shock, and is composed of kaolinite, quartz, montmorillonite, and illite. Pelagic Foraminifera date the matrix as Paleocene(?) or early Eocene(?).

Exotic blocks within the olistostrome include limestone, serpentinite, andesite, pillow volcanics, and tuffaceous rocks. The longest dimensions of the blocks range from 1 cm to 1.5 km, and the known ages range from pre-Paleocene to middle Eocene. Thus, the exotic blocks are older, the same age as, and younger than their enclosing matrix.

The matrix probably was once a marine tuffaceous unit of Paleocene–early Eocene age which was deposited rapidly and retained enough water to liquify spontaneously and to begin moving rapidly down a gentle submarine basin slope following an earthquake shock during the late-middle Eocene. Younger more competent overlying rocks were ruptured and incorporated into the moving mass, while older rocks were torn loose from the submarine slope. The entire unit was emplaced essentially instantaneously into a marine sedimentary sequence.

Although olistostromes record catastrophic events in the geologic record, they do not necessarily imply a major orogeny. However, most described olistostromes were formed during times of tectonic activity. The middle Eocene was such a time in the Puerto Plata area.

INTRODUCTION

During the course of field mapping of some 700 sq km in the vicinity of Puerto Plata, Dominican Republic (*see* Fig. 1 and Nagle, 1966, 1971), a major problem of geologic interpretation centered upon a mudstone unit incorporating a unique assemblage of boulders and blocks of various sizes and rock types. The San Marcos olistostrome was not at first recognized as a unit. During the initial stages of the investigation attention was centered on the blocks, and as more were examined, it became increasingly difficult to construct a reasonable map, no matter what sort of fold or fault patterns were invented. In the summer of 1961, H. H. Hess (oral commun.) suggested that this chaos was an allocthonous unit and that the blocks were exotic. In the following months, field work indicated that this was so.

The Puerto Plata area is located along the extension of the axial zone of the Puerto Rico trench which dies out 250 km east of Puerto Plata. Negative isostatic gravity anomalies extend beyond the topographic expression of the trench through the Puerto Plata area. The area is currently tectonically active (Fig. 2); two linear zones of epicenters intersect at roughly 70°10′ W., 19°30′ N., a point 50 km southeast of the mapped area. One of these zones continues along the south slope of the Puerto Rico trench, indicating the occurrence of a fault or fault zone along that line which may connect with known fault zones to the west in Hispaniola and Cuba. Faults of one type or another have been suggested for one or both slopes of the Puerto Rico trench by many authors, although past and present motions on these faults have been debated (*see,* for example, Bunce and Fahlquist, 1962; Glover, 1967; Chase and Hersey, 1968; Monroe, 1968; Molnar and Sykes, 1969; Bracey and Vogt, 1970; Nagle, 1970). Focal mechanism solutions from fault plane data (Molnar and Sykes, 1969) indicate current underthrusting nearly perpendicular to the Lesser Antillean portion of the West Indies arc but almost parallel to the Puerto Rico trench trend in the area north of the Virgin Islands to Hispaniola

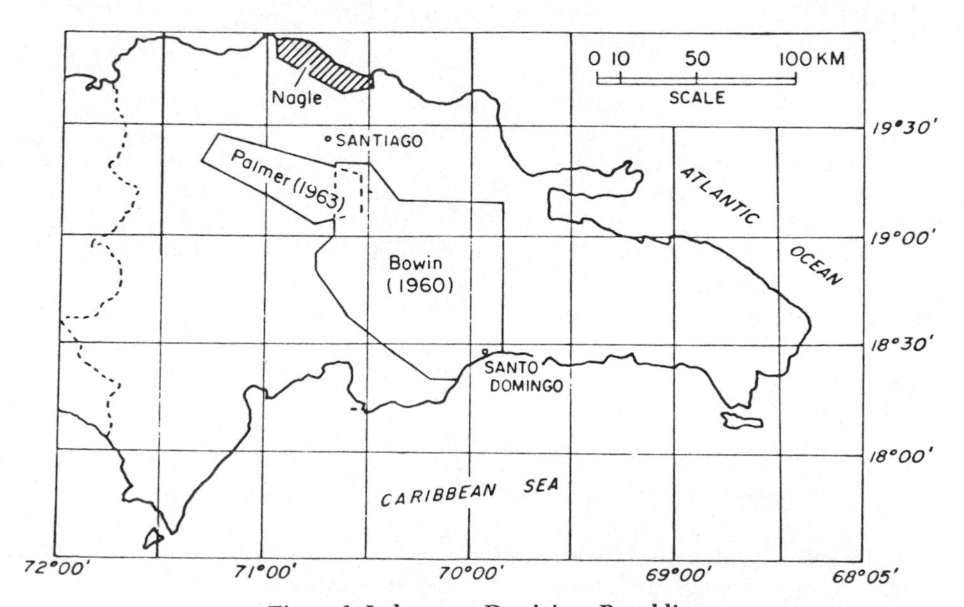

Figure 1. Index map, Dominican Republic.

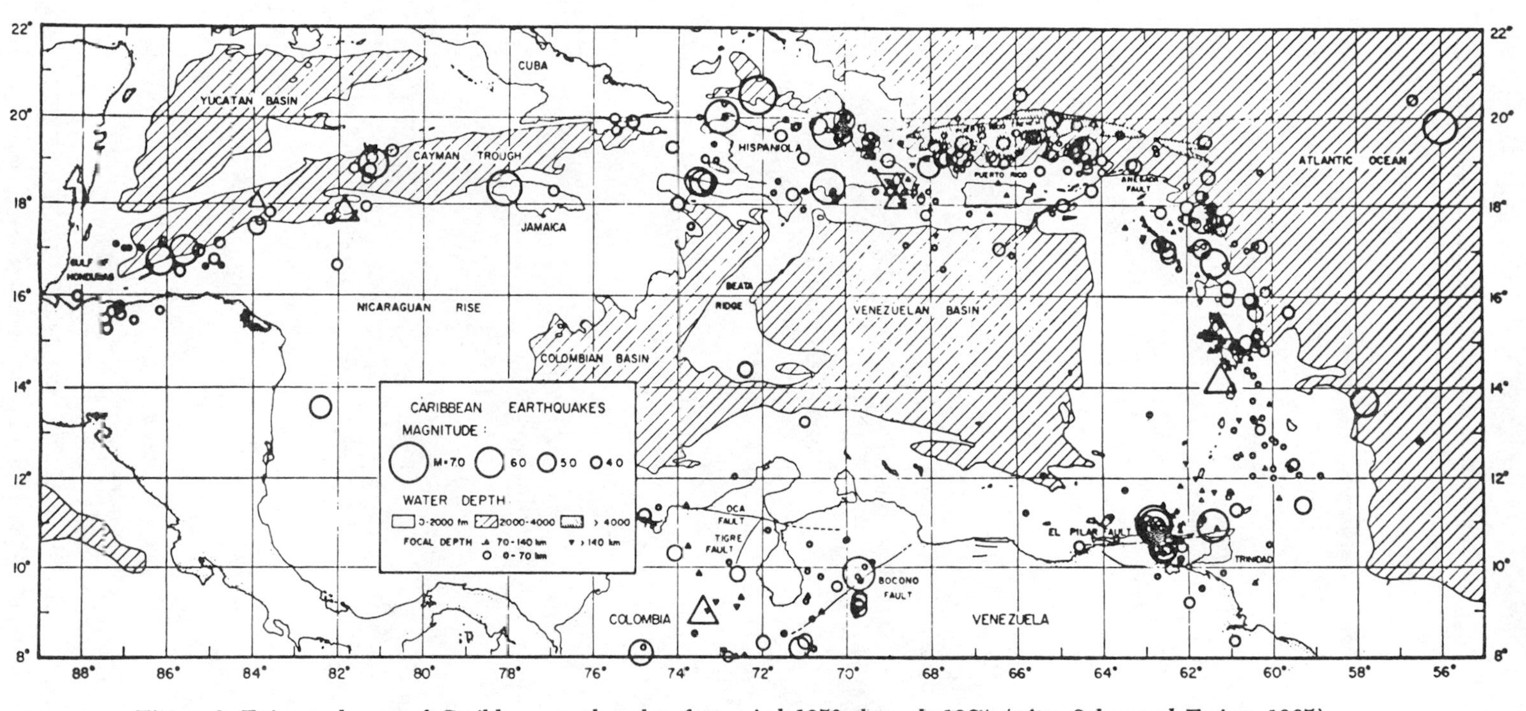

Figure 2. Epicentral map of Caribbean earthquakes for period 1950 through 1964 (*after* Sykes and Ewing, 1965).

(Fig. 3). Tectonic activity, especially vertical movements, can be documented from middle Eocene to the present in the Puerto Plata region, and I have suggested elsewhere (Nagle, 1971) that the rocks and tectonic history indicate that this region is an upfaulted sliver of the south slope of the Puerto Rico trench.

The San Marcos olistostrome is interpreted as a chaotic submarine gravity slide, containing blocks older, contemporaneous with, and younger than an unindurated clay-size matrix, the whole mass of which was emplaced during a single catastrophic event. There is no evidence for later tectonic remobilization.

Hsu (1968) made an important contribution to the subject by pointing out that two different sets of processes could lead to stratal disruption and mixing and suggests that the term *olistostrome* be restricted to sedimentary deposits as suggested by Flores (1955) and that the term *melange* be used to refer to those deposits which have resulted from tectonic mixing of consolidated rocks under an overburden after sedimentation. Hsu details several diagnostic characteristics of each type.

Olistostromes very similar to the San Marcos have been described in many parts of the world: Timor (Audley-Charles, 1965), Sicily (Marchetti, 1957), and New Zealand (Kear and Waterhouse, 1967).

Although neither the term olistostrome nor melange has been applied previously to describe gravity slide deposits in the Caribbean, one or the other of the terms might be applicable to certain units in Barbados (Senn, 1940; Saunders, 1968), Trinidad (Kugler, 1953; Kugler and Saunders, 1967; Barr and Saunders, 1968), Venezuela (Renz and others, 1955; Bushman, 1958; Bell, 1967), Cuba (Thayer and Guild, 1947; Kozary, 1968), and Puerto Rico (Glover, 1967). Other similar deposits around the world are listed in Horne (1969). Submarine slide and slump deposits on the present continental shelves are reviewed by Moore and others (1970).

STRATIGRAPHIC AND STRUCTURAL POSITION OF THE OLISTOSTROME

In the stratigraphic sequence for the Puerto Plata area (Fig. 4) the oldest rocks are serpentinites which represent basement upon which the Los Caños Formation

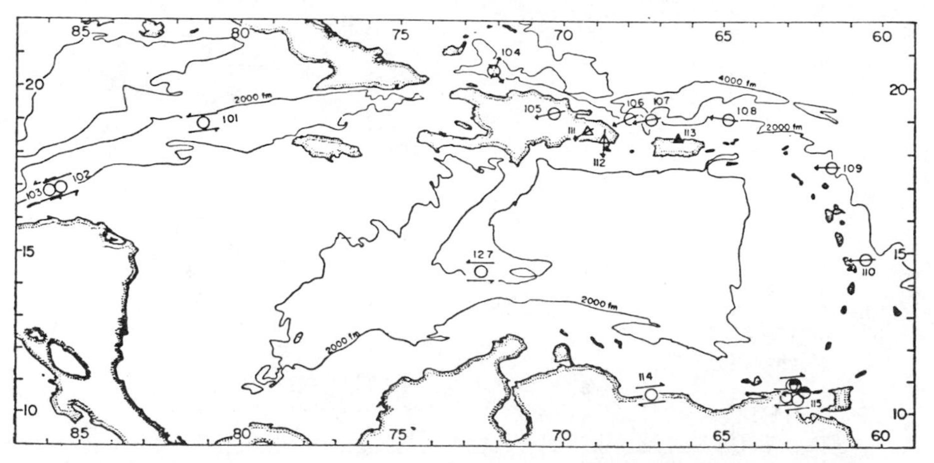

Figure 3. Azimuths of slip vectors for earthquake mechanisms in the Caribbean region. Earthquakes deeper than 100 km shown as triangles, those shallower than 100 km as circles (*after* Molnar and Sykes, 1969).

(pre-Paleocene) was deposited. The serpentinites occur as infaulted bodies in the Los Caños Formation and are probably detached fragments of the oceanic crust. The Los Caños Formation and serpentinites, as well as their gabbro, rodingite, and metamorphic associates, occur together near the core of the major anticlinal structure in the area.

The Los Caños Formation consists of andesite flows and tuffs, with spilites and keratophyres near the base. Relic glassy textures indicate the formation is submarine in origin. The exact age of the Los Caños Formation is unknown, but it is overlain unconformably by the Paleocene Imbert Formation. Its thickness is estimated at several kilometers.

Unconformably above the Los Caños Formation is a 1-km-thick succession of fine-grained graded calcareous tuffs which grade upward to vitric andesite and dacite tuffs, with rare interbedded green cherts and thin white aphanitic limestones.

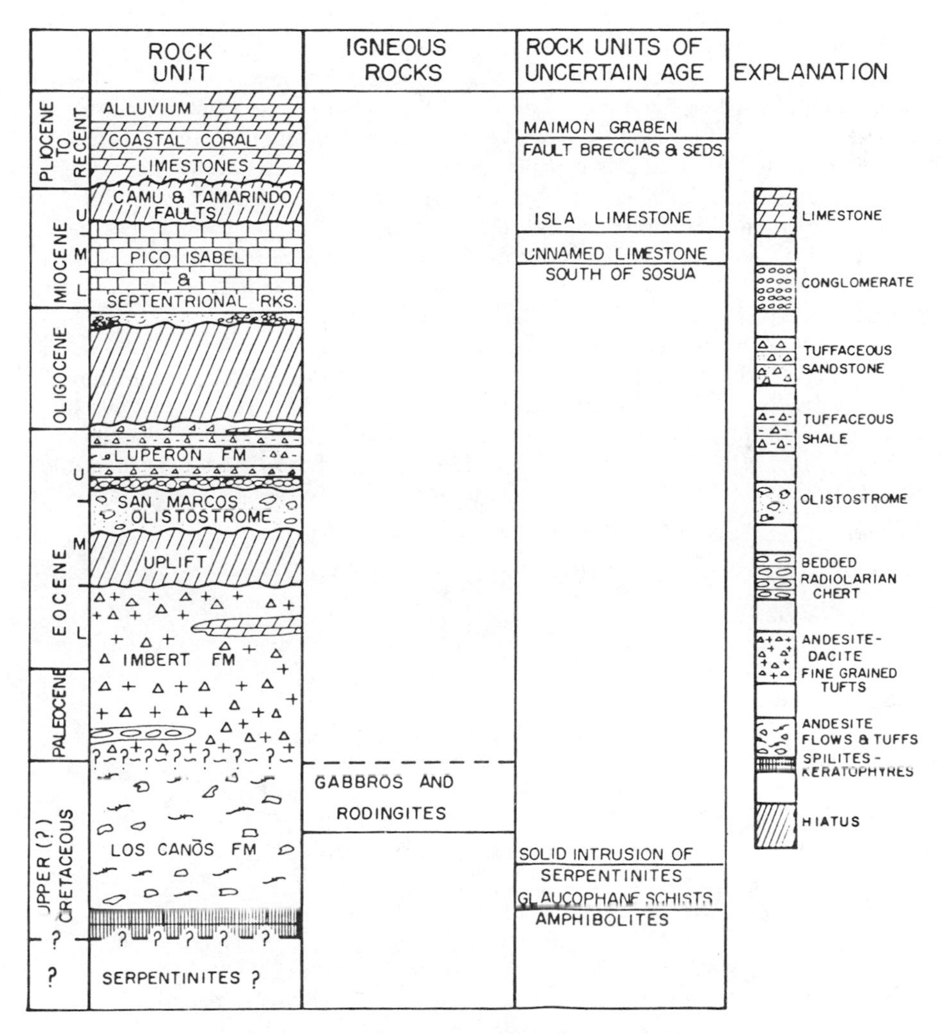

Figure 4. Stratigraphic sequence for the Puerto Plata area.

This marine succession, the Imbert Formation, has been dated as Paleocene to early Eocene.

The only known middle Eocene rocks, tuffaceous limestones, occur as exotic blocks in the San Marcos olistostrome. No middle Eocene rocks were found in situ. Upper Eocene rocks are present in the Luperon Formation, a 1-km-thick unit consisting of a coarse basal conglomerate and an alternating sequence of calcareous tuffaceous graded bedded sandstone and shales with rare interbedded bioclastic limestone. The Luperon Formation contains the first recognizable stream-worn detritus from the older formations in the area. It rests unconformably upon the Imbert Formation, but it is nowhere in contact with the San Marcos deposits.

The dominant structural feature of the area is a broad faulted anticline, 5 to 8 km wide and 25 km long, which trends N. 55° W., parallel to the north coast of the Dominican Republic, and plunges to the northwest. The San Marcos olistostrome is presently distributed over some 90 sq km within the core of this major anticline.

The contact relations of the San Marcos to adjoining units are puzzling. If the stratigraphic interpretation presented here is correct, the San Marcos unit should be or should have been enclosed within a middle Eocene section. I have been unable to document or disprove the occurrence of Middle Eocene rocks at either contact.

The age and nature of the rocks beneath the olistostrome are unknown. In the area west of Puerto Plata, the San Marcos rests unconformably(?) on Los Caños rocks, whereas to the south it is in fault contact with middle Miocene rocks of the Cordillera Septentrional. The San Marcos thins to the northwest and east of Puerto Plata where it is covered by recent alluvium.

There are no middle Eocene sediments now on top of the olistostrome. Either these sediments were removed by later slumping, faulting and erosion, or more likely the Puerto Plata area was generally rising during the middle Eocene and there was little or no post–San Marcos middle Eocene deposition.

Since contact relations of the San Marcos deposits with other units are unknown, nondiagnostic, or subject to alternative interpretations, I have interpreted the history of the San Marcos olistostrome primarily on the basis of known stratigraphy outside the area of olistostrome exposure and on the basis of fossils present in the olistostrome.

SAN MARCOS OLISTOSTROME

General Description

The hummocky topographic expression of the blocks within the San Marcos is a particularly outstanding feature of this deposit visible in the field and in aerial photographs (Fig. 5). A similar appearance in aerial photographs of other regions may indicate the presence of a similar unit.

The olistostrome has a minimum thickness of 300 m. However, this estimate is based only upon topographic exposures. There is no bedding preserved within the matrix, the lower contact was not found, and an upper contact with suspected late-middle Eocene or early-upper Eocene rocks was never proven. The oldest dated rocks now on top of the San Marcos are the late Oligocene–Miocene rocks of the Pico Isabel Formation, but these rocks are suspected to be an allochthonous block which slid over the olistostrome sometime in the Late Miocene from the south, where rocks of the same age and rock type were uplifted during that time. There

Figure 5. Aerial photograph of Puerto Plata area, D. R. San Marcos olistostrome in center portion of photograph; some olistoliths (exotic blocks) marked by arrows, many others visible. The city visible is Puerto Plata. Pico Isabel de Torres is the area outlined immediately south of Puerto Plata. Scale 1:60,000.

is some thinning of the San Marcos unit to the east as well as dilution of blocks in this direction, but no apparent thinning or dilution to the north or south.

Olistostrome Matrix

Fifty samples of the San Marcos matrix were collected over a distance of 20 km along the road running northwest about 4 km south of Pico Isabel. These samples represent the freshest matrix known in the area. The matrix is unconsolidated clay-sized sediment, sometimes streaked with red hematite or brown iron oxides, and it lacks bedding or directional features except for occasional swirling of layers in the immediate vicinity of a block. No blocks larger than 3 to 4 m were found surrounded entirely by the matrix, primarily because outcrops along this road and in stream valleys are limited to about 4 m in height.

The ratio of matrix to exotic blocks varies from about 1:1 to 5:1. Nowhere is the volume of blocks greater than that of the matrix.

Since there is no bedding or marker horizon in the San Marcos matrix, it is impossible to determine what part of the section was sampled. Samples in the west were collected from higher topographic levels than those in the east.

I believe all matrix samples collected are at least partially weathered. None were collected any deeper than 4 m from the present topographic surface; all samples were moist with ground water, and had the texture of wet clay. There is no evidence that the matrix was ever lithified. It is, however, undergoing weathering and probably represents the total product of source material, weathering in its present location, and possibly decomposition of the smaller enclosed blocks. Intense weathering has probably helped to eliminate any primary sedimentary features which survived initial churning.

X-ray diffraction traces of the less than 2 micron fraction indicate that the San Marcos matrix is composed of a mixture of quartz, kaolinite, montmorillonite and (or) chlorite and illite. The strongest peaks on all traces, other than those of quartz, are the 7 Å and 3.5 Å peaks of kaolinite. Three samples were heated to 550° C for several hours to separate kaolinite from chlorite. The 7Å and 37.5°2θ peaks disappeared, indicating kaolinite rather than chlorite. No samples were glycolated so that the 14 Å peak distinction of montmorillonite from chlorite is impossible, although the heat test mentioned above suggests montmorillonite rather than chlorite.

Portions of the coarse fractions (0.25 mm to 1 mm) of 20 samples viewed in thin section consisted mainly of turbid globs of clay(?) minerals and iron oxides, and rare unaltered broken crystals of plagioclase, probably volcanic ash remnants. No carbonate fragments were noted in the coarse fraction, nor did carbonate appear in the x-ray traces of the finer portion. This must indicate strong leaching or deep-sea deposition.

Thirty samples disaggregated in water in an ultrasonic vibrator and sieved through no. 60 (0.25 mm), no. 230 (0.62 mm) and no. 325 (0.044 mm) sieves were searched for organic remains. Only four of the samples yielded fossils. P. J. Bermudez examined this collection and found only members of the genus *Globigerina* and genus *Globorotalia* in addition to unidentified radiolaria. In his opinion (1962, written commun.) the fauna represents a marine deep-water environment and is of early Eocene–Paleocene age, although he notes that the collection "is insufficient to determine the age." Having no data to the contrary, I have tentatively assigned an early Eocene–Paleocene age to the matrix.

Olistostrome Blocks

The olistostrome blocks have sharp boundaries with the finer matrix, are invariably angular, show inconsistent strikes and dips and varied rock types, and range in size from 1 cm to 1.5 km in the longest dimension. Outcrops of blocks occur on the slopes of hills, on the tops of hills, or as boulders in stream valleys. Any recognizable pattern of outcrops or rock type is completely lacking. No geologic parameter is continuous from one locale to the next. Wise and Bird (1964) describe the blocks of the *argille scagliose* of the Apennines as a "chaotic littering of the landscape like so many popcorn boxes after a Saturday matinee." This analogy applies to the San Marcos blocks.

The most common rocks forming blocks of all sizes are gray to black tuffs and gray or buff tuffaceous limestones. Also rather common are blocks of andesite; pillow volcanics; veined, fractured, and recrystallized limestones; serpentinites; and rodingites. Approximately 30 different rock types have been recognized in thin sections. Some are definitely from the Los Caños and Imbert Formations and the older rocks (serpentinites and gabbros), whereas others have never been seen in place (for example, the recrystallized and veined limestones). The few paleontologic dates the blocks have yielded indicate a range in clasts from the Late Cretaceous(?) to the middle Eocene, excluding one doubtful Miocene age (FN-245-D) which I interpret as a fragment from a block which slumped over the olistostrome in late Miocene time forming Pico Isabel de Torres (*see* Fig. 4).

The olistolith (exotic block) dates are listed in Table 1. Except for sample *Bowin #202A* which is a gray limestone, all others listed are tuffaceous limestones. Textures of these rocks indicate that they were calcareous muds and calcarenites. None have grains coarser than 1 mm. In addition to organic and calcite fragments and irregular patches of calcite, volcanic rock, plagioclase, quartz, and pyroxene fragments are present in all samples.

Blocks or Basement?

Some rocks crop out continuously over distances of 1.5 km in the area of the San Marcos Formation. In all cases these rocks are serpentinites with associated rodingites or volcanic rocks, which in thin section look like Los Caños rocks.

These outcrops could be large blocks within the olistostrome. However, they cannot be seen completely in three dimensions so that they may also be bedrock poking through the olistostrome, representing basement irregularities that were

TABLE 1. PALEONTOLOGIC AGES OF SAN MARCOS OLISTOLITHS*

Sample No.	Age	Comments
Bowin 202-A (P. Bronniman)	Upper Cretaceous, Maestrichtian, or Eocene with reworked Upper Cretaceous	*Discocyclina* sp. (fragment?) *Rotalid* (large form nondescript) *Vaughanina cubensis* (fragments) *Orbitocyclina minima* *Calcisphaerula innominata* *Sulcoperculina* cf. *vermunti* *Kathina* sp.
FN 63CC (P. J. Bermúdez)	Probably Paleocene	*Actinosiphon barbadensis* (Vaughan) *Operculinoides* sp. (?) fragment *Ecology:* marine, shallow water (reef or near reef)
FN 245-D (W. S. Cole)	Possibly lower Miocene	Mostly planktonics—one or two larger Foraminifera suggestive of *Miogypsina*
FN 181-C (W. S. Cole)	Definitely Eocene	discocyclinid fragments *Eoconuloides* (?) . . . if so suggestive of middle Eocene
FN 121 (W. S. Cole)	Eocene	without question Eocene; could be middle or upper; would have to have more material to be certain
FN 111 (W. S. Cole)	Eocene	See comments under FN 121

* The name of the investigator who examined the fossils follows in parenthesis after the sample number. The comments and age determinations are those of the investigator listed.

never covered by the olistostrome or that have been stripped of their cover. Large areas around these outcrops are uncontaminated by blocks of other rock types, a feature not common to the olistostrome elsewhere, and they occur mainly in the east where the olistostrome appears to thin out.

On the other hand, there is no independent evidence that the base of the olistostrome has been discovered and there are no outcrops of the Imbert Formation (also older than the San Marcos) within the olistostrome which can be anything other than exotic blocks. On this basis, then, the large outcrops are considered to be blocks, rather than basement in place.

I can make no ordered pattern out of the distribution of the various types of blocks as Hsu (1968) suggests should be the case for melange units.

Origin of the Olistostrome

Any hypothesis of origin of the San Marcos olistostrome involves considerable speculation. The nonstratified and unconsolidated nature of the matrix, relative ages of the blocks and matrix, and lack of evidence for tectonic remobilization all indicate that some mechanism of subaqueous gravity sliding was most probably involved.

There is also no evidence for mud or salt diapirs or mud volcanism. Mud volcanism such as that described by Higgins and Saunders (1967) in Trinidad had been considered a likely alternative. However, in the Puerto Plata area there is no typical fault pattern usually associated with doming; the blocks in the San Marcos are larger than any described products of mud volcanism or mud diapirs; and, significantly, there are exotic blocks which are older than the matrix while those in mud volcanoes are invariably contemporaneous or younger than the matrix. Finally, folding usually associated with mud volcanoes is not present in the Puerto Plata area. All major rock deformation here can be explained by vertical tectonics; that is, tilting. normal faulting, slumping, and gravity sliding—not typical of mud volcano tectonics.

The few dates from the San Marcos matrix are tentatively accepted as indicating the time of deposition of the original unit which later became the matrix of the olistostrome. These Paleocene-Eocene dates are the time equivalent of Imbert tuffaceous deposition. Most likely the matrix represents a former tuffaceous unit within the Imbert Formation which has gone through a post-sliding alteration sequence of volcanic ash plus some montmorillonite and chlorite changing to kaolinite, a process which may still be continuing. That kaolinite can be formed from montmorillonite at low temperatures is suggested by the field work of Altschuler and others (1963) and the experimental work of Poncelet and Brindley (1967). Mackenzie and Mitchell (1966) point out that kaolinite or halloysite along with sesquioxides are generally the final products of rock alteration under humid tropical conditions irrespective of starting rock types. Hay (1959, 1960) reports that late Pleistocene andesitic subaerial ash deposits on St. Vincent, British West Indies, have weathered to halloysite, allophane, hydrated ferric oxide, and iddingsite.

One important feature of the San Marcos olistostrome was that it must have retained a large volume of pure water as other sediments rapidly buried it (Bredehoeft and Hanshaw, 1968), or it acquired excess water from a source layer at depth, perhaps its own montmorillonite (Hanshaw and Bredehoeft, 1968). That high water pressures may be maintained by unconsolidated sediments to burial depths of 2,100 to 5,000 m (more than twice the load necessary on top of the postulated tuffaceous layer in this case) eventually resulting in slumping has recently been documented by the work of Dickey and others (1968). These not unlikely conditions, in addition to probable random orientation of clay-sized particles (Liebling and Kerr, 1965) and location in an area of known tectonic activity set the stage for instability and mass flow, awaiting only a triggering mechanism such as earthquake shock or a series of shocks.

Shock-induced marine slumps in historical times are not uncommon; a good review is provided by Andresen and Bjerrum (1967) and by Morgenstern (1967), who lists several submarine slumps occurring on slopes of 3 to 20° following earthquake shocks of magnitude 6.75 to 8.5. Both figures are reasonable for the Puerto

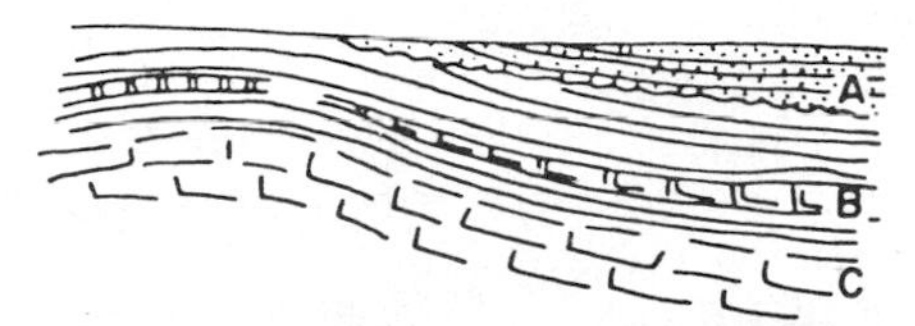

1. MIDDLE EOCENE TIME, PRIOR TO EARTHQUAKE SHOCK

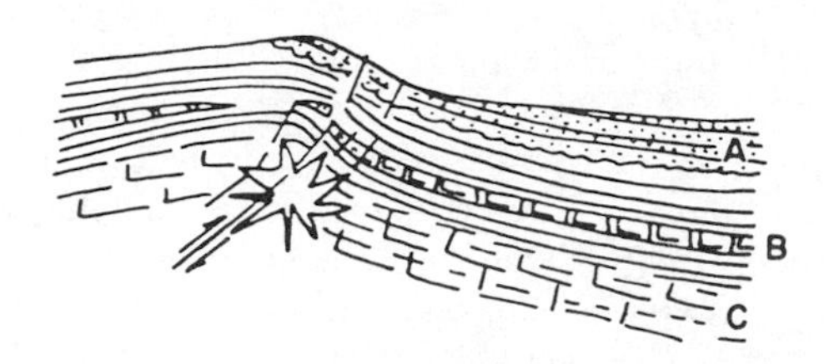

2. INSTANT OF EARTHQUAKE SHOCK.

3. POST EARTHQUAKE SHOCK SLUMPING.

EXPLANATION:

A. OVERBURDEN

B. TUFFACEOUS (?) UNIT IN IMBERT FM.

C. LOWER IMBERT FORMATION AND OLDER ROCKS

(THRUST FAULT IN 2,3 IS HYPOTHETICAL)

Figure 6. Sequence of events (from top to bottom) leading to production of olistostrome and olistoliths (adapted *from* Marchetti, 1957).

Plata area. It should be noted, however, that not all marine sediments on sloping surfaces are unstable and prone to liquid flow by shocking, a point made by Dill (1969) and reviewed by Morgenstern (1967).

The clay matrix of San Marcos Formation liquefies under uncontrolled laboratory conditions. Several of the matrix samples were placed in mason jars and varying amounts of water were added up to the saturation point of the sample. Liquefaction of the sample was caused by shaking or jarring. The degree of liquification brought about by this means increased with increasing water content.

The exotic blocks in the olistostrome are older than, contemporaneous with, and younger than their surrounding matrix; the youngest known are of middle Eocene age. This is taken as the probable time of shocking and consequent sliding, slipping taking place on a stratigraphically lower Paleocene–lower Eocene horizon. Flow was turbulent, presumably destroying the original stratification. Blocks younger than the matrix were incorporated as overlying layers ruptured and were swept into the moving mass. Older blocks from the Imbert Formation, the Los Caños Formation, and the peridotites were torn loose from a submarine slope where they had formed promontories in the sea floor (Fig. 6).

The San Marcos need not have traveled far and probably is near its original source. Where that is, is conjectural, although most likely it is only a few kilometers to the west where many of the rock types which form exotic blocks are in place.

Olistostromes do not necessarily indicate major orogenesis in progress. More likely they indicate shock-induced failure and catastrophic slumping of unstable sedimentary accumulations or perhaps sea-floor deposits scraped up at plate boundaries. However, many olistostromes are associated with tectonic movements, and the San Marcos deposit is no exception. The emplacement of the San Marcos probably was triggered by movement on the Camu fault, a major northwest-trending high-angle fault running across the entire mapped area, a fault which has been active since at least the middle Eocene. This fault strikes parallel to the south slope of the Puerto Rico trench.

ACKNOWLEDGMENTS

The writer gratefully acknowledges financial support from the National Science Foundation, Grant No. GA-14217, and the Woodrow Wilson Fellowship Foundation while at Princeton University; and current research support fron NSF Grant GA-13531 at the University of Miami Rosenstiel School of Marine and Atmospheric Science.

Helpful criticism of the manuscript was given by F. B. Van Houten and A. G. Fischer.

REFERENCES CITED

Altschuler, A. Z., Dwornik, E. J., and Kramer, H., 1963, Transformation of montmorillonite to kaolinite during weathering: Science, v. 141, p. 148–152.

Andresen, A., and Bjerrum, L., 1967, Slides in subaqueous slopes in loose sand and silt, *in* Richards, A. F., ed., Marine geotechnique: Urbana, Univ. Illinois Press, p. 221–239. p. 221–239.

Audley-Charles, M. G., 1965, A Miocene gravity slide from eastern Timor: Geol. Mag., v. 102, no. 3, p. 267–276.

Barr, K. W., and Saunders, J. B., 1968, An outline of the geology of Trinidad: Trans. Fourth Caribbean Geol. Conference, p. 1–10.
Bell, J. S., 1967, Geology of the Camatagua area, Estado Aragua, Venezuela [Ph.D. thesis]: Princeton, Department of Geology, Princeton Univ., 282 p.
Bracey, D. R., and Vogt, P. R., 1970, Plate tectonics in the Hispaniola area: Geol. Soc. America Bull., v. 81, p. 2855–2860.
Bredehoeft, J. D., and Hanshaw, B. B., 1968, On the maintenance of anomalous fluid pressures: I. Thick sedimentary sequences: Geol. Soc. America Bull., v. 79, p. 1097–1106.
Bunce, E. T., and Fahlquist, D. A., 1962, Geophysical investigation of the Puerto Rico trench and outer ridge: Jour. Geophys. Research, v. 67, p. 3955–3972.
Bushman, J. R., 1958, Geology of the Barquisimeto area, Venezuela [Ph.D. thesis]: Princeton, Department of Geology, Princeton Univ., 169 p.
Chase, R. L., and Hersey, J. B., 1968, Geology of the north slope of the Puerto Rico trench: Deep-Sea Research, v. 15, p. 297–317.
Dickey, P. A., Shriram, C. R., and Paine, W. R., 1968, Abnormal pressures in deep wells of southwestern Louisiana: Science, v. 160, p. 608–615.
Dill, R. F., 1969, Earthquake effects on fill of Scripps Submarine Canyon: Geol. Soc. America Bull., v. 80, p. 321–328.
Flores, G., 1955, Definition of olistostrome: Fourth World Petrol. Congress, Sec. 1, p. 122.
Glover, L., III, 1967, Geology of the Coamo area, Puerto Rico; with comments on Greater Antillean volcanic island arc-trench phenomena [Ph.D. thesis]: Princeton, Princeton Univ., 363 p.
Hanshaw, B. B., and Bredehoeft, J. D., 1968, On the maintenance of anomalous fluid pressures: II. Source layer at depth: Geol. Soc. America Bull., v. 79, p. 1107–1122.
Hay, R. L., 1959, Origin and weathering of Late Pleistocene ash deposits on St. Vincent, B.W.I.: Jour. Geology, v. 67, p. 65–87.
—— 1960, Rate of clay formation and mineral alteration in a 4000-year-old volcanic ash soil on St. Vincent, B.W.I.: Am. Jour. Sci., v. 258, p. 354–368.
Higgins, G. E., and Saunders, J. B., 1967, Report on 1964 Chatham Mud Island, Erin Bay, Trinidad, West Indies: Am. Assoc. Petroleum Geologists Bull., v. 51, no. 1, p. 55–64.
Horne, G. S., 1969, Early Ordovician chaotic deposits in the central volcanic belt of northeastern Newfoundland: Geol. Soc. America Bull., v. 80, p. 2451–2464.
Hsu, K. J., 1968, Principles of melanges and their bearing on the Franciscan-Knoxville Paradox: Geol. Soc. America Bull., v. 79, p. 1063–1074.
Kear, D., and Waterhouse, B. C., 1967, Onerahi chaos-breccia of Northland: New Zealand Jour. Geology and Geophysics, v. 10, no. 3, p. 629–646.
Kozary, M. T., 1968. Ultramafic rocks in thrust zones of northwestern Oriente Province, Cuba: Am. Assoc. Petroleum Geologists Bull., v. 52, no. 12, p. 2298–2317.
Kugler, H. G., 1953, Jurassic to Recent sedimentary environments in Trinidad: Assoc. Suisse des Geol. et Ing. du Petrole Bull., v. 20, no. 59, p. 27–60.
Kugler, H. G., and Saunders, J. B., 1967, On Tertiary turbidity-flow sediments in Trinidad, B.W.I.: Asoc. Venezolana Geología, Minería y Petróleo Bol. Inf., v. 10, no. 9, p. 241–259.
Liebling, R. S., and Kerr, P. F., 1965, Observations on quick clay: Geol. Soc. America Bull., v. 76, p. 853–878.
Mackenzie, R. C., and Mitchell, B. D., 1966, Clay mineralogy: Earth-Sci. Rev., v. 2, p. 47–91.
Marchetti, M. P., 1957, The occurrence of slide and flowage materials (olistostromes) in the Tertiary series of Sicily: Internat. Geol. Cong., 20th, Mexico 1956, Comptes Rendus sec. 5, p. 209–225.
Molnar, P., and Sykes, L. R., 1969, Tectonics of the Caribbean Middle America regions from focal mechanisms and seismicity: Geol. Soc. America Bull., v. 80, p. 1639–1684.
Monroe, W. H., 1968, The age of the Puerto Rico trench: Geol. Soc. America Bull., v. 79, p. 487–494.
Moore, T. C., Jr., van Andel, Tj. H., Blow, W. H., and Heath, G. R., 1970, Large submarine slide off northeastern Continental Margin of Brazil: Am. Assoc. Petroleum Geologists Bull., v. 54, p. 125–128.
Morgenstern, N. R., 1967, Submarine slumping and the initiation of turbidity currents, *in*

Richards, A. F., ed., Marine geotechnique: Urbana, Univ. Illinois Press, p. 189–220.
Nagle, F., 1966, Geology of the Puerto Plata area, Dominican Republic [Ph.D. thesis]: Princeton, Princeton Univ., 171 p.
—— 1970, Serpentinites and metamorphic rocks at the Caribbean boundaries: Geol. Soc. America, Abs. with Programs (Ann. Mtg.), v. 2, no. 7, p. 633.
—— 1971, Geology of the Puerto Plata area, Dominican Republic, relative to the Puerto Rico Trench: Fifth Caribbean Geol. Conf. Trans., Geol. Bull. No. 5, Queens College Press, p. 79–84.
Poncelet, G. M., and Brindley, G. W., 1967, Experimental formation of kaolinite from montmorillonite at low temperatures: Am. Mineralogist, v. 52, p. 1161–1173.
Renz, O., Lakeman, R., and van der Meulen, E., 1955, Submarine sliding in western Venezuela: Am. Assoc. Petroleum Geologists Bull., v. 39, p. 2053–2067.
Saunders, J. B., 1968, Field trip guide, Barbados: Fourth Caribbean Geol. Conf. Trans., p. 443–450.
Senn, A., 1940, Paleogene of Barbados and its bearing on history and structure of Antillean-Caribbean region: Amer. Assoc. Petroleum Geologists Bull., v. 24, no. 9, p. 1548–1610.
Sykes, L. R., and Ewing, M., 1965, The seismicity of the Caribbean region: Jour. Geophys. Research, v. 70, no. 20, p. 5065–5074.
Thayer, T. P., and Guild, P. W., 1947, Thrust faults and related structures in eastern Cuba: Am. Geophys. Union Trans., v. 28, no. 6, p. 919–930.
Wise, D., and Bird, J. M., 1964, International Field Institute, Italy, 1964: Geotimes, v. 9, no. 5, p. 12–15.

MANUSCRIPT RECEIVED BY THE SOCIETY MARCH 29, 1971

CONTRIBUTION NO. 1471 FROM THE UNIVERSITY OF MIAMI ROSENSTIEL SCHOOL OF MARINE AND ATMOSPHERIC SCIENCE, MIAMI, FLORIDA 33149

10

Reprinted from pages 447-450 and 454-466 of *Geol. Soc. London Jour.*
133:447-466 (1977)

The Moni Mélange, Cyprus: an olistostrome formed at a destructive plate margin

A. H. F. ROBERTSON

SUMMARY

Detailed mapping has shown that the Moni Mélange of southern Cyprus consists of two separate allochthonous components: a variety of Triassic to Cretaceous sedimentary and volcanic rocks representative of an original continental margin sequence, and substantial sheets of serpentinite derived from Upper Cretaceous Troodos-type oceanic crust. All these allochthonous rocks were emplaced by sliding into an *in situ* host matrix of Upper Cretaceous bentonitic clays and radiolarian siltstones of deep water hemipelagic facies. The postulated geotectonic setting involves a phase of late Cretaceous northward subduction of Troodos-type oceanic crust beneath a continental margin now located in southern Turkey. Subduction culminated in the Maastrichtian in a major collision of oceanic crust with the adjacent continental margin. This triggered off major oceanward gravity sliding of continental margin rocks. The collision was also associated with deep tectonic slicing of Troodos oceanic crust with the result that sheets of partly serpentinized ultramafic rocks were emplaced by sliding into the host sediments of the mélange from the opposite direction to the other olistoliths. Soon afterwards there was a brief period of submarine erosion, then local tectonic movements ceased, suggesting an end to subduction in the area.

1. Introduction

THE SIGNIFICANCE OF OLISTOSTROMES and tectonic *mélanges* associated with ophiolites in the Alpine-Mediterranean Tethys is now widely appreciated as a key to geotectonic evolution. Despite this, there have been few detailed field descriptions of olistostromes or other mélanges. Consequently, the position of the Moni Mélange to the south of the Troodos Massif must be of considerable significance. Here it is shown that the Moni Mélange consists of several distinct mappable allochthonous components, principally olistoliths derived from a Triassic to Cretaceous continental margin and serpentinite sheets which originated as Troodos-type oceanic basement. All these allochthonous rocks were rapidly emplaced in the Maastrichtian by gravity sliding into an Upper Cretaceous sequence of hemipelagic clays and siltstones. The field relations are also important in reconstructions of the geotectonic setting of the Troodos Massif in the late Cretaceous.

2. Mélange nomenclature

From a widespread debate about the definitions of mélanges and olistostromes (e.g. Hsu 1968, 1974, Abbate *et al.* 1970, Hoedemaeker 1973, Dimitrijevic & Dimitrijevic 1974), a consensus has now emerged that mélanges are chaotically deformed bodies characterized by tectonically mixed rocks set in a pervasively sheared fine grained matrix which may be ophiolitic. In contrast olistostromes, although also chaotically deformed, are typically stratigraphically ordered bodies which were produced by gravity sliding in a poorly consolidated state. The admixed olistoliths generally occur in a matrix which may range from structureless sediments to graded turbidites or even boulder beds.

Complicating factors arise when tectonic mélanges and olistostromes are gradational, or when a later tectonic fabric has been imprinted on an original olistostrome. Likewise, the Moni Mélange fails to fit the straightforward definitions of either a tectonic mélange or an olistostrome. As will be discussed, although the olistoliths were emplaced by gravity sliding, the matrix of the Moni Mélange lacks the internal stratigraphical organization which is characteristic of a typical olistostrome.

3. Previous interpretations

The entire igneous and sedimentary cover of the Troodos Massif, termed the 'Trypa Group', was once regarded as autochthonous (Henson *et al.* 1949). The south Troodos mélange was then interpreted as an erosional remnant of the 'Trypa Group' which disintegrated during contemporaneous volcanism to form a chaotic mixture of blocks in the underlying bentonitic clays.

Subsequently, on the basis of detailed mapping, Pantazis (1967) established the allochthonous nature of many of the south Troodos rocks to which he gave the name Moni Mélange after a village in the area. Pantazis interpreted the mélange as a chaotic mixture of exotic rocks embedded in bentonitic clays of his 'Moni Formation'. He also described other rocks, including various Triassic limestones, quartzose sandstones and siltstones which he believed to rest *in situ* on the Troodos volcanic basement. If true, this would hold important implications for the ocean floor interpretation of the Troodos Massif.

In general, the Moni Mélange as now defined crops out along the south margin of the Troodos Massif from Cape Dolos in the east to near Yerasa in the west, a distance of about 30 km (Fig. 1). Maximum total thickness of the mélange, about 200 m, is reached in the Pendakomo to Pareklisha areas. When traced westwards into the foothills of the Troodos Mountains in the Limassol Forest, the mélange thins progressively to around 75–100 m, as in the Akrounda and Phinikaria areas (Fig. 1); westwards near Apsiou, the mélange disappears entirely, leaving an upward, unbroken sedimentary sequence above the Troodos lavas.

4. The matrix of the Mélange

The matrix of the Moni Mélange is essentially a para-autochthonous sequence of non-calcareous clays and siltstones. These sediments form part of the Kannaviou Formation, an *in situ* sequence of Upper Cretaceous volcanogenic clays, siltstones and sandstones which crop out around the circumference of the Troodos Massif (Lapierre 1968*a*, 1975, Robertson & Hudson 1974).

The Kannaviou sediments of the Moni Mélange, which are non-calcareous, have been dated as Campanian using Radiolaria, especially *Dictyomitra multicostata* which Mantis (1970) took as a fossil diagnostic of the Campanian. However, identical dictyomitrids have also been found in the upper levels of the sequence of

Paphos District (Fig. 1) in which Maastrichtian planktonic foraminifera are also found. Thus it is probable that all the Kannaviou sediments, including those of the Moni Mélange, range from Campanian well into the Maastrichtian.

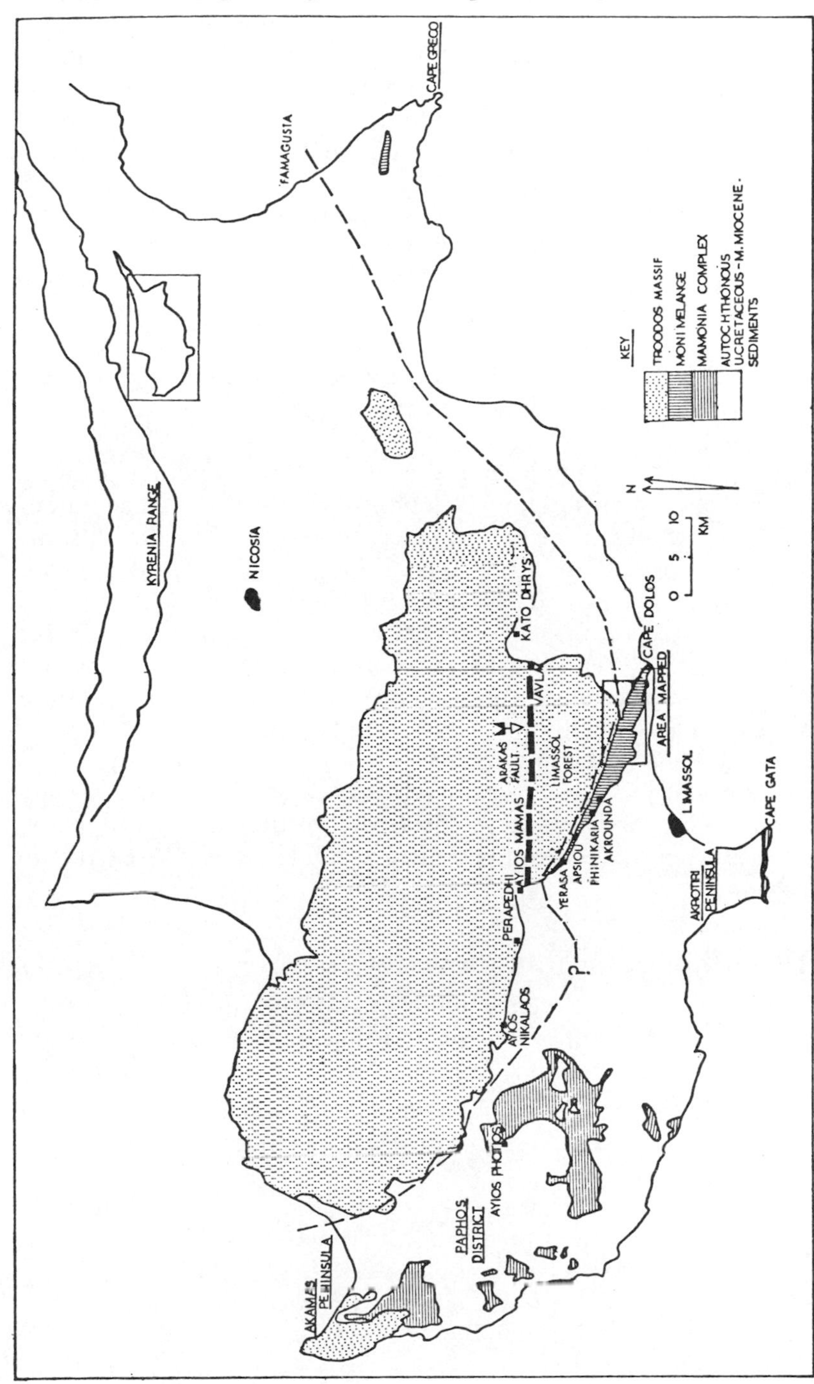

Fig. 1. Outline geological map of S Cyprus to show the locations of allochthonous rocks, particularly the Moni Mélange and the Mamonia Complex. The broken line marks the approximate northern limit of the emplacement of allochthonous rocks in the Maastrichtian.

Where the Kannaviou sediments have locally escaped deformation in the mélange zone (Fig. 2), the original sequence can be established (Fig. 3). Where the mélange is thickest, as in the Monagroulli area, the base of the sequence comprises up to 30 m of bentonitic clays. Then there is an incoming of up to 100 m of thinly bedded white or pale grey radiolarites and radiolarian siltstones, generally 2–4 cm thick. These are well exposed south of the Limassol to Nicosia road (Fig. 2). Above, there is a return to up to 30 m of bentonitic clays which pass upwards over several centimetres into pale grey Maastrichtian chalks of the Lower Lefkara Formation (Fig. 3).

When traced westwards into the Limassol Forest, the Kannaviou sediments thin drastically to an approximately 10 m sequence of bentonitic clays which grade progressively upwards over about 25 m into grey marls of the Lower Lefkara Formation.

North of the Moni Mélange, as at Vavla and Asgata (Fig. 3), deposits of up to 10 m are found in hollows in the surface of the Troodos pillow lavas. These clays are overlain by Maastrichtian marls and chalks without any sedimentary break or deformation, an observation highly relevant to the direction of emplacement of the Moni Mélange.

Overall, it is clear that the Kannaviou deposits of the Moni Mélange are relatively deep water pelagic sediments. The restriction of the Kannaviou sediments immediately north of the mélange zone to non-calcareous bentonitic clays ponded in hollows in the surface of the Troodos lavas is an indication that this area was then topographically relatively elevated. To the south, in the main mélange zone, a much greater thickness of Kannaviou sediments accumulated in deeper water. Deposition of these sediments beneath the carbonate compensation depth is likely in view of their non-calcareous nature which contrasts with their contemporaneous counterparts in Paphos district. Unfortunately, the depth of the carbonate compensation depth is unknown in the Troodos ocean which may have been relatively small and possibly isolated from wider oceanic areas. Interestingly, the gradual transition of clays or marls to the northwest of the mélange zone may imply that this area was then a topographical high located close to the carbonate compensation depth.

5. Allochthonous components of the Mélange

Detailed mapping reveals that the allochthonous rocks of the Moni Mélange can be grouped into several categories on the basis of lithology, size, distribution and tectonic state:

(i) Olistoliths of clastic, calcareous and siliceous sedimentary rocks and basic volcanics—rocks all known from the Mamonia Complex of Paphos District in SW Cyprus.

(ii) Olistoliths of quartzose sandstone, limestone, and siltstone which are generally larger and less deformed than the other olistoliths; these rocks are unknown elsewhere in Cyprus.

(iii) Localized sheets of serpentinite which are unrelated to the olistoliths.

(iv) A distinctive 'small-olistolith' mélange discontinuously developed above the main mélange.

[*Editor's Note:* Figure 2 has been omitted.]

(A) OLISTOLITHS OF MAMONIA ROCKS

Olistoliths of almost all the rocks found in the Mamonia Complex of SW Cyprus have been identified in the Moni Mélange. Although the Mamonia Complex is itself severely disrupted, being intermediate between intact nappes and mélange, Lapierre (1968*b*, 1975) has reconstructed a coherent stratigraphical succession: in generalized upward stratigraphical sequence, Middle and Upper Triassic flysch (*grés à végeteaux*), Upper Triassic pelagic limestone (Halobia limestones) locally interbedded with alkaline volcanics (Mamonia lavas), followed by sequences of mid-Jurassic to Lower Cretaceous radiolarian mudstones and cherts which are locally interbedded with quartzose sandstones (Akamas sandstones). Other lithologies found in both the Mamonia Complex and the Moni Mélange which do not fit this basic succession include Triassic reef limestones (Petra tou Romiou Formation) and various metamorphic rocks.

The olistoliths of Mamonia rocks in the mélange range up to 1·5 km in diameter; the average is 1–10 m. In general the most resistant lithologies, for example the Akamas sandstones, Petra tou Romiou limestones and the Mamonia volcanics form numerous large olistoliths, whereas the more readily disrupted radiolarian cherts and thinly bedded Triassic sandstones are normally found as small olistoliths scattered chaotically through the matrix of the mélange.

Excellent exposures of olistoliths of Mamonia rocks crop out north of the Limassol to Nicosia road, south of Pareklisha, south of Ayios Thykhonas (Fig. 2), and near Akrounda and Phinikaria in the Limassol Forest.

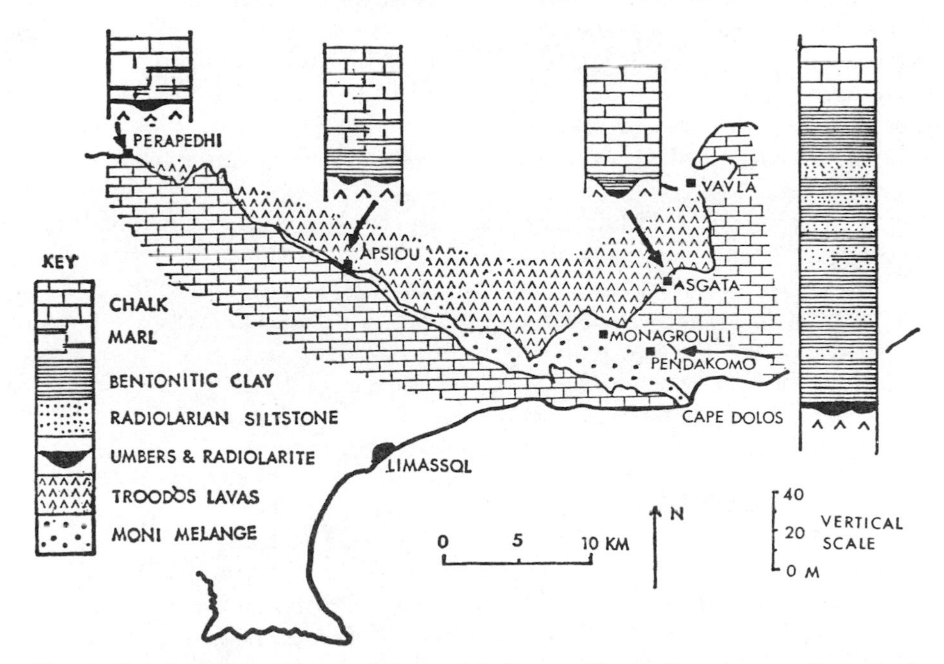

FIG. 3. Stratigraphic columns of the autochthonous Upper Cretaceous sediments of the Kannaviou Formation along the south margin of the Troodos Massif. These sediments form the matrix of the Moni Mélange.

Significantly, Pantazis (op. cit.) interpreted a substantial body of internally brecciated Petra tou Romiou limestone and Mamonia lavas north of the Limassol to Nicosia road (near milepost 42) as an inlier of Triassic rocks. Here this body is reinterpreted as one of the largest olistoliths in the mélange.

(B) OLISTOLITHS OF NON-MAMONIA ROCKS

A feature of the Moni Mélange is the occurrence of numerous olistoliths of uncemented quartzose sandstones, siltstones and limestones which are not known in the Mamonia Complex of SW Cyprus. The significance of these olistoliths is increased by their unusually large size, relatively undeformed state and their preferential distribution along the northern margin of the mélange close to the Troodos lavas, factors which earlier suggested that these rocks were *in situ.*

(i) Olistoliths of Pareklisha Sandstone

Substantial bodies of pale yellow, uncemented, quartzose sandstone within the mélange zone were named the *Pareklisha Sandstone* by Pantazis (1967). At the type locality south of Pareklisha village (Fig. 2) the rock is a friable yellow to orange orthoquartzite, generally massive except for occasional stringers of quartz pebbles. Numerous spherical calcite- or iron oxide-cemented concretions reach 1·7 m in diameter. Elsewhere, the Pareklisha sandstones are locally cross-bedded, occasionally with interbeds of kaolinite-rich or bituminous siltstone.

South of Pareklisha the sandstones are not deformed and rest directly on the Troodos pillow lavas with an apparently undisturbed sedimentary contact. This led Pantazis (op. cit.) to believe that these rocks were *in situ.* However, close examination of the type locality reveals thin intercalations of red-purple scaly clay along the lava-sandstone contact. Also, not far to the west (Fig. 2) another large mass of Pareklisha Sandstone forming a low hill again rests directly on the Troodos lavas to the north; to the south this sandstone is underlain by typical Moni Mélange.

Petrographically, the Pareklisha sandstones are orthoquartzites, distinguished by a strongly bimodal size distribution of quartz grains. The larger grains which range from 0·4 to 2·1 mm in diameter (average 0·5 mm) are rounded to subangular with a matrix of much smaller interlocking quartz grains on average 0·1 mm in diameter. There are also occasional grains of alkali feldspar, epidote, hornblende, muscovite, zircon and tourmaline, an assemblage indicative of either a predominately low grade metamorphic or granitic provenance. Several samples of the sandstone also contain shell fragments and occasional echinoderm plates.

(ii) Olistoliths of Monagroulli Siltstone

A variety of siltstones, also some impure limestones and fine grained sandstones, crop out in the area south of Monagroulli village. This distinctive assemblage is here collectively termed the *Monagroulli Siltstone.* A substantial body of these sediments forms a low cliff 1 km south of Monagroulli (Fig. 2) in which the strata are inclined at up to 25° to the south. This was Pantazis' (op. cit.) type locality of his 'Tuffaceous Limestone Member'. This he regarded as *in situ*, the upper part of the 'Moni Clay' sequence which was thought to rest on the Pareklisha Sandstone.

Here mapping has established that all the Monagroulli siltstones are olistoliths of the Moni Mélange.

Pantazis' type locality lies within a giant olistolith of siltstone, over 1 km in diameter, in which an approximately 30–40 m sequence was pieced together. Near the base of the sequence exposed in this olistolith, the siltstones are interbedded with impure calcilutites and calcarenites. These are redeposited bioclastic limestones which contain rounded shell fragments, occasional benthonic foraminifera, Radiolaria, echinoderm plates, possible algal debris and intraclasts as well as scattered quartz grains. Upwards there is an incoming of increasingly thinly bedded siltstone and fine grained sandstone which show grading, channelling and intraformational slumping. Significantly, the size distribution of the quartz grains in these sandstones is closely comparable to the Pareklisha sandstones.

Other siltstone olistoliths which crop out south of Monagroulli (Fig. 2) include a 17 m sequence of rhythmically alternating finely laminated siltstones and more massive siltstones which range from 0·3–0·5 m in thickness. These massive siltstones show ripple cross-lamination, small scale channelling and flaser bedding, features indicative of deposition in a current-swept environment.

Several kilometres east of Monagroulli River a low hill is formed of the largest single olistolith of Monagroulli Siltstone, over 1 km in diameter. Frequent drastic changes of dip and strike within this olistolith indicate extreme internal disruption. When the Monagroulli siltstones are traced along their general direction of strike, the huge olistoliths disappear leaving only small olistoliths scattered through the mélange.

(C) THE SERPENTINITE SHEETS

A striking feature of the Moni Mélange is the occurrence of substantial sheets of serpentinite which crop out only in the areas south and southwest of Pareklisha (Fig. 2). Although previously regarded as olistoliths (Pantazis 1967), the serpentinites can be traced laterally as sheets up to 5 km long and 40 m thick. The upper surfaces of these sheets always exhibit a blocky clinker-like appearance reminiscent of the upper surface of a glacier. Significantly, there are no signs of any admixing of serpentinite with the adjacent mélange clays.

(D) THE SMALL-OLISTOLITH MÉLANGE

In some areas the typical Moni Mélange is overlain by deposits of a strikingly different form of mélange in which small olistoliths, mostly less than 10 cm in diameter, of all the other allochthonous rocks, except serpentinite, are strewn through a matrix of grey Kannaviou clay and red-brown silt containing kaolinite. This 'small-olistolith' mélange reaches a maximum thickness of around 20 m in the main mélange zone between Pendakomo and Monagroulli, but disappears completely when traced towards the Limassol Forest.

6. Provenance of the olistoliths

The majority of the olistoliths in the Moni Mélange are formed of rocks also known from the Mamonia Complex and the Antalya nappes of Turkey (e.g. Brunn *et al.*

1971, Juteau 1975); all form part of an original Triassic to Lower Cretaceous continental margin sequence.

The provenance of both the Pareklisha sandstones and the Monagroulli siltstones is less obvious as these rocks have not been identified in the Mamonia Complex or in Turkey. The Pareklisha sandstones are petrographically similar to the Akamas sandstones of the Mamonia Complex in which both facies possess a similar bimodal size distribution of quartz grains. In contrast, the Akamas sandstones are extremely tough due to silica cementation, whereas the Pareklisha sandstones are friable without cement. The Akamas sandstones in the Mamonia Complex of Paphos District first appear as thin interbeds of Upper Jurassic red radiolarian mudstones and radiolarites; then they thicken upwards in the Lower Cretaceous to form beds of massive structureless orthoquartzite up to 10 m thick. These sandstones show evidence of emplacement by mass flow into a relatively deep water sequence of hemi-pelagic sediments (Robertson & Woodcock, unpublished work). In contrast the Pareklisha sandstones reach greater thicknesses of up to 50 m in a single olistolith; their locally crossbedded and pebbly nature, with other sedimentary features, is suggestive of marine deposition on a stable continental shelf, probably in a deltaic environment.

The various Monagroulli Siltstone facies including the thinly bedded and massive siltstones, the impure sandstones and the redeposited bioclastic limestones, probably all also form part of a relatively shallow water continental shelf association. Although these sediments have not been accurately dated in Cyprus, comparable sediments have been discovered in the Baër-Bassit region of northern Syria, where they are of Lower to Mid-Cretaceous age (Delaune-Mayère, pers. comm. 1976). Indeed, a Lower to Mid-Cretaceous age for the Monagroulli siltstones and the Pareklisha sandstones is consistent with their greater degree of induration relative to the matrix sediments which are Upper Cretaceous.

In summary, taking account of all the available data, the Pareklisha sandstones and the Monagroulli siltstones are interpreted as relatively proximal facies of the stratigraphically higher parts of a continental margin sequence. Prior to emplacement in the Maastrichtian, the sedimentary-volcanic sequences of the Mamonia Complex and the olistoliths of the same rocks in the Moni Mélange probably represented the stratigraphically and structurally deeper levels of the same continental margin. Significantly, deposition of the Pareklisha sandstones and Monagroulli siltstones during the early to Mid-Cretaceous would fill the only remaining major gap in the Triassic to Recent stratigraphy of Cyprus.

7. Emplacement of the Mélange

(a) Directions of emplacement

The orientation of bedding in the olistoliths of the Moni Mélange has been measured in an attempt to infer the directions of emplacement. This is complicated by Miocene folding in the area around the southern margin of the Limassol Forest. To avoid these difficulties, the structural data were obtained from the eastern area of the mélange zone where the sedimentary cover has remained intact until the Pleistocene or later.

In the area mapped, most long axes of the olistoliths are orientated to the present NNW–SSE; bedding, where visible, is generally steeply inclined to the present SSW (Fig. 4A, B). In contrast, a group of the largest olistoliths of Pareklisha sandstones and Monagroulli siltstones which are situated close to the Troodos lavas near Pareklisha Village are orientated nearer to the present E–W. The strata in these olistoliths are generally more gently inclined than the strata within the Mamonia olistoliths (Fig. 4A, B).

Studies of the orientation of olistoliths in Italian olistostromes have demonstrated that the direction of motion lies in the opposite direction to the dip within most olistoliths. As in pebbles in a stream, the olistoliths are imbricated with the dip in the 'upstream' direction of flow (Görler & Reutter 1968, Görler 1975). Accordingly, the direction of movement of olistoliths of Mamonia rocks of the Moni Mélange is believed to have been from the present SW, a conclusion supported by the regional geological setting (see below). The presently more E–W strike of many of the large olistoliths of Pareklisha sandstone and Monagoulli siltstone is complicated by their frequent proximity to the Troodos volcanic basement.

In sharp contrast to all the olistoliths, the substantial sheets of serpentinite are internally displaced by numerous, mostly north-dipping, small thrusts and slickensides. This situation is well exposed in a small quarry 200 m beyond where the road from the coast forks to Pareklisha and Pyrgos (Fig. 2). At this locality the

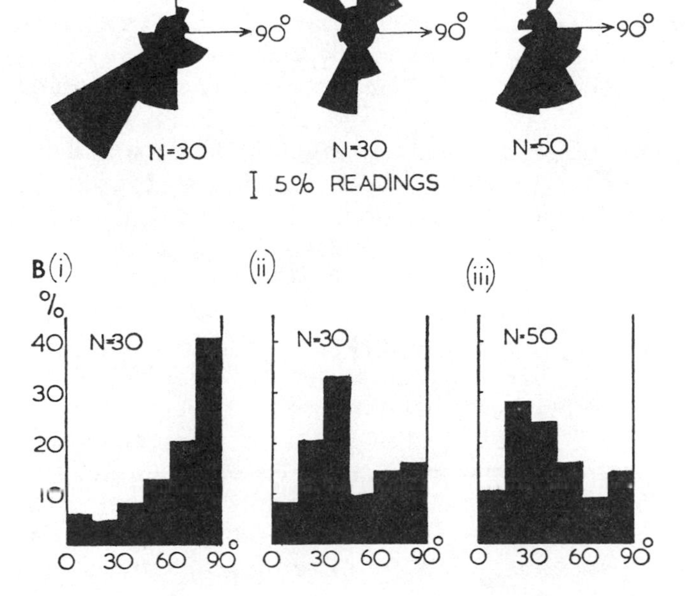

FIG. 4

Orientations of olistoliths in the Moni Mélange. Directions of maximum dip (Fig. 4A) and angles of maximum dip (Fig. 4B) of the strata within:

(i) olistoliths of Mamonia rocks:

(ii) para-autochthonous Kannaviou sediments:

(iii) olistoliths of Pareklisha Sandstone and Monagroulli Siltstone.

serpentinites can be seen in thrust contact with Kannaviou clays which are locally scaly. Close to the contact these clays have been deformed into numerous small south-facing asymmetrical folds which are indicative of southward-directed movement. This evidence of derivation of the serpentinite sheets from the present north is strongly supported by detailed structural analysis of similar serpentinite sheets in the Mamonia Nappes of Paphos District (Woodcock & Robertson, unpublished work).

Alternatively, Lapierre (1970, 1972, 1975) concluded that all the Mamonia rocks including both the Mamonia Complex and the Moni Mélange were all derived from the present NE. This interpretation would require that the Mamonia rocks were all transported over the Troodos Massif to reach their present position. However, there is no evidence of any Mamonia rocks ever having existed along the north margins of the Troodos Massif (Fig. 1). In particular, Mamonia rocks have not been encountered in deep bore holes in the Mesaoria Plain north of Troodos. Also, whereas the south margin of the Troodos Massif is severely deformed associated with emplacement of the allochthonous rocks, the northern margin of Troodos shows no signs of any deformation which would have resulted from southward overthrusting of an entire continental margin. Significantly, north of the Moni Mélange, as at Valva and Asgata (Fig. 3), the Campanian to Tertiary sedimentary sequences are unbroken. This alone would appear to preclude derivation of the Mamonia rocks from the present north.

(B) MODE OF EMPLACEMENT

The Moni Mélange and the Mamonia nappes represent exposed portions of a much larger area of allochthonous rocks. Jurassic red cherts crop out south of the Moni Mélange, in the Akrotiri Peninsula, south of Limassol (Morel 1960), and there are small scattered outcrops of Mamonia rocks near Cape Greco in the Famagusta area (Fig. 1). Consequently, it appears that the Tertiary chalks are underlain by allochthonous rocks throughout much of southern Cyprus (Fig. 1).

Co-emplacement of all the allochthonous rocks is favoured by similarities in gross structural organization of both the Moni Mélange and the Mamonia Complex (Fig. 5). Both bodies include similar continental margin lithologies and sheets of serpentinite derived from Troodos-type oceanic basement; both are overlain by the distinctive 'small-olistolith' mélange.

On the basis of the field data, the olistoliths of the Moni Mélange must have been emplaced as a result of tectonic disruption of a continental margin, followed by a major phase of gravity sliding into the unconsolidated clays and silts of the Kannaviou Formation. The precise mechanism of this initial tectonic break-up is not clear from the Moni Mélange alone. However, it is significant that the olistoliths range from scattered blocks in a clay matrix to structurally more coherent zones, as seen SE of Pareklisha. There, although internally jumbled, the olistoliths form a series of relatively laterally continuous sheets which dip towards the present SSW (Fig. 2). These sheets are structurally intermediate between the adjacent mélange and the more substantial thrust sheets found in the Mamonia Complex of SW Cyprus. Hence it is likely that the initial disintegration of the continental margin sequence first gave rise to a series of relatively undeformed sheets which were then jumbled to form the olistoliths during the extensive gravity sliding which ensued.

The orientations of most of the olistoliths was dominantly controlled by sliding down the regional palaeoslope. However, the emplacement of the larger olistoliths of Pareklisha Sandstone and Monagroulli Siltstone which are floored by lavas must have been influenced by the local palaeotopography of the Troodos basement. The preferential E–W orientation of the strike of these olistoliths, relative to the dominant NW–SE strike of the other olistoliths may be an indication that the Troodos basement still sloped to the south, as earlier, in the Campanian, when the Kannaviou sediments were first deposited. Also the position of many of the olistoliths of Pareklisha Sandstone and Monagroulli Siltstone at the highest structural levels of the Moni Mélange, and their relatively undeformed state, implies that these olistoliths were among the latest to have been emplaced. This is consistent with their sedimentological interpretation as relatively proximal portions of the original continental margin sequences.

In contrast, the serpentinites were emplaced as large sheets derived from the present north. Their internal structure is suggestive of movement as a serpentinite flow or 'glacier'. However, unlike the various serpentinite gravity flows described by Lockwood (1971), the serpentinites show no signs of admixing with the adjacent bentonitic clays. Instead a multi-phase emplacement of the serpentinites is implied by the field relations. First, the deep-seated Troodos ultramafic rocks must

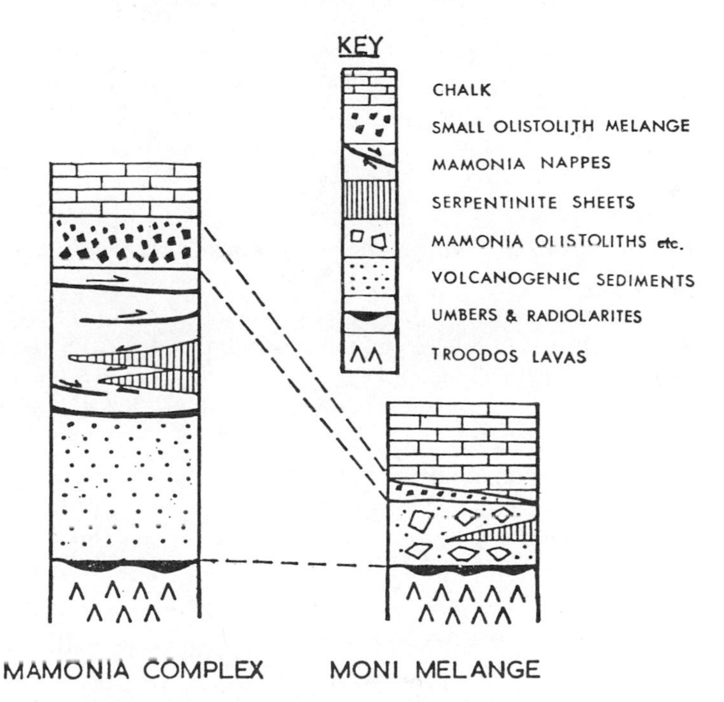

FIG. 5. Generalized sections correlating the Moni Mélange with the Mamonia Complex of SW Cyprus.

have been mobilized, presumably as serpentinite which flowed upwards to reach the ocean floor. This material then flowed some distance southwards over the ocean floor to produce the flow features; finally, portions of the serpentinite slid rapidly into their present positions without admixing with the adjacent clays. The exact location of the source of the serpentinite sheets is uncertain but it is likely to have been from the major E–W trending Arakapas fault belt or from the NW–SE trending Yerasa fault belt located further south (Fig. 1). The Arakapas fault belt has been interpreted as a fossil transform fault in which serpentinite was already present at high structural levels prior to the Maastrichtian emplacement of the Moni Mélange (Simonian 1975). Partly serpentinized untramafic rocks were probably expelled from one or other, or both, of these fault zones during the formation of the Moni Mélange. This material then slid some distance over the Troodos lavas coming to rest within the host sediments of the mélange.

This leaves the 'small-olistolith' mélange to be explained. Recent study has shown that its counterpart above the Mamonia Complex of SW Cyprus is essentially a debris flow formed by rapid submarine partial erosion and redeposition of areas of the Mamonia Complex soon after its emplacement. This submarine erosion which is suggestive of rapid uplift, also affected the Moni Mélange but on a much smaller scale. Subsequently, still in the Maastrichtian, the entire area was blanketed with pelagic chalks and marls of the Lefkara Formation.

8. Regional tectonic setting

There is firm palaeomagnetic evidence of a 90° anticlockwise rotation of the Troodos Massif since the ocean floor genesis in the Upper Cretaceous (Moores & Vine 1971). When exactly this rotation took place is not yet clear, but it is likely to have accompanied the final emplacement in the Miocene of the Kyrenia Range of northern Cyprus. Hence, with respect to the late Cretaceous position of Cyprus (Fig. 6):

(i) The olistoliths of the Moni Mélange came from a Mesozoic continental margin originally located to the NW.
(ii) The serpentinite sheets of the Moni Mélange represents slivers of Troodos-type oceanic crust which were derived locally from the E.
(iii) The matrix of the mélange is a deep water sequence of Upper Cretaceous hemi-pelagic clays and radiolarian siltstones which were deposited directly on Troodos oceanic crust.
(iv) All the allochthonous rocks were emplaced by a short-lived episode of gravity sliding in the Maastrichtian, followed by the deposition of pelagic marls and chalks without further deformation.

Critically, the olistoliths of all the continental margin rocks, of which some are as old as late Triassic, were emplaced directly on to Upper Cretaceous oceanic crust. To explain this the Troodos Massif might represent either the proximal margins of a late Cretaceous ocean, or the more axial portion of an older but extremely slow-spreading ocean basin; alternatively, the juxtaposition of continental margin material with oceanic crust might have resulted from crustal shortening involving the subduction of oceanic lithosphere. This last possibility is strongly favoured by evidence from the Mamonia Complex (Robertson & Woodcock, unpublished work) which shows that the continental margin was initiated in the Upper Triassic associated with alkaline magmatism, followed by sedimentation on an essentially passive continental margin until the late Cretaceous. Generation of oceanic crust was presumably going on throughout the Jurassic and much of the Cretaceous, with the implication that the Troodos Massif is likely to represent the axial portion of a long-lived and hence relatively wide ocean, rather than a small ocean basin. For example it is unlikely to represent a marginal sea formed behind an island arc in the late Cretaceous, as has recently been proposed (Pearce 1974, Smewing *et al.* 1975).

Taking account of the palaeorotation of Cyprus and the geometrical relationships of the Moni Mélange, it follows that the postulated subduction of Troodos-type oceanic crust was originally directed towards the NW beneath a continental margin now located in southern Turkey (Fig. 6). Significantly, neither the Moni Mélange nor the Mamonia nappes show much in common with subduction

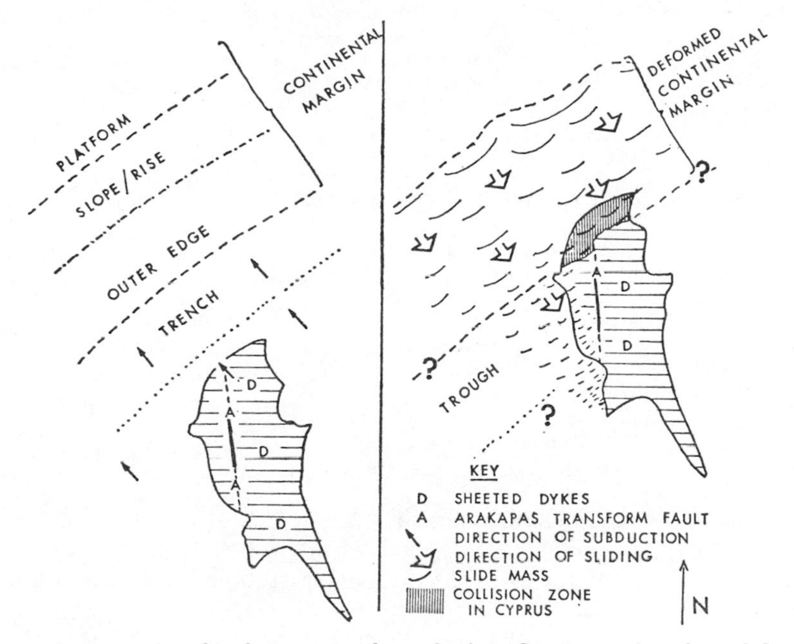

FIG. 6. Interpretative sketch maps to show the late Cretaceous location of Cyprus relative to a continental margin to the original NW, now represented in Cyprus by the Moni Mélange and the Mamonia Complex.

(A) A period of late Cretaceous steady-state subduction is postulated to bring Upper Cretaceous Troodos oceanic crust close to a continental margin to the NW. At this time the Arakapas fault belt formed a series of then W-facing elevated fault scarps.

(B) In the Maastrichtian there was a major collision of Troodos oceanic crust with a distal part of the continental margin, now probably located outside Cyprus. This collision triggered off major oceanward sliding of portions of the continental margin to form the Mamonia Complex and the Moni Mélange.

complexes, which are attributed to extended periods of steady-state subduction (e.g. Karig & Sharman 1975). Specifically, the brief period of emplacement of all the allochthonous rocks and the evidence of deep tectonic slicing of oceanic crust involving the emplacement of ultramafic rocks to high structural levels are not features of steady-state subduction, as seen, for example, in the Java trench at the present time (Beck & Lehner 1974).

In contrast, the gravity emplacement of the Moni Mélange and the Mamonia Complex took place during a brief period of the Maastrichtian, accompanied by tectonic intercalation of sheets of partly serpentinized ultramafic rocks from the adjacent Troodos oceanic crust. Although the exact nature of this event is still unclear, it may have involved a forceful collision of the then recently active Troodos ocean ridge with the adjacent continental margin. This collision, which was probably most intense outside the present area of Cyprus, may have been sufficient to bring subduction to an end in the area. Whatever the exact mechanism, the pre-existing trench in which Troodos oceanic crust was presumably

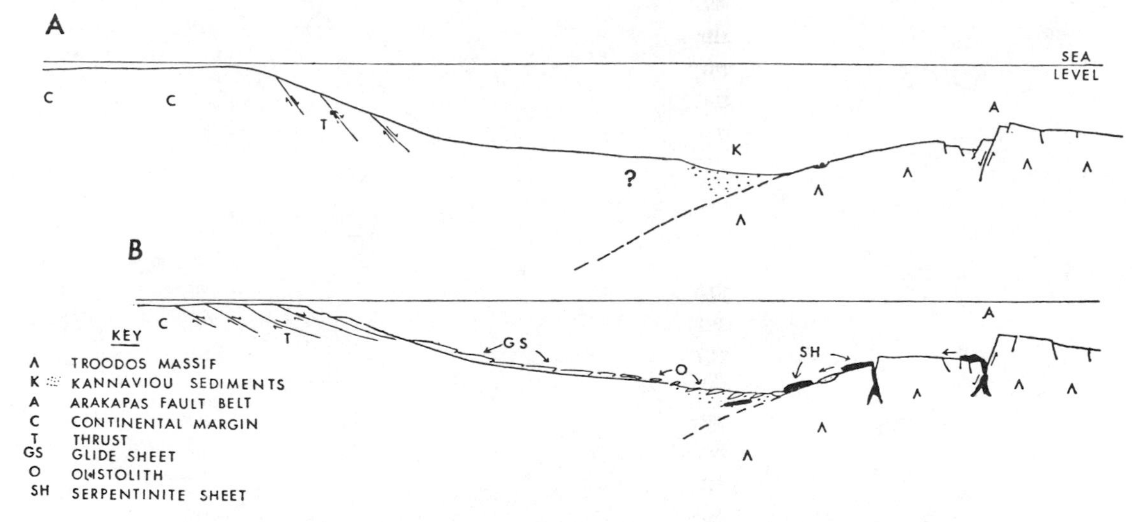

FIG. 7. Diagrammatic cross-section through Moni Mélange (not to scale).

(A) During a period of postulated late Cretaceous steady-state subduction, the Troodos Massif was carried towards a subduction zone dipping northwards beneath a continental margin then located to the NW. During the Campanian and Maastrichtian the Kannaviou sediments were deposited on topographically low-lying areas on the ocean floor.

(B) A major collision centred away from the area of the Moni Mélange then triggered off major oceanward sliding of portions of the continental margin as relatively large semi-intact sheets. The furthest travelled most distal portions of these sheets then disintegrated to form olistoliths which slid into the unconsolidated hemi-pelagic sediments deposited on the Troodos lavas. Simultaneously, large bodies of partly serpentinized ultramafic rocks were mobilized from pre-existing fault zones which are attributed to a pre-existing transform fault system. The serpentinite slid westwards as sheets becoming embedded in the host sediments and olistoliths of the Moni Mélange. Afterwards, there was a brief period of submarine erosion, then onset of calcareous pelagic sedimentation.

undergoing earlier northwards subduction, was then completely eliminated in SW Cyprus by the emplacement of the Mamonia Complex as large portions of the continental margin which slid into the trench. In contrast, the Moni Mélange was located to the original S, further from the main collision zone (Fig. 6). For a brief period a trough still remained in this area between the deformed continental margin and a portion of topographically elevated Troodos oceanic crust (possibly part of the ridge itself) which remained, unsubducted, to the original E (Fig. 7). This trough which was possibly located close to the oceanward margin of the pre-existing trench, was then rapidly filled by sliding of both olistoliths from the continental margin to the NW and sheets of serpentinite from the topographically elevated oceanic crust to the E (Figs. 6, 7).

In summary, the olistoliths of the Moni Mélange are taken to represent the most distal and hence furthest travelled portions of essentially the same oceanward gravity slides as form the more intact thrust sheets and nappes of the Mamonia Complex which were emplaced nearer the source area to the NW. Soon after the emplacement of the Moni Mélange there was a pulse of uplift giving rise to rapid submarine erosion to form the distinctive 'small-olistolith mélange', then the area was blanketed by latest Cretaceous marls and chalks of pelagic facies. Significantly, the absence of any signs of deformation after deposition of these chalks implies that subduction in this area had come to an end with the emplacement of the allochthonous rocks.

9. Conclusions

Mapping has shown that the Moni Mélange of southern Cyprus consists of two chief allochthonous components, olistoliths which originated in the late-Cretaceous disintegration of a Mesozoic continental margin, and serpentinite sheets derived from Troodos-type oceanic crust.

Taking account of post-Cretaceous rotation of Cyprus, the original geotectonic setting is believed to have involved a period of late-Cretaceous subduction of Troodos ocean floor beneath a continental margin then located to the NW of Cyprus in southern Turkey. In the Maastrichtian this subduction culminated in a major collision of oceanic crust with the continental margin thus triggering off major oceanward gravity sliding of a variety of continental margin rocks. At the same time, the adjacent Troodos oceanic crust was deeply tectonically sliced which allowed the passage of sheets of partly serpentinized ultramafic rocks directly on to the ocean floor. The emplacement of all the allochthonous rocks was followed by a brief period of rapid submarine erosion, then the area was blanketed by typically open marine, calcareous, pelagic sedimentation.

Finally, the Moni Mélange must be distinguished from a typical olistostrome. As an example, the olistostromes of the Italian Apennines (e.g. Abbate *et al.* 1970, Görler & Reutter 1968, Görler 1976) developed as a series of internally stratigraphically ordered, giant, submarine debris flows, said to have been active over millions of years. The very much more rapid emplacement of the Moni Mélange as a series of huge submarine slides explains the lack of the internal stratigraphical order which is characteristic of fully developed olistostromes.

ACKNOWLEDGEMENTS. This work was supported by an N.E.R.C. Studentship at the University of Leicester. I thank Dr J. D. Hudson for his advice and encouragement, and Dr N. H. Woodcock, M. A. Naylor, and Dr J. E. Dixon for their help with the manuscript.

References

ABBATE, E., BRTOLOTTI, V. & PASSERINI, P. 1970. Development of the Northern Apennines Geosyncline-olistostromes and olistoliths. *Sediment. Geol.* **4**, 521–57.

BECK, R. H. & LEHNER, P. 1974. Oceans, new frontiers in exploration. *Bull. Am. Ass. Petrol. Geol.* **58**, 376–94.

BRUNN, J. H., DUMONT, J. F., GRACIANSKY, P. CH. de., GUTNIC, M., JUTEAU, T., MARCOUX, J., MONOD, O. & POISSON, A. 1971. Outline of the geology of the western Taurides. *In* Campbell, A. S. (ed.) *Geology and History of Turkey.* Tripoli, 225–55.

DIMITRIJEVIC, M. D. & DIMITRIJEVIC, M. N. 1974. Internal fabric of Mesozoic bodies in the mélange around Nova Varos: a case for gravity tectonics. *Bull. scient. Cons. Acads RPF Yougosl.* **19**, 70–2.

GÖRLER, K. 1976. The determination of former mudflow-directions in olistostromes. *9th Int. Congr. Sediment.* 1975 Nice.

—— & REUTTER, K. J. 1968. Entstehung und Merkmale der olistostrome. *Geol. Rdsch.* **57**, 484–514.

GRACIANSKY, P. C. 1972. *Recherches géologiques dans le Taurus Lycien.* Thesis, University of Paris-Sud, 571 pp.

HENSON, F. R. S., BROWNE, R. V. & MCGINTY, J. 1949. A synopsis of the stratigraphy and geological history of Cyprus. *Q. Jl geol. Soc. Lond.* **105**, 1–41.

HOEDEMAEKER, PH. J. 1973. Olistostromes and other delapsional deposits, and their occurrence in the region of Moratalla (Prov. of Murcia, Spain). *Scr. geol.* **19**, 1–207.

HSÜ, K. J. 1968. Principles of mélanges and their bearing on the Franciscan-Knoxville paradox. *Bull. geol. Soc. Am.* **79**, 1063–74.

—— 1974. Mélanges and their distinction from olistostromes. *In* Dott, R. H. & Shaver, R. H. (eds) *Modern and Ancient Geosynclinal Sedimentation. Soc. Econ. Palaeont. Mineral. Spec. Publ.* **19**, 321–33.

JUTEAU, T. 1974. *Les ophiolites des Nappes d'Antalya (Taurides occidentales, Turquie). Pétrologie d'un fragment de l'ancienne croûte océanique téthysienne.* Doctoral thesis, University of Nancy, 692 pp.

KARIG, D. E. & SHARMAN, G. F. III. 1975. Subduction and accretion in trenches. *Bull. geol. Soc. Am.* **86**, 377–89.

LAPIERRE, H. 1968a. Découverte d'une série volcano-sédimentaire probablement d'âge Crétacé supérieure au S.W. de l'île de Chypre. *C.r. hebd. Séanc. Acad. Sci., Paris* **D266**, 1817–20.

—— 1968b. Nouvelles observations sur la série sédimentaire de Mamonia (Chypre). *C.r. hebd. Séanc. Acad. Sci., Paris* **D267**, 32–5.

—— 1970. Découverte de plusieurs phases orogénétiques Mésozoiques au Sud de Chypre. *C.r. hebd. Séanc. Acad. Sci., Paris* **D270**, 1876–8.

—— 1972. *Les formations sédimentaires et éruptives des nappes de Mamonia et leurs relations avec le massif de Troodos (Chypre).* Doctoral Thesis, University of Nancy, 420 pp.

—— 1975. *Les formations sédimentaires et éruptives des nappes de Mamonia et leur relations avec le massif du Troodos. Mem. Soc. géol. Fr.* **123**, 132 pp.

LOCKWOOD, J. P. 1971. Sedimentary and gravity slide emplacement of serpentinite. *Bull. geol. Soc. Am.* **82**, 919–36.

MANTIS, M. 1970. Upper Cretaceous-Tertiary foraminiferal zones in Cyprus. *Scientific Research Centre of Cyprus, Epithris (Cyprus Research Centre)* **3**, 227–41.

MOORES, E. M. & VINE, F. J. 1971. The Troodos Massif, Cyprus and other ophiolites as oceanic crust: evaluation and implications. *Phil. Trans. R. Soc.* **A268**, 433–66.

MOREL, S. W. 1960. The Geology and Mineral Resources of the Apsiou-Akrotiri Area. *Mem. geol. Surv. Dep. Cyprus* **7**, 51–88.

PANTAZIS, TH. M. 1967. The geology and mineral resources of the Phamakas-Kalavassos area. *Mem. geol. Surv. Dep. Cyprus* **8**, 120 pp.

PEARCE, J. A. 1975. Basalt chemistry used to reconstruct past tectonic environments on Cyprus. *Tectonophysics* **25**, 41–69.

ROBERTSON, A. H. F. & HUDSON, J. D. 1974. Pelagic sediments in the Cretaceous and Tertiary history of Cyprus. *In* Hsü, K. J. & Jenkyns, H. C. (eds) *Pelagic sediments: on land and under the Sea. Spec. Publs. Int. Ass. Sedimentologists* **1**, 403–36.

SIMONIAN, K. 1975. The geology of the Arakapas Fault Belt area, Troodos Massif, Cyprus. Thesis (Ph.D.) Open University. (unpubl.) 151 pp.

SMEWING, J. D., SIMONIAN, K. O. & GASS, I. G. 1975. Metabasalts from the Troodos Massif, Cyprus: genetic implication deduced from petrology and trace element composition. *Contrib. Mineral. Petrol.* **51**, 49–64.

Received 2 June 1976; revised typescript received 10 December 1976.

ALASTAIR HARRY FORBES ROBERTSON, Grant Institute of Geology,
University of Edinburgh, Edinburgh. EH9 3JW.

Part III

CALEDONIAN MÉLANGES

Editor's Comments
on Papers 11 Through 14

The Caledonian mélanges provided the source of the term *mélange* (Greenly, 1919). He wrote (summary due to Hsü, 1968, p. 1064):

> The essential characters of an autoclastic mélange may be said to be the general destruction of the original junctions, whether igneous or sedimentary, especially bedding, and a shearing down of the more tractable material until it functions as a schistose matrix in which fragments of the more obdurate rocks float as isolated lenticles of phacoids. . . . It is developed on a grand scale. . . . In a country of this kind, the larger masses are found to be arranged as trains of lenticles overlapping *en echelon*. . . . Upon a map, the masses that can be separated out appear as if embedded in a homogeneous and structureless matrix. But this country rock is itself built up of interdigitating lenticular bodies; and, could we take in the region at a glance, it would present itself as a mélange of torn and sheared lenticular masses of all sizes, from such as are two or three miles in length to the smallest the eye can see, of spilite, lava, diabase, quartzite, limestone, jasper and grit, floating in an undifferentiated but schistose body that is a weft of all the more easily deformable elements, itself pervaded throughout in the same lenticular structure.

Greenly described a Paleozoic mélange as well as the Precambrian Gwna mélange, a mixture of all local Ordovician rock units with the

Gwna Group. His concept was one "of a tectonically derived chaotised body, of immense scale, related to thrusting, including both ophiolitic and sedimentary material—derived from a number of superimposed units and even bridging unconformities."

A number of authors have contributed to the redefinition of Anglesey, in terms of stratigraphy and structures, and its mélanges (Shackleton, 1956, 1969; Wood, Paper 11; Maltman, 1975; Barber and Max, 1979). In Paper 11, Wood noted that the Gwna Mélange is developed on a microscopic to megascopic scale and that it includes oolitic limestone, dolomite, orthoquartzite, graywacke, pelite, chert, jasper, basic pillow lava, tuff, and hyaloclastite. He believed that Greenly intended no genetic connotation of the term *mélange,* otherwise he would not have qualified it with *autoclastic.* It may be observed here that, in any case, the olistostrome (gravity slide) interpretation, favored by later authors, was unknown at the time when Greenly was writing—he has a simple choice between eruptive fragmentation, as in an agglomerate, and tectonic fragmentation. Wood favored the olistostrome interpretation, regarding it as a submarine slide deposit of sedimentary origin, though commonly affected by secondary tectonic deformation. He recorded a ghost stratigraphy but did not describe the matrix in any detail. Later authors (Barber and Max, 1979) mention a fine-grained matrix and "purple slate in a low state of deformation and metamorphism." Wood rejected the possibility of an origin in tectonic boudinage, mainly because of the low level of metamorphism. However, clearly the matrix is at least of slate grade, and more recent investigations on the Franciscan (Paper 25) and the sedimentary mélange of the Makran, Iran (Paper 21) suggest that not even this low grade is necessary for tectonic boudinage to be a possibility. Other criteria invoked by Wood include lack of deformation in limestone ooliths, angularity of blocks, lack of evidence of ductile flow in the matrix, and flat-lying attitude of the mélange body. He rejected volcanic mudflow origin because of the large extent of the mélange development. He also stressed the role of lithification contrasts between arenaceous and argillaceous components. Some doubt remains whether the case for gravity slide origin in a soft sediment environment has been unequivocably established, and the suggestion was lately made (Bachman, 1980) that Wood may have confused tectonic deformation of semilithified sediments with soft rock sedimentary structures. There is a need for a very detailed fabric study here of the type carried out by Cowan (Paper 25).

The assemblage was again described by Wood in discussion to the paper by Barber and Max (1979) as a "series of olistostromes and interslide members . . . interbedded with distal turbidites, proximal turbidites and tuffs. . . with some normal contacts . . . easterly directed and becoming progressively younger to the northwest." There is an association with disrupted ophiolites and blueschists, and the latter association would suggest a paleo-subduction zone environment (Coleman, 1972). This was favored by Wood, though Robertson (discussion of paper by Maltman, 1975) questioned the validity of such a conclusion.

The possibility that the mélange is Cambrian rather than Precambrian was suggested by Barber and Max (1979), who cited evidence of contained Cambrian fossils. The Gwna Mélange is, if Wood's interpretation is correct, an olistostrome mélange containing considerable ophiolitic block content.

The other regions where mélanges have been described from the Caledonian orogen are Newfoundland and the Appalachians, and three papers dealing with these mélanges have been included (Papers 12-14). Williams and Talkington (Paper 12) give a review of the occurrence of ophiolitic mélanges in the Appalachian orogen, which they relate to the closure and destruction of the Iapetus Ocean. It is interesting that they suggest that some ophiolitic mélanges may simply be dismembered parts of ophiolitic sequences; there is a possible analogy here with the Makran and Oman (Papers 21 and 22) where a chaotic character indistinguishable from the Coloured Mélange apparently has been produced on a rather local scale simply by dislocation related to extreme tectonic compression. The Dunnage mélange, the most important and best documented of the Appalachian mélanges, is not of this type. It is essentially non-ophiolitic, though locally it contains ophiolitic slabs and exotic blocks, and is probably a lateral equivalent of the Carmanville ophiolitic Mélange. This review points above all to the need for detailed factual descriptions of all these Appalachian mélanges; probably there is a wide spectrum of mélanges of diverse origin represented. In such descriptions, there is a need to separate factual observation from conjecture or theorization—something not always achieved under the euphoria of the new global tectonics.

In their detailed account of the Dunnage mélange, Hibbard and Williams (Paper 13) commence with a discussion of the usage of the term *mélange*. This is so important and so applicable to the philosophy of this volume that it is stressed here:

> It is obvious from a review of the literature and communications between geologists that a common understanding of what is meant by the term mélange is urgently needed. In view of the recent barrage of definitions and usages of the term, whether authors revert back to its original use in geological literature or concoct revised or new definitions of the word, the present authors feel that only a descriptive usage of the term can save it from oblivion. Thus the term is used in the present study to

> denote a chaotic, heterogeneous assemblage of unsorted blocks set in a fine grained matrix, the larger blocks being of outcrop size; modifiers such as tectonic or olistostromal may be prefixed to designate mode of genesis.

This is, in effect, what the editor of this volume has concluded is the only practical solution to the problem of usage—that is he would only insert the words *nongenetic* before *descriptive* and the word *characterizing* before *modifiers* to achieve an entirely satisfactory definition for the term *mélange*.

The Carmanville Mélange was considered by Kennedy and McGonigal (1972) to be either tectonic (a zone of gravity sliding) or an olistostrome. There are overlying slumped turbidities that may or may not be deformed in the same sliding event, according to Pajari, Pickerill, and Currie (Paper 14). The matrix of the mélange is pyritic black shale or silt, believed to be derived from disintegration of thixotropic, unconsolidated sediment. The fragments within it are commonly rounded and range from granules up to several kilometers diameter. Some were tectonically deformed before incorporation. Some, lithologically identical to other unmetamorphosed fragments nearby, are metamorphosed—such relationships being attributed to dynamothermal metamorphism during mélange formation. Locally, the entire mélange is regionally metamorphosed, up to garnet-cordierite-andalusite grade, and it may also exhibit superimposed tectonic deformation.

The basalt in the mélange occurs as huge rafts, which are shown to be allochthonous. The ultramafic blocks and some plagiogranite fragments come from the Gander River ultramafic belt to the southeast. The mélange is believed, from evidence of contained Cambrian orthoquartzite fragments, to have sampled the whole continental rise and prism, and back to the shelf (including the Avalon zone and the Gander River belt to the northwest of it). However, much of the debris in the slides is believed to be of local provenance.

Pajari and his associates relate this mélange to obduction of oceanic crust and overlying volcanics at the convergent plate margin bordering the Iapetus Ocean. Again, the main doubt that arises is whether this mélange does represent submarine gravity sliding (a sedimentary process) in an unloaded system or tectonic deformation of a loaded (i.e., buried), partially consolidated sequence. The authors use the term *ophiolitic mélange* following Gansser (Paper 1) and remark that most such mélanges have a sedimentary matrix, though some have a serpentinite matrix. This observation is broadly true and the usage adopted is justified; yet it must be remarked here that the serpentinite mélanges and the block-to-block mélanges, devoid of any matrix, may have a far greater extent worldwide than is apparent from the meager references in the literature. Such mélanges occur, particularly, in remote areas of Iran, Afghanistan, Pakistan and the Himalayas, and the USSR and have, in many cases, only been the subject of brief descriptions.

All these descriptions of Caledonian mélanges contain reference to slaty matrixes, yet some doubt remains as to whether such matrixes are related to soft rock, superficial processes of submarine gravity sliding—indeed the authors cited, other than Wood, all stress the role of tectonic processes in mélange formation, as well as olistromal processes. The development of mélanges along very extensive sectors of geotectonic sutures (destructive plate margins) is emphasized by Williams and Talkington (Paper 12), and such a scale of development is certainly not consistent with small-scale, localized olistostromes, despite the thinness of some of the mélange sheets. The Dunnage Mélange is related by Hibbard and Williams (Paper 13) to thixotropic activation and superficial slumping to a trench slope deposit overlying an accretionary prism within an extensive geotectonic zone—a gravity slide of regional dimensions—and the Carmanville Mélange (Paper 14) is probably an extension of it, but there would seem to be room for future evaluation of the role of olistostromal and tectonic deformation in these Caledonian mélanges.

REFERENCES

Bachman, S. B., 1980, The Coastal Belt of the Franciscan, Northern California: Youngest Phase of Subduction, in *Trench and Fore-Arc Sedimentation and Tectonics in Modern and Ancient Subduction Zones, Abstract Volume,* Geological Society of London and British Sedimentological Research Group, p. 3.

Barber, A. J., and M. D. Max, 1979, A New Look at the Mona Complex (Anglesey, North Wales), *Geol. Soc. London, Jour.* **136:**407-424.

Coleman, R. G., 1972, Blueschist Metamorphism and Plate Tectonics, *Internat. Geol. Congress 24th, Proc.* **2:** 19-26.

Greenly, E., 1919, The Geology of Anglesey, *Great Britain Geol. Survey Mem. 1,* 980p.

Hsü, K. J., 1968, Principles of Mélanges and Their Bearing on the Franciscan-Knoxville Paradox, *Geol. Soc. America Bull.* **79:**1063-1074.

Kennedy, M. J., and M. H. McGonigal, 1972. The Gander Lake and Davidsville Groups of Northeastern Newfoundland: New Data and Geotectonic Implications, *Canadian Jour. Earth Sci.* **8:**452-469.

Maltman, A. J., 1975, Ultramafic Rocks of Anglesey: Their Non-Tectonic Emplacement, *Geol. Soc. London, Jour.* **131:**593-606.

Shackleton, R. M., 1956, Notes on the Structures and Relations of the Precambrian and Ordovician Rocks of Southwestern Lleyn (Caernarvonshire), *Liverpool and Manchester Geol. Jour.* **1:**400-409.

Shackleton, R. M., 1969, The Precambrian of North Wales, in *The Precambrian and Lower Paleozoic Rocks of Wales,* A. Wood, ed., University of Wales Press, Cardiff, pp. 1-22.

11

OPHIOLITES, MELANGES, BLUESCHISTS, AND IGNIMBRITES: EARLY CALEDONIAN SUBDUCTION IN WALES?

DENNIS S. WOOD
University of Illinois, Urbana

INTRODUCTION

Since 1831, when Adam Sedgwick in company of the youthful Charles Darwin commenced his geological investigations, North Wales has been a classical region. The remarkable variety of its geology is above all proportion to its size, but for many well described regions some interpretations that now may be offered in the light of new developments in geological thought have been either delayed or overlooked. The associations within a late Precambrian eugeosynclinal sequence that includes pillowed olivine-basalt lavas, cherts, jaspers, mélanges, serpentinites, gabbros, and blueschist metamorphic rocks provide adequate scope for speculation. When the responsible processes are considered in a space-time context together with regional structure, subsequent effusive ignimbrite activity, and with the overall aspect of the lower Paleozoic Welsh Basin, a unified view of this region unfolds that may prove to be an ideal small-scale model for one of the more important crustal processes.

Edward Greenly (1919) provide most of our modern knowledge of the Precambrian of Anglesey. Marshall Kay is one of few geologists to have been shown this region by Greenly. It is, therefore, appropriate that a symposium to honor Professor Kay and to recognize his contribution to geology should refer to this fascinating region.

THE PRECAMBRIAN (MONIAN) ROCKS OF NORTH WALES

Precambrian rocks occur in four main regions of the island of Anglesey and in one area in the southwestern promontory of Caernarvonshire, the Lleyn Peninsula. The area of outcrop is approximately 180 square miles in Anglesey and 35 square miles in Caernarvonshire. The stratigraphic thickness is at least 35,000 ft. Equivalent rocks are present in Ireland on Carnsore Head, County Wexford.

Of the earliest investigators, Henslow (1822) referred to Anglesey as containing "the earliest stratified rocks," Sedgwick (1838, 1843) regarded the "metamorphic slates" of Anglesey and southwest Caernarvonshire as coeval and older than the Paleozoic systems; and Sharpe (1846) recorded the presence of jaspers and serpentines. The term "Monian System" was introduced by Blake (1887, 1888a), who described the importance of lavas and large-scale breccias and discovered glaucophane in Anglesey (Blake, 1888b). Gabbros and "spheroidal basalts" were recorded by Raisin (1893), the basalts being recognized as spilitic pillowed lavas by Dewey and Flett (1911). The most detailed descriptions of the Monian of Caernarvonshire are by Matley (1913, 1928) and Greenly, whose 25 years of research and mapping of the whole of Anglesey were published in 1919.

Stratigraphy.—The lower part of the pertinent sequence is best preserved on Holy Island at the western extremity of Anglesey. It consists in upward succession of turbidite graywackes interbedded with pelites (South Stack Series); thick orthoquartzites (Holyhead Quartzite); turbidite graywackes, pelites and thin quartzites (Rhoscolyn Beds); and a thick finely laminated flysch group of pelites and semipelites (New Harbour Group).

The uper part of the sequence, which is best seen in northern and southeastern Anglesey and in southwest Caernarvonshire, consists of basic to acid tuffs and volcanic conglomerates (Skerries Group) and is overlain by an extremely varied assemblage not less than 12,000 ft thick (Gwna Group). The Gwna contains oolitic limestones, dolomites, orthoquartzites, graywackes, pelites, cherts, jaspers, basic pillow lavas, tuffs, and hyaloclastites. Much of the Gwna Group is a mélange on scales ranging from microscopic to megascopic. Above the Gwna Group at the top of the succession is the Fydlyn Group of unlayered acidic volcanic rocks and tuffs. The stratigraphic terms are, with one exception, those assigned by Greenly, but the stated order of succession is almost the reverse of his. Greenly's interpretation of the Anglesey structure invoked enormous recumbent folds with westerly vergence. This interpretation, together with his discovery in the Holyhead Quartzite of jasper fragments, which he presumed to be derived from the Gwna Group, led him to conceive of the succession in the reverse order from that given here. Shackleton (1953, 1954) used sedimentary structures and applied the innovative concept of facing (Shackleton, 1957) to demonstrate the true sequence.

Provenance of sediments.—The graywackes of the lower Monian of Holy Island are in graded beds as much as 20 ft thick. They are best seen at South Stack (204823)[1] and at Rhoscolyn (258753–264750). The great thickness of many units and the observable consistency over hundreds of yards, imply that the sediments accumulated in relatively deep water. Channeling and grooving show that they were emplaced by currents that flowed along a northeast-south west trough; the sense of flow is not known. Associated pelites slumped down slopes which inclined to the northwest. Hence, axial transport and lateral supply from the southeast were involved. This graywacke facies is uncommon, being interbedded with extremely pure quartzites. In 5 miles, from Holyhead to Rhoscolyn, a single quartzite, at least 800 ft thick, splits into three thinner units, one of which is 120 ft thick and devoid of internal stratification. Wood (1960) found gigantic flute casts as much as 17 ft long and 3 ft deep at the base, indicating flow from the northeast (see fig. 1). The upper few feet of this quartzite is cross bedded and indicates that reworking currents also flowed from the northeast. The quartzites were interpreted as relatively deep water sand-flow deposits.

The mélange.—The mélange of the Gwna Group is perhaps the most extraordinary and spectacular aspect of the North Wales Precambrian. It is best seen on Wylfa Head (353946–358945) and at Cemaes Bay (369938) in northern Anglesey and on Bardsey Island (120220) some 2 miles off the southwest coast of Caernarvonshire. Matley (1913) gave the first clear description of the mélange from Bardsey Island. Matley and, subsequently, Greenly (1919) both regarded the mélange as a tectonic breccia as did Sir Edward Bailey (personal communication, 1961). Matley used the terms "crush-breccia" and "crush-conglomerate" before Greenly introduced the term "autoclastic mélange." Greenly intended no genetic implication for "mélange"; otherwise, he would not have needed to prefix it. For the term to carry

[1] Good field examples are referred to in the text by six-figure National Grid References of the Kilometer grid of the Ordnance Survey of Great Britain. All are in the 100-kilometer square SH and are located on the 1-inch-to-1-mile sheets 106 (Anglesey), 107 (Snowdon), and 115 (Pwllheli).

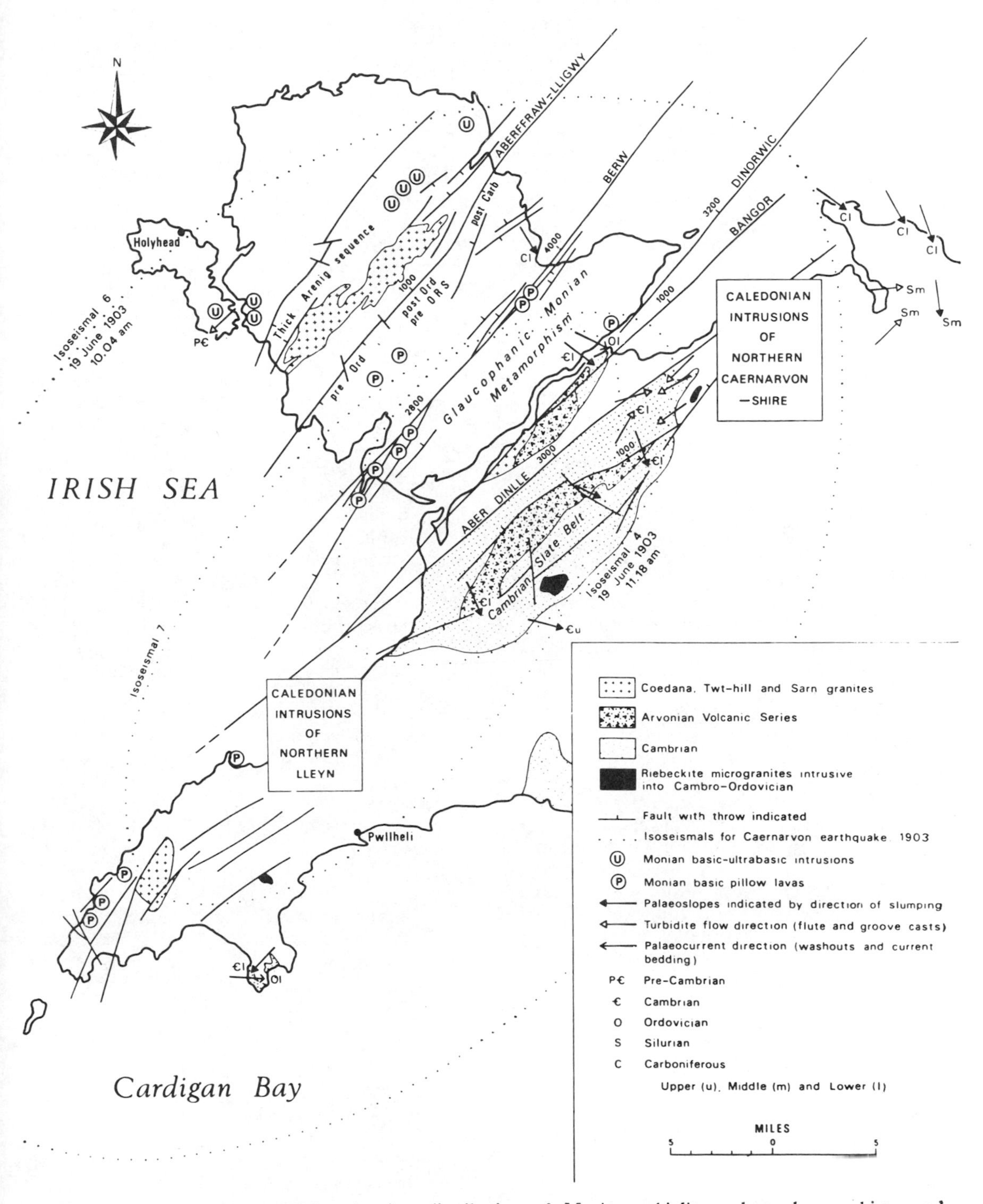

FIG. 1.—Map of North Wales showing distribution of Monian ophiolites, glaucophane schists, and Arvonian ignimbrites together with major acidic intrusions and Caledonian faults of the region and sedimentary transport directions.

the connotation of either a purely tectonic or purely sedimentary origin would be very unfortunate. The Gwna mélange is appreciably deformed in many places, but Shackleton (1956, 1969) and the writer believe it to be ultimately of sedimentary origin, probably the result of submarine sliding. The mélange has regional extent, for, as Greenly (1919, p. 66) noted, "the Gwna Group as a whole is usually in the condition of an Autoclastic General Mélange in which all the members of the group are involved."

In most places the mélange shows a convincing ghost stratigraphy, the individual blocks of one lithology being separated by distances ten to twenty times greater than the block size. It may be argued that tectonic boudinage could account for such a relationship. For such to be the case, the ductility contrasts between the materials and the mean ductility of all the materials involved would need to be impossibly high in view of the very low greenschist level of metamorphism. Lithologies having contrast as great as those in the mélange occur elsewhere in Anglesey without such associated disruption. Furthermore, the ooliths of the limestones lack penetrative deformation in most examples; nor does the matrix of the mélange show ductile flow. Many separated blocks within an individual ghosted unit are very angular. More cogently, much of the mélange is still relatively flat lying. For boudinage to account for the structure would require that the whole of the mélange outcrop had been subjected to a vertical principal compressive force acting perpendicularly to the bedding and under an enormous confining pressure. The undoubted principal deformation features of the Monian, both above and below the mélange, are steep structures, not easily reconcilable with intervening flat structures. The penetrative deformation in the mélange is everywhere less than that characteristic of the deeper levels of the Monian stratigraphy. In many places the only tectonic structure is a rather crude cleavage, which cuts across the intervals containing separated blocks of the same lithology, as at Porth Cadwaladr (362664) in southern Anglesey. Sedimentary contacts with undisturbed sediments are present at the top and bottom of the mélange. The undisturbed beds consist of strongly variable, well-layered materials, which would have responded to strong deformation with distinct ductility contrast and would have resulted in disharmonic relationships. Well-exposed lower contacts occur at Porth Cadwaladr, Anglesey (362664) and well-exposed upper contacts at Braich y Pwll, Caernarvonshire (136258).

Though effects of tectonic deformation can be evaluated and removed, the mélange still remains as a highly disturbed stratigraphic unit of regional extent and nontectonic origin. The ghost stratigraphy and lack of materials exotic to the Gwna Group preclude the mélange from being a tillite. The great areal extent and paucity of volcanic materials preclude it from being a volcanic mudflow. The remaining possibility is that the mélange is an olistostrome. Much complication may pertain because the materials were in varying stages of lithification at the time of their initial movement, arenitic members being more completely lithified than argillites but less lithified than carbonate members. Near Porth Gwylan (214367), the sense of rotation of highly angular blocks of arenite and limestone and the deformation of enclosing layered matrix suggest that relative movement of the stratigraphically higher material was toward the west.

The ophiolites.—The term "ophiolite" may be used in the original strict sense (Steinmann, 1905) to include commonly serpentinized mafic and ultramafic rocks ranging from spilite and basalt to gabbro and periodotite. In view of Steinmann's emphasis upon association of ophiolite with radiolarian cherts, the Steinmann Trinity of serpentinite, spilite, and radiolarian chert came into prominence (Hess, 1955; Bailey and McCallien, 1960), and both "ophiolite" and "Steinmann Trinity" have been used synonymously. Dewey and Bird (1970) have urged that the term "ophiolite" be "restricted to a full sequence of ultramafic rocks, gabbro dike complexes, pillow lava and chert, a sequence that, if fully developed, almost certainly represents oceanic mantle and crust." Any of these usages would apply to the Precambrian of North Wales. The distribution of mafic-ultramafic intrusives and pillow lavas is shown in figure 1.

Mafic and ultramafic intrusives, best developed in the southern part of Holy Island (267772) and the adjacent part of the main island of Anglesey, consists of a variety of largely serpentinized ultramafic rocks and altered gabbros emplaced prior to the regional late Precambrian deformation and metamorphism. The serpentinites were not emplaced cold, so to speak, because they preserve good igneous textures and have caused appreciable contact effects in the enclosing sediments. The ultramafics may be small layered bodies. The majority of lithic types were dunite and harzburgite, but pyroxenite, lherzolite, and minor amounts of wehrlite were also present (Wood, 1960). The dunites, formerly consisting of 90 percent olivine and 10 percent orthopyroxene and accessories, and harzburgites have been replaced by serpentine minerals. Olivine and orthopyroxene did not survive in the lherzolites but clinopyroxene did. The serpentinization of ferromagnesian components of the peridotites and pyroxenites ranges from 10 percent for dunite and harzburgite to about 85 percent for lherzolite and to 65 to 60 percent for pyroxenites. The serpentine minerals are dominantly lizardite and antigorite. Carbonate metasomatism of the serpentinites has formed rocks that, except for surviving magnetite and chromite-picotite, are wholly dolomite. In some cases the nature of both the preexisting serpentine minerals and the earlier ferromagnesian minerals can be recognized. The carbonation ranges from the mere presence of isolated dendritic dolomite rhombs to complete replacement. Dunite and harzburgite serpentinites may be found in all conditions of carbonation from zero to 100 percent (Wood, 1960). The associated gabbros are equally strongly altered, pyroxene being replaced by tremolite and the remaining feldspar being oligoclase. The original feldspar altered to epidote-group minerals, talc, prehnite, scapolite, and calcic garnet. Remaining problems with regard to interpretation of structural form, mode of emplacement, and mineralogical and textural changes should be clarified by the current mapping and petrographic work of A. J. Maltman.

The Monian pillow lavas are exceptionally well preserved and have their finest expressions in the Newbrough Forest and Llandwyn Island area of Anglesey (391635) and on the headland west of Porth Dinllaen (276420) and at Dinas Bach (158293) in the Lleyn Peninsula of Caernarvonshire. The lavas, at least 2,000 ft thick, are spilitic albite basalts containing either original olivine or its alteration products. The basalts are generally vesicular and variolitic, much fresh augite remains, and the albite seems to be original. Except that alumina (17.5 percent) is enriched at the expense of total iron (9.5 percent), the lavas have the chemical character of typical oceanic basalts. The interpillow interstices and the central vacuoles of many pillows are occupied by jasper or carbonate. The pillowed sequences are interbedded with tuffs, basic hyaloclastites, dolomite, dolomite breccias, and massive ferruginous cherts. The pillow lavas have been replaced by carbonate so extensively that some are reduced virtually to ellipsoidal dolomites. Greenly (1919, p. 84) recognized these as metasomatic pseudomorphs after spilitic agglomerate. Every stage in the replacement of complete pilows can be recognized in the Newbrough Forest (395641), commencing with the interpillow spaces and vacuoles being filled with dolomite and ending with a thin relict film of epidote and hematite marking the outermost skin of the former pilows. Some axinite and prehnite are present as minor metasomatic products.

[*Editor's Note:* Material, including Figure 2, has been omitted at this point.]

INTERPRETATIONS AND ANALOGIES

In view of the association of a full ophiolite suite with mélanges together with a steeply inclined zone of blueschists seen to pass below the almost contemporaneous edge of an initiating sedimentary basin, an analogy between North Wales on a small scale and western California on a larger scale can be drawn.

The North Wales relationships may be interpreted as the result of late Precambrian (Monian) deposition in a trench above a southeasterly descending subduction zone, which generated its own largely subaerial volcanic arc and gave rise to the Arvonian ignimbrites. The scale is comparable with that envisaged by Marshall Kay (1972) in his interpretation of the Dunnage Mélange of northeastern Newfoundland as the result of early Paleozoic subduction. That this involved the collision of two major crustal plates is not implied, but at least one of the plates was a small oceanic one (e.g., Dewey and Bird, 1970, p. 2641, fig. 12 A-D). It may be that limited subduction, perhaps along a series of contemporaneous *en echelon* zones of restricted linear extent, could have been a common occurrence during pre-Mesozoic time.

A limited and slow continuation of subduction could have accompanied the development of the Welsh Paleozoic basin during the Ordovician. The mid-Ordovician volcanicity of Snowdonia, the Caledonian alkali-rich intrusions, which are located at either end of the Padarn Pericline (fig. 1), and the riebeckite microgranite intrusions of Caernarvonshire could also be genetically related. The associated Late Ordovician copper mineralization of North Wales together with the mineralization of the Dolgelly gold belt could be ascribed to the same fundamental subduction.

It is equally possible that the Monian was partly consumed beneath continental crust, depressing the continental crustal margin to initiate the Welsh Basin. In either event, the older Monian rocks are comparable with the Franciscan Sequence of the California Coast Ranges, and the younger Welsh Basin rocks are comparable with the Great Valley Sequence. The model is similar to that proposed by Ernst (1970, fig. 3) for the Franciscan and explains the near synchroneity of glaucophane metamorphism and ignimbrite extrusion on the edge of the Welsh Basin and deposition of lowest Paleozoic sediments within the basin. The time separation between the oldest Monian sediments and youngest lower Paleozoic rocks of the Welsh Basin is greater than that between oldest Franciscan and youngest Great Valley rocks and may be accounted for by supposing a smaller amount of plate consumption and a slower rate of subduction for North Wales. Also, later sedimentation within the Welsh Basin may not have been directly related to events that delineated the basin's northwestern margin and provided its early sediment source.

The timing of late Precambrian metamorphism together with extrusive and intrusive activity in Wales, dated within the range of 610 to 580 my, finds parallels in the Appalachian region of North America. Certain stratigraphic similarities of basal Cambrian relationships between Wales and the Appalachian region have been noted previously (Wood, 1969). Events of the same age as those discussed here for North Wales occurred throughout the length of the Appalachian region. The rocks, relationships, and timing of events, which most closely simulate those of North Wales, are to be found in the Avalon Peninsula of eastern Newfoundland. As has been pointed out (e.g., Rodgers, 1972, p. 514), the volcanics and sediments of the Avalon Peninsula belong to a belt "whose north-eastern extension is to be sought in south-western Wales and Shropshire and perhaps also in southeastern Ireland and Anglesey."

The Harbour Main Group of the Avalon Peninsula consists of varied sediments and largely subaerial volcanics, which include ignimbrites. These are intruded by the Holyrood Granite, which in turn is unconformably overlain by the Lower Cambrian. The Holyrood Granite has yielded a rubidium-strontium isochron age of 574 ± 11 my (McCartney and others, 1966), which is in reasonable agreement with a re-calculated feldspar age of 565 ± 42 my obtained by Fairbairn (1965). Somewhat similar ages have been obtained farther south, where Hills and Dasch (1969) reported an age of 610 my for the crystallization of the Stony Creek Granite of southeastern Connecticut and where Fairbairn and others, (1967) recorded an isochron age of 559 ± 28 my for the Dedham Granodiorite of southeastern Massachusetts. The latter intrusion is overlain by the Lower Cambrian. In Virginia and North Carolina, Glover and others, (1971) have obtained lead ages of 620 my on zircons from the late Precambrian volcanics of the Virgilina Synclinorium and a minimum age of 570 my for the Roxboro Granodiorite, which is intrusive into the synclinorium.

There can be no doubt that the late Precambrian events of North Wales are part of a widespread major episode involving thick sedimentary accumulation, volcanicity, metamorphism, deformation, and plutonism. It is becoming clear for most parts of the Appalachian-Caledonian zone where relationships have not been obscured by Ordovician (Taconic), late Silurian (main Caledonian), or Acadian events, that there is a widely recognizable earlier event dated from 610 to 570 my. It is termed Avalonian in North America, Monian in Britain, and Cadomian in Brittany. The Avalonian-Monian event marked the initiation of the Appalachian-Caledonian mobile belt, and as such it deserves to stand as one of the four important activity phases of the Eocambrian and early Paleozoic. In Wales, the earliest phase' may have been characterized by small-plate tectonics, so to speak.

REFERENCES

Bailey, E. B., and McCallien, J. W., 1960, Some aspects of the Steinmann Trinity: mainly chemical: Geol. Soc. London Quart. Jour., v. 111, p. 365–395.

Blake, J. F., 1887, Introduction to the Monian System of rocks: British Assoc. Rept. (1886), p. 669.

———, 1888a, The Monian System of rocks: Geol. Soc. London Quart. Jour., v. 44, p. 463–547.

———, 1888b, The occurrence of a glaucophane-bearing rock in Anglesey: Geol. Mag., ser. 3, v. 5, p. 125–127.

Cowie, J. W., 1964, The Cambrian Period: Geol. Soc. London Quart. Jour., v. 120s, p. 225–228.

Davison, C., 1908, On some minor British earthquakes of the years 1904–1907: Geol. Mag., v. 45, p. 296–309.

———, 1924, A history of British earthquakes: Cambridge, England, Cambridge Univ. Press, 416 p.

Dewey, H., and Flett, J. S., 1911, Some British pillow-lavas and the rocks associated with them: Geol. Mag., v. 8, p. 202–209, 241–248.

Dewey, J. F., and Bird, J. M., 1970, Mountain belts and the new global tectonics: Jour. Geophys. Research, v. 75, p. 2625–2647.

Ernst, W. G., 1970, Tectonic contact between the Franciscan Mélange and the Great Valley Sequence—Crustal expression of a late Mesozoic Benioff Zone: *ibid.*, v. 75, p. 886–901.

Fairbairn, H. W., 1965, Personal communication in McCartney and others, 1966.

———, and others, 1967, Rb-Sr age of granitic rocks of southeastern Massachusetts and the age of the Lower Cambrian at Hoppin Hill: Earth and Planetary Sci. Letters, v. 2, p. 321–328.

Fitch, F. J., and Miller, J. A., 1964, Age of the paroxysmal Variscan Orogeny in England: Geol. Soc. London Quart. Jour., v. 120s, p. 159–173.

———, ———, and Brown, P. E., 1964, Age of the Caledonian Orogeny and metamorphism in Britain: Nature, v. 203, p. 275–278.

———, ———, and Meneisy, M. Y., 1963, Geochronological investigations on rocks from North Wales: *ibid.*, v. 199, p. 449–451.

———, and others, 1969, Isotopic age determinations on rocks from Wales and the Welsh borders, *in* Wood, A. (ed.), The Precambrian and lower Palaeozoic rocks of Wales: Cardiff, Univ. Wales Press, p. 23–45.

Glover, L., Sinha, A. K., and Higgins, M. W., 1971, Virgilina phase (Precambrian and Early Cambrian?) of the Avalonian Orogeny in the central Piedmont of Virginia and North Carolina (abs.): Geol. Soc. America Abs. with Programd, v. 3, p. 581–582.

Greenly, E., 1919, The geology of Anglesey: Geol. Survey Great Britain Mem., 2 v.

Henslow, J. S., 1822, Geological description of Anglesea: Cambridge Philos. Soc. Trans., v. 1, p. 359–452.

Hess, H. H., 1955, Serpentines, orogeny and epeirogeny, *in* Poldervaart, A. (ed.), The crust of the earth: Geol. Soc. America Special Paper 62, p. 391–408.

Hills, F. A., and Dasch, E. J., 1969, Rb-Sr evidence for metamorphic remobilization of the Stony Creek Granite, southeastern Connecticut (abs.): Geol. Soc. America Special Paper 121, p. 136–137.

Jones, O. T., 1956, The geological evolution of Wales and the adjacent regions: Geol. Soc. London Quart. Jour., v. 111, p. 323–350.

Kay, M., 1972, Dunnage Mélange and lower Paleozoic deformation in northeastern Newfoundland: 24th Internat. Geol. Cong., Sec. 3, p. 122–133.
Ksiakiewicz, M., 1960, Pre-orogenic sedimentation in the Carpathian Geosyncline: Geol. Rundschau, v. 50, p. 8–31.
McCartney, W. D., and others, 1966, Rb-Sr age and geological setting of the Holyrood Granite, southeast Newfoundland: Canadian Jour. Earth Sci., v. 3, p. 947–957.
Matley, C. A., 1913, The geology of Bardsey Island: Geol. Soc. London Quart. Jour., v. 29, p. 514–533.
———, 1928, The Precambrian complex and associated rocks of southwestern Lleyn: *ibid.*, v. 84, p. 440–504.
Moorbath, S., and Shackleton, R. M., 1966, Isotopic ages from the Precambrian Mona Complex of Anglesey, North Wales, Great Britain: Earth and Planetary Sci. Letters, v. 1, p. 113–117.
Nataraj, T. S., 1967, Glaucophanic metamorphism in Anglesey (thesis); Leeds, England, Univ. Leeds, 103 p.
Raisin, C. A., 1893, Variolite of the Lleyn and associated volcanic rocks: Geol. Soc. London Quart. Jour., v. 44, p. 145–165.
Rodgers, J., 1972, Latest Precambrian (post-Grenville) rocks of the Appalachian region: Am. Jour. Sci., v. 272, p. 507–520.
Sedgwick, A., 1838, Synopsis of the English stratified rocks inferior to the Old Red Sandstone: Geol. Soc. London Proc., v. 2, p. 675–685.
———, 1843, An outline of the geological structure of North Wales: *ibid.*, v. 4, p. 212–224.
Shackleton, R. M., 1953, The structural evolution of North Wales: Liverpool and Manchester Geol. Jour., v. 1, p. 261–297.
———, 1954, The structure and succession of Anglesey and the Lleyn Peninsula: British Assoc. Adv. Sci., v. 11 (41), p. 106–108.
———, 1956, Notes on the structure and relations of the Precambrian and Ordovician rocks of southwestern Lleyn (Caernarvonshire): Liverpool and Manchester Geol. Jour., v. 1, p. 400–409.
———, 1957, Downward-facing structures of the Highland Border: Geol. Soc. London Quart. Jour., v. 113, p. 361–392.
———, 1969, The Precambrian of North Wales, in Wood, A. (ed.), The Precambrian and lower Palaeozoic rocks of Wales: Cardiff, Univ. Wales Press, p. 1–22.
Sharpe, D., 1846, Contributions to the geology of North Wales: Geol. Soc. London Quart. Jour., v. 2, p. 283–316.
Steinmann, G., 1905, Gelogische Beobachtungen in den Alpen: Naturf. Gesell. Freiburg, Ber., v. 16, p. 18–67.
Wood, D. S., 1960, The geology and structure of the Rhoscolyn district, Holy Island, Anglesey (thesis): Liverpool, England, Univ. Liverpool, 65 p.
———, 1969, The base and correlation of the Cambrian rocks of North Wales, *in* Wood, A. (ed.), The Precambrian and lower Palaeozoic rocks of Wales: Cardiff, Univ. Wales Press, p. 47–66.

12

Reprinted from pages 1-11 of *North American Ophiolites,* Oregon Dept. Geology and Mineral Industries Bull. 95, 1977

DISTRIBUTION AND TECTONIC SETTING OF OPHIOLITES AND OPHIOLITIC MÉLANGES IN THE APPALACHIAN OROGEN

H. Williams and R. W. Talkington

[*Editor's Note:* Plate I, a large foldout map, has been omitted because reduction would make it illegible.]

ABSTRACT

Ophiolites of highly allochthonous character and associated ophiolitic melange occur along the western margin of the Appalachian Orogen, i.e. the ancient continental margin of eastern North America (Humber Zone). The Bay of Islands Complex is a typical example of a transported ophiolite, and ophiolitic melanges such as the Milan Arm, Second Pond and Coachman's Melanges are all interpreted as related to ophiolite obduction. The melanges vary in structural style from undeformed thin sheets in the west to polydeformed and metamorphosed in the east.

Small mafic-ultramafic bodies in eastern parts of the Humber Zone occur along the entire length of the orogen. Some of these are blocks in ophiolitic melange and others are sited at major structural contacts. They are affected by the full range of deformations and metamorphism related to the destruction of the ancient continental margin of eastern North America. Most are probably dismembered parts of ophiolite suites, and their occurrence from Newfoundland to Alabama implies a similar structural history for western parts of the system.

In central parts of the Appalachian Orogen, ophiolites that represent vestiges of an ancient Iapetus Ocean (Dunnage Zone) occur along the Baie Verte-Brompton Line and they form the basement to island arc sequences farther east. Mafic-ultramafic complexes along eastern parts of the Dunnage Zone in the north may be blocks in melange, and some may represent diapiric intrusions.

Relationships in the northern Appalachians imply a major suture within the exposed parts of the southern Appalachians, either at the Brevard Zone or somewhere between the Brevard Zone and the Carolina Slate Belt.

Appalachian ophiolites are Late Cambrian and Early Ordovician in age, where dated, and they were emplaced and deformed during the Ordovician Taconic Orogeny. Silurian or younger ophiolites are unknown and the stratigraphic record for this period indicates largely terrestrial conditions without the presence of important continental margins or major oceans.

INTRODUCTION

The interpretation of the on-land ophiolite suite of rock units as oceanic crust and mantle has brought new and considerable interest to the study of ophiolitic rocks in orogenic belts. Because the ophiolite suite of rock units relate in a total way to their place and mode of generation, their presence provides an important adjunct to the interpretation of the regional geology of the orogen in which they are found. In ancient orogenic belts, ophiolite suites occur as highly allochthonous structural slices transported across former continental margins, e.g. western Newfoundland, Oman, Zagros, Himalayas, or as imbricated and deformed slices in central parts of orogens, e.g. northeastern Newfoundland, Ballantrae, Indus Suture, and Pindus Zone. Ophiolitic melanges (Gansser, 1974), or chaotic rocks that contain blocks derived from the ophiolite suite of rock units are associated with allochthonous ophiolites (obduction) or they are related to the destruction of oceanic crust by its descent at oceanic trenches (subduction). The distribution and structural setting of ophiolitic melanges in orogenic belts are in most cases therefore as important to tectonic syntheses as the presence of the ophiolite suite itself.

The common occurrence of mafic-ultramafic rocks in the Appalachian Orogen was first pointed out by Hess (1939; 1955), and since then there have been numerous attempts to explain their origin. Most occurrences are interpreted now as ancient oceanic crust and mantle (Moores, 1970; Stevens, 1970; Church and Stevens, 1971; Dewey and Bird, 1971). Some are part of well-preserved ophiolite suites, e.g. Bay of Islands Complex, others are dismembered ophiolites, e.g. Advocate Complex, and still others form blocks in ophiolitic melange, e.g. Coachman's Melange. In a few cases, local examples are interpreted as intrusions (Chidester and Cady, 1972; Kennedy and Phillips, 1971) or mantle diapirs situated above subduction zones (Stevens and others 1974; Kean, 1974).

Although Hess suggested that there were two ultramafic belts in the Appalachian Orogen, new data require a revision of this view and indicate a wide spectrum of occurrences in a variety of different tectonic settings. Recent geologic syntheses of the Appalachian Orogen are based upon the recognition of contrasting tectonic-stratigraphic zones across the system. Nine zones are defined in the Canadian Appalachians (Williams and others 1972; 1974) and these have been amalgamated for purposes of broad correlation into five zones that are extrapolated along the full length of the system from Newfoundland to Alabama (Williams, 1976). From west to east these zones are given local names in the northern Appalachians, viz: Humber, Dunnage, Gander, Avalon, Meguma (plate 1). Ophiolitic rocks are restricted almost entirely to the Humber and Dunnage Zones.

The model for the development of the Appalachian Orogen follows the suggestion of Wilson (1966) and involves the generation and destruction of a late Precambrian - early Paleozoic Iapetus Ocean (Dewey, 1969, Bird and Dewey, 1970; Stevens, 1970; McKerrow and Cocks, 1977; etc.). The Humber Zone records the development and destruction of an Atlantic type continental margin of eastern North America (Williams and Stevens, 1974). It contains the best examples of transported complete ophiolite suites (e.g. Bay of Islands Complex; Williams, 1973) as well as a variety of ophiolitic melanges in various states of structural complexity. The Dunnage Zone represents the former site of the Iapetus Ocean. In places it exhibits well-developed ophiolite suites (e.g. Betts Cove Complex, Upadhyay and others 1971; Thetford Mines ophiolites, Laurent, 1975), commonly overlain by thick island arc volcanic sequences. As well, it contains melanges that possibly relate to subduction (e.g. Dunnage Melange, Kay, 1976; Fournier Complex, N. Rast, pers. comm. 1975), and in the northeast a belt of mafic-ultramafic complexes that are either blocks in melange or mantle diapirs. The Gander and Avalon Zones developed upon continental crust and lay to the east and southeast of the Iapetus Ocean. Neither contains well-preserved examples of a complete ophiolite. The Meguma Zone may represent the eastern continental margin of an ancient ocean that lay to the east of the Avalon Zone (Schenk, 1971). Like the Gander and Avalon Zones, it too is devoid of ophiolite suites.

In the Canadian Appalachians, the boundary between the Humber and Dunnage Zones is marked by the occurrence of ophiolites in a steep structural belt. These ophiolites can be traced as discontinuous bodies from Baie Verte, Newfoundland, to Brompton Lake, Quebec. Accordingly, the steep ophiolite zone has been termed the Baie Verte-Brompton Line (St. Julien and others 1976). It is an important structural junction in the Northern Appalachians and its ophiolites are host to the asbestos deposits that make this zone the world's richest asbestos belt. In places where the ophiolitic rocks are absent, the Humber-Dunnage boundary is marked by faults that separate Humber Zone metamorphosed clastics (west) and Dunnage Zone less metamorphosed volcanic rocks (east). In other places, the boundary is hidden by Silurian and younger cover rocks, e.g. Gaspé Peninsula (plate 1).

South of the Canada-United States border, the projection of the Baie Verte-Brompton Line is marked by a zone of small isolated ultramafic bodies in Vermont that extends all the way southward to Staten Island, New York. Farther South, the Baltimore Gabbro Complex (Crowley, 1969) of Maryland lies at or near the continuation of the same structural zone, and the zone may be marked by local occurrences of ophiolitic melange in the James River Synclinorium (Brown, 1976). From there, it projects between the Grenvillian basement rocks of the Blue Ridge and Sauratown Mountains, and appears to continue farther south along the Brevard Zone (Hatcher, 1972). In the northern Appalachians, the Baie Verte-Brompton Line marks an ancient continent-ocean interface and it is the most westerly possible root zone for allochthonous ophiolites found farther west. If this same structural zone continues southward, as proposed, the Brevard Zone of the southern Appalachians is an important suture marking the site of the former Iapetus Ocean.

The distrubution and tectonic setting of ophiolitic rocks in the Humber and Dunnage Zones form the basis of the discussion that follows. No attempt is made to describe each individual ophiolite occurrence. Instead, a description of the regional geology and extrapolations along the length of the system for each kind of ophiolite occurrence is followed by a description of a type example. Most examples are taken from the northern Appalachians because of a greater familiarity to the authors and because southerly examples are treated by other contributions to this volume (see papers by Laurent and Morgan).

OPHIOLITES AND OPHIOLITIC MELANGES OF THE HUMBER ZONE

The Humber Zone consists of a crystalline Grenvillian basement overlain by a thick clastic sequence with associated volcanic rocks, and a prominent Cambro-Ordovician carbonate sequence. Ophiolites of highly allochthonous nature occur in western parts of the zone where they overlie relatively undeformed parts of the autochthonous carbonate sequence. These ophiolites are associated with transported sedimentary rocks and collectively they constitute allochthons emplaced during Middle Ordovician.

The carbonate sequence of the Humber Zone and correlative coarse limestone breccias in overlying structural slices are interpreted as bank and bank foot deposits, respectively, formed at the ancient continental margin of eastern North America (Rodgers, 1968). Underlying clastics that rest on Grenvillian gneisses formed a rise prism at the margin (Williams and Stevens, 1974), and transported ophiolites represent oceanic crust and mantle that lay farther east (Church and Stevens, 1971).

Examples of allochthonous ophiolites in westerly parts of the Humber Zone include the White Hills Peridotite of the Hare Bay Allochthon, the Bay of Islands Complex of the Humber Arm Allochthon, and the Mount Albert ophiolite of the Shick Shock Mountains (Williams, 1975). The Baltimore Gabbro Complex of Maryland occurs in a similar structural position but lies nearer the east boundary of the zone.

Ophiolitic melanges occur across the Humber Zone and they are particularly well-developed in western Newfoundland. Their formation is attributed to the transport of ophiolites from their root zone at the Baie Verte-Brompton Line, across the rise prism and carbonate bank successions, to their present positions. The best exposed and most extensive melanges form integral parts of the Humber Arm and Hare Bay Allochthons, e.g. Companion and Milan Arm Melanges (Williams, 1975), and comparable examples, though lacking ophiolitic blocks in most places, are associated with transported sedimentary rocks in Taconic klippen all the way southward to Harrisburg, Pennsylvania. These melanges are mainly thin subhorizontal sheets between other transported rocks that collectively lie above the carbonate sequence. Ophiolitic melanges occur also at the east margin of the carbonate terrane in Maryland and in western White Bay, Newfoundland, e.g. Second Pond Melange (Williams, 1977a). Farther east, ophiolitic melanges are associated with clastics of the rise prism at the eastern margin of the Humber Zone (e.g. Coachman's Melange, Williams, 1977b).

Deformation and metamorphism increase from west to east across the Humber Zone. Melanges associated with Taconic-type allochthons above the carbonate terrane have been little deformed since formation. These near the present easternmost exposures of the carbonate sequence vary from polydeformed and metamorphosed in Maryland to locally deformed and mildly metamorphosed in Newfoundland. Ophiolitic melanges at the eastern margin of the Humber Zone are everywhere polydeformed and metamorphosed and now bear little resemblance to occurrences farther west.

Small mafic-ultramafic bodies are common throughout the full length of the Appalachian Orogen in the belt of deformed clastic rocks at the eastern margin of the Humber Zone. Although locally interpreted as intrusions, some are clearly blocks in ophiolitic melange and other isolated bodies occur at structural contacts. All are affected by the full range of deformation and metamorphism that accompanied the destruction of the ancient continental margin of eastern North America. Examples in easterly parts of the Fleur de Lys Supergroup in Newfoundland are thought to occur at structural discontinuities (Williams and others, 1977), the Pennington Dike and nearby ultramafic bodies of the Eastern Townships of Quebec are sited at structural contacts (Pierre St. Julien, pers. comm. 1975), the continuous string of small ultramafic occurrences from Vermont to Staten Island marks a zone of nappes and imbricate slices (Barry Doolan, pers. comm. 1976), and many of the small ultramafic bodies of the eastern Blue Ridge from Virginia to Alabama may owe their presence to structural emplacement.

If these small mafic-ultramafic occurrences are intrusions, there is no apparent mechanism or obvious reason for their emplacement into an undeformed rise prism of clastic sediments. More likely, they represent blocks in melange and dismembered ophiolite at structural contacts. Their widespread occurrence within the deformed and metamorphosed clastic terrane at the eastern margin of the Humber Zone implies a similar early tectonic history for the full length of the western part of the Appalachian System.

The distinction between the Humber and Dunnage Zones is subtle in places where deformed ophiolitic melanges, which are incorporated structurally within the Humber Zone clastics, are juxtaposed with ophiolite suites along the Baie Verte-Brompton Line. The latter are a natural part of the Dunnage Zone, but the melanges are grouped in places with the Humber Zone clastic rocks and considered a natural part of local stratigraphic successions. This situation exists at the Baie Verte-Brompton Line in Newfoundland and it may be a common circumstance elsewhere. Stratigraphic studies in metamorphic terranes that include ophiolitic melanges or small ultramafic bodies of possible ophiolitic parentage should be made with extreme caution, as experience has shown that major structural disruptions have gone unnoticed in the polydeformed and metamorphosed rocks immeidately west of the Baie Verte-Brompton Line (Williams and others 1977; Williams, 1977b; Pierre St. Julien, pers. comm. 1976; Barry Doolan, pers. comm. 1976).

Allochthonous complete ophiolite suites: The Bay of Islands Complex

The Bay of Islands Complex affords an excellent example of an allochthonous complete ophiolite suite that forms the highest structural slice of a composite allochthon in the western part of the Humber Zone. It is represented in four separate massifs, which from south to north are Lewis Hills, Blow Me Down, North Arm Mountain, and Table Mountain (pl. I). All lie in the same structural position and either represent separate transported bodies or erosional remnants of a once continuous slice. Two of the massifs (Blow Me Down and North Arm Mountain) display a completely developed ophiolite stratigraphy, but all four include the basal ultramafic unit.

The sequences of ophiolite units in the three northernmost massifs are disposed in synclines with northeast-trending subhorizontal axes and moderately to steeply dipping limbs. The present tectonic base of each massif is subhorizontal so that the ophiolite

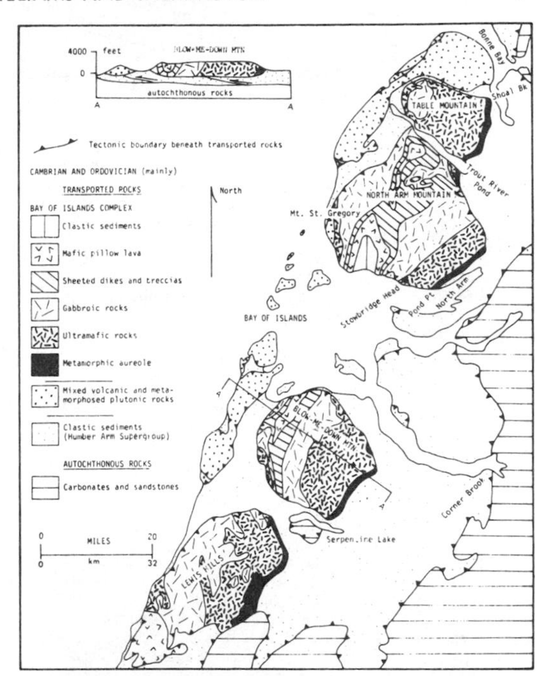

Figure 1: Geologic setting of the Bay of Islands Complex, Western Newfoundland (after Williams and Smyth, 1973).

units are structurally truncated at depth in much the same way as they are truncated at their top by the present erosional surface (see cross-section AA, fig. 1). A contact between the stratigraphic base of the ophiolite sequence and a dynamo-thermal aureole of supracrustal rocks, now frozen into the folded ophiolite slices, is interpreted to represent the actual zone of obduction where the hot oceanic plate moved across the continental margin. The contacts of latest emplacement of the structural slices are marked now by thin zones of shale melange with sedimentary, volcanic, and sparse gabbro and serpentinite blocks. These are the result of mass wastage and tectonic mixing that accompanied gravity sliding.

Trondhjemite from the Bay of Islands Complex has been dated isotopically at 504 m.y. $\pm$ 10 (Mattinson, 1976) and amphiboles from its dynamo-thermal aureole at 460 to 470 m.y. (Dallmeyer and Williams, 1975; Archibald and Farrar, 1976). The former gives the time of generation of the ophiolite suite as Late Cambrian. The latter indicates an Early Ordovician age for initial displacement and agrees with the sedimentologic evidence of ophiolite detritus in Lower Ordovician sedimentary rocks deposited during ophiolite transport (Stevens, 1970). An upper limit to the time of final emplacement is given by the Middle Ordovician age of the neo-autochthonous Long Point Formation (Bergström and others 1974). For more detailed descriptions the reader is referred to Smith (1958) and Malpas (this volume).

Ophiolitic melanges of the Humber Zone

The Milan Arm Melange (Williams, 1975) of northern Newfoundland is the clearest example of an ophiolitic melange that forms an integral part of a Taconic-type allochthon above the Humber Zone carbonate sequence. It structurally overlies autochthonous rocks in some places and intervening transported clastics of the rise prism (Maiden Point Formation) in other places. It is in turn structurally overlain by the ophiolitic White Hills Peridotite. The melange is similar in most respects to all of the shale melanges that separate structural slices of the west Newfoundland allochthons, however, it has a much wider variety of exotic blocks; some up to a kilometer or more across and resembling the largest slices that make up the west Newfoundland allochthons.

Its commonest blocks are serpentinized peridotite, mafic volcanic rocks, amphibolites, foliated gabbro, greywacke, diorite, and exceptionally coarse grained pyroxenite and hornblendite associated with tonalite and hornblende-biotite schist. Nephrite blocks are also known locally (R.K. Stevens, pers. comm. 1976). Most of these rock types can be matched directly with rocks in nearby structural slices, but a few are of unknown origin.

Many of the amphibolite, gabbro, and diorite blocks are encased in a relatively thin, hard rind of light grey calc-silicate alteration products (rodingite). The tough and resistant alteration halos form coastal wave-washed outcrop surfaces where the matrix shales are eroded. In some examples the rodingite alteration halos are surrounded in turn by a thinner serpentinite coating, implying that the rodingite represents an alteration zone between mafic rocks and serpentinite. These blocks appear therefore to have been once immersed in serpentinite or serpentinite melange, so that they are recycled where they now occur in a shale matrix.

Local serpentinite and gabbro blocks in melanges at the base of the west Newfoundland allochthons imply that the sequences of transported slices were emplaced as already-assembled allochthons (Stevens and Williams, 1973).

The recently recognized Coachman's Melange (Williams, 1977b) is an example of a polydeformed and metamorphosed ophiolitic melange that occurs at the eastern margin of the Newfoundland Humber Zone. It is closely associated with psammitic schists of the rise prism (Fleur de Lys Supergroup) and it has been affected by the full range of deformations recognized in nearby rocks. The melange occurs in a multitude of narrow zones that rarely exceed more than 50 m in structural thickness. If all occurrences represent a complexly folded single unit, exceedingly tight isoclines of more than 3 km amplitude affected easternmost local exposures of the rise prism.

The Coachman's Melange has a black pelitic matrix with conspicuous deformed and recrystallized ultramafic blocks now represented by bright green actinolite-fuchsite schist. Sedimentary blocks with ill-defined outlines are common everywhere, and in some places large serpentinized ultramafic blocks, foliated gabbro blocks, and marble are also known.

Actinolite-fuchsite schist occurs in lenses from 10 cm to 3 m in length and rarely more than 1 m in width. They exhibit minor folds and folded schistosities identical to structures in the surrounding schistose matrix and nearby psammitic schists. Pale green actinolite crystals are locally 2-4 cm in length, set in a fine-grained fuchsite-carbonate matrix. An ultramafic origin for these rocks is indicated by their mineralogy and because larger nearby ultramafic blocks are recrystallized to similar mineral assemblages at their margins. Interior parts of large ultramafic blocks are in places brecciated, and this feature predates both serpentinization and incorporation into the melange.

Recognition of the Coachman's Melange and an appreciation of its complex structural history bears upon one of the major problems of northern Appalachian geology, i.e. the timing of deformation and metamorphism within the rise prism in relation to the time of generation of nearby ophiolite suites and the time of their transport across an ancient continental margin. As is the case with other worldwide examples of ophiolitic melanges, the Coachman's Melange implies transport of oceanic crust across the rise prism represented by the local Fleur de Lys Supergroup. This transport is equated most reasonably with the emplacement of highly allochthonous ophiolites in western parts of the Humber Zone from an initial position to the east of the Baie Verte-Brompton Line. Similar structural histories for both the Coachman's Melange and nearby parts of the Fleur de Lys Supergroup indicate that the rise prism was undeformed at the time of melange formation and initial ophiolite displacement. This conclusion leads to a simple model for the place of origin and time of transport of ophiolitic complexes in western Newfoundland compared to the time of deformation and metamorphism in the intervening Fleur de Lys terrane (fig. 2). As well, it explains the marked structural contrasts between the Fleur de Lys Supergroup and nearby ophiolite suites, while implying a mechanism for deformation and metamorphism through ophiolite transport and structural loading at the ancient continental margin.

Ophiolitic melanges comparable to those at Coachman's Harbour, Newfoundland are unknown elsewhere in the Appalachian Humber Zone. Other occurrences are predicted because of structural similarities along the length of the system.

OPHIOLITES AND OPHIOLITIC MELANGES OF THE DUNNAGE ZONE

Ophiolites and ophiolitic melanges are represented in the Dunnage Zone from Newfoundland to Virginia. Farther south, rocks typical of the Dunnage Zone are absent and the Humber Zone is bordered eastward by crystalline rocks of the Inner Piedmont (Hatcher, 1972). Relationships in the northern Appalachians predict a major suture in the southern Appalachians, either at the Brevard Zone or somewhere between this zone and Avalon Zone equivalents of the Carolina Slate Belt.

The most prominent ophiolite occurrences in the Dunnage Zone are found at its western margin along the Baie Verte-Brompton Line. In Newfoundland, examples can be traced from Baie Verte of the Burlington Peninsula, e.g. Advocate and Point Rousse Complexes (Williams and others 1977) to Glover Island of Grand Lake. From there, the Baie Verte-Brompton Line is ill-defined, but it is probably coincident with the Cape Ray Suture (Brown, 1973) farther south, and transported ophiolites at Cape Ray presumably root in this zone. In mainland Canada, volcanic rocks and deformed mafic-ultramafic rocks of the Fournier Complex, New Brunswick may lie at or near the Baie Verte-Brompton Line Farther west, the line is

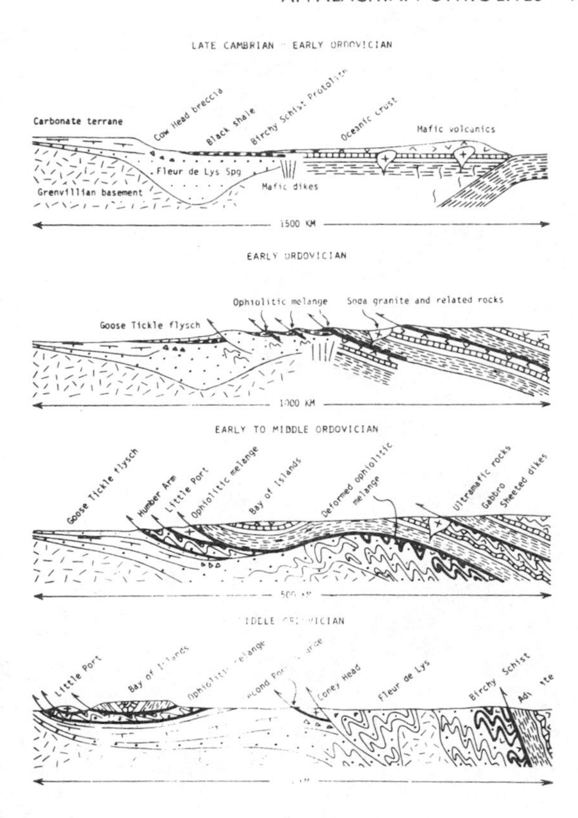

Figure 2: Model for the development of ophiolitic melange and the transport of ophiolites across the Humber Zone (after Williams, 1977b).

covered by Upper Ordovician rocks of the Metapedia Belt and Silurian rocks of the Gaspé Synclinorium. Where the Baie Verte-Brompton Line reappears in the Eastern Townships of Quebec, it is marked by major ophiolite occurrences along most of its length.

The Baie Verte-Brompton Line is a zone of intense deformation and ophiolites along its length are in most places thrust-imbricated and penetratively deformed. The ophiolite suites are mainly steeply-dipping to overturned, but their stratigraphic sections face eastwards. In Newfoundland, the Baie Verte-Brompton Line is a zone of marked gravity gradient with negative anomalies on its western continental side, and positive anomalies across oceanic rocks to the east (Miller and Deutsch, 1976).

East of the Baie Verte-Brompton Line, separate ophiolite occurrences are common throughout central parts of the Dunnage Zone. These form the basement to thick volcanic arc sequences that are locally dated as Lower Ordovician, thus defining an upper age limit for the underlying ophiolite suites. In Newfoundland, the Betts Cove Complex is probably the best known example, but sheeted dikes occur throughout the central volcanic terrane (Strong, 1972) and gabbros and ultramafic rocks occur locally at South Pond, Brighton and Gull Island of Cape St. John. In Maine, mafic-ultramafic rocks along the southern margin of the Chain Lakes massif are interpreted as ophiolites (G.M. Boone, pers. comm. 1975). Their occurrence and relationships to nearby crystalline rocks are still poorly understood.

Olistostromes that contain ophiolitic blocks occur along the east margin of the Thetford Mines ophiolite belt, e.g. St. Daniel Formation (St. Julien and Hubert, 1975), and megaconglomerates with outcrop size gabbro and penetratively deformed and altered ultramafic blocks occur along the east margin of the Advocate Complex in Newfoundland. The significance and time of deposition of these rocks is still poorly understood, but in Newfoundland, deposition post-dates earliest deformations in nearby ophiolites and predates Silurian volcanism. Farther south in the New England Appalachians, small metamorphosed ultramafic bodies in sulphidic schists of the Partridge Formation may represent blocks in olistostrome rather than small intrusions.

The Dunnage Melange (Kay, 1976), which lies to the east of the central Newfoundland island arc terrane, is not itself ophiolitic but a similar melange, 20 km eastward at Carmanville, contains sparse ultramafic blocks. The Carmanville melange may represent a subsurface continuation of the Dunnage. The Dunnage Melange has been interpreted as an oceanic trench fill (Dewey and Bird, 1971; Williams and Hibbard, 1976; Kay, 1976). If so, then ophiolitic melange would be expected in this position.

A prominent belt of mafic-ultramafic complexes along the eastern margin of the Dunnage Zone in Newfoundland, and east of the Dunnage Melange, remains poorly understood in present models for the development of the Appalachian Orogen. Some occurrences contain ultramafic rocks, gabbros and volcanic rocks that collectively are reminiscent of an ophiolite suite, e.g. Pipestone Pond (Kean, 1974). Others are mainly clinopyroxenite bodies in structural contact with surrounding dark shales, e.g. Gander River Belt, and still others may represent differentiated intrusions of mantle derivation, e.g. Great Bend of Gander River (Stevens and others 1974). At one locality on the north shore of Gander Lake, penetratively deformed ultramafic rocks are unconformably overlain by conglomerates of probable Caradocian age, thus defining an upper age limit for some of these occurrences.

Transported ophiolites of the Humber Zone, ophiolites at the Baie Verte-Brompton Line, and ophiolites that form the basement to island arc sequences across the Dunnage Zone may all relate to a single cycle of ophiolite generation (as summarized in Figure 2). This simple view is contrasted with an earlier interpretation that relates each ophiolite belt to an equal number of small ocean basins that formed, at least in part, after deformation and metamorphism of the Humber Zone continental rise prism (Dewey and Bird, 1971; Kennedy, 1975; Kidd, in press).

Ophiolites at the Baie Verte-Brompton Line

Ophiolites along the Baie Verte-Brompton Line are bounded to the west by polydeformed and metamorphosed clastic rocks of the Humber Zone, and they are bounded to the east by olistostromes and volcanic sequences. Several occurrences in Newfoundland and Quebec are overlain by thick volcanic sequences, which are similar to volcanic arc sequences found above ophiolites farther east. The Advocate and Point Rousse Complexes along the Baie Verte-Brompton Line

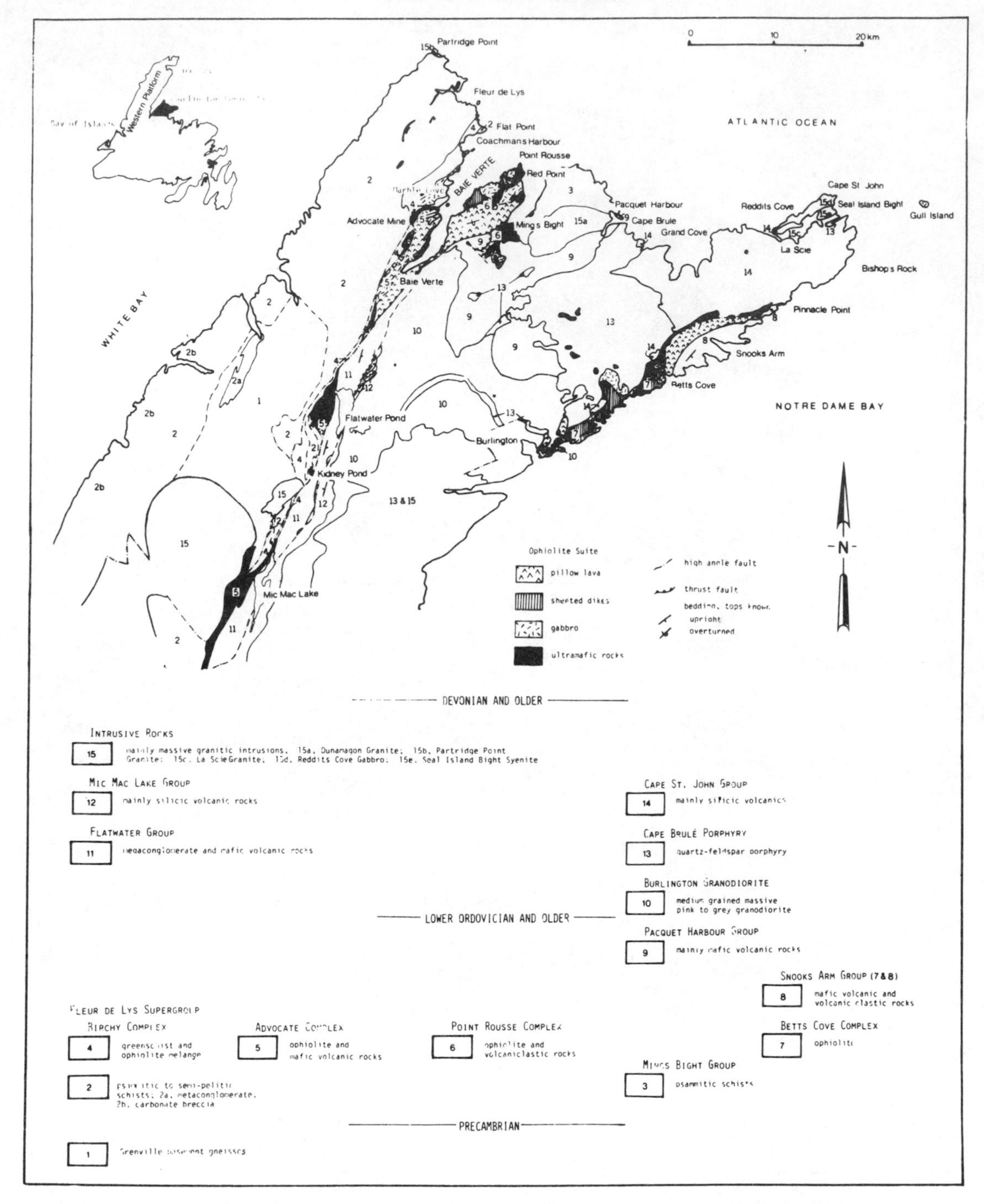

Figure 3: Distribution of ophiolites at and near the Baie Verte-Brompton Line, Burlington Peninsula, Newfoundland (After Williams and others, 1977).

in northern Newfoundland are taken as a type example and their distribution and relationships to nearby groups are summarized in Figure 3.

The Advocate Complex can be traced from Baie Verte 60 km southward beyond Mic Mac Lake. It has a steep northeast-trending foliation in most places and it is cut by numerous steep northeast-trending shear zones that repeat its rock units. A large ultramafic body at its northwest margin near Baie Verte consists of intensely foliated and fractured serpentinites that are host to the Advocate asbestos deposit. Its gabbros vary from massive to intensely foliated and some are distinctive white-altered rocks with green fuchsitic smears, e.g. those at Marble Cove. Sheeted dikes are best exposed 1 km northeast of Baie Verte and well-preserved pillow lavas occur nearby.

Black and green slates, black pebbly slates and chaotic black slaty melange are also present in the Advocate Complex. Some of these rocks occur at tectonic contacts so that they are structurally comingled with the segmented ophiolite. Their chaotic character results, in part, from tectonic processes that accompanied imbrication of the ophiolite suites. Others may be depositional breccias. Some of these chaotic shale zones in the Advocate Complex resemble nearby chaotic rocks and black schists of the Coachman's Melange, so that all appear to mark significant structural dislocations.

The Point Rousse Complex is made up of several distinct structural blocks with their tectonic boundaries marked by foliated serpentinite or foliated carbonate-talc-fuchsite alterations of ultramafic rocks (Norman and Strong, 1975). Three separate blocks on Point Rousse Peninsula contain southeast-facing, overturned sections of gabbro, sheeted dikes and pillow lava. The most complete section occurs south of Red Point where a northwest-dipping, southeast-facing sequence of gabbro, sheeted dikes and pillow lava is followed by local chert beds and a thick section of volcaniclastic rocks, all southeast-facing. Locally, on the western side of the Point Rousse Peninsula, pillow lavas of the complex are intensely deformed and converted to greenschists. Southward thrusting of gabbros above the greenschists postdates the development of a steep foliation in the mafic volcanic rocks.

The Point Rousse Complex is less altered and deformed than the Advocate Complex, except in local zones of intense deformation. Farther east, the Betts Cove Complex is even less deformed. The regional distribution of ophiolite complexes across the Burlington Peninsula and the pattern of their deformation suggest westward imbrication of east facing ophiolitic suites with pervasive intense deformation in lower levels (Coachman's Melange, Advocate Complex) and less intense deformation and fewer deformed zones higher in the structural pile (Point Rousse Complex, Betts Cove Complex).

Ophiolites beneath volcanic arc sequences

The Betts Cove Complex (fig. 4) is the clearest example of an ophiolite suite that forms the basement to a volcanic arc sequence. It consists of a basal ultramafic member overlain transitionally by a poorly developed gabbroic member, in turn overlain by a sheeted dike complex that consists of practically 100% mafic dikes (Upadhyay and others 1971). The sheeted dike complex is faulted against nearby mafic volcanic rocks, but locally the contact is gradational

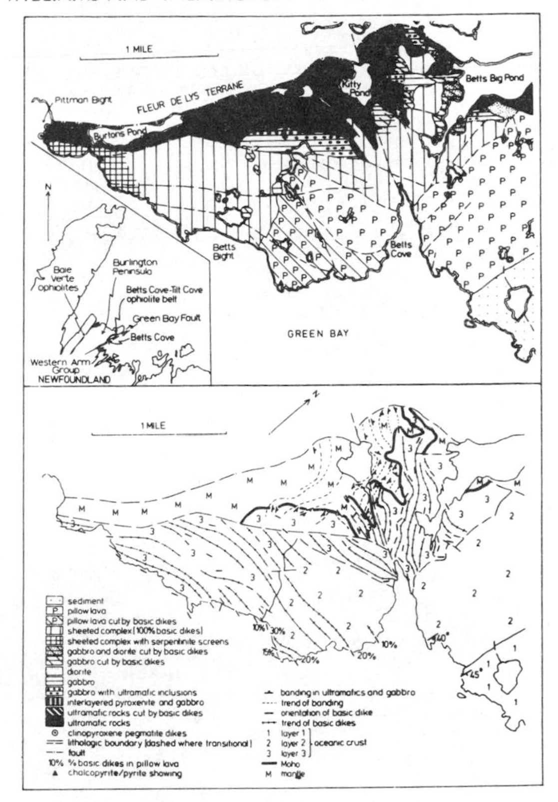

Figure 4: Geology of the Betts Cove Complex and its comparison to models of oceanic crust and mantle (after Upadhyay and others 1971).

across a narrow zone that shows a large decrease in the percentage of dikes over a short distance. Overlying Lower Ordovician rocks of the Snooks Arm Group are nearly 4 km thick and comprise a conformable sequence of pillow lavas, cherts, argillites, andesitic pyroclastic rocks and immature volcanic sediments. A lowermost pillow lava unit constitutes the upper part of the Betts Cove Complex so that a completely conformable transition exists from the ultramafic member of the ophiolite complex to the top of the thick overlying mainly volcanic succession.

Nearby correlatives of the Snooks Arm Group show a lithological evolution from lowermost pillow lavas upward through marine cherts and turbidites into pyroclastic rocks and volcaniclastic sedimentary rocks capped by limestone and subareal tuffs. The overall deep to shallow water lithic change is accompanied by geochemical changes in the volcanic stratigraphy from low potassium tholeiites at the base to calc-alkaline low-silica andesites toward the top that show progressive enrichment in Al_2O_3 and K_2O and a decrease in CaO and MgO (Kean and Strong, 1975). Accordingly, the Dunnage Zone volcanic sequences are interpreted as analogues to modern island arcs and the relationships at Betts Cove leave little doubt that volcanic rocks of the Dunnage Zone, at least in westerly exposures, directly overlie oceanic crust.

The prominent island arc volcanism of the Dunnage Zone continued into the Middle Ordovician, indicating that island arcs were evolving in the Dunnage Zone during the emplacement of allochthonous ophiolites across the Humber Zone. A cessation to Ordovician volcanism and the deposition of Caradocian black shales across the entire Dunnage Zone coincides exactly with the time of final emplacement of allochthonous ophiolites such as the Bay of Islands Complex in the Humber Zone.

Melanges, olistostromes and megaconglomerates of the Dunnage Zone

The Dunnage Melange, though devoid of ophiolitic components, is included in this analysis because of its similarity and possible subsurface continuity with melange at Carmanville that locally contains ultramafic and gabbroic blocks. As well, it is a prominent feature of the Dunnage Zone and its interpretation as an oceanic trench deposit bears heavily upon models proposed for the development of the northern Appalachians.

The melange is a strikingly heterogeneous deposit composed of blocks of mainly clastic sedimentary and mafic volcanic rocks enveloped in a dark scaly shale matrix. It is well-exposed along the rugged coast and clusters of islands of the Bay of Exploits, where it extends for 40 km along strike with a maximum outcrop width of 10 km (fig. 5). Its clasts vary in size from granules and cobbles to boulders and huge blocks up to a kilometre in diameter, thus producing a chaotic mosaic that contrasts sharply with nearby stratified volcanic and sedimentary rocks. Most blocks are indigeneous to nearby volcanic arc sequences and they can be matched with formations of the Exploits and Summerford Groups (Williams and Hibbard, 1976). Shale is much more important in the melange than in nearby terranes.

The Dunnage Melange overlies and interdigitates with the New Bay Formation of the Exploits Group in the southwest, and it has an apparent ghost stratigraphy comparable to that of the Exploits Group. Its matrix is Tremadocian (Hibbard and others 1977) and the melange is overlain by Caradocian black shales toward the northwest. These are succeeded by greywackes and Silurian conglomerates that are coarser and of shallower water deposition higher in the stratigraphic section. The sequence of units above the melange can be viewed therefore as representing the gradual sedimentary infilling of a marine trough, or an upward shoaling sequence built upon a melange basement.

A variety of small intrusions that are localized within the melange terrane are rare or absent in surrounding country rocks. These are mainly quartz-feldspar porphyries and related rocks, which in places contain numerous small mafic and ultramafic inclusions. Dated isotopically as Early to Middle Ordovician and exhibiting relationships suggesting contemporaneity with melange formation (Williams and Hibbard, 1976), the intrusions imply a direct magmatic linkage with deeper parts of the crust. Mafic and ultramafic inclusions in the porphyries indicate that the melange is underlain by a mafic-ultramafic substrate.

West of the Dunnage Melange, the main volcanic sequences of Notre Dame Bay are interpreted as island arc accumulations built upon oceanic crust. These are bordered to the southeast by mixed sedimentary and volcanic rocks of the Exploits Group, that interdigitate farther southeastward with the Dunnage Melange. The melange is interpreted therefore to occupy a fore arc area, based on the geographical distribution of these ancient elements and their similarity to that outlined for modern volcanic arcs (Dickinson, 1974; Karig, 1974). There is no evidence that the Dunnage Melange was ever buried in an actual subduction zone. It is most reasonably considered therefore as a trench-slope deposit that overlies an accretionary prism, analogous to the positioning of some modern melanges with respect to the arc and oceanic trench (Seely and others 1974; Karig and Sharman, 1975).

A zone of megaconglomerates and olistostromal melange with local ophiolitic blocks occurs along the east margin of the Thetford Mines ophiolite belt of Quebec. These rocks are known as the St. Daniel Formation and they rest conformably on basic volcanic rocks of the ophiolite suite. The unsorted rocks contain fragments of greywacke, quartz arenite, shale, siltstone, volcanic rocks, and outsize serpentinite blocks, all set in a dark grey to red and green shale matrix. The age of the St. Daniel Formation is unknown, more than that it is overlain by the Middle Ordovician Beauceville Formation.

The St. Daniel shale-melange facies has been interpreted as an offshore oceanic deposit equivalent to clastics of the Humber Zone rise prism, i.e. Rosaire and Caldwell Groups of Quebec (St. Julien and Hubert, 1975). It has been interpreted also as a subduction related melange (Laurent, 1975).

In Newfoundland, unsorted shale matrix megaconglomerates occupy a similar position to the St. Daniel Formation where they lie to the east of the Advocate Complex (included in Flatwater Group of Figure 3). These locally contain gabbro blocks up to tens of metres in diameter, a variety of sedimentary and volcanic clasts, granodiorite pebbles, altered and deformed ultramafic blocks, and rare semipelitic schist blocks. Deposition of the Newfoundland examples followed deformation in nearby ophiolites and deformation in sedimentary rocks of the rise prism to the west, i.e. Fleur de Lys Supergroup. These examples are interpreted therefore as coarse slump conglomerates derived mainly from deformed ophiolitic rocks at a destroyed continental margin. They are therefore not thought to be correlative with sediments of the rise prism or connected with subduction, as has been suggested for the Quebec examples.

Dismembered ophiolites or mantle diapirs at the eastern margin of the Dunnage Zone

Mafic-ultramafic bodies along the east side of the Dunnage Zone in Newfoundland (Gander River Belt) have been interpreted as blocks in melange or as mantle diapirs related to subduction and intruded into the country rocks. One occurrence on the north shore of Gander Lake is overlain by conglomerate of presumed Middle Ordovician age, indicating an upper time limit for the age of some of these bodies.

The largest body at Pipestone Pond is approximately 16 km long and 5 km wide, and it is composed mainly of pyroxenite, gabbro, diorite and serpentinized equivalents. It is faulted against metasediments to the east and it is followed westward by volcanic rocks that may be an integral part of the plutonic complex.

A nearby occurrence at Great Bend of Gander

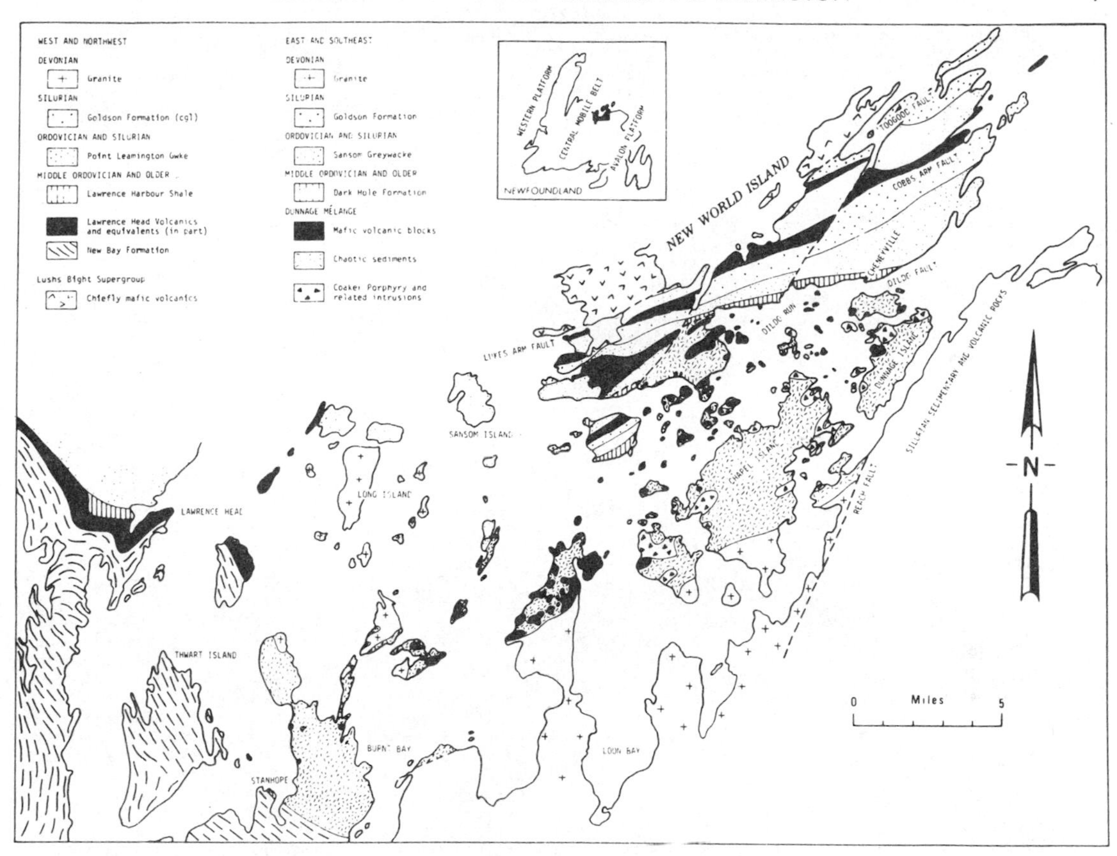

Figure 5: Internal characteristics and regional setting of the Dunnage Melange (after Williams and Hibbard, 1976).

River is composed of dunite and peridotite with an outer zone of gabbro, anorthosite and minor serpentinite. It is circular in shape and surrounded by hornfels of probable Silurian age. This body is most likely an intrusion (Kean, 1974).

Numerous circular, oval and elongate ultramafic bodies occur along the shores of Gander Lake and extend northward along the east side of Gander River. These range in size from a kilometre to less than a metre in diameter and extend over a distance of 60 km. They are mainly pyroxenites and gabbros with minor amounts of peridotite. Although described as intrusions, most contacts are tectonic.

MAFIC-ULTRAMAFIC ROCKS OF THE SOUTHERN APPALACHIANS

In the Southern Appalachians (plate I), the greatest concentration of small mafic-ultramafic bodies lies to the west of the Brevard Zone. These occurrences are equated with similar occurrences farther north throughout the eastern parts of the Humber Zone. East of the Brevard Zone, small ultramafic bodies within the crystalline terrane of the Inner Piedmont are of unknown significance and have no northern counterparts, except perhaps for occurrences in metamorphosed easterly parts of the Dunnage Zone in Newfoundland. Farther east, in metamorphosed parts of the Carolina Slate Belt, small ultramafic bodies occur near the Georgia-South Carolina state line and in the Raleigh Belt of eastern North Carolina. All of these occurrences are of unknown significance and many of them may represent metamorphosed small intrusions rather than oceanic ophiolites.

APPALACHIAN OPHIOLITES AND THE AGE OF IAPETUS

Stratigraphic analysis of the Humber Zone (Williams and Stevens, 1974) indicates rifting and the development of an ancient continental margin of Eastern North America in the late Precambrian with a stable margin existing well into the Ordovician. Destruction of this margin began in the Lower Ordovician and it was completed by Middle Ordovician. Ophiolite suites of the Appalachian Orogen are all of Late Cambrian to Early Ordovician age, where dated. Whether or not the preserved ophiolites represent the crust of a wide Iapetus Ocean, or a sampling of the floor of a related small ocean basin, remains debatable.

The Silurian and later development of the Appalachian Orogen presents a picture entirely different from that which prevailed during its

Cambrian and Ordovician development. Ophiolites of Silurian or younger age are unknown in the Appalachian Orogen and it is impossible to restore the stratigraphy of possible oceans and margins for the Silurian and later periods. An unconformity beneath Silurian rocks across the Humber Zone and westerly parts of the Dunnage Zone indicates the destruction of an earlier Ordovician continental margin and ocean basin. Elsewhere in central areas of the Dunnage Zone, where the stratigraphic record is complete, marine Ordovician shales pass upward into Silurian conglomerates and continental volcanics and red beds.

The common view that a Silurian or Devonian Iapetus Ocean closed in the Devonian to produce the Acadian orogeny (Dewey, 1969; McKerrow and Ziegler, 1971; Schenk, 1971; McKerrow and Cox, 1977) is based more on the premise that orogeny is the result of moving plates and closing oceans, rather than on the stratigraphic record.

ACKNOWLEDGEMENTS

We wish to thank W.R. Church, J.F. Dewey, B. Doolan, R.D. Hatcher, J. Malpas, R.K. Stevens, P. St. Julien and D.F. Strong who through discussions helped extrapolate zones of mafic-ultramafic occurrences along the Appalachian System. The first author also wishes to acknowledge financial support of his field work on ophiolites in the Northern Appalachians through the National Research Council of Canada, the Canadian Department of Energy, Mines and Resources, and an Izaak Walton Killam Special Senior Research Scholarship. The second author wishes to acknowledge financial support of his studies through a Memorial University of Newfoundland Fellowship.

REFERENCES CITED

Archibald, D.A., and Farrar, E., 1976, K-Ar ages of amphiboles from the Bay of Islands ophiolite and the Little Port Complex, Western Newfoundland, and their geological implications: Can. Jour. Earth Sci., v. 13, p. 993-996.
Bergström, S.M., Riva, John, and Kay, Marshall, 1974, Significance of conodonts, graptolites, and shelly faunas from the Ordovician of western and north-central Newfoundland: Can. Jour. Earth Sci., v. 11, p. 1625-1660.
Bird, J.M. and Dewey, J.F., 1970, Lithosphere plate-continental margin tectonics and the evolution of the Appalachian Orogen: Geol. Soc. America Bull., v. 81, p. 1031-1060.
Brown, P.A., 1973, Possible cryptic suture in southwest Newfoundland: Nature, Phys. Sci., v. 245, p. 9-10.
Brown, W.R., 1976, Tectonic melange in the Arvonia Slate District of Virginia (abs.): Geol. Soc. America Abs. with Programs, v. 8, no. 2, p. 142.
Chidester, A.H., and Cady, W.M., 1972, Origin and emplacement of Alpine-type ultramafic rocks: Nature Phys. Sci., v. 240, p. 27-31.
Church, W.R., and Stevens, R.K., 1971, Early Paleozoic ophiolite complexes of Newfoundland Appalachians as mantle-ocean crust sequences: Jour. Geo-Phys. Research, v. 76, p. 1460-1466.
Crowley, W.P., 1969, Stratigraphic evidence for a volcanic origin of part of the Bel Air belt of Baltimore Gabbro Complex in Baltimore County, Maryland (abs.): Geol. Soc. America, pt. 1, p. 10.
Dallmeyer, R.D., and Williams, Harold, 1975, $^{40}Ar/^{39}Ar$ release spectra of hornblende from the Bay of Islands metamorphic aureole, western Newfoundland: their bearing on the timing of ophiolite obduction at the Ordovician continental margin of eastern North America: Can. Jour. Earth Sci., v. 12, p. 1685-1690.
Dewey, J.F., 1969, The evolution of the Appalachian/Caledonian Orogen: Nature, v. 222, p. 124-129.
Dewey, J.F., and Bird, J.M., 1971, Origin and emplacement of the ophiolite suite: Appalachian ophiolites in Newfoundland: Jour. Geophys. Research, v. 76, p. 3179-3207.
Dickinson, W.R., 1974, Plate tectonics and sedimentation: Society Econ. Paleont. and Miner. Special Pub. no. 22, p. 1-27.
Gansser, A., 1974, The ophiolitic melange, a world wide problem on Tethyan examples: Eclogae Geologicae Helvetiae, v. 67, p. 479-507.
Hatcher, R.D., Jr., 1972, Developmental model for the Southern Appalachians: Geol. Soc. America Bull., v. 83, p. 2735-2760.
Hess, H.H., 1939, Island arcs, gravity anomalies and serpentinite intrusions: Internat. Geol. Cong. 17th, Moscow 1937, Rept. 17, v. 2, p. 763-783.
Hess, H.H., 1955, Serpentinites, orogeny and epeirogeny: Geol. Soc. America Spec. Paper 62, p. 391-408.
Hibbard, J.P., Stouge, S., and Skevington, D., 1977, Fossils from the Dunnage Melange, north central Newfoundland: Can. Jour. Earth Sci., v. 14, p. 1176-1178.
Karig, D.E., 1974, Evolution of arc systems in the Western Pacific: Ann. Rev. Earth and Plan. Sci., v. 2, p. 51-75.
Karig, D.E., and Sharman, G.F., 1975, Subduction and accretion in trenches: Geol. Soc. America Bull., v. 86, p. 337-389.
Kay, Marshall, 1976, Dunnage Melange and subduction of the Protacadic Ocean, Northeast Newfoundland: Geol. Soc. America, Spec. Paper 175, 49p.
Kean, B.F., 1974, Notes on the Geology of the Great Bend and Pipestone Pond ultramafic bodies: Newfoundland Department of Mines and Energy, Min. Dev. Div., Report of Activities, p. 33-42.
Kean, B.F., and Strong, D.F., 1975, Geochemical evolution of an Ordovician island arc of the central Newfoundland Appalachians: Am. Jour. Sci., v. 275, p. 97-118.
Kennedy, M.J., 1975, Repetitive orogeny in the northeastern Appalachians - new plate models based upon Appalachians examples: Tectonophysics, v. 28, p. 39-87.
Kennedy, M.J., and Phillips, W.E.A., 1971, Ultramafic rocks of Burlington Peninsula, Newfoundland, in A Newfoundland Decade. Proc. Geol. Assoc. Canada, v. 24, p. 35-46.
Kidd, W.S.F., (in press), The Baie Verte Lineament, Newfoundland: ophiolite complex floor and mafic volcanic fill of a small Ordovician marginal basin: Maurice Ewing Volume.
Laurent, Roger, 1975, Occurrences and origin of the ophiolites of Southern Quebec, Northern Appalachians: Can. Jour. Earth Sci., v. 12, p. 443-455.
Mattinson, J.M., 1976, Ages of zircons from the Bay of Islands ophiolite complex, Western Newfoundland: Geology, v. 4, p. 393-394.
McKerrow, W.S., and Cocks, L.R.M., 1977, The location of the Iapetus Ocean suture in Newfoundland: Can. Jour. Earth Sci., v. 14, p. 488-495.
McKerrow, W.S., and Ziegler, A.M., 1971, The Lower Silurian paleogeography of New Brunswick and adjacent areas: Jour. Geology, v. 79, p. 635-646.
Miller, H.G., and Deutsch, E.R., 1976, New gravitational evidence for the subsurface extent of oceanic crust in north-central Newfoundland: Can. Jour. Earth Sci., v. 13, p. 459-469.

Moores, E., 1970, Ultramafics and orogeny, with models of the U.S. Cordillera and the Tethys: Nature, v. 228, p. 837-842.
Norman, R.E. and Strong, D.F., 1975, The geology and geochemistry of ophiolitic rocks exposed at Ming's Bight, Newfoundland: Can. Jour. Earth Sci., v. 12, p. 777-797.
Rodgers, John, 1968, The eastern edge of the North American continent during the Cambrian and Early Ordovician, in Zen, E-an, White, W.S., Hadley, T. B. and Thompson, J.B., Jr. (eds.), Studies of Appalachian Geology: northern and martime: Wiley-Interscience, New York, p. 141-149.
Schenk, P.E., 1971, Southeastern Atlantic Canada, Northwestern Africa and continental drift: Can. Jour. Earth Sci., v. 10, p. 1218-1251.
Seely, D.R., Vail, P.R., and Walton, G.G., 1974, Trench slope model, in Burk, C.A. and Drake, C.L. (eds.), The Geology of Continental Margins: Springer-Verlag, New York, p. 249-260.
Smith C.H., 1958, Bay of Islands Igneous Complex, western Newfoundland: Geol. Survey Canada, Memoir 290, 132p.
Stevens, R.K., 1970, Cambro-Ordovician flysch sedimentation and tectonics in West Newfoundland and their possible bearing on a proto-Atlantic ocean, in Lajoie, J. (ed.), Flysch sedimentology in North America: Geol. Assoc. Canada Spec. Paper No. 7, p. 165-177.
Stevens, R.K., Strong, D.F., and Kean, B.F., 1974, Do some Eastern Appalachian ultramafic rocks represent mantle diapirs produced above a subduction zone? Geology, v. 2, p. 175-178.
Stevens, R.K., and Williams, Harold, 1973, The emplacement of the Humber Arm Allochthon, Western Newfoundland (abs.): Geol. Soc. America, Northeast Section Annual Meeting Program, Allentown, Pennsylvanic, v. 4, No. 2, p. 222.
Strong, D.F., 1972, Sheeted diabases of Central Newfoundland: new evidence for Ordovician seafloor spreading: Nature, v. 235, p. 102-104.
St. Julien, P., and Hubert, C., 1975, Evolution of the Taconian orogen in the Quebec Appalachians. Am. Jour. Sci., v. 275A, p. 337-362.
St. Julien, Pierre, Hubert, C., and Williams, Harold, 1976, The Baie Verte-Brompton Line and its possible tectonic significance in the Northern Appalachians (abs.): Geol. Soc. America Abs. with Programs, v. 8, p. 259-260.
Upadhyay, H.D., Dewey, J.F., and Neale, E.R.W., 1971, The Betts Cove ophiolite complex, Newfoundland: Appalachian oceanic crust and mantle, in A Newfoundland Decade: Proc. Geol. Assoc. Canada, v. 24, p. 27-34.
Williams, Harold, 1973, Bay of Islands Map-Area, Newfoundland. Geol. Survey Canada Paper 72-34, 7p.
Williams, Harold, 1975, Structural succession, nomenclature and interpretation of transported rocks in Western Newfoundland: Can. Jour. Earth Sci., v. 12, p. 1874-1894.
Williams, Harold, 1976, Tectonic-stratigraphic subdivision of the Appalachian orogen (abs.): Geol. Soc. America Abs. with Programs, v. 8, no. 2, p. 300.
Williams, Harold, 1977a, The Coney Head Complex. Another Taconic allochthon in west Newfoundland: Am. Jour. Sci. (in press).
Williams, Harold, 1977b, Ophiolitic melange and its significance in the Fleur de Lys Supergroup, northern Appalachians: Can. Jour. Earth Sci., v. 14, p. 987-1003.
Williams, Harold and Hibbard, J.P., 1976, The Dunnage Melange, Newfoundland: Geol. Survey Canada, Paper 76-1A, p. 183-185.
Williams, Harold, Hibbard, J.P., and Bursnall, J.T., 1977, Geologic setting of asbestos-bearing ultramafic rocks along the Baie Verte Lineament, Newfoundland: Geol. Survey Canada, Paper 77-1, pt. A, p. 351-360.
Williams, Harold, Kennedy, M.J., and Neale, E.R.W., 1972, The Appalachian structural province, in Price, R.A. and Douglas, R.J.W. (eds.), Variations in tectonic styles in Canada: Geol. Assoc. Canada Spec. Paper 11, p. 181-261.
Williams, Harold, Kennedy, M.J., and Neale, E.R.W., 1974, The northeastward termination of the Appalachian orogen, in Nairn, A.E.M. and Stehli, F.G. (eds.), The Ocean Basins and Margins, v. 2, the North Atlantic: Plenum, New York, p. 79-123.
Williams, Harold, and Smyth, W.R., 1973, Metamorphic aureoles beneath ophiolite suites and Alpine peridotites: Tectonic implications with west Newfoundland examples: Am. Jour. Sci., v. 273, p. 594-621.
Williams, Harold, and Stevens, R.K., 1974, The ancient continental margin of Eastern North America, in Burk, C.A. and Drake, C.L. (eds.), the geology of continental margins: Springer-Verlag, New York, N.Y., p. 781-796.
Wilson, J.T., 1966, Did the Atlantic close and then reopen? Nature, v. 211, p. 696-681.

Department of Geology
Memorial University of Newfoundland
St. John's, Newfoundland, Canada

13

Reprinted from page 993 of *Am. Jour. Sci.* **279:**993-1021 (1979)

REGIONAL SETTING OF THE DUNNAGE MELANGE IN THE NEWFOUNDLAND APPALACHIANS

JAMES HIBBARD* and HAROLD WILLIAMS**

ABSTRACT. The Dunnage Melange consists mainly of clastic sedimentary and mafic volcanic blocks set in a dark shale matrix. It outcrops in the Dunnage Zone of the Newfoundland Appalachians, the former site of the Paleozoic Iapetus Ocean. Regional tectonic elements of this zone indicate that the melange formed on the flank of a Lower Ordovician island arc complex; thus many workers believe that it formed in a trench and marks a major suture.

To the southwest, the melange overlies and interdigitates with the gabbro-infested Ordovician New Bay Formation; near the contact and only there, the melange contains gabbro blocks. Similarly, in the northwest portion of the melange, the largest mafic volcanic blocks define a wide zone that is correlative westward with the Lawrence Head Volcanics of the Exploits Group. Along its northwest border the melange is locally conformable with overlying graywacke and Caradocian black shale, though in most places the contact is modified by faulting. These relations imply that the Dunnage Melange is a chaotic equivalent of the nearby Exploits Group, and age relations suggest correlation with similar nearby groups on New World Island.

A suite of intrusions ranging from quartzo-feldspathic porphyries in the northeast to gabbros in the central portions occur as dikes and stocks in the melange. Field relationships and isotopic ages indicate intrusion during or immediately following melange formation.

The chaotic aspect of the Dunnage Melange distinguishes it as a lithostratigraphic unit. This structural character is primary and formed as a result of thixotropic activation and surficial slumping of a complex Ordovician volcanic-sedimentary pile. The present attitude of the melange is controlled by later, post-lithification structures.

Formation of the melange appears to have heralded a major change in the tectonic regime of central Newfoundland. It marks the change from active arc volcanism to arc degradation and basinal infilling. These phenomena are probably linked to events along a continent-ocean interface that lay to the east of the melange. The polarity of subduction along this interface is largely conjectural; thus the paleogeographic siting of the Dunnage Melange with respect to arc-trench elements remains equivocal. The limited extent of the melange within the Appalachian Orogen shows that the processes controlling its formation were local, perhaps reflecting local irregularities in the ancient eastern margin of Iapetus.

REVIEW†

The Dunnage mélange was termed a *slump breccia* by Patrick (1956). Horne (1969) conjectured that it is a few thousand meters thick; both this and the Carmanville Mélange (≃1,000m) are an order thicker than the upper limit suggested for individual olistostromes (Abbate, Bortolotti, and Passerini, Paper 5). Horne also suggested that the mélange has normal stratigraphic field relations. Hibbard and Williams note that despite the consistent dips and north facing, there is a doubt whether the mélange

* Department of Mines and Energy, 95 Bonaventure Avenue, St. John's, Newfoundland, Canada

** Department of Geology, Memorial University of Newfoundland, St. John's, Newfoundland, Canada

†This review was prepared by the volume editor.

does not represent a series of structural repetitions. Locally the mélange is of greenchist metamorphic facies, due to thermal effects associated with Devonian batholiths. The matrix is characteristically composed of black and green, pyritic, smeared shale. Blocks within it are up to hundreds of meters across. Manganese nodules with cupriferous cores are locally present in the shale matrix, which displays fluidal banding and attenuation, in contrast to the brittle fracture and disruption of quartz-rich siltstone interbeds. Cleavage is developed preferentially where large clasts are present. Sedimentary blocks account for more than half of the content, and clastic sediments predominate, chert and carbonate being only of local occurrence. The blocks are of random distribution in the matrix. Carbonate blocks contain both Cambrian and Arenig fossils, whereas the matrix contains Tremadocian fossils. The presence of younger blocks in an older matrix is attributed to some form of structural comingling. The larger basic volcanic inclusions form a distinct and wide stratigraphic zone, and smaller blocks of similar material, occur to the southeast of it. Massive, pillowed, and ropy lava; agglomerate, pillow breccia; lapilli tuff; and both angular and splashbombs occur among the basic volcanic blocks—the latter inset in the shale matrix and testifying to comtemporaneous volcanic activity. Gabbro, diorite, and dunite occur locally and appear to be disrupted sill material of the New Bay Formation. The mélange has conformable contacts with bordering units; at the lower boundary it interdigitates with the New Bay Formation, and at the top it has a normal sedimentary contact with overlying graywacke, shale, and siltstone. It has a consistent stratigraphic position over its entire length. The mélange is attributed to an earlier phase of soft rock deformation and a later phase of hard rock deformation. Lack of shearing precludes hard rock deformation for the main phase, according to the authors (but again the possibility arises of tectonic deformation of a partially consolidated sedimentary sequence). Dewey (1969) favored tectonic origin for this mélange, and Kay (1970) was uncertain whether it is sedimentary or tectonic (Hsü, Paper 2). The gabbroic and volcanic inclusions are believed by the authors to be derived from original layers within the mélange matrix, and thus they are not truly exotic. The deformation is attributed to thixotropic sensitivity of the matrix to stress, and the mélange is regarded as a failed sedimentary klippe (see editorial comments on Paper 5). It is related to the ancient continental margin of the Iapetus Ocean, but the paleogeographic situation is considered to be equivocal—that is, it may represent a fore-arc or rear-arc basin; if the former, an analogy is drawn with extensive slope deposits in recent volcanic arcs, especially the Bassein slide (Moore, Curray, and Emmel, 1976).

REFERENCES

Dewey, J. F., 1969, Evolution of the Appalachian/Caledonian Orogen, *Nature* **222:**124–129.

Horne, G. S., 1969, Early Ordovician Chaotic Deposits in the Central Volcanic Belt of Northeastern Newfoundland, *Geol. Soc. America Bull.* **80:**2451–2464.

Kay, M., 1970, Flysch and Bouldery Mudstone in Northeastern Newfoundland, *Geol. Assoc. Canada Spec. Paper 7,* pp. 155-164.

Moore, D. G., J. R. Curray, and F. J. Emmel, 1976, Large Submarine Slide (Olistostrome) Associated with Sundra Arc Subduction Zone, Northeast Indian Ocean, *Marine Geology* **21:**211-226.

Patrick, T. O. H., 1956, Comfort Cove, Newfoundland: Map with Notes, *Canada Geol. Survey Paper 55,* 31p.

14

Reprinted from *Canadian Jour. Earth Sci.* **16:**1439–1451 (1979)

The nature, origin, and significance of the Carmanville ophiolitic mélange, northeastern Newfoundland[1]

G. E. PAJARI AND R. K. PICKERILL
Department of Geology, University of New Brunswick, Fredericton, N.B., Canada E3B 5A3
AND
K. L. CURRIE
Geological Survey of Canada, 601 Booth Street, Ottawa, Ont., Canada K1A 0E8
Received March 20, 1978
Revision accepted March 8, 1979

The Carmanville ophiolitic mélange of northeastern Newfoundland forms an olistostrome within a thick succession of Middle to Upper Ordovician flyschoid shale, siltstone, and greywacke. The olistostrome consists of sedimentary, volcanic, and ultramafic olistoliths ranging in size from granules to several kilometres. The matrix appears to have been derived entirely by disaggregation and disintegration of hydroplastic and thixotropic sediments. The matrix was sufficiently fluid for turbulent motion to occur in the lower parts of the olistostrome, yet viscous enough to produce alignment of fragments and a pervasive cleavage, as well as drag folds in the surrounding hydroplastic sediments. The olistoliths have been drawn from a stratigraphic column hundreds, if not thousands, of metres thick, and an area many kilometres across. A small proportion of the olistoliths were deformed and metamorphosed prior to incorporation in the olistostrome.

The Carmanville ophiolitic mélange is tentatively correlated with the Dunnage mélange to define the northern margin of a sheet of ocean floor and island arc obducted toward the southeast onto an accreting continental rise in Llanvirnian–Llandeilian time. Olistostrome formation within this unstable pile commenced in post-Arenigian time, and may have continued intermittently until Llandoverian time. The olistostrome and surrounding rocks were further deformed, metamorphosed, and intruded by granitoid plutons during Silurian and Devonian time.

Erratum

Two of the photographs have been rotated relative to the directions given in the captions. The corrected directions are given in italics: "Fig. 4. A sedimentary melange unit (*at top*) . . ." and "Fig. 5. A melange unit (*right of the pen*) . . . ."

Introduction

Greenly (1919) introduced the term 'mélange' to describe a rock unit near Gwna, Anglesey, consisting of fragments of very diverse size, shape, state of deformation, and lithology enclosed in a sedimentary matrix. Similar rocks form a significant part of many orogenic belts. According to current plate tectonic theory, mélange yields important information on sedimentation and tectonism during orogeny (Dewey and Bird 1970; Hsü 1971, 1974; Dickinson 1973). These units comprise an integral part of several Cambro–Ordovician successions in the Newfoundland Appalachians. In

[1]Canadian contribution number 7 to the International Geological Correlation Program, Caledonide Orogen Project 27.

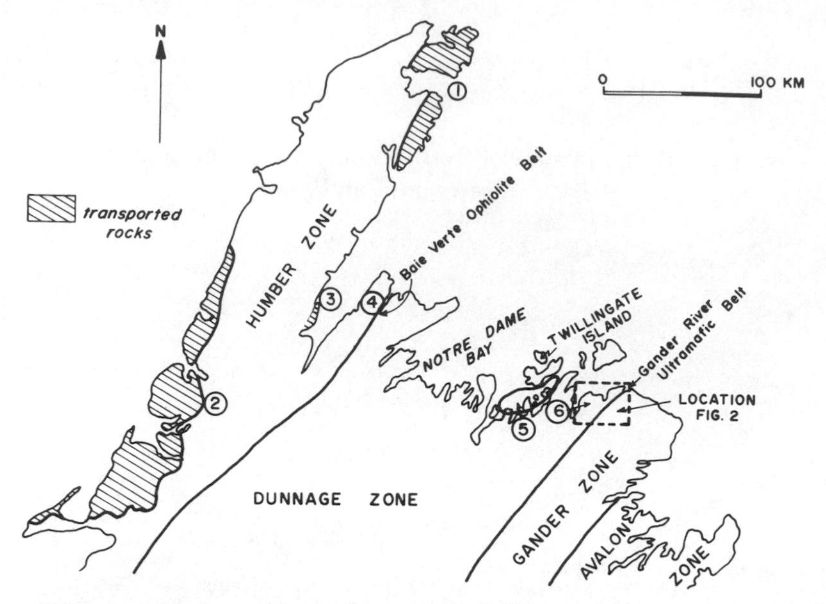

FIG. 1. Northern and western Newfoundland illustrating the location of known mélanges: (1) Hare Bay (see Stevens 1970); (2) Humber Arm (see Stevens 1970); (3) Second Pond mélange (see Williams 1977*b*); (4) Fleur de Lys mélange (see Williams 1977*a*); (5) Dunnage mélange (see Kay 1976), and (6) Carmanville ophiolitic mélange. The tectonic divisions of Newfoundland are after Williams *et al*. (1974) and Williams (1975, 1977*a*, *b*).

western Newfoundland, mélange is found associated with transported allochthons (localities 1 and 2 of Fig. 1) and with the complex western margin of the Dunnage Zone (localities 3 and 4 of Fig. 1). These occurrences have been interpreted in terms of westward transport of oceanic crust across an initially undisturbed continental rise prism onto continental crust (Humber Zone) (Stevens 1970; Williams 1977*a*,*b*). Horne (1969) interpreted the celebrated Dunnage mélange in northeastern Newfoundland (locality 5 of Fig. 1) as a tectonized olistostrome shed from the scarp of a normal fault. Subsequent workers have suggested that this mélange formed at or near an oceanic trench associated with a northwest-dipping subduction zone (Bird and Dewey 1970; Hibbard and Williams 1975; Kay 1976). More recently Hibbard and Williams (1979) have interpreted the mélange as an olistostrome formed by submarine slumping of Cambrian and Lower Ordovician sedimentary and mafic volcanic rocks. The mélange near Carmanville (locality 6 of Fig. 1) has not previously been interpreted in terms of plate tectonic theory. Kennedy and McGonigal (1972, p. 457) stated that "the significance of the melange zone ... is presently obscure. It may be tectonic and represent a zone of gravity sliding, ... or it may be an olistostrome." This paper describes the nature and origin of the mélange around Carmanville as revealed by detailed mapping (Currie and Pajari 1977; Pickerill *et al*. 1978*a*; Currie *et al*. 1979*a*), and comments on its significance for the tectonic history of northeastern Newfoundland.

Geological Setting

The geological history of the Carmanville region involves a series of documentable events that began before the initial stages that closed the Iapetus Ocean during early Ordovician, and that continued through into Devonian and possibly Carboniferous times. We briefly outline below a model based on the geological relationships in the Carmanville area before we present the pertinent details of the stratigraphic record (see also Currie *et al*. 1979*a*, *b*).

In Cambrian and early Ordovician times, the southeastern part of the Carmanville map-sheet (Fig. 2, unit 1) represented the sedimentologically active rise prism of a continent lying to the southeast. To the northwest there existed ocean floor and incumbent volcanic arcs of the Iapetus Ocean (Fig. 3*a*), the remnants of which now constitute the Dunnage Zone (Fig. 1). The events that involved the closing of the Iapetus resulted in the emplacement of the allocthonous sheet of ultramafic and mafic volcanic rocks (units A and B, Fig. 3) onto the

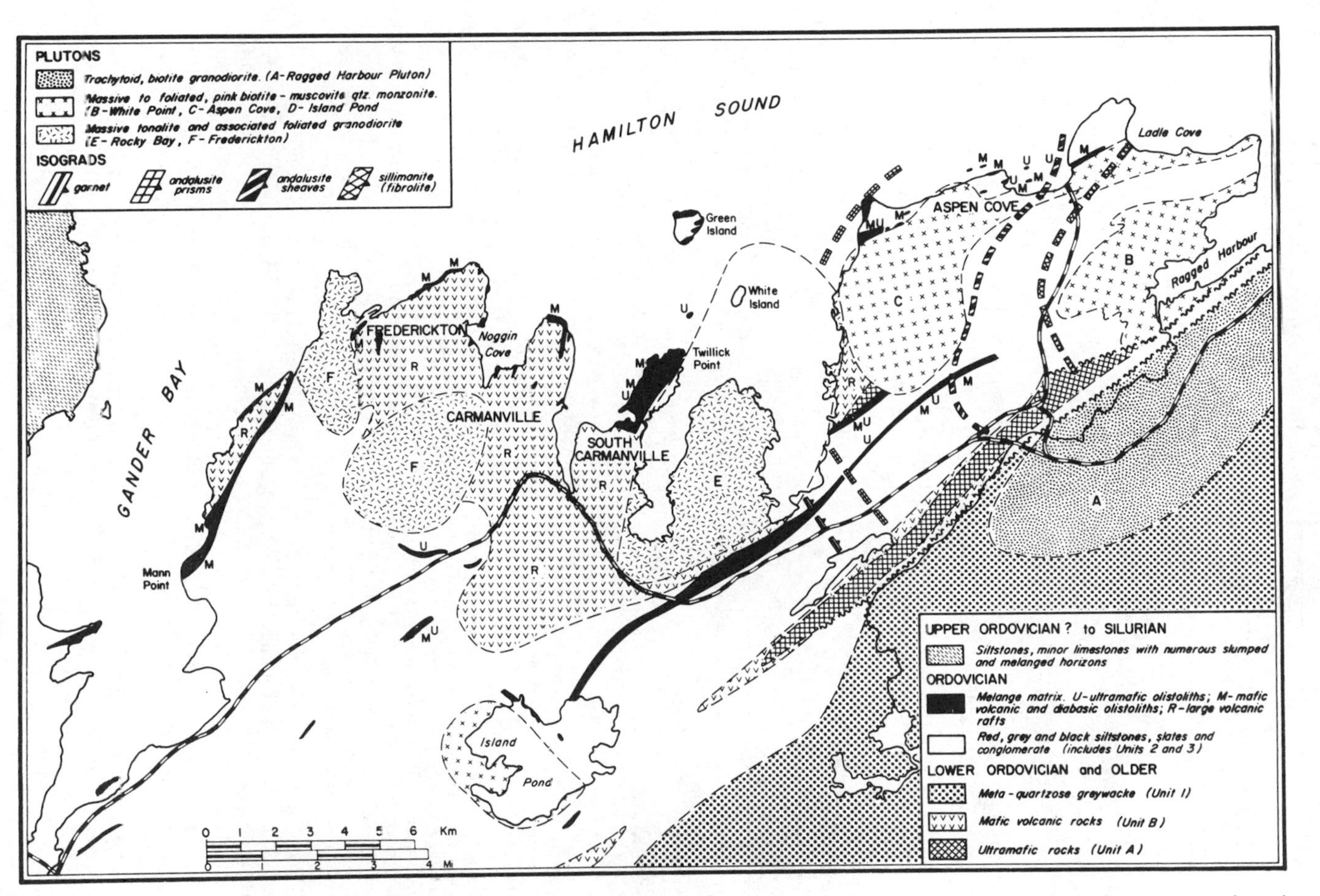

FIG. 2. Generalized geological map of the Carmanville area (after Williams 1963; Pickerill *et al.* 1978*a*). The stippled ornament designates the continental terrane.

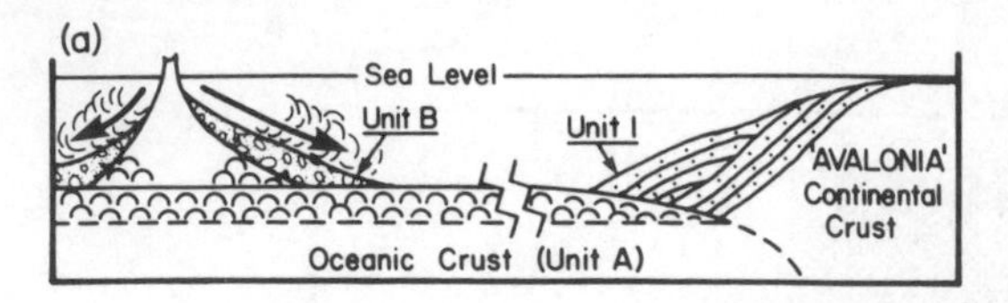

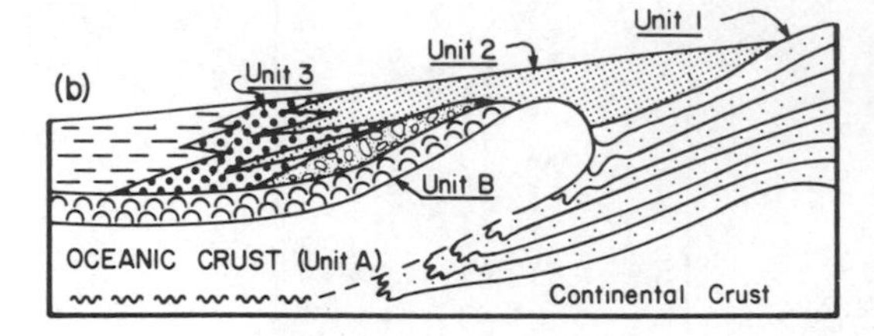

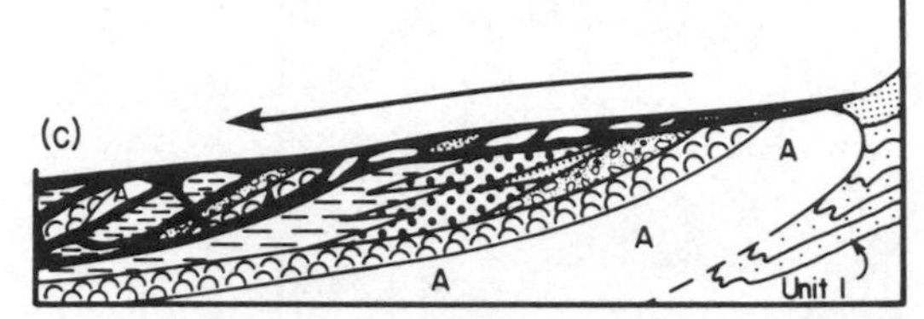

FIG. 3. Schematic representation of the tectonostratigraphic evolution of the Carmanville area. (*a*) Pre-Llandeilian. Ocean floor (unit A) and incumbent volcanic arc (unit B) material existed to the northwest whilst the continental rise prism sequence (unit 1) of the Avalonian continent existed immediately to the southeast. (*b*) Latest Llanvirnian and earliest Llandeilian. Slabs of units A and B have now been obducted onto the toe of the continental rise prism partially interrupting normal sedimentation processes and resulting in complex and localized facies distributions (represented here schematically by units 2 and 3). The dashed ornament represents the progradational mass-emplaced (essentially turbiditic) sediments of the post-Llanvirnian – Middle Ordovician basinal sediments to the west. (*c*) Post-Caradocian. The sedimentary pile was involved in one or more submarine gravity slides (olistostromes) resulting in the formation of the Carmanville ophiolitic mélange. The mélange matrix (black) comprised the mobile medium (lubricant) along which the large rafts moved and contains numerous small olistoliths (not illustrated). Note: the diagrams are not drawn to scale and incorporate a large vertical exaggeration.

sediments of the continental rise prism in pre-Llandeilian time. This sheet was fragmented during emplacement and remnants now constitute the 150 km linearly disposed but discretely isolated slabs of the Gander River ultramafic belt. The largest of these slabs within the Carmanville map-area (Shoal Pond slab) lies southwest of Ragged Harbour and is approximately 15 km in length (Fig. 2). The emplacement of these slabs partially interrupted normal sedimentation processes on the continental rise prism resulting in complex and localized facies distributions.

Sedimentation continued through to the Caradocian and possibly into the Silurian. The sedimentary pile was further involved in one or more submarine gravity slides, which formed two distinctly different olistostromal horizons. The exact timing of these slides is unknown but faunal evidence (see below) suggests they were post-Caradocian.

Although sedimentation was essentially continuous from pre-Ordovician through the Ordovician and possibly into the Silurian, the resultant sedimentary sequence cannot be given a single stratigraphic descriptor owing to the presence within the sequence of clearly allocthonous slabs. We therefore avoid previously published nomenclature (Gander Lake Group, Gander Group, Davidsville Group) as this distorts the observed relationships between rock units. For a more detailed discussion of the inherent stratigraphic nomenclatural problems the reader is referred to the discussions of Kennedy and McGonigal (1972), Jenness (1972), Brückner (1972, 1978), Pajari and Currie (1978), and Pickerill *et al.* (1978*b*).

Major tectonic deformation, metamorphism, and plutonism commenced in this sedimentary terrane in Late Silurian time and continued through the Devonian, possibly terminating in the Carboniferous (Bell *et al.* 1977; Pajari and Currie 1978). The rocks underwent a complex deformational history with at least four periods of deformation observable locally (Kennedy and McGonigal 1972; Pickerill *et al.* 1978*a*). However, because strain was very markedly inhomogeneously distributed, the bulk of the rocks now exposed east of Gander Bay and south of Hamilton Sound face northwest and exhibit only a single penetrative cleavage. The metamorphic grade of the rocks is low in the western part of the area, but rises to the east, so that virtually all of unit 1 (Fig. 2) is at biotite or higher grade metamorphism, and east of Carmanville the other units (Ordovician of Fig. 2) are at sillimanite grade with localized development of anatectic melt (Pajari and Currie 1978).

The Stratigraphic Record

The oldest sedimentary unit (unit 1, Figs. 2 and 3) comprises a thick monotonously uniform silty greywacke rich in quartz, indicating an ultimately continental source for the sediment. Primary sedimentary structures occur rarely in this unit, and most of the presently observed foliation is of tectonic origin.

A faulted contact between unit 1 and a slab of ultramafic and mafic rocks (units A and B) is well exposed on the southeastern edge of the Shoal Pond slab. This fact and the observations that the footwall sedimentary rocks are intensely deformed and the slabs are unconformably overlain by sediments rich in ultramafic and mafic debris and (or) red siltstones and shales, constitute the evidence for their allocthonous character. The unconformity above these slabs has been noted elsewhere in the Gander River ultramafic belt by Kennedy (1975, 1976) and Blackwood (1978).

The red sediments have proved to be a reliable marker horizon, traceable for tens of kilometres. At their base, they locally contain a thin fossiliferous carbonate member, which has yielded a highly characteristic conodont assemblage defining an age on the Llanvirnian–Llandeilian boundary (G. Nowlan, written communication, 1978; cf. Blackwood 1978). This important datum fixes a minimum age for the emplacement of the allocthonous ultramafic and mafic slabs.

A gradational sequence of sediments between the greywackes of unit 1 and the red shales of unit 2 can be examined in detail along the bush roads southeast of Island Pond in the interval between the Shoal Pond slab to the northeast and an unnamed slab to the southwest (Fig. 2). The sequence here exhibits neither an unconformity nor a fault (Currie *et al.* 1979*a*), and therefore, the gap between the ultramafic slabs constituted a primary site of continuous sedimentation whilst the allochthonous slabs were 'topographic highs' concurrently shedding sedimentary debris.

Unit 1 is therefore older than the Llanvirnian–Llandeilian boundary. We assume that the Arenigian or Llanvirnian age obtained by McKerrow and Cocks (1977) from a horizon near the top of Jenness' (1963) lower Gander Lake Group represents the probable age of the top of unit 1.

The allocthonous slabs, consisting of ultramafic, mafic intrusive, and volcanic rocks, and minor plagiogranite, shed coarse debris to the southwest, producing a belt of relatively coarse clastic rocks (unit 3) up to at least 1000 m thick. Interbedded with and overlying this facies, pebble and cobble horizons up to 10 m thick occur in a matrix of fissile black shale. These horizons contain debris from the slabs as well as sediment transported from elsewhere. The coarse horizons fine to the southwest, where they grade into finely bedded siltstone with numerous thin beds of granule conglomerates and sandstones. It is evident, therefore, that shortly after the emplacement of the allocthons, the depositional environment adjacent to the slabs was simultaneously receiving sediments from a proximal and a more distal source.

The units are overlain conformably by hundreds of metres of finely laminated greyish black siltstone and shale consisting mainly of beds 5–10 mm thick grading from fine siltstone at the base to discontinuous shaly tops, with local thin seams of shale flake conglomerate, or granule beds. We interpret this unit to have accumulated by distal deposition

FIG. 4. A sedimentary mélange unit (to the left) parallel to bedding in an otherwise coherently bedded sequence (to the right). The mélange contains both hydroplastically deformed and more lithified olistoliths. Cleavage is parallel to elongation of olistoliths and bedding. (Location: northeastern side of Green Island.) Map folder is 30 cm long. Arrowhead points north.

from turbidity currents. More proximal turbidite facies occur in the southwestern part of the mapped area, where both partial and complete Bouma sequences as well as coarse conglomerates are observed, indicating the presence of a large submarine fan complex. These sediments were largely derived from a mafic provenance, but the presence of quartz–feldspar sandstones and the existence of well-rounded orthoquartzite pebbles in conglomerates attest to a sediment contribution from a continental source. Caradocian fossils have been recovered from these sediments south of Mann Point (Bergström *et al.* 1974).

The geographic extent of the sedimentary rocks of this unit, as well as of those lying between this unit and unit 1, is shown as the unshaded area on Fig. 2. The sediments of this unit have been involved in a slide event of distinctive characteristics that incorporated, in a limited way, all the older lithologies described above. It is also evident that mass movements of the sedimentary pile started after Caradocian time.

The sedimentary rocks that lie to the northwest of Gander Bay (Upper Ordovician to Silurian, Fig. 2) have yielded specimens of the tabulate coral *Favosites cf. favosus* attributable to a Llandoverian age (Williams 1972). These sediments have undergone extensive soft rock slumping giving rise to a unit hundreds or thousands of metres thick in which details of the sedimentary bedding and other primary structures and stratigraphic relationships have been destroyed over large areas. As we have not identified a regionally distinct break in the sedimentary record within the map-area west of Gander Bay (if indeed this will ever be possible) we assume that sedimentation continued into Silurian time. However, we do make a distinction between the younger slumped sediments northwest of Gander Bay (Fig. 2), which appear to have involved the mechanical failure of large portions of the sedimentary pile to give rise to incoherent masses of sediment over extensive areas, and the mélange event, which affected the older rocks to the southeast. This underlying mélange (hereafter termed Carmanville ophiolitic mélange) is characterized by the movement of coherent or semicoherent blocks, and rafts along mud-lubricated slide planes identified as mélange matrix in Fig. 2. It is this latter mélange that forms the subject of this paper.

Whether a single event was responsible for the mélange and the overlying slumped terrain is not known, but tentatively we view the two terrains to be geologically distinct.

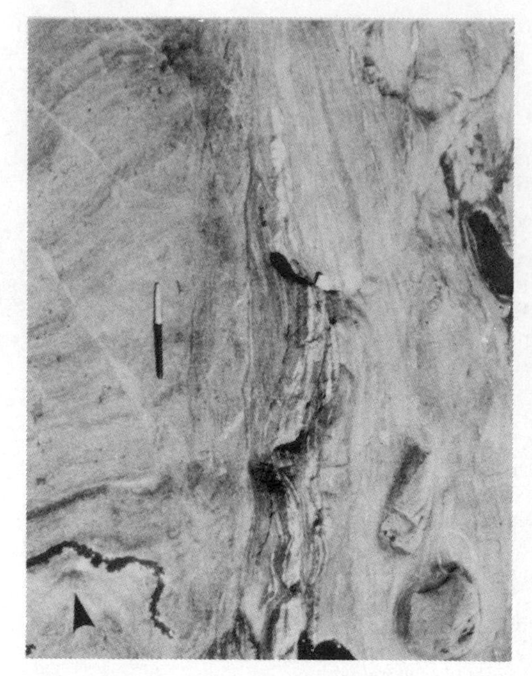

FIG. 5. A mélange unit (above the pen) intruding a coherently bedded succession at a high angle (see Fig. 10 for location). Length of the pen is 15 cm. Arrowhead points north.

The Carmanville Ophiolitic Mélange

General Characteristics

The mélange consists of sedimentary, mafic volcanic or volcaniclastic and ultrabasic clasts, blocks, and rafts (henceforth all included in the term olistolith) enclosed within a matrix, commonly pyritiferous, of black shale and silt. The mélange matrix (Fig. 2) comprises less than 10% of the terrain affected by the slide event. This matrix, which acted both as a carrier for the olistolith load and as a lubricant during movement of the largest olistoliths, was derived solely from the disaggregation of thixotropic, unconsolidated sediment.

Numerous discrete mélange horizons can be mapped, and these vary from a few centimetres to tens of metres in width and a few tens of metres to more than 10 km in length. Other occurrences of unmappable amounts of clast-bearing black matrix in isolated outcrops suggest that many more horizons may be present. The more extensive horizons represent major slide planes, whereas many of the smaller ones are matrix-filled dilational fractures, presumably connected to some more extensive source of mobile muddy or silty matrix. Most of the continuous horizons strike roughly parallel to the regional structure of the sedimentary rocks (Fig. 4), but cross-cutting mélange also occurs (Fig. 5). The boundaries of the mélange can only be defined by the geographical extent of the matrix horizons (Fig. 2).

The bulk of the olistoliths consist of flyschoid sedimentary rocks. The remainder comprise ultramafic rocks and an intimately associated volcanogenic suite including pillowed lavas, hyaloclastites, volcaniclastic sediments, and limestone. The dimensions of the olistoliths range from several kilometres down to granules, that is, to the smallest size in which the polymict character of the olistolith load can be megascopically identified. A few of the sedimentary, volcanic, and ultramafic olistoliths had been tectonically deformed and metamorphosed prior to incorporation in the matrix.

Like other sedimentary rocks in this region, the mélange was subsequently regionally metamorphosed and intruded by granitoid plutons.

Description of the Olistoliths

Sedimentary olistoliths occur throughout the mélange, but are the sole constituent on Green Island. Mafic volcanic olistoliths generally occur throughout but have not been observed on Green Island (Fig. 2). Their abundance is extremely variable in any single locality. Ultramafic olistoliths occur mainly in the northeastern part of the olistostrome, where their concentration may reach 10% of the olistolith load. The easily defined olistoliths are those surrounded by mélange matrix. Although numerically most of the olistoliths are smaller than 20 m in diameter, some large but unidentified sedimentary rafts probably also exist.

Basalts and associated olistoliths occur mainly in huge rafts bordered, and presumably underlain, by black shale matrix. The evidence indicating the allocthonous character of these rafts, which range up to 11 km by 7 km in extent, can be summarized as follows.

(1) The rafts are invariably surrounded by black shale matrix, regardless of the character of the other nearby rocks. Detailed mapping around the volcanic raft on the east side of Gander Bay has shown it to be completely surrounded by matrix. Good coastal exposures indicate that the raft exposed in the Carmanville district is similarly surrounded by matrix on its northern side; the extent of this matrix on the southern boundary is unclear because of exposure difficulties.

(2) The rafts are fractured and intruded by matrix dykes up to 30 m in width, some of which can be seen to wedge out within the volcanic rocks.

(3) The olistolith load of the dykes is variable in both concentration and species, but invariably includes sedimentary fragments.

(4) Younging directions (northeasterly) and fold trends (commonly north-northwesterly to northeasterly) are consistent within single rafts, but vary between rafts, and are anomalous compared with the surrounding sedimentary terrane in which the beds trend northeast and young to the northwest.

These observations demonstrate that the rafts are allocthonous masses that at one time reposed on a fluidized shaly medium.

The volcanic rocks exhibit two major lithofacies: (a) poorly sorted volcanic conglomerates characterized by abundant amygdaloidal clasts up to boulder size, associated with lesser amounts of volcaniclastic sandstone and minor limestone beds; and (b) essentially nonvesicular pillowed lavas and associated hyaloclastites. The former lithofacies constitutes a thick succession of resedimented debris, grain, and turbidite flows, and the latter an environment of pillow formation with *in situ* granulation and disaggregation (R. K. Pickerill *et al.*, in preparation). The differing physical characters of the lithofacies reflect differences in their environment of generation, most probably differences in depth and slope of the paleosurface. Their present juxtaposition results from mass transport of shallow-water basaltic debris into a deep-water environment characterized by the formation of pillow lavas and hyaloclastites prior to the obduction event.

The ultramafic olistoliths are lithologically identical to the rocks of the Gander River ultramafic belt (Jenness 1963), that is, websterite, lherzolite, peridotite, and dunite, now largely altered to antigorite and amphibole. Ultramafic olistoliths up to 40 m in diameter are dispersed among volcanic and sedimentary olistoliths in the lower horizons of the mélange. In one case an ultramafic olistolith 20 m in diameter occurs in a 23 m wide mélange horizon within bedded sediments that are continuously exposed and otherwise undisturbed for more than 100 m on both sides. Small leucocratic plagiogranite and quartz–albite porphyry bodies intrude individual ultramafic and volcanic olistoliths of the pillowed lava – hyaloclastite facies. Plagiogranite pebbles and granules occur throughout the sedimentary succession and were presumably derived from this same source. Diabase dykes, occasionally locally quite common, and gabbro occur in the ultramafic slabs and as rare isolated olistoliths.

We consider the ultramafic and volcanic olistoliths to derive from the allocthonous slabs (units A and B of Fig. 2). The character of the olistolith load indicates that these slabs originally consisted of an ultramafic layer surmounted by a submarine volcanic edifice similar to a volcanic island arc. In areas to the south of those studied by us, the Gander River belt retains this character (Jenness 1963; Williams 1963). The formation of mélange removed fragments of these slabs up to several kilometres in diameter and transported them several kilometres.

Sedimentary olistoliths consist of shale, interbedded shale and siltstone, and sandstone derived from the underlying sedimentary rocks. Small olistoliths (< 1 m in diameter) of carbonate material are occasionally seen in the lower part of the mélange. Olistoliths up to 150 m in length have been identified, but the distinction between large(r) olistoliths and lithologically identical surrounding rocks can only be made with certainty where the surrounding mélange matrix can be observed. Hence, much larger but unrecognizable sedimentary olistoliths may be present.

Most olistoliths exhibit rounding (Fig. 6), hydroplastic deformation (Fig. 7), and (or) disaggregation, but resistant and more lithified beds fractured in brittle fashion. In larger bedded olistoliths these characteristics may occur internally (Fig. 8), their distribution and character controlled by differing competency of the various layers.

In several locations sedimentary rocks can be continuously traced from a coherent well-bedded

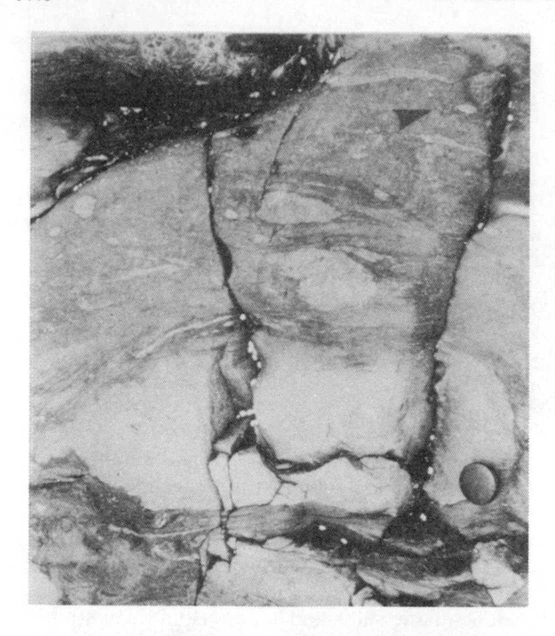

FIG. 6. Sedimentary olistoliths enclosed within a shale matrix. Note the variability of shapes resulting from the incorporation of hydroplastic to brittle olistoliths (see Fig. 10 for location). Lens hood is 6 cm in diameter. Arrowhead points north.

FIG. 8. Hydroplastic and brittle deformation within a more coherently bedded succession. The small displacements and shale intrusions are parallel to cleavage (see Fig. 10 for location). Coin diameter is 2 cm. Arrowhead points north.

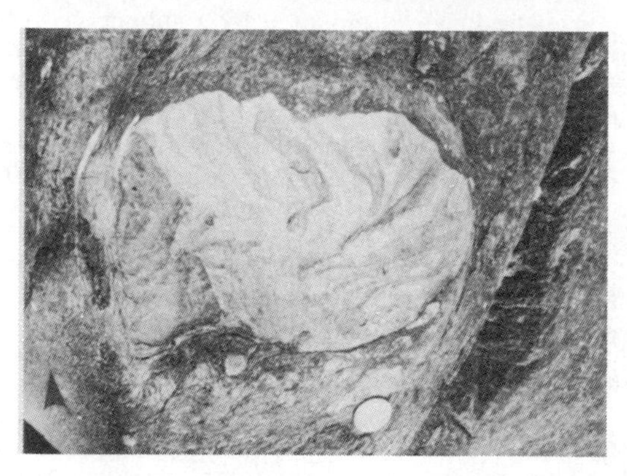

FIG. 7. An olistolith exhibiting internal hydroplastic deformation enclosed within a cleaved shale matrix. The cleavage and flow pattern of elongated fragments wrap around the olistolith (see Fig. 10 for location). Coin diameter is 2 cm. Arrowhead points north.

FIG. 9. Hydroplastically deformed sediment becoming progressively more disaggregated to the right of photo. (Location: eastern Green Island.) Width of photo is 4 m. Arrowhead points north.

section, through hydroplastically deformed beds, to completely disaggregated material (Fig. 9), and finally to discrete olistoliths enclosed within a thixotropic matrix, which exhibits a cleavage and a preferred dimensional orientation of olistoliths in the cleavage plane. This cleavage commonly lies at a high angle to the bedding in surrounding undisturbed beds (Fig. 10). Sharp contacts have been observed between mélange matrix and sedimentary rocks in many localities.

Bedded sedimentary olistoliths greater than 10 m in size, each separated or truncated by matrix, locally occur stacked in a succession so that bedding and facing directions are consistent. The stratigraphic succession may have been preserved because the matrix was sufficiently fluid that it did not

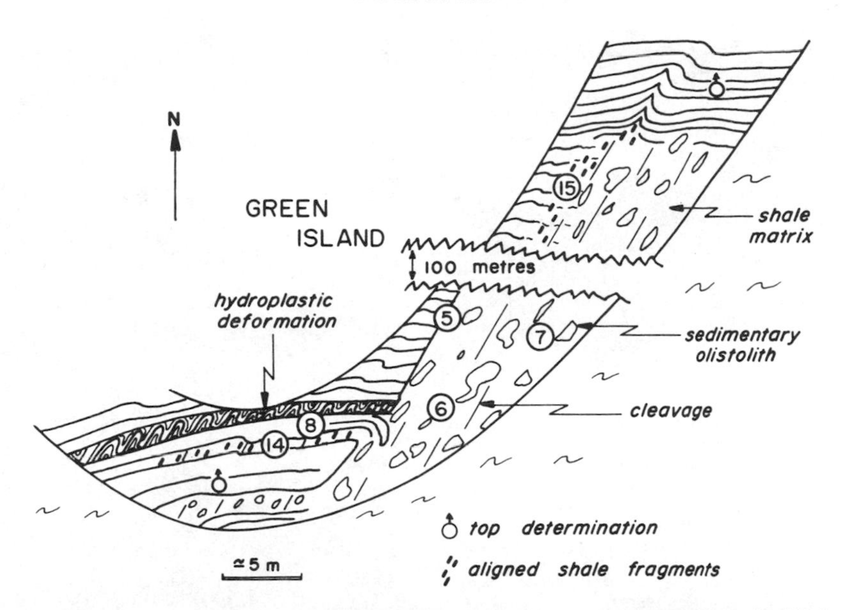

FIG. 10. Schematic diagram of the relationship between bedded sedimentary successions and intrusive black shale. Note that the alignment of cleavage, olistoliths, and fold axes is common throughout the exposure. Circled numbers refer to figures in this paper. The thickness of bedding and size of olistoliths and shale fragments are not to scale. (Location: southwestern corner of Green Island.)

exert a significant force on these large olistoliths. However, other observations, for example at Mann Point, show that sedimentary olistoliths of comparable size were rotated through at least 180°, since they now face downward. Small olistoliths have all undergone rotation.

Metamorphosed and deformed olistoliths with histories of deformation and metamorphism differing from adjacent lithologically identical olistoliths are occasionally observed in the mélange. Several large olistoliths of layered volcanogenic schist occur associated with normal mafic volcanic olistoliths. The layered metamorphic fabric has been folded at least twice and later shattered by small shears and breccia zones, many of which are filled by quartz veins and pods. Large sedimentary olistoliths exhibiting three or even four episodes of folding, as well as shattering and intense quartz veining, occur associated with the volcanogenic schists. We have also observed pillowed lava olistoliths containing metamorphic garnet and a recrystallized groundmass. At two locations, on the east side of Noggin Cove and on Twillick Point, veined and deformed olistoliths are intruded by mélange matrix, which contains boudined and disrupted quartz veins (Fig. 11), and are then incorporated into still mobile, perhaps younger, mélange matrix. Therefore metamorphism, deformation, and quartz-vein emplacement were active during mélange formation. Dynamothermal metamorphism in similar environments has been described by Moores (1969), Reinhardt (1969), Church and Stevens (1971), and Williams and Smyth (1973).

Regionally metamorphosed mélange outcrops south of Aspen Cove (Fig. 2). We have traced the mélange continuously into garnet–cordierite– andalusite grade, and observed isolated localities at sillimanite grade. The mélange in the contact aureole of the Rocky Bay pluton has been severely deformed and metamorphosed by the intrusive event, producing an intricately plicated, hornfelsed mélange (Fig. 12).

Character of the Matrix

The matrix of the olistostrome exhibits features that establish its former intrusive and dynamic character. A cleavage and cleavage-parallel alignment of small olistoliths is universal. The latter feature is particularly well-exhibited on Green Island (Fig. 10). At the southwestern point of the island is exposed a 20 cm mélange layer in otherwise coherently bedded sedimentary olistoliths up to 7 cm in length (Fig. 13), none of which derive from immediately adjacent beds. Along the upper edge of the mélanged layer, elongate shale fragments have been forcefully injected into the overlying hydroplastic beds (Figs. 14, 15). These in-

FIG. 11. Quartz-bearing black shale matrix irregularly intruding the contact between volcanogenic schist (background) and metasedimentary schist (foreground). The white areas in the matrix (e.g., under hammer) are quartz that are somewhat fractured. (Location: northwestern end of peninsula between Noggin Cove and Carmanville.) Arrowhead points north.

FIG. 12. Remobilized metasedimentary olistoliths and quartz in black slate in the contact aureole of the Rocky Bay pluton. Contrast this photograph with the quartz-bearing 'older' matrix, which was not affected by contact metamorphism, in Fig. 11. (Location: 800 m west of Twillick Point.) Coin diameter is 2 cm. Arrowhead points north.

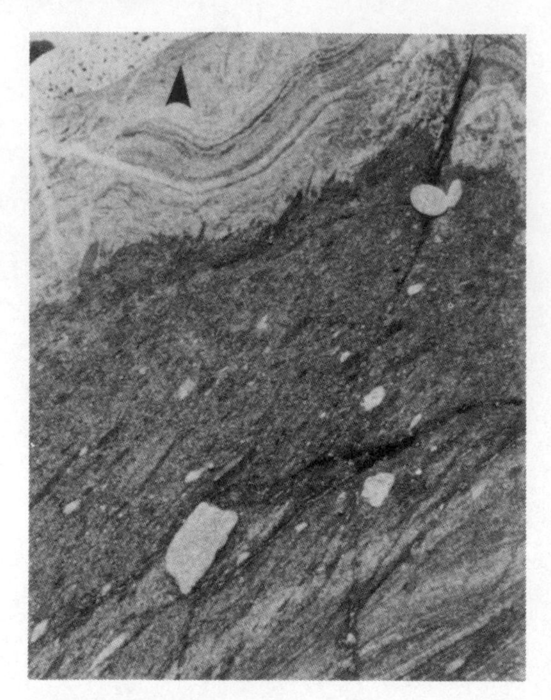

FIG. 13. A 20 cm thick olistostrome horizon in an otherwise bedded sedimentary succession. Note the parallelism of elongate olistoliths, which also parallels cleavage (not observable in photo). None of the small olistoliths were derived from adjacent beds (see Fig. 10 for location). Coin diameter is 2 cm. Arrowhead points north.

jected fragments lie parallel to the cleavage, the preferred orientation of the olistoliths, and the cleavage in a mélange 'dyke' that intrudes the bedded sequence a few metres to the east (Fig. 10). The axial planes of minor folds in the coherent beds are also parallel to this direction, although the cleavage is only weakly developed outside the mélange horizon. These observations show that the mélange matrix was sufficiently viscous that it could inject fragments and sustain differential stresses sufficient to align the fragments and produce small folds (Fig. 8) in the surrounding rocks.

In the upper horizons of the mélange the facing directions of separated sedimentary blocks remain consistent over distances of at least 800 m. The fragment motion was therefore almost a pure translation with little or no rotation. In the lower part of the mélange, blocks of similar size have been rotated at least 180°, since they are now inverted. These differences may have been due to higher pressures, with resulting higher viscosity and more turbulent flow in the lower parts of the slide. Regardless of position in the matrix, the transport of the very largest sedimentary olistoliths was by sliding, with limited rotation about horizontal axes.

The shale that formed the matrix of the mélange may have represented a single depositional unit, which was forced under high hydrostatic pressures into fractures formed as the sedimentary pile became unstable.

Evolution of the Mélange around Carmanville

The sedimentary rocks immediately below the southeasternmost matrix horizons consist of resedimented turbidity and inertia flows with locally developed manganese nodules, characteristics that also apply to much of the sedimentary rocks of the mélanged terrain. Unlike the continentally derived sediment of unit 1, this sediment consisted mainly

FIG. 14. A close-up of the upper edge of the olistostrome horizon in Fig. 13. Note the injection of shale fragments into the then soft but still coherent overlying bed. Coin diameter is 2 cm. Arrowhead points north.

FIG. 15. Shale fragments injected into bedded sediments, which were still plastic at the time the black shale matrix was mobilized (see Fig. 10 for location). Coin diameter is 2 cm. Arrowhead points north.

of mafic volcanic detritus, with a minor but ubiquitous component of plagiogranite, mafic, and ultramafic granules. However, the distinctive quartz-rich continentally derived sediment continued to be supplied in subordinate amount.

The stratigraphy and structure of the Carmanville region, discussed above and by Currie *et al.* (1979*a*), show that the source of the volcanic and ultramafic material must have been from the so-called Gander River ultramafic belt, and that the change in sediment type correlates with the emplacement of these rocks by obduction. Despite the drastic nature of this emplacement the sedimentological regime appears to have continued little changed, except for the shedding of clastic debris from the allocthonous slabs. The nature of the rocks beneath the mélange strongly suggests that this area continued to receive some sediments from distal turbidity flows originally initiating on the continental rise to the east. Eventually, however, possibly due to further tectonic movement, much of this continental rise prism became unstable and the variably lithified sediments were involved in submarine slides (olistostromes) on a major scale. We have recovered undeformed orthoquartzite pebbles from the mélange whose nearest source appears to be in the Cambrian Random Formation (Jenness 1963) well within the Avalon Zone, suggesting that the slides directly or indirectly sampled the whole of the continental rise prism and back onto the shelf. However, at any point, much of the debris in the slides was of local origin, leading to the observed crude zonation of the mélange between ultramafic, volcanic, and sedimentary olistoliths. A major effect of such large slides was to dissect the ultramafic belt, undoubtedly already badly fractured during transport and emplacement, so that the volcanic rocks originally part of the belt are, in the Carmanville area, found some kilometres to the northwest. The ultimate source of this ultramafic–volcanic association lies beyond the scope of this paper, but its present geologic configuration clearly shows that the Gander River ultrabasic belt forms part of an obducted slab, now partly buried by later sediments. The association of mélange-bearing ultramafic and volcanic rocks with obducted slabs has been observed world-wide (Gansser 1974).

Subsequent to formation and emplacement, the mélange was deformed and metamorphosed. The essentially posttectonic Rocky Bay pluton yields a latest Silurian or early Devonian age (Pajari and Currie 1978). Thus, most of the deformation is probably of Silurian age, although metamorphism probably continued into the Devonian. Dean and Strong (1977) have shown that regional thrust faults with southeasterly transport directions were reactivated during Devonian time in the Moretons Harbour area, 30 km to the northwest. Some of the deformation in this area may be contemporaneous with these thrusts. We point out that if the obduction and formation of mélange indicate oceanic closure and continental collision, then the metamorphism and plutonism considerably postdate these events, contrary to opinions expressed by McKerrow and Cocks (1977).

Discussion

Gansser introduced the term 'ophiolitic mélange' "in order to distinguish it from purely sedimentary mélange of the olistostrome type or purely tectonic mélange ..." (Gansser 1974, p. 482) and to emphasize the world-wide importance of ophiolite-bearing mélange in relation to ophiolitic suites. Essentially an ophiolitic mélange is a mixture of ophiolitic and sedimentary blocks set in either a shale–sandstone matrix or a matrix of tectonically sheared serpentinite. Gansser (1974, p. 483) in his original definition stated that the base of an ophiolitic mélange should have a tectonic contact. Some ophiolitic mélanges possess a sheared serpentinite matrix, and appear truly tectonic, which is hardly surprising since they typically occur at ancient convergent plate boundaries in association with oceanic material obducted across continental rocks. Most ophiolitic mélanges, however, possess a sedimentary matrix, and a wide variety of variably (plastically) deformed indigenous and exotic blocks suggestive of an origin by surficial mass wastage. We therefore view most examples of ophiolitic mélanges as olistostromal mixtures of tectonically displaced rocks (cf. Williams 1977*a*). The Carmanville mélange clearly exhibits these characteristics. We therefore propose that the most appropriate stratigraphic descriptor for this series of olistostromes is the Carmanville ophiolitic mélange.

The Dunnage mélange exhibits striking similarities to the Carmanville ophiolitic mélange, from which it is separated by a few kilometres of Silurian sedimentary and volcanic rocks (see Horne 1969; Hibbard and Williams 1975; Williams and Hibbard 1976; Kay 1976). The Dunnage mélange lacks ultramafic olistoliths and contains a greater proportion of carbonate olistoliths and matrix relative to olistoliths than the Carmanville ophiolitic mélange. Both sample deep water, mainly turbiditic sediments. The Dunnage mélange involves Middle Cambrian to Arenigian sedimentary rocks (Hibbard *et al.* 1977). If the correlation of the orthoquartzite pebbles in the Carmanville ophiolitic mélange is correct, it samples Lower Cambrian to Caradocian sediments. The age of the two mélanges may differ slightly, since the sediments in the Carmanville ophiolitic mélange contain Caradocian fossils, whereas the Dunnage mélange is overlain by the Caradocian Dark Hole Formation (Hibbard and Williams 1979). It seems probable that some of the black shale in the matrix of the Carmanville ophiolitic mélange may be correlative with the Dark Hole Formation.

We believe that the Carmanville and Dunnage mélanges form a single zone, probably slightly time-transgressive. Since we have shown that formation of the Carmanville ophiolitic mélange was closely associated with obduction of oceanic crust and superincumbent volcanic rocks, we assume that the formation of the Dunnage mélange, on the southern and eastern flanks of an oceanic island arc complex (Strong and Payne 1973; Hibbard and Williams 1979), was also linked to this event.

Acknowledgements

We wish to thank Hank Williams, Ben Kennedy, Les Fyffe, Bill Poole, and Lawson Dickson for critically reviewing earlier versions of the manuscript. Darlene Pajari, Sherri Townsend, and Mike Oudemans are also thanked for their technical assistance during its preparation. Two of us (G.E.P. and R.K.P.) acknowledge the support of National Research Council of Canada grants. This paper represents the Canadian contribution to IGCP Publication No. 7.

BELL, K., BLENKINSOP, J., and STRONG, D. F. 1977. The geochronology of some granitic bodies from eastern Newfoundland and its bearing on Appalachian evolution. Canadian Journal of Earth Sciences, **14**, pp. 456–476.

BERGSTRÖM, S. M., RIVA, J., and KAY, M. 1974. Significance of conodonts, graptolites and shelly faunas from the Ordovician of western and north-central Newfoundland. Canadian Journal of Earth Sciences, **11**, pp. 1625–1660.

BIRD, J. M., and DEWEY, J. F. 1970. Lithosphere plate – continental margin tectonics and the evolution of the Appalachian Orogen. Geological Society of America Bulletin, **81**, pp. 1031–1060.

BLACKWOOD, R. F. 1978. Northeastern Gander zone, Newfoundland. Mineral Development Division, Newfoundland Department of Mines and Energy, Report of Activities for 1977, pp. 72–79.

BRÜCKNER, W. D. 1972. The Gander Lake and Davidsville Groups of northeastern Newfoundland: new data and geotectonic implications: Discussion. Canadian Journal of Earth Sciences, **9**, pp. 1778–1779.

——— 1978. Carmanville map-area, Newfoundland; the northeastern end of the Appalachians—Discussion. *In* Current Research, Part C. Geological Survey of Canada, Paper 78-1C, pp. 127–129.

CHURCH, W. R., and STEVENS, R. K. 1971. Early Paleozoic ophiolite complexes of the Newfoundland Appalachians as mantle-oceanic crust sequences. Journal of Geophysical Research, **76**, pp. 1460–1466.

CURRIE, K. L., and PAJARI, G. E. 1977. Igneous and metamorphic rocks between Rocky Bay and Ragged Harbour, northeastern Newfoundland. *In* Report of Activities, Part A. Geological Survey of Canada, Paper 77-1A, pp. 341–346.

CURRIE, K. L., PAJARI, G. E., and PICKERILL, R. K. 1979*a*. Tectono-stratigraphic problems in the Carmanville area, northeastern Newfoundland. *In* Current Research, Part A. Geological Survey of Canada, Paper 79-1A, pp. 71–76.

CURRIE, K. L., PICKERILL, R. K., and PAJARI, G. E. 1979*b*. Plate tectonic models of the southeastern margin of the Iapetus Ocean in northeastern Newfoundland. To be published.

DEAN, P. L., and STRONG, D. F. 1977. Folded thrust faults in Notre Dame Bay, Central Newfoundland. American Journal of Science, **277**, pp. 97–108.

DEWEY, J. F., and BIRD, J. M. 1970. Mountain belts and the new global tectonics. Journal of Geophysical Research, **75**, pp. 2625–2647.

DICKINSON, W. R. 1973. Widths of modern arc-trench gaps proportional to past duration of igneous activity in associated magmatic arcs. Journal of Geophysical Research, **78**, pp. 3376–3389.

GANSSER, A. 1974. The ophiolitic mélange, a world-wide problem on Tethyan examples. Ecologae Geologicae Helvetiae, **67**, pp. 479–507.

GREENLY, E. 1919. The geology of Anglesey. Memoirs of the Geological Survey of Great Britain, England and Wales. 980 p.

HIBBARD, J. P., and WILLIAMS, H. 1975. The Dunnage mélange, northeast Newfoundland. Geological Society of America, NE Section, Abstracts with Programs, Vol. 7, pp. 781–782.

——— 1979. Regional setting of the Dunnage mélange in the Newfoundland Appalachians. American Journal of Science, **279**. In press.

HIBBARD, J. P., STOUGE, S., and SKEVINGTON, D. 1977. Fossils from the Dunnage mélange, north-central Newfoundland. Canadian Journal of Earth Sciences, **14**, pp. 1176–1178.

HORNE, G. S. 1969. Early Ordovician chaotic deposits in the central volcanic belt of northeastern Newfoundland. Geological Society of America Bulletin, **80**, pp. 2451–2464.

HSÜ, K. 1971. Franciscan mélanges as a model for eugeosynclinal sedimentation and underthrusting tectonics. Journal of Geophysical Research, **76**, pp. 1162–1170.

——— 1974. Mélanges and their distinction from olistostromes. *In* Modern and ancient geosynclinal sedimentation. *Edited by* R. H. Dott, Jr. Society of Economic Paleontologists and Mineralogists, Special Publication 19, pp. 321–333.

JENNESS, S. E. 1963. Terra Nova and Bonavista map-areas, Newfoundland. Geological Survey of Canada, Memoir 327. 184 p.

——— 1972. The Gander Lake and Davidsville Groups of northeastern Newfoundland: new data and geotectonic implications: discussion. Canadian Journal of Earth Sciences, **9**, pp. 1779–1781.

KAY, M. 1976. Dunnage mélange and subduction of the Protacadic Ocean northeast Newfoundland. Geological Society of America, Special Paper 175. 49 p.

KENNEDY, M. J. 1975. Repetitive orogeny in the northeastern Appalachians—new plate models based upon Newfoundland examples. Tectonophysics, **28**, pp. 39–87.

——— 1976. Southeastern margin of the northeastern Appalachians: Late Precambrian orogeny on a continental margin. Geological Society of America Bulletin, **87**, pp. 1317–1325.

KENNEDY, M. J., and MCGONIGAL, M. H. 1972. The Gander Lake and Davidsville Groups of northeastern Newfoundland: new data and geotectonic implications. Canadian Journal of Earth Sciences, **8**, pp. 452–459.

MCKERROW, W. S., and COCKS, L. R. M. 1977. The location of the Iapetus Ocean suture in Newfoundland. Canadian Journal of Earth Sciences, **14**, pp. 488–495.

MOORES, E. M. 1969. Petrology and structure of the Vourinous ophiolite complex, northern Greece. Geological Association of America, Special Paper 118. 74 p.

PAJARI, G. E., and CURRIE, K. L. 1978. The Gander Lake and Davidsville Groups of northeastern Newfoundland: a re-examination. Canadian Journal of Earth Sciences, **15**, pp. 708–714.

PICKERILL, R. K., PAJARI, G. E., CURRIE, K. L., and BERGER, A. R. 1978*a*. Carmanville map-area, Newfoundland; the northeastern end of the Appalachians. *In* Current research, Part A. Geological Survey of Canada, Paper 78-1A, pp. 209–216.

——— 1978*b*. Carmanville map-area, Newfoundland; the northeastern end of the Appalachians—Reply. *In* Current research, Part C. Geological Survey of Canada, Paper 78-1C, pp. 129–132.

REINHARDT, B. M. 1969. On the genesis and emplacement of ophiolites in the Oman mountains geosyncline. Schweizerische Mineralogische und Petrographische Mitteilungen, **49**, pp. 1–30.

STEVENS, R. K. 1970. Cambro-Ordovician flysch sedimentation and tectonics in west Newfoundland and their possible bearing on a Proto-Atlantic Ocean. *In* Flysch sedimentology in North America. *Edited by* J. Lajoie. Geological Association of Canada, Special Paper, No. 7, pp. 165–177.

STRONG, D. F., and PAYNE, J. G. 1973. Early Paleozoic volcanism and metamorphism of the Moreton's Harbour area, Newfoundland. Canadian Journal of Earth Sciences, **10**, pp. 1363–1379.

WILLIAMS, H. 1963. Botwood map-area, Newfoundland. Geological Survey of Canada, Preliminary Series, Map 60-1963.

——— 1972. Stratigraphy of Botwood map-area, northeastern Newfoundland. Geological Survey of Canada, Open File 113. 103 p.

——— 1975. Structural succession, nomenclature and interpretation of transported rocks in western Newfoundland. Canadian Journal of Earth Sciences, **12**, pp. 1874–1898.

——— 1977*a*. Ophiolitic mélange and its significance in the Fleur de Lys Supergroup, northern Appalachians. Canadian Journal of Earth Sciences, **14**, pp. 987–1003.

——— 1977*b*. The Coney Head Complex: another Taconic allochthon in west Newfoundland. American Journal of Science, **277**, pp. 1279–1295.

WILLIAMS, H., and HIBBARD, J. P. 1976. The Dunnage mélange, Newfoundland. *In* Report of Activities, Part A. Geological Survey of Canada, Paper 76-1A, pp. 183–185.

WILLIAMS, H., and SMYTH, W. R. 1973. Metamorphic aureoles beneath ophiolitic suites and Alpine peridotites: tectonic implications with west Newfoundland examples. American Journal of Science, **273**, pp. 594–621.

WILLIAMS, H., KENNEDY, M. J., and NEALE, E. R. W. 1974. The northeastward termination of the Appalachian orogen. *In* The ocean basins and margins, Vol. 2. *Edited by* A. E. M. Nairn and F. G. Stehli. Plenum Press, New York, NY, pp. 79–123.

Part IV

ALPINE MÉLANGES OF THE DINARIDS AND HELLENIDS

Editor's Comments on Papers 15, 16, and 17

15 **DIMITRIJEVIĆ and DIMITRIJEVIĆ**
Olistostrome Mélange in the Yugoslavian Dinarides and Late Mesozoic Plate Tectonics

16 **CELET et al.**
Volcano-sedimentary and Volcano-detritic Phenomena of Mesozoic Age in Dinarid and Hellenic Ranges: A Comparison

17 **MERCIER and VERGELY**
Abstract from and a Review of *Les Mélanges ophiolithiques de Macédoine (Grèce): Décrochements d'Age anté-crétacé supérieur*

This group of three papers concerns mélanges in the Balkan region, in the Alpine orogen. Dimitrijević and Dimitrijević (Paper 15) reinterpret the so-called Diabase Hornstein Formation as an olistostrome mélange connected with a destructive plate margin active to the late Mesozoic. The age of the mélange is uncertain, and it may either be entirely of late Cretaceous age or include both late Jurassic and late Cretaceous mélanges. The matrix is a chaotic assemblage of dark gray, commonly argillaceous sandstone and silt, unstratified and structureless, with no visible sedimentary breaks. The enclosed fragments are lumps, blocks, and irregular plates of turbiditic sandstone, diverse limestones (including oolitic, reefal, and pelagic types), radiolarian red chert, and basic ultramafic igneous rocks. The limestones include Triassic, Jurassic, and Cretaceous rocks, and Permian limestones and conglomerates also are represented. The amount of matrix is variable and, locally, almost absent. Blocks are up to hundreds of meters in diameter and display some concentration in horizons, especially chert blocks. There is evidence of deformation of the sedimentary blocks prior to lithification, and the red cherts are tightly folded. The mélange commonly overlies Triassic limestone and may also interdigitate with it or be overlain by it (in the form of klippen). Ophiolitic rocks represent a considerable part of this mélange. Some ultrabasic rocks have amphibolite selvages and marginal contact metamorphic zones up to 100 m thick, though geophysical evidence suggests that most are rootless bodies. Some ultrabasic bodies were apparently emplaced into the mélange as warm crystal mush, but others are slabs of oceanic crust or diapiric introductions.

The authors believe that mass transport was the prime agency of mélange formation, accompanied by disruption and occurring prior to consolidation. It may have commenced as early as the middle Triassic. The mélange is believed to have samples of the continental margin (shelf), deep sea sediments of the lower trench slope and ocean floor, and the oceanic crust itself. It is believed to include shelf (oolite, reefal limestone), continental shelf (turbidites), oceanic sea (radiolarite, chert, micrite), and oceanic crust (mafic, ultramafic) rocks. Only part of the mafic and ultramafic rock content was so derived, part being intruded into the mélange later. The authors state that the older indurated blocks slid in as olistolites, but they seem to ignore the possibility of lithification differentials allowing contrasting responses to strain in a simultaneously deposited sequence. They favor the olistostrome model rather than the tectonic models of Pamir (1944) and Mercier and Vergely (Paper 17), considering tectonization to have been an important late event, connected with final collision, subduction, and tectonic and gravity nappe formation—not the primary cause of nappe formation. They believe that the process of mélange formation may have been prolonged and may have occupied the entire period of active plate consumption. One must question whether an olistostrome interpretation is entirely satisfactory to explain a mélange that incorporates such diverse components, gleaned from such an immense paleogeographic spectrum over such a long period; the concept of tectonic deformation and fragmentation of semilithified sediments would explain the soft rock deformations in the sedimentary fragments and would allow a holistic interpretation of such mammoth-scale mélanges in terms of the immense compressional stresses operating at a destructive plate boundary. Such an alternative model would allow the stresses to act on both hard rocks and any semiconsolidated sequences caught up in the process. It would eliminate the problem, noted by the authors, that this mélange is too large to be formed by single submarine gravity slide, yet they can delimit no internal boundaries.

Celet et al. (Paper 16) discussed mélanges of the Dinarids and Hellenids, where Celet (1976) proposed the term *mélange de type volcano-sedimentaire*. The Upper Jurassic volcano-detrital mélange is distinguished from a Triassic volcano-sedimentary series of radiolarites, pelites, sandstones, volcanics, and so on, which includes calc-alkaline lavas and also some alkalic lavas. The Jurassic mélange is unstratified, disordered, and of chaotic aspect. It has a pelitic, sandy matrix with radiolarites wrapping around the exotic blocks, the lithologies of which include various limestones (Permian, Triassic, and so on), sandstone, and radiolarites. Eruptive rocks such as basic and ultrabasic magmatic rocks (gabbro, peridotite, serpentinite), diabase, and spilitic basalt (including pillow lavas) are abundant among the blocks and derive from ophiolitic suites. Metamorphic rock fragments (amphibolite, mica schist, gneiss) are also represented. This mélange has complex relationships to overlying and underlying units, but it occupies a distinct stratigraphic position and may overlie or underlie shelf limestones and flyschoid or pelagic sediments.

Tectonization was superimposed and resulted in increased chaotization, but the authors do not believe that it was the fundamental cause of the mélange character. Layers of overthrusting in the mélange are of latest Jurassic to earliest Cretaceous age, and the authors suggest that the mélange-forming episode may have occupied a relatively short time period prior to this. The blocks of basic and ultrabasic rocks are very bulky and are clearly derived from the ophiolite nappe, whereas Triassic limestone xenoliths come from series outcropping in the inner zones (also probably radiolarites, sandstones, and undated crystalline limestones). Metamorphic components are of obscure origin but may come from a Paleozoic basement or from contemporaneous metamorphic suites (i.e., coexistent Jurassic-Cretaceous metamorphic zones). Evidence of internal structure led the authors to reject a purely tectonic origin for this mélange and to invoke olistostrome origin with subsequent tectonic modification. This conclusion again meets the possible objection that the soft rock deformations observed could stem from tectonic deformation of partially consolidated sedimentary sequences. Models relating to (1) subduction on the edge of an island-arc or marginal sea and (2) prior obduction of oceanic crust and its disintegration and redeposition in a sedimentary basin are suggested.

Wigniolle (1977) described a Jurassic mélange from the Iti Mountains, Greece, and he noted that not even a feeble metamorphism has affected the pelitic and siliceous, locally gritty matrix. The mélange is of similar character to those described by Celet et al. (Paper 16), and he supports their interpretation. Zimmerman (1972) described a mélange from the Vourinos Complex, northern Greece, with a matrix of metasiltstone and meta-tuff, now mainly chlorite-sericite phyllite. Exotic blocks of mafic volcanics, marble, radiolarite, and minor serpentinite are set in this commonly structureless matrix, which, however, locally preserves folded bedding laminations. The blocks were lithified prior to inclusion, whereas the matrix remained plastic. Zimmerman does not discuss the mode of origin of this mélange.

Mercier and Vergely (Paper 17) described an ophiolitic mélange of pre-late Cretaceous age from Macedonia. This important paper is in

French and includes numerous detailed sketches of the internal structure of the mélange. Because of the technical problems presented, it could not be presented in translation or reproduced in facsimile. It is represented here by the English language abstract and a summary prepared by the editor of this volume. The mélange is considered to be purely tectonic in origin, due to a peculiar type of deformation subsequent to the isoclinal synmetamorphic deformation of late Jurassic-early Cretaceous age and prior to another deformation of late Cretaceous-early Tertiary age and related to dextral transcurrent dislocation of the region. The authors suggest that the colored mélange of Cyprus and Asia embraces a large number of mélange occurrences and probably embraces mélanges of diverse types, including olistostromes, eruptive mixtures, and tectonic mélanges related to both overthrusting and transcurrent dislocation. They suggest that similar studies of deformations in detail would be required to classify the large number of ophiolitic mélange occurrences in Asia and central Europe, as yet not studied in any detail.

The nature of the *prasinite* matrix of this mélange is obviously critical, and it seems that further study including geochemical study is needed. At present, it is unclear whether it was basic igneous rock, a tuff, a sediment, or a mixture of these.

REFERENCES

Celet, P., 1976, A propos de mélange de type "volcano-sedimentaire" de l'Iti (Grèce méridionale), *Soc. Géol. France Bull.* **18:**299-307.

Pamir, H. N., 1944, Une ligne séismogène en Anatolie septentrionale, *Istanbul Univ. Fen Fakültesi Mecmuasi Rev.* **9.**

Wigniolle, E., 1977, Données nouvelles sur la géologie du massif de l'Iti (Grèce continentale), *Soc. Géol. Nord Annales* **97:**239-251.

Zimmerman, J., 1972, Emplacement of the Vourinos Ophiolitic Complex, Northern Greece, *Geol. Soc. America Mem. 132,* pp. 225-239.

15

OLISTOSTROME MÉLANGE IN THE YUGOSLAVIAN DINARIDES AND LATE MESOZOIC PLATE TECTONICS[1]

M. D. DIMITRIJEVIĆ AND M. N. DIMITRIJEVIĆ
Rudarsko-geološko-metalurški fakultet, LMGK, University of Beograd, Beograd, Djušina 7; and Zavod za geološka i geofizička istraživanja, Beograd, Karadjordjeva 48

ABSTRACT

The "Diabas-Hornstein Formation" of Hammer occurs in the Ophiolite belt and in the outer and central Vardar zone of the Inner Dinaric region, characterizing this area as an "eugeosyncline." We interpret this formation as a mélange of olistostrome origin. The mélange originated in deep troughs over a plate-consuming margin by accumulation of sediments from the oceanic bottom, continental rise, and continental shelf, with intermixing of the oceanic crust and mantle ultramafic material as mechanical fragments, cold diapiric bodies, or warm crystal mush. It marks the boundary between the Serbo-Macedonian continental element and part of the Tethys, a margin active also during the Hercynian orogeny. The Ophiolite belt ocean is explained as a secondary ocean, opened by differences in velocity between the marginal and inner parts of the Dinaric microcontinent (probably in the Middle Triassic), and possibly closed without a regular Benioff zone. The Upper Cretaceous-Tertiary magmatic activity beyond this complex ophiolitic scar is explained as a high-temperature result of the processes in the consumed plate.

The peculiar "Diabas-Hornstein Formation" of Hammer (1921) is characteristic of the Inner Dinaric region. It occurs in three belts: (1) the Ophiolite belt ("zone of Mesozoic ophiolites, chert, and the Gossau-type Upper Cretaceous" of Kossmat [1924]), in places up to 50 km wide; (2) the outer Vardar zone (Kopaonik-West Serbian zone); and (3) the central Vardar zone. In a broad sense, these last two belts belong to the Vardar zone (fig. 1). In the classical geotectonic view, this formation has been regarded as a typical deposit with basic "initial volcanism," which gave rise to the subdivision of the Dinarides (the southern Alpine branch in Yugoslavia) into the Outer Dinarides, corresponding to a miogeosyncline, and the Inner Dinarides, corresponding to a eugeosyncline (Carte téctonique internationale de l'Europe 1964). The "Diabas-Hornstein Formation" is interpreted here as a typical olistostrome mélange, connected with a plate-consuming margin active until the Late Mesozoic. The terms "olistostrome" and "olistolite" are used here as defined by Flores (1955): (1) "By olistostromes we define those sedimentary deposits occurring within normal geologic sequences that are sufficiently continuous to be mappable, and that are characterized by lithologically or petrographically heterogeneous material, more or less intimately admixed, that were accumulated as a semifluid body. They show no true bedding, except for possible large inclusions of previously bedded material." (2) "The name olistolith . . . is applied to the masses included as individual elements within the binder" [of the olistostrome].

The sedimentology, geological setting, and interpretation of this formation are presented here.

AGE OF THE MÉLANGE

The age of the mélange has been doubtful for a long time, and still is in some areas (for a bibliography, see Bešić 1970). Although Katzer (1906) had characterized this formation as "Ophiolite-jasper beds" of Jurassic age, classical workers considered it to be Triassic, mainly because of the Middle Triassic fauna of limestones found in the mélange, the superpositional relationships with the Middle and Upper Triassic observed to make its roof, and the

[1] Manuscript received June 19, 1971; revised September 15, 1972.

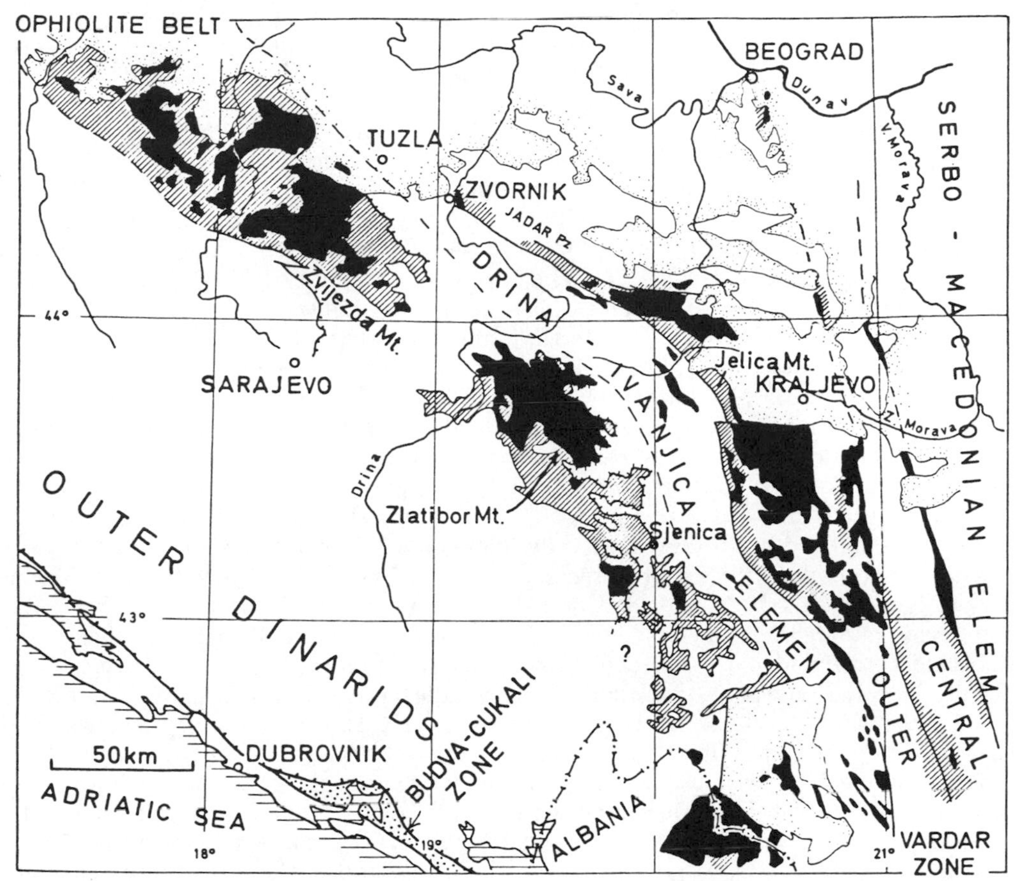

Fig. 1.—Distribution of the mélange (*hachured*) and ophiolites (*black*) in the Ophiolite belt and the Vardar zone.

base of the mélange in places represented by Upper Permian and Lower Triassic strata. New mapping has shown that the mélange in the Ophiolite belt overlies Triassic, as well as somewhat younger strata of possible Middle Jurassic age, and is unconformably overlain by the Uppermost Jurassic. In the Jelica Mountains west of Kraljevo in the Outer Vardar zone, the mélange develops upward from rudist limestones, and is of Upper Cretaceous age. There are still no final data on the age of the mélange in the whole Outer Vardar zone; it may be Upper Cretaceous only, or there may be both Upper Cretaceous and Upper Jurassic mélange formations.

LITHOLOGY

Many investigators have tried to find continuous horizons and consistent lithofacies in the mélange, searching for a normal superpositional succession in the column. However, the mélange has few layers which can be followed for more than a few tens or hundreds of meters, and show coherent succession.

The most typical and widespread rocks of the mélange are the dark gray, frequently argillaceous sandstones and silts which form its matrix. These rocks are entirely chaotic, without stratification, graded bedding, or visible breaks in sedimentation, and have the appearance of sediment-flow and similar mass-transport deposits.

Within this matrix are lumps, blocks, and irregular or platy bodies composed of greenish and gray turbiditic sandstone (mostly subgraywacke and graywacke), basic rocks (mostly melaphyre and diabase, frequently coated with a calcareous envelope), ultramafic rocks, red radiolaritic chert, limestones of various ages and facies (Triassic, Jurassic, and in the Jelica Mountains even Cretaceous), dark sandstones similar to the arenites of the matrix itself, cherty limestone, oolitic limestone, Permian limestone and conglomerate, etc. The quantity of matrix is extremely variable: the rocks are sometimes almost without it, although rarely, and grade up to types with blocks loosely disseminated in an abundant sandy matrix. Blocks of these rocks have diameters from a few centimeters to tens, and probably even hundreds, of meters. Some of these blocks are undoubted olistolites, frequently located along definite horizons. Smaller clasts and blocks of sandstone and chert are frequently folded, smoothed, striated, rounded, or twisted in a way suggesting deformation before lithification was complete.

The red chert horizons are the most persistent but can only rarely be traced more than 1 km. They sometimes display tight folds, probably due to submarine sliding (the Ovčar-Kablar gorge of the Zapadna Morava river, just northwest of the Jelica Mountains.

RELATIONS OF TRIASSIC LIMESTONE TO MÉLANGE

In the whole Ophiolite belt, the mélange is almost always associated with limestone of Triassic age. Exposures show (1) the mélange overlying the limestone, which is then the exposed base of the mélange, sometimes driven tectonically into a position much higher than the original; (2) the intimate intermixing of the two units, generally explained as imbricate structure or lateral transitions (R. Jovanović 1957), but mostly showing olistolite/host relations; or (3) Triassic limestones overlying the mélange, which had previously been regarded as indicating its Triassic age, but correspond to "tectonic" klippen, as parts of a gravity nappe which represents a large part of the northeastern boundary of the Ophiolite zone. Around Sjenica, Triassic limestones definitely overlie the mélange over a vast area. This allochthony was recorded on the 1:200,000 map of SR Serbia by Milovanović and Ćirić (1968), and later described by Rampnoux (1969) as the "Pešter nappe."

The northeastern contact of the mélange is almost everywhere covered by Triassic limestones, which we interpret as a gravity nappe, and the original relations between this formation and its floor cannot be observed. Very interesting relations have been observed along the Zvijezda Mountains, northeast of Sarajevo, on the southwestern contact of the mélange. There, the base of the mélange is represented by Middle Triassic limestones, overlain mostly by a unit of very variable thickness and peculiar lithology, informally named by Bosnian geologists "the marly unit" and here called "the Zvijezda formation." Three of the contacts observed will be briefly described; they are all situated about 35 km northeast of Sarajevo (fig. 2):

Stupari locality; Tuzla-Sarajevo road.—The lowermost unit is composed of gray to reddish Middle Triassic limestones, with incipiently nodular beds separated by films and thin (up to a few centimeters) interlayers of reddish argillaceous sediments. They grade upward into 5–6 m of well-stratified, light gray micrites with conspicuous greenish lenses of mm-cm dimensions, where Middle Triassic conodonts have been found. The uppermost bed of the micritic sequence is divided into cake-like bodies, but without signs of plastic deformation. The mélange follows with typical appearance.

Stavnja River.—The lower exposed unit, corresponding in the vertical position to the micrites at Stupari, consists of a very well stratified alternation of greenish marls, grey siltstones, and red chert in beds 2–20 cm thick, with less frequent layers of marly micrite. The

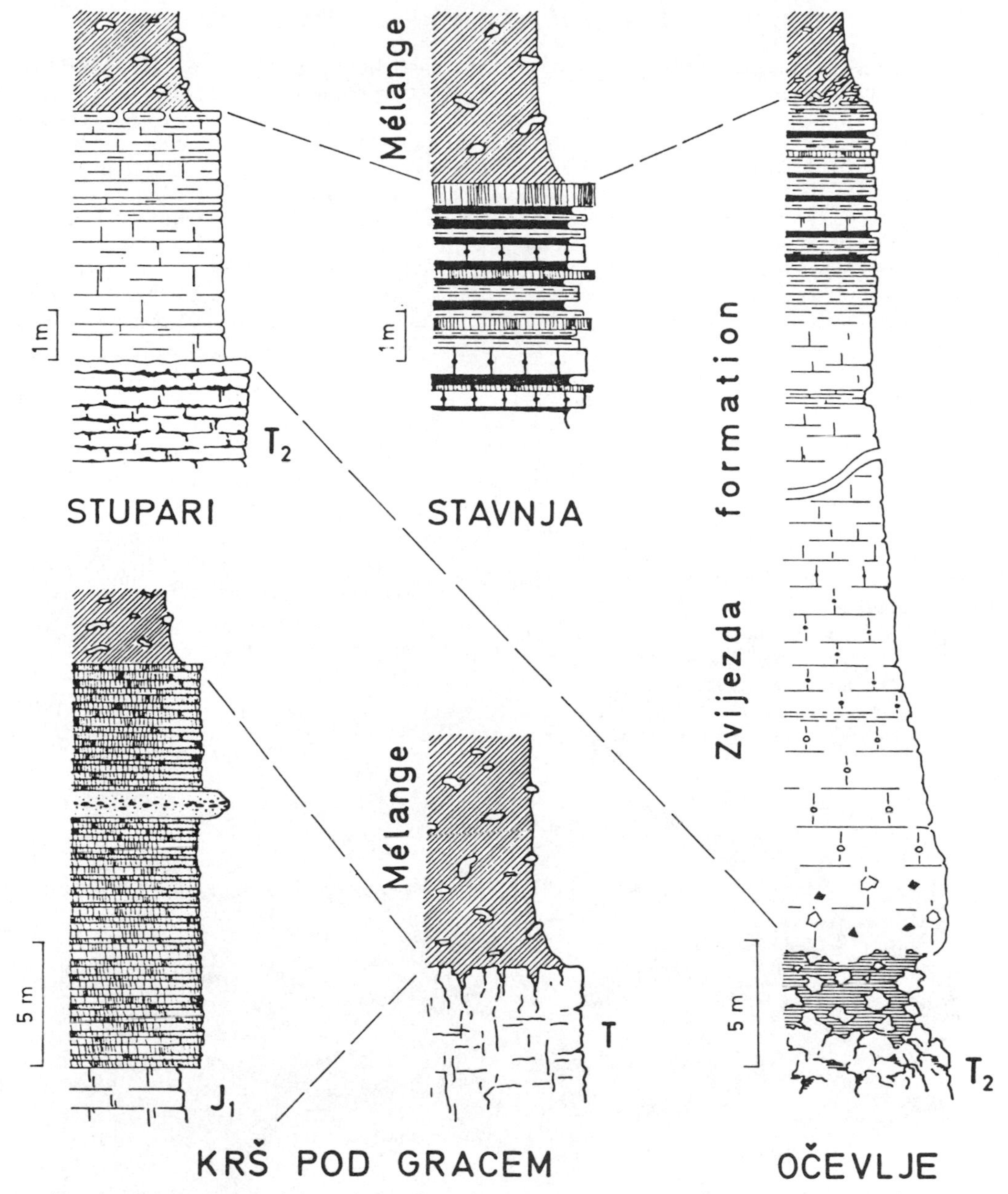

FIG. 2.—Local columns of the mélange, the Zvijezda formation, and the base along the Zvijezda Mountain (Stupari, Stavnja, and Očevlje localities), and at Krš pod Gracem near Sjenica.

unit ends with a chert bed 50 cm thick, overlain by very schistose mélange. No fossils have so far been found in these sediments, representing here the Zvijezda formation, and it does not show any kind of sedimentary or penetrative tectonic deformations.

Očevlje.—The section begins with brecciated Middle Triassic limestone of the Han Bulog type (*rosso ammonitico*). The last 3 m of these limestones represent a conspicuous breccia with red binder, where a break in sedimentation seems to be present. They are overlain by the Zvijezda formation, beginning here with 3.5 m of light gray calcirudite with chert fragments, grading upward into 6 m of calcarenites in about 15 cm-thick beds displaying in places a rough lamination. The grain size and bed thickness decrease upward, and the sequence shows a transition into several decameters of poorly bedded, marly, light gray to slightly purplish micrites. They have only a few laminated horizons, up to 0.5 m thick. The thickness of the unit could not be accurately measured because of several local faults. The uppermost part of the sequence, more than 8 m thick, consists of a finely laminated alternation of grey siltstone and marl with some calcarenite and chert beds. Upward, in a zone about 1 m thick, this sequence shows a transition into mélange; the beds are more and more discontinuous, with increasing deformations in a semilithified state and grade into typical mélange. This transition is shown by figures 3–6. Figure 3 represents the undisturbed Zvijezda formation several meters below the transition zone; figure 4, the lower part of the transition zone; figure 5, the upper part of the transition zone, and figure 6 the typical mélange about 2 m above the transition zone.

We interpret the Zvijezda formation, occurring between Middle Triassic limestones and the mélange, as the remainder of the primary complex from which the largest part of the mélange was derived by intimate mixing in the semiconsolidated state. At the Očevlje section, the Zvijezda formation and the mélange have the same lithologic composition, only the latter is not stratified; the general appearance of Zvijezda formation is exactly what one would expect of the original, pre-mélange complex on the basis of the lithological examination

FIG. 3.—Undisturbed Zvijezda formation, a few meters below the transition zone. Očevlje locality, Zvijezda Mountain.

FIG. 4.—Lower part of the transition zone between the undisturbed Zvijezda formation and the mélange. Očevlje locality, Zvijezda Mountain.

FIG. 5.—Upper part of the transition zone between the undisturbed Zvijezda formation and the mélange. Očevlje locality, Zvijezda Mountain.

FIG. 6.—Typical mélange, a few meters above the transition zone. Očevlje locality, Zvijezda Mountain

of the mélange rocks. The Middle Triassic conodonts from the micrites in the Stupari section may indicate the beginning of primary sedimentation even from this period (D. Veljković, personal communication, 1972).

Different relations have been observed at Krš pod Gracem, a locality described and interpreted by Ledebur (1941), Ćirić (1954), Z. Jovanović (1963), and others. The relationships between the mélange and its base vary there within a distance of 20 m. In a little gully below the road, the base is represented by Liassic limestones overlain by about 1 m of sandstone and microconglomerate with multiple graded bedding, followed by several meters of red and green chert and the mélange. For about 20 or more meters along the contact, the mélange directly overlies karstified Triassic limestones, filling even its karst cavities.

OPHIOLITES IN THE MÉLANGE

The association of ophiolites and mélange is characteristic (Steinmann 1906, 1927), but has been explained in the Dinarides in various ways. The opinion prevailed that serpentinites have always a cold tectonic contact with the surrounding strata; that they are ancient rocks, foreign in relation with the adjacent rocks; that amphibolites and schists, so frequent along their margins, represent the Paleozoic or even older base transported tectonically with diapiric serpentinites into the present position. It seems clear that the ultramafic rocks represent an integral part of the mélange, and that contacts of large masses are frequently magmatic. That is conspicuous around the Zlatibor massif, where all transitions can be traced from the unchanged mélange through green schists to amphibolites and gneisses immediately at the contact. Metamorphic gradients are always very steep. The contact zone is seldom much more than 100 m thick, the metamorphics being located alongside the lower surface of the ultramafics. Geophysical prospecting (Milovanović and Mladenović 1966–1967) has shown that serpentinites lie largely as rootless plates.

SUMMARY OF FEATURES

From these data the following can be concluded:

1. The matrix (arenites and siltstones, mostly) was formed by a type of mass transport with turbulent particle movements. They are usually retransported prior to full consolidation from their original sedimentary environments, and torn into slabs and blocks, or completely intermixed and rearranged. During their first sedimentation, they had the characteristics of turbidites and miogeoclinal sediments, the deposition of which began probably even in the Middle Triassic. The quantities of sediments involved are too extensive to be explained by one submarine slump, although no breaks in sedimentation have been observed.

2. The fragments and blocks in this rock association correspond to deposits of highly different environments: shelf (oolitic and reef limestones), continental rise (turbidites and silt), oceanic sea (radiolaritic chert and micrites), and oceanic crust (mafic and ultramafic rocks).

3. Mafic and ultramafic rocks occur as mechanical fragments, but ultramafic rocks also form large bodies with magmatic contacts. The emplacement of these bodies is later than the deposition of at least one part of the mélange.

These features show an environment where the products of various sedimentary areas, including oceanic bottom, continental rise, and shelf, were accumulated; where the oceanic crust had the opportunity to participate in deposition; and where a force was present, able to stir the accumulated deposits throughout the deposition of mélange.

FEASIBLE INTERPRETATION

We think that only the plate tectonics model adequately explains all the facts observed. The mélange environment can best be explained by the existence of a deep oceanic trough along a plate-consuming continental margin. In this trough have been accumulated:

1. Miogeoclinal deposits of the passive (Dinaric) continental margin; they are now mostly reworked into the mélange, except for micrites, siltstones, cherts, and calcarenites described from the Zvijezda formation and possible similar deposits.

2. Deep-sea sediments conveyed by the oceanic plate; they represent the lower part of the sedimentary apron covering continental margins, and sediments of the oceanic bottom (red radiolaritic chert; marly micrite; and biomicrite with pelagic microfauna now present as olistolites; others intermixed into the mélange).

3. Rocks of the oceanic crust itself.

4. Synchronous deposits of the active continental margin, almost completely reworked before lithification into the mélange, including Jurassic pelmicrites, pelsparites, biosparites, and pseudo-oosparites as ingredients which still can be identified.

5. Older, indurated rocks of the active margin, slid into the trough as olistolites, including Triassic sparites and Permian conglomerate.

All these ingredients were mixed during the mélange development, many of them prior to their complete lithification. Older, indurated rocks gave rise to olistolites and clasts. The mixing mechanism was perhaps controlled by simultaneous differential vertical and horizontal movements in the trough, which caused constant agitation of the mélange by a type of transport intermediate between sliding and turbulent suspensional conveyance.

We think that this olistostrome model explains the origin of the mélange much better than the tectonic models (Greenly 1919; Pamir 1944; Blumenthal 1948; Bailey and McCallien 1950; Mercier and Vergely 1972); tectonization of the mélange is an important late event, connected with the final collision of continental blocks, subduction, and accompanying "tectonic" and "gravity" nappe formation, but not the primary cause of mélange organization.

It is interesting that part of the ultramafic material in the form of large bodies was warm enough to produce active magmatic contacts with a metamorphic rim (Zlatibor). Such features can be explained by fracturing of the descending oceanic slab in the buckling zone (see Savage 1969), and injection of sufficiently warm crystal mush from below. Thus the ultramafic materials were emplaced into the mélange (1) as large bodies of warm crystal mush, (2) as blocks of the oceanic mantle mechanically intermixed with the mélange, and (3) as diapiric lenses and bodies, mechanically transported through the mélange and its roof (see Milovanović and Karamata 1957).

Geological data seem to be sufficient for such a general explanation, but details are not clear as yet. At the present stage of our knowledge, only speculations on several points can be made.

M. D. Dimitrijević (1971, 1972) interpreted the Serbo-Macedonian massif in the Late Paleozoic to be an active margin of a microcontinent in the Tethys. This element was probably separated from Paleoeurope and possibly included also the Panonian basement and the Rhodope massif. At this time, the Dinarides were separated from this continental element, being situated somewhere in the Tethys. In front of the Serbo-Macedonian margin, a subduction zone was active, with a period of very intense plate consumption during the Late Carboniferous-Early Permian. This activity produced conspicuous low-to-intermediate pressure metamorphism in the Serbo-Macedonian massif, and was followed by tearing and spatial rearrangement of continental elements. The Panonian fragments have been possibly torn off and rotated, and the continental margin during the Mesozoic received a new form corresponding roughly with the Vardar zone. Further on, this margin seems to deflect northwest and west, including possibly Fruška Gora Mountain and part of Slavonia. The western extension of this margin is not known, being covered by the Tertiary of the Panonian basin, and probably torn off along the Zagreb line. To the west it seems to have its continuation in the Periadriatic line (Insubric-Pusteria).

This margin was active at least during the Jurassic, with a final collision of the Dinaric and Serbo-Macedonian elements in the Late Jurassic and in the Vardar zone in the Late Cretaceous. It had possible branches along sutures of more ancient, and at that time already rearranged, continental elements of the Hercynian continental margin (Apuseni?).

The details regarding the internal anatomy of the collisional zone are obscure. Particularly puzzling is the existence of at least two mélange zones, separated by Paleozoic blocks—namely the Ophiolite zone in the southwest and the Vardar zone in the east. They are separated by a belt of Paleozoic rocks overlain by a thin Permian and Triassic cover. These strata—the Drina and Ivanjica Paleozoic—correspond rather more to the synchronous formations in the central Dinaric region than to the analogous deposits in the Vardar zone (the Jadar Paleozoic and others). They must thus be regarded as part of the Dinaric element. The Drina-Ivanjica element does not represent an island arc and it is presumably too narrow to be regarded as an independent continental element. The available geological data permit only hypothetical explanations. A possible hypothesis is that the "Zvornik ocean" (which has its suture in the Vardar zone) represented the main Tethys branch, separating the Dinaric and the Serbo-Macedonian elements. The Ophiolite belt ocean is a secondary sea, opened by differences in velocity of the Dinaric margin during its drift toward the Serbo-Macedonian massif (fig. 7). It would thus represent a kind of marginal sea created in the passive continental margin with a crust of oceanic type. If further evidence shows an Upper Cretaceous age for the whole mélange of the Outer Vardar zone, the Ophiolite belt would represent a subduction system closed in the Late Jurassic, and the Outer Vardar zone a system closed in the Late Cretaceous. If at least one part of the mélange in the Outer Vardar zone is Upper Jurassic, it would be difficult to envisage two synchronous subduction zones

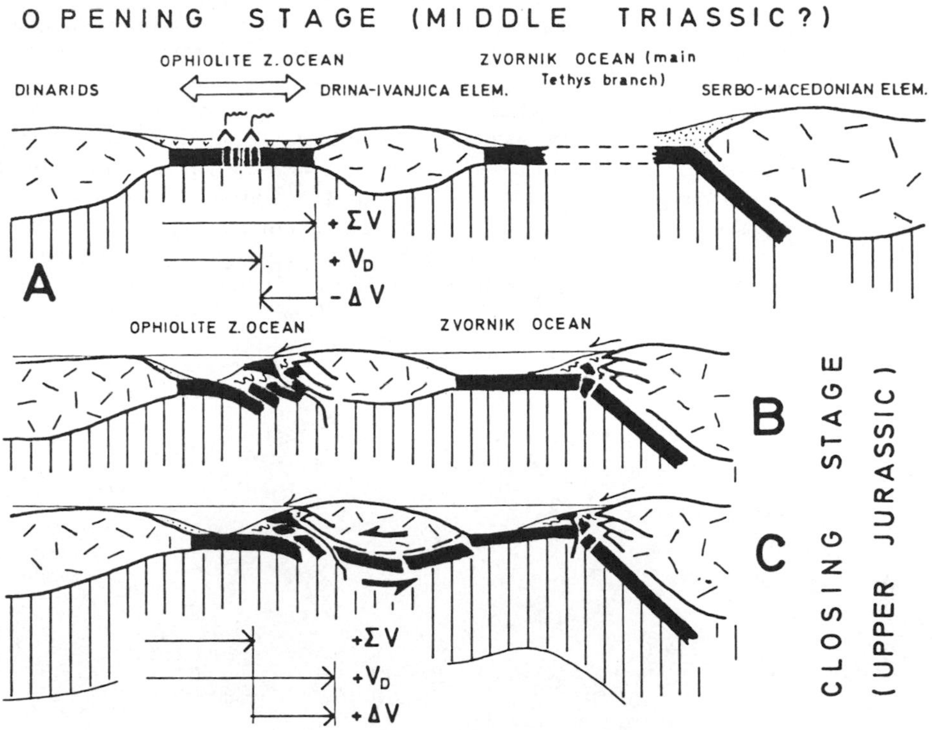

FIG. 7.—Models for the opening and closing history of the Ophiolite belt ocean. *A*, Opening stage: the general velocity of the Dinaric microcontinent, defined as the velocity of its outer margin ($+\Sigma V$) is higher than the velocity of the Dinarides proper ($+V_D$); the resulting difference in velocity ($-\Delta V$) opens the Ophiolite belt ocean. Closing stage: *B*, Hypothesis of the ocean destruction without a Benioff zone; *C*, Hypothesis of the ocean destruction by gliding of the oceanic crust under the Drina-Ivanjica element. The velocity difference of the proper Dinarides and the microcontinental margin is positive and closes the ocean.

separated by only several tens of kilometers and new hypotheses should be invoked. If this proves so, the Ophiolite belt micro-ocean could have been destroyed either (*a*) by fracturing without a regular Benioff zone, with accumulation of sediments and piling of the oceanic crust in a trench; or (*b*) by gliding of the fractured oceanic crust under the Drina-Ivanjica element and joining the Vardar subduction zone (fig. 7, *B* and *C*). In both hypotheses, the Ophiolite trench would represent a feature included in the broad Benioff system of the Mesozoic Vardar zone, but not situated directly over the plate-consuming belt. It is interesting that a few kilometers from the Zvornik suture southward, a thin zone of ultramafic rocks occurs, which could represent an injection of oceanic crust rocks from below the Paleozoic.

The age of opening of the Ophiolite belt ocean in this hypothesis is questionable. The products of this event could be represented by the Ladinian "Porphyrite-Chert Formation," whose igneous activity was similar to the rift-type volcanism. This formation is thickest along the southeastern border of the Ophiolite belt and along the Budva-Cukali zone, which could represent a rift in Middle Triassic. The development of the Uppermost Paleozoic and Triassic permits the assumption that this micro-ocean was opened even earlier (M. Mujičević, personal communication, 1972).

Duration of deposition of the mélange represents another problem. Theoretically, the sedimentation might have been simultaneous with the whole period of the plate-consuming activity, when there was a trench connected with the Benioff system. Geological data are still unsatisfactory. In several places, some of which are described here, the Upper Jurassic mélange of the Ophiolite belt seems to grade from the ?Middle Jurassic; the Upper Cretaceous mélange developed in the Jelica Mountains from breccias containing fragments of rudist limestones. This would imply a rather short period of deposition. It is, nevertheless, not easy to say for such a formation where it is "conformable" and where mechanically transported and piled up over older or lateral deposits entangled in the broad trough environment.

In the Ophiolite belt no metamorphism has been observed as yet, but Mesozoic strata of the Vardar zone show a high-pressure metamorphism. This metamorphism, as well as jadeite-glaucophane schists in the Drina Paleozoic area immediately along the Zvornik suture, can be ascribed to the effects of the plate-consuming conti-

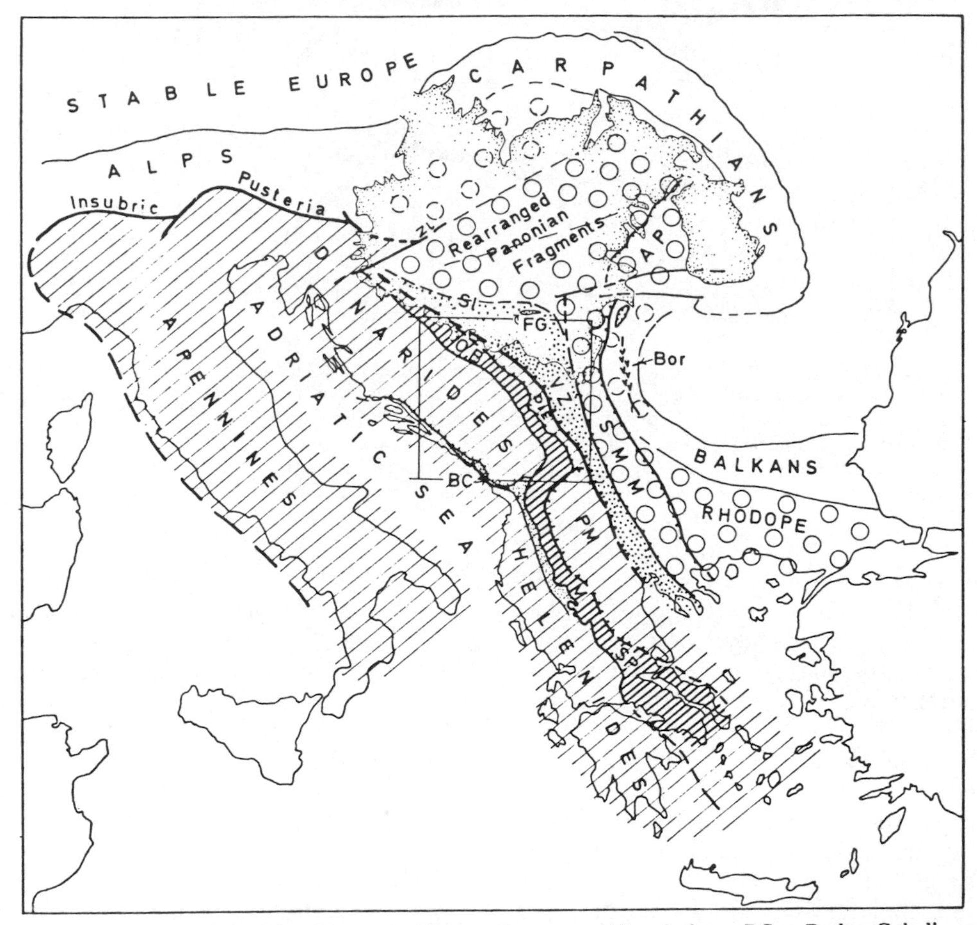

FIG. 8.—Relations of the Dinarides and adjacent elements. Abbreviations: *BC* = Budva-Cukali zone; *OB* = Ophiolite belt (*M* = Mirdita zone in Albania, *SP* = Subpelagonian zone in Greece); *DIE* = Drina-Ivanjica element; *PM* = Pelagonian massif; *VZ* = Vardar zone; *Sl* = Slavonia; *FG* = Fruška Gora Mountain; *Zl* = Zagreb line; *SMM* = Serbo-Macedonian massif; *AP* = Apusseni Mountains. *Circles* = Hercynian continental blocks; *broken circles* = Hercynian blocks reworked in the Alpine movements. Framed area covered by fig. 1.

nental margin. High-temperature effects are recognizable in abundant Tertiary granitoids and andesitic volcanism all over the area east and northeast of the Ophiolite belt. This activity started in the Late Cretaceous with the opening of the Bor fissure with submarine volcanism of the linear type, analogous in position and origin to the gigantic belt of Eocene volcanics situated in Iran beyond the Upper Cretaceous Coloured Mélange belt. Another effect of the collision is represented by large gravity movements of deposits from the Drina-Ivanjica zone over the Ophiolite belt. Bulging of the Serbo-Macedonian massif due to thermal processes could also be responsible for a more recent nappe tectonics in Eastern Serbia, which ceased as late as the Middle Miocene (compare with the model of Dewey 1969).

As has been stated before, many of these tentative explanations are speculative and conceived only as a working hypothesis; the relationships in the Mediterranean are much more complicated than along margins of large continental plates. It seems, however, highly probable that the mélange—the "Diabas-Hornstein Formation"—marks off a boundary between formerly separated continental elements, and shows their collisional scar; that the Balkan Peninsula—as part of a very complicated belt situated south of the "stable Europe"—represents an assembly of originally independent blocks which drifted together and finally collided in the Upper Mesozoic (as suggested previously by Carey 1959; Dewey and Bird 1970; Hsü 1971); and that the "eugeosynclinal" features of the Inner Dinarides can be plausibly explained by the collision of continental margins, marked by a mélange of olistostrome origin.

ACKNOWLEDGMENTS.—We are greatly indebted to Stanislav Dzulynski, Krakow; J. Dewey, Albany, New York; K. J. Hsü, Zurich, and Radu Dimitrescu, Jaşi, for stimulating discussions and reading the manuscript. We also wish to thank our colleagues who acquainted us with the terrains they were mapping, especially M. Mojićević, Sarajevo; and A. Ćirić, Beograd.

REFERENCES CITED

BAILEY, E. B., and MCCALLIEN, W. J., 1950, The Ankara-Mélange and the Anatolian Thrust: Nature, v. 166, p. 938–940.

BEŠIĆ, Z., 1970, Le développment de la formation Diabase-radiolarite dans les Dinarides: Inst. géol. Montenegro, Titograd, Bull. géol., v. 6, p. 243–281.

BLUMENTHAL, M., 1948, Un aperçu de la géologie des chaines nord-anatoliennes entre d'Ova de Bolu et le Kizilermak inférieur: Ankara, M.T.A. Pub., B., No. 13.

CAREY, S. W., 1959, The tectonic approach to continental drift, *in* Continental drift, a symposium: Univ. Tasmania, Geology Dept., p. 177–355.

Carte téctonique internationale de l'Europe, 1:2,500,000, 1964, Moscou.

ĆIRIĆ, B., 1954, Neka zapažanja o dijabaz-rožnaćkoj formaciji Dinarida: Vesnik Zavoda za geol. i geof. istraživanja NR Srbije, Beograd, p. 31–88.

DEWEY, J. F., 1969, Continental margins: a model for the transition from Atlantic type to Andean type: Earth and Planetary Sci. Letters, v. 6, p. 189–197.

———, and BIRD, J. M., 1970, Mountain belts and the new global tectonics: Jour. Geophys. Research, v. 75, p. 2625–2647.

DIMITRIJEVIĆ, M. D., 1971, Variscijski metamorfizam u aksijalnom delu Balkanskog poluostrva (mogućnosti nove genetske interpretacije): Srpsko geološko društvo, Comptes rendus, p. 115–124.

——— 1972, Herzynian metamorphism in the axial part of the Balkan Peninsula, 4: Aegaeis-Symposium, Hannover; and Deutsche geol. Gesell., v. 123, p. 329–335.

FLORES, G., 1955, Discussion, *in* BENEO, E., Les résultats des études pour la recherche pétrolifère en Sicilie: World Petrol. Cong., 4th, Rome, Proc., sec. IA2, p. 120–121.

GREENLY, E., 1919, The geology of Anglesey: p. 1–980.

HAMMER, W., 1921, Die Diabashornsteinschichten, *in* AMPFERER, O. and HAMMER, W., eds., Ergebnisse der geologischen Forschungsreisen im Westserbien: Akad. Wiss. Wien Denkschr., math. naturf. Kl., v. 98, p. 45–56.

HSÜ, K. J., 1971, Origin of the Alps and western Mediterranean: Nature, v. 233, p. 44–48.

JOVANOVIĆ, R., 1957, Pregled razvoja mezozojika i

neki novi podaci za stratigrafiju i tektoniku FNRJ: Kongres geologa Jugoslavije, 2d, Sarajevo, p. 38-63.

Jovanović, Ž., 1963, Prilog poznavanju stratigrafskog položaja krečnjaka i Dijabaz-rožnačke formacije Krša Gradac (Sjenica): Zapisnici SGD za 1960, i 1961, godinu, Beograd.

Katzer, F., 1906, Geologische Uebersichtskarte von Bosnien-Hercegovina, 1:200,000: Erstes Sechtelblatt, Sarajevo.

Kossmat, F., 1924, Geologie der zentralen Balkanhalbinsel, die Kriegsschauplätze 1914–1918 Geol. dargestelt: v. 12, p. 1-198.

Ledebur, K., 1941, Stratigraphie und Tektonik Jugoslaviens zwischen Lim und Ibar: Neues Jahrb. Geologie u. Paläontologie, Beil., v. 85, Abt. B., p. 463-506.

Mercier, J., and Vergely, P., 1972, Les mélanges ophiolitiques de Macedonie (Grece): Dechrochements d'age anté-crétacé superieur, 4: Aegaeis-Symposium, Hannover: Deutsche geol. Gesell., v. 123 (in press).

Milovanović, B., and Ćirić, B., 1968, Carte géologique de la R.S. de Serbie, 1:200,000: Zavod za geol. i geof. istraživanja, Beograd.

———, and Karamata, S., 1957, O dijapirizmu serpentina: Vesnik Zavoda za geol. i geof. istraživanja, Beograd, v. 13, p. 7-28.

———, and Mladenović, M., 1966-1967, Quelques résultats des recherches géologiques-géophysiques dans la zone ophiolotique des Dinarides: Vesnik Zavoda za geol. i geof. istraživanja, Beograd, v. 24–25, ser. A., p. 7-24.

Pamir, H. N., 1944, Une ligne séismogène en Anatolie septentrionale: Rev. Fac. Sci., Istanbul, v. 9, pt. 3.

Rampnoux, J. P., 1969, Sur la géologie du Sandjak: mise en evidence de la nappe de Pešter (confins serbo-monté-négrins, Yougoslavie): v. 9, p. 872–881.

Savage, J. C., 1969, The mechanism of deep-focus faulting: Tectonophysics, v. 8, p. 115-127.

Steinmann, G., 1906, Geologische Beobachtungen in den Alpen: die schardtsche Ueberfaltungstheorie und die geologische Bedeutung der Tiefseeabsätze und der ophiolitischen Massengesteine: Naturf. Gesell. Freiburg Ber., v. 16, p. 1-49.

——— 1927, Die ophiolitischen Zonen in den mediterranen Kettengebirgen: Internat. Geol. Cong., 14th, Madrid, rept. 2, p. 638-667.

16

Reprinted from pages 35-46 of *International Symposium on the Structural History of the Mediterranean Basins, Split, Yugoslavia, 1976,* B. Biju-Duval and L. Montadert, eds., Editions Technip, Paris, 1977

VOLCANO-SEDIMENTARY AND VOLCANO-DETRITIC PHENOMENA OF MESOZOIC AGE IN DINARID AND HELLENIC RANGES : A COMPARISON

P. CELET [1], J.-P. CADET [2], J. CHARVET [1] and J. FERRIERE [1]

I. MAIN FEATURES

Mesozoic dinaric and hellenic series are frequently characterized during Trias and Jurassic by associations of magmatic elements with marine sedimentary formations. These phenomena are important because they reflect fundamental crustal or sub-crustal transformations linked with the evolution of the whole range.

However, despite apparent similitudes, these episodes are very different in their lithological associations and B. CIRIC clearly established in 1954 the distinctions between these formations.

The association of middle triassic magmatic and sedimentary rocks is widespread in Yugoslavia where it is known under the name of "Porphyrit-Radiolarite Formation" (CIRIC 1954) ; it's a volcano-sedimentary formation with equivalents in Albania and Greece.

The association of upper Jurassic, known in the Dinarids as the "Diabase-radiolarite formation " (CIRIC 1954) is of a volcano-detritic nature and mainly made of volcanoclastic and terrigenous elements associated with radiolarites. More recently the terms of "Olistostrome Melange" (DIMITRIJEVIC and DIMITRIJEVIC 1973), "Mélange de type volcano-sedimentaire" (CELET 1976, CELET et al 1976) have been proposed. To account for the chaotic aspect of the polygenetic material included in this type of deposit, we'll take the general term of "mélange" for the volcano-detritic formation of the upper Jurassic.

Although there exist similar and subcontemporary formations in more inner zones (i. e. Vardar zone, Mercier et Vergely 1972), we won't talk about them at present, our purpose being essentially to collect the characters which draw together or differenciate these volcanogenic formations after having recalled their display. At last we we'll attempt to situate these manifestations in their geotectonic situation.

II. DISTRIBUTION AND LOCALISATION OF TRIASSIC AND JURASSIC VOLCANOGENIC FORMATIONS

The witnesses of middle triassic and upper Jurassic magmatism are widespread in the dinaro-hellenic ranges and mainly in the areas we study here (Fig. 1).

1. <u>The volcano-sedimentary triassic formation</u> is found as well in inner as in outer areas (BEBIEN et al 1975) :

- Budva trough, High-Karst platform, Bosnian trough, Serbian set, and Golija rise in the

(1) Laboratoire de Géologie Dynamique, Université des Sciences et Techniques de Lille, B.P. 36, 59650 Villeneuve d'Ascq, France

(2) Laboratoire de Géologie, Université d'Orléans, 45045 Orléans Cédex, France

Yugoslavian Dinarids.

- Mirdita zone (N.E. of Tirana) in Albania.
- Pindus zone (unit of Megdovas), Vardoussia subzone (west of Parnassus) Maliac zone (Othrys), Pelagonian zone (Attica, Euboea, Argolis (?), Northern Pindus in Greece.

This large scattering underlines the ubiquity of this middle Triassic phenomenon.

2. The upper jurassic volcano-detritic "melange" is essentially linked with the ophiolitic field. Its outcropping most of the time can be found together with those of the ophiolitic nappe in the Serbian zone (northern Montenegro and Bosnia, southern Serbia) in Golija zone, Mirdita zone or Maliac, Pelagonian zone (Geranées, Iti Kallidromon, Western Othrys and Pindus). However this relation is not constant ; in some areas, the "mélange" is separated from the ophiolitic suite (Ceotina, Durmitor in Yugoslavia, central Othrys in Greece).

So in the dinaric branch of the alpine system, there is a geographical differenciation in the distribution of the volcanic phenomena during Trias and Jurassic. This opposition is still more apparent in the composition of the formations themselves.

III. STRATIGRAPHIC POSITION, PETROGRAPHICAL COMPOSITION AND SEDIMENTARY ENVIRONMENT.

The age and deposit conditions of the volcanic material can be analysed thanks to the stratigraphic context. Here are the main components.

A. TRIASSIC VOLCANO-SEDIMENTARY SERIES

1. In the Dinarides, in all paleogeographic zones, a volcano-sedimentary formation is known during middle Trias. It is constituted by the association of sedimentary (Radiolarites, pelites, sandstones) and volcanic elements (flows and projections). The substratum of this series is made of anisian neritic limestones crowned by nodulous red limestones Han Bulog facies) of upper anisian age (Fig. 2, II et III), except in the Budva zone where there is a thick flysch during the Anisian (Fig. 2 - I). It is capped by Ladinian and upper Triassic limestones in the zones with having a meaning of ridge and by chert limestones in the zones that will evolve into troughs.

A preliminary analysis of the volcanic elements and their repartition (BEBIEN, BLANCHET et al 1975, 1976) allows to distinguish four sets : rhyolitic tuffs and lavas, andesites, diabases and basalts ; the rhyolitics tuffs can be found in all zones, andesite, of calco-alkaline nature, are mainly known in the outer zones and in the serbian zone south of the Sarajevo transverse sector whereas basalts are limited to the serbian zone north of the same transverse sector and to the bosnian zone.

In short, the middle triassic volcano-sedimentary formation of the Dinarides, excorting the differenciation of the paleogeographical zones, is characterized by its wide repartition and varied volcanic phenomena mainly of calco-alkaline nature.

2. In the Hellenides, the triassic volcano-sedimentary outcrops are scarcer and more scattered ; it seems that this is due to tectonic phenomena (shearings and slides often occuring at the basis of the nappes).

However there seems to be some differenciations corresponding to two types of volcanic floods.

a. An alkaline series (HYNES, 1974) predominently lavic with various interstratifications : fine tuffs, breccias and hyaloclastites, very often pillow-lavas and slaggy spilitic lavas in alternance with radiolarites, siliceous pelites and platy limestomes with cherts Conodonts or Halobia (Pindus) dating the top from Carnian-Norian. Such outcroppings can be observed, up to now in Othrys in the area of Neochorion (series of Garmeni and Loggistion, FERRIERE 1972-1974) and in the northern Pindus (area of Panagia, TERRY, 1971).

We must equally mention a series of breccias and effusive rocks with pillow lavas flows of spilitic to basaltic nature, petrographically fairly similar to the preceding one but situated in Arkanania at the front of the Pindus nappe and thrust over the ionian flysch (unit of Megdovas, FLEURY, 1976). At the top, this volcano-sedimentary series turn to an Ammonitico-Rosso of upper Scythian to Carnian age and Halobia limestones.

b. Another volcanic formations with pyroclastites, rhyolitic and dacitic light green tuffs ("Pietra verde" type) can be found in Attica, Parnis,(CLEMENT, 1968) central Euboea (GUERNET, 1971, KATSIKATSOS, 1971) also associated with sedimentary rocks (pelites, sandstones, radiolarites). In Parnis this volcano-sedimentary formation lies over ladinian cherts limestomes and microbreccias and bear carnian chert limestones.

3. Remark.

In Albania (Mirdita zone, N.E of Tirana) a Triassic volcanism has been noticed (A. PAPA, 1970). It seems similar to the Ladinian-Carnian one of the Dinarides and Hellenides.

4. Conclusion

In Dinarides and Hellenides, the triassic volcanism is bonded with the paleogeographical differenciation of the isopic zones this evolution can be followed up to the end of the Jurassic (cf. AUBOUIN et al, 1970) and its first stages-with notably the middle Trias volcano-sedimentary formation - recall the rifting known in the continental margins but it raises the problem of the paleoceanographic organisation of the Dinarides and Hellenides (cf conclusion).

B. JURASSIC VOLCANO-DETRITIC MELANGE

1. Nature

Despite great similarities of facies and petrology, the hellenic and Dinaric volcano-detritic formations are located in various stratigraphical and tectonic surroundings. They display constant characteristics similar to the classical "melanges" (cf BAILEY et al 1950, HSU 1971 et 1974 etc) : absence of stratification, disordered and chaotic aspect of the formation, pelitico-sandy matrix sometimes with radiolarites wrapping various exogenous elements such as blocks of limestones of different nature and age (Permian with Neoschwagerina, upper Trias with Conodonts,etc) pieces of quartzitic sandstone, Radiolarites. The eruptive matter, very abundant is mainly composed of fragments belonging to the ophiolitic cortege : basic and ultrabasic magmatic rocks from gabbros to peridotites and serpentines ; veins stones (dolerites, diabases) effusive rocks (basaltes or spilites with pillow lavas structure ; metamorphic rocks with diverse meaning (amphibolite, gneiss mica schists) are associated there as well as granitic blocks (Iti).

The recents works, in Yugoslavia (BLANCHET, 1970, 1975, CADET, 1970-1976, CHARVET, 1970, RAMPNOUX, 1970-1973) as in Greece give precise details about the composition and different aspects of this formation and we send the reader back to these publications (CELET, 1976).

The volcano-detritic formation constantly lies upon radiolarites, often with pelitic intercalations ; flyschoïd series separate these formations in some cases. The radiolarites and the matrix of the mélange are seldom dated, however Protopeneroplis of Dogger - Malm have been reported (FERRIERE, 1974 ; TERRY, 1975 ; WIGNIOLLE, 1975) together with neritic microfauna of Kimmeridgian-Portlandian (CADET,1967 ; RAMPNOUX, 1970). So, if the age of these formations is still imprecise, its normal stratigraphic support's is better known :

a. In the Dinarides, the melange and its sole lies upon, - either Dogger-Malm condensed or pelagic carbonated series, their chert-limestones progressively turn to pelites and radiolarites (Rogatica, Bosnia, fig 3, III) or are capped by alternances of microbreccias silstones and radiolarites,

- or neritic limestones topped by liassic ammonitico-rosso and, locally, by oxfordian limestones (Romanija, Bosnia, fig 2, IV).

b. In the Hellenides, the mélange lies conformably on various substrata, in Argolis (fig. 2, V) the transition between the norian Megolodonte limestones and a reduced terrigenous series occurs through a Lias-Dogger ammonitico-rosso as in the Yugoslavia Golija zone.

In Othrys the mélange lies either upon neritic series in the pelagonian zone or upon pelagic series in the maliac zone (fig. 3, VI) (FERRIERE, 1974 et 1976).

In Iti (NW of Parnassus) (fig. 3, VII)and in the Kallidromon, the pelagonian limestones series including two horizons of bauxites at different stages of middle-upper Jurassic, dated at the top by Cladocoropsis ends with limestones getting richer in silica and progressively passing to radiolarites and the flyschoïde detrital layers of the mélange basis.

Theses examples shows that the volcano-detritic phenomena normally take place in the sedimentary evolution of the time and is part of it. The subsequent tectonisation doubtlessly increased the chaotic aspect of these formation but is not their fundamental cause. The layers overthrusting the melange can be dated back to Tithonian or Beriasian in the place where they have been spared by the ophiolite overthrusts or others. Keeping in mind the upper or late Jurassic age of the substratum and the thickness of the volcano-detritic deposits, one realizes the brevity and the extent of the event in the history of a range, showing at this time a very important tectonic activity and a major phase in the geosynclinal evolution of the Dinarides and Hellenides.

2. Conclusions

Thus the mélange appears as a stratigraphic formation closely bounded to its substratum in relation with the ophiolitic complex whose material is inherited for a great part and where the resedimentation play a great role. Nevertheless a few differences between Dinarides and Hellenides can be noted, particularly :

- in the lithological composition of the deposits, the Dinaric volcano-detritic seems richer in volcanic products whereas the granitic elements are more abundant in the Hellenides ;
- in the nature of the infra-radiolaritic substratum better diversified in the Dinarides where a transition between the serbian basin in the S W and the Golija ridge in the N E, can be observed in central Bosnia. The same tendency can be noted in Greece in the Iti, on the pelagonian platform, where the carbonate deposits rise to upper jurassic while an oceanization begins in the maliac zone;
- in the early age of the limestone cover (fig. 3, IV) or the flysch like cover (BLANCHET et al, 1970), lateral passage to Bosnian flysch, (fig. 3, III) in the Dinarides;
- finally in the greater thickness of the mélange in Yugoslavia than in Greece.

IV. GENERAL CONSIDERATIONS ABOUT THE ORIGIN AND GENESIS OF THE DEPOSITS

The nature of the material and its stratigraphic context show a contrast between the triassic volcano-sedimentary manifestations and the upper jurassic volcano-detritic phenomena which leads to contemplate an origin and a genetic meaning specific to each of these episodes.

1. Triassic volcanism

The vast platform with a detritic sedimentation during lower Trias and a neritic one during Anisian was first differenciated during upper Anisian in the Dinarides. Marks of this differenciation can be observed in the inner zones as well as in the outer ones (cf BEBIEN et al 1975 for the Dinarides). We can suppose that these strains have fostered a system of fractures that caused the rise of the magma and started the volcanism with its diverse manifestations (explosive volcanism or lava flows above or under the sea). A rifting-like segmentation of a thinned (continental ?) crust can be considered in the case of marginal areas in the adriatic or apulian subplate (cf DERCOURT 1970, DEWEY et al 1973 etc). The first model (fig. 4, A) shows up such a fragmentation with neritic platform and the beginning of subsiding basins some of whick will turn short and become shallow troughs while others will evolve toward genuine oceanic basins. The development of such opening processes consistent with the alkaline characteristic of triassic lavas has been suggested to situate the origin of the oceanic crust begetting the ophiolitic nappes in eastern Mediterranea (HYNES, 1974, LAPIERRE et ROCCI 1976). However, in the Dinarides, the chemistry of the volcanic products seems rather to indicate a calco-alkaline tendency inconsistent with the generalisation of the former model. The facts as they have been observed in these areas would rather induce us to imagine the existence of a subduction zone on the edge of the plate, admitting the possibility of an interplate opening system according to a second model (fig. 4 B). But the geochemical knowledge of the volcano-sedimentary products is too rudimentary to allow us to propose now a simple explanation of these phenomena or to choose between either of these hypothesis.

Anyhow, the triassic volcanism is the beginning of an important stage in the geodynamic evolution of the dinaro-hellenic system.

2. Volcano-detritic mélange

The nature and heterogeneity of the melange raise the problem of the origin of the accumulated material.

The blocks of basic and ultrabasic rocks, often bulky and very frequent, doubtlessly belong to the ophiolitic complex and it is sometimes difficult to distinguish them from the ophiolitic nappe itself when the slabs are big.

As for the limy xenolithes, some of them of Triassic age come from series outcropping in inner zones. But the origin of numerous chips (non dated cristalline limestones, radiolarites, sandstones, greywackes) is still unknown, but probally situated mainly in inner zones.

The origin of the metamorphic elements is much more difficult to determine ; and they can have been torn from the paleozoic basement or from series metamorphised during jurassico-cretaceous tectonic phases.

The setting of such a material must take into account the data of the sedimentology about the relations between the wrapped blocks and their matrix. The blunt aspect of the elements, indeed dissociated and fractured previously to their deposit, their surface coated with pelitic and clayey material, the close union of the interfaces, the scarceness of shearing contacts and the absence of coherent net of fractures being generalised into the whole of the formation lead us to discard the hypothesis of the merely tectonic origin of a series previously stratified or of a tectonic mélange (DIMITRIJEVIC 1973, GANSSER 1974, HSU 1974). This induces us to consider the deposit of this exogenous blocks in a basin with a pelitic sedimentation and several supplies to that basin, itself submitted to a subsequent tectonization. The phenomena of resedimentation would have occured following early deformations phases having affected the Dinaric and Hellenic ranges at the upper Jurassic and early Cretaceous either under the shape of olistostrome, olistolith and xenolith (CELET et al 1976).

Considering the narrow association of the ophiolites and the volcano-detritic mélange, the explanatory schemas must account for their origin and situation. If they represent a slab of oceanic crust tectonically settled during upper Jurassic (cf DERCOURT 1970, BERNOUILLI and LAUBSCHER 1972 etc), several genetic models can be proposed.

- The mélange could result from the transportation and accumulation in a subsident basin of a material originating in a subduction trench on the edge of an island-arc or a marginal sea (Fig. 5 A).

- The volcano-detritic mélange could have been formed following obduction with dilaceration of two crusts and redeposit of the products in a basin and or upon platform (Fig. 5 B) situated at the front of the system.

- The two mechanisms, obduction and subduction, can be combined in a more complex model (Fig. 5 C) enabling us to explain the coexistence of such varied elements inside the terrigenous matrix.

So the volcano-detritic mélange reflects the instability of the sedimentary environment.The hypothesis of a sedimentary complex with olistostromes already considered for Hellenids (CELET 1976, CELET et al 1976, FERRIERE 1976, Othrys, NAYLOR et al 1976 Vourinos), and Dinarids (DIMITRIJEVIC 1973) fits better in the paleogeographic and tectonic frame of the Dinaric branch of the alpine range.

Conclusions

These triassic and jurassic events are highlights in the history of the chain. With the volcano-sedimentary formation of the middle Trias, we witness a revolution whose consequences are fundamental with the birth of the genuine mesozoic alpine paleogeography which will progressively be diversified. This stage, contemporary of the opening period of the tethysian ocean, directly leads to the late Jurassic upsetting. A progressive strain dynamics is developped and creates the first nappes, notably the ophiolites overthrust. The volcano-detritic deposits are the advanced witnesses and testify of the violence of the phenomenon. The following flyschs are a more remote consequence.

AUBOUIN J., BLANCHET R., CADET J.P., CELET P., CHARVET J., CHOROWICZ J., COUSIN M. et RAMPNOUX J.P., 1970. Essai sur la Géologie des Dinarides. Bull.Soc. Géol. Fr., 7, XII, 1060-1095.

BAILEY E.B. et Mc CALLIEN, 1950. The Ankara mélange and the Anatolian thrust. Nature, 166, 938-940.

BEBIEN J., BLANCHET R., CADET J.P., CHARVET J., CHOROWICZ J., LAPIERRE H. et RAMPNOUX J.P., 1975. La série "Porphyrit-Radiolarite" élément de reconstitution du cadre géotectonique des Dinarides au Trias. 3ème réunion des Sc. de la Terre, Montpellier (ronéot.), 31.

BERNOUILLI D. and LAUBSCHER H., 1972. The palinspastic problem of the Hellenides. Eclogae geol. Helv., 65, 1, 107-118.

BLANCHET R., 1974. De l'Adriatique au Bassin pannonique. Essai d'un modèle de chaîne alpine. Mém. Soc. Géol. Fr., LIII, 120, 171 p

BLANCHET R., DURAND-DELGA M., MOULLADE M. et SIGAL J., 1970. Contribution à l'étude du Crétacé des Dinarides internes : la région de Maglaj, Bosnie (Yougoslavie). Bull. Soc. Géol. Fr., 7, XII, 1003-1009.

CADET J.P., 1976. Contribution à l'étude géologique des Dinarides : les confins de la Bosnie-Herzégovine et du Montenegro. Essai sur l'évolution alpine d'une paléomarge continentale. Thèse, ronéot. Orléans, 450 p.

CELET P., 1976. A propos du mélange de type "volcano-sédimentaire" de l'Iti (Grèce méridionale). Bull. Soc. Géol. Fr., XVIII, (7), 299-307 (Col. ORSAY, 1975).

CELET P., CLEMENT B. et FERRIERE J., 1976. La zone béotienne en Grèce. Implications paléógéographiques et structurales. Eclogae géol. Helv., 69/3, 577-599.

CELET P., FERRIERE J. et WIGNIOLLE E., 1976. Le problème de l'origine des blocs exogènes du mélange de type volcano-sédimentaire au Sud du Sperchios et dans le Massif de l'Othrys (Grèce). Bull. Soc. Géol. Fr. (colloque de la Méditerranée, Montpellier), sous presse.

CHARVET J., 1970. Aperçu géologique des Dinarides au méridien de Sarajevo (Bosnie). Bull. Soc. Géol. Fr., XII, (7), 968-1002.

CHARVET J., 1973. Sur les mouvements orogéniques du Jurassique-Crétacé dans les Dinarides de Bosnie Orientale. C.R. Acad. Sc. Paris, 276 (D), 257-259.

CIRIC B., 1954. Einige Betrachtungen über die Diabas-Hornstein Formation der Dinariden Bull. Serv. Geol. Serbie, XI, 31-88.

CIRIC B., 1975. The geological composition and tectonic structure of the outer Dinaric Alps in the Montenegro. Acta geologica (res. anglais), VIII, 41, 317-345, Zagreb (en Serbo-croate).

CLEMENT B., 1968. Observations sur le Trias du Patseras et du Parnis en Attique C.R. somm. S.G. Fr., 9, 332-334.

COLEMAN R.G., 1971. Origin and emplacement of the Ophiolite suite : Appalachian ophiolites in New-foundland. Journ. Géophys.Res., 5,14, 1212-1222.

DERCOURT J., 1970. L'expansion océanique actuelle et fossile ; ses implications géotectoniques. Bull. Soc. Geol. Fr., XII, (7), 261-309.

DERCOURT J., CELET P., COTTIN J.Y., DEWEVER P., FERRIERE J., GRANDJACQUET C., HACCARD D. et WIGNIOLLE E., 1976. Importance d'une tectonique Jurassique supérieur sur les marges de la plaque d'Apulie (Hellenides et Apennins ligures). Bull. Soc. Geol. Fr. (Colloque de la Méditerranée, Montpellier), sous presse.

DEWEY J.F., PITMAN W.E., RYAN W.B. and BONNIN J., 1973. Plate tectonics and the evolution of the Alpine system. Geol. Soc. Ann. Bull., 84 (10), 3137-3180.

DIMITRIJEVIC M.D. and DIMITRIJEVIC M.N.,1973. Olistostrome mélange in the Yougoslavian Dinarides and late mesozoïc plate tectonics. Journ. Geol. 81, 328-340.

FERRIERE J., 1974. Etude géologique d'un secteur des zones helleniques internes subpélagonienne et pélagonienne (massif de l'Othrys - Grèce continentale). Importance et signification de la période orogénique anté-crétacé supérieur. Bull. Soc. Geol. Fr. XVI, (7), 543-562.

FERRIERE J., 1976. Sur la signification des séries du Massif de l'Othrys (Grèce continentale orientale) : la zone isopique maliaque. Ann. Soc. Geol. Nord, XCVI, 121-134.

FLEURY J.J., 1976. Unité paléogéographique originale sous le front de la nappe du Pinde Olonos : l'Unité du Megdovas (Grèce continentale), C.R. Ac. Sc. Paris, 282, D, 25-28.

GANSSER A., 1974. The Ophiolis mélange, a world-wide Problem on Tethyan examples. Eclog. Geol. Helv. 67 (3), 479-507.

GUERNET Cl., 1971. Etudes géologiques en Eubée et dans les régions voisines (Grèce). Mém. Lab. Geol. 1, Fac. Sc. Paris, 395 p.

HSÜ J.K., 1971. Franciscan mélange as a model for eugeosynclinal sedimentation and underthrusting tectonics. Journ. Geophys. Res., 76, 5, 1162-1170.

HSÜ J.K., 1974. Melanges and their distinction from olistostromes. Soc. Eco. Paleon. Mineral., spec. pub. (M.Kay), 19, 321-333.

HYNES A.J., 1974. Igneous activity at the birth of an ocean in Eastern Greece, Canad. Journ. Earth Sc., 11, 842-853.

KARIG D.E. and SHARMAN G.F., 1975. Subduction and accretion in trenches. Geol. Soc. Amer. Bull., 86, 377-389.

KATSIKATSOS G., 1971. Les formations triasiques de l'Eubée centrale. An. Geol. Pays Hellen., 22, 62-76.

LAPIERRE H. et ROCCI G., 1976. Le volcanisme alcalin du Sud-Ouest de Chypre et le problème de l'ouverture des régions téthysiennes au Trias. Tectonophysics, 30, 299-313.

MAXWELL J.C., 1974. Anatomy of an Orogen. Geol. Soc. Amer. Bull. 85, 1195-1204.

MERCIER J. et VERGELY P., 1972. Les mélanges Ophiolitiques de Macédoine (Grèce) : décrochements d'âge anté-crétacé supérieur. z. deutsch. Geol. Ges., 123, 469-489.

NAYLOR M.A. and HARLE T.J., 1976. Paleogeographic significance of rocks and structures beneath the Vourinos ophiolite, Northern Greece. Journ. Geol. Soc. Lond., 132, 667-675.

PAPA A., 1970. Conceptions nouvelles sur la structure des Albanides (présentation de la carte tectonique de l'Albanie au 500.000e). Bull. Soc. Geol. Fr., XII, (7), 1096-1109.

RAMPNOUX J.P., 1970. Regards sur les Dinarides internes yougoslaves (Serbie méridionale et Monténégro oriental) : stratigraphie, évolution paléogéographique et magmatique. Bull. Soc. Geol. Fr., XII, (7), 948-966.

RAMPNOUX J.P., 1973. Contribution à l'étude géologique des Dinarides : un secteur de la Serbie méridionale et du Montenegro oriental (Yougoslavie). Mém. Soc. Geol. Fr., LII, 119 100 p.

TERRY J., 1971. Sur l'âge triasique de laves associées à la nappe ophiolitique du Pinde Septentrional (Epire et Macédoine, Grèce). C.R. Somm. S.G. Fr., 7, 384-385.

TERRY J., 1975. Echo d'une tectonique jurassique : les phénomènes de resédimentation dans le secteur de la nappe des ophiolites du Pinde septentrional (Grèce). CR. Somm. S.G. Fr., 2, 49-51.

TURKU I. et NDOJAJ I. GJ., 1973. Mbi disa karakteristika gjeologo-petrgrafike te vulkanizmit mesozoik ne Shqiperi. Permbledhye Studimesh, 2, 23-33 (résumé en français).

VERRIEZ J.J., 1976. Sur les formations volcaniques basiques d'Atalanti (Locride Grèce). Bull. Soc. Geol. Fr., XVIII, (7), 293-298.

WIGNIOLLE E., 1975. Contribution à l'étude géologique de la région centrale du Massif de l'Iti (Grèce continentale). Diplôme Et. Approf. Lille, 94 p.

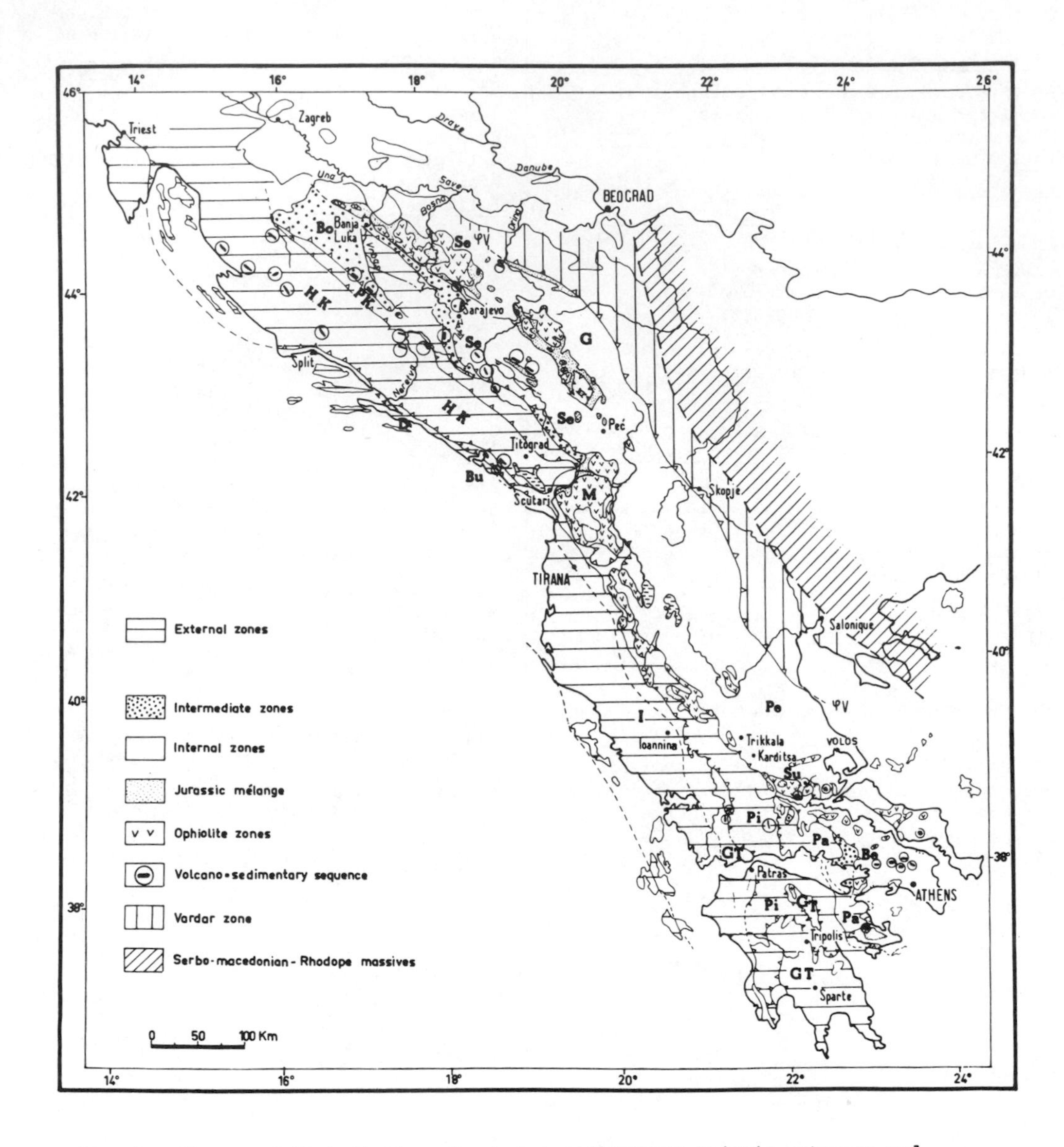

Fig. 1 - Place and distribution of volcano-sedimentary triasic outcrops and of volcano-detritic Jurassic mélange in Dinarides and Hellenides.
External zones. I : Ionian zone ; D : Dalmatian zone ; GT : Gavrovo-Tripolitza zone ; Bu : Budva-Cukali zone ; Pi : Pindus zone ; HK : High-Karst zone ; Pa : Parnassian zone.
Intermediate zones. Bo : Bosnian zone ; Be : Beotian zone.
Internal zones. Se : Serbian zone ; M : Mirdita zone ; Su : "Subpelagonian" (Maliac) zone ; G : Golija zone ; Pe : Pelagonian zone.

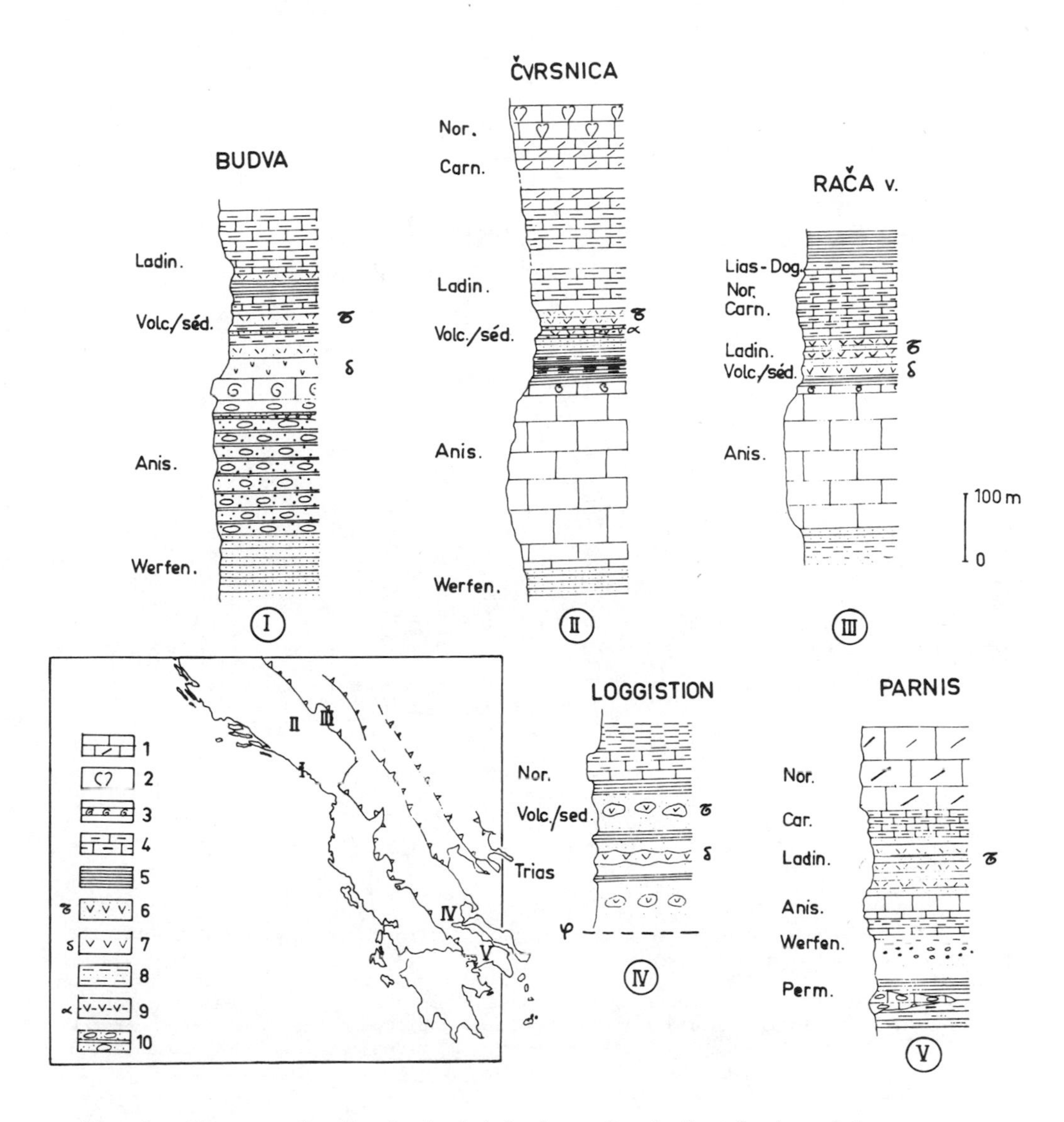

Fig. 2 - Diagrams of a few typical triasic series in Yugoslavia and Greece. 1) limestones and dolomitic limestones ; 2) Megalodonte limestones ; 3) Ammonitico-Rosso ; 4) Cherty limestones and cherts ; 5) radiolarites and siliceous pelites ; 6) pyroclastic rocks ; 7) dolerites ; 8) spilites ; 9) diabases ; 10) triasic detritic rocks (sandstones, conglomerates). Dinarides : I Budva zone ; II : High-Karst zone ; III : Internal bosnian zone. Hellenides : IV : Maliac zone ; V : Pelagonian zone.

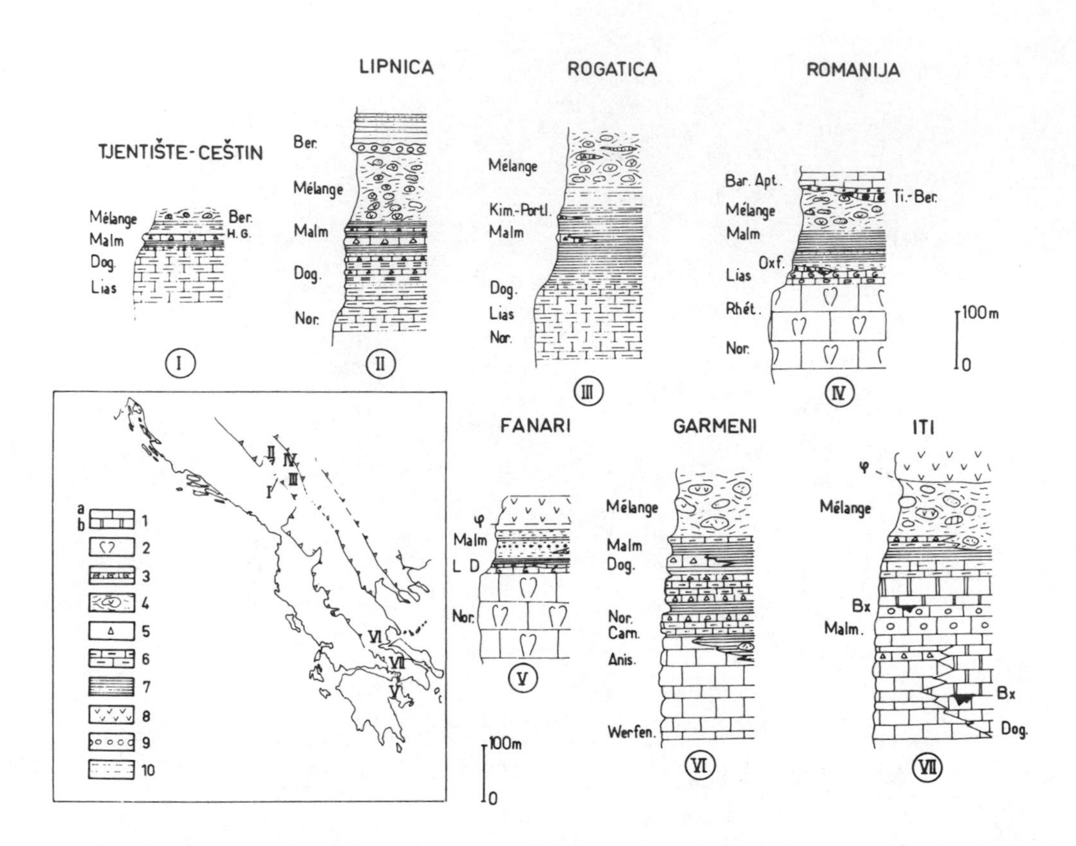

Fig. 3 - Profiles showing different stratigraphic environments of the volcano-detritic melange in Yugoslavia and in Greece.
1) Massive and oolithic limestones (a) ; reef limestones (b) ; 2) Megalodonte limestones ; 3) Ammonitico-Rosso ; 4) Mélange ; 5) Breccias and microbreccias ; 6) Pelagic cherty limestones ; 7) Radiolarites and siliceous pelites ; 8) Ophiolites ; 9) conglomerates ; 10) Siltstones, sandstones and associated rocks. Dinarides : I to III : Serbian zone, IV : Golija zone. Hellenides : V to VII : Internal zones (V : Argolis, VI : Othrys, VII : Locride).

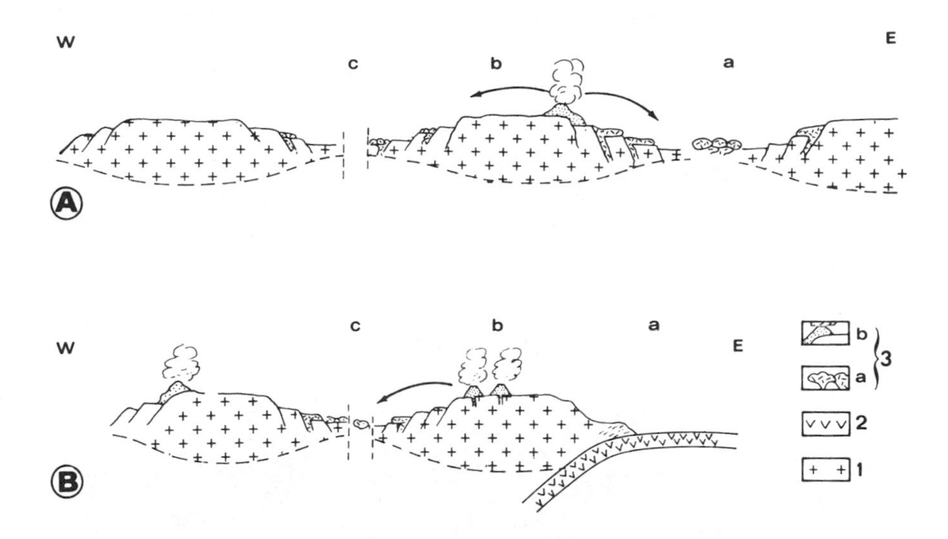

Fig. 4 - Hypothesis about the genesis of volcano-sedimentary triasic events of Dinarides and Hellenides.
1) Continental crust ; 2) Oceanic crust ; 3) effusive materials ; 3 a) mainly submarine ; 3 b) partly outward.

Remarks : The diagrams are not paleogeographic surveys on precisely determined zones but only illustrations of possible working of settlements. The correlations (a,b,c,) given as indications, are pure speculations and can be modified according to the chosen geotectonic patterns and the considered fields.

Possible equivalences : In a platform-basin or insular arc marginal sea system, b) could illustrate anyone of the neritic zones affected by triasic volcanism (Golija or High-Karst in Dinarides, Pelagonian or Parnassian in Hellenides). If b) corresponds to the Pelagonian zone or Golija, c) would be equivalent to Maliac zone and a) to Vardar zone in Greece.

A) Platform breakage that corresponds to a strong paleogeographic transformation after Anisian time and differentiation of the main deep basins ("troughs").

This hypothesis fits to the alkalin type of zone triasic Hellenide lavas.

B) Craton breakage and ocean crust subduction.

Besides necessary breakage to explain facies changes one or several subduction zones are thought over. If the process is able to explain the calcalkaline nature of some triasic lavas, more especially in Yugoslavia, there remains to solve the quick subduction disappearance.

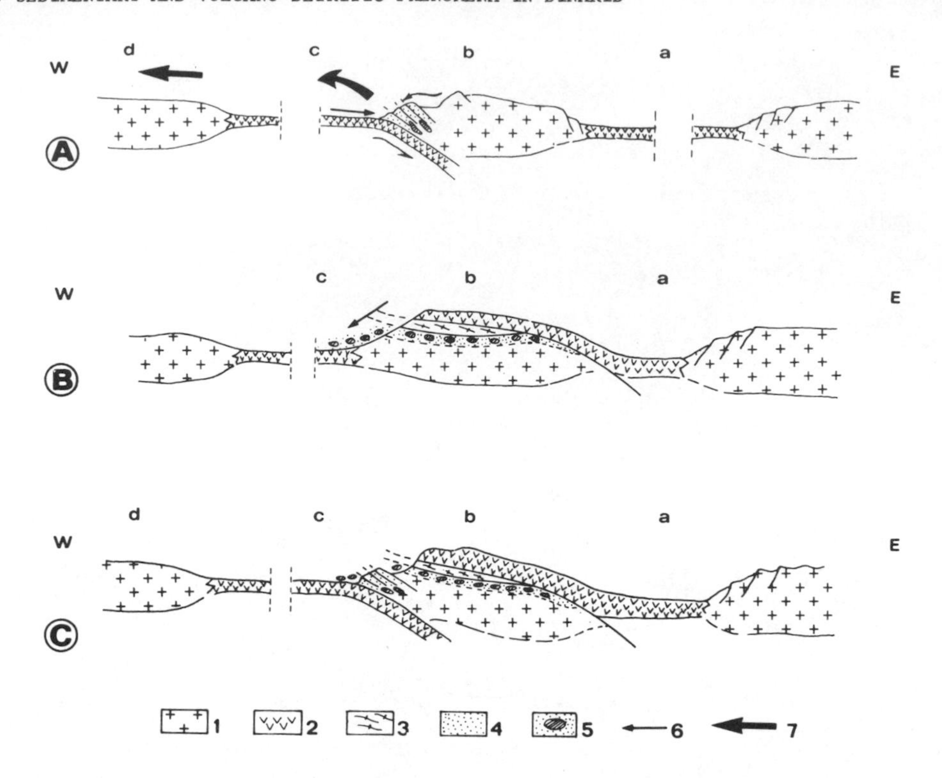

Fig. 5 - Different models of volcano-detritic upper Jurassic mélange working out in Dinarides and Hellenides.
1) Continental crust ; 2) Oceanic crust ; 3) Amphibolites ; 4) Detritic rocks ; 5) Blocks (peridotites and others in the terrigenous matrix) ; 6) Origin and transport of the detritic material ; 7) Tectonic reworking and resedimentation of the detritic formations.
N.B. : Same remarks as fig. 4 and some possible equivalences in the B or C hypothesis considered below.

A) A melange genesis at a subduction zone level (accretionary prisms).

This origin, frequently set forth leads to consider a melange formation without its own basement. Yet, such basements exist and make think of a tectonic reworking followed with resedimentation at least in certain places. In the Hellenides case, the process implies a melange deposit in the internal zones (a : Peonias subzone ; b : Païkon subzone ; c : Almopias subzone ?) and a westward conveying of the materials.

B) Formation of olistostrome melange through destruction of the ophiolitic nappe and the gathered sheets.

That is an intermediate stage where the front of the ophiolitic nappe is destroyed but yet it already covers a melange newly formed during the tectonic shifting of the nappe. There can be also set forth resedimentation of chaotic previously built formation and the "mélange" deposit over various substrata.

C) Simultaneous action of the two phenomena : obduction, subduction and redepositing.

That schema enables to explain "Mélange" without any own basement (accretionary prism) and others with one (olistostromes lying on triasico-jurassic unbroken series).

Nevertheless, such patterns can be seen through a tectonic "décollement".

17

Reprinted from pages 470–471 of *Deutsch. Geol. Gesell. Zeitschr.* **123:**469–489 (1972)

LES MÉLANGES OPHIOLITHIQUES DE MACÉDOINE (GRÈCE): DÉCROCHEMENTS D'AGE ANTÉ-CRÉTACÉ SUPÉRIEUR

J. Mercier and P. Vergely

[The ophiolitic melanges of prae-uppercretaceous age in Macedonia (Greece)]

Abstract: "Ophiolitic Mélanges" are chaotic mega-breccias consisting of blocks of sedimentary, metamorphic or ophiolitic rocks floating in an ophiolitic matrix which often includes serpentinite bodies; in Macedonia these "Mélanges" constitute a NNW—SSE oriented zone measuring 4 km in width and 40 km in length.

The "ophiolitic Mélanges" constitute several distinct tectonic units thrusted towards the SW. The nature of their matrix favours the definition of distinct lithological units corresponding to two types of "Mélange": "Mélanges" containing autochthonous blocks and "Mélanges" containing exotic blocks (fig. 1).

The matrix of the "Mélanges" is composed of basic ortho-greenschists (prasinites) or basic tuffs (tuffites), while the blocks include marbles, amphibolites, mica schists, gneiss, and ophiolites. In the field (fig. 1) these blocks are aligned in a NE—SW direction parallel to the direction of the system consisting of reverse-faulted flexure-folds (phase M) which affect these blocks (fig. 4 B). It is evident therefore that these blocks are not olistoliths incorporated within lava or deposited in sediment; had this been the case their distribution would have been haphazard.

These "ophiolitic Mélanges" have a tectonic origin. A comparison between the type of deformation (fig. 2 and 3) affecting the "Mélanges" (blocks and matrix) and that affecting adjacent units lacking "Mélanges" has shown that a specific deformation is peculiar to the "Mélanges" (phase M). This latter deformation is posterior to an isoclinal, syn-metamorphic deformation (JE 1) of Upper Jurassic—Lower Cretaceous age, and anterior to another deformation (CT 1) of Upper Cretaceous—Lower Tertiary age. The structures (M) peculiar to the "ophiolitic Mélanges" (fig. 4) are shear-fractures (normal, reverse, and strike-slip faults) which have caused the rupture and fragmentation of blocks within the "Mélange". Their study shows that the "ophiolitic Mélanges" of Macedonia are the result of a transcurrent fault of considerable magnitude.

The authors conserve to original meaning of the term "Mélange" defined by GREENLY (1919): the "Mélanges" are tectonic in origin; a separate nomenclature therefore should be utilised for "Coloured Mélanges", which include volcanic agglomerates, magmatic xenoliths, and olistostromes.

The "Coloured Mélanges" in Cyprus, Syria and in the Zagros Mountains, are volcanic and sedimentary complexes which include volcanic agglomerates thrusted into closely spaced tectonic units. The "Ankara Mélange" (Turkey) of Pre-Upper Cretaceous age is related possibly to a zone sibly to a zone of transcurrent faults similar to the "ophiolitic Mélanges" of Macedonia.

REVIEW*

The original definition of the term *mélange* was "tectonic" (Greenly, 1919). However, three interpretations of mélange now exist, two tectonic and the third nontectonic.

1. Tectonic breccias related to major overthrust (Pamir, 1944; Blumenthal, 1948; Bailey and McCallien, 1950, 1953). Moores (1970) incorporated the same concept in plate tectonic theory, suggesting that slices of oceanic plates were actually incorporated in mélanges that would follow fossil subduction zones (Benioff zones). Ernst (1970) believed that the chaotic deformation of the Franciscan mélange occurred as it was subducted

*This review was prepared by the volume editor from a translation of the French text.

beneath the American continent, though he did not exclude the possibility that at least part of the mélange is of sedimentary origin.

2. The ophiolitic olistostrome model as suggested by Gansser (1955, 1959), is very widely accepted. The ophiolites are supposed to have been poured out from the Jurassic to the early Cretaceous and then modified late in the late Cretaceous, with formation of mélanges of olistostrome type following definitions of the Apennine olistostromes by Italian geologists (Rigo de Righi and Cortesina, 1964; Boccaletti, Bortolotti, and Sagri, 1966). An unfortunate ambiguity was introduced between this concept and the earlier one of Dubertret (1939, 1953), who proposed a volcanic origin for ophiolites of Hatay (Syria), which were supposed to have spread out as immense expanses of submarine agglomerates—an idea also adopted by Kundig (1956) and constituting a second nontectonic interpretation.

3. Ophiolitic mélanges related to major dextral transcurrent dislocations (Mercier, 1966, 1973) based on, but modifying, the model of Brunn (1960, 1961). The volcanic interpretation was rejected because rarity of thermal metamorphic effects and the complete lack of blocks in the massive ophiolites, contrasting with their presence in the crushed ophiolites. Relation to nappes was also rejected on the basis of field evidence.

In the Macedonian mélanges, the matrix is either *tuffaceous* or *prasinitic* (greenschist). The tuffaceous matrix contains only carbonate blocks that were once normally interstratified with the basic tuffs before fragmentation. The prasinitic matrix contains blocks of marble, gneiss, and amphibolite that, prior to their fragmentation, belonged to assemblages independent of the volcanic matrix, which was presumably derived from some sort of basic rock. They are thus exotic. Blocks are up to 1 km long, the largest being of marbles of various types and schistose, and in some cases isoclinally folded. They correspond to pelagonian marbles exposed to the east and west of the mélange belt. They are traversed by reverse and normal faults, and contacts with the matrix are invariably planes of rupture. Smaller blocks are composed of marble, amphibolite, micaschist, and gneiss. The blocks have been stirred up after rupture by some mechanism not as yet understood. The flow schistosity of the matrix is not a plastic deformation related to the rupture and stirring; it is a superimposed, later phenomenon. The larger blocks comprise regular trains along structural trends and cannot be olistoliths derived from the substrate and incorporated in an emission of volcanic and ophiolitic character because, in that case, the blocks would have a turbulent, not a regular, arrangement.

Study of the deformations shows that the mélange does not result from any phase of tangential deformation known in this region but from a deformation postdating the late Jurassic–earliest Cretaceous synmetamorphic deformation and prior to the latest Cretaceous–early Tertiary synmetamorphic deformation. The deformations within the blocks are believed to be critical and to establish the origin of the mélange.

Three chronological stages succeeded one another:

1. *Rupture:* Extension (normal faults) and compression (reverse faults) combined to produce dislocation. Flexure is also evident, but is rarely

simple, being associated mainly with reverse faults on planes of rupture.
2. *Fragmentation:* This occurs on planes of fracture within blocks. Initially of minor significance, in zones of intense deformation it comes to affect a greater volume of material. It stems from further development of the rupture pattern, and little blocks become detached and entrained along rupture planes. One can follow all the progressive stages from rupture of large blocks to the formation of little blocks.
3. *Stirring:* The mechanism is as yet not understood.

The deformation is related to a series of gigantic dextral dislocations in this region, trending at N 120°–140°. The mélange zone is up to 5 km wide and 40 km long. The probable time of this dislocation was between the late Jurassic and late Cretaceous.

Broader implications are discussed, particularly to the Coloured Mélange of the Near East. The authors favor a return to the usage of Greenly (1919) and Bailey and McCallien (1953), attributing the fragmentation and stirring to purely tectonic causes. They distinguish tectonic mélanges; volcanic agglomerate, breccia, coarse tuffs that can carry sedimentary rock fragments; magmatic rocks with sedimentary enclaves; and sedimentary breccias and olistostromes carrying blocks of ophiolitic material. Despite superficial similarities, they believe the term mélange should be restricted to the tectonic mélanges. They believe that there are such mélanges related to nappe emplacement [the tuff-matrix mélanges with limestone inclusions of Macedonia, described by them; the similar Mammonia mélange of Cyprus, variegated like the Coloured Mélange of the Iranian-type area, but composed of serpentinites in which there are layers of carbonate sediments, neither fragmented or stirred (Grunau, 1965; Lapierre, 1968); and a similar mélange of Antalya, Turkey (Marcoux, 1970)]. All these consist of scales of carbonate rocks interstratified with basic lava. The scales are crowded, but regularly arranged. Such mélanges could be related to the sole of a powerful nappe, and detailed tectonic study should distinguish it from the type of tectonic mélange described by Mercier and Vergely.

That volcanic agglomerates can resemble a tectonic ophiolitic mélange is accepted. However, such cases—for example, Metalikon, Macedonia (Mercier, 1966*a, b*), and Antalya, Turkey (Marcoux, 1970)—are quite clear from their structure.

Magmatic enclaves can also similarly simulate a tectonic mélange. The example is cited of a nontectonized harzburgite with carbonate enclaves at Neyriz, Iran (Ricou, 1971). This is due to hot intrusion of ultrabasic magma into carbonate sediment as shown by the contact skarns. Ricou believed that such evidence could establish a magmatic origin for colored mélange, but it is quite clear that at Nain (Davoudzadeh, 1969), there is no thermal metamorphism, only cold emplacement of the ultrabasics can be accepted and there is invariably tectonic deformation of the ultrabasic matrix.

Two branches of the colored mélange zone of the Near East are suggested (following Gansser, 1959; Grunau, 1965). The southern branch includes the Mammonia, Antalya, Bassit (Syria), East Turkey, Kurdistan, and Neyriz. Purely tectonic origin is suggested, related to regional sliding. Ricou (1971) admitted purely tectonic origin. These mélanges occur at the sole of the ophiolite nappe and comprise a chaotic assemblage of diverse formations churned up by it.

The northern branch includes the Ankara pre-late Cretaceous-Eocene mélange, and it could be of similar origin to the Macedonian mélange described in this paper. This zone disappears towards the east, but mélange occurs again in the Elburz zone, also around the Lut massif, especially in Iranian Baluchistan, and again to the east in the Himalayas. Classification of these mélanges, of great interest, could only be established after detailed tectonic study. This study of the Macedonian mélange could lead to further studies of similar type and to precise tectonic study and classification of all the other mélanges of the Alpine chain, in Asia and central Europe, the geotectonic character of which remains as yet to be established.

REFERENCES

Bailey, E. B., and W. J. McCallien, 1950, The Ankara-Mélange and the Anatolian Thrust, *Nature* **166:**938–940.

Bailey, E. B., and W. J. McCallien, 1953, Serpentine Lavas, the Ankara-Mélange and the Anatolian Thrust, *Royal Soc. Edinburgh Trans.* **62**(Part II):403–442.

Blumenthal, M., 1948, Un apercu de la géologie des chaînes nord-anatoliennes entre l'Ova de Bolu et le Kizilermak inférieur, *Ankara (Turkey), Maden tekik ve arama enstitüsü yayinlarindan,* ser. B, **13.**

Boccaletti, M., V. Bortolotti, and M. Sagri, 1966, Ricerche sulle ofioliti delle Catene Alpine. I: Osservazio ni sull' Ankara Mélange nella zona di Ankara, *Soc. Geol. Italaliana Boll.* **85:**485–508.

Brunn, J. H., 1960, Les zones helléniques internes et leur extension. Réflexions sur l'orogénèse alpine. *Soc. Géol. France Bull.* **2:**470–486.

Brunn, J. H., 1961, Les sutures ophiolithiques. Contribution á l'étude des relations entre phénomènes magmatiques et orogéniques. *Rev. Géographie Phys. et Géologie Dynam.* **4:**181–202.

Davoudzadeh, M., 1969, Geologie und Petrographie des Gebietes nördlich von Nain, Zentral-Iran. *Zürich (ETH) Geol. Inst. Mitt., new ser., No. 98,* 91p.

Dubertret, L., 1939, Sur la génèse et l'âge des roches vertes syriennes, *Acad. Sci. Comptes Rendus* **209:**763.

Dubertret, L., 1953, Géologie des Roches Vertes du Nord-Ouest de la Syrie et du Hatay (Turquie), *Mus. Natl. Histoire Nat. Notes et Mém. Moyen-Orient* **6:**5–179.

Ernst, W. G., 1970, Tectonic Contact between the Franciscan Mélange and the Great Valley Sequence. Crustal Expression of a Late Mesozoic Benioff Zone, *Jour. Geophys. Research* **75:**886–901.

Gansser, A., 1955, New aspects of the Geology of Central Iran, *World Petroleum Congress, 4th, Rome, Proc. Section 1,* pp. 179–300.

Gansser, A., 1959, Ausseralpine Ophiolith Problem, *Eclogae Geol. Helvetiae* **52:**659–679.

Greenly, E., 1919. The Geology of Anglesey, *Great Britain Geol. Survey Mem. 1,* 980p.

Grunau, H. R., 1965, Radiolarian Cherts and Associated Rocks in Space and Time, *Eclogae Geol. Helvetiae* **58:**157–208.

Kündig, E., 1956, The Position in Time and Space of the Ophiolites with Relation to Orogenic Metamorphis, *Geologie en Mijnbouw,* new ser., **18:**106–114.

Lapierre, H., 1968, Nouvelle observations sur la série sédimentaire de Mamounia (Chypre), *Acad. Sci. Comptes Rendus* **267:**32–35.

Marcoux, J., 1970, Age carnien de termes effusifs du cortege ophiolithique des nappes d'Antalya (Taurus lycien oriental, Turquie), *Acad. Sci. Comptes Rendus, ser. D,* **271:**285–287.

Mercier, J., 1966, Paléogéographie, orogénèse, métamorphisme et magmatisme des zones internes des Hellénides en Macédoine (Grèce): vue d'ensemble, *Soc. Géol. France Bull.* **8:**1020–1049.

Mercier, J., 1973, Etude géologique des zones internes des Hellénides en Macédoine centrale (Grèce), *Annales Géol. Pays Helléniques* **20:**1-586. (In November 1966, 200 copies of a preprint edition were sent out to various geological laboratories in Europe and it was filed with the archives of the Centre National de la Recherche Scientifique in Paris as number A.O. 1142.)

Moores, E., 1970, Ultramafics and Orogeny, with Models of the U.S. Cordillera and the Tethys, *Nature* **228:**837-842.

Pamir, H. N., 1944, Une ligne séismogène en Anatolie septentrionale, *Istanbul Univ. Fen Fakültesi Mecmuasi Rev.* **9.**

Ricou, L. E., 1971, Le métamorphisme au contact des péridotites de Neyriz (Zagros interne, Iran): développement de skarns à pyroxéne, *Soc. Géol. France Comptes Rendus* **1:**43.

Rigo de Righi, M., and A. Cortesini, 1964, Gravity Tectonics in Foothills Structure Belt of Southeast Turkey, *Am. Assoc. Petroleum Geologists Bull.* **48:**1911-1937.

Part V

ALPINE MÉLANGES OF TURKEY, IRAN, AND OMAN

Editor's Comments on Papers 18, 19, and 20

This group of papers concerns mélanges in Turkey. The paper by Bailey and McCallien is a classic in the record of mélanges, and the shorter account is included here (Paper 18), but the full account (Bailey and McCallien, 1953) includes much additional detail. These authors emphasized the importance of the Steinmann (1905, 1927) trinity of decomposed submarine basalts (spilitic pillow lavas), serpentinite, and radiolarite as worldwide and age-long characteristic of geosynclinal deposition. It is the modern consensus of opinion that spilites are secondary hydrothermal alteration derivatives of basalts, mainly tholeiites. These authors also believed that the serpentinites originated as lava flows. Though the reality of ultrabasic lava flows is nowadays established (Gass, 1958; Viljoen and Viljoen, 1971; McCall and Leishman, 1971; Hallberg and Williams, 1972), they are mainly related to Archean suites and, though ultrabasic flows may be represented in ophiolitic mélanges, the greater part of the ultrabasic content is probably not of volcanic origin. Their study of the fragments and the detailed structure of the mélange led them to attribute it to tectonic deformation of Taurus rocks beneath the Anatolian thrust—that is, to the sole deformation beneath a powerful nappe.

Graciansky (Paper 19) describes a mélange in Lycia, Turkey. This paper, illustrated by several excellent diagrams, is in French, and for technical reasons it could not be reproduced here either in translation or in facsimile. It is instead represented by a summary prepared from a translation by the volume editor. This mélange is not more that 200–300 m thick, but its internal structure reveals that it stems from a combination of early synsedimentational deformations (gravity sliding) and superimposed tectonic deformations, introducing some of the exotic material at the sole of an overriding peridotite nappe, traversed by dolerite dikes. Graciansky realizes that this mélange may be atypical, but the paper is extremely important as it shows how exotic slices can enter a mélange at the sole of a powerful nappe by purely tectonic agencies.

Hall (Paper 20) described an ophiolitic mélange of tectonic origin from the Taurus suture zone of Turkey. This is one of several papers on the subject (Hall and Mason, 1972; Hall, 1978; Hall, 1980), and in the last paper he has modified his original ideas on the basis of more detailed petrological study. The mélange is tectonic and consists of predominantly igneous rocks of ophiolitic association. It is overriden by pre-Alpine metamorphic rocks of the Bitlis massif on a low angle, northward-dipping zone of imbrication. In this mélange, the matrix material is so intricately mixed with the enclosed blocks that it is not practicable to map each component separately, and five contrasting rock associations have been mapped. Hall defined five such associations: serpentine, greenschist, chromite, sediment, and marble. Blocks in the first are up to 300 m diameter and include serpentinite, gabbro, amphibolite, and rodingite. The matrix is schistose serpentinite. This association is a mélange in its own right but also hosts large blocks up to several kilometers across of the greenschist association, which consists of unpillowed alkalic basalt and radiolarite, probably both of oceanic provenance (and presumed late Cretaceous age on the basis of indirect evidence). It is interesting that the only matrix defined in this mélange is the serpentinite in the serpentinite association; it seems likely that there was no continuous matrix as in the Franciscan mélange or olistostrome mélanges described from the Appenines and others (Papers 5 to 15). This mélange appears to be closer to the block-to-block mélanges of Iran (Paper 21) but with a higher grade of metamorphism and more penetration of serpentinite so that it actually forms a fragmental matrix to other blocks in at least part of the mélange. The chromite association is a marker zone characterized by podiform chromitite, and the sedimentary association is a minor development of unmetamorphosed flysch, mudstone, and pebbly conglomerate along thrust contacts between the serpentinite association and the greenschist association. The marble association, made up of blocks up to 1 km diameter of pure white marble similar to Permian marbles of the Bitlis massif, mainly occurs structurally below the mélange but also as small blocks in the mélange.

The internal structure of the mélange is a series of subparallel thrust slices. A reasonable interpretation is that large blocks of greenschist have been sheathed by the serpentinite association and that other associations occur mainly at contacts between the two. The probable late Cretaceous age for the metabasites and cherts indicates that some Alpine metamorphism affects the mélange, and as none affects the pre-Permian rocks of the Bitlis massif nearby, such metamorphism must have occurred far from the crustal level indicated by the present position of the mélange. The metamorphism is consistent with low temperature and high fluid pressure. The serpentinite was the lubricant of the mélange, a process believed to be related to failure along internal shear surfaces during the formation of the mélange. The juxtaposition of pre-Permain and Alpine metamorphic rocks probably occurred after this internal organization became the imbricate zone of contact is oblique to the dip of the major thrust slices within the mélange. This juxtaposition produced mylonites and cataclastites in the pre-Permian rocks and local formation of serpentinite breccia in the mélange.

Ophiolites are now usually regarded as fragments of oceanic crust and upper mantle tectonically emplaced in orogenic belts. These basalts are anomalous as they are alkalic, not tholeiitic. Volcanic island or sea mount provenance is suggested (Hall, 1980, later suggested also possible marginal basin provenance). Subduction is believed to have been established by the late Cretaceous. Some authorities have proposed that mélanges are actually formed as oceanic floor is carried down beneath the overriding plate (Hamilton, 1969; Hsü, Paper 3), and by disruption, fragmentation and mixing of the material occurring there, and offscraping the subduction zone on to the leading edge of the overriding plate then taking place. There is certainly some similarity between the Taurus and Franciscan mélanges (Maxwell, 1970), but blueschist metamorphism is not recognized, though the unusually low thermal gradient indicated by the greenschist metamorphism is consistent with a subduction zone environment. Though the absence of blueschists could have been due to lack of detailed petrographic examination, a later detailed study (Hall, 1980) showed that the metamorphism is on the borderline between *epidote-amphibolite* and blueschist facies, and some indications were found that the absence of crossite might be related to some compositional control. In the same paper, Hall recognized that the gabbro in the mélange had not undergone the same metamorphism as the volcanic rocks and cherts and had only suffered a greenschist metamorphism, which he attributed to a sub-ocean floor event, related to the depth of penetration of hydrothermal fluids, and thus not affecting the picrite. In this later paper, Hall proposed a separation of the gabbro from the basalt, which suffered the high pressure metamorphism during subduction; this led him to attribute the mélange formation to a later phase (Miocene, according to his table). There is some doubt whether there is really a need for spatial separation of the gabbro and basalt as suggested, and it is not clear what happened to the gabbro now forming the blocks while the basalts were being subducted. Perhaps some form of compositional and fabric control was exerted on the metamorphism rather than a spatial separation. If it were so, then the mélange formation need not have been a later and quite separate phase. In this respect, perhaps the last word has not been said on this mélange, and there remains room for a model that more closely relates mélange formation to subduction in the Cretaceous and offscraping of slabs. The rather similar Coloured Mélange of the Makran, occupying the same position beneath and extension of the Bitlis pre-Alpine metamorphics, appears to have been formed as a mélange in the Cretaceous–early Paleocene and to comprise offscraped slabs related to pre-Eocene subduction (Paper 21).

Hall discussed the origin of the *Orbitoides* reefal limestone component in the serpentinite association and the peculiar fact that it contains serpentinite clasts. He noted the fact that serpentinites are unknown in the older pre-Permian assemblage, but such rocks are difficult to date and it is difficult to be certain that they are really absent, despite lack of dating evidence. They may also be small bodies and so overlooked. Various possible origins are discussed, including turbidite deposition in a trench and analogies with the top of the scarp of a fracture zone along a large transform fault (Romanche) or an atoll provenance.

Timing of events in the mélange history is difficult, and in the original paper, a late Maastrichtian to Miocene age for the mixing of blocks, including the serpentinite and high pressure metamorphic assemblages, was suggested. This mixing was related to diapiric movement upwards, during the final compressive phase, of the serpentinite, probably up the oblique zone of weakness provided by the subduction zone.

Hall invoked late emplacement by gravity slide tectonics in a subaqueous environment and concluded that the last Miocene phase of compression, bringing the mélange to its present situation high in the crust, did nothing to complicate the internal structure of the mélange, probably because the blocks were insulated by the serpentinite surrounding them, which took up the strain.

He briefly considered the ophiolitic wildflysch zone of the Foothills zone to the south of a belt of normal, late Cretaceous to Paleocene flysch. Olistostromes had previously been recognized in this zone (Rigo de Righi and Cortesini, 1964). It differs from the mélange that formed his principal theme in the higher proportion of sedimentary rocks, much lesser degree of metamorphism (only local, low-grade dynamothermal), and structural position. He noted the suggestion of Oxburgh (1974) that oceanic sediments are not mechanically coupled to the underlying lithosphere and are thus likely to be offscraped and to accumulate in the trench zone. Based on this idea, he related the sedimentary-rock-dominated character of this mélange to the fact that it represents trench material, not subducted, but thrust or slid under gravity from the trench during a phase of uplift. In his later paper (Hall, 1980, Fig. 2), he separated this into olistostromes to the south and mélange in a thrust sheet to the north. Obviously, this mélange and the olistostromes are suitable subjects for further more detailed study of the same excellent quality as those of the northern mélange; and the sedimentary facies of matrix and contained block, their age relations as established by very detailed foraminiferal and radiolarian determinations, and detailed description and analysis of the internal deformation structures are all aspects that would be profitable. Despite the advance in our knowledge of mélanges provided by these studies, yet a feeling exists that here lies an area with great potential for further study.

REFERENCES

Bailey, E. B., and W. J. McCallien, 1953, Serpentine Lavas, the Ankara Mélange and the Anatolian Thrust, *Royal Soc. Edinburgh Trans.* **62**(part II):403–442.

Gass, I. G., 1958, Ultrabasic Pillow Lavas from Cyprus, *Geol. Mag.* **95:**241–251.

Hall, R., 1978, Pyroxenes of Basic Rocks and Rodingites from an Ophiolite Mélange, Southeastern Turkey, *Mineralog. Mag.* **42:**511–512.

Hall, R., 1980, Unmixing a Mélange: The Petrology and History of a Disrupted and Metamorphosed Ophiolite, South East Turkey, *Geol. Soc. London Jour.* **137:**195–206.

Hall, R., and R. Mason, 1972, A Tectonic Mélange from the East Taurus Mountains, Turkey, *Geol. Soc. London Jour.* **128:**395–398.

Hallberg, J., and D. A. C. Williams, 1972, Archaean Mafic and Ultramafic Rock Associations on the Eastern Goldfields Region, Western Australia, *Earth and Planetary Sci. Letters* **15:**191–200.

Hamilton, W., 1969, Mesozoic California and the Underflow of the Pacific Mantle, *Geol. Soc. America Bull.* **80:**2409–2430.

McCall, G. J. H., and J. Leishman, 1971, Clues to the Origin of Archean Eugeosynclinal Peridotites and the Nature of Serpentinisation, *Geol. Soc. Australia Spec. Pub.* **3:**281–300.

Maxwell, J. C., 1970, The Mediterranean, Ophiolites and Continental drift, in *The Megatectonics of Continents and Oceans,* H. Johnson and B. C. Smith, eds., Rutgers University, New Brunswick, N. J., pp. 167–193.

Oxburgh, E. R., 1974, The Plain, Man's Guide to Plate Tectonics, *Geol. Assoc. London Proc.* **85:**299–357.

Rigo de Righi, M., and A. Cortesini, 1964, Gravity Tectonics in Foothills Structure Belt of Southeast Turkey, *Am. Assoc. Petroleum Geologists Bull.* **48:**1911–1937.
Steinmann, G., 1905, Geologische Beobachtungen in den Alpen, *Naturf. Gesell. Freiburg Ber. 16,* pp. 1–49.
Steinmann, G., 1927, Die ophiolitischen Zonen in den mediterranen Kettengebirgen, *Internat. Geol. Congress, 14th, Espagne, Proc.,* pp. 637–668.
Viljoen, M. J., and R. P. Viljoen, 1971, The Geology and Geochemistry of the Lower Ultramafic Unit of the Onverwacht Group and a Proposed New Class of Igneous Rocks, *Geol. Soc. South Africa Spec. Pub.* **2:**55–85.

18

Reprinted from *Nature* **166:**138–140 (1950)

THE ANKARA MÉLANGE AND THE ANATOLIAN THRUST

By SIR EDWARD BAILEY, F.R.S., and PROF. W. J. McCALLIEN

ONE of us (W. J. M.) was professor of geology in the University of Ankara from 1944 until 1949. During this period opportunities were taken of exploring many parts of Turkey. Welcome financial assistance was received from the Royal Society through the British Council: and invaluable geological guidance was obtained from the findings of previous workers (for example, Chaput[1]). Fortunately, a very useful summary of these findings is available in the eight sheets of the Turkish Geological Survey of the country, 1 : 800,000, issued during 1941–46 by the Maden Tetkik ve Arama Enstitüsü (Institute of Mining Study and Research), which corresponds roughly with the Department of Scientific and Industrial Research in Britain.

One result has already appeared, namely, an announcement of the widespread occurrence of pillow lavas[2]. Apparently the name pillow lava has not previously been employed for Turkish examples, though some of them are among the clearest in the world. In many cases these lavas occur with serpentine and radiolarian cherts in what is known as the green-rock or ophiolite association. Thus the recognition of pillow lavas completes for Turkey the Steinmann trinity—pillow lava, serpentine, radiolarite. Steinmann's realization of this trinity as a world-wide, age-long characteristic of geosynclinal deposition ranks among the most exciting achievements of geological research (cf. ref. 3, p. 1718). The pillow lavas of Turkey are decomposed basic rocks, of the general type called spilite by Flett and others in Britain. Another feature of Turkish geology, well illustrated in the Ankara district along with Steinmann's trinity, is the broken, block condition of pre-Tertiary rocks, holding as a general feature over an extremely wide area. The resultant complication is so great that it was decided, if possible, to arrange a joint excursion to investigate the phenomenon in considerable detail. The adventure has materialized through the support of the Carnegie Trust, the University College of the Gold Coast, and the Maden Tetkik ve Arama Enstitüsü. Our work was much facilitated by our enrolment as temporary members of the Turkish Geological Survey.

We worked for rather more than a month from two centres, the modern Turkish capital, Ankara, and the ancient Hittite site, Alaca Höyük, excavated by the Turkish Historical Society since 1935. Alaca Höyük lies a dozen kilometres west-north-west of the town of Alaca and 160 km. east-north-east of Ankara. Our facilities included 'jeep' transport, which enabled us to wander everywhere. The broken rocks, consisting mainly of greywackes, limestones and the Steinmann trinity, are in great disarray; yet, with the help of fossils, mostly of the micro order, and found quite commonly both in the limestones and the cherts, the Geological Survey map has diagrammatically divided the area into Palæozoic and Mesozoic, with local refinements such as Permo-Carboniferous, Lias, Oolite and Jurasso-Cretaceous. We established undoubted, important recurrences of rock types at more than one horizon; but broadly speaking, we are satisfied that the main greywackes are Devonian, the main limestones Permo-Carboniferous in an inclusive sense, and the main developments of Steinmann's trinity, Mesozoic. It had been impossible for one of us (W. J. McC.) to undertake detailed mapping while at the University of Ankara, since, quite properly under existing conditions, a foreigner attempting such a task is subjected to endless questioning. On the other hand, a research student, Oğuz Erol, starting in 1947 and working under close supervision, has produced a valuable thesis the title of which when translated reads "A Study of the Geology and Geomorphology of the Region S.E. of Ankara in Elma Dağ and its Surroundings"; and this is illustrated with a careful geological map (not as yet published) on the scale of 1 : 100,000. For our present purpose it is only necessary to say that Erol firmly established a broken, block structure for the greywacke–limestone complex of the wide district he described, in many cases separating individual blocks of limestone on his map; and that he found numerous microfossils in limestones, some of which have been referred by palæontologists to the Middle Permian. We had the great advantage of Erol's company and guidance during one week of our visit.

It is thought that it may be helpful to furnish a preliminary note on our results, as they bear closely upon current discussions of such topics as the Simme

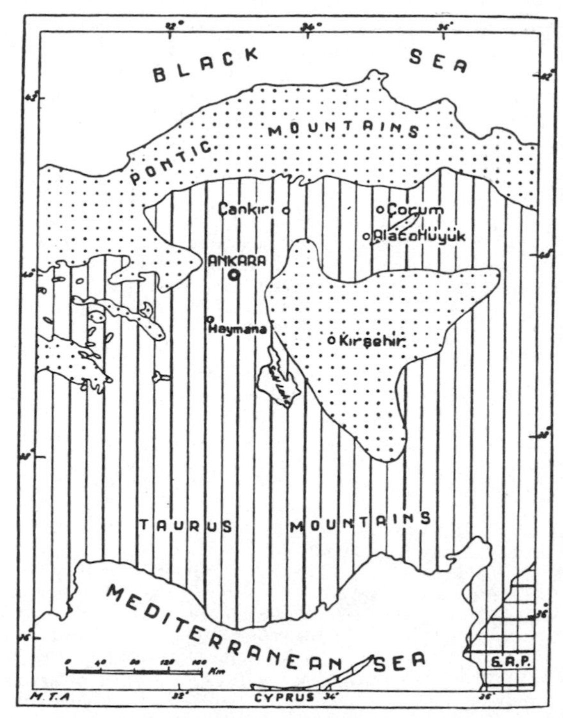

First sketch of Anatolian thrust separating Pontic and Taurus rocks, stripped of Tertiary and Quaternary cover. S.A.P. = Syro-Arabian Platform

Nappe of the Alps, the Ligurian Nappe of the Apennines, and the Radiolarite Nappe of the Zagros Mountains in Iran.

(1) Ankara is situated upon exposures of pre-Tertiary rocks obscured to some extent by Neogene cover. These exposures, rimmed by Eocene or Oligocene, are due to what may be called the Ankara Anticlinal, and run in a general north-easterly direction, continuing thus to the foot of the Pontic (Black Sea) Mountains past Çankiri, the latter 100 km. north-east of the capital. Our personal knowledge of the exposures extends from near Haymana, 60 km. south-west of Ankara, to about half-way between Ankara and Çankiri. The width of the belt at Ankara is about 60 km. In all parts of this anticlinal country which we have seen, the pre-Tertiary rocks are in the condition of a mélange, with the possible exception of some of the greywacke, which, however, only gives poor exposures. We know definitely, from our own experience, of some 6,000 sq. km. of mélange as exposed along the Ankara anticlinal, and it is clear that this is only a small fraction of an immense whole. For example, Blumenthal[4] in a map showing part of the pre-Tertiary beyond Çankiri reserves a tablet for "Mésozoique à faciès tectonique brouillé" (see also under heading 4 below). We propose to call this broken mixed zone the Ankara mélange, out of respect for the descriptions that E. Greenly and C. A. Matley have given of the somewhat similar Gwna mélange in Anglesey and the adjoining Lleyn peninsula of Wales[5].

(2) The Ankara mélange is composed of fragments and masses of the component rocks already cited, ranging exceptionally up to kilometres in length. The most widespread and conspicuous blocks are made of white-weathering limestone, bare of soil. Search invariably shows debris of associated rock, greywacke or spilite as the case may be, as soon as one steps off a bare limestone patch, no matter how bizarre its outline. Moreover, numerous exposures of sheared contacts and of discordant dips show that the blocks are tectonic. Some, of course, are composite, consisting, for example, of limestones with accompanying interbedded spilites. The block fragmentation of the Steinmann trinity is also self-evident wherever soil has been washed away. It is advertised by sharp colour and structural contrasts among the component rocks.

(3) The general elongation of component blocks throughout the Ankara mélange tends to be north-easterly. Though contacts are greatly sheared, the interiors of limestone blocks are often massive and there are numerous exposures of undeformed pillow lavas. Where shearing and folding are prominent a crude cleavage is general. The greywackes and accompanying shales are much more broken down and sheared than the limestones.

(4) A group of anticlinal outcrops centred about Alaca Höyük shows the Ankara mélange reappearing from under the Eocene. They leave no doubt that the Ankara mélange continues without change of style, not only eastwards as far as Alaca Höyük, but also far beyond this ancient site and modern village.

(5) Although the breaking and mixing of the components of the mélange are wonderful to behold, it is evident that it is not complete, since there is a marked tendency for certain types of rock to maintain an irregular association. For example, large areas, predominantly formed of the Steinmann trinity, have been picked out, more or less faithfully, on existent maps. As these areas generally adjoin the Eocene cover, it is certain that at the time of its formation the upper portion of the mélange was formed from what was previously the upper portion of the affected stratigraphical column.

(6) The mélange in places assumes an appearance of *schuppen struktur*, but not with sufficient regularity to warrant deductions regarding the direction of the movement responsible for its production.

(7) Eocene conglomerate-breccias rest on the mélange and present a patent contrast between submarine and subterranean accumulation. The Eocene has derived much of its material from the mélange, including occasional limestone blocks measuring tens of metres, and obviously distributed by tunamis.

(8) The Eocene is sharply folded with dips ranging to vertical, and this adds to the difficulty of interpreting the attitude of blocks contained in the mélange.

(9) Among the serpentines and radiolarites of the Steinmann trinity, we found in five independent exposures variations of the following succession: serpentine; some few metres of calcitized serpentine veined by rather crystalline sedimentary-looking limestone; a small thickness of the same limestone, well bedded; thick radiolarian cherts. These left us in no doubt that many of the serpentines of Turkey are submarine lava flows. Such a conclusion would probably have been reached long ago in this or other regions, if it had not been for the large size of the original olivine crystals, a size that proclaims plutonic growth. Fortunately, N. L. Bowen, as was emphasized at a recent meeting of the Geological Society when he received the Wollaston Medal, has argued on

laboratory and field experience that the great bulk of the crystallization of peridotites, from which serpentines have been derived, was completed before the crystal-concentrate plus lubricant-liquid undertook its last journey.

(10) Obviously the radiolaria of the Steinmann trinity have fed on silica-water, supplied not only by submarine, pillow, spilite lavas, as Dewey and Flett have so well maintained[6], but also by submarine, peridotite flows.

(11) There are features of the Steinmann trinity, such as a frequent association of interbedded fine-grained greywacke, which suggest that Steinmann was wrong in assuming great depth of sea at the time of its formation. On the other hand, until peridotite or serpentine lavas are found among terrestrial accumulations, it is plausible to argue that considerable external hydrostatic pressure is required to maintain the lubricant necessary for their outpouring.

(12) We have mentioned under heading (4) how the Ankara mélange, exposed in the Ankara anticlinal, reappears eastwards in the Alaca Höyük group of anticlinals. The width of separating Early Tertiary outcrop occupying the intervening synclinal is, in this traverse, 60 km. The synclinal outcrop diminishes greatly northwards past Çankiri ; but it increases correspondingly southwards and eastwards past Kirşehir, after which the synclinal may be named.

(13) A remarkable feature of the Kirşehir synclinal is that its Early Tertiary outcrop is in very many places interrupted by great spreads of marble and crystalline schists intruded by granite or diorite. These rocks are clearly pre-Tertiary : they pass under the adjoining Tertiary and have yielded debris to the same. Similar rocks are found occasionally in direct contact with the assemblage characteristic of the Ankara mélange. We shall speak of these crystallines as the Kirşehir crystallines.

(14) The highly accentuated synclinal disposition of the Kirşehir crystallines with reference to the Ankara mélange shows that the former constitute part of a great *nappe de recouvrement* covering the latter. In the anticlinal areas the Early Tertiaries rest preferentially on the mélange, because the anticlines originated with concomitant erosion before the Early Tertiaries started to accumulate. In the synclinal areas the Early Tertiaries rest preferentially on the Kirşehir crystallines, because here erosion has spared extensive masses of the *nappe de recouvrement.*

(15) We believe that the Kirşehir nappe is most probably a Pontine nappe, and that the thrust at its base, which we suggest be called the 'Anatolian thrust', is essentially a boundary between Pontic (Black Sea) and Taurus (Mediterranean) pre-Tertiary rocks, which differ in facies. The Ankara mélange we interpret as Taurus rocks beneath the Anatolian thrust ; the Taurus Mountains as a continuation of similar rocks bulldozed towards the south.

(16) The horizontal displacement postulated along the Anatolian thrust amounts to about 350 km. This is greater than the minimum displacement measured along other thrusts as yet described. The Törnebohm thrust in Scandinavia has a minimum displacement of about 130 km.

(17) We are conscious that our conclusions differ considerably from Kober's conception[7] of a *Zwischengebirge* between the Pontic and Taurus Mountains, and also from Paréjas's interpretation[8] of transversal anticlinals and synclinals in the same region.

In conclusion, we wish to express our thanks to Prof. H. N. Pamir, director-general of Maden Tetkik ve Arama Enstitüsü, and Dr. Recep Egemen, director of the Geological Branch of the same, for their constant help during our sojourn in Turkey.

[1] Chaput, E., Carte géologique de la région d'Angora, 1 : 135,000. (University of Istanbul, 1930). "Voyages d'études géologiques et géomorphogéniques en Turquie" (Paris, 1936).
[2] McCallien, W. J., *Bull. Turk. Geol. Soc.*, **2**, 1 (1950).
[3] Bailey, E. B., *Bull. Geol. Soc. Amer.*, **77**, 1713 (1936).
[4] Blumenthal, M., *M.T.A. Pub.*, B, No. 13 (1948) : see Fig. 2.
[5] Greenly, E., "The Geology of Anglesey" (*Mem. Geol. Surv.*, 1919).
[6] Dewey, H., and Flett, J. S., *Geol. Mag.*, 202 (1911).
[7] Kober, L., "Das alpine Europe" (Berlin, 1931).
[8] Paréjas, E., *Rev. Fac. Sci. Univ. Istanbul*, B, 5, fasc. 3–4 (1940).

19

Reprinted from page 555 of *Rev. Géographie Phys. et Géologique Dynam.* **15**:555–566 (1973)

LE PROBLÈME DES « COLOURED MELANGES » A PROPOS DE FORMATIONS CHAOTIQUES ASSOCIÉES AUX OPHIOLITES DE LYCIE OCCIDENTALE (TURQUIE)

Pierre Charles de GRACIANSKY

ABSTRACT. — The *coloured melange* in Western Lycia (SW Turkey) is included in a 200-300 m thick sheet which is overlain by hudge masses of thrusted ultrabasics and lying on a wildflysch formation of Upper Cretaceous to lowermost Eocene age. All these units are making part of a nappe pile which has been thrust onto autochtonous formations of Lower Miocene age.

The coloured melange is mainly composed of alkali-basalt pillow lavas, volcanic tuffs, radiolarites, pink cherty limestones with globotruncanids, graded bedded calcarenites and breccias that are associated in a very chaotic way with blocks of crystalline schists, dolerites (rodingitized at places) and serpentinites.. These rock types don't form continuous stratas or bodies but are divided into blocks of various sizes, from some kilometres to a few centimetres, which are embedded into a matrix of serpentinite, or volcanic sandstone, or calcarenite.

It may be deduced from structural considerations as from field detailed observations that these various rock types have been disturbed successively (1) by penecontemporaneous reworking of lavas, volcanic sandstone and lime muds (2) by later tectonic events.

So the Lycian *coloured melange*, according to the term coined by A. Gansser (1959), has to be distinguished from melanges *sensu stricto* which are of tectonic origin only, as defined by E. Greenly (1919).

REVIEW*

Giant breccias of great longitudinal extent (tens of kilometers at the least) occur in the Alpine orogenic belts of Europe and Asia on variable scales of thickness (up to kilometers). These are characterized by chaotic structure, fragmentation (of varied type and on varied scale) of component elements, and great lithological variety of components, including ophiolitic igneous rocks, sediments (radiolarite, limestone, graywacke), and metamorphic rocks, producing variegated color in outcrop. These have been termed *colored mélanges* (Gansser, 1959). Mercier and Vergely (1972*a*, and 1972*b*) provided a valuable classification system. Previously, ideas concerning such assemblages were confused. There are two groups:

1. *Tectonic:* These include mélanges of Macedonia (Brunn, 1960; Mercier and Vergely, 1972*a*, 1972*b*) related to major transcurrent faults and mélanges of the Ankara region (Bailey and McCallien, 1950) related to overthrust nappes.
2. *Nontectonic:* These include bodies of eruptive origin and sedimentary olistostromes containing ophiolitic fragments in an argillaceous or tuffaceous matrix.

Later have come interpretations related to plate tectonics theory (e.g., Temple and Zimmerman, 1969). These are syntheses that could explain

*This review was prepared by the volume editor from a translation of the French text.

several features: synsedimentary deformation, tectonic dislocation, common metamorphic zones of high pressure type, spatial association with fresh sediments, diverse ages of sediments, metamorphic rock association including eclogite and upper mantle(?) material (Ernst, 1970; Dickinson, 1970; Coleman, 1971). Similarities in appearance due to a form of geological convergence has resulted inthe term *mélange* being applied to bodies variously interpreted on the basis of often well conducted studies. Thus, to consider that all mélanges indicate the trace of fossil subduction zones is to confound the proposition with the converse. In some cases, unicausal explanation does not account for all the deformations observed in a mélange. This is the case for the Lycian mélange in Asia Minor. It merits the name colored mélange, but here processes of separation of blocks, generating dislocation, have acted on material that has undergone prior dispersion processes contemporaneous with sedimentation. In the Lycian example, there is a threefold sequence: (1) a peridotite nappe, the most stretched and continuous of the allochthonous units, overlies (2) a complex of slices including the Coloured Mélange (Graciansky, 1967, 1968, 1972), which in turn overlies (3) the autochthon, of which the youngest components are Miocene. The intermediate complex of slices consists of carbonates beneath (Permian-late Cenomanian), overlain by flysch (passing up into wildflysch, Senonian-Maastrichtian), and overlain in turn by a giant breccia, with tuffitic or serpentinitic matrix and containing carbonates, radiolarites, alkalic basalts (pillow lavas), volcanic grits, metamorphic rocks, and tholeiitic dolerites. The nappe of peridotites is cut by tholeiitic dolerite dikes that came from beneath. The contact at the base of this nappe is consistently tectonic, whatever the nature of the underlying formation. The contact between the giant breccia and the wildflysch has to be studied closely in every case.

The contact between the peridotite nappe and the mélange beneath is indicated by the appearance of abundant dolerite and crystalline schist bodies localized at the sole. They are sheathed in serpentinite. The dolerite dikes have a different petrographic character to basic igneous rocks in the mélange and thus can be readily distinguished. They have not been so tectonically deformed. In the interior of the peridotite, their trace is still preserved. At the edge of the peridotite mass and at the ends of the veins they are invariably enveloped by serpentinite and become disordered. Some are totally detached from the main mass and inserted into the adjacent mélange. These arrangements give a picture of mechanical fragmentation undergone by the dolerite dikes at the base of the peridotite nappe during transport. The greater part of the dolerite blocks in the mélange have been mechanically torn from the base of the peridotite nappe as it moved over the mélange. The chemical mobility has been sufficient to locally effect alteration to white rodingite with idocrase in vesicles, the dolerite being thus transformed in addition to the serpentinization. The origin of the slices of metamorphic rock at the base of the peridotite nappe is unknown; they cannot be related to any known outcrop of basement in Asia Minor. Amphibolite, quartzite, gneiss, micaschist, marble and its skarns, of epidote amphibolite facies, comprise these slices. Their presence strengthens the belief that the peridotite is allochthonous, resting as it does on unmetamorphosed rocks. The threads of tholeiitic dolerite and crystalline schist in the mélange must owe their presence to purely tectonic processes. They invariably occupy the abnormal contact of the peridotite with the underlying sequence—with other components of the mélange and with Permian-Mesozoic carbonates of the intermediate complex, or even with the autochthon.

The contact of the mélange with the wildflysch beneath appears to be normal, sedimentary, and progressive (gradational). The wildflysch carries basalt and radiolarite blocks of the same character as those in the mélange, bonded by a gritty or silty cement. The upward passage is simply a change in the bonding matrix, from mainly quartzose and argillaceous below to tuffaceous above. This transition occurs within a thickness of 5 m. It is concluded, however, on the basis of diverse considerations of age and facies, that the lavas and the sedimentary rocks associated with them, which comprise the mélange, have been mobilized as a transported nappe that, toward the end of the Cretaceous, overwhelmed the basin in which the wildflysch was being deposited. This nappe is called the *nappe of diabases* (Graciansky, 1968) and is formed of colored mélange. It is 200–300 m thick. The contact above it is mechanical, but that beneath, separating it from the wildflysch, is believed to be sedimentary.

The relationships between the dolerite, crystalline schist, and components of the mélange (lava, carbonates) are manifestly abnormal and related to the nappe emplacement. This has also resulted in interstratification of limestones, disrupted into shreds, in the pillow lavas. This is a common effect beneath a nappe. However, relationships between the lavas and sediments of the mélange, where still preserved, reveal the submarine environment of the pillows, which are rarely entire, all variations being seen from whole pillows, through fissured pillows to split off fragments (the effect of fragmentation being due to thermal expansion; see Vaugnat, 1967). Pillows and fragments are set in a hyaloclastite or tuff matrix. Gulleying and sorting of clasts are well seen in suitable outcrops. Interstitial enclaves included in the basalts include limestone-carrying fossils, which establish the age of the eruptions as late Cretaceous in most cases. Other sediments directly overlie the pillow basalt, jaspers, or micrites that are encrusted on them; they contain detritic diabase grains and planktonic microfossils (radiolaria, foraminifera). They represent a descending suspension of fine particles between the lava eruptions. Other sediments are also developed. A full sequence consists of carbonate-silex conglomerates at the base, oolite with rudists, hydrozoons, sponges and echinoderms above, and fine red limestones with red or black chert bands, containing radiolaria and *Globotruncana*, at the top. The upper facies resembles the *rosso ammonitico*. Commonly only the calcarenites and micrites are represented, not the full sequence. These are again interstratified in the breccias. Reworking structures in the volcanic grits and synsedimentary folds and sorting in the limestones show that the sea floor was characterized by slopes and relief. Seamounts or the margins of emergent areas represent very shallow zones, close to the water surface, where the organisms such as rudists and sponges grew and oolites formed prior to reworking in the pyroclastic formations.

Blocks of carbonate occur in the matrix of most of the diabasic elements, up to meters scale, dispersed in distinct horizons. These are angular, rounded, or pebble-like, and surfaces are dented and coated with volcanic sand. Their shape is asymmetric and strongly tapered, suggesting reworking, the sliding of shreds of unconsolidated calcareous slime and their insertion among the diabasic blocks. The largest carbonate blocks, of up to kilometer scale, representing a pile of coarser carbonates, threaded at their extremities into the tuff matrix, passing into swarms of other blocks of the same facies, but smaller and grouped in distinct horizons. The outer surfaces may be parallel, oblique or normal to the internal stratification. Where not destroyed by friction, the surface is a green chloritic or red hematitic crust, on which there is a coating of volcanic sand. Blocks are in some cases separated by sharp angled fractures; here the surfaces are particularly bare and the volcanic sand matrix gets partly into the silt. These are late fractures. The existence of fracture surfaces recutting the internal stratification of carbonate masses coated with chlorite and hematite suggests strongly that the masses, already partially diagenized, have undergone breaking and separation in the submarine environment itself.

Comparisons are drawn with other regions. Similar assemblages are known from both orogenic and nonorogenic environments. The white bioclastic limestone of the Franciscan mélange of California, carrying *Globotruncana* (Laytonville type), is remarkably similar to limestones described here, and the blueschist association (Bailey et al., 1964) recalls Lycia. The Triassic Kunose Group in the Sambosan Range of Kyushe, Japan (Kanmera, 1969) also presents similarities. The resedimentation phenomena that characterize the sediments in these chaotic formations are present in formations sheltered from any orogenic perturbation—for example, in the Italian Dolomites (Cros, 1967*a*, 1967*b*, 1968, 1971; Leonardi, 1967). These formations occur in the form of very rounded bodies with contoured lobes. It is easy to reconcile them with the different form of the bodies that have undergone a phase of tangential tectonics, allowing the lithological variety and competence contrast in the rocks. Mistakes have occured due to overregular and linear paleogeographic reconstructions of the basins.

The conclusion is reached that the Lycian mélange is compatible with the modified definition of Gansser (1959) rather than the strict tectonic definition of Greenly (1919). It differs from the typical mélange because of:

1. Lack of great thickness of the order of kilometers (the Lycian mélange is no more than 200–300 m thick);
2. Lack of prasinitic matrix (in the Lycian mélange the matrix is tuffaceous or locally calcareous);
3. Lack of superimposed schistosity (no trace in the Lycian mélange);

4. Intense perturbation of original structures (the structure of the Lycian mélange is only chaotic on the coarse scale, the detailed structure including the petrographic structure being preserved in entire slabs of up to kilometer scale).

During the Cretaceous, there was a fine sedimentation of carbonate and radiolarite silt at the time of emplacement of the submarine pillow lavas that filled up most of the delicate cavities. Relief in the sea floor allowed the local deposition also of reefal carbonates. Before the termination of diagenesis, blocks and calc-sands descended, undergoing sorting and resedimentation toward the deeper zones, where they were interstratified with find muds resulting from continuing pelagic sedimentation and with diabasic breccias. During the same sedimentation, collapse and submarine sliding occurred periodically, upsetting the regularity of the stratification. This is evidence of unstable bottom conditions related perhaps to volcanism, the manner of emplacement of the flows, and pyroclastic rocks.

At the beginning of the Eocene, mechanical deformation proper commenced. An initial phase of movement must have brought the volcanic-sedimentary complex to rest on the nearby continental margin or the wildflysch, already deposited. Probably the great peridotite nappe, carrying crystalline schist shavings at its base, lost at its sole serpentinite slices and dolerite—these being intercalated in the mass of volcanic rocks beneath as displaced scales. Later, in the Miocene, further movements introduced new deformations into the edifice.

Thus, diverse processes, both synsedimentary and tectonic, have superimposed their effects in the elaboration of the chaotic structure of the Lycian colored mélange.

REFERENCES

Bailey, E. H., W. D. Irwin, and D. J. Jones, 1964, Franciscan and Related Rocks and Their Significance in the Geology of Western California, *California Div. Mines Geol. Bull. 183,* 177p.

Bailey, E. B., and W. J. McCallien, 1950, The Ankara-Mélange and the Anatolian Thrust, *Nature* **166:**938–940.

Brunn, J. H., 1960, Les zones helléniques internes et leur extension. Réflexions sur l'orogénèse alpine, *Soc. Géol. France Bull.* **2:**470–486.

Coleman, R. G., 1971, Plate Tectonic Emplacement of Upper Mantle Peridotites along Continental Edges, *Jour. Geophys. Research* **76:**1212–1230.

Cros, P., 1967*a*, Hypothèses sur la génèse de bréches triasiques dans les Dolomites italiennes. *Acad. Sci. Comptes Rendus,* ser. D, **264:**793–796.

Cros, P., 1967*b*, Origine de certains blocs calcaires de cipit du Trias dolomitique, *Acad. Sci. Comptes Rendus,* ser. D, **264:**556–559.

Cros, P., 1968, Enchaînements de faciès au Ladinien dans le massif du Latemar (Dolomites italiennes), *Acad. Sci. Comptes Rendus,* ser. D, **266:**313–316.

Cros, P., 1971, Glissements sous-marins et passages de faciès dans le Carnien. Dolomies de Braies (Italie du Nord), *Soc. Géol. France Bull.* **13:**57–66.

Dickinson, W. R., 1970, The new global tectonics, 2nd Penrose Conference, *Geotimes,* **15:**18–22.

Ernst, W. G., 1970, Tectonic Contact between the Franciscan Mélange and the Great Valley Sequence. Crustal Expression of a Late Mesozoic Benioff Zone, *Jour. Geophys. Research* **75:**886–901.

Gansser, A., 1959, Ausseralpine Ophiolith Problem. *Eclogae Geol. Helvetiae* **52:**659–679.

Graciansky, P. Ch. de, 1967, Existence d'une nappe ophiolitique à l'extrémité occidentale de la chaîne sud-anatolienne; relations avec les autres unités charriées et avec les terrains autochtones (Province de Mugla, Turquie), *Acad. Sci. Comptes Rendus,* ser. S, **264:**2876–2879.

Graciansky, P. Ch. de, 1968, Stratigraphie des unités superposées dans le Taurus Lycien et place dans l'arc dinaro-taurique, *Ankara (Turkey), Maden tekik ve arama enstitüsü Bull.* **71:**42–62.

Graciansky, P. Ch. de, 1972, Recherches géologiques dans le Taurus Lycien occidental, *Fac. Sci. Thèse Univ. Paris-Sud (Orsay) no 896.*

Greenly E., 1919, The Geology of Anglesey, *Great Britain Geol. Survey Mem. 1,* 980p.

Kanmera, K., 1969, Litho and Bio-Facies of Permo-Triassic Geosyncline Limestone of the Sambosan Belt in Southern Kyushu, *Palaeont. Soc. Japan Spec. Paper 14,* pp. 13–40.

Leonardi, B., 1967, *Le Dolomiti Geologia dei monti tra Isarco e Piave,* 2 vols., Consiglio Nazionale delle Ricerche and Giunta Provinciale di Trento, Trento.

Mercier, J., and P. Vergeley, 1972*a*, Les mélanges colorés de la zone s'Almopias (Macédoine, Grèce), *Soc. Géol. France Comptes Rendus* **2:**70–73.

Mercier, J., and P. Vergely, 1972*b*, Les Mélanges ophiolitiques de Macédoine, Grèce: Décrochements d'âge ante'crétacé supérieur, *Deutsch. Geol. Gesell. Zeitschr.* **123:**469–490.

Temple, P., and J. Zimmerman, 1969, Tectonic Significance of Alpine Ophiolites in Greece and Turkey, *Geol. Soc. America Programs Ann. Meeting, 1969/7,* pp. 221–222.

Vuagnat, M., 1967, Les coussins éclatés du Lago Nero (Massif du Mentgenèvre, province de Turin) et le problème des brèches ophiolitiques, *Soc. Physique et Histoire Nat. Genève Compte Rendu,* N. S., **1:**163–167.

20

Reprinted from *Geol. Soc. America Bull.* **87:**1078–1088 (1976)

Ophiolite emplacement and the evolution of the Taurus suture zone, southeastern Turkey

ROBERT HALL* *Department of Geology, University College, University of London, London, WC1E 6BT, Great Britain*

ABSTRACT

The interior of the southeastern Taurus Mountains of Turkey is occupied by an extensive area of metamorphic rocks known as the Bitlis Massif, previously regarded as pre-Permian in age. An ophiolitic mélange occurs within the Bitlis Massif, and detailed mapping has shown that the mélange can be subdivided and an internal structure recognized. It contains components of Late Cretaceous age, some of which have been metamorphosed, indicating an Alpine metamorphic event. It is proposed that the Bitlis Massif is a composite structural entity consisting of a northern area of pre-Permian metamorphic rocks and a southern zone of Alpine rocks that have undergone high-pressure–low-temperature metamorphism.

The Bitlis Massif is thrust southward over an ophiolite-flysch complex, which is also thrust southward over sedimentary rocks of the Arabian foreland. The ophiolite-flysch complex is here divided into three zones, one of which is an ophiolitic-wildflysch zone correlated with ophiolitic gravity slides occurring further west in the southeastern Taurus Mountains.

In the tentative plate-tectonic model for the evolution of a section of the southeastern Taurus Mountains presented here, two episodes of ophiolite emplacement are recognized. It is suggested that the ophiolitic-wildflysch represents trench mélanges that were not subducted but were thrust out of the trench zone because of uplift associated with the final phase (Late Cretaceous) of subduction. The metamorphosed ophiolitic mélange is thought to represent successfully subducted mélange emplaced during the final phase (Miocene) of continental collision. *Key words: ophiolites, mélanges, emplacement, Taurus foldbelt, Turkey.*

INTRODUCTION

The Middle East is well known for the large areas of ophiolitic rocks that crop out extensively in the mountains of the Alpine chain. These ophiolitic rocks have presented many problems with regard to their structural position, their age, and the mechanisms of their emplacement, and they have been regarded at various times as both autochthonous and allochthonous (see reviews by Gansser, 1959, and Ricou, 1971). In the light of modern theories of global tectonics, ophiolitic rocks of mountain chains have been interpreted as fragments of oceanic lithosphere (see, for example, Dewey and Bird, 1971), and this has led to renewed investigation of their geology. Particular attention has been given to the stratiform ophiolite complexes of the Middle East (Cyprus, Hatay, and Oman), partly because of their excellent exposure but mainly because of their obvious similarities to models of the structure of present-day oceanic lithosphere. These complexes are currently providing much information on processes occurring at *constructional* plate margins. However, ophiolitic mélanges are more characteristic of the ophiolite zones of the Middle East, and although detailed studies of their geology have been made (for example, Bailey and McCallien, 1953), they have been somewhat neglected in recent years in comparison to the stratiform ophiolites. This is unfortunate, because it seems likely that detailed studies of ophiolitic mélanges may provide many insights into the processes occurring at *destructional* plate margins (as has been shown by Hamilton, 1969, and Haynes and McQuillan, 1974). I discuss here the geology of an ophiolitic mélange and the general structure of the ophiolite zone in the southeastern Taurus Mountains of Turkey. I have attempted to interpret this work and previous studies of the southeastern Taurus Mountains in the light of modern theories of global tectonics.

REGIONAL SETTING

The Taurus foldbelt is the southern of the two major foldbelts forming the Anatolian sector of the Alpine-Himalayan mountain chain. The interior of the eastern Taurus Mountains (Fig. 1) is occupied by an extensive area of metamorphic rocks known as the Bitlis Massif, which extends approximately 300 km eastward from Elaziğ to Hakkari and is approximately 40 km wide from north to south. This appears to be part of a belt of metamorphic rocks which can be traced from the Zagros Mountains of Iran in the east through the Bitlis Massif westward into the Malatya-Pütürge Massif. This belt is not continuously exposed, and the Bitlis Massif appears to be an arched structure, plunging eastward and westward under Cretaceous and Paleocene rocks (Altinli, 1966).

The metamorphic rocks of the Bitlis Massif have not been systematically studied, but the limited information available (Yilmaz, 1971; Boray, 1973) suggests that they are mainly within the greenschist and amphibolite facies. A variety of lithologies has been recorded, including phyllite, chlorite schist, mica schist, garnet mica schist, amphibolite, quartzite, and marble, and locally these rocks are intruded by granitic rocks. Altinli and others (1964) estimated that the thickness of the metamorphic rocks exposed in the deeply eroded valleys of the massif exceeds 1,000 m. No fossils have been found in the metamorphic rocks, and their age is not yet clearly established. Most Turkish workers have considered them to be pre-Permian (Altinli and others, 1963, 1964), and Precambrian ages have been suggested (Peyve, 1969). Rigo de Righi and Cortesini (1964) have suggested that some of the rocks of the adjacent Malatya-Pütürge Massif may be metamorphosed Mesozoic sediments, representing deeper parts of the Taurus eugeosyncline, and Kamen-Kaye (1971), following Ketin (1966), showed the Bitlis Massif as "presumed early Alpine" metamorphic rocks. However, some of the areas of metamorphic rocks considered by Ketin to be "early Alpine" or "middle Alpine" clearly have a long and complex pre-Mesozoic history (see Van der Kaaden, 1971), and in the Bitlis Massif there are good geologic grounds for considering that some of the metamorphic rocks are pre-Permian in age. Tolun (1953) has shown that marbles of Permian age rest unconformably on metamorphic rocks at a few localities within the Bitlis Massif, although generally the contacts of the Permian marbles with the metamorphic rocks are tectonic. In addition, the geochronological work of Yilmaz (1971) in the region of

* Present address: Department of Geology, University College Galway, Galway, Ireland.

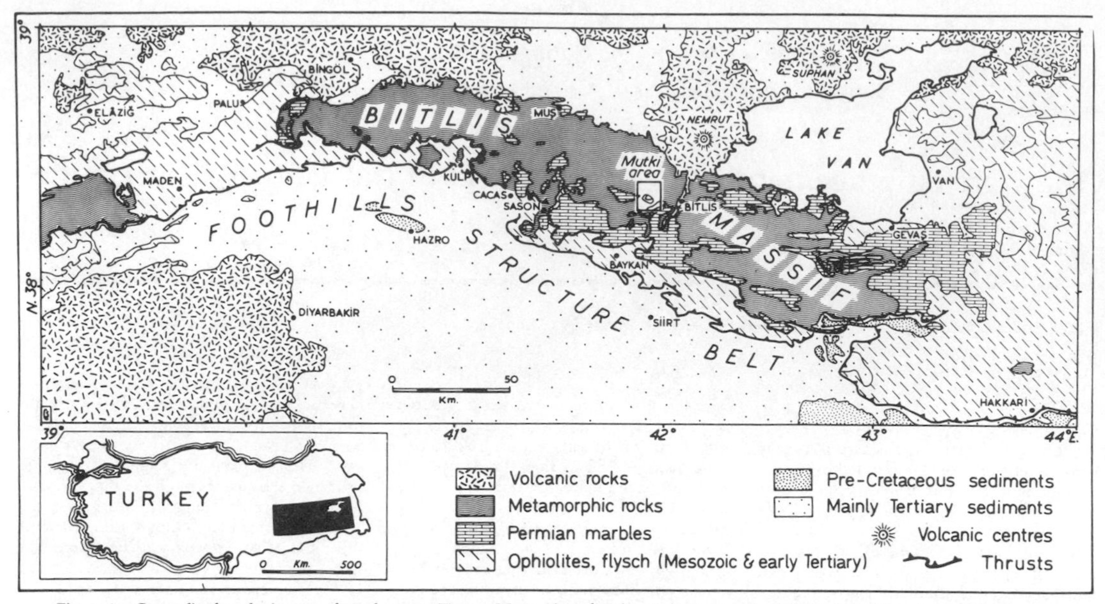

Figure 1. Generalized geologic map of southeastern Taurus Mountains, showing main structural subdivisions and location of Mutki area.

Cacas has established that some of the rocks were metamorphosed and intruded by granite in Paleozoic time, and this suggests that some were deposited during late Precambrian or early Paleozoic time.

The northern edge of the Bitlis Massif is not clearly seen because it is overlain unconformably by extrusive and intrusive rocks (basalt, andesite, and dolerite) and by sedimentary rocks, all of Neogene age. Faulting of the margin has been suggested because Neogene lacustrine limestones of the Muş area are exposed at an altitude 700 m lower than similar limestones found on the peaks of the Bitlis Massif to the south (Altinli, 1966). Ternek (1953) has suggested that further east the contact between metamorphic and Tertiary rocks is a thrust, because near Gevaş, metamorphic rocks overlie rocks of Late Cretaceous and Paleocene age, and the metamorphic rocks apparently were thrust northward.

To the south, the margin of the Bitlis Massif is clearly tectonic. Metamorphic rocks of the Bitlis Massif are in contact with an ophiolite-flysch complex referred to by Rigo de Righi and Cortesini (1964) as the Elaziğ gravity nappe. This contact is considered to be a northward-dipping overthrust (Altinli, 1966). The rocks of the ophiolite-flysch complex are mainly Late Cretaceous and Paleocene in age but occasionally include components of Late Jurassic age (Altinli and others, 1963). The complex comprises a variety of rocks, including rocks of the ophiolite suite (serpentinite, basalt, and radiolarite) and flysch, and occasional pink limestone and conglomerate. Altinli and others (1964) noted that the upper part of this complex often resembles the wildflysch of the Alps.

South of this zone is a second, approximately parallel tectonic line where the ophiolite-flysch complex is in contact with the Foothills structure belt. Rigo de Righi and Cortesini (1964) referred to this as the Maden-Bitlis tectonic line and described it as a 600-km-long arcuate front, concave toward the south. They considered it to be the outcrop of a regional gliding surface, whereas other workers (Ilhan, 1967) described it as an overthrust. The surface is nearly horizontal or gently dipping toward the north, and whereas the Miocene shales immediately beneath this surface are affected by only minor tectonism, the allochthonous complex above is highly tectonized. Rigo de Righi and Cortesini (1964) considered that the allochthonous complex has a minimum southward displacement of 15 km in the region of Maden, where an inlier of Miocene shale is observed, with the displacement on the gliding surface increasing eastward.

The Foothills structure belt is a zone of sedimentary rocks affected by foreland-type folding and thrusting. South of this zone, approaching the undisturbed shelf sequences of the Arabian platform, an almost continuous stratigraphic sequence, with minor breaks, can be traced from lower Paleozoic to Miocene (Temple and Perry, 1962). The Paleozoic sequence consists largely of siltstone, shale, and sandstone resting on a series of volcanic and clastic sedimentary rocks of possible Precambrian age. The Mesozoic sequence is formed predominantly of carbonate rocks, with subordinate evaporite and clastic rocks. In this area the sequence continues with only minor breaks through to rocks of Miocene age, but within the Foothills structure belt the sequence is interrupted by the Besni and Kevan gravity slides. These gravity slides, which Rigo de Righi and Cortesini (1964) regarded partly as thrust sheets and partly as olistostromes, contain fossils ranging in age from Jurassic to Late Cretaceous and are chaotic assemblages that include rocks of the ophiolite suite, limestone, cherty limestone, and variegated shale. They are thought to represent part of the Taurus eugeosynclinal sequence thrust onto the shelf sequence in Late Cretaceous time and are overlain by a varied and much-interrupted sequence of shelf sedimentary rocks.

Rigo de Righi and Cortesini (1964) interpreted the regional geology in terms of geosynclinal theory, but recently the ophiolite-flysch complex has been reinterpreted as the outcrop of a Tethyan suture (Smith, 1971; Dewey and others, 1973) and the Bitlis Massif regarded as the crystalline basement of the overriding Anatolian continental fragment (Peyve, 1969). The work described here began with the aim of discovering more about the internal structure of the Bitlis Massif and its relationship to the Mesozoic and Tertiary rocks to the south.

OPHIOLITIC MÉLANGE

Reconnaissance work within the Bitlis Massif demonstrated the presence of a tectonic mélange, consisting predominantly of ophiolitic rocks, near the village of Mutki (Hall and Mason, 1972). Subsequently, a detailed study (Hall, 1974) has been made of the petrology and structural relations of this mélange.

Structurally above and to the north of the mélange, the metamorphic rocks of the Bitlis Massif comprise a series of metasedimentary and metabasic rocks intruded by a synmetamorphic granite. Petrographic and geochemical study of these rocks indicates that they have undergone regional metamorphism from epidote-amphibolite facies to amphibolite facies and later were intruded by a number of postmetamorphic mafic dikes. No direct evidence is available in the Mutki area for the age of these rocks, but structural evidence indicates that they are pre-Alpine; similarities between these rocks and those of the Bitlis area (Boray, 1973) and the Cacas region (Yilmaz, 1971) strongly suggest a pre-Permian date for the metamorphism. The contact of the pre-Permian metamorphic rocks with the mélange is inclined at a low angle northward and appears to be a zone of imbrication. Mylonites have been developed locally along planes of movement associated with this imbrication zone.

Within the mélange itself, widely varied lithologies are represented. These include ultramafic rocks with associated chromites and calcium-rich metasomatic rocks, metagabbros and amphibolites, metabasalts and radiolarites, and a variety of sedimentary rocks. Although no pillow lavas have been observed, the ophiolitic character of the assemblage is apparent, and mafic and ultramafic rocks constitute approximately 90 percent of the outcrop area of the mélange. Matrix material and enclosed blocks of different rock types are in general so intricately mixed that it is not practicable to map each lithological component separately. However, I have been able to distinguish a number of different types of *rock association*, which I have used as the fundamental units of my mapping within the mélange (Hall, 1973). These mappable units are empirically defined; they are the consistently associated lithologies of blocks and matrix occurring throughout the mélange, and they correspond to the mélange units or tectono-stratigraphic units of Hsü (1968). Five rock associations have been defined in mapping the mélange.

Rock Associations

Serpentinite Association. Internally, the Serpentinite association has the character of a tectonic mélange and consists of a matrix of schistose serpentinite, enclosing blocks of massive serpentinite, picritic gabbro, metagabbro, unfoliated amphibolite, and calcium-rich metasomatic rocks (rodingites). Detailed petrographic and geochemical work suggests that the amphibolites (which have no oriented fabric) are the recrystallized equivalents of the metagabbros, and that this recrystallization occurred before the blocks were incorporated in the ultramafic matrix. The evidence for the timing of this recrystallization is indirect and is based on the failure of many of the mafic blocks to reach textural equilibrium. Many very small blocks (less than 10 m across) still contain relict igneous minerals (pyroxene and plagioclase), whereas most of the larger blocks (more than 100 m across) appear to be completely recrystallized amphibolites. If the blocks had been metamorphosed while incorporated in the ultramafic rocks, one would expect that the smaller blocks would be more likely to achieve textural equilibrium, whereas the cores of the larger blocks would be more likely to show disequilibrium textures. The reverse is true, and the blocks show none of the features that would be expected if they had been further disrupted after metamorphism. Therefore, the most plausible explanation of these observations is that the gabbro blocks underwent amphibolitization before their incorporation in the ultramafic rocks. Serpentinization of the ultramafic rocks followed, accompanied by considerable calcium metasomatism that affected all the tectonic blocks and resulted in the formation of the calcium-rich reaction zones.

The tectonic blocks vary considerably in size and shape. They range from angular blocks as much as 300 m across to small slivers and angular fragments less than 0.5 m across. Among the most significant of the tectonic blocks are several of fossiliferous marble. The marble contains angular fragments of detrital serpentinite (as much as 20 cm across) and chrome spinel, and the fossil assemblage includes abundant *Orbitoides*. This assemblage has been described by Meriç (1973), who concluded that it is clearly of late Maestrichtian age. However, not only is the Serpentinite association a mélange in its own right, but it also acts as a matrix for the very large blocks (which range from one to several kilometres across) of the Greenschist association.

Greenschist Association. The Greenschist association consists of a series of basalts and radiolarites that have undergone regional metamorphism and are wholly or partly recrystallized in the glaucophanitic-greenschist transitional facies (Winkler, 1967). Pillow structures are absent, but relict igneous textures and igneous pyroxenes are preserved in many of the samples, which usually have a poor foliation. Locally, the foliation is well developed, and the basalts have recrystallized as greenschists (hornblende + albite + chlorite + quartz + sphene + opaques ± calcite ± epidote ± biotite) or as sodic amphibole–bearing greenschists (crossite + albite + phengite + chlorite + quartz + sphene + opaques ± calcite ± epidote). In the schists, the development of sodic amphibole appears to be dependent on the bulk-rock iron oxide ratios, as Ernst and others (1970) have demonstrated for similar rocks from the Sanbagawa terrain. Whole-rock analyses and microprobe analyses of relict igneous clinopyroxenes suggest that the rocks of the Greenschist association represent a series of mafic volcanic rocks of alkaline affinities, together with associated radiolarites, and are probably of oceanic origin.

The cherts are too recrystallized for identification of the radiolaria. However, Boray (1973) has described similar cherts associated with sodic amphibole–bearing metabasalts, which can be correlated with the Greenschist association, from a thrust zone in the Bitlis area 20 km to the east. In these cherts, the radiolaria have been identified tentatively as *Dictyomitra*, which is of Late Cretaceous age. It is reasonable to propose a Late Cretaceous age, therefore, for some, at least, of the cherts and associated metabasalts of the Greenschist association.

Chromite Association. The Chromite association is recognized as a separate rock association because it provides a distinctive structural marker horizon within the mélange. It is separated from the Serpentinite association by tectonic contacts, although tectonism appears to have been slight and to postdate the mélange emplacement. The Chromite association includes podiform chromite bodies (as much as 20 m long) that are incorporated in a variety of silica-carbonate rocks ranging from pure silica to pure carbonate extremes. Petrographic and x-ray diffraction work on the nonopaque phases and chemical analyses of the chrome spinel (R. Wood, 1974, personal commun.) indicate that the podiform chromites were originally incorporated in serpentinite that has suffered low-temperature CO_2 metasomatism of the type discussed by Greenwood (1967) and Barnes and others (1973). Relict serpentinite textures are well preserved in rocks totally composed essentially of magnesite, dolomite, and quartz. Structural evidence suggests that CO_2 metasomatism was a late event, postdating the emplacement of the mélange in its present position. (Currently active travertine springs suggest that such alteration may be occurring at depth today.)

Sediment Association. The Sediment association includes flysch, mudstones, and pebbly conglomerates, all of which are present as thin slivers along the thrust contacts between the Serpentinite and Greenschist associations. Therefore, because of their structural position, the rocks of the Sediment association have an importance in mapping which is much greater than their areal extent.

Marble Association. The Marble association consists of blocks of pure white marble, highly recrystallized and unfossiliferous, with subordinate chlorite schist. The blocks range in size from a few hundred metres to approximately 1 km across. In general, the marbles appear to be structurally below the rest of the mélange, although a few smaller marble blocks are present within the mélange as well. These marbles closely resemble the pure white marble that crops out extensively throughout the Bitlis Massif and is known to be Permian in age; although no fossil evidence is available in the Mutki area, a Permian age for the white marbles of the Mutki area has been tentatively accepted.

Structure and Metamorphism

Mapping of the rock associations indicates that internally the mélange is formed of a series of subparallel thrust slices. The combination of incomplete exposure and the tectonic characteristics of serpentinite, which occurs as discontinuous bodies with sudden variations in thickness, makes it difficult to obtain a detailed picture of the overall structure. However, an interpretation of the mélange as large blocks of the Greenschist association enclosed in a sheath of the Serpentinite association and of the other rock associations as restricted mainly to contacts between these two associations seems most reasonable (Fig. 2).

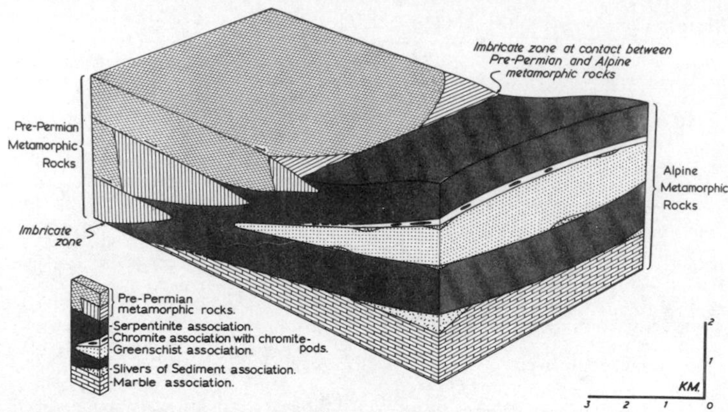

Figure 2. Postulated three-dimensional form of ophiolitic mélange and nature of contact with pre-Permian metamorphic rocks. Structure is much oversimplified in diagram, and effects of postemplacement faulting and relief have been ignored.

The probable Late Cretaceous age for the metabasites and associated cherts of the Greenschist association identifies an Alpine metamorphic event affecting some of the mélange rocks. Study of the mineral compositions and mineral parageneses in the pre-Permian rocks indicates that they were unaffected by the Alpine event, and this evidence, together with the absence of pre-Permian metamorphic rocks from the mélange, indicates that the Alpine metamorphism and deformation occurred far from the high crustal level that is the present position of the mélange. This metamorphism and deformation occurred under conditions of low temperature and high fluid pressure ($T \sim 350°C$, $P \approx P_{H_2O} \sim 4.5$ kb), and serpentinite has acted as a lubricant, probably by failure along numerous internal shear surfaces, during the internal organization of the mélange. The juxtaposition of the pre-Permian and the Alpine metamorphic rocks probably occurred after this internal organization had been achieved, because the imbricate zone of contact between them is oblique to the dip of the major thrust slices within the mélange. The juxtaposition has been accompanied by the local formation of mylonites and cataclasites in the pre-Permian rocks and the local formation of serpentinite breccias in the mélange.

STRUCTURE OF THE MASSIF AND OPHIOLITE-FLYSCH COMPLEX

The discovery of pre-Permian and Alpine metamorphic rocks in an area of the Bitlis Massif previously thought to consist only of pre-Permian metamorphic rocks raised the question of the relative importance of Alpine metamorphic rocks in the Bitlis Massif and their structural relations with other rocks. In examining these questions, the word "massif" may mislead because it appears that for many authors, there is an implication that the Bitlis Massif is a single structural entity. With a few exceptions (for example, Rigo di Righi and Cortesini, 1964), this has led to the assumption that all the metamorphic rocks of the Bitlis Massif are the same age (Altinli, 1966; Ketin, 1966). This interpretation would lead one to regard the area of ophiolitic rocks around Mutki as part of the ophiolite-flysch complex, and to interpret the structure as a window through the Bitlis Massif (Peyve, 1969).

The general interpretation of Peyve's cross section (Fig. 3) that the Bitlis Massif is thrust southward over the ophiolite-flysch complex is accepted by all geologists who have worked in the region. Yet, while the southern contact of the Bitlis Massif is regarded as a thrust, the massif is supposed to plunge eastward and westward under the same ophiolite-flysch complex (Altinli, 1966). If it is assumed that the structural relations have been correctly observed, it is clear that the Bitlis Massif cannot be a

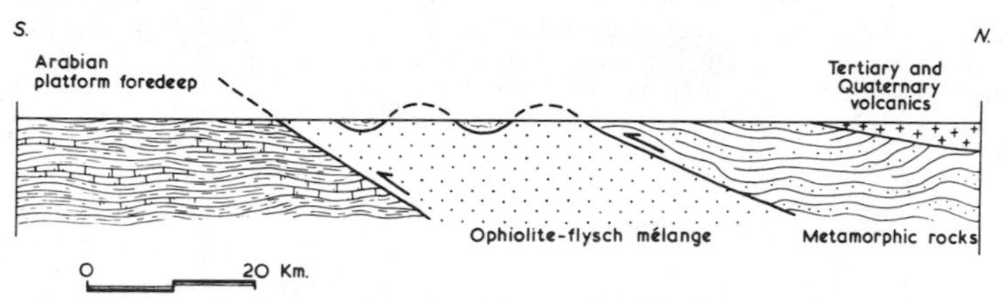

Figure 3. Generalized cross section across Bitlis Massif, ophiolite-flysch complex, and Arabian foreland in Bitlis region (after Peyve, 1969).

single structural entity, because in the south the metamorphic rocks are structurally *above* the ophiolite-flysch complex, whereas in the east and west they are structurally *below* the same complex. This conclusion is supported by examination of the 1:500,000-scale geological maps (Altinli and others, 1963, 1964), which show that metamorphic rocks must be both structurally above and structurally below the ophiolite-flysch complex.

I suggest that these large-scale structural relations are best explained by supposing the Bitlis Massif to be a number of major slices separated by thrusts. Local support for this view is provided by the detailed studies of the Bitlis (Boray, 1973) and Mutki (Hall, 1974) areas, but in order to test this suggestion and estimate the extent of Alpine metamorphic rocks in the Bitlis Massif, reconnaissance work was carried out by R. Mason and myself along the Bitlis River to the south of the Mutki area.

The Bitlis River gorge cuts deeply into the Bitlis Massif and runs about 50 km southwestward from Bitlis to Baykan. The gorge cuts across all the major structural units of the area, from the Quaternary volcanic rocks of Nemrut in the north to the Foothills structure belt in the south. Our survey indicated that in the region south of Mutki, between the Mutki mélange and the sedimentary sequences of the Foothills belt, there are no metamorphic rocks of pre-Alpine age except as exotic fragments in the flysch sequences. The survey indicated also that the ophiolite-flysch complex, which is structurally between the Mutki mélange and the Foothills belt, can be divided into three parts (Fig. 4): two wildflysch zones separated by a thick sequence of monotonous folded flysch that is generally unfossiliferous. In the structural succession, the upper wildflysch zone is immediately beneath the Mutki mélange. It consists of a matrix of flysch containing numerous large blocks of Permian marble (some are at least 1 km across, and others are probably much larger) with smaller boulders and pebbles of metamorphic rocks and marble. The Permian marbles have previously been regarded as tectonic intercalations in the flysch (Altinli and others, 1964), and in order to explain the field relations it has been suggested that some of these marbles are of Tertiary and not Permian age (P. Ibbotson and others, Turkey Mineral Research and Exploration Inst. unpub. rept.). However, sedimentary contacts between the marble blocks and the enclosing flysch can be observed, and I suggest that these marble blocks were derived from preexisting outcrops of Permian age and incorporated in the flysch sequence as olistoliths. Subsequent tectonism has resulted in the resemblance to tectonic inclusions.

Structurally below this wildflysch zone is the thick sequence of folded flysch, and below this is the second wildflysch zone. This zone differs from the zone of wildflysch with marble boulders because it contains a high proportion of ophiolitic rocks, and although blocks of marble are present, they appear to have been incorporated by a tectonic rather than a sedimentary mechanism. Therefore, this ophiolitic-wildflysch is not regarded as a repetition of the upper wildflysch zone. It is proposed here that the ophiolitic-wildflysch zone is the equivalent of the Kevan gravity nappe of Rigo de Righi and Cortesini (1964) and represents the root zone of the ophiolitic olistostromes emplaced on the Arabian foreland sequences. The olistostromes, which are exposed in the Foothills belt further west, are covered by extensive upper Tertiary deposits in the area south of Baykan.

The reconnaissance suggests that the Alpine metamorphic rocks of the Mutki mélange are present as a narrow slice structurally above the ophiolite-flysch complex to the south (Fig. 4). The reason for this arrangement is discussed below. I believe that this arrangement of pre-Permian metamorphic rocks structurally above Alpine metamorphic rocks may apply in many areas of the Bitlis Massif. Therefore, I propose that the Bitlis Massif is a composite structural entity consisting of a northern area of pre-Permian metamorphic rocks arranged as a series of thrust slices, separated by thrusts marked by dynamic metamorphism associated with Alpine tectonism, and a southern zone of Alpine metamorphic rocks that have undergone high-pressure–low-temperature metamorphism. This proposal is supported by the only other occurrence of sodic amphibole reported from the Bitlis Massif; it is also at the southern margin of the massif, in the area of Kulp (Van der Kaaden, 1966).

INTERPRETATION OF OPHIOLITIC MÉLANGE

Formation

The association of chert and metabasic and ultramafic rocks in the mélange complies with current definitions of the ophiolite suite (Dewey and Bird, 1971; Moores and Vine, 1971; Penrose Field Conference, 1972), and at present ophiolites are often regarded as fragments of ocean crust and upper mantle tectonically emplaced in orogenic belts (see, for example, Vine and Hess, 1971; Dewey and Bird, 1971). It is tempting to suggest that the Mutki ophiolitic mélange represents such a fragment. The regional geologic environment is appropriate; the serpentinites are of the alpine

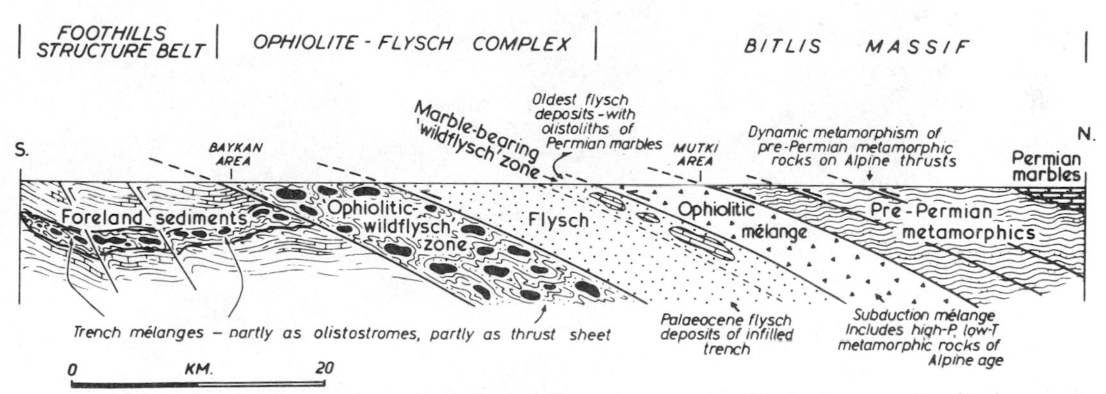

Figure 4. Main structural subdivisions of Bitlis Massif and ophiolite-flysch complex recognized in Bitlis gorge between Mutki and Baykan areas. Interpretation of main units shown in italics.

type; their constituent serpentine minerals (lizardite and chrysotile), highly tectonized character, and lack of a contact aureole, and the presence of podiform chromites are all features typically associated with alpine-type ultramafites (Thayer, 1960; Wyllie, 1967), and such ultramafites are commonly regarded as slices of upper mantle depleted by partial melting (Dietz, 1963; Moores, 1969, 1970; Maxwell, 1970; Davies, 1971). Greenbaum (1972) suggested that podiform chromite bodies may be an original feature of a layered sequence developed at the ocean-crust–upper-mantle interface. However, volcanic rocks formed at accreting plate margins normally have tholeiitic affinities (Engel and others, 1965; Miyashiro and others, 1970) and a characteristic trace-element chemistry (Pearce and Cann, 1971, 1973), whereas the metabasic rocks of the Mutki area have alkaline affinities and a quite different trace-element chemistry. I speculate that the Mutki rocks represent a volcanic island or seamount (rather than true ocean floor) underlain by upper mantle. This explains the chemical character of the mafic igneous rocks and the association of rocks of such character with alpine-type serpentinites and podiform chromites. It could also, by postulating a sub–ocean-floor metamorphic event, explain the observation that recrystallization of gabbroic blocks in the Serpentinite association appears to have occurred before their incorporation in the ultramafic rock. The age of the cherts in the Greenschist association suggests that an oceanic area existed to the south of the Bitlis Massif in Late Cretaceous time.

The regional evidence (see Plate-Tectonic Model, below) suggests that a subduction zone had been established by Late Cretaceous time. Many authors have proposed that mélanges consisting of chaotic assemblages including ophiolitic rocks are formed as ocean floor is carried down beneath plate edges in subduction zones (Hamilton, 1969; Hsü, 1971). In these mélanges, deformation is characterized by disruption, fragmentation, and mixing, and they are thought to be formed as material is scraped off the subducted plate onto the leading edge of the overriding plate. There are many resemblances in the style of deformation, nature of included lithologies, and type of metamorphism between the Mutki mélange and the Franciscan mélanges, which are interpreted as subduction mélanges (Maxwell, 1974). The metamorphism of the Greenschist association, although not reaching the unusual combination of high pressures and low temperatures of metamorphism of blueschist terranes (for example, the Franciscan), does seem to reflect an unusually low geothermal gradient, and the most likely place for such metamorphic conditions is a subduction zone. Thus, contrary to the view of Van der Kaaden (1966, p. 65) that "the Taurus geosyncline was never very deep in most of its parts and the *PT* conditions required for the formation of lawsonite-glaucophane or the glaucophanitic-greenschist facies were not reached. . .," I suggest that the apparent absence of high-pressure metamorphism in the eastern Taurus Mountains is due to the lack of detailed petrographic observations in the region.

The exotic blocks of *Orbitoides*-bearing marble occurring in the Serpentinite association also support the hypothesis of a subduction zone to the south of the Bitlis Massif.[1] The presence of clasts of serpentinite in these marbles indicates that some of the ultramafic rocks were serpentinized at this time, since no pre-Mesozoic ultramafic bodies are known in the Bitlis Massif. The angularity and size of the serpentinite clasts also indicate that rapid erosion and deposition must have occurred. However, *Orbitoides* is known to indicate relatively shallow-water environments and the fossil assemblage, which includes corals, suggests shallow, warm-water conditions such as might prevail on a continental shelf. Erosion of serpentinite near such an environment and the effect of deposition in shallow water, where current and wave activity is high, would be expected to break down large clasts quickly to produce fine rounded serpentinite fragments or serpentinite dust. Such material is absent in the blocks, but in the larger blocks a sedimentary banding can be observed and in these bands, and in smaller, apparently unlaminated blocks, a preferred orientation of microfossils is present, suggesting deposition by strong currents. All these features are explained by deposition from turbidity currents, which would carry loose material from the continental shelf onto the ocean floor or into a trench. Trenches are among the several oceanic areas from which serpentinite has been dredged (Gresens, 1970), and Fisher and Engel (1969) reported serpentinite in dredge hauls from the nearshore flank of the Tonga trench. Thus, if a turbidity current were to carry material from a continental shelf into a trench in which serpentinite was exposed, it might erode the serpentinite, which would then be deposited almost immediately as assorted angular fragments on the trench bottom. It is noteworthy that Aykulu and Evans (1974) have described very similar serpentinite-bearing and chrome spinel–bearing fossiliferous limestones of Senonian age from an area of Upper Cretaceous and Paleocene rocks at the western end of the Bitlis Massif. They noted that these limestones contain features that indicate that much of the material was deposited from turbidity currents.

[1] W.S.F. Kidd (1975, personal commun.) has suggested the alternative hypothesis that the *Orbitoides* marble was formed at the base of a large transform-fault–fracture-zone scarp, and the limestone was derived either from a reef on top of the scarp (similar to the Romanche fracture zone) or from one or more separate oceanic atolls. This would form a good sediment trap and enable a thick limestone sequence to accumulate well before disruption in the trench.

Emplacement

Timing of events in the mélange is difficult to determine without isotopic dates and further paleontological evidence. Even the simplest interpretation suggests that mixing, incorporation of blocks, serpentinization and metasomatism, and the high-pressure metamorphism of the Greenschist association all overlapped considerably in time. All were closely related events, and the regional evidence (see below) suggests that they occurred between late Maestrichtian and Miocene time. Significantly, the study of the mélange has indicated that the serpentinites have played a major role both in promoting metamorphism and metasomatism and in deformation and emplacement of the mélange. The field evidence suggests that the development of equilibrium mineral assemblages in the metabasalts of the Greenschist association has been facilitated by shearing at contacts with serpentinites. Calcium metasomatism of the metagabbro blocks of the Serpentinite association appears to have occurred as a direct result of serpentinization of the ultramafic body.

Most important, however, is the role of serpentinite during deformation and emplacement. Raleigh (1967) has shown that at temperatures between 300° and 600°C, serpentinite is considerably weakened by the generation of a high pore-fluid pressure, and he suggested that this weakening greatly facilitates the emplacement of large ultramafic bodies. A marginal envelope of serpentinite into which water has penetrated is weakened in relation to cooler surrounding country rock, and during compression and rapid sedimentary loading, the pore pressure may equal or exceed the overburden pressure and allow solid emplacement. A similar mechanism may have operated during the emplacement of the mélange. Mineralogical work on the Greenschist association rocks indicates that the fluid pressure during metamorphism was close to overburden pressure. During the final phase of compression of the eastern Taurus orogen, fluid pressures in the subduction mélange may have increased rapidly, causing the serpentinites to be suddenly weakened. Given suitable tectonic stress gradients, they may have begun moving upward, rather in the manner of a diapir, although in this case the movement was probably not vertical but inclined, since the obvious zone of weakness (and probable access) was likely to have been the dipping surface of the fossil subduction zone. Thus, the sheath of serpentinite, acting as a lubricant, may have allowed the emplacement of the relatively large blocks of the more competent metabasites enclosed in it and facilitated much of the mixing. The action of the serpentinites in this way explains the absence of cataclastic phenomena in the mélange and also its sheathlike structure (described above; see Fig. 2).

INTERPRETATION OF OPHIOLITIC-WILDFLYSCH ZONE

The allochthonous units of the Foothills belt have many features in common with the mélange of the Mutki area. Rigo de Righi and Cortesini (1964) divided these allochthonous rocks into two parts: the Besni olistostrome, which is present in the Foothills belt sedimentary sequence, and the Kevan gravity nappe, which represents the roots of the olistostrome. The lower, ophiolitic-wildflysch zone of the Baykan area is here proposed as the equivalent of the Kevan gravity nappe, and hence it may be regarded as part of the Foothills belt allochthon. The Kevan and Besni gravity slides have been differentiated by Rigo de Righi and Cortesini (1964) on the basis of their emplacement characters only and together have been divided into three main units. However, only one of the three units recognized consists predominantly of ophiolitic rocks; the two remaining units (Perdeso and Hezan) consist mainly of sedimentary rocks with subordinate exotic blocks of igneous rocks. The Perdeso unit, which is largely a mass of tectonized, variegated shale, is considered to be the shaly matrix of the slide complex. The allochthonous complex is overlain by shallow-marine sedimentary rocks and continental deposits, some of which appear to have been deposited during the emplacement of the allochthon. The chief differences between these allochthonous units and the Mutki mélange are in the proportions of igneous and sedimentary rocks, degree of metamorphism, and structural position. The Mutki mélange is now structurally *above* the flysch zone, whereas the Foothills belt allochthon is structurally *below*. The Mutki mélange consists predominantly of igneous and metaigneous rocks, whereas in the Foothills allochthon, igneous rocks are less abundant than sedimentary rocks; also, the Mutki mélange contains rocks metamorphosed under high-pressure–low-temperature conditions, whereas the Foothills allochthon is affected only locally by low-grade dynamothermal metamorphism.

I suggested above that the Mutki mélange represents a subduction mélange; I propose that the Foothills belt allochthon represents trench deposits that were not subducted. The differences between them have arisen because the Mutki mélange was subducted and therefore subjected to tectonism and metamorphism, whereas the Foothills belt allochthon was thrust, or slid, under gravity from the trench during a phase of uplift and was therefore not subducted. Several authors (for example, Oxburgh, 1974) have suggested that since oceanic sediments are not mechanically coupled to the underlying lithospheric plate, they are likely to be scraped off and will accumulate in the trench during subduction of the plate. Therefore, the subducted material will con-

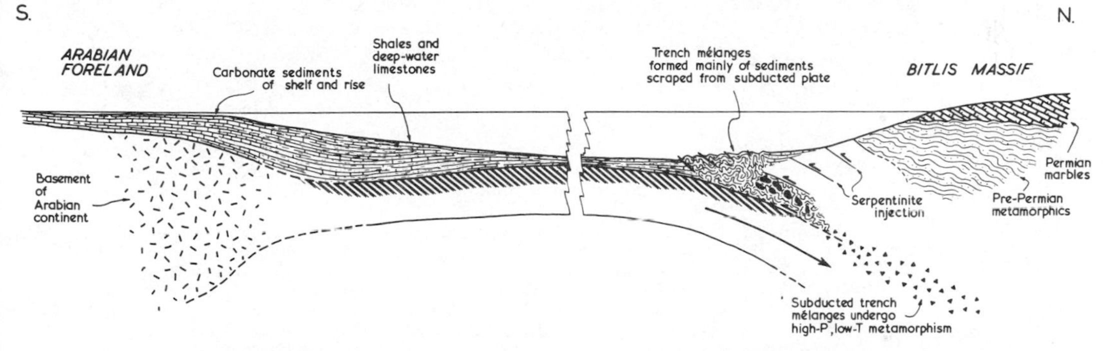

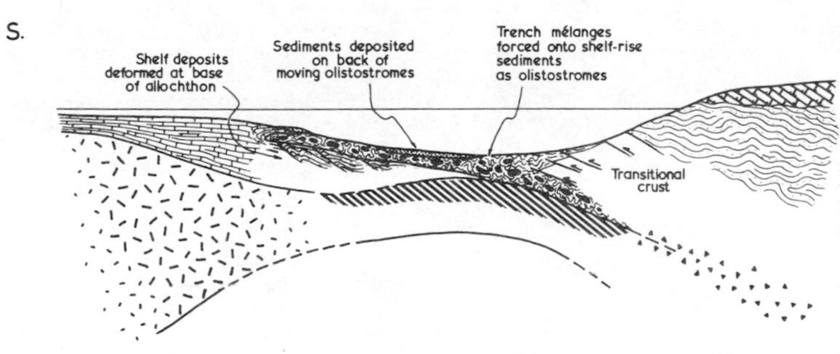

Figure 5. Cross sections illustrating main stages (discussed in text) in evolution of suture zone in southeastern Taurus Mountains. Note that although B shows olistostrome moving uphill, this is a consequence of constructing oversimplified diagrams with limited information. It is likely that temporary troughs formed during this period of compression (compare with Taconic region), but our knowledge of sedimentary history of Foothills belt is as yet inadequate to distinguish such features. Rectangle outlined in D indicates approximate position of present-day suture zone in Bitlis gorge shown in Figure 4.

sist predominantly of igneous and metamorphic rocks (the Mutki mélange), whereas the trench deposits will consist predominantly of sedimentary rocks (the Foothills belt allochthon).

PLATE-TECTONIC MODEL

At present, the regional geology of Turkey is poorly known, and attempts to interpret the geology in terms of the Mesozoic and Tertiary history of plate movements are restricted to general syntheses of the Alpine system (Smith, 1971; Dewey and others, 1973). McKenzie (1970) has shown that the present-day seismicity of the Mediterranean region can be interpreted in terms of motion between a complex pattern of small plates, and Dewey and others (1973) have attempted to identify the microplate pattern that existed during the evolution of the Alpine system. However, the complex distribution of ophiolitic mélanges and metamorphic massifs suggests that even this relatively sophisticated plate-tectonic model is oversimplified for the Anatolian region. Knowledge of the regional geology is not yet sufficient to give a three-dimensional account of the development of the eastern Taurus Mountains; it is probable that the history of plate interaction varied considerably in both space and time in the orogen. Therefore, I propose an essentially two-dimensional model for the Mesozoic and Tertiary history of the plate interaction for the Bitlis-Baykan section of the southeastern Taurus. Schematic cross sections illustrating the main stages in the development of the model discussed below are given in Figure 5.

Permian to Late Cretaceous Time

In continental reconstructions for the Permian Period, the position of present-day Turkey is difficult to determine. Most authors (Dietz and Holden, 1970; Smith, 1971; Dewey and others, 1973) have considered that a large oceanic area (Tethys) separated Eurasia and Africa-Arabia, and the older massifs of Turkey are generally assigned to one or another of these continental areas. Little can be said about the position of the pre-Permian metamorphic rocks of the Bitlis Massif at this time. However, it should be noted that a simple fit of the Bitlis Massif to the present-day Arabian foreland basement is not possible. The pre-Permian rocks of the Bitlis Massif appear to have been metamorphosed during early Paleozoic time (?Silurian; Yilmaz, 1971), whereas on the Arabian foreland an almost complete unmetamorphosed sequence of sedimentary rocks ranging in age from Miocene to fossiliferous Middle Cambrian rests on probable Precambrian volcanic rocks (Flügel, 1971; Van der Kaaden, 1971).

During late Paleozoic time the Bitlis Massif was submerged, and limestones of Permian age were deposited extensively. These are probably represented in the Mutki area by the pure-white recrystallized marbles of the Marble association. Elsewhere in the Bitlis Massif, the limestones are occasionally fossiliferous, and the faunal assemblages indicate a warm shallow-water environment at this time. No Mesozoic rocks are found deposited in the area of the Bitlis Massif, but on the Arabian foreland, continuous carbonate deposition commenced in Late Triassic time and continued until the end of the Jurassic. If the Bitlis Massif were north of the Arabian foreland at this time, it is possible that subsequent erosion has removed all traces of sediments deposited on it, although an equally plausible alternative is that the Bitlis Massif was a continental area from Permian time onward (Tolun, 1960).

At some time prior to the Late Cretaceous (Fig. 5A), oceanic development began between the Bitlis Massif and the Arabian foreland. In the Mutki area the only evidence for the age of the oceanic rocks is the date obtained from the radiolarian cherts indicating deposition in Late Cretaceous time. Dewey and others (1973) considered that oceanic development occurred in this region during the Late Triassic, but there is no evidence in southeastern Turkey to support this date. Altinli and others (1963) noted that the lower age limit of blocks in the ophiolite-flysch complex is Late Jurassic, and Rigo de Righi and Cortesini (1964) postulated that radiolarite and limestone of Jurassic age occurring as blocks in the allochthonous units of the foreland indicate deep-water conditions, and they suggested that igneous activity took place in the mobile belt from Late Jurassic to Late Cretaceous time. Cordey (1971, p. 319) noted that "the late Jurassic or early Cretaceous saw the onset of a major marine transgression with carbonate deposition extending over all of south-east Turkey." Thus, the limited evidence indicates that an ocean was established in the region at some time around Late Jurassic to Early Cretaceous.

The nature of the oceanic area is uncertain. Dewey and others (1973) implied that a continuous oceanic area existed between the Bitlis region and the Oman region, and for the Oman region Glennie and others (1973) have estimated a minimum distance

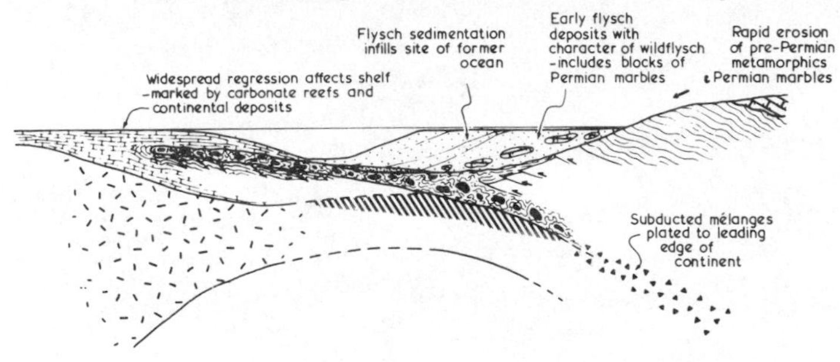

C. Palaeocene : site of former ocean filled by flysch deposits.

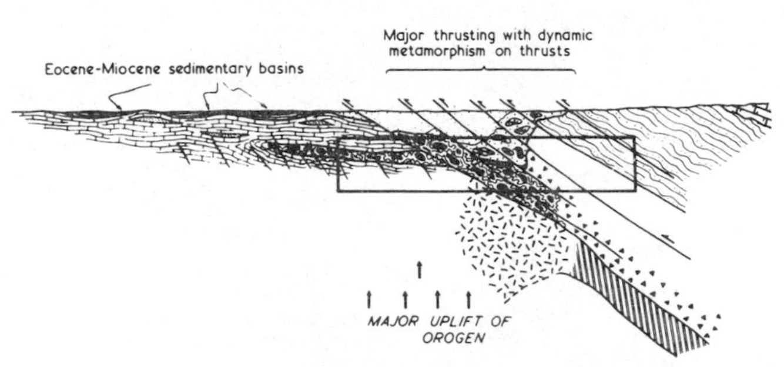

D. Final phase of continental collision.

Figure 5. *(continued).*

between the spreading center and the Arabian foreland of 400 to 1,200 km. Other work hints at the complexity of this oceanic area. For example, Takin (1972) has suggested that small microcontinents, separated by oceanic areas, are today represented by the areas of metamorphic rocks and the "colored mélange" of Iran: This hypothesis of microcontinents is particularly appealing when applied to Turkey, because it offers a very fruitful working model for interpreting the areas of metamorphic massifs surrounded by ophiolitic mélanges that are common in Turkey. With this model, the discontinuous metamorphic massifs of the eastern Taurus Mountains may be regarded as continental fragments separated from the Arabian foreland and situated within the oceanic area stretching between the Bitlis and Oman regions.

The evidence from the Mutki area and the Foothills belt indicates that a subduction zone, situated on the southern edge of the Bitlis Massif, had developed by Late Cretaceous time (Fig. 5A).

Late Cretaceous Time

The dating of the end of subduction is difficult, but the evidence to be considered is as follows: (1) None of the few datable components in the Mutki mélange are younger than late Maestrichtian. (2) The ophiolite-flysch complex to the south contains components that range in age from Late Jurassic to Paleocene. Flysch sedimentation appears to have begun in Late Cretaceous time (Altinli and others, 1963, 1964). (3) The ophiolitic gravity slides of the Foothills belt contain components that range in age from Late Jurassic to Late Cretaceous. They are overlain by sedimentary rocks of Late Cretaceous age, and the gravity slides are considered to have been emplaced in Campanian time (Rigo de Righi and Cortesini, 1964). (4) A regional regression, representing the end of the Upper Cretaceous–lower Tertiary sedimentary sequence in the Foothills belt, occurred in Eocene time. The regression is marked by local carbonate reefs and appears to be related to extensive uplift to the north and east (Rigo de Righi and Cortesini, 1964). (5) A profound unconformity separates Eocene and Upper Cretaceous sedimentary rocks in the Palu area at the western end of the Bitlis Massif. The unconformity is regarded as Maestrichtian-Paleocene in age and is correlated with major uplift, folding, and thrusting in that region (Aykulu and Evans, 1974).

This evidence indicates that a major tectonic event occurred during Late Cretaceous and early Tertiary time (Fig. 5B); I consider this event to be the cessation of subduction and uplift of the trench zone associated with the beginning of the collision between the Arabian landmass and the Bitlis Massif. The limited data suggest that the timing of this collision may have varied along the orogenic belt, being earlier in the west, but this impression may be due to the paucity of information in this complex suture zone. However, although more detailed studies are required to clarify this postulated evolution, it appears that the cessation of subduction occurred rather earlier in this part of the Alpine system than the Pliocene date suggested by Dewey and others (1973).

Late Cretaceous to Paleocene Time

Rapid uplift was not restricted to the trench zone; it also affected the Bitlis Massif and the Foothills belt and resulted in the filling of the ocean site by flysch (Fig. 5C). I propose that the rapid erosion accompanying uplift of the massif caused the deposition of an early wildflysch containing large exotic blocks of the Permian marble cover and pebbles of the underlying metamorphic rocks. The wildflysch zone is now structurally beneath the subduction mélange of the Mutki area and is probably the oldest part of the Upper Cretaceous–Paleocene flysch sequence. Later, as erosion and uplift slowed, the flysch sequences became monotonous, and they contain no exotic components. Finally, as the trough representing the former ocean was filled, flysch sedimentation ceased (Paleocene time), and the shallowing of the trough is reflected further south by the regional regression affecting the shelf sequences. Continental red beds and shallow-marine carbonate reefs form the lower Eocene deposits of the foreland.

During the remainder of Eocene and Oligocene time, shallow-water carbonates and evaporites were formed in the foreland area, but toward the end of the Oligocene some tectonism occurred (Rigo de Righi and Cortesini, 1964) which subsequently controlled the early Miocene sedimentation. Although little orogenic activity appears to have taken place during this period, it seems likely that metamorphism, metasomatism, and serpentinization occurred at this time in the Mutki mélange.

Miocene to Holocene Time

The final phases of continental collision in the eastern Taurus Mountains took place during Miocene time (Fig. 5D). They are marked by the ending of marine sedimentation, the thrusting of the Bitlis Massif over the flysch trough, and the thrusting of both onto the foreland sedimentary rocks of the Foothills belt. The date of this tectonic event is late Miocene and is fixed by the abundant evidence of lower Miocene sedimentary rocks overthrust by older rocks from the north. Rigo de Righi and Cortesini (1964) considered the southern tectonic line (the Maden-Bitlis line) to be a regional gliding surface rather than an overthrust. This surface, which separates the ophiolite-flysch complex from the Foothills belt, is nearly horizontal, and the Miocene shales beneath are affected by only minor tectonism. However, the upper tectonic surface, which separates the pre-Permian and Alpine metamorphic rocks in the Mutki area, must be a thrust. Overthrusting was accompanied by imbrication and dynamic metamorphism of the pre-Permian rocks.

Uplift and erosion of the orogen during Pliocene and Quaternary time have produced extensive thick deposits of continental clastic rocks in the region and have exposed the suture. The present-day plate motion in the area is still mainly convergent (McKenzie, 1970) and is marked by earthquake and volcanic activity. Although the centers of Nemrut and Suphan have erupted in historical times, they are now dormant, and present-day volcanic activity is mainly restricted to hot springs, many of which are producing CO_2 and depositing travertine. It is not yet possible to say whether these springs represent the continuation of the CO_2 metasomatism that resulted in the silica-carbonate rocks of the Mutki area.

CONCLUSIONS

This study suggests that the Bitlis Massif is a complex structural entity that includes both pre-Permian and Alpine metamorphic rocks. High-pressure–low-temperature metamorphic rocks of Alpine age are probably restricted to the southern edge of the Bitlis Massif, and they are thought to be subduction mélanges plated to the edge of the continental area represented by the Bitlis Massif. It seems likely that mapping of these mélanges will be more successful if *rock associations*, rather than lithological units, are used as fundamental mappable units. Tracing the distribution of these mélanges and their relationship to the ophiolite-flysch complex to the south should help in the recognition of major tectonic events in the suture zone and provide important evidence of the space and time relations of subduction and continental collision in the southeastern Taurus region.

The early history and character of the oceanic area south of the Bitlis Massif are still very poorly known, but this study indicates that two significant tectonic events, both marked by the emplacement of ophiolitic mélange, can be identified during the later phases of ocean development and continental collision. The first of these events occurred in Late Cretaceous time: ophiolitic mélange was emplaced southward by gravitational gliding tectonics, probably in a subaqueous environment, and its emplacement appears to be associated with the end of subduction of oceanic lithosphere in the region. The major uplift that occurred at this time has been documented in many other parts of the orogenic belt from Cyprus to Oman and is marked in a variety of ways. The Mamonia nappes, composed of sedimentary rocks of predominantly continental-margin facies (Robertson and Hudson, 1974), were emplaced in Cyprus; chaotic ophiolitic nappes were emplaced in south-eastern Turkey and in Iran (Gansser, 1959; Ricou, 1971), while in Oman a sequence of nappes, including a well-ordered ophiolite nappe (Reinhardt, 1969; Glennie and others, 1973), was emplaced. If, as suggested above, the southern Tethys were a discontinuous oceanic area (see also Takin, 1972; Robertson and Hudson, 1974), it is possible that this tectonic activity marks the change in relative motions between major plates (Dewey and others, 1973) that had unique manifestations in each ocean basin. Therefore, the limits and extent of these ocean basins must be more fully defined before regional and tectonic correlations along the orogenic belt can be made with any certainty. The second major tectonic event in the southeastern Taurus indicated by this study occurred in Miocene time. This is considered to be the final phase of continental collision and is marked by the emplacement of ophiolitic mélange, including high-pressure–low-temperature metamorphic rocks, at high levels in the crust. The ophiolites were emplaced by thrusting, in which serpentinites played an important role.

ACKNOWLEDGMENTS

The work reported here was carried out as part of a doctoral thesis at University College London (UCL). Financial support during the work was provided by the Royal Society, the Natural Environment Research Council, and the Department of Geology, UCL. I am grateful to the Institute of Mineral Research and Exploration (MTA), Ankara, for much assistance. I thank R. Mason, who supervised this work, and M. K. Wells for their advice, criticism, and encouragement. I also thank other members of the academic and technical staff of the Department of Geology, UCL. I am grateful to my many Turkish friends for much assistance, especially to E. Aydin, A. Boray, A. Erdoğan, and F. Oktay for discussions and help.

REFERENCES CITED

Altinli, I. E., 1966, Geology of eastern and south-eastern Anatolia, Part II: Turkey Mineral Research and Explor. Inst. Bull., v. 67, p. 1–22.

Altinli, I. E., Pamir, H. N., and Erentöz, C., 1963, Explanatory text of the geological map of Turkey — Erzurum: Ankara, Turkey Mineral Research and Explor. Inst., 131 p.

——1964, Explanatory text of the geological map of Turkey — Van: Ankara, Turkey Mineral Research and Explor. Inst., 90 p.

Aykulu, A., and Evans, A. M., 1974, Structures in the Iranides of south-eastern Turkey: Geol. Rundschau, v. 63, p. 292–305.

Bailey, E. B., and McCallien, W. J., 1953, Serpentine lavas, the Ankara mélange and the Anatolian thrust: Royal Soc. Edinburgh Trans., v. 62, p. 403–442.

Barnes, I., O'Neil, J. R., Rapp, J. B., and White, D. E., 1973, Silica-carbonate alteration of serpentine: Wall rock alteration in mercury deposits of the California Coast Ranges: Econ. Geology, v. 68, p. 388–398.

Boray, A., 1973, The structure and metamorphism of the Bitlis area, south-east Turkey [Ph.D. thesis]: London, London Univ., 233 p.

Cordey, W. G., 1971, Stratigraphy and sedimentation of the Cretaceous Mardin formation in south-eastern Turkey, *in* Campbell, A. S., ed., Geology and history of Turkey: Petroleum Explor. Soc. Libya, 13th ann. field conf., p. 317–348.

Davies, H. L., 1971, Peridotite-gabbro-basalt complex in eastern Papua: An overthrust plate of oceanic crust and mantle: Australia Bur. Mineral Resources, Geology and Geophysics Bull. 128, 48 p.

Dewey, J. F., and Bird, J. M., 1971, Origin and emplacement of the ophiolite suite: Appalachian ophiolites in Newfoundland: Jour. Geophys. Research, v. 76, p. 3179–3206.

Dewey, J. F., Pitman, W. C., III, Ryan, W.B.F., and Bonnin, J., 1973, Plate tectonics and the Alpine system: Geol. Soc. America Bull., v. 84, p. 3137–3180.

Dietz, R. S., 1963, Alpine serpentinites as oceanic rind fragments: Geol. Soc. America Bull., v. 74, p. 947–952.

Dietz, R. S., and Holden, J. C., 1970, Reconstruction of Pangea: Break-up and dispersion of the continents, Permian to present: Jour. Geophys. Research, v. 75, p. 4939–4957.

Engel, A. E., Engel, C. G., and Havens, R. G., 1965, Chemical characteristics of oceanic basalts and the upper mantle: Geol. Soc. America Bull., v. 76, p. 719–734.

Ernst, W. G., Seki, Y., Onuki, H., and Gilbert, M. C., 1970, Comparative study of low-grade metamorphism in the California Coast Ranges and the outer metamorphic belt of Japan: Geol. Soc. America Mem. 124, 276 p.

Fisher, R. L., and Engel, C. G., 1969, Ultramafic and basaltic rocks dredged from the near-shore flank of the Tonga trench: Geol. Soc. America Bull., v. 80, p. 1373–1378.

Flügel, H. W., 1971, Palaeozoic rocks of Turkey, *in* Campbell, A. S., ed., Geology and history of Turkey: Petroleum Explor. Soc. Libya, 13th ann. field conf., p. 211–224.

Gansser, A., 1959, Ausseralpine Ophiolithprobleme: Eclogae Geol. Helvetiae, v. 52, p. 659–680.

Glennie, K. W., Boeuf, M.G.A., Hughes Clark, M. W., Moody-Stuart, M., Pilaar, W.F.H., and Reinhardt, B. M., 1973, Late Cretaceous nappes in the Oman Mountains and their geologic evolution: Am. Assoc. Petroleum Geologists Bull., v. 57, p. 5–27.

Greenbaum, D., 1972, Magmatic processes at ocean ridges: Evidence from the Troodos Massif, Cyprus: Nature Phys. Sci., v. 238, p. 18–21.

Greenwood, H., 1967, Mineral equilibria in the system MgO-SiO_2-H_2O-CO_2, *in* Abelson, P. H., ed., Researches in geochemistry, Vol. 2: New York, John Wiley & Sons, p. 542–567.

Gresens, R. L., 1970, Serpentinites, blueschists, and tectonic continental margins: Geol. Soc. America Bull., v. 81, p. 307–310.

Hall, R., 1973, Rock association mapping: Cong. Earth Sciences, 50th anniversary Turkish Republic, Proc. (in press).

——1974, The structure and petrology of an ophiolitic mélange near Mutki, Bitlis province, Turkey [Ph.D. thesis]: London, London Univ., 351 p.

Hall, R., and Mason, R., 1972, A tectonic mélange from the Eastern Taurus Mountains, Turkey: Geol. Soc. London Jour., v. 128, p. 395–398.

Hamilton, W., 1969, Mesozoic California and the underflow of Pacific mantle: Geol. Soc. America Bull., v. 80, p. 2409–2430.

Haynes, S. J., and McQuillan, H., 1974, Evolution of the Zagros suture zone, southern Iran: Geol. Soc. America Bull., v. 85, p. 739–744.

Hsü, K. J., 1968, Principles of mélanges and their bearing on the Franciscan-Knoxville paradox: Geol. Soc. America Bull., v. 79, p. 1063–1074.

—1971, Franciscan mélanges as a model for eugeosynclinal sedimentation and under-thrusting tectonics: Jour. Geophys. Research, v. 76, p. 1162–1170.

Ilhan, E., 1967, Toros-Zagros folding and its relation to Middle East oil-fields: Am. Assoc. Petroleum Geologists Bull., v. 51, p. 651–667.

Kamen-Kaye, M., 1971, A review of depositional history and geological structure in Turkey, *in* Campbell, A. S., ed., Geology and history of Turkey: Petroleum Explor. Soc. Libya, 13th ann. field conf., p. 111–137.

Ketin, I., 1966, Tectonic units of Turkey: Turkey Mineral Research and Explor. Inst. Bull., v. 66, p. 23–34.

Maxwell, J. C., 1970, The Mediterranean, ophiolites, and continental drift, *in* Johnson, H., and Smith, B. L., eds., The megatectonics of continents and oceans: New Brunswick, N.J., Rutgers Univ., p. 167–193.

——1974, Anatomy of an orogen: Geol. Soc. America Bull., v. 85, p. 1195–1204.

McKenzie, D. P., 1970, Plate tectonics of the Mediterranean region: Nature, v. 226, p. 239–243.

Meriç, E., 1973, Sur la presence d'un petit affleurement du Maestrichtien supérieur au S de Mutki (Bitlis-Turquie): Istanbul Univ. Rev. Fac. Sci., sec. B., v. 38, p. 49–51.

Miyashiro, A., Shido, F., and Ewing, M., 1970, Crystallisation and differentiation in

abyssal tholeiites and gabbro from mid-oceanic ridges: Earth and Planetary Sci. Letters, v. 7, p. 361–365.
Moores, E. M., 1969, Petrology and structure of the Vourinos ophiolitic complex of northern Greece: Geol. Soc. America Spec. Paper 118, 74 p.
——1970, Ultramafics and orogeny, with models of the U.S. Cordillera and the Tethys: Nature, v. 228, p. 837–842.
Moores, E. M., and Vine, F. J., 1971, The Troodos Massif, Cyprus and other ophiolites as oceanic crust: Evaluation and implications: Roval Soc. London Philos. Trans., ser. A., v. 268, p. 443–466.
Oxburgh, E. R., 1974, The plain man's guide to plate tectonics: Geol. Assoc. London Proc., v. 85, p. 299–357.
Pearce, J. A., and Cann, J. R., 1971, Ophiolite origin investigated by discriminant analysis using Ti, Zr, and Y: Earth and Planetary Sci. Letters, v. 12, p. 339–349.
——1973, Tectonic setting of basic volcanic rocks determined using trace element analyses: Earth and Planetary Sci. Letters, v. 19, p. 290–300.
Penrose Field Conference, 1972, Report of conference on ophiolites: Geotimes, v. 17, no. 12, p. 24–25.
Peyve, A. V., 1969, Oceanic crust of the geologic past: Geotectonics, v. 4, p. 210–223.
Raleigh, C. B., 1967, Experimental deformation of ultramafic rocks and minerals, *in* Wyllie, P. J., ed., Ultramafic and related rocks: New York, John Wiley & Sons, p. 191–199.
Reinhardt, B. M., 1969, On the genesis and emplacement of ophiolites in the Oman Mountains geosyncline: Schweizer. Mineralog. u. Petrog. Mitt., v. 49, p. 1–30.
Ricou, L. E., 1971, Le croissant ophiolitique péri-arabe: Une ceinture de nappes mises en place au crétacé supérieur: Rev. Géographie Phys. et Géologie Dynam., v. 13, p. 327–349.
Rigo de Righi, M., and Cortesini, A., 1964, Gravity tectonics in Foothills structure belt of south-east Turkey: Am. Assoc. Petroleum Geologists Bull., v. 48, p. 1911–1937.
Robertson, A.H.F., and Hudson, J. D., 1974, Pelagic sediments in the Cretaceous and Tertiary history of the Troodos Massif, Cyprus, *in* Hsü, K. J., and Jenkyns, H. C., eds., Pelagic sediments: On land and under the sea: Internat. Assoc. Sedimentologists Spec. Pub. 1, p. 403–436.
Smith, A. G., 1971, Alpine deformation and the oceanic areas of the Tethys, Mediterranean, and Atlantic: Geol. Soc. America Bull., v. 82, p. 2039–2070.
Takin, M., 1972, Iranian geology and continental drift in the Middle East: Nature, v. 235, p. 147–150.
Temple, P. G., and Perry L. J., 1962, Geology and oil occurrence, southeast Turkey: Am. Assoc. Petroleum Geologists Bull., v. 46, p. 1596–1612.
Ternek, Z., 1953, Geological study southeastern region of Lake Van: Geol. Soc. Turkey Bull., v. 2, p. 28–32.
Thayer, T. P., 1960, Some critical differences between alpine-type and stratiform peridotite-gabbro complexes: Internat. Geol. Cong., 21st, Copenhagen 1960, sec. 13, p. 247–259.
Tolun, N., 1953, Contributions à l'étude géologique des environs du sud et sud-ouest du Lac de Van: Turkey Mineral Research and Explor. Inst. Bull., v. 44/45, p. 77–112.
——1960, Stratigraphy and tectonics of southeastern Anatolia: Instanbul Univ. Rev. Fac. Sci., sec. B., v. 25, p. 203–264.
Van der Kaaden, G., 1966, The significance and distribution of glaucophane rocks in Turkey: Turkey Mineral Research and Explor. Inst. Bull., v. 67, p. 36–67.
——1971, Basement rocks of Turkey, *in* Campbell, A. S., ed., Geology and history of Turkey: Petroleum Explor. Soc. Libya, 13th ann. field conf., p. 191–210.
Vine, F. J., and Hess, H. H., 1971, Sea-floor spreading, *in* Maxwell, A. E., ed., The sea, Vol. 4: New York, Wiley-Interscience, p. 587–622.
Winkler, H.G.F., 1967, Petrogenesis of metamorphic rocks (2nd ed.): Berlin, Springer-Verlag, 237 p.
Wyllie, P. J., 1967, Review, *in* Wyllie, P. J., ed., Ultramafic and related rocks: New York, John Wiley & Sons, p. 403–416.
Yilmaz, O., 1971, Etude pétrographique et géochronologique de la région de Cacas [Ph.D. thesis]: Grenoble. France, Grenoble Univ., 230 p.

Manuscript Received by the Society May 28, 1975
Revised Manuscript Received October 27, 1975
Manuscript Accepted January 12, 1976

Editor's Comments on Papers 21 and 22

21 **McCALL**
Mélanges of the Makran, Southeastern Iran

22 **LIPPARD et al.**
Mélanges Associated with the Semail Ophiolite in the Northern Oman Mountains Allochthon, Southwest Arabia

The Coloured Mélange of the Iranian region is well known (Gansser, 1955). There are numerous descriptions of mélanges from Iran, but unfortunately few of these are in a form suitable for facsimile reproduction, being contained in long regional reports, in theses, and as incidental paragraphs in papers dealing with subjects otherwise not relevant to this volume. A special summary review of the Makran colored mélanges (there are two types) and the sedimentary mélange from the same mountain belt is presented as Paper 21. It has been prepared by the volume editor from seven reports compiled by himself (McCall, in press) and shortly to be published by the Geological and Mineral Survey of Iran. Amplifying this summary review, brief comments are given here on the principal publications covering this and other developments in Iran, available in the older literature.

Huber (1952) initially described and illustrated colored mélange occurrences near Minab, during a reconnaissance survey of the western Makran related to the oil search, and Gansser (1955), who was in the same team of geologists but working in the Bandar Abbas sector, later defined the term in an international publication. The essential features—pillow lavas, andesite, serpentinite, other ultrabasic rocks, gabbro, amphibolite, red radiolarian chert, pink Globotruncana limestone, sandstone, siltstone, shale, minor breccia and conglomerate, exotic reefal limestone and metamorphic rock bodies, in a chaotic assemblage—were then established for this variegated mappable zone, which looks as if the outcrops have been splashed with all the colors from a paint box. Ricou (1971*a, b;* 1974) described the mélange occurrences from Neyriz in the Zagros sector. He summarized the mélange there as a ground up zone where one finds chaotically assembled and strongly tectonized shreds that appear to be torn off from a diverse series: shreds of formations of the Pichakun nappe, a series of interbedded radiolarites, pelagic limestones, and lava—without doubt originally near Pichakun—blocks of the Megalodon limestone and fusulinid limestone, blocks of metamorphic rocks (marbles, amphibolites, biotite schists), and serpentinite. There is also a thick sheet, several kilometers wide, of peridotite and gabbro associated with crystalline limestone with contact skarns (such contact metamorphism is generally unknown in the Coloured Mélange). Ricou attributed the chaotization to tectonic agencies, to the effect of the overriding by the ultrabasics and gabbros, churning up an intermediate layer between them and the allochthon. He ruled out magmatic or sedimentary origin for the mélange. The mélange is overlain by Maastrichtian rudistic limestone cover, and the mélange within the autochthon beneath must have been formed by the late Cretaceous (Campanian). It is important to note that this occurrence, together with those of Kermanshah to the northwest and Oman to the south, is now believed to represent the southwestern convergent margin of the Southern Tethys Ocean, whereas the Makran occurrences represent the northeastern convergent margin (McCall and Kidd, 1980; see also Section 14 on plate tectonic aspects by Kidd and McCall in *Report on East Iran Project Area No. 1* [McCall, in press]).

The following three descriptions are all from colored mélanges that appear to represent neither the southwestern margin of Southern Tethys nor the northeastern margin (represented by the Coloured Mélange complex discussed in Paper 21, developed through the Makran); rather, they appear to represent a pattern of riftlike marginal basins, subsidiary to the main Southern Tethys Ocean. Davoudzadeh (1969) described a 40 X 12 km colored mélange zone from Nain in central Iran. Peridotite and serpentinite are associated with less common pyroxenite and diabase, rodingite occurs in the ultrabasic rocks, and limestone and radiolarite form blocks from meter to hundreds of meters in maximum dimension. There are some large peridotite masses up to several kilometers in maximum dimension. The red *Globotruncana* limestone is of Campanian-Maastrichtian age and contains thin chert layers. Davoudzadeh recognized Lower Eocene flysch and *Nummulite-Alveolina* limestones caught up in the mélange as youngest components. Such young components have also been noted from mélanges in other areas by Huber (1978). These occurrences led Davoudzadeh to attribute the mélange to Eocene and later tectonic disturbances, but such inclusions may be a minor, late tectonic introduction subsequent to mélange formation, which seems, in the case of the Coloured Mélange, to have occurred not later than Campanian (in the mélanges of the southwestern margin of Southern

Tethys) or early Paleocene (in the case of the mélanges of the northeastern margin of Southern Tethys). Davoudzadeh sugested that the ophiolitic components are of Paleocene-early Eocene age, but it seems more likely that they are no younger than early Paleocene. An unfortunate aspect of this study is the failure to discriminate between diabase (of minor intrusions) and pillow basalts ("all fine to medium grained green basic rocks were designated as diabases"). No discussion of the mechanism of mélange formation was given, but a tectonic interpretation and cold emplacement of the ultrabasic bodies were favored.

Sabzehei (1974) described colored mélanges from Esfandaghegh. He initially listed certain overall facts about colored mélanges in Iran. Of particular interest is the statement that most such zones are situated in the midst of cratonic blocks or platforms. This reflects the unusual riftlike or marginal basin situation of many of these zones. A second interesting point is that deformation in the Zagros was considered to be essentially of late Cretaceous age, whereas the development in other zones continued after the Eocene. This is true in a sense—there was major tectonic deformation in the Makran and central Iran in the Miocene, for example—but it is important to distinguish the actual mélange formation in the colored mélanges from superimposed tectonic effects including the incorporation of minor tectonic slivers of Cenozoic sediments. Sabzehei noted that at Esfandaghegh the Cenozoic rocks are faulted against the mélange, mostly consist of Eocene flysch, and are partly imbricated with the mélange. This is exactly the relationship seen in the Makran, except that there are also unconformities. It seems likely that mélange formation occurred prior to the Eocene, despite these imbrications, though some purely tectonic mélanges (for example, those at Sabzevar described by Lensch, Mihm, and Alavi-Tehrani, 1977) may not have adopted mélange character until the Cenozoic. Sabzehei recorded great ultrabasic bodies associated with cumulate sequences including olivine, pyroxene, and plagioclase cumulates and small bodies of serpentinite. The mélange assemblage otherwise consists of *Globotruncana* limestone, red radiolarite and red argillite, graywacke, shale, basic pillow lavas with micrite containing *Globotruncana* in the interstices, and diabase feeders to the lavas. This assemblage has undergone high pressure metamorphism, and glaucophanitic assemblages are present. He recorded limestone lenses up to 100 m long and beds of limestone up to 2 km long. The intimate association of volcanics, radiolarites, pelagic limestones, and argillites is regarded by him as a sort of matrix, an intimately associated sequence. Serpentinite may well be the only introduced component (other than minor exotics) introduced as slices on planes of movement. The minor exotic blocks (Jurassic sediments, metamorphics) are believed to have been introduced after solidification, like the ultrabasic introductions; they are, however, related to very late post-Eocene tectonic deformations. No evidence of synsedimentary deformations was found and, though they cannot be excluded utterly, olistostrome mechanisms were not favored by Sabzehei, who followed Ricou (1971*a,* 1974) in accepting a tectonic origin.

Lensch et al. (1977) described tectonic mélanges from Sabzevar (in another rift zone or marginal basin, see McCall and Kidd, 1980), which exist only in planes of tectonic motion—that is, a fine mixture of ophiolitic rocks with isolated larger fragments, mainly in zones following the regional strike, is produced. This description matches the description of the $KPe^{cm/ma}$ mélange in minor developments in the Inner Makran rift zone or marginal basin (Paper 21). However, there may have been limited subduction in these zones (McCall and Kidd, 1980), and not all the mélanges of such zones need necessarily represent such simple tectonic deformation zones in ophiolite sequences.

The Makran mélanges (Paper 21) include the main Coloured Mélange complex. Though the semantics of description differ and the metamorphism is very weak and entirely static hydrothermal, the model erected is not dissimilar to that of Sabzehei (1974), though this mélange is that of the main northeastern subduction zone of Southern Tethys. The cognate sequence of pillow lavas, radiolarites; micrites, and turbidites appears locally to be sequential and little disturbed. Serpentinite exotics have been introduced, with increasing chaotization, lubricating the planes of separation, and again there are large ultrabasic bodies including cumulates. The most surprising aspect is the anomaly of paleontological ages between the radiolarites and *Globotruncana* limestones, which is not understood. There are also early Paleocene globigerinid oozes represented (packed biomicrites). The pelagic sediments and pillow lavas appear to form a sedimentary-volcanic sequence, despite the age anomalies. Reefal limestones and metamorphic rocks are exotics, also introduced. The mélange is essentially a block-to-block mélange, and all the evidence indicates tectonic deformation, related to subduction, prior to the Eocene as the origin, though again one cannot rule out some ancillary olistostrome involvement. The mélange appears to form a series of offscraped slabs.

The later sedimentary mélange situated to the south and west is also related to subduction, but in the late middle Miocene. It is called sedimentary because of its dominant flysch content, but the origin is tectonic, due to the boudinage and disruption of competent beds and rafts in the flysch, plastic deformation of the incompetent beds, dislocation of the stratigraphy, and upward protrusion of exotic blocks, up to kilometers in maximum dimension, from the substratum (mainly the older Coloured Mélange). The mélange is related to schuppen structures affecting an accretionary prism of the older Coloured Mélange, capped by flysch, during renewed subduction.

A minor development of a special type of colored mélange, formed in the tectonic movement planes from the ophiolites of the Inner Makran rift zone or marginal basin, is also recorded, constituting yet a third type of mélange in the Makran.

The mélanges of the Oman have not been previously covered in a single comprehensive treatment, and a special review has been prepared for this volume by Lippard et al. (Paper 22). There are three mélanges of quite distinct character. The lower Hawasina mélange, which contains no ophiolite fragments, is considered to be largely sedimentary in origin on the evidence of contained debris flows (olistostromes), but it has suffered subsequent tectonic modification overprinting the primary structure. It is related to Cretaceous subduction. The second, Semail mélange, has a sheared serpentinite matrix, is clearly of tectonic origin from the field evidence, and is connected to the emplacement of the ophiolite nappe, having a sole relationship to it. The mechanism suggested resembles that described by Graciansky (Paper 19) from Lycia. It was formed in the late Cretaceous. The third, Batinah mélange, is enigmatic; it overlies the Semail ophiolite and is partly of block-to-block mélange character. It is suggested that it formed by protrusion through fault zones in the late stage of ophiolite emplacement. It includes ophiolite and exotic components. The authors accept that this mélange may be the subject of a revised interpretation consequent on further studies ongoing at present. The review illustrates the diversity of origin of mélanges, even in a relatively small area, and the foolishness of attempting to refer attribution of all mélanges to a single genetic process, and of even attributing a single occurrence necessarily to a single process.

REFERENCES

Davoudzadeh, M., 1969, Geologie und Petrographie des Gebeites nördlich von Nain, Zentral-Iran, *Zürich (ETH) Geol. Inst. Mitt., new ser., No. 98,* 91p.

Gansser, A., 1955, New Aspects of the Geology of Central Iran, *World Petroleum Congress, 4th, Rome, Proc. Section 1,* pp. 279–300.

Huber, H., 1952, *Geology of the Western Coastal Makran Area,* Iranian Oil Company Report, GR 91B, filed at National Iranian Oil Co., Tehran.

Huber, H., 1978, *Geological Map of Iran, 1:1,000,000: with Back Notes,* National Oil Company, Tehran.

Lensch, G., A. Mihm, and N. Alavi-Tehrani, 1977, Petrography and Geology of the Ophiolite Belt, North of Sabzevar, Khorassan (Iran), *Neues Jahrb. Mineralogie Abh.* **131:**156–178.

McCall, G. J. H., and R. G. W. Kidd, 1980, The Makran, Southern Iran: The Anatomy of a Convergent Plate Margin Active from the Cretaceous to the Present Day, in *Trench and Fore-Arc Sedimentation and Tectonics in Modern and Ancient Subduction Zones, Abstract Volume,* Geological Society of London and British Sedimentological Research Group, London, pp. 21–22.

McCall, G. J. H., in press, *Explanatory Text of the Minab Quadrangle Map, 1:250,000; Explanatory Text of the Taherui Quadrangle Map, 1:250,000; Explanatory Text of the Fannuj Quadrangle Map, 1:250,000; Explanatory Text of the Pishin Quadrangle Map, 1:250,000; Report on East Iran Project Area No. 1; Note File on the Southern Half of the Nikshahr Quadrangle, 1:250,000; Note File on the Southern Half of the Saravan Quadrangle, 1:250,000,* Geological and Mineral Survey of Iran.

Ricou, L. E., 1971*a,* Le croissant ophiolithique péri-Arabe: une ceinture des nappes mises en place au Crétacé supérieur, *Rev. Géographie Phys. et Géologie Dynam.* **13:**327–350.

Ricou, L. E., 1971*b,* Le métamorphisme au contact des peridotites de Neyriz (Zagros interne), Iran, *Soc. Géol. France Bull.* **13:**146–155.

Ricou, L. E., 1974, L'étude géologique de la région de Neyriz (Zagros Iranien) et l'evolution structurale des Zagrides, Thèse Doct. État, Paris.

Sabzehei, M., 1974, Les mélanges ophiolitiques de la région d'Esfandagegh (Iran meridionale). Etude pétrologique et structurale interprétation dans le cadre Iranien, Thèse Doct. État, Grenoble.

21

This is an original article written expressly for this Benchmark volume

MÉLANGES OF THE MAKRAN, SOUTHEASTERN IRAN

G. J. H. McCall

The Makran has been geologically mapped on a regional scale only recently. The reports on this mapping (McCall, in press) are awaiting publication and the two note files (McCall, in press) are awaiting incorporation of additional mapping of the northern halves of the Nikshahr and Saravan quadrangles. This article has been prepared on the basis of the three regional reports on the Minab, Taherui, and Fannuj quadrangles and on the comprehensive report on the East Iran Project Area No. 1.

The six quadrangles that comprise this region, and the outcrop areas of the three separate mélanges that occur within it, are shown in Figure 1. The mélanges are defined in Table 1.

COLORED MÉLANGE COMPLEX

In the mapping of the Makran region, this complex was treated as a lithostratigraphic identity, although no formal stratigraphy was erected. This was on account of the large areas in which it appears to be normally sequential, with consistent facings of pillows and inward dip despite the manifest lensing and shearing. The correctness of such a usage remains debatable, and possibly it is better regarded as a structural unit. Certainly, despite the very few reversals of pillow facings, it may be a repetitive pile of imbrics rather than a throughgoing sequence. It forms a mappable zone more that 150 km long and up to 20 km wide, with faulted boundaries, except for local unconformable boundaries with overlying Eocene sediments. It occurs in a schuppen terrain of blocks bounded by steep inward-dipping reverse faults and separates geotectonic zone no. 3 (Bajgan complex, overlying Dur-Kan complex: Paleozoic metamorphics and Permian to Paleocene shelf carbonates above) from the Eocene flysch on the other, southern and western, boundary, but it appears to pass under the Eocene and younger Oligocene-Miocene flysch to the south as a wide accretionary prism formed before their deposition (evidence of inclusions in dislocated flysch and sedimentary melange). The scale of the Coloured Mélange development here is abnormally large. If the mélange is a through-going sequence, it would have a thickness of at least tens of thousands of meters. It is, furthermore, known to occur within a zone at least 70 km wide, though it does not actually outcrop extensively in the southern part of this zone where it is obscured by younger cover and has only limited protrusions through this cover.

The mélange consists of a cognate volcano-sedimentary assemblage [this is what Sabzehei (1974) refers to with some justification as a matrix, but this term is confusing, and it has been avoided here]. Pillowed basalts, mainly vesicular olivine basalts, only affected by static hydrothermal metamorphism characteristic of ocean floor suites (rarely more than zeolite facies),

Figure 1. Diagram showing colored and sedimentary mélange zones in the Makran, Iran (inset shows area of Makran mapped in the recent geological survey).

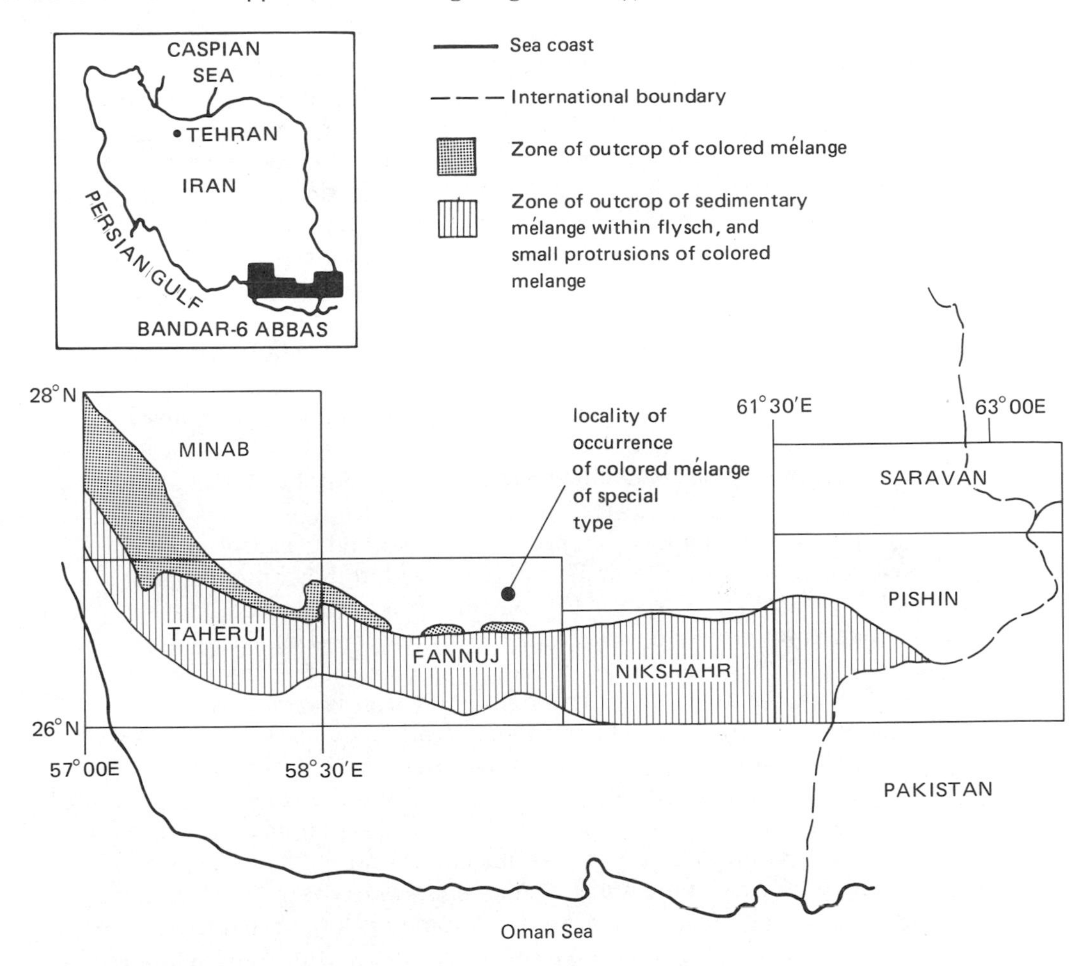

Table 1. Mélanges of the Makran, Southeastern Iran

Mélange	*Time of Formation*	*Geotectonic Zone*
Colored Mélange complex	Late Cretaceous-early Paleocene	No. 4, trench
Sedimentary mélange	Late middle Miocene	Nos. 5 and 6, flysch (a little in No. 7, Miocene, neritic)
Colored mélange—a special type (KPe$^{cm/ma}$) derived from ophiolitic complexes in localized zones of tectonic movement, related probably to reverse faulting	unknown	No. 2, Inner Makran spreading zone (riftlike marginal basin)

are intimately associated with red radiolarian chert and argillite; pink biomicrite with *Globotruncana* microfaunas (and white recrystallized limestone derived from it); massive, rather structureless distal turbidite sandstone; associated siltstone and shale; limey and phyllitic conglomerates and breccias; rare packed biomicrites (globigerinid oozes) of early Paleocene age; and volcanic (hyaloclastite) breccias, andesite, andesitic tuff, trachyte, rhyolite, and rhyodacite and associated welded tuffs. The mélange can be divided into a sedimentary and basic association in mapping purely on the basis of dominance of rock types: elsewhere it is neutral and not so differentiated. The principal exotic component is serpentinite derived from a variety of peridotite types—dunite, harzburgite, wehrlite, lherzolite, websterite—and the same rock types also occur in large masses up to 17 km long within the mélange (the largest of these are the Sorkhband and Rudan ultrabasics). These are grossly layered and include minor pyroxenite and gabbro layers, but cumulate tectures are only present in associated finely layered chromitites, having elsewhere apparently been obliterated by the ubiquitous overprint of a tectonite fabric. These bodies are dismembered ophiolites.

Trondhjemite is a minor component, also derived from ophiolite sequences. There is some amphibolite, derived from gabbro, fringing smaller serpentinite bodies in the mélange. These are tectonically introduced shreds and lenses, along the planes of movement. They do not form a continuous matrix to blocks of other lithologies except in rare cases, but increase in numbers of such shreds and lenses is accompanied by increased chaotization of the mélange. Clearly they had a lubricating role. Rodingite selvages are not uncommon to these bodies. The mélange is essentially a tectonic block-to-block mélange, and the edges of serpentinite and other bodies in it are sheared, but the cores are largely undeformed. There is no true matrix in the sense of a continuous host to blocks of other lithologies, and certainly not an argillaceous matrix as in Franciscan mélanges. Blocks may be up to several kilometers long; the largest are of limestone and chert, and these may also occur as definite strata more than 10 k long in the mélange, where it has remained largely sequential.

Other exotics include rare Albian reefal *Orbitolina*—limestones and metamorphic blocks of phyllite, schist, and amphibolite. Both these types are very similar to the rocks of the Bajgan and Dur-Kan complexes of geotectonic zone no. 3 (carbonate fore-arc) to the north and east, and may be either olistostromes or tectonically spalled off bodies derived from that source, or else the limestones may be part of the cognate ocean floor suite, but derived from atoll situations, though the metamorphics could not be so derived.

The passage from incipient (sequential) mélange to fully chaotic mélange is traced from the inner to the outer side of the mélange zone. The mélange includes Paleocene packed biomicrites so its formation must have at least extended into the Paleocene. It is unconformably overlain locally by Eocene limestones of shelf facies and flysch-type sediments, so it was formed prior to the Eocene. The ages obtained from its internal sedimentary components are anomalous and baffling. For example, the pink biomicrite contains *Globotruncana* microfaunas of the expected age range, mainly Campanian-Maastrichtian, but in rare cases as old as Cenomanian, Turonian, Coniacian, or Santonian. These biomicrites occur interstitial to some pillow lavas, which must be of the same age. The discovery of early Paleocene packed biomicrites was surprising, but not so surprising as the fact that the radiolarites contain Pliensbachian to Coniacian radiolarite microfaunas, despite the fact that they seem to be intimately related in normal sedimentary sequences to the *Globotruncana* limestones.

The recrystallization of the pink biomicrites to white limestones may have occurred during plastic deformation in the solid state. The mélange is considered to be tectonic and to probably represent offscraped slabs, related to subduction that was active during the late Cretaceous and early Paleocene. There is no evidence at all of sedimentary olistostrome involvement, other than that mentioned before, but such gravity sliding processes in a soft medium could have contributed to and have had all traces obliterated by tectonism. Certainly such processes were not the major agency in mélange chaotization.

This mélange is the main mélange of the northeast convergent margin of the Southern Tethys ocean.

SEDIMENTARY MÉLANGE

This is the "zone of schuppen and blocks" of Harrison et al. (1935, 1936) and Huber (1952). The existence of a second mélange occupying a zone of irregular developments to the south of the Coloured Mélange zone, over a strike length of more than 200 km, and in places more than 20 km wide, caused some confusion in mapping until it was realized that this is a quite distinct, later mélange, chaotized no earlier than the late middle Miocene, the age of the youngest components—and probably then, as younger sequences never display such mélange developments. It is a tectonic mélange, and the name is given on account of the dominant sedimentary rock content, not to indicate a sedimentary mode of origin. It occurs in steeply penetrative zones within the flysch outcrop belts. These zones are bounded by faults, tectonized contacts, or structural gradations into dislocated flysch. It is regarded as a structural unit, like the dislocated flysch.

The sedimentary components account for at least 75% of the mélange and commonly more. They are mostly of flysch, but locally this mélange is also developed from Miocene neritic sediments. In the north the flysch host is Eocene, but farther south it is Oligocene-Miocene or Miocene. The mélange zones have a manifest spatial relationship to the steep reverse faults, bounding the schuppen and dipping to the north and east. These relate to a phase of movement that had its climax in the late middle Miocene. The structure of the mélange is essentially a combination of dislocation and exotic block protrusion. The two processes are quite distinct, except that dislocation of rocks with an appreciable incompetent shale content provides an easy passage for upward protruded blocks.

The dislocation process is a more intense expression of the process elsewhere seen in dislocated flysch. Large areas of Eocene flysch nearby are characterized by plastic flow of the shaley matrix and brittle fracture of the competent sandstone beds to form boudins and rafts, accompanied by protrusion of exotic blocks. This structure is similar to that of the sedimentary mélange described by Cowan (1974, 1980) from the Franciscan mélanges of California, but in California no exotic blocks are protruded and the deformation apparently acted on a semiconsolidated sequence, whereas in the Makran it appears to have acted on a consolidated sequence. In both dislocated flysch and the dislocated host of the sedimentary mélange, it is the 1:1 shale-sandstone flysch that preferentially suffers dislocation. Sandstone-dominated sequences accommodate by chevron folding and faulting, whereas shale dominated sequences accommodate by small-scale contortion. This is a matter of the optimum competence contrast being achieved in the 1:1 sequence. The shaley matrix is green, shiny, and subphyllitic. Similar dislocation also occurs in Oligocene-Miocene and Miocene flysch, and in some Miocene neritic sequences, but is rarely associated with exotic block protrusion. The dislocation process is mainly related to reverse faulting, but also some relation exists to the early stages of the synchronous folding—to the progression from buckle type to similar folds involving tightening up and transposition. Dislocation also tends to show a preferential association with the subdued and appressed anticlinal zones in this syncline-dominated region (which displays the reverse of the Zagros anticline-dominated style).

The dislocated flysch differs from the sedimentary mélange in that whereas the stratigraphy can still be followed through the former, despite the dislocation, full chaotization has affected the sedimentary mélange, and it has been obliterated. This mélange produced low country of irregular hummocks of flysch, with scattered abrupt spines and strike elongated ridges formed by the protruded exotics. These are mainly of colored mélange lithologies and are believed to be derived from an older, pre-existing accretionary prism of colored mélange offscraped slabs, which formed the foundation for the Cenozoic flysch through deposition and later Miocene neritic deposition.

Pink *Globotruncana* limestone, calcilutite, sandstone, shale, radiolarite, siltstone and chert, pillow basalt, serpentinite, peridotite, troctolite, gabbro, and listvenite are the main Cretaceous lithologies. Paleocene and Eocene pelagic sediments and volcanics, and shallow water limestones, are also common. Blocks may be up to several kilometers diameter. There are some large rafts of colored mélange in the mélange. Blocks are commonly rounded, but angular boundaries are also recorded. The small boudins of sandstone in the mélange are typically rounded, though angular surfaces are again recorded; and the large rafts of sandstone typically have angular boundaries.

It is rare to encounter sedimentary mélange free from exotics, but a few such areas were mapped. In one locality, mélange was mapped with a host of virtually undislocated shale, and it is clear from this that dislocation is not a prerequisite for exotic block protrusion. In the same area, near Morton within the Pishin quadrangle, exotics map in trains, strung along the major reverse faults like a line of pips and standing up as spines. This clearly demonstrates the relationship between protrusion and reverse faulting, though more commonly the exotics are randomly dispersed in the sedimentary mélange zones and do not show such strong structural control on their distribution. Trains of exotics of the same lithology (particularly pelagic limestone) are quite common and suggest that some of the exotics at least are derived from continuous older strata.

Micropaleontological studies have been unrewarding on the matrix but rewarding on both exotics and sandstone boudins or rafts. The exotics mainly contain *Globotruncana* microfaunas, but Paleocene and Eocene planktonic microfaunas are also common. There are rare early Cretaceous *Orbitolina* bearing shelf limestone exotics, and more common Paleocene algal and *Lockhartia* bearing shelf limestone exotics; also Eocene *Nummulites* and *Alveolina* bearing shelf limestone exotics. Calc-wacke and shales contain Oligocene benthonic and planktonic microfaunas, and Miocene microfaunas of both types occur in sandstones, reefal limestones, and pelagic marls.

The three-dimensional geometry of the sedimentary mélange zones remains little understood, and the exact mechanism of protrusion of exotic blocks remains unknown, though the dislocation process is well understood. Destabilization of sequences carrying considerable pore water under seismic influences has been suggested as a possible mechanism for exotic block protrusion.

The dislocated flysch has been previously mapped as *wildflysch* in photogeological mapping (Huber, 1978), a terminology that implies olistostrome origin (Dzulynski and Walton, 1965:189). Neither the dislocated flysch nor the sedimentary mélange has any features indicative of such origin, and there is much evidence to the contrary in the penetrative nature of the zones, their obvious relationship to reverse faulting and associated folding, the trains of exotic blocks strung along faults, and the lack of any soft rock deformation structures. The subphyllitic nature of matrix indicates chaotization in a buried rock mass, not an open, sea floor gravity slide. The most emphatic evidence is to be found in the lack of any stratigraphic contacts with other sedimentary units. Olistostromes must be formed more or less at one time and therefore should stratigraphically overlie other formations, normally or unconformably. They should also be sealed conformably or unconformably by other formations of younger age that are not chaotically disturbed (though they may, of course, have nothing superincumbent). Such relationships we never see in the Makran. There is also a lack of systematic relationship to the paleogeography—olistostromes should have systematic relationships to the paleoslopes. There is also a limit to the size of blocks that can be transported downslope in submarine gravity slides, and many of the exotic blocks, kilometers in maximum dimensions, may well be too large to have been so displaced. There is probably a general scale constraint on olistostromes (see Papers 5 and 6), and the sedimentary mélange is developed on a truly immense scale.

Clearly this is not a trench mélange. All the evidence suggests that it was produced during renewed rapid subduction in a zone shifted well out oceanward from the site of the earlier Mesozoic subduction: the zone affected by chaotization was covered by quite shallow platform seas at the time of chaotization; the flysch trough had been entirely filled up; there was an enormously thick accretionary prism of older colored mélange offscraped slabs covered by successive flysch developments and some terminal neritic sediments; and the chaotization related to the subduction affected not a trench assemblage dominated by ocean-floor lavas and sediments but a compound accretionary prism. A final suggestion that may be offered is that the process of exotic block protrusion is really the process of offscraping of material that is resistant to the process of subduction. The dislocation is a reflection of the extreme compression operative close to zones of active subduction, accentuated in a highly resistant, sediment-dominated pile.

COLORED MÉLANGE OF SPECIAL TYPE ($KPe^{cm/ma}$)

Minor developments of chaotic rocks occur near Fannuj, within the ophiolite sequences of the Inner Makran spreading zone (geotectonic zone no. 2). These developments appear to be simply due to dislocation in zones of movement of rocks of the Mokhtarabad and Remeshk complexes, which together form an ophiolite sequence within this zone. These rocks are of either Jurassic or early Cretaceous to early Paleocene age. The component rocks—pillow lavas, cherts, pelagic and reefal limestone, turbidites, diabase, and ophiolitic plutonics—are similar to those of the Coloured Mélange complex, and these chaotic bodies closely resemble the Coloured Mélange complex. The distinction is due to the fact that they are developed on the inward side of the continental shelf limestone covered zone (Bajgan-Dur-Kan, geotectonic zone no. 3). They may in fact reflect an exactly similar process of chaotization because there are blueschists nearby in geotectonic zone no. 2—in the Deyader complex—and such developments are commonly supposed to indicate fossil subduction zones (Coleman, 1972). It is likely that there was limited subduction affecting the ancillary rift-like marginal basins within and around the Lut continental nucleus, and that colored mélange was produced in such zones, though in no way related to the main oceanic trench zone at the northeastern margin of the Southern Tethys Ocean.

REFERENCES

Coleman, R. G., 1972, Blueschist Metamorphism and Plate Tectonics, *Internat. Geol. Cong., 24th, Proc.* **2:**19-26.

Cowan, D. S., 1974, Deformation and Metamorphism of the Franciscan Subduction Zone Complex Northwest of Pacheco Pass, California, *Geol. Soc. America Bull.* **85:**1623-1634.

Cowan, D. S., 1980, Deformation of Partly Dewatered and Consolidated Franciscan Sediments, Piedras Blancas and San Simeon, California, in *Trench and Fore-Arc Sedimentation and Tectonics in Modern and Ancient Subduction Zones, Abstract Volume,* Geological Society of London and British Sedimentological Research Group, p. 7.
Dzulynski, S., and E. K. Walton, 1965, *Sedimentary Features of Flysch and Greywackes,* Developments in Sedimentology, vol. 7, Elsevier, Amsterdam, London, New York, 274p.
Harrison, J. V., and N. L. Falcon, 1935–1936, *Geological Map of the Coastal Makran,* unpublished, filed at the Geological Society of London.
Harrison, J. V., N. L. Falcon, A. Allison, J. A. Hunt, P. B. Maling, and R. J. C. McCall, 1935–1936, *Geological Map of the Landward Makran,* unpublished, filed at the Geological Society of London.
Huber, H., 1952, *Geology of the Western Coastal Makran Area,* Iran Oil Company Report, GR 91B, filed at the National Iranian Oil Company, Tehran.
Huber, H., 1978, *Geological Map of Iran, 1:1,000,000 with Backnotes,* National Iranian Oil Company, Tehran.
McCall, G. J. H., in press, *Explanatory Text of the Minab Quadrangle Map, 1:250,000; Explanatory Text of the Taherui Quadrangle Map, 1:250,000; Explanatory Text of the Fannuj Quadrangle Map, 1:250,000; Explanatory Text of the Pishin Quadrangle Map, 1:250,000; Report on East Iran Project Area No. 1; Note File on the Southern Half of the Nikshahr Quadrangle, 1:250,000; Note File on the Southern Half of the Saravan Quadrangle, 1:250,000,* Geological and Mineral Survery of Iran. (Nine regional maps on 1:100,000 scale of parts of the Minab, Taherui, and Fannuj quadrangles [Now-Dez, Kahnuj, Minab, Qal'eh Manujan, Dur-Kan, Dar Pahn, Avartin, Rameshk, Ramak] with extended side notes prepared by G. J. H. McCall and M. S. Peterson are now published. These maps, available at the Geological Survey of Iran, show the mélange discussed in Paper 21.)
Sabzehei, M., 1974, *Les mélanges ophiolitiques de la région d'Esfandageh (Iran meridionale). Etude pétrologique et structurale interprétation dans le cadre Iranien,* Thèse, Doct. État, Grenoble.

22

This is an original article written expressly for this Benchmark volume

MÉLANGES ASSOCIATED WITH THE SEMAIL OPHIOLITE IN THE NORTHERN OMAN MOUNTAINS ALLOCHTHON, SOUTHWEST ARABIA

S. J. Lippard, G. M. Graham, J. D. Smewing, and M. P. Searle

Abstract

The Semail ophiolite in the northern Oman Mountains is structurally over- and underlain by mélanges. It is possible to subdivide the underlying mélange into lower and upper components. The lower is largely sedimentary in origin, lacks ophiolite material, and was probably formed in a Cretaceous subduction zone complex. The upper mélange is tectonic in origin with an ophiolitic (sheared serpentinite) matrix and was formed during ophiolite emplacement in the Upper Cretaceous. A third, more enigmatic, mélange overlies the Semail ophiolite. It contains a mixture of ophiolite and exotic components and appears to have formed by the protrusion of material through fault zones in the ophiolite during the later stages of its emplacement.

INTRODUCTION

The Oman Mountains form part of the late Mesozoic to Tertiary thrust and fold belt that extends around the northern edge of Arabia (Falcon, 1967; Ricou, 1968; Stocklin, 1968; Murris, 1980). The mountains are largely allochthonous in origin and dominated by a large (700 km by 50 km areal extent) overthrust mass of Upper Cretaceous oceanic lithosphere (the Semail ophiolite) (Reinhardt, 1969; Coleman, 1977; Smewing et al., 1977). The Semail ophiolite is structurally underlain by an extensive mélange, which in turn overlies a strongly imbricated thrust stack of basinal sediments—the Hawasina mélange (Lees, 1928). These comprise part of a Mesozoic continental margin that was emplaced onto the northwestern edge of the Arabian continent, together with the Semail ophiolite, in the late Cretaceous (Glennie et al., 1974). Although the interpretation of the mélange has been an integral part of all recent models of the emplacement of the Oman Mountains allochthon (Glennie et al., 1974; Gansser, 1974; Stoneley, 1975; Dewey, 1976; Brookfield, 1977; Gealey, 1977; Welland and Mitchell, 1977), it has received no previous detailed study.

There have been several different interpretations of the Oman mélange, partly because of the difficulty of distinguishing structures produced by sedimentary and tectonic processes. In addition, it has frequently been assumed that only one process was responsible for its formation; for example, Wilson (1969) considered it to be entirely sedimentary in origin. In contrast, Glennie et al. (1974) proposed that the Oman mélange was tectonic, "a mechanical mixture formed when the Oman Exotics and the

Semail Nappe were emplaced on the uppermost units of the Hawasina" (Glennie et al., 1974, p. 180). In a similar way, Gansser (1974) interpreted the entire Oman thrust belt, excluding the Semail ophiolite, as an ophiolitic mélange, similar in character to the colored mélanges of Iran and Pakistan, and specifically related to the process of ophiolite obduction. In a later model of ophiolite emplacement, Welland and Mitchell (1977) recognized olistostromes, and hence a sedimentary component to the mélange, and proposed that it formed as a subduction zone complex.

During the course of the present study, it has become apparent that the mélanges in the Oman Mountains are not the result of a single process, but formed by a variety of sedimentary and/or tectonic processes. Three structurally and lithologically distinct mélanges are recognized (Table 1).

Table 1 Subdivision of the Oman Mélanges

Name	*Position*	*Components*	*Origin*	*Age*
Hawasina mélange	Top of Hawasina thrust sheets	Dominantly sedimentary, olistostromes; blocks of Hawasina sediments, exotic limestones, volcanics; shale/mudstone matrix.	Debris flows, gravity sliding, and tectonic mixing; trench infill?	Cretaceous (?Lower)
Semail mélange	Base of Semail nappe	Sheared serpentinite matrix containing blocks of metamorphics, "Exotic" limestones, Hawasina sediments, volcanics and cherts.	Tectonic, synophiolite emplacement; downward protrusion of serpentinite.	Upper Cretaceous
Batinah mélange	Overlies Semail	Mainly "block on block" mélange. Blocks of Semail, Hawasina, "Exotic" limestones, volcanics, serpentinite and metamorphics. Matrix occasionally sedimentary or serpentinite.	Tectonosedimentary; formed of material transported from beneath the ophiolite, perhaps by protrusion along fault zones.	Upper Cretaceous

These are informally named the Hawasina, Semail, and Batinah mélanges. The first two underlie the Semail; the third, although possessing many features in common with the other two, overlies it. The general lithostratigraphic and structural setting of the Oman Mountains is shown on Figures 1 and 2.

HAWASINA MÉLANGE

The Hawasina thrust sheets comprise a pile of redeposited turbiditic and fine grained pelagic sediments formed between the Permian and Upper Cretaceous (Cenomanian) (Glennie et al., 1974). In general, the high units show progressively deeper water and more distal facies. Glennie et al. (1974) proposed that the Oman Tethys or Hawasina Ocean could be palinspastically reconstructed by stretching out the thrust sheets in a northeasterly direction. From this they concluded that the ocean had a width of between 400 km and 1,200 km and that, although part of the Hawasina was deposited on the outer part of the continental slope and rise, the greater part was formed in the ocean basin itself. A recent sedimentological and structural reassessment of the type area of the Hawasina in the central Oman Mountains (Graham, 1980) has shown that only the most distal units are truly oceanic and that most of the sediments were deposited on marginal block-faulted continental crust.

The uppermost Hawasina units are structurally overlain by the mélange that consists of unsorted megabreccias and conglomerates up to 1,500 m, but usually around 25–50 m thick. The blocks include all the lithological types of the Hawasina, together with volcanics and a minor amount of

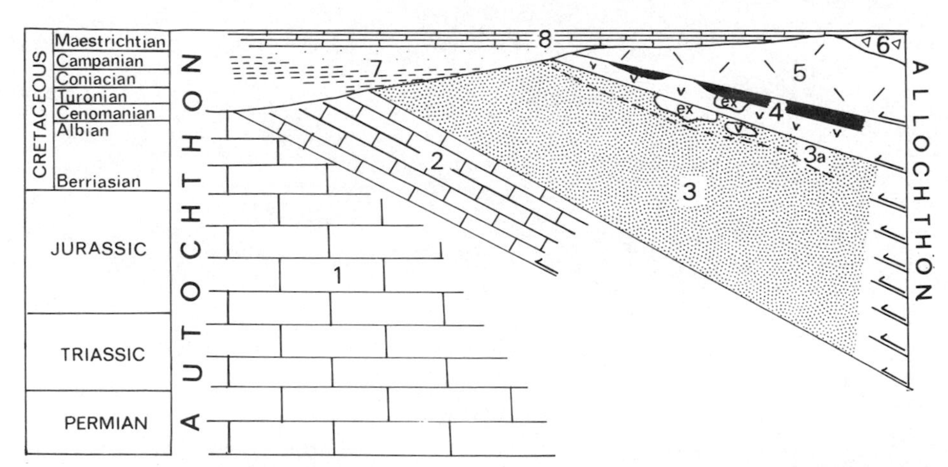

Figure 1 Tectonostratigraphic relations in the Oman Mountains (modified after Glennie et al., 1974). 1: Hajar Supergroup, autochthonous shelf limestones; 2: Sumeini Group, parautochthonous shelf-edge limestones; 3: Hawasina Series, allochthonous basinal sediments; 3a: Hawasina mélange; 4: Haybi complex (Allochthonous); 5: Semail ophiolite (Allochthonous); 6: Batinah complex; 7: Muti and Juwiza Formations (Neoautochthon); 8: Simsima Formation (Neoautochthon); v: Haybi volcanics; ex: Exotic limestones metamorphic sheet: —: thrusts (diagrammatic).

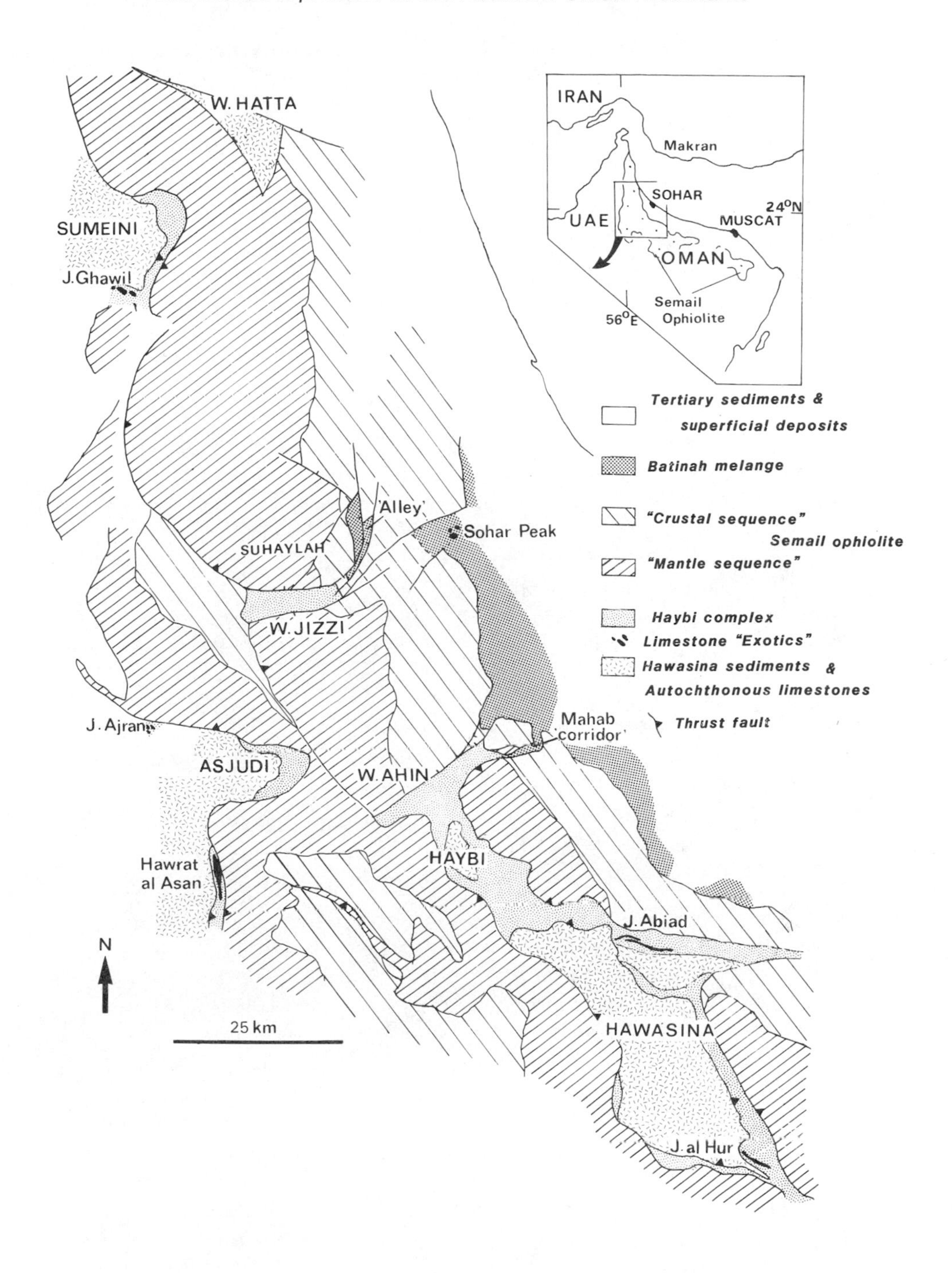

Figure 2 Simplified geological map of the northern Oman Mountains (Open University Project mapping, 1975–1980)

basic igneous and metamorphic rocks. The largest blocks are mountain size masses of Permian and Triassic exotic reefal limestones (Lees, 1928; Glennie et al., 1974). The Hawasina mélanges lack definite ophiolite-derived material. Some of the deposits have a recognizable matrix that is mainly pelitic, a sheared red siliceous mudstone or shale, and forms up to 30% of the rock mass that may be matrix supported. In places, there are recognizable beds, 1-10 m thick, that are occasionally graded and interpreted as debris flows or olistostromes. The matrix, which is interpreted as a contemporaneously formed pelagic sediment, has, in samples from the Dibba area, United Arab Emirates, yielded Berriasian to Cenomanian radiolaria (P. Tippit, pers. comm.). The olistostromes may locally be interbedded with normal marine pelagic sediments, cherts, and fine-grained limestones.

The exotic limestones are a prominent and distinctive feature of the Oman Mountains, often forming steep-sided, isolated peaks, or in the case of the largest (Jebel Kawr, which has an areal extent of 600 km^2 and a thickness of 900 m) whole mountain masses. They were called the Permian klippen by Lees (1928) and the Oman exotics by Glennie et al. (1974). The last named authors showed that they are predominantly of Permian and Upper Triassic ages. Detailed studies show that they are only partly truly reefal, that they are Bahama-type carbonate build-ups that include oolitic and bioclastic sands and carbonate muds (Graham, 1980). Despite their shallow water origin, the exotics occupy an anomalous position in the upper part of the allochthon above the Hawasina thrust sheets. Many occur as blocks in the Hawasina mélange, others appear to have a primary volcanic substrate and may have formed as cappings to volcanic islands at the site of ocean basin rifting in the early Mesozoic off the continental margin (Glennie et al., 1974; Searle et al., 1980).

SEMAIL MÉLANGE

The Haybi complex tectonically overlies the Hawasina in northern Oman, the name being used informally for the rocks that lie between the Hawasina and the base of the Semail nappe (Searle et al., in press) (Fig. 1). It comprises a mixture of rocks including Trias volcanics [the Haybi volcanics (Searle et al., 1980)], some of the largest exotic limestone masses, and thin pelagic sediments. Tectonically overlying these is a discontinuous unit of metamorphic rocks that was formed by the dynamothermal metamorphism of the sole rocks of the ophiolite slab during the early stages of its displacement from the ocean lithosphere (Searle and Malpas, 1979). Much of the Haybi complex, particularly its upper part, is incorporated in a tectonic mélange with a largely serpentinite matrix. Because of its obvious genetic connection with the overlying ophiolite, this is called the Semail mélange.

The Semail mélange is characterized by a matrix of pervasively sheared serpentinite in which is included a variety of blocks that, in some cases, can be seen in the process of being intruded and broken up by the serpentinite. Slickensided surfaces to the blocks are found, as are discolored outer zones caused by metasomatic "rodingitization" by the serpentinite. Blocks of metamorphic rocks are most common, particularly near the top, where they can be seen to have been derived by the break-up of the metamorphic sheet. The blocks vary in size from a few meters up to several kilometers, and in the case of some of the largest, the original upward metamorphic zonation from greenschist to amphibolite facies rocks can be seen. The main rock types are amphibolites, quartzites, marbles, and pelitic schists. Lower down, the serpentinite incorporates blocks of Haybi volcanics and unmetamorphosed sediments, including exotic limestones and Hawasina lithologies. The maximum thickness of the Semail mélange in Oman is about 500 m, and it is confined to the immediate base of the Semail ophiolite. However, in the Dibba area, United Arab Emirates, serpentinite mélanges penetrate deeper into the thrust pile and are found as lenticular bodies along thrust planes between sheets of Hawasina sediments and parautochthonous shelf-edge facies limestone.

BATINAH MÉLANGE

Along the eastern edge of the northern Oman Mountains, the Semail ophiolite is overlain by a group of rocks that have many structural and lithological similarities with the Haybi and Hawasina complexes that underlie it. At their maximum development, west of Sohar (Fig. 2), these rocks are about 3 km thick and comprise a lower mélange unit (the Batinah mélange) overlain by largely intact masses of Hawasina-type sediments. The mélange consists of a generally chaotic mixture of the following: serpentinite; bedded cherts and limestones; volcanics, mainly basic pillow lavas; and exotic limestones, metamorphics, and dolerites and gabbros. It rests, without major structural break, on the top of the Semail, often on the locally uppermost lavas or pelagic sediments overlying them. The lowermost exotic material appears as blocks set in a pelagic sediment matrix and mixed with Semail lithologies. Higher up sequence, the mélange is composed entirely of exotic material and usually has a block-on-block structure with no matrix, although locally there are sheets and masses of serpentinite that include blocks of sediment and other lithologies. Locally the mélange cuts down along fault zones in the ophiolite—for example, in the alley, north of Wadi Jizzi (Smewing et al., 1977) and in the Mahab corridor, south of Wadi Ahin (Fig. 2)—and rests on sheeted dikes and gabbros.

GENESIS OF THE OMAN MÉLANGES

The Hawasina mélange contains recognizable debris flow deposits (olistostromes) and is clearly partially sedimentary in origin, although many of its primary characters have been tectonically overprinted during emplacement. A deep, rapidly subsiding trough off the continental margin, perhaps a subduction zone trench, is suggested by its nature and tectonic position. This possibility is supported by the presence of volcanics of island-arc type occurring above the mainly Triassic Haybi volcanics and exotic limestones in the Haybi complex (Searle et al., 1980). From the presence of interbedded cherts bearing Cretaceous radiolarites, these are probably contemporaneous with the Hawasina mélange and may have formed to the northeast of the subduction zone (Fig. 3). The lack of

Stage 1 ?Lower Cretaceous - Initiation of a NE-facing subduction zone off the Oman continental margin

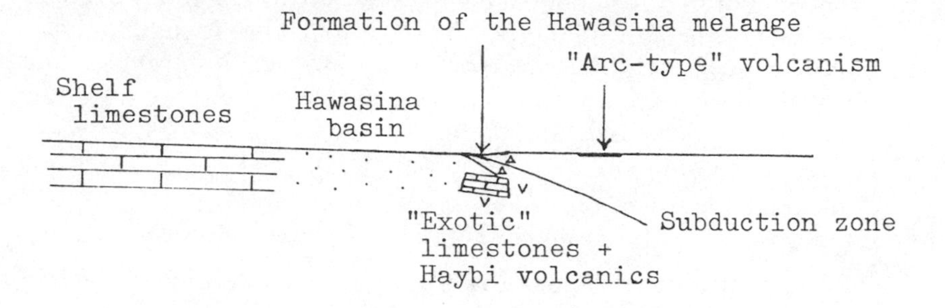

Stage 2 Turonian (ca 90 Ma) - Initial displacement of the Semail Ophiolite

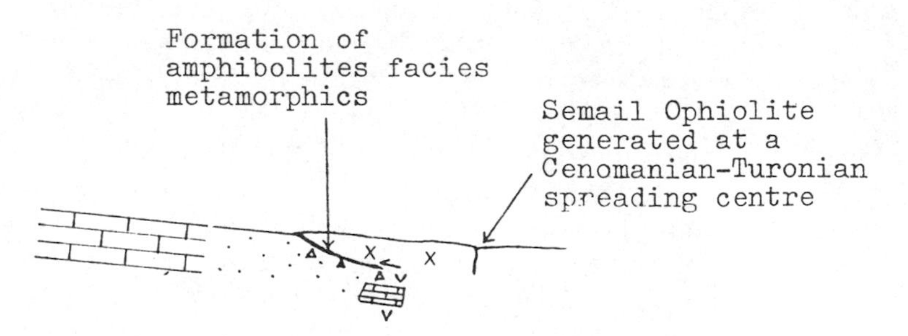

Stage 3 Turonian-?Campanian - Emplacement of the Semail Ophiolite across the continental margin. Detachment and emplacement of Hawasina thrust sheets

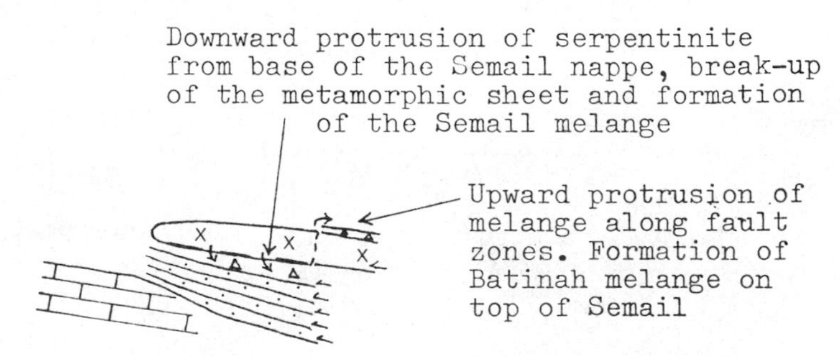

Figure 3 Schematic cross-sectional development of the Oman allochthon with particular reference to mélange formation

ophiolite material in the Hawasina mélange indicates that it formed prior to ophiolite emplacement in the Upper Cretaceous (Turonian-Campanian) and that it predates both the Semail and Batinah mélanges.

The Semail mélange is a true ophiolitic mélange (Gansser, 1974). There seems little doubt that it was produced by the emplacement of the Semail ophiolite, particularly the downward penetration of serpentinite from the ultramafic base of the ophiolite slab and the break-up of the sole rocks. Both it and the underlying Hawasina mélange were further disrupted by later imbrication and thrusting during the final phases of emplacement. As a result, the contact between the two is not always clearly defined.

Because of its position above the Semail ophiolite, the Batinah mélange is the most difficult to interpret. Most its components can be matched with the Hawasina and Haybi complexes beneath the ophiolite. In particular, the presence of subophiolite metamorphics shows that they were originally overriden by it. A possible explanation for its formation is that, during the later stages of emplacement, the ophiolite began to break up into a number of blocks along major fault zones [e.g., Wade Jizzi, Wadi Ahin (Fig. 2)]. This process was aided by the upward protrusion of serpentinite and low density sediments from beneath the sheet. This material was finally emplaced by gravity sliding from high points close to the fault zones where it broke through onto the surface. This process can be seen at an arrested stage in the Hahab corridor south of Wadi Ahin (Fig.2) where it is possible to trace continuously the mélange from its position above to beneath the ophiolite. However, this explanation of the Batinah mélange is inadequate to explain the overlying Hawasina-type sediment sheets and may have to be modified in the light of more detailed studies of the whole Batinah complex currently being undertaken (Woodcock and Robertson, in preparation).

ACKNOWLEDGMENTS

This work forms part of the Open University Oman Ophiolite Project, directed by Professor I. G. Gass, and funded by the Overseas Development Administration (ODA) and the Natural Environment Research Council (NERC). We have benefited from discussion with Dr. N. H. Woodcock and Dr. A. H. F. Robertson (universities of Cambridge and Edinburgh).

REFERENCES

Brookfield, M. E., 1977, The Emplacement of Giant Ophiolite Nappes: 1. Mesozoic-Cenozoic Examples, *Tectonophysics* **37:**247–303.

Coleman, R. G., 1977, *Ophiolites: Ancient Oceanic Lithosphere?* Springer-Verlag, New York, 229p.

Dewey, J. F., 1976, Ophiolite Obduction, *Tectonophysics,* **31:**93–120.

Falcon, N. L., 1967, The Geology of the North-east Margin of the Arabian Basement Shield, *Adv. Sci.* **24:**31–42.

Gansser, A., 1974, The Ophiolite Mélange: A Worldwide Problem on Tethyan Examples, *Eclogae Geol. Helvetiae* **67:**479–507.

Gealey, W. K., 1977, Ophiolite Obduction and Geologic Evolution of the Oman Mountains and Adjacent Areas, *Geol. Soc. America Bull.* **88:**1183–1191.

Glennie, K. W., M. G. A. Boeuf, M. W. Hughes-Clark, M. Moody Stuart, W. F. H. Pilaar, and B. M. Reinhardt, 1974, Geology of the Oman Mountains, *Koninkl. Nederlands Geol. Mynbouw. Genoot. Verh. Part 1,* 423p.
Graham, G. M., 1980, Evolution of a Passive Margin and Nappe Emplacement in the Oman Mountains, in *Proceedings of the International Ophiolite Symposium, Nicosia,* A. Panayiotou, ed., Republic of Cyprus Ministry of Agriculture and Natural Resources, Geological Survey Department, pp. 414–423.
Lees, G. M., 1928, The Geology and Tectonics of Oman and Parts of South East Arabia, *Geol. Soc. London Jour.* **84:**585–670.
Murris, R. J., 1980, Hydrocarbon Habitat of the Middle East, *Canadian Soc. Petroleum Geol.Mem.* 6, pp. 765–799.
Reinhardt, B. M., 1969, On the Genesis and Emplacement of Ophiolites in the Oman Mountains Geosyncline, Schweizer. *Mineralog. u. Petrog. Mitt.* **49:**1–30.
Ricou, L. E., 1968, Sur le mise en place au Crétacé supérieur d'importantes nappes à radiolarites dans les Zagros (Iran), *Acad. Sci. Comptes Rendus* **267:**2272–2275.
Searle, M. P., and J. Malpas, 1979, The Structure and Metamorphism of Rocks Beneath the Semail Ophiolite of Oman (abstract), *EOS (Am. Geophys. Union Trans.)* **60:**962.
Searle, M. P., S. J. Lippard, J. D. Smewing, and D. C. Rex, 1980, Volcanic Rocks Beneath the Semail Ophiolite Nappe in the Northern Oman Mountains and Their Tectonic Significance in the Mesozoic Evolution of Tethys, *Geol. Soc. London Jour.* **137:**589–604.
Smewing, J. D., K. O. Simonian, I. M. Elboushi, and I. G. Gass, 1977, Mineralised Fault Zone Parallel to the Oman Ophiolite Spreading Axis, *Geology* **5:**534–538.
Stocklin, J., 1968, Structural History and Tectonics of Iran: A Review, *Am. Assoc. Petroleum Geologists Bull.* **52:**1229–1258.
Stoneley, R., 1975, On the Origin of Ophiolite Complexes in the Southern Tethys Region, *Tectonophysics* **25:**303–322.
Welland, M. J., and A. H. Mitchell, 1977, Emplacement of the Oman Ophiolite: A Mechanism Related to Subduction and Collision, *Geol. Soc. America Bull.* **88:**1081–1088.
Wilson, H. H., 1969, Late Cretaceous Eugeosynclinal Sedimentation, Gravity Tectonics and Ophiolite Emplacement in the Oman Mountains, South East Arabia, *Am. Assoc. Petroleum Geologists Bull.* **53:**626–671.

Part VI

MÉLANGES OF THE CIRCUM-PACIFIC ZONE

Editor's Comments on Papers 23, 24, and 25

The mélanges of California have provided some of the basic literature on the subject of mélanges (for example, see Papers 2 and 3), and only a selection of the very substantial output of literature on the subject from this region can be reproduced here. Saleeby (Paper 23) described an older mélange, not the renowned Franciscan mélange. He noted that distinctive mappable rock assemblages characterized by the inclusion of exotic and native fragments and blocks within a pervasively deformed more tractable material (i.e., mélanges) are widespread in numerous belts in California, but hitherto no serpentinite-matrix mélange as described from the Tethyan realm and the USSR has been recorded from this region, the mélanges described being characterized by inclusion of blocks of various igneous, metamorphic, and sedimentary rocks in a pelitic/psammitic matrix. He described a serpentinite-matrix mélange, an ophiolitic mélange *sensu stricto* occupying a belt 57 miles long along the west side of the Sierra Nevada, in the foothills. It is the southern part of the Kings-Kaweah ophiolite belt. The essential character is of tectonic blocks, up to 20 km long and elongated parallel to the regional trend, surrounded by serpentinite, which tends to form areas of low-lying nonoutcrop, whereas the blocks tend to produce knobby outcrops. The serpentinite matrix includes both serpentinite derived from a variety of peridotite types, and also sedimentary serpentinite, with visible clastic structure—debris flow and so on. Some of the sedimentary serpentinite may in reality be protrusion breccia, but much is truly sedimentary; tectonic processes have obliterated the primary structure in much of the matrix, and it is suggested that a large part of it may be sedimentary. Blocks include chert, ophicalcite, silica-carbonate rocks, basalt, gabbro, metabasite, and serpentinized peridotite. The mélange is locally affected by contact metamorphism related to a variety of Cretaceous intrusions. The mélange units mapped, defined by clusterings of blocks (following the method of Hsü, 1968), parallel the regional trend and have primary contacts preserved. It is clear that a deep ocean floor stratigraphic succession of the Kings River ophiolite is represented by this sequence of units The ophiolitic rocks were highly disrupted and internally mixed to form the mélange on the deep ocean floor, it is believed. The mélange contains only rocks of the deep ocean domain.

Saleeby provides an original explanation for this mélange. Unlike most modern authorities, he relates the mélange not to plate convergence along a fossil continental margin but to a mid-ocean spreading center traversed by a large fracture zone. It was later displaced, not to a convergent plate margin situation but to a locus of arc magmatism. The age of the ophiolitic rocks of the displaced oceanic lithosphere comprising the mélange is latest Paleozoic to earliest Mesozoic. There is some superimposed later deformation, but the main deformation is believed to have occurred before displacement from the oceanic domain. The remnants of continental margin rocks are believed to be olistostromes; the deformation in these is not attributed to tectonic processes. A latest Triassic–middle Jurassic age is inferred for these rocks.

Saleeby has presented an excellent description of a tectonic ophiolitic mélange, with ancillary olistostrome developments of another age and quite unrelated to the mélange-forming process. His model interpretation is excellently presented and convincing, though it may be that alternative models will be forthcoming from other authorities; this is a very recent paper and must surely stimulate argument. It is possible to envisage alternative models in which the mélange formation occurs in the deep ocean domain, but not at a spreading center or on a major fracture zone—certainly the present-day understanding of ophiolites allows them to be developed in a wide range of deep ocean situations, and the need to invoke a spreading center situation in the sense of a principal mid-ocean ridge is not apparent.

Blake and Jones (Paper 24) discussed the origin of the classic Franciscan mélanges, which they believe cannot be explained by oversimple models of subduction of oceanic crust and mantle material swept from mid-ocean ridge across the ocean and down the subduction zone at the convergent plate margin. The mélanges occur at the boundaries of imbricate thrust sheets at several distinct structural horizons. They consist of blocks of graywacke, greenstone, chert, serpentinite, together with the isolated so-called knockers of high grade blueschists and eclogite, set in a matrix of sheared and quartz-veined mudstone and minor sandstone. Except for the high grade schists, the blocks are similar to, but more deformed than, the normal sequence immediately above the Coast Range thrust at the base of the Great Valley Sequence. The mélanged sequence is interpreted as a distal or seaward extension of the basal part of the Great Valley Sequence; the associated greenstone, chert, and serpentinite in tectonic blocks are interpreted as being derived from the underlying oceanic crust and upper mantle. They are dated on radiolarite and so on as of Tithonian-Valanginian age. The sediments were shed from the ancestral Sierran and Klamath Ranges, first by river action and then by turbidity currents, filling the trench and accumulating in quiet periods, and formed into mélanges in repeated episodes of more active subduction, when lawsonite and jadeite bearing meta-graywackes were also produced. The underlying oceanic crust and upper mantle that supplied the greenstone, chert, and serpentinite blocks is believed to be either from interarc basin or marginal sea situations. The authors favor east-dipping subduction zone geometry. They believe that subduction occurred possibly from the early Cretaceous to the Tertiary. The blueschist and eclogite tectonic blocks are related to earlier subduction and are believed to have been tectonically mixed with the other rocks of the mélanges during the subsequent subduction.

Of particular interest are the reasons given for rejecting a provenance in an East Pacific Rise-type mid-ocean ridge for the oceanic component of the mélange—namely, the abnormally thin, heterogeneous nature of the oceanic crust involved in the mélanges; the unusual abundance of quartz keratophyres and albite granite; the lack of sheeted dikes; and the lack of a thick pelagic cover are all considered to be much more consistent with an interarc basin or marginal sea provenance. The andesite-dacite-rhyolite-dominated igneous suite, despite the high soda and low potash content (which may be secondary), favors such a provenance.

An interesting detail in this paper is the mention of sedimentary serpentinites in the Great Valley Sequence. Such sedimentary serpentinites appear to have a major development in the mélange described by Saleeby (Paper 23) and have also been described from flysch sequences in the Makran of Iran (McCall, in press), where a harzburgite conglomerate unit, apparently shed off peridotite imbrics synsedimentationally protruded up reverse faults and rapidly eroded on the floor of a flysch basin, is recorded. Another interesting detail is the presence of radiolarites as an intrinsic part of a turbidite sequence; too many geologists at once relate radiolarites and pelagic biomicrites to deep ocean floor abyssal environments, whereas in fact, they are quite common intercalations in the turbidites of the more distal parts of the trench slope.

The authors rejected sedimentary slide processes for the formation of the mélange on the basis of evidence of only very restricted mixing (limited to the contacts). They instead related the mélange primarily to tectonic processes. Their model is not very different to that proposed in Paper 21 in that the ocean crust material is believed to have been derived from quite close to the continental margin, not the spreading center. The Coloured Mélange of the Makran, in particular, has an anomalous abundance of andesitic, rhyolitic, and even trachytic components, also plagiogranites, and appears to be mainly calc-alkaline. However, that mélange is a true ophiolitic mélange, in many ways more similar to the Kaweah mélange (Paper 23), and the younger sedimentary mélange of the Makran is much more like the Franciscan mélanges in that it has a dominant matrix of turbiditic sediments.

Cowan (Paper 25) described the deformation in a tectonic mélange within the Franciscan of California. This Garzas mélange is considered to be typical of the Franciscan mélanges in general; he described a specific area near Pacheco Pass. Like most mélanges it contains exotic inclusions not derived from adjacent units. He limits the exotics to those clearly metamorphosed prior to incorporation (including the knockers of glaucophane-lawsonite schist and albite-chlorite greenstone) but admits that there is a difficulty in determining whether other blocks are exotic or part of the local sequence—this is especially the case with chert and greenstone. Cowan used the terms *native* and *exotic* (following Hsü, 1968) in the same way that others (e.g., Paper 21) have used *cognate* and *exotic*. He was mainly concerned with the deformations in the mélange that he believed to be related to a pervasive shearing of consolidated rock bodies and the production of tectonic inclusions of all sizes immersed in a pervasively sheared, generally fine-grained matrix. There are no soft rock deformation structures, and though accepting the possibility of deformation of olistostromes (the exotics being inherited from submarine slumps), he found this hypothesis unsatisfactory because of no evidence of such soft rock processes and an apparent spatial relationship between the highest grade rocks and the most tectonically disturbed zones. He preferred to invoke a tectonic mixing of rocks derived from older terrains—quite dense blocks once deeply buried being uplifted. The mélanges are regarded as the structural equivalent of faults. The process of their formation involved shearing and brittle fracturing of already consolidated rock bodies.

Cowan noted the distinction of Hsü (1968) between broken formations and mélanges, one that distinguished the disruption of competent beds in the native fraction and the introduction of exotic blocks. He adopted the methodology of listing the important questions to be answered, an excellent approach to discussion of mélanges. In this case they are:

What was the nature of the mélange prior to deformation?
How were the exotic inclusions introduced?
What do the observed structures tell us about the deformation environment?

In a later paper (Cowan, 1978), which cannot be reproduced here because of space limitations, the San Simeon mélange, also of the Franciscan, was described—the mélange previously studied in detail by Hsü (1969). The two successive deformations recognized are not considered to have produced the chaotization. Cowan again lists the critical questions, in this case:

How did the lithologic heterogeneity originate?
What produced the nonbedded chaotic fabric?

He lists the possible modes of origin as olistostrome mechanism, postdepositional deformation, or a combination of both. Essentially this mélange consists of a matrix of black argillite, quite devoid of internal layers or bedding, and is both heterogeneous and lacks sorting in respect to the component blocks. Graywacke blocks are dominant (75%), and of the rest greenstone is most abundant, while chert, metamorphics, serpentinite, other ultramafics, diabase, and gabbro are minor. Late Jurassic radiolarite has been recorded. The scale of blocks is rather small (up to 10 m). Cowan could find no evidence of a tectonic mixing process such as he favored in Paper 25. He believed that the predeformational fabrics are consistent with an olistostrome origin (though he did not give any very detailed account of these fabrics) and suggested that a pebble or boulder bearing mudstone or diamictite underwent submarine gravity sliding as one or more olistostromes. He related the olistostrome to a trench slope situation, and the subsequent tectonic deformation to a subduction zone environment, suggesting recycling as subduction continued. In this paper Cowan seems to be suggesting that this mélange, though the subject of Hsü's classic studies (1968, 1969), is atypical of the Franciscan as a whole.

Cowan (1980) discussed the San Simeon mélange again, together with the mélange at Piedras Blancas, in an abstract of a paper shortly to be published. This is so important that it is summarized here. However, how much this requires modification of the earlier interpretation (1978) is uncertain. He concentrates on the deformation processes and once again is concerned with postdepositional premetamorphic deformation, in which the sandstone suffered extension of the layers, local disruption, and adopted the pattern of oblate ellipsoids. Accommodation on outcrop and hand-specimen scale was by ductile pinch-and-swell behavior and boudinage, extreme necking leading to extension fracturing and rare brecciation, and formation of parallel shear fractures oblique to bedding. On the smaller scale, intergranular flow and slip are observed, but there is a lack of microfracturing, granulation, and cataclasis even close to microfaults. The mudstone is megascopically ductile and was transformed by flow, by distributed displacements on penetrative, anastomizing surfaces of slip to a scaly clay. Cowan made the suggestion that appreciable pore fluids were present in the sediments during deformation and that the essential process involved was deformation of partially dewatered and consolidated sediments. The progressive deformation was coaxial on a mesoscopic scale, suggesting gravitational collapse and spreading on a trench slope or offscraping and accretion at the base of the inner wall. The bulk strain and strain history are incompatible with either simple uniaxial compaction or fault zones characterized by progressive simple shear. He seems to be saying that either an olistostrome-type process or a tectonic offscraping process related to subduction could have produced the deformation and chaotization, at least of the native components.

Once again, the possibility of confusion of soft rock submarine slump deformations and tectonic deformations of not fully dewatered and consolidated sediments in the rock mass in the subduction zone environment is raised.

REFERENCES

Cowan, D. S., 1978, Origin of Blueschist-bearing Chaotic Rocks in the Franciscan Complex, San Simeon, California, *Geol. Soc. America Bull.* **89:**1415–1423.

Cowan, D. S., 1980, Deformation of Partly Dewatered and Consolidated Franciscan Sediments, Piedras Blancas and San Simeon, California, in *Trench and Fore-Arc Sedimentation and Tectonics in Modern and Ancient Subduction Zones, Abstract Volume,* Geological Society of London and British Sedimentological Research Group, pp. 21–22.

Hsü, K. J., 1968, Principles of Mélanges and Their Bearing on the Franciscan-Knoxville Paradox, *Geol. Soc. America Bull.* **79:**1063–1074.

Hsü, K. J., 1969, Preliminary Report and Geological Guide to Franciscan Mélanges of the Morro Bay-San Simeon Area, California, *California Div. Mines and Geology Spec. Pub. No. 35,* 46p.

McCall, G. J. H., ed., in press, *Explanatory Text of the Minab Quadrangle Map, 1:250,000; Explanatory Text of the Taherui Quadrangle Map, 1:250,000; Explanatory Text of the Fannuj Quadrangle Map, 1:250,000; Explanatory Text of the Pishin Quadrangle Map, 1:250,000; Report on East Iran Project Area No. 1; Note File on the Southern Half of the Nikshahr Quadrangle, 1:250,000; Note File on the Southern Half of the Saravan Quadrangle, 1:250,000,* Geological and Mineral Survey of Iran.

23

Reprinted from pages 29-31, 34-42, and 44-46 of *Geol. Soc. America Bull.*
90(Part I):29-46 (1979)

Kaweah serpentinite mélange, southwest Sierra Nevada foothills, California

JASON SALEEBY *Division of Geological and Planetary Sciences, California Institute of Technology, Pasadena, California 91125*

[*Editor's Note:* Figures 2 and 7 have been omitted.]

ABSTRACT

Prebatholithic rocks of the southwest Sierra Nevada foothills contain a 125-km-long northwest-trending disrupted and metamorphosed ophiolite belt. Much of the belt consists of a serpentinite-matrix mélange in which the ophiolitic material is dispersed. The mélange matrix consists of schistose serpentinite derived from tectonitic peridotite and sedimentary serpentinite. The tectonic blocks consist mainly of peridotite, gabbro, basalt, chert, ophicalcite, and silica-carbonate rock and are as long as several kilometres. Within the tectonic blocks relict primary features such as bedding, pillows, dikes, and cumulate layering remain. The mélange represents disrupted and internally mixed oceanic lithosphere of latest Paleozoic to possibly earliest Mesozoic age.

Outcrop mapping of the mélange reveals clustering of blocks into several lithologic associations. The associations are defined as mélange units, which appear to represent the vestiges of once-intact ocean-floor sections. Three types of mélange units have been recognized: (1) peridotite-gabbro units; (2) gabbro-basalt units; and (3) peridotite-chert-basalt units. The first two units represent crust and upper mantle whose deformation and metamorphic history began at the site of ocean-floor genesis. The third unit represents abnormal crust composed of ultramafic protrusions and interbedded sedimentary serpentinite, ophicalcite, chert, and pillow lava. Protrusive activity also started at the site of ocean-floor genesis and continued for an extended time after sea-floor–spreading transport of the ophiolite belt away from the genesis site. Genesis was at an oceanic spreading center that was cut by a transverse fracture zone. Ocean-floor mélange developed along the fracture zone by the combined effect of protrusive acitivity and wrench faulting. Emplacement of the fracture-zone complex (ophiolite belt) resulted from large-scale wrench faulting that truncated the ancient continental margin and juxtaposed the complex against the modified margin.

During transport to the continental margin, a chert-argillite oliostostrome complex was shed across the ophiolite belt. The olistostromes carried limestone blocks with fauna exotic to North America. Once in the vicinity of the continental margin, the ophiolite belt served as basement for continent-derived submarine-fan deposits and island-arc volcanic rocks, both of Late Triassic to Middle Jurassic age. Deformation of these strata along with their ophiolitic basement continued along the older fracture zone trends. The strata now exist as highly deformed depositional remnants above serpentinite mélange.

INTRODUCTION

Distinctive mappable rock assemblages characterized by the inclusion of exotic and native fragments and blocks within a pervasively deformed more tractable material are referred to as mélanges (Hsü, 1968; Berkland and others, 1972). The widespread occurrence of mélange belts throughout California has been well documented (Hsü, 1969, 1971; Blake and Jones, 1974; Cashman, 1974; Cowan, 1974; Maxwell, 1974; Duffield and Sharp, 1975). These mélange belts are characterized by the inclusion of blocks of various igneous, metamorphic, and sedimentary rocks in a pelitic-psammitic matrix. Serpentinite-matrix mélanges such as those found in the Soviet Union and Tethyan realm (Gansser, 1955, 1974; Peyve, 1969; Knipper, 1973; Makarychev and Visnevsky, 1973) have not been reported in North America. Serpentinite-matrix mélanges that include mainly ophiolite assemblage rocks are important in that they probably reveal how sediment-poor oceanic basement behaves during intense disruption and emplacement along once-active plate junctures.

The Kaweah serpentinite mélange forms the southern half of the Kings-Kaweah ophiolite belt, which comprises part of the western wall of the southern Sierra Nevada batholith (Fig. 1). The mélange is exposed almost continuously for 57 km along the regional trend of the ophiolite belt. The isolated exposures of disrupted ophiolite that occur farther north, between the Kaweah and Kings Rivers (Fig. 1), and serpentinite mélange bounding ophiolite slabs of the Kings River area (Saleeby, 1978) indicate that the mélange extends along the entire length of the 125-km-long Kings-Kaweah ophiolite belt. In this paper I describe the mélange, discuss the relationship between the mélange and the Kings River ophiolite, discuss the origin of the mélange, discuss the relationships between the mélange and highly deformed continental margin rocks that appear to rest depositionally above it, and briefly cover the tectonic settings of mélange development. A comprehensive tectonic model of mélange development and emplacement of the Kings-Kaweah ophiolite belt into its present position has been presented in Saleeby (1977b).

Segments of the central and southern parts of the mélange have been mapped in reconnaissance and studied petrographically by Durrell (1940) and Plafker (1956), respectively. The remainder of the belt has been mapped only in reconaissance (Jennings and Strand, 1966). It was not recognized as a mélange in those studies.

GEOLOGIC SETTING

The mélange is exposed at the extreme western edge of the Sierran foothills in steep-sided grassy hills. The tectonic inclusions weather out as jagged slope and hilltop outcrops, whereas the serpentinite matrix forms grassy slopes and smooth, less conspicuous outcrops. In the Yokohl Valley and Tule River areas occur remnants of continental-margin rocks that were apparently deposited across the mélange (Fig. 1). The continental-margin assemblage consists of a chert-argillite olistostrome complex, quartz-rich clastic rocks, and basalt-andesite volcanic rocks. The metamorphic rocks that bound the eastern margin of the entire Kings-Kaweah ophiolite belt (Fig. 1) consist primarily of schist, quartzite, and marble derived from quartz-rich clastic rocks, chert, and limestone.

Voluminous plutons ranging in composition from olivine-hornblende gabbro to biotite granodiorite invaded the Kaweah mélange between 125 and 102 m.y. ago (Saleeby, 1975). The resulting contact metamorphism of the mélange is mainly in the hornblende-hornfels facies, but locally albite-epidote hornfels facies and pyroxene hornfels facies assemblages are present. Metamorphic recystallization is in some places incomplete: thus, some evidence of the original mineralogy of the ophiolitic protoliths is available. In addition, where metamorphic recrystallization is complete, earlier textures and structures are commonly preserved. Thus, the ophiolitic protoliths have been readily deduced from field, petrographic, mineralogical, and chemical data. Much of the discussion below is in terms of the protoliths. Contact-metamorphic mineral assemblages along with prebatholith metamorphic and primary mineral assemblages are listed in Table 1.

The Kings-Kaweah ophiolite belt represents a tectonic suture in the Earth's crust (Saleeby, 1977a, 1977b). The suture apparently extends at least 400 km to the north, partly as the Sierran foothills fault system (Clark, 1960; Saleeby and others, 1978). The term "suture" is used here as a zone of joining. As discussed in the references cited above, the foothill suture joins latest Paleozoic to early Mesozoic fossil oceanic lithosphere to older North American continental lithosphere. The Kings-Kaweah ophiolite belt represents the deepest structural level of the foothill suture now exposed. Thus, the history of ophiolite genesis and deformation is critical for an understanding of the entire foothill suture. The discussion of ophiolite petrogenesis presented for the ophiolitic slabs of the Kings River area (Saleeby, 1978) is applicable along the entire length of the ophiolite belt. In addition, structural features inherited from the earliest stages of ophiolite deformation are well displayed in the Kings River slabs and are therefore discussed at length in Saleeby (1978).

STRUCTURE OF THE MELANGE

The internal structure of the mélange is clearly shown by outcrop maps. These maps, in which exposures as small as 5 m in diameter are represented, have been prepared for the entire mélange. Figure 2 shows a geologic map for part of the mélange in the area of Yokohl Valley and Round Valley. The geologic map was made in the field on an outcrop base map that had previously been prepared. This area was chosen for the geologic map because (1) a wide range of mélange features are well displayed; (2) the largest area with the lowest grade of contact metamorphism occurs, and (3) relationships with the continental-margin assemblage are well displayed. The following description applies to the entire mélange, but the Yokohl Valley area is referred to frequently in the discussion of critical relationships.

The mélange is composed of coherent to semicoherent blocks within a pervasively deformed matrix. Blocks of chert, ophicalcite, silica-carbonate rock, basalt, gabbro, metabasite, and serpentinized

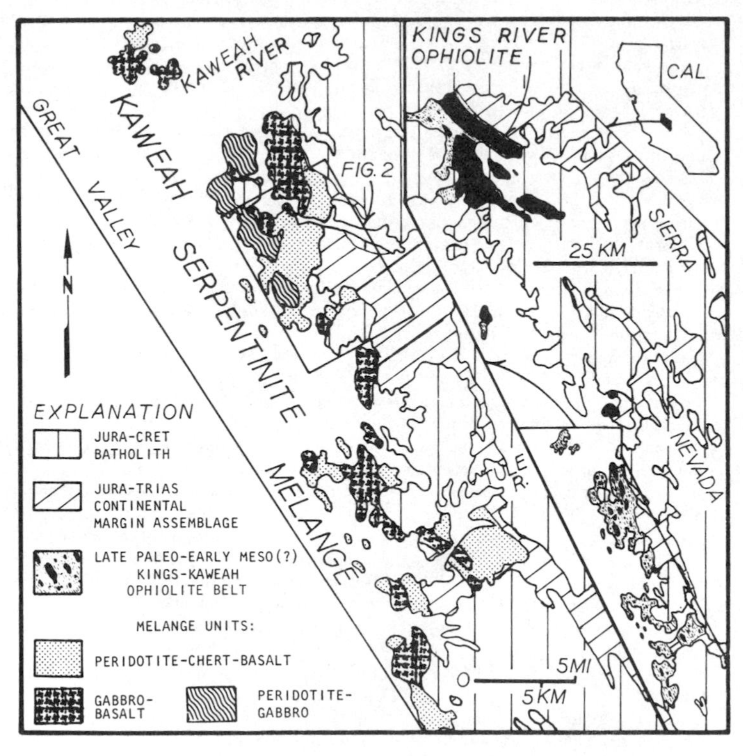

Figure 1. Location map of Kings-Kaweah ophiolite belt, showing large tectonic slabs and blocks, and mélange unit map of Kaweah serpentinite mélange.

peridotite are exposed as knobby outcrops (Fig. 3, A and B). Schistose serpentinite and deformed sedimentary serpentinite form less-resistant outcrops and underlie the areas of nonexposure. As shown in Figure 2 and photogaphs A through D of Figure 3, the tectonic blocks are surrounded by serpentinite or by areas of nonexposure. This is true along the entire length of the mélange, and wherever the nonexposed areas have been opened by roadcuts, canal cuts, or prospects, they are invariably underlain by serpentinite. Thus, the mélange blocks are suspended within a serpentinite matrix.

Ophiolite rock fragments within the mélange matrix range from grit size to large blocks and slabs. The largest fragments of the entire ophiolite belt occur in the Kings River area, where tectonic slabs as long as 20 km contain mappable intervals of the ophiolite sequence (Saleeby, 1978). Large blocks and fault slices occur mainly as isolated exposures between the Kings and Kaweah Rivers (Fig. 1). These are intermediate in size between the Kings River slabs and the Kaweah mélange blocks. Most of the mélange blocks are between 25 and 500 m in length and are elongated parallel to both the schistosity of the matrix and the regional trend of the ophiolite belt.

Mélange Matrix

A mélange matrix can be defined as any pervasively deformed rock material that encloses coherent blocks that are generally less deformed. The matrix of the Kaweah mélange is composed almost entirely of serpentinite or silica-carbonate rock derived from serpentinite. The matrix is commonly schistose, but in many instances it has been recrystallized to granoblastic textures as a result of contact metamorphism. The dip of the schistosity rarely diverges more than 10° from vertical, and it generally strikes parallel to the trend of the ophiolite belt, which is about N30°W. Local deviations in the attitude of the schistosity occur at all scales where the schistosity wraps around tectonic blocks and where crosscutting plutons deflect the serpentinite wall rocks. Small-scale folds and kinks also occur locally in the matrix schistosity. These invariably have steep-plunging axes and sometimes have asymmetries suggestive of a dextral sense of motion. Small remnants of nonschistose serpentinite occur commonly and show various stages of conversion to the schistoste matrix. They are usually ellipsoidal in shape, with long axes lying in the schistosity plane of the matrix. As these remnants decrease in size, they commonly show strong flattening in the plane of the matrix and occasional vertical elongation.

Significant zones of what appear to be sedimentary serpentinite also occur in the matrix. These rocks are distinguished from most other ultarmafic rocks of the mélange by several factors: (1) they have relict clastic textures (Fig. 4, A and B); (2) they occasionally have relict bedding features; (3) they occasionally contain rounded clasts of ophicalcite and chert; (4) they occur primarily with ophicalcite and chert; and (5) they are always enriched in talc and commonly enriched in carbonate and chlorite. These rocks are usually easily distinguished in the field from tectonitic serpentinized peridotite and have mesoscopic features most similar to accumulations of sedimentary serpentinite examined in the California Coast Ranges (see locations described by Dickinson, 1966; Cowan and Mansfield, 1970; Moiseyev, 1970; Lockwood, 1971). They are, thus, interpreted as sedimentary accumulations of serpentinite; however, some of them may represent protrusion breccias. Relict bedding features suggest that deposition was primarily by debris-flow mechanisms; however, deposition by turbidity currents and settling of fine suspended serpentinitic mud also appears to have occurred. The friable sedimentary serpentinites commonly grade abruptly into schistose serpentinite. The detrital components were apparently easily obliterated during schistosity development. Thus, a significant amount of the matrix could have been derived from sedimentary serpentinite that has since lost its diagnostic features.

Alteration of the serpentinite matrix to silica-carbonate rock is common. There are a number of stages in the development of the alteration. Patchy growth of magnesite occurs in the incipient stage, followed by progressively more pronounced segregation of magnesite into veins that run both parallel and at high angles to the schistosity. Talc is also a common constituent that manifests a siliceous shift in the rock's bulk composition. In the advanced stages of the alteration all primary features are lost, and a residue of jasper,

TABLE 1. PRE–SIERRA NEVADA BATHOLITH PROTOLITHS OF KAWEAH SERPENTINITE MELANGE AND THEIR CORRESPONDING CONTACT-METAMORPHIC DERIVATIVES

Protolith	Contact-metamorphic derivative
Radiolarian chert	Metaquartzite
Impurities:	
Metalliferous	Iron and manganese oxide minerals
Ultramafic	Chlorite + tremolite
Volcanic	Hornblende ± plagioclase ± epidote
Argillaceous	Biotite
Ophicalcite	Dirty marble (calcite ± dolomite ± talc ± antigorite ± chlorite ± quartz magnesite)
Basalt (plagioclase ± clinopyroxene ± olivine phyric)	Mafic hornfels (green hornblende + plagioclase ± epidote or clinozoisite)
Diabase (mafics: clinopyroxene ± brown hornblende or olivine)	Mafic hornfels (green hornblende + plagioclase ± epidote or clinozoisite)
Gabbro and plagioclase clinopyroxenite (mafics: clinopyroxene ± brown hornblende or olivine)	Mafic hornfels (green hornblende + plagioclase ± epidote or clinozoisite)
Wehrlite and olivine clinopyroxenite	Tremolite ± Mg − chlorite ± antigorite hornfels
Harzburgite-dunite	Antigorite ± talc ± metamorphic olivine hornfels
Mafic greenschist tectonite (actinolitic amphibole + albite + epidote ± chlorite ± calcite ± sericite ± quartz)	Mafic hornfels (green hornblende + plagioclase ± epidote)
Mafic amphibolite tectonite (brown or green hornblende + plagioclase ± garnet ± epidote ± clinopyroxene)	Mafic hornfels (green hornblende + plagioclase ± epidote or clinozoisite)
Schistose and sedimentary serpentenite (chrysotile ± Mg chlorite ± tremolite ± calcite	Silica carbonate rock (magnesite + jasper ± talc) or antigorite ± talc ± tremolite hornfels

spinel, iron oxide, and traces of various ore minerals remain. The alteration zones apparently resulted from metasomatic processes driven by Cretaceous contact metamorphism (Saleeby, 1975). Segments of the mêlange containing significant amounts of sedimentary serpentinite, ophicalcite, and chert appear to have been particularly susceptible to silica-carbonate alteration. The ophicalcites are thought to have served as a carbonate source for the metasomatic processes. Jasper-magnesite alteration of the matrix is a distinctly different process from the earlier process that formed the silica-carbonate mêlange blocks. The silica-carbonate mêlange blocks will be discussed in conjunction with the ophicalcite blocks.

The schistose serpentinite matrix was derived both from a sedimentary serpentinite protolith and from a peridotite protolith. The peridotite mêlange blocks appear to be the incompletely digested remnants of the peridotite protolith. Structural relations between the matrix and the peridotite blocks are thus critical for an understanding of the matrix.

Peridotite Blocks

The peridotite blocks consist of harzburgite, dunite, and wehrlite. All of the peridotite blocks are at least 50% serpentinized, and many of them are completely serpentinized. Some of the larger blocks, such as the one underlying Elephant Back (Fig. 2), contain

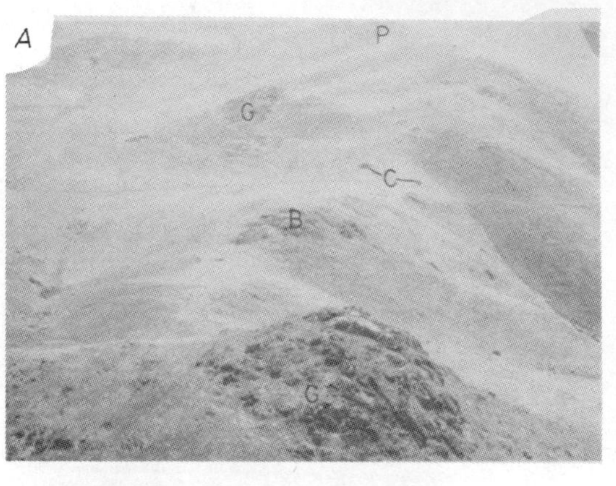

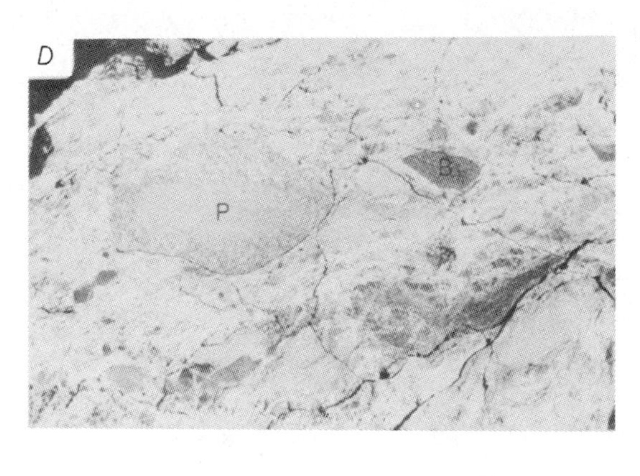

Figure 3. Some important structural features of Kaweah serpentinite mélange. A: Blocks of chert (C), basalt (B), gabbro (G), and peridotite (P) weathering out with matrix underlying grassy slopes. B: Chert block (C) with schistose matrix (S) wrapping around its margins. C: Basalt block (B) with schistose matrix wrapping around its margins. D: Fragments of serpentinized peridotite (P) and basalt (B) floating in schistose serpentinite; field of view is about 1 m long.

gabbro and amphibolite derived from gabbro. Several different structural features occur within the peridotite blocks. The earliest of these features is a steep-dipping foliation that locally contains a downdip lineation. The foliation is defined by deformed olivine and pyroxene grains and occasional aggregates of chrome spinel. The foliation in many instances grades into the matrix schistosity (Fig. 3, E) and in some instances is cut by it. Mafic inclusions within the larger peridotite blocks have overall shapes, foliations, and lineations that parallel the structure of the peridotite host. The structural features introduced above are best displayed within the slabs of the Kings River area. A structural analysis of those features is presented in Saleeby (1978).

Throughout peridotite blocks and most commonly along block margins, structural and textural features are progressively obliterated by several later structural features. The most common is a cleavage or shear fabric and associated serpentine schistosity that grades into the matrix schistosity (Fig. 3, E). Another important feature is the development of a blocky fracture system in which schistose domains of serpentinite are woven through the serpentinized peridotite autoclasts. Contact-metamorphic textures are superimposed over the blocky fracture systems and in many instances obscure and probably obliterate the systems. Thus, the blocky fracture systems are in general a prebatholith feature. The blocky fractured serpentinized peridotites also grade into the mélange matrix by obliteration of the autoclasts and development of a penetrative serpentine schistosity. There are thus three observed pathways for schistose matrix development: (1) obliteration of sedimentary serpentinite, (2) peridotite foliation grading into matrix schistosity, and (3) blocky fractured serpentinized peridotite grading into schistose serpentinite. Pathways 1 and 3 are sometimes difficult to distinguish from one another.

Evidently, the peridotite blocks are vestiges of larger peridotite masses that are now mostly altered to serpentinite matrix. These larger peridotite masses were probably similar at one time to the large peridotite slabs of the Kings River ophiolite (Saleeby, 1978). The wehrlitic peridotite blocks yield a tremolite-bearing serpentin-

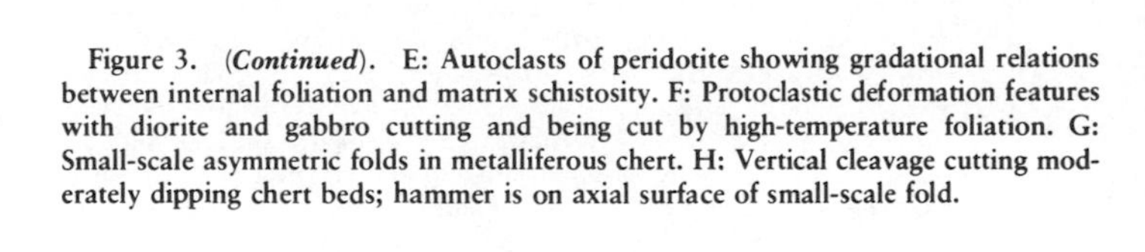

Figure 3. (***Continued***). E: Autoclasts of peridotite showing gradational relations between internal foliation and matrix schistosity. F: Protoclastic deformation features with diorite and gabbro cutting and being cut by high-temperature foliation. G: Small-scale asymmetric folds in metalliferous chert. H: Vertical cleavage cutting moderately dipping chert beds; hammer is on axial surface of small-scale fold.

Figure 4. Some important sedimentary features of Kaweah mélange. A: Sedimentary serpentinite with subangular to subrounded clasts of serpentinized peridotite within serpentinitic sand-mud matrix; large clast of ophicalcite occurs in lower left. B: Ophicalcite–sedimentary serpentinite debris flow with blocks and clasts of altered peridotite and ophicalcite within carbonate-serpentinite mud matrix. C: Chaotic folds of probable soft-sediment origin in metalliferous chert; note quartz-filled tension fractures, which are the only hard-rock structures in this chert block. D: Chert clast conglomerate with matrix of metalliferous chert; crude bedding runs subparallel to hammer handle, with faint cleavage crossing at high angle.

ite matrix, whereas harzburgite-dunite blocks yield a serpentinite matrix that lacks a Ca-bearing phase.

Mafic Blocks

Mafic mélange blocks consist of gabbro, pyroxenite, diabase, basaltic pillow lava, breccia and tuff, massive basalt, various metamorphic tectonites, and mafic detrital rock. Relict primary structures such as cumulate layering, dikes, pillows, and bedding are locally preserved. Relict igneous textures in gabbroic rocks are commonly preserved, whereas relict phenocrysts are usually the only preserved textural feature in basaltic rocks. Most of the smaller blocks (100 m or less in length) are monolithologic. Many of the larger blocks contain complex mixtures of the rock types listed above.

The large heterolithologic blocks are of particular interest. These consist of chaotic mixtures of intrusive and extrusive rock. The internal structures are so complex that they usually cannot be accurately mapped. The blocks contain numerous domains that either cut one another or are intergradational. Gabbroic, diabasic, and rare leucocratic rocks commonly have contradictory intrusive relations and sometimes show chaotic protoclastic-type deformation features (Fig. 3, F). They also commonly grade into amphibolite tectonite. In some instances, domains of pillow lava are surrounded by mafic intrusive rock and/or mafic tectonite of greenschist or amphibolite facies. Chert rarely occurs in association with the pillow-lava domains. Domains of mafic breccia also occur in the complex blocks. In some instances the breccias can be recognized as clearly intrusive or sedimentary, but in many instances their origin cannot be resolved.

A variety of deformation features occur within the mafic blocks. Chaotic mixtures of different intrusive rocks, foliation surfaces, and ductile faults having contradictory relationships with one another are considered primary features, suggesting that tectonic deformation commenced during igneous petrogenesis. Dense clusters of veins and dikelets also occur, indicating intense fracturing during igneous petrogenesis. These clusters, along with a significant number of dikes, commonly have steep dips and strike at high angles to the regional structural trends. Metamorphic foliation surfaces are developed to various degrees in some blocks but are lacking in many blocks. The foliation is defined by transposition of primary features, and metamorphic recrystallization in greenschist to amphibolite facies. In some instances there is a strong downdip lineation within the foliation plane. A distinct line cannot be drawn between the metamorphic tectonites and the protoclastically deformed igneous rocks. In addition to the high-temperature features, intense fracture systems are common in mafic blocks. The fracture systems usually consist of a longitudinal shear fabric and transverse tension fractures. The fracture systems occur in blocks with or without the high-temperature features.

The shear and tension fracture systems appear to be the surfaces along which the blocks were broken apart and dispersed within the mélange matrix. Some blocks have been frozen into incipient stages of being rafted apart, as shown by the presence of schistose serpentinite veins that were injected along the fracture systems. Some blocks have been pulled apart along transverse tension fractures like large boudins.

Blocks that have a penetrative metamorphic foliation usually sit concordantly within the structure of the matrix. This is not due to metamorphism of the block and matrix together following tectonic mixing, because blocks of different metamorphic and textural grade occur in proximity to one another. The concordant relationship between the internal structure of the tectonic blocks and the structure of the matrix appears to be the result of two processes: (1) The mafic tectonites originally formed within a peridotite tectonite host, and the peridotite has since been converted into schistose serpentinite; this is shown by ultramafic blocks that contain mafic inclusions and are in intermediate stages of conversion into schistose serpentinite. (2) The mafic tectonites fractured preferentially into elongate blocks along foliation surfaces and then the blocks became oriented concordantly with the matrix structure. Similar relationships are common with bedded chert blocks.

Chert Blocks

Chert blocks are a common constituent of the mélange. Relict primary features are commonly preserved within them. These consist of highly recrystallized radiolaria tests, bedding, and soft-sediment folds and breccias (Fig. 4, C and D). Some beds are graded according to radiolaria test size. Metalliferous impurities are a common constituent of the chert blocks. These consist of fine disseminations and discrete interbeds of primarily Fe and lesser Mn oxide. Interbeds of chlorite-tremolite also occur in cherts associated with sedimentary serpentinite and ophicalcite. These impurities appear to have been derived from fine ultramafic detritus. Rarely, thick interbeds of ophicalcite occur within chert, and chert clasts occur commonly within ophicalcite (Fig. 5, A). Argillaceous impurities occur locally in some of the larger chert blocks. Volcanic impurities also occur in cherts that are associated with pillow lava and mafic tuff. In a number of instances the cherts occur interbedded with pillow lava, although the chert blocks are most commonly associated with peridotite, sedimentary serpentinite, and ophicalcite.

Longitudinal shear and transverse tension fractures (Fig. 4, C) are common features in chert blocks. Ductile deformation features are also common. These consists of flattening of shear foliations, boudinaging of relict bedding or the block itself, and tight steep-plunging folds. The ductile deformation features usually occur in chert blocks that have been faulted into peridotite block clusters where there is little matrix. In these instances, the chert blocks are considered to have undergone ductile deformation during and following tectonic mixing with the peridotite.

Chert blocks are occasionally kinked about vertical axes. The kinks usually show a dextral sense of motion. Asymmetric folds also displaying a dextral sense of motion occur within the chert blocks (Fig. 3, A). Rare structures displaying sinistral motion have also been observed. Similar kinks, asymmetric folds, and crenulations have been observed in foliated mafic blocks, but the chert blocks appear to display them much better than the mafic blocks.

Ophicalcite Blocks

Calcite- and dolomite-cemented breccias with clasts composed primarily of serpentinized peridotite and lesser chert occur as tectonic blocks and as diffuse zones within the mélange matrix. Basalt and gabbro clasts also occur rarely in these ophicalcites. The breccias contain poorly sorted, sand to boulder size, subangular to subrounded clasts that commonly show partial replacement by carbonate. In most cases the carbonate matrix is recrystallized to marble, but rare vestiges of a micritic protolith have been observed. The ratio of clasts to matrix is highly variable, with both clast- and matrix-supported rocks common and intergradational. Nearly pure beds of sedimentary serpentinite and limestone also occur. Composite clasts occur frequently, indicating that much of the material has undergone multiple cycles of brecciation and cementation. Bedding relationships with chert and sedimentary serpentinite indicate that many of the ophicalcites were derived from sedimentary serpentinites (Fig. 4, B). In several instances the ophicalcites occur in fissured peridotite along with patches of silica-carbonate rock. In a couple of these instances it appears as though chert and ophicalcite were resting depositionally above highly fissured peridotite, with the ophicalcite–silica-carbonate assemblage extending down into the fissures. The ophicalcites also occur adjacent to pillow lavas that occur with sedimentary serpentinite, and in one instance ophicalcite was observed as the matrix of a basaltic breccia.

Most matrix-supported ophicalcites occur as monolithologic mélange blocks. When small chert clasts occur, and when highly fissured and veined, the ophicalcite blocks are indistinguishable from silica-carbonate blocks. The chert-carbonate blocks are not the same as the jasper-magnesite ± talc rocks that occur as an alteration product of the mélange matrix near Cretaceous plutons. Most clast-supported ophicalcites occur as diffuse zones in the mélange matrix along with sedimentary serpentinite. The ophicalcites are interpreted as a product of chemical interaction between ultramafic rock and percolating ocean water and/or hydrothermal fluids. In many instances the ultramafic rock appears to have been sedimentary serpentinite. It appears that the finer grained components of these deposits were the most susceptible to alteration. In some instances, highly fractured serpentinized peridotite was the protolith. The occurrence of ophicalcite zones in sedimentary serpentinite near pillow lava suggests that hydrothermal fluids related to ocean-bottom volcanism was at least locally significant.

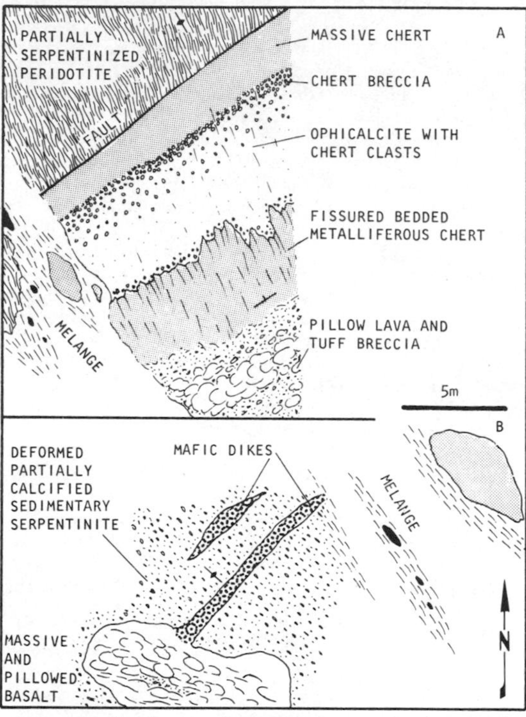

Figure 5. Sketch maps of some important primary and deformational features of peridotite-chert-basalt mélange units. A: Sequence of basalt, chert, and ophicalcite fissured along northwest trend and cross-faulted against foliated ultramafic block. These relations in turn are truncated by thoroughly mixed mélange with northwest trends. B: Nondeformed mafic feeders for pillow lava cutting across structure in deformed sedimentary serpentinite. These relations are also truncated by northwest-trending mélange.

Mélange Units

Outcrop maps of the mélange reveal patterns defined by clustering of blocks of similar rock types. The larger the average block size the better defined the clusters are. For example, along the western margin of the Yokohl Valley area (Fig. 2), long clusters of peridotite and gabbro blocks occur. East of the northern cluster occurs a cluster composed of gabbro, mafic dike rock, pillow lava, and rare chert (Figs. 1, 2). South and east of these mafic-ultramafic clusters occurs a complex cluster containing mainly sedimentary serpentinite, ophicalcite, pillow basalt, and chert. The block clusters are designated as mélange units (after Hsü, 1968). The units are named after the main blocks that occur within them. A mélange unit map is shown in Figure 1. As shown on the map, the mélange units are elongate parallel to the trend of the ophiolite belt and are interdigitated in a complex fashion. Contacts between the units are at some localities sharp and at others gradational over distances of as much as 200 m. The internal structure of most units is dominated by deformation features with regional trends. However, structures oriented at high angles to the regional trends occur within a number of units and actually bound some units (Fig. 5, A). This is particularly well displayed in the Yokohl Valley area (Fig. 2). Three types of mélange units are recognized.

1. Peridotite-gabbro units consist of dunite, harzburgite, wehrlite, pyroxenite, gabbro, and amphibolite tectonite derived from gabbro. These units occur along the west side of Yokohl Valley (Fig. 1) and in the Kings River area adjacent to the peridotite-gabbro slabs (Saleeby, 1977b, Pl. 1). Relict primary intrusive contacts and cumulate depositional contacts occur locally within the blocks of these units. The matrix schistosity in these units is usually intergradational with the peridotite foliation. The mafic tectonites are usually situated concordantly within this structure. Blocky fracture systems are occasionally developed along the margins of peridotite blocks.

2. Gabbro-basalt units consist mainly of the rock types discussed above in the section on mafic mélange blocks. An additional component of these units is small chert blocks, which never exceed 20 m in thickness when measured normal to bedding. The gabbro-basalt units occur along the entire length of the mélange. They commonly contain the large heterolithologic blocks that were discussed above. The large blocks commonly have monolithologic mafic and chert blocks scattered around their margins. Relict primary contacts in these units occur only in the large blocks, as discussed above. The smaller mafic blocks commonly have a strong shear fabric, which appears to represent the structure along which the smaller blocks were dislodged from the larger blocks. The matrix in these units is usually schistose serpentinite without vestiges of the ultramafic protolith.

3. Peridotite-chert-basalt units consist mainly of serpentinized peridotite and chert with variable amounts of pillow basalt,

ophicalcite, and metabasite. These are the most commonly occurring mélange units. Some significant points concerning these units are that (1) in contrast to the peridotite-gabbro units, the peridotite blocks of these units commonly contain the blocky fracture systems; (2) significant zones of the mélange matrix appear to have been derived from sedimentary serpentinite as well as blocky fractured serpentinized peridotite; (3) the pillow lavas of these units are without an intrusive complex; furthermore, the rare feeders found for these pillow lavas cut sedimentary serpentinite and serpentinized peridotite (Fig. 5, B); (4) the pillow lavas of these units are the only ones along the entire ophiolite belt that contain interpillow limestone (marble) and vesicular rinds; (5) pillow lavas of these units are commonly nondeformed, in contrast to pillow lavas of the gabbro-basalt units; and (6) vestiges of depositional contacts appear to exist between chert, sedimentary serpentinite, ophicalcite, and pillow lava.

Some of the larger blocks of chert ± pillow lava in the peridotite-chert-basalt units have noteworthy relationships with the rest of the mélange. These large blocks appear to be highly deformed depositional remnants that lie above previously deformed and partially mixed ophicalcite, sedimentary serpentinite, blocky fractured serpentinized peridotite, chert, and pillow lava. This relationship is displayed in the central part of the Yokohl Valley area (Fig. 2). Here, the contact between the large chert-basalt block and the rest of the mélange cuts irregularly across mélange structure along much of its length, suggesting a folded and faulted depositional contact. Part of the contact consists of longitudinal faults along which blocks of chert and basalt were dislodged from the main block and mixed part way into mélange. Much of the bedding within this block strikes at high angles to the structure of the mélange. The chert bedding is cut by a spaced cleavage that locally becomes penetrative, yielding a siliceous slate. The cleavage is axial planar to steeply plunging to moderately plunging folds (Fig. 3, H). This cleavage is coplanar to the adjacent and underlying matrix schistosity. The intensity of deformation manifested by the structure in the adjacent mélange cannot be matched by the cleavage and faults that cut the large chert-basalt block. It is thus suggested that the petrogenesis of the large block commenced after its substrate had been partially mixed by tectonic and possibly sedimentary processes. Signs of soft sediment deformation are well displayed in chert of the large block. This, coupled with the fact that the cleavage and faults that cut the large block are coplanar with the matrix schistosity, indicates that the mixing processes that probably commenced prior to the genesis of the block continued during and after its genesis. Another possible depositional remnant of chert above blocky fractured peridotite and ophicalcite–silica-carbonate rock occurs east of Round Valley (Fig. 2).

All of the mélange units contain vestiges of primary contacts between their various components. The units are thus interpreted as dismembered remnants of once-intact rock assemblages composed of their constituent blocks. As discussed below, this interpretation carries important implications about the nature of the processes that formed the mélange and the environment in which these processes operated.

MELANGE AS DISRUPTED OPHIOLITE

Distinctive rock assemblages having the succession of ultramafic tectonites and cumulates, mafic plutonic rocks, pillow lava, and pelagic sedimentary rocks are referred to as ophiolites (Dewey and Bird, 1971; Church, 1972; Coleman, 1977). Most ophiolites represent fragments of oceanic crust and upper mantle that were detached from oceanic lithosphere and emplaced along continental edges by plate-margin processes. The Kaweah serpentinite mélange is composed almost entirely of rocks typical of the ophiolite assemblage. Furthermore, it contains all of the significant components of the ophiolite assemblage in approximately the right relative proportions. The mélange is thus interpreted as an internally mixed ophiolite.

The ophiolitic rocks of the Kaweah serpentinite mélange are equivalent to the Kings River ophiolite. This is shown by the following facts. (1) The serpentinite mélange can be traced to the Kings River area, where it encloses the large ophiolitic slabs. (2) Mesoscopic structural features of the Kings River slabs and Kaweah mélange blocks are similar. (3) Petrographic features of the Kings River slabs and Kaweah mélange blocks are similar. (4) Palinspastic reconstructions of gabbro-basalt and peridotite-gabbro mélange units to less-deformed states yield hypothetical slabs that are similar to the Kings River slabs (Saleeby, 1977b). (5) U-Pb zircon ages on rare diorite and plagiogranite differentiates from the Kings River slabs and Kaweah mélange blocks are similar. These ages indicate that the igneous rocks of the Kings-Kaweah ophiolite belt originated in latest Paleozoic to possibly earliest Mesozoic time (Saleeby, 1977b, 1978). (6) K-Ar ages on mafic metamorphic tectonites from the Kings River slabs and Kaweah mélange blocks have similar minimum-age spreads.

The Kings River ophiolite contains vestiges of a deep ocean floor stratigraphic succession (Saleeby, 1978). The Kaweah serpentinite mélange thus represents highly disrupted and internally mixed deep ocean floor. As discussed below, the relative chronology of petrogenetic and deformation events within the mélange suggest that the mélange itself also formed on the deep ocean floor.

MELANGE DEVELOPMENT

In this section structural and petrologic data are used to develop a model for the petrogenetic and deformational history of the mélange. Because the mélange contains only the elements of the ophiolite suite, it is considered to have formed in the oceanic domain. A large oceanic fracture zone appears to be the most suitable tectonic environment for ocean-floor mélange development (Saleeby, 1977b, 1978). The northwest-trending structures of the mélange are believed to represent the fracture-zone trends, whereas the cross structures are believed to have been inherited from a rifting axis oriented at a high angle to this trend. The sequential block diagrams of Figure 6 show diagrammatically how this model applies to the Yokohl Valley area.

Deformation of the mélange's ophiolitic protolith commenced during ophiolite genesis. This is demonstrated in the slabs of the Kings River area and in peridotite-gabbro and gabbro-basalt mélange units, all of which contain protoclastic deformation features intimately related to metamorphic tectonite features. The complex transitions that the deformed igneous rocks go in and out of to form greenschist and amphibolite facies tectonites suggest that the prebatholith metamorphic rocks originated as metaigneous rocks at the site of ophiolite genesis. In zones where contact metamorphism by the batholith is at its lowest mineralogic and textural grades, vestiges of earlier greenschist facies assemblages occur primarily in volcanic and hypabyssal protoliths, whereas protoliths containing amphibolite facies assemblages occur primarily in gabbroic rocks. This is consistent with the patterns observed in ocean-floor metamorphism that occurs at the site of ocean-floor genesis

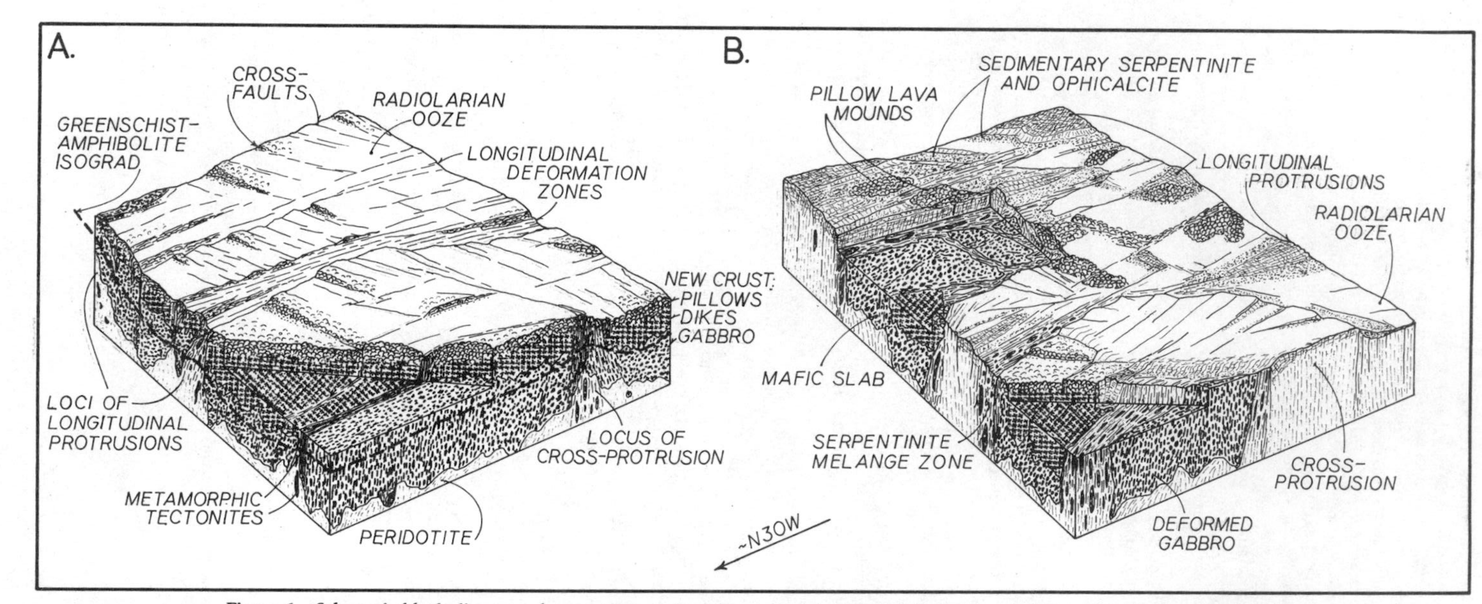

Figure 6. Schematic block diagrams showing early and middle stages of mélange deformation. A: Disruption of new oceanic crust along longitudinal deformation zones and cross-faults, with localized development of greenschist to amphibolite facies tectonites. B: More advanced stages of ocean-floor disruption, with continued longitudinal and cross-faulting joined by ultramafic protrusive activty. Pelagic sedimentation and volcanism is shown continuing throughout these stages of disruption. Sedimentary serpentinites and ophicalcites form along protrusive uplifts. Later stages of deformation would result in configuration similar to that displayed in Yokohl Valley area. Direction arrow refers to present orientation of structures.

(Miyashiro and others, 1971; Miyashiro, 1972; Spooner and Fyfe, 1973; Fox and others, 1976).

Diagram A in Figure 6 shows the syngenetic deformation and metamorphism of the ophiolitic rocks that now occur in gabbro-basalt and peridotite-gabbro mélange units. Deformation is shown as being concentrated in some zones, diffuse in some zones, and lacking in other zones. The deformation modes include brittle fracturing and shearing, ductile and protoclastic flow, and metamorphic recrystallization. Mesoscopic field data suggest that the deformation trends that prevailed during this stage were roughly the same as the trend of the entire ophiolite belt. Exceptions to this pattern occur in (1) zones of cross-fracturing, with concomitant emplacement of dikes and veins, (2) zones of cross-faulting, as shown within and along some margins of mélange units, and (3) zones of protoclasis where deformation features take on chaotic forms and orientations.

The critical process in the formation of the serpentinite mélange was the upward emplacement of ultramafic rock from mantle levels to upper-crustal levels. The common occurrence of blocky fractured serpentinized peridotite, sedimentary serpentinite, and ophicalcite and the common association of chert with various ultramafic rocks favors the upward emplacement of ultramafic material rather than the downfaulting of crustal mafic rocks to mantle depths. Upward flow of ultramafic material started during the igneous petrogenesis of the mafic rocks and continued throughout mélange development. This is shown in the mafic-ultramafic slabs of the Kings River area (Saleeby, 1978) and at numerous locations along the mélange where small metagabbro-amphibolite inclusions occur within peridotite blocks. In these instances, steep downdip lineations indicate intense episodes of both vertical extension and slip. These lineations formed along with and within the plane of high-temperature foliations that commonly pervade peridotite and mafic blocks. It thus appears that while the mafic crustal rocks were syngenetically deformed, portions of hot upper mantle were flowing and slipping upward to crustal levels. The upward emplacement of ultramafic rock occurred primarily along the regional structure; however, emplacement along cross-faults also occurred.

The attendant disruption of the young oceanic crust probably facilitated migration of ocean water into deeper levels of the crust. This resulted in progressive serpentinization of the ascending ultramafic rocks and probably accelerated the protrusion process. The blocky fractured serpentinized peridotite blocks may be the signature of expansion and autobrecciation during diapiric rise and serpentinization (Coleman and Keith, 1971). A continuum is envisaged between thermal-tectonic rise and expansive rise related to serpentinization. This is suggested by the fact that in ultramafic rocks the high-temperature foliation surfaces, blocky fracture systems, and serpentine schistosities each have intergradational relationships with one another.

The surface expressions of ultramafic protrusive activity are sedimentary serpentinite, ophicalcite, and ultramafic detritus within chert. Soft-sediment deformation in chert and multiple cycles of brecciation, sedimentation, and cementation in sedimentary serpentinite and ophicalcite indicate that continued protrusion and faulting sustained an unstable depositional environment on the ocean floor. In addition to clastic and biogenic sedimentation operating during ultramafic protrusive activity, ocean-floor igneous activity continued. This is shown by pillow lavas and breccias that occur with sedimentary serpentinite, ophicalcite, and chert. The pillow lavas probably formed as local mounds. Some may have been built on protrusive highs, as suggested by their high vesicularity and interpillow limestone.

The critical stage of mélange development characterized by the disruption of mafic crust by ultramafic protrusions, the surfacing of protrusions that shed sedimentary serpentinites, and continued ocean-floor volcanism is depicted in diagram B of Figure 6. It appears that pelagic sedimentation preceded, continued throughout, and followed the main pulse of protrusive activity. The earliest formed cherts appear to have been mixed with ultramafic rock prior to and during deposition and penecontemporaneous deformation of the later formed chert. The disrupted crust beneath the later formed cherts consisted of mafic slabs interwoven with narrow mélange zones, ultramafic protrusions, sedimentary serpentinite, and local pillow-lava mounds. The fact that pelagic sedimentation progressed through this stage of mélange development, coupled with the absence of continental-margin rocks as blocks or matrix within the mélange, indicates an oceanic origin for the mélange.

The ensuing deformation that affected the disrupted ocean floor depicted in diagram B of Figure 6 led to a structural configuration that is close to that shown in Figure 2. This deformation consisted of (1) flattening in the plane of the matrix schistosity, as shown by transverse tension fractures within many mélange blocks and boudinaging of some mélange blocks, (2) longitudinal faulting and further tectonic mixing, as shown by the faulted and mixed margins of the large chert ± basalt blocks that appear to be depositional remnants, (3) cleavage development and folding about mainly steep axes also shown in these large blocks, and (4) continued serpentinization and protrusion of ultramafic material, as shown by protrusions and sedimentary serpentinites that occur within the continental margin assemblage. The degree to which these features formed during and after sedimentary overlap of the continental-margin assemblage is difficult to resolve. As discussed below, faulting, folding about steep-plunging axes, and regional flattening also affected the continental-margin rocks along with their serpentinite mélange basement. However, significant tectonic mixing is limited to the mélange and thus predates sedimentary overlap by the continental-margin assemblage.

CONTINENTAL-MARGIN SEDIMENTATION AND DEFORMATION

Continental-margin assemblage rocks are in tectonic contact with the ophiolite belt along its eastern margin (Fig. 1) and occur as depositional remnants along much of the serpentinite mélange. In this section the petrologic character, age, and possible correlatives of the assemblage are discussed briefly, and its structural relations with the serpentinite mélange are covered in more depth. Petrologic data in relation to regional tectonics and paleogeography are covered in Saleeby and others (1978).

Metamorphic Rocks East of the Ophiolite Belt

Remnants of continental-margin assemblage rocks lying east of the ophiolite belt occur in numerous roof pendants in the drainages of the Kings, Kaweah, and Tule Rivers (Saleeby and others, 1978). The present contact between these rocks and the ophiolite belt is a vertical fault or fault zone with little or no tectonic mixing between the two. It is significant that mélange of the ophiolite belt along the fault rarely contains inclusions of the continental-margin assemblage. This indicates that tectonic mixing of the ophiolite belt predates the present contact.

Quartz-mica (± andalusite) schist, quartzite, marble, and calc-silicate rock constitute the majority of this part of the assemblage.

The protoliths of these metamorphic rocks appear to have been primarily siliceous-argillite, quartzose to subarkosic flysch, chert, massive quartzose sandstone, limestone, and calcareous sandstone. Felsic volcanic and lesser mafic volcanic and hypabyssal rocks are also present at several localities. Tectonic deformation generally increases in intensity toward the ophiolite belt. Deformation in these rocks consists of (1) soft-sediment folding, slumping, and chaotic mixing, (2) regional tectonic flattening parallel to the ophiolite belt, (3) faulting and folding along regional trends, (4) local intense ductile folding, flattening, and boudinaging during batholith emplacement, and (5) crenulation and hot fracturing of the structures that developed from the processes listed above.

Late Triassic and Early Jurassic fossils have been recovered at several localities (Christensen, 1963; Jones and Moore, 1973; Saleeby and others, 1978). The uniformity in general lithologic character and deformation that these rocks show suggests that these ages are applicable for the entire assemblage east of the ophiolite belt. These rocks constitute the main part of the Kings sequence of Bateman and Clark (1974) and the southern continuation of the eastern unit of the Calaveras Complex (Schweickert and others, 1977). Regional tectonic relations and petrochemical studies of the Sierran region suggest that they were deposited for the most part on continental basement situated east of the foothill suture (Saleeby, 1977b).

Depositional Remnants above Serpentinite Mélange

Remnants of continental-margin rocks lying depositionally above serpentinite mélange consist of a complex mixture of primarily hemipelagic, submarine volcanic, and continent-derived clastic rocks. The mixed nature of these rocks appears to have resulted from olistostrome processes. However, penetrative deformation, resulting in a flattening foliation and slaty cleavage, and localized faulting have obscured primary relations. In Figure 2 lithologic zones are shown for part of the large remnant that occurs in the area of Yokohl and Round Valleys. Mappable marker beds and blocks are also shown to help elucidate the complex structure of these rocks. Much of the discussion below is based on relations observed in this area. The rocks of this area consist of the six components described below.

1. Mud is a ubiquitous component that occurs in cherty, tuffaceous, clastic, and chaotically intermixed rocks. The mud is now present as white mica and biotite and as andalusite near contacts with the batholith.

2. Radiolaria tests are another significant component occurring in chert beds or mixed with mud to form siliceous argillite. The tests have been highly recrystallized by deformation and contact metamorphism, which has so far prohibited paleontological dating.

3. Quartzsose to subarkosic silt and sand occur in graded sandstone beds and in mudstone as part of a flysch sequence (Fig. 7, A). The sandstone also occurs as cobbles and blocks as much as 150 m in diameter in chaotic deposits. In zones of intense contact metamorphism, the sandstones have been recrystallized to quartz-mica schists that are difficult to distinguish from metamorphic derivatives of argillaceous chert.

4. Limestone occurs as lenses and irregular to rounded blocks as much as 500 m in diameter and as well-rounded cobbles, all in chaotic deposits (Fig. 7, B). Recrystallization to marble is commonly complete. However, relict pisolites are quite common, and fossil bryozoans, crinoid stems, and fusilinids have been recovered. The fusilinids are significant in that they indicate a Late Permian age for the limestone blocks, and they are also of a fauna that is exotic to North America (Saleeby and others, 1978). It is emphasized that the age of the limestone blocks is a maximum age for the chaotic deposits in which they occur.

5. Volcanic rocks ranging in composition from basalt to andesite occur as discrete units consisting of breccia, tuff-breccia, pillow lava, and hypabyssal rock. Silicic tuff also occurs as thin beds and as an admixture in quartzitic deposits. The volcanic rocks have been interpreted as the products of arc magmatism (Saleeby and Sharp, 1977; Saleeby and others, 1978).

6. Ophiolitic rocks occur as chaotic deposits within the Yokohl-Round Valley remnant. These occur as local chlorite-serpentine mud layers, in deposits containing the components listed above, and as mappable olistostrome units. The olistostromes consist primarily of chert and ophicalcite blocks and clasts within a sedimentary serpentinite–pebbly mudstone matrix (Fig. 7, C). The olistostromes were probably shed from protrusive highs in the mélange basement. This is favored over simple fault-bounded highs because the olistostrome assemblage consists of the earlier protrusion assemblage of the serpentinite mélange and generally lacks peridotite, gabbro, and basalt.

Chaotic rocks of the depositional remnants are not considered tectonic in origin. Exceptions to this may occur along faults where shear-fracture fabrics have developed. However, there is no evidence of widespread brittle shearing with the formation of phacoids of resistant material within a milled-down matrix. Chaotic mixing by mass movement in a nonlithified to semilithified state is favored by the following facts. (1) Vestiges of disrupted beds have chaotic forms that include roll-up or ball structures, complex interpenetration structures, and pull-apart structures that have inconsistent relationships with definite tectonic structures. (2) There are no traces of the tectonically eroded fragments of chert, limestone, or sandstone in the matrix of these rocks that would be expected if mixing had been tectonic in origin. Instead, the matrix material is siliceous argillite or argillite, regardless of the clast content. (3) Tectonic structures of the chaotic deposits commonly have mesoscopic relationships with chaotic structures, demonstrating that chaotic mixing predated tectonic deformation. A common example is where the regional cleavage cuts across limbs of chaotic folds, ball structures, and long axes of clasts (Fig. 7, D). (4) Shallow-water limestone and deeper water radiolarian chert have been mixed together in these rocks. Downward transport of the limestone blocks into a deeper water environment is suggested by the presence of these blocks in chert-argillite diamictites that are apparently interbedded with thin, relatively continuous chert horizons that could not have been transported far from their site of origin. It is also significant that most of the features discussed above have been reported from large-scale olistostromes of late Cenozoic age that have slid off both stable and active continental margins (T. C. Moore and others, 1970; Embley, 1976; Jacobi, 1976; D. G. Moore and others, 1976).

The continental-margin assemblage was deformed during and after its deposition; this is manifest by tectonic deformational features, intraformational breccias, and olistostromes shed from basement uplifts. The dominant tectonic structural feature that developed was a vertical-flattening foliation parallel to the matrix schistosity of the underlying serpentinite mélange. The flattening foliation is best displayed in distorted diamictite clasts (Fig. 7, E). In many instances there is a cleavage associated with the foliation which is axial planar to moderately plunging to mainly steeply plunging folds. Almost all of the folds observed are symmetric.

Open to moderately tight varieties are most commonly observed; isoclinal varieties are rarely observed. This is not taken to indicate that isoclinal folds are rare, since bedding surfaces have been rotated into the plane of the cleavage. In the Yokohl Valley–Round Valley area, preferences in facing directions have not been observed; this is consistent with the symmetric form of the folds. In the Tule River area, large well-bedded intervals tend to face west (Fig. 7, A), but the observed folds of that region are also symmetric.

Longitudinal faults have also cut these rocks. These faults are parallel to the regional fabric and are characterized by abrupt changes in lithology, the development of shear-fracture fabric, the presence of schistose serpentinite slivers, and the presence of felsic dikes of probable Cretaceous age. The serpentinite slivers appear to have been protruded along the faults from the serpentinite mélange basement. This is well displayed where the fault that extends northward from Oat Canyon passes into the serpentinite mélange near Yokohl Valley (Fig. 2). Here, sedimentary serpentinite of the mélange can be traced into the fault zone, where it thins to discontinuous schistose slivers within the fault zone. Some of the longitudinal faults line up with the rare slivers of terrigenous rock that have been faulted into the mélange.

Absolute-age constraints on the depositional remnants place them between latest Permian and Jurassic. However, a Triassic to Middle Jurassic age is inferred for these rocks from the following relations. (1) The flysch sequence can be traced across the foothill suture where it occurs in fossiliferous strata of Late Triassic to Early Jurassic age. (2) Quartzitic to subarkosic clasts and blocks that occur in olistostrome deposits also occur as clasts, blocks, and intact units east of the suture. (3) The apparent intrusive equivalents of the arc-type volcanic rocks have yielded Middle Jurassic zircon ages (Saleeby, 1975; Saleeby and Sharp, 1977). The structure, composition, exotic limestone fauna, and age constraints of the depositional remnants suggest that they may be correlative with the western belt of the Calaveras Formation exposed along the main segment of the foothill metamorphic belt (Clark, 1964; Duffield and Sharp, 1975). Possible correlatives within the main belt of the Calaveras Formation (or Complex) (Clark, 1964; Schweickert and others, 1977) may also exist.

Significance of Depositional Remnants Relative to Serpentinite Mélange Development

Recognition of continental-margin rocks as depositional remnants above serpentinite mélange is significant from two standpoints: (1) it identifies the basement upon which the continental-margin rocks were deposited, and (2) it puts constraints on the age, tectonic setting, and relative intensity of the later stages of mélange deformation, since the depositional remnants were deformed along with their mélange basement. Important criteria used for the recognition of these rocks as depositional remnants may be summarized as follows. (1) Tectonic deformation of the depositional remnants is much less intense than in the mélange basement and in general does not involve intense tectonic mixing. (2) The depositional remnants are in contact with the mélange in zones that can be interpreted as the ophiolite belt's highest structural level, as shown by the predominance of chert, ophicalcite, and sedimentary serpentinite. (3) Except for longitudinal fault contacts, contacts between the depositional remnants and the mélange cross the pervasive planar structures of the region and have the appearance of open to tight folds (Fig. 2). These apparent folds have geometries similar to those of the mesoscopic folds in the depositional remnants, and both apparent and observed folds share the same axial surface, which is the regional flattening plane. The highly irregular contact, thus, has the appearance of a transposed depositional contact. (4) The intrusive equivalents to the volcanic component of the depositional remnants intrude the mélange basement rocks. (5) Ophiolitic debris ranging from fine detritus to large slide blocks occurs within the depositional remnants. This debris had to have been shed from basement exposures. (6) Serpentinite slivers that occur along faults within the depositional remnants were apparently emplaced along vertical protrusion pathways from the underlying serpentinite mélange basement.

Most of these criteria taken alone do not dictate the interpretation presented above. However, all of these criteria taken together seem to necessitate such an interpretation.

The structural configuration of the depositional remnants can be used to deduce the final stages of mélange deformation. These appear to be (1) regional flattening along the pre-existing trends of the ophiolite belt, (2) faulting along the same regional trend, and (3) remobilization and protrusion of sedimentary serpentinite and mélange matrix along the vertical faults within the overlying strata. This is consistent with observations within the mélange that show that in some locations the matrix schistosity appears to represent a plane of flattening and that discontinuities within and between some mélange units line up with mappable faults in the depositional remnants (Fig. 2).

From the preceding discussion it is apparent that the pre-existing structural trends within the serpentinite mélange basement controlled the trends along which the younger overlying strata were deformed. Chaotic soft-sediment mixing within the younger strata is believed to have been induced primarily by basement tectonics. It is suggested that much of the tectonic flattening and folding of the younger strata occurred prior to and during lithification. The effects of flattening are best displayed in slaty rocks in which mud was an important component. Massive volcanic units composed of breccia, pillow lava and shallow-level intrusions, and irregular to rounded limestone olistoliths commonly show no signs of flattening. Since the massive volcanic units were lithified at the time of their emplacement and the limestone olistoliths were introduced into the argillaceous rocks in a lithified state and since tectonic flattening occurred prior to high-temperature metamorphism, the ductility contrast displayed between the flattened and nonflattened rocks can be best explained if deformation commenced prior to complete lithification of the slaty rocks. In contrast to this pattern, slaty rocks, massive volcanic units, and limestone olistoliths all have shear-fracture fabrics in the vicinity of mappable faults. This suggests that a significant component of the faulting occurred after or during the late stages of regional flattening.

Because constraints on the age of the depositional remnants are at present poorly defined, the precise timing of late-stage deformations of the serpentinite mélange basement rocks are not well known. The inferred age of Triassic to Middle Jurassic for the depositional remnants suggests that intense tectonic mixing of ophiolite basement rocks ceased at some time in the Triassic and that regional flattening, folding, longitudinal faulting, and late-stage protrusive activity continued through Middle Jurassic time. These late-stage deformations are probably in part related to the Late Jurassic Nevadan orogeny, which is considered a distinct orogenic pulse of regional extent (Bateman and Clark, 1974). In contrast to this view, the views set forth above suggest that the Nevadan orogeny in this region was the culmination of a deformational regime that was controlled by basement tectonics and that this regime existed throughout early Mesozoic and possibly back into latest Paleozoic time. The later stages of this regime recorded in the continental-margin assemblage occurred within the locus of Jurassic island-arc volcanism. It must be emphasized, however, that sedimentation during this stage was dominated by North American–derived detritus and by reworked hemipelagic deposits. Thus, the tectonic and paleogeographic settings of late-stage mélange deformation are considered to have been an intra-arc deformation zone, with the volcanic arc built across the foothill suture (Saleeby and others, 1978). The deformation zone followed the pre-existing structure of the suture. The link between deep ocean-floor tectonics that yielded serpentinite mélange and the continental-margin regime outlined above are discussed in Saleeby (1977b, 1978).

TECTONIC SETTING OF OCEAN-FLOOR MELANGE DEVELOPMENT

The ophiolite-mélange association is generally accepted as the signature of plate convergence along fossil continental margins (Dewey and Bird, 1970, 1971; Coleman, 1971; Hsü, 1971). This view is not adopted for the Kings-Kaweah ophiolite belt. In the section on mélange development the point was made that tectonic mixing of the ophiolite belt commenced at a site of ocean-floor genesis and that the mixing processes operated primarily in the oceanic realm without the involvement of continental-margin rocks. Continental-margin rocks were deposited across ocean-floor mélange following significant tectonic mixing. Furthermore, once the ophiolitic mélange was in the continental-margin environment, it coincided with the locus of arc magmatism, not the locus of structural imbrication and high pressure/temperature metamorphism — the signature of the subduction process.

An alternative to a continental-margin subduction mode of mélange development is presented in Saleeby (1977b). Here the Kings-Kaweah ophiolite belt is considered to have originated at a mid-ocean spreading center where cut by a large transverse fracture zone. Mixing of ocean floor to form serpentinite mélange resulted from combined protrusion and wrench tectonics. Ophiolite emplacement was primarily a result of large-scale wrench faulting that truncated the southwest continental margin and transported the fracture-zone complex into the continental-margin regime, where it subsequently served as an exotic basement terrane.

Structural features of the mélange permit a hypothetical kinematic pattern consisting of dextral wrench movements oriented along the regional trend of the ophiolite belt and rifting with an axis oriented at high angles to the regional trend (Fig. 6). Relationships between structures with regional trends and the less prevalent cross structures indicate that they formed simultaneously. This can be seen at a wide range of scales (Figs. 2, 5) where northwest-trending structures cut and are cut by the cross structures. Perhaps the cross structures formed in response to both ridge-crest faulting and protrusion of mantle-derived blocks. The interaction of cross-faulting with protrusion and longitudinal wrenching is believed to have greatly facilitated mélange development.

The structural and petrologic complexities of oceanic fracture zones have just recently begun to emerge from marine geological and geophysical studies. Metamorphic tectonites, ultramafic protrusions, sedimentary serpentinites, ophicalcites, offridge volcanism, and complex patterns in structural overprinting abound in these zones (Aumento and others, 1971; Bonatti and others, 1971, 1973, 1974; Melson and Thompson, 1971; Melson and others, 1972; Thompson and Melson, 1972; Bonatti and Honnorez, 1976; Fox and others, 1976; DeLong and others, 1977; Schreiber and Fox, 1977). Perhaps significant strips of modern oceanic basement are composed of serpentinite mélange formed along fracture zones.

ACKNOWLEDGMENTS

Special thanks go to C. A. Hopson for his guidance and criticism in the early stages of this study. Field assistance from W. S. Leith, W. D. Sharp, S. E. Goodin, C. J. Busby, and L. F. Drake is gratefully acknowledged. This study was supported in part by a Geological Society of America Penrose Bequest Grant and by National Science Foundation Grant EAR77-08691.

REFERENCES CITED

Aumento, F. Loncarevic, B. D., and Ross, D. I., 1971, Hudson geotraverse: Geology of the Mid-Atlantic Ridge at 45°N: Royal Society of London Philosophical Transactions, ser. A, v. 268, p. 623–650.

Bateman, P. C., and Clark, L. D., 1974, Stratigraphic and structural setting of the Sierra Nevada batholith, California: Pacific Geology, v. 8, p. 79–89.

Berkland, J. D., Raymond, L. A., Kramer, J. C., and others, 1972, What is Franciscan?: American Association of Petroleum Geologists Bulletin, v. 56, p. 2205–2302.

Blake, M. C., and Jones, D. L., 1974, Origin of Franciscan mélanges in northern California, *in*, Dott, R. H., Jr., and Shaver, R. H., eds., Modern and ancient geosynclinal sedimentation: Society of Economic Paleontologists and Mineralogists Special Publication 19, p. 345–357.

Bonatti, E., and Honnorez, J., 1976, Sections of the Earth's crust in the equatorial Atlantic: Journal of Geophysical Research, v. 81, p. 4104–4116.

Bonatti, E., Honnorez, J., and Ferrara, G., 1971, Peridotite-gabbro-basalt complex from the equatorial Mid-Atlantic Ridge: Royal Society of London Philosophical Transactions, ser. A, v. 268, p. 385–402.

Bonatti, E., Honnorez, J., and Gartner, S., Jr., 1973, Sedimentary serpentinites from the Mid-Atlantic Ridge: Journal of Sedimentary Petrology, v. 43, p. 728–735.

Bonatti, E., Emiliani, C., Ferrara, G., and others, 1974, Ultramafic-carbonate breccias from the equatorial Mid-Atlantic Ridge: Marine Geology, v. 16, p. 83–102.

Cashman, P. H., 1974, Mélange terrane in the western Paleozoic and Triassic subprovince, Klamath Mountains, California: Geological Society of America Abstracts with Programs, v. 6, p. 153.

Christensen, M. L., 1963, Structures of metamorphic rocks at Mineral King, California: University of California Publications in Geological Sciences, v. 42, p. 159–198.

Church, W. R., 1972, Ophiolite: Its definition, origin as oceanic crust, and mode of emplacement in orogenic belts, with special reference to the Appalachians, *in* Irving, E., ed., The ancient oceanic lithosphere: Canada Department of Energy, Mines and Resources Earth Physics Branch Publication 42, p. 71–85.

Clark, L. D., 1960, Foothills fault system, western Sierra Nevada, California: Geological Society of America Bulletin, v. 71, p. 483–496.

——1964, Stratigraphy and structure of part of the western Sierra Nevada metamorphic belt, California: U.S. Geological Survey Professional Paper 410, 70 p.

Coleman, R. G., 1971, Plate tectonic emplacement of upper mantle peridotites along continental edges: Journal of Geophysical Research, v. 76, p. 1212–1222.

——1977, Ophiolites, ancient oceanic lithosphere?: New York, Springer-Verlag, 240 p.

Coleman, R. G., and Keith, T. E., 1971, A chemical study of serpentinization — Burro Mountain, California: Journal of Petrology, v. 12, p. 311–328.

Cowan, D. S., 1974, Deformation and metamorphism of the Franciscan subduction zone complex northwest of Pacheco Pass, California: Geological Society of America Bulletin, v. 85, p. 1623–1634.

Cowan, D. S., and Mansfield, C. F., 1970, Serpentinite flows on Joaquin Ridge, southern Coast Ranges, California: Geological Society of America Bulletin, v. 81, p. 2615–2628.

DeLong, S. E., Dewey, J. F., and Fox, P. J., 1977, Displacement history of oceanic fracture zones: Geology, v. 5, p. 199–202.
Dewey, J. F., and Bird, J. M., 1970, Mountain belts and the new global tectonics: Journal of Geophysical Research, v. 75, p. 2625–2647.
——1971, Origin and emplacement of the ophiolite suite: Appalachian ophiolites in Newfoundland: Journal of Geophysical Research, v. 76, p. 3179–3266.
Dickinson, W. R., 1966, Table Mountain serpentinite extrusion in California Coast Ranges: Geological Society of America Bulletin, v. 77, p. 451–472.
Duffield, W. A., and Sharp, R. V., 1975, Geology of the Sierra foothills mélange and adjacent areas, Amador County, California: U.S. Geological Survey Professional Paper 827, p. 1–30.
Durrell, C. D., 1940, Metamorphism in the southern Sierra Nevada northeast of Visalia, California: University of California Department of Geological Sciences Bulletin, v. 24, p. 1–118.
Embley, R. W., 1976, New evidence for occurrence of debris flow deposits in deep sea: Geology, v. 4, p. 371–374.
Fox, P. J., Schreiber, E., Rowlett, H., and others, 1976, The geology of the Oceanographer fracture zone: A model for fracture zones: Journal of Geophysical Research, v. 81, p. 4127–4128.
Gansser, A., 1955, New aspects of the geology in central Iran: World Petroleum Congress, 4th, Rome 1955, Proceedings, sec. 1, p. 279–300.
——1974, The ophiolite mélange, a world-wide problem on Tethyan examples: Eclogae Geologicae Helevetiae, v. 67, p. 479–507.
Hsü, K. J., 1968, Principles of mélanges and their bearing on the Franciscan-Knoxville paradox: Geological Society of America Bulletin, v. 79, p. 1063–1074.
——1969, Preliminary report and geologic guide to Franciscan mélanges of the Morro Bay–San Simeon area, California: California Division of Mines and Geology, Special Publication 35, 46 p.
——1971, Franciscan mélanges as a model for eugeosynclinal sedimentation and underthrusting tectonics: Journal of Geophysical Research, v. 76, p. 1162–1170.
Jacobi, R. D., 1976, Sediment slides on the northwestern continental margin of Africa: Marine Geology, v. 22, p. 157–173.
Jennings, W. C., and Strand, R. G., 1966, Geologic map of California: Fresno and Bakersfield sheets: California Division of Mines and Geology, scale 1:250,000.
Jones, D. L., and Moore, J. G., 1973, Lower Jurassic ammonite from the south-central Sierra Nevada, California: U.S. Geological Survey Journal of Research, v. 1, p. 453–458.
Knipper, A. L., 1973, Serpentinitic mélange of the Lesser Caucasus: U.S.S.R. Academy of Science, International Symposium on Ophiolites in the Earth's Crust, Abstracts, p. 71–89.
Lockwood, J. P., 1971, Sedimentary and gravity-slide emplacement of serpentinite: Geological Society of America Bulletin, v. 82, p. 919–936.
Makarychev, G. I., and Visnevsky, Y. S., 1973, Excursion 2: U.S.S.R. Academy of Sciences International Symposium on Ophiolites in the Earth's Crust, Excursion Guidebook, p. 106–116.
Maxwell, J. C., 1974, Anatomy of an orogen: Geological Society of America Bulletin, v. 85, p. 1195–1204.
Melson, W. G., and Thompson, G., 1971, Petrology of a transform fault zone and adjacent ridge segments: Royal Society of London Philosophical Transactions, ser. A, v. 263, p. 423–442.
Melson, W. G., Hart, S. R., and Thompson, G., 1972, St. Paul's Rocks, equatorial Atlantic: Petrogenesis, radiometric ages, and implications on sea-floor spreading, *in* Shagam, R., and others, eds., Studies in Earth and space sciences: Geological Society of America Memoir 132, p. 241–272.
Miyashiro, A., 1972, Pressure and temperature conditions and tectonic significance of regional and ocean-floor metamorphism, *in* Ritsema, A. R., ed., The upper mantle: Amsterdam, Elsevier, p. 141–159.
Miyashiro, A., Shido, F., and Ewing, M., 1971, Metamorphism in the Mid-Atlantic Ridge near 24° and 30°N: Royal Society of London Philosophical Transactions, ser. A, v. 268, p. 589–603.
Moiseyev, A. N., 1970, Late serpentinite movements in the California Coast Ranges: New evidence and implications: Geological Society of America Bulletin, v. 81, p. 1721–1732.
Moore, D. G., Curray, J R.,and Emmel, F. J., 1976, Large submarine slide (olistostrome) associated with Sunda arc subduction zone, northeast Indian Ocean: Marine Geology, v. 21, p. 211–226.
Moore, T. C., Jr., Van Andel, T. H., Blow, W. H., and others, 1970, Large submarine slide off northeastern continental margin of Brazil: American Association of Petroleum Geologists Bulletin, v. 54, p. 125–128.
Pevye, A. V., 1969, Oceanic crust of the geologic past: Geotectonics, v. 4, p. 210–224.
Plafker, G., 1956, Geology of the southwest part of the Kaweah 30′ quadrangle [M.S. thesis]: Berkeley, University of California.
Saleeby, J. B., 1975, Structure, petrology and geochronology of the Kings-Kaweah mafic-ultramafic belt, southwestern Sierra Nevada foothills, California [Ph.D. thesis]: Santa Barbara, University of California, 286 p.
——1977a, Fieldtrip guide to the Kings-Kaweah suture, southwestern Sierra Nevada foothills, California: Geological Society of America Cordilleran Section 73rd Annual Meeting, 47 p.
——1977b, Fracture zone tectonics, continental margin fragmentation, and emplacement of the Kings-Kaweah ophiolite belt, southwest Sierra Nevada, California, *in* Coleman, R. G., and Irwin, W. P., eds., North American ophiolites: Oregon Department of Geology and Mineral Industries Bulletin 95, p. 141–160.
——1978, Kings River ophiolite, southwest Sierra Nevada foothills, California: Geological Society of America Bulletin, v. 89, p. 617–636.
Saleeby, J. B., and Sharp, W. D., 1977, Jurassic igneous rocks emplaced along the Kings-Kaweah suture, southwestern Sierra Nevada foothills, California: Geological Society of America Abstracts with Programs, v. 9, p. 494–495.
Saleeby, J. B., Goodin, S. E., Sharp, W. D., and others, 1978, Early Mesozoic paleotectonic-paleogeographic reconstruction of the southern Sierra Nevada region, *in* Howell, D. G., and McDougall, K., eds., Mesozoic paleogeography of the western United States: Society of Economic Paleontologists and Mineralogists Pacific Section Publication, p. 311–336.
Schreiber, E., and Fox, P. J., 1977, Density and *P*-wave velocity of rocks from the FAMOUS region and their implication to the structure of the oceanic crust: Geological Society of America Bulletin, v. 88, p. 600–608.
Schweickert, R. A., Saleeby, J. B., Tobisch, O. T., and others, 1977, Paleotectonic and paleogeographic significance of the Calaveras Complex, western Sierra Nevada, California, *in* Stewart, J. H., and others, eds., Paleozoic paleogeography of the western United States: Society of Economic Paleontologists and Mineralogists Pacific Section Publication, p. 381–394.
Spooner, E.T.C., and Fyfe, W. S., 1973, Sub–sea-floor metamorphism, heat and mass transfer: Contributions to Mineralogy and Petrology, v. 42, p. 287–304.
Thompson, G., and Melson, W. G., 1972, The petrology of the oceanic crust across fracture zones in the Atlantic Ocean: Evidence of a new kind of sea-floor spreading: Journal of Geology, v. 80, p. 526–538.

MANUSCRIPT RECEIVED BY THE SOCIETY JULY 26, 1976
REVISED MANUSCRIPT RECEIVED DECEMBER 1, 1977
MANUSCRIPT ACCEPTED JANUARY 20, 1978

24

ORIGIN OF FRANCISCAN MELANGES IN NORTHERN CALIFORNIA

M. C. BLAKE, JR., AND DAVID L. JONES
U. S. Geological Survey, Menlo Park, California

ABSTRACT

In northern California, chaotic Franciscan mélange occurs beneath the overlying ophiolite and Great Valley Sequence. Identical mélanges occur to the west, separating well-bedded, coherent Franciscan units that differ markedly in age. Detailed studies in several places indicate that these mélanges mark the boundaries of imbricate thrust sheets, and they appear to occur at several discrete structural horizons.

The mélange comprises blocks of graywacke, greenstone, chert, serpentinite, and isolated so-called knockers of high-grade blueschist and eclogite set in a matrix of sheared and quartz-veined mudstone and minor sandstone. Except for the blocks of high-grade schist, these rocks are similar to, but more deformed than, the orderly sedimentary, volcanic, and other rocks that occur immediately above the Coast Range thrust at the base of the Great Valley Sequence. Unlike the other Franciscan units, the mélanges contain relatively abundant fossils, mainly *Buchia,* radiolarians, and dinoflagellates. Significantly, all of these fossils are of Tithonian to Valanginian age.

We suggest, on the basis of similarity of lithology and fossil content, that the matrix of the mélanges represents a distal, or seaward, portion of the basal sediments of the Great Valley Sequence and that the abundant greenstone, chert, and serpentinite found as tectonic blocks within the mélanges were derived from the underlying oceanic crust and upper mantle.

Formation of the mélanges must be related to multiple subduction of separate plates, the mélange being generated repeatedly from the ultramafic-mafic-chert and *Buchia*-bearing shale and minor graywacke sequence that constitutes the oldest rocks of the Coast Ranges. This process of subduction probably began in the Early Cretaceous and continued into the Tertiary, as Eocene fossils have been found recently in deformed Franciscan (coastal belt) rocks structurally below the mélanges.

The tectonic blocks of high-grade blueschist and eclogite were formed during an earlier period of subduction, then embedded in serpentinite and carried westward by flow in the upper mantle. During subsequent subduction, the serpentinite and embedded blocks of schist were tectonically mixed with the overlying volcanic rocks, chert, graywacke, and fossiliferous shale.

INTRODUCTION AND GEOLOGIC BACKGROUND

A popular concept of the Franciscan assemblage is a mélange comprising blocks of ultramafic rocks, gabbro, diabase, volcanic rocks, radiolarian chert, limestone, graywacke, and metamorphic rock. According to plate tectonic models proposed by several geologists, late Mesozoic oceanic crust was generated at a midocean ridge and was subsequently covered by radiolarian oozes formed on abyssal plains, and limestone formed on rises and seamounts as it was swept toward the continent by ocean-floor spreading. As this oceanic plate encountered a postulated trench marking the western margin of the North American plate, it was further covered by clastic sediments. "Mixing of these rocks under the trench and along the Benioff zone, produced the mélanges" (Hsü, 1971, p. 1168). This hypothesis may be correct for some parts of the Franciscan but does not appear to fit known geologic data for the Coast Ranges north of San Francisco. Published mapping of the Coast Ranges by Brown (1964a), Rich (1971), Swe and Dickinson (1970), Fox and others (1973), Berkland (1969, 1972), Sims and others (1973), Blake and others (1971 and in preparation), and Suppe (1972) and unpublished reports by geologists of the California Department of Water Resources indicate that the Franciscan rocks comprise a number of subparallel, north-northwest trending lithologic belts. These include (1) relatively well-bedded graywacke and mudstone with interbedded chert and volcanic rocks; (2) metagraywacke and metachert and metagreenstone, characterized by a faint to pronounced metamorphic fabric and development of new metamorphic minerals such as lawsonite, jadeite, and glaucophane; and (3) massive arkosic sandstone and mudstone alternating with thin-bedded flyschlike sandstone and mudstone. The contacts between these distinctive lithologic units are shear zones defined by mélanges made up of blocks and slabs of graywacke, greenstone, chert, serpentinite, and their metamorphosed equivalents set in a matrix of dark, sheared, and quartz-veined mudstone and minor sandstone. Structurally overlying all these lithologic belts is a discontinuous sheet of ultramafic rocks, gabbro, diabase, pillow lavas, and chert that in turn is overlain by the sedimentary rocks of the Great Valley Sequence of Late Jurassic and Cretaceous age. The major outcrop area of these upper plate rocks is along the west side of the Great Valley; they also occur as scattered tectonic outliers (klippen) west of the valley. The Coast Range thrust marks the base of this upper plate.

DESCRIPTION OF LITHOLOGIC BELTS

Franciscan assemblage.—The structurally highest Franciscan unit in northern California is informally named the Yolla Bolly belt after the Yolla Bolly Mountains in the northern part of the area mapped (fig. 1). Rocks of this belt are predominantly graywacke and mudstone but include significant interbedded radiolarian chert and volcanic rocks.

The medium- to coarse-grained graywacke occurs in beds that range in thickness from centimeters to tens of meters but most commonly are about 30 cm to 1 m thick. The darker mudstone is generally subordinate, forming interbeds only a few centimeters thick. Other rocks present are mappable beds of pebble conglomerate and pebbly to bouldery mudstone. The mudstone locally is several hundred meters thick. These rocks contain these abundant sedimentary structures that suggest deposition by turbidity currents: graded beds, sole markings, and small-scale current-ripple laminations.

Radiolarian chert in a number of localities is clearly interbedded with coarse-grained graywacke and mudstone without associated volcanic rocks. In the Yolla Bolly area (fig. 1), three chert beds, each more than 50 m thick, have been mapped along strike for more than 15 km and extend northwest and southeast far beyond the mapped area (Blake, 1965).

Volcanic rocks in the Yolla Bolly belt range in composition from quartz keratophyre to basalt; tuff is particularly abundant. Pillow structures are extremely rare and may be restricted to bouldery mudstone intervals. Only in a very few places are volcanic rocks and radiolarian cherts found together; this suggests that in this area there is no genetic relation between them.

Fossils are extremely rare in this belt. Several collections of *Buchia* of Late Jurassic and Early Cretaceous age are known from the pebbly and bouldery mudstone beds; one collection of Lower Cretaceous *Buchia* is known from coarse-grained, foliated, and metamorphosed metagraywacke containing blueschist minerals.

Exotic rocks such as high-grade glaucophane schist, amphibolite, and eclogite are unknown from this belt, and serpentinite is extremely rare.

The Yolla Bolly belt is continuous along the west side of the Great Valley north of Stonyford, where it is overlain by the ophiolite and sedimentary rocks of the Great Valley Sequence (fig. 1). It extends north at least as far as the Pickett Peak Quadrangle (Irwin and others, in press) and possibly into southwestern Oregon (Blake and others, 1967), where it is structurally overlain by the older rocks of the Klamath Mountains province. Detailed study in the Yolla Bolly area (Blake, 1965, and unpublished data, 1973) indicates that the belt includes a number of mappable lithologic subunits upon which a regional metamorphic fabric is superposed. This fabric ranges from faint in the south and west to pronounced in the east and north, increasing in intensity toward the thrust fault that marks the upper boundary of the belt. The part of this belt texturally highest in grade has been named the South Fork Mountain Schist (Blake and others, 1967).

The lower contact of the Yolla Bolly belt has been mapped in detail near its southern end by Brown (1964a), who clearly showed that the metamorphic rocks (Brown's phyllonite unit) structurally overlie, above a low-angle fault contact, a mélange, termed a friction carpet by Brown. Later work by geologists of the California Department of Water Resources (unpublished mapping by Michael Dwyer, James Vantine, and others) extended Brown's contact to the north and west, where the friction carpet has been given the informal structural term, Skunk Rock mélange.

Several metagraywacke units lie west of the Skunk Rock mélange. Some of these are probably klippen of the Yolla Bolly belt, but at least one differs both lithologically and paleontologically, informally called the Hull Mountain belt (fig. 1) for the prominent mountain that it underlies. To the west is another metagraywacke belt, informally referred to as the Mount Sanhedrin belt, which consists of a number of other graywacke-rich Franciscan units, each bounded by mélange. These units, including the bounding mélanges, are largely unnamed, and very little is known concerning their age or metamorphic state except that they generally lack a metamorphic fabric and contain the mineral assemblage quartz-albite-pumpellyite-chlorite-white mica-celadonite ± aragonite, but not lawsonite. One of these belts, along the Eel River near English Ridge (fig. 1), yielded palynomorphs of late Early Cretaceous (Albian) age (J. O. Berkland, written commun., 1972).

The westernmost mapped unit is the coastal belt Franciscan, long known to contain fossils of Late Cretaceous age (Bailey and others, 1964). Rocks of the coastal belt differ from other Franciscan units in that the sandstones are notably more arkosic and contain very little lithic volcanic or chert detritus. The southernmost known occurrence of this unit is in the San Francisco Bay area. Fossils found in the coastal belt are of Late Cretaceous (Turonian and Campanian), Paleocene, and Eocene ages (fig. 1). Metamorphic grade appears to be lower than in the other Franciscan rocks, laumontite being abundant and prehnite-pumpellyite being locally present. Because of the widespread later faulting related to the San Andreas system, all rocks in the western part of the map area are fragmented, and an adequate stratigraphic or structural sequence in these youngest known Franciscan rocks has not yet been established. Near Cloverdale (fig. 1, west-central), however, the coastal belt structurally underlies a chaotic mélange unit and is isoclinally folded and sheared, suggesting that it was involved in deformation of the same kind of, but less intense than, deformation seen along the Coast Range thrust to the east (Blake and others, 1971).

Great Valley Sequence.—Rocks of the Great Valley Sequence differ from those of the Franciscan assemblage in many ways (see Bailey and others, 1964, and Bailey and Blake, 1969). The most pronounced differences are the great continuity of individual sandstone and shale beds, the broad open style of deformation, the striking lack of small-scale crumpling, folding, or faulting, and the lower sand/shale ratio. The base of the Great Valley Sequence consists of ophiolite and minor amounts of chert, but in many places these basal rocks have been faulted out and the sedimentary rocks of the Great Valley Sequence are in fault contact with sheared serpentinite or metamorphosed Franciscan rocks. Overlying the ophiolite is 16,000 m or more of mudstone, sandstone, and conglomerate ranging in age from Late Jurassic (mid-Tithonian) to latest Cretaceous. Sedimentary rocks low in the sequence are dominantly dark mudstone and some thin-bedded sandstone, including basaltic sandstone and tuffs (Brown, 1964a), pebbly mudstone, and conglomerate that locally aggregate 5,000 m in thickness.

Graded bedding and sole markings are common in these strata, suggesting deposition by turbidity currents. Current-direction indicators, such as flute casts, groove casts, and small-scale crossbedding, are abundant along the west edge of the northern half of the Great Valley. They show remarkably uniform, nearly north to south transport through the entire sequence except for a local reversal in Turonian time (Ojakangas, 1968). Their abundance and consistency suggest that the currents were longitudinal and that through much of the time of deposition this area was the central part of a northward trending trough in which the Great Valley Sequence accumulated.

In addition to significant differences in the sand/shale ratio between the coeval older Franciscan rocks (Yolla Bolly belt) and the *Buchia* beds of the Great Valley Sequence (sandstone/mudstone ratio 3:1 in Yolla Bolly belt as compared with 1:3 in Great Valley Sequence), there are important compositional differences. The petrology of the sandstone of the Great Valley Sequence in northern California has been detailed by Ojakangas (1968) and Dickinson and Rich (1972), and similar studies have been made in the southern Coast Ranges (Gilbert and Dickinson, 1970). These studies indicate that sediments of the Great Valley Sequence fall into several distinctive petrologic units that closely approximate mapped stratigraphic units. The oldest petrofacies (Tithonian and Neocomian) consists of quartz-poor and feldspatholithic sandstones containing very little potassium feldspar and mica. These rocks are considered by most workers to have been derived largely from the Upper Jurassic volcanic and associated plutonic rocks of the western Sierra Nevada, but our own studies suggest that much of the volcanic detritus near the base was derived from the underlying ophiolite sequence.

That ophiolite was locally present at the surface during the filling of the late Mesozic basins is proven by the extensive development of what may be called sedimentary serpentine interbedded with Lower Cretaceous strata of the Great Valley Sequence. These unusual sedimentary beds, composed almost entirely of serpentine detritus, locally contain fossils. Although they have been known for many years, their significance seems to have been largely overlooked. They are nowhere adequately described but are mentioned by Taliaferro (1943, p. 206–207), Bailey and others (1964, p. 164), Lawton (1956), Brice (1953, p. 25), Averitt (1945, p. 73), and Moiseyev (1966).

Undoubted Sierran detritus appears near the Jurassic-Cretaceous boundary in the form of conglomerates containing granitic pebbles and cobbles dated as Late Jurassic (Irwin, 1966). Composition of Cretaceous sandstones younger than Neocomian shows a good correlation with the dated intrusive episodes in the Sierra Nevada. There seems to be little doubt that the batholiths and related calcalkaline volcanoes were the source of most of the Cretaceous sediments (Dickinson and Rich, 1972).

West of the Great Valley, a number of iso-

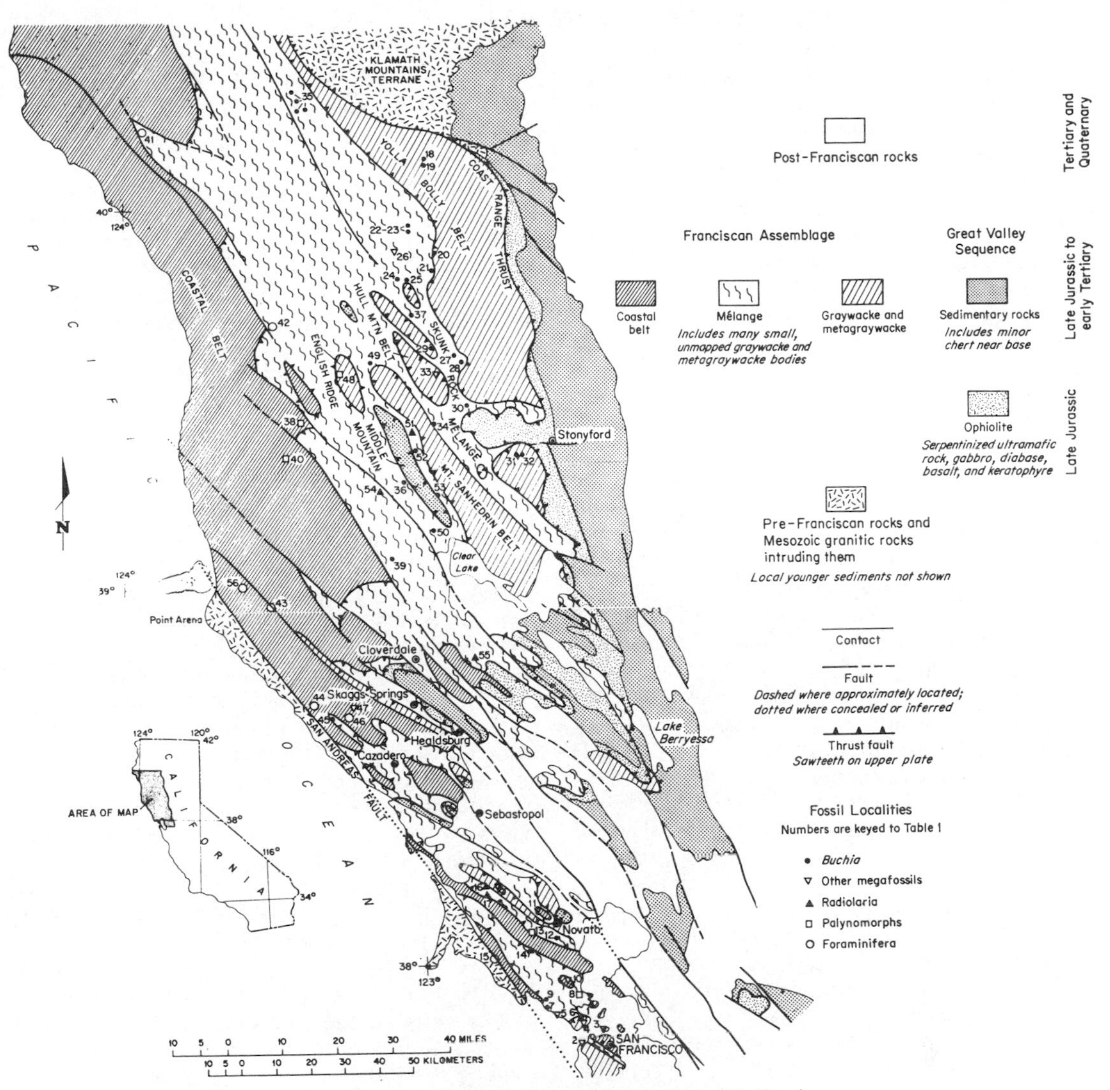

FIG. 1.—Geologic sketch map showing major units and fossil localities within Franciscan assemblage in northern California.

lated outliers of similar sedimentary rocks have been considered to be klippen of a once-continuous thrust sheet of the Great Valley Sequence (Irwin, 1964; Bailey and others, 1964). These rocks are similar to those along the west side of the Great Valley and contain the same fossils. They differ from them, however, by being more conglomeratic and by containing more abundant potassium feldspar. The sandstones and conglomerates, present at Healdsburg (Gealey, 1951), near Cazadero (Irwin, 1964), Novato (fig. 1, no. 11; Berkland, 1969), and another belt running from the mouth of the Russian River to south of Sebastopol (Blake and others, 1971) contain pebbles and cobbles of andesite, dacite, rhyolite, and their plutonic equivalents, all containing abundant quartz and potassium feldspar. Biotite from a rhyolite cobble near Novato has been dated at 138 my (Berkland, 1969). Vesicular volcanic rocks are intercalated within the lower mudstones south of Cloverdale in the eastern part of the Skaggs Springs 15-minute quadrangle (unpublished mapping by E. H. Bailey) and along Atherton Avenue in the Novato 7½-minute quadrangle (Berkland, 1969). These differences suggest that the western outliers were derived from a different source than the *Buchia* beds along the west side of the Great Valley, perhaps from a former remnant volcanoplutonic arc that lay to the west. This possibility is discussed further in the section dealing with our proposed sedimentary-tectonic model.

The sedimentary petrology of the Franciscan sandstones has not been studied in as much detail as has that of the Great Valley Sequence, largely because of the widespread secondary alteration and low-grade metamorphism that have affected these rocks. Nevertheless, studies made [Soliman, 1965; Swe and Dickinson, 1970; and Bailey and others, 1964, p. 31 (has a list of older studies)] indicate that the Franciscan sandstones (graywackes) are generally coarser grained and more quartzose than sandstones of similar age from the Great Valley Sequence. Chemical analyses of the older Yolla Bolly belt graywackes (Blake, unpublished data, 1973) show that these rocks contain 65 to 70 percent SiO_2 and, except for uncommonly high Na_2O:K_2O, are similar in composition to the granitic rocks of the Klamath and Sierran areas (Bateman and Dodge, 1970; Hotz, 1971). According to most workers, this coarser plutonic detritus somehow was carried across the intervening Great Valley basin and deposited in a trench. This puzzling relation is discussed further in the section dealing with sedimentary and tectonic models.

AGE AND LITHOLOGY OF MÉLANGES AND THEIR SIMILARITY TO THE GREAT VALLEY SEQUENCE

Unlike the other Franciscan units, the mélanges contain relatively abundant fossils. These include megafossils such as *Buchia,* which occur mainly in limy concretions and lenses within the sheared mudstone matrix, as well as microfossils such as spores and pollen, also in limy concretions, and Radiolaria in blocks and slabs of chert. We have plotted on the geologic sketch map (fig. 1) all known Franciscan fossil localities in this part of northern California. Paleontologic data for these fossils are given in table 1. Most of these localities, including those for the radiolarian cherts, are in mélanges. It is significant that fossils at these localities are of Tithonian to Valanginian age. We have found no evidence that the process of mélange formation involved widespread mixing[1] of rocks of widely differing series, for example, of the Upper Jurassic and Upper Cretaceous Series. Rather, the mélanges appear to have formed by grinding up a predominantly shale sequence that included much chert and volcanic rocks, all with a very narrow age span of Tithonian to Valanginian—mélanges of other ages may occur in other parts of the Coast Ranges.

Other significant aspects to be considered are the lithologic character of the mélanges and their structural and metamorphic histories. As first pointed out by R. D. Brown, Jr. (1964b), in his study of the Stonyford Quadrangle, the mélange (Brown's "friction carpet") consists of sedimentary, volcanic, and ultramafic rocks, which, although chaotically deformed, are otherwise indistinguishable from the orderly sedimentary, volcanic, and other rocks that occur immediately above the thrust contact at the base of the Great Valley Sequence. Subsequent work has supported this observation, for over large areas of northern California many outcrops of the relatively unmetamorphosed mélange matrix can be distinguished in the field from the basal part (*Buchia*-bearing beds) of the Great Valley Sequence only by their being harder, more sheared, and cut by quartz veins. Our studies in the San Francisco Bay area indicate that the sandstone (graywacke) within the mélanges is poorer in quartz and richer in volcanic detritus than in the other Franciscan units. In this, our study agrees with earlier petrographic studies of the lower part of the Great Valley Sequence (Ojakangas, 1968; Dickinson and Rich, 1972; Gilbert and Dickinson, 1970).

[1] Along the contact between the coastal belt and mélange, blocks of Upper Cretaceous foraminiferal limestone have locally been tectonically mixed with the overlying mélange.

PROPOSED SEDIMENTARY AND TECTONIC MODEL

Following an episode of accelerated spreading and related deformation, metamorphism, and volcanoplutonic activity in the Late Jurassic but prior to middle Tithonian time, new oceanic crust was generated west of the North American continental margin. Preliminary studies suggest that this new crust was not formed at a spreading ridge such as the Mid-Atlantic Ridge or the East Pacific Rise, for the new crust is:

(a) abnormally thin in places (Bailey and others, 1970);

(b) highly heterogeneous in character and contains a much greater volume of quartz kera-

TABLE 1.—PALEONTOLOGIC DATA FOR FOSSIL LOCALITIES SHOWN ON GEOLOGIC MAP

Locality no.	Location	Fossils	Unit	Age	Reference
1	Tiburon Peninsula	*Buchia*	Unnamed mélange	Valanginian	Blake and others, 1973
2	San Francisco	*Douvilleiceras*	Unnamed Franciscan sandstone	Albian	Same
3	Alcatraz Island	*Inoceramus*	Coastal belt(?) Franciscan	Cretaceous	Same
4	South Marin headlands	*Mantelliceras*	Unnamed Franciscan sandstone	Cenomanian	Same
5	South Marin coast	Belemnite	Unnamed mélange	Hauterivan	Same
6	Tamalpais Valley	Radiolaria	Chert at base of unnamed sandstone unit	Early Cretaceous(?)	Same
7	Muir Beach	*Buchia*	Unnamed mélange	Late Jurassic(?)	Same
8	Tiburon Peninsula	Palynomorphs	Unnamed mélange	Early Cretaceous(?)	Same
9	Mount Tamalpais	*Buchia*	Unnamed mélange	Valanginian	Same
10	San Rafael	*Inoceramus*	Coastal belt(?) Franciscan	Campanian	Same
11	Novato	*Inoceramus*	Coastal belt(?) Franciscan	Campanian	Same
12	San Rafael	*Buchia*	Unnamed mélange	Valanginian(?)	Same
13	Bolinas	Foraminifera	Coastal belt(?) Franciscan	Cenomanian	Same
14	Fairfax	Radiolaria	Unnamed mélange	Tithonian and Berriasian	Blake and others, 1973
15	San Rafael	Palynomorphs	Unnamed mélange	Early Cretaceous	Same
16	Salmon Creek	Radiolaria	Unnamed mélange	Tithonian and Berriasian	Same
17	Salmon Creek	Radiolaria	Unnamed mélange	Tithonian and Berriasian	Same
18	Yolla Bolly	*Buchia*	Yolla Bolly belt	Early Cretaceous	Irwin, 1957
19	Yolla Bolly	*Buchia*	Yolla Bolly belt	Late Jurassic	Blake, 1965
20	Doll Ridge	*Buchia*	Yolla Bolly belt	Late Jurassic	Jones and Blake, in preparation
21	Anthony Peak	*Buchia*	Skunk Rock mélange	Early Cretaceous	Ghent, 1963
22–26	Leech Lake Mountain	*Buchia*, snail	Skunk Rock mélange	Late Jurassic and Early Cretaceous	Suppe, in press
27–28	Black Butte Creek	*Buchia*	Skunk Rock mélange	Early Cretaceous	Jones and Blake, in preparation
29	Calamese Rock	*Buchia*	Unnamed metagraywacke unit	Early Cretaceous	Same
30	Bowery Flat	*Buchia*	Skunk Rock mélange	Early Cretaceous	Irwin, 1957
31–32	Stonyford	*Buchia*	Yolla Bolly belt(?)	Late Jurassic and Early Cretaceous	Brown, 1964b
33	Hull Mountain	*Inoceramus*	Hull Mountain belt	Cenomanian(?)	Jones and Blake, in preparation
34	Lake Pillsbury	*Buchia*	Skunk Rock(?) mélange	Late Jurassic	Same
35	Pickett Peak Quadrangle (four separate localities)	*Buchia*	Unnamed mélange	Late Jurassic and Early Cretaceous	Irwin and others, in press
36	Near Middle Mountain	*Buchia*	Unnamed mélange	Late Jurassic	J. O. Berkland, written commun., 1972
37	Edsel Ridge	*Buchia*	Skunk Rock mélange	Early Cretaceous	Jones and Blake, in preparation
38	Brooktrails	Palynomorphs	Coastal belt Franciscan	Eocene	Jones and Blake, in preparation
39	Ukiah	*Buchia*	Unnamed mélange	Late Jurassic	Same
40	Willits and Ft. Bragg (12 separate localities)	Palynomorphs and coccoliths	Coastal belt Franciscan	Paleocene and Eocene	James Berkland, written commun., 1972; O'Day and Kramer, 1972
41	Garberville	Foraminifera	Coastal belt Franciscan (tectonic block)	Cenomanian	Bailey, Irwin, and Jones, 1964
42	Laytonville Limestone	Foraminifera	Coastal belt Franciscan (tectonic block)	Cenomanian	Same
43	Ornbaun Limestone	Foraminifera	Coastal belt Franciscan	Cenomanian	Same
44	Annapolis Limestone	Foraminifera	Coastal belt Franciscan	Cenomanian	Same
45	Cazadero	*Buchia*	Unnamed mélange	Late Jurassic	Same
46	Cazadero	Foraminifera	Coastal belt Franciscan	Cenomanian	Same
47	Skaggs Springs	*Inoceramus*	Coastal belt Franciscan	Turonian	Same
48	English Ridge (three separate localities)	Palynomorphs	Unnamed graywacke unit	Albian	James Berkland, written commun., 1972
49	Eden Valley	*Buchia*	Unnamed mélange	Tithonian to Valanginian	Bailey, Irwin, and Jones, 1964
50	Witter Springs	*Buchia*	Unnamed mélange	Tithonian to Valanginian	James Berkland, written commun., 1972
51	Eel River	Radiolaria	Unnamed mélange	Tithonian and Berriasian	Same
52	Bucknell Creek	*Buchia*	Unnamed mélange	Tithonian	Same
53	Middle Creek	*Buchia*	Unnamed mélange	Tithonian	Same
54	Potter Valley	Radiolaria	Unnamed mélange	Tithonian	Same
55	The Geysers	Radiolaria	Unnamed mélange	Tithonian and Berriasian	Same
56	Alder Creek	Foraminifera, palynomorphs	Coastal belt Franciscan	Upper Cretaceous	Berkland, 1964

tophyre and albite granite than has been reported from the present active ridge crests;

(c) generally lacks abundant dikes comparable to the sheeted dike complexes described elsewhere and attributed to formation at a spreading ridge crest; and

(d) lacks the thick pelagic cover that would be expected if it formed in midocean far to the west of the continental mass (Scholl and Marlow, this volume).

A more reasonable environment for generation of new crust is an interarc basin or marginal sea near the present Great Valley, although an inferred remnant arc to the west has not been identified. One possible solution to this dilemma has already been suggested by Karig (1972, p. 1065). He pointed out that the siliceous and intermediate igneous rocks, which locally make up much of the Great Valley ophiolite, may be related to arc volcanism rather than representing midocean ridge material. Chemical analyses of the California ophiolite (M. C. Blake, Jr., and E. H. Bailey, unpublished data, 1972) indicate that many of these rocks are close to andesite, dacite, and rhyolite in composition except for their extremely high soda-to-potash ratios, which may be due to later metasomatism.

Because of the lack of reliable radiogenic data and the great area covered by younger sediments in the Great Valley itself, the exact sequence of events and even the polarity of the inferred subduction zones are still highly speculative. An east-dipping zone is favored because this seems to fit best the overall tectonic style of the Pacific coast region. For example, possible fossil subduction zones in the Klamath Mountains all appear to have dipped to the east (Irwin, 1966, fig. 6, p. 32), and K_2O contents in Mesozoic granitic plutons appear to increase eastward at some places, extending into Nevada (Dickinson, 1970; Hotz, 1971; Bateman and Dodge, 1970; Evernden and Kistler, 1970).

According to the most recent sedimentary and tectonic models (Bailey and Blake, 1969; Hamilton, 1969; Ernst, 1970; Dickinson, 1971; and many others) and beginning in mid-Tithonian time, vast quantities of sediment were derived from the uplifted ancestral Sierran and Klamath mountains, were carried into the Great Valley basin by west-flowing rivers, then redeposited by longitudinal turbidity currents. At the same time, somewhat coarser grained sediments were periodically carried across the Great Valley basin in submarine canyons and dumped into the Franciscan trench. During periods of relative quiescence, thick lenses of radiolarian chert accumulated in the trench. Concurrent with sedimentation but probably taking place during several distinct episodes or pulses, the trench material was subducted to form high-pressure lawsonite and jadeite-bearing metagraywackes as well as forming the mélanges.

Recognition that the mélanges of the northern Coast Ranges formed from a restricted stratigraphic succession that belongs to the lowest part of the upper Mesozoic sedimentary sequence of the Coast Ranges imposes definite constraints on interpretation of the mode of origin and significance of these remarkable tectonostratigraphic units as well as on the previously described model. We interpret their significant features as follows:

(1) The matrix of the mélanges is thought to represent a distal, or seaward, portion of the so-called Knoxville Formation on the basis of similarity of lithology and fossil content with the Knoxville at the base of the Great Valley Sequence.

(2) The abundant greenstone, chert, and serpentinite found as tectonic blocks within the mélanges probably were derived from the immediately underlying oceanic crust and upper mantle.

(3) This oceanic crust and serpentinite probably was not formed at an oceanic ridge lying far to the west, but may have formed in an interarc basin or a marginal sea.

(4) Some or all the siliceous to intermediate igneous rocks, which intrude and overlie mafic portions of the ophiolite, may represent the products of island arc volcanism west of the marginal basin.

(5) The tectonic blocks, or knockers as they have been called, of high-grade blueschist and eclogite that occur in the mélanges were formed during an earlier period of subduction that deformed the Galice and Mariposa Formations 150 to 160 my ago, that is, at about the time the ophiolite was formed (Lanphere, 1971).

(6) Because the fossiliferous (*Buchia*-bearing) mélanges in places are in contact with much younger rocks (Late Cretaceous and younger), yet do not contain abundant blocks of these younger rocks, it seems unlikely that they could have formed through sedimentary sliding, for such a process would lead to an intimate mixing of rocks of widely differing ages throughout the mélange. Only along the sheared contacts between mélange and Franciscan graywacke or metagraywacke units is much mixing of rock types observable.

Formation of the mélanges must involve multiple subducted plates, the mélange being generated repeatedly from the ultramafic, mafic, chert, and *Buchia*-bearing shale sequence that constitutes the oldest rocks of the Coast Ranges. The mélanges could have formed in two ways: (1) by tearing up and shearing out the subducted plate composed of these rocks or (2) by abrading the base of a previously subducted ophiolite and shale sequence by passage of another younger plate beneath the overlying older rocks. According to the paleontologic evidence, this process of subduction probably began in late Valanginian or Hauterivian time. It continued through the Eocene, as shown by the recent discovery of Eocene fossils in deformed Franciscan rocks (coastal belt) lying structurally below a typical mélange.

In order to fit these petrologic, paleontologic, tectonic, and radiogenic data into a reasonable geologic history, we have prepared a series of diagrams of hypothetical models based on the plate tectonic model and in particular on the recent studies by Karig (1972) in the southwest Pacific: The first diagram (fig. 2a) shows conditions during Tithonian to Valanginian time, which is considered to be about the same time as the Yosemite intrusive epoch (148 to 132 my, Evernden and Kistler, 1970). We suggest in our highly generalized cartoon that the basal part of the Great Valley Sequence was deposited in an interarc basin above an east-dipping subduction zone. The entire ophiolite at the base of the Great Valley Sequence probably formed prior to the deposition of the Knoxville Formation, but this is not clear. The older Franciscan sediments (Yolla Bolly belt and probably the Tithonian to Valanginian metagraywacke and chert of the Diablo Range) are seen as formed in an arc-trench gap rather than in a trench environment. This concept is partly based on radiogenic ages on blueschists, indicating that the type III metabasalts and metacherts of Cazadero (Lee and others, 1964; Suppe and Armstrong, 1972) and other areas were being subducted at about the time that the Knoxville and older Franciscan beds were being deposited. The diagram (fig. 2a) shows a possible remnant volcanoplutonic arc that is inferred to have rifted away from the Mariposa-Amador arc at the site of the present Sierra Nevada during the initial formation of the ophiolite-floored Great Valley basin. This arc would have been in the right position to provide the coarse-grained volcanoplutonic detritus to the westernmost Great Valley Sequence, seen in klippen today north of San Francisco, and also to the older Franciscan graywackes. The problem here is that the inferred arc is now completely gone and if once present must have been subsequently eroded away or subducted. [A similar remnant arc seems to be required to explain the geologic data recently presented by Ross, Wentworth, and McKee (1973). Their data indicate that the Late Cretaceous Gualala Formation of Weaver (1943) was derived from a granitic source area on the west and oceanic crust to the east. Prior to Tertiary offset on the San Andreas Fault, the inferred Gualala basin was believed to lie near the extreme southern margin of the Great Valley, west of the Sierra Nevada Batholith.)]

During the Early Cretaceous (fig. 2b) a pronounced change occurred in the tectonic regime as the subduction zone flattened and began to consume the older Franciscan rocks deposited in the former arc-trench gap. This event is inferred from radiogenic dating of Sierran plutonic rocks (the Huntington Lake epoch of Evernden and Kistler, 1970) and from the dating of blueschists in the Yolla Bolly and other areas (Suppe and Armstrong, 1972). The change in dip of the subduction zone to near horizontal is inferred from the present geometry of the Coast Range thrust (Bailey and others, 1970), from the wide east-west extent of calcalkaline plutonic rocks of this age, and the apparent lower P-T conditions inferred in the mineral assemblages of the 110 my blueschists as compared with the older ones. Detailed studies of the Great Valley Sequence in northern Sacramento Valley (Jones, Bailey, and Imlay, 1969; Jones and Irwin, 1971) indicate that the North American plate was being deformed by tear faults related to absolute westward movement of the Klamath Mountains relative to the Great Valley Sequence at the time the older Franciscan sediments were being underthrust along the oceanic plate. The timing of these events coincides closely with a pronounced acceleration in plate motion inferred from magnetic studies in both the Pacific and Atlantic (Larson and Chase, 1972; Larson and Pitman, 1972). It has recently been proposed by Hyndman (1972) that, if the absolute motion of *both* plates is convergent, there would be a pronounced flattening of the 45° or greater dip in most presently active Benioff zones.

It is at this time that the development of the mélanges is believed to have commenced. The oldest *in situ* Franciscan unit is the Yolla Bolly belt, which, as described, directly underlies both the older rocks of the Klamath Mountains province and, to the south, the Great Valley Sequence, including the basal ophiolite. Following subduction of the Yolla Bolly belt, the rapid westward motion of the overriding North American plate apparently resulted in periodic underthrusting of part of the Great Valley Sequence and underlying ophiolite. This material represents the tectonic mélange that marks the

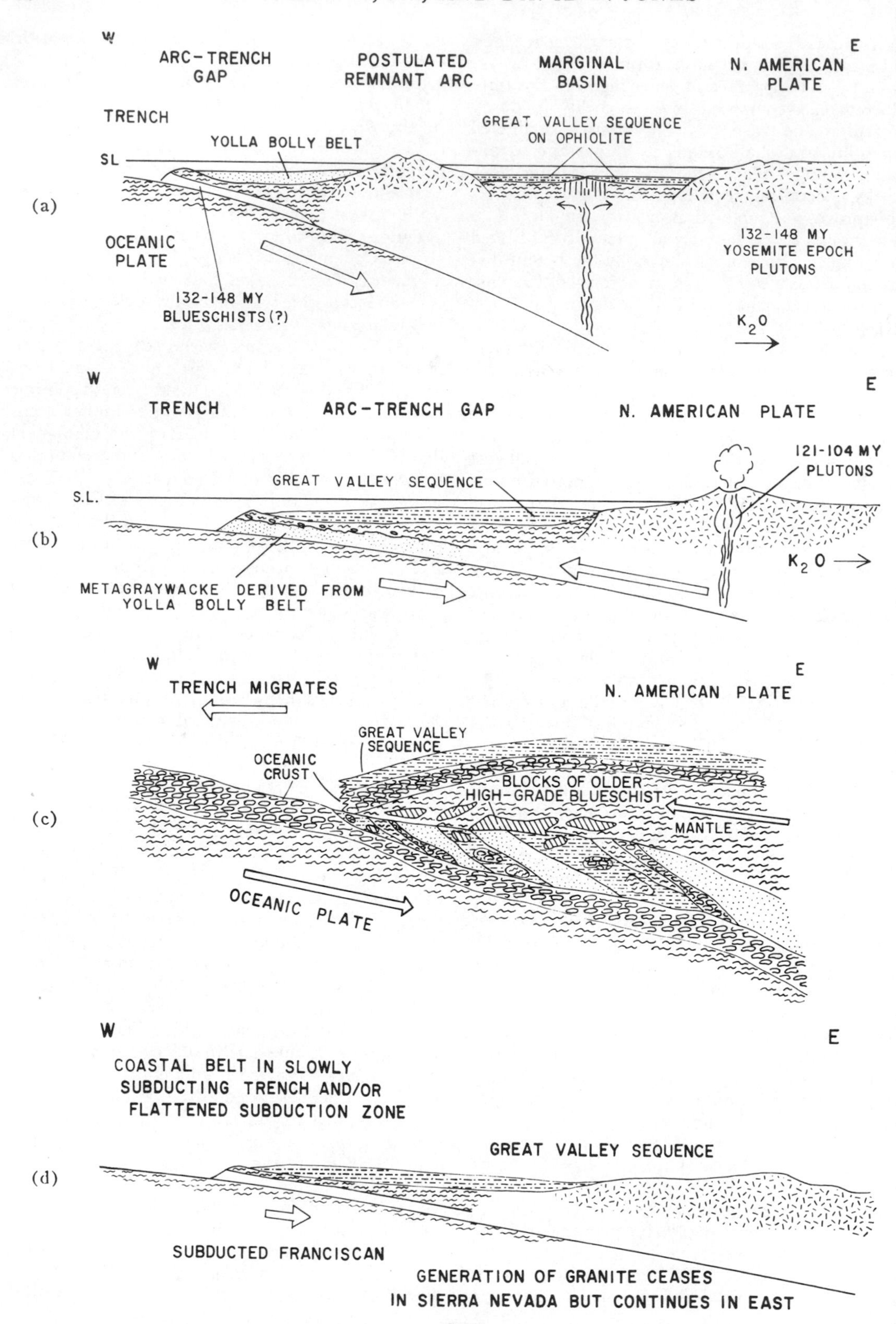

W
E
ARC-TRENCH GAP
POSTULATED REMNANT ARC
MARGINAL BASIN
N. AMERICAN PLATE
TRENCH
GREAT VALLEY SEQUENCE ON OPHIOLITE
YOLLA BOLLY BELT
SL
(a)
OCEANIC PLATE
132-148 MY YOSEMITE EPOCH PLUTONS
132-148 MY BLUESCHISTS (?)
K_2O
W
E
TRENCH
ARC-TRENCH GAP
N. AMERICAN PLATE
121-104 MY PLUTONS
GREAT VALLEY SEQUENCE
S.L.
(b)
K_2O
METAGRAYWACKE DERIVED FROM YOLLA BOLLY BELT
W
E
TRENCH MIGRATES
N. AMERICAN PLATE
GREAT VALLEY SEQUENCE
OCEANIC CRUST
BLOCKS OF OLDER HIGH-GRADE BLUESCHIST
(c)
MANTLE
OCEANIC PLATE
W
E
COASTAL BELT IN SLOWLY SUBDUCTING TRENCH AND/OR FLATTENED SUBDUCTION ZONE
GREAT VALLEY SEQUENCE
(d)
SUBDUCTED FRANCISCAN
GENERATION OF GRANITE CEASES IN SIERRA NEVADA BUT CONTINUES IN EAST

boundaries between successive Franciscan subducted plates. The previously subducted high-grade metamorphic rocks, such as eclogite, glaucophane-epidote gneiss, and amphibolite, were moved up and westward as part of the upper plate (embedded in the serpentinized basal ophiolite) and were resubducted to form the so-called polyphase knockers (Coleman and Lanphere, 1971). Our proposed model for these structures is shown in more detail in figure 2c. Eventually, the subduction slowed or nearly ceased, possibly as a result of a decrease in plate motions about 80 my ago as inferred from the lack of calcalkaline volcanoplutonic activity as well as from the lack of evidence for blueschist formation younger than that age (fig. 2d). That subduction continued through the Eocene is shown by the presence of deformed Eocene Franciscan rocks (coastal belt). No granitic plutons of this age are known from the Sierra Nevada Batholith, although Laramide and younger events farther east are known to be of this same age. Apparently, then, some kind of change in tectonic activity occurred about 80 my ago, when subduction slowed or when some other related activity took place to cause igneous activity to be shifted far to the east and high-pressure metamorphism apparently to be replaced by a shallower phenomenon. This change possibly was brought about by an even greater flattening of the subduction zone in such a manner that the underthrust rocks in California did not attain the depths necessary for generation of either blueschists or calcalkaline magmas.

ACKNOWLEDGMENTS

We are greatly indebted to numerous individuals for providing geologic and paleontologic data. For their generous premission to use unpublished data, we thank James Berkland, Appalachian State University; Salem Rice, California Division of Mines and Geology; and M. J. Dwyer and James Vantine, California Department of Water Resources. For identifying the palynomorphs and radiolarians, we thank respectively W. R. Evitt, Stanford University, and E. A. Pessagno, Jr., University of Texas. We acknowledge much valuable criticism and review from our colleagues of the U.S. Geological Survey and from the participants at the "Conference on Modern and Ancient Geosynclinal Sedimentation," Madison, Wisconsin, November 10–11, 1972.

Publication has been authorized by the Director, U.S. Geological Survey.

REFERENCES

Averitt, Paul, 1945, Quicksilver deposits of the Knoxville district, Napa, Yolo, and Lake Counties, California: California Jour. Mines and Geology, v. 41, no. 2, p. 65–89.

Bailey, E. H., and Blake, M. C., Jr., 1969, Tektonicheskoe razvitiye zapadnoy kalifornii v pozdnem mezozoe (pts. 1 and 2) (Tectonic development of western California during the late Mesozoic): Geotektonika, no. 3, p. 17–30; no. 4, p. 24–34 (in Russian).

———, ———, and Jones, D. L., 1970, On-land Mesozoic oceanic crust in California Coast Ranges: U.S. Geological Survey Prof. Paper 700-C, p. 70–81.

———, Irwin, W. P., and ———, 1964, Franciscan and related rocks and their significance in the geology of western California: California Div. Mines and Geology Bull. 183, 177 p.

Bateman, P. C., and Dodge, F. C. W., 1970, Variations of major chemical constituents across the central Sierra Nevada Batholith: Geol. Soc. America Bull., v. 81, p. 409–420.

Berkland, J. O., 1964, Notes on the geology of the Alder Creek area near Point Arena, California: Calif. Div. Mines and Geology Min. Inf. Service, v. 17, p. 139–141.

———, 1969, Geology of the Novato Quadrangle, Marin County, California (M.S. thesis): San Jose, California, San Jose State Coll., 146 p.

———, 1972, Paleogene "frozen" subduction zone in the Coast Ranges of northern California, *in* Tectonics: 24th Internat. Geol. Cong., Montreal, 1972, Repts. Sec., no. 3, p. 99–105.

Blake, M. C., Jr., 1965, Structure and petrology of low-grade metamorphic rocks, blueschist facies, Yolla Bolly area, northern California (Ph.D. thesis): Stanford California, Stanford Univ., 91 p.

———, Irwin, W. P., and Coleman, R. G., 1967, Upside-down metamorphic zonation, blueschist facies, along a regional thrust in California and Oregon, *in* Geological Survey Research, 1967: U.S. Geol. Survey Prof. Paper 575-C, p. 1–9.

←

Fig. 2.—Hypothetical cross sections in northern California showing progressive changes in plate tectonic regime and development of Franciscan mélanges. *a,* Tithonian to Valanginian: formation of hypothetical marginal ocean basin at present site of the Great Valley; *b,* Early Cretaceous; acceleration of westward movement of North American plate leads to overriding of trench and flattening of subduction zone. *c,* Schematic diagram showing formation of imbricate mélange and metagraywacke units. *d,* Late Cretaceous to post-Eocene; note very much flattened subduction zone.

———, AND OTHERS, 1971, Preliminary geologic map of western Sonoma county and northernmost Marin County, California: *ibid.*, open-file report.
———, AND ———, 1973, (in preparation) Preliminary geologic map of Marin County, California: *ibid.*, map, scale 1:62,500.
BRICE, J. C., 1953, Geology of the Lower Lake Quadrangle, California: California Div. Mines Bull. 166, 72 p.
BROWN, R. D., JR., 1964a, Geological map of the Stonyford Quadrangle, Glenn, Colusa, and Lake Counties, California: U.S. Geol. Survey Min. Inv. Field Studies Map MF-279, scale 1:48,000.
———, 1964b, Thrust-fault relations in the northern Coast Ranges, California: *ibid.*, Prof. Paper 475-D, p. 7–13.
COLEMAN, R. G., AND LANPHERE, M. A., 1971, Distribution and age of high-grade blueschist, associated eclogites, and amphibolites from Oregon and California: Geol. Soc. America Bull., v. 82, p. 2397–2412.
DICKINSON, W. R., 1970, Relations of andesites, granites, and derivative sandstones to arc-trench tectonics: Rev. Geophysics and Space Physics, v. 8, p. 813–860.
———, 1971, Clastic sedimentary sequences deposited in shelf, slope, and trough settings between magmatic arcs and associated trenches: Pacific Geology, v. 3, p. 15–30.
———, AND RICH, E. I., 1972, Petrologic intervals and petrofacies in the Great Valley Sequence, Sacramento Valley, California: Geol. Soc. America Bull., v. 83, p. 3007–3024.
ERNST, W. G., 1970, Tectonic contact between the Franciscan mélange and the Great Valley Sequence, crustal expression of a late Mesozoic Benioff zone: Jour. Geophys. Research, v. 75, p. 886–902.
EVERNDEN, J. F., AND KISTLER, R. W., 1970, Chronology of emplacement of Mesozoic batholithic complexes in California and western Nevada: U.S. Geol. Survey Prof. Paper 623, 42 p.
FOX, K. F., AND OTHERS, 1973, Preliminary geologic map of eastern Sonoma, Napa, and Solano Counties, California: *ibid.*, Map MF-483, scale 1:62,500.
GEALEY, W. K., 1951, Geology of the Healdsburg Quadrangle, California: California Div. Mines Bull. 161, p. 7–50.
GHENT, E. D., 1963, Fossil evidence for maximum age of metamorphism in part of the Franciscan Formation, northern Coast Ranges, California: *ibid.*, Special Rept. 82, p. 41.
GILBERT, W. G., AND DICKINSON, W. R., 1970, Stratigraphic variations in sandstone petrology, late Mesozoic Great Valley Sequence, southern Santa Lucia Range, California: Geol. Soc. America Bull., v. 81, p. 949–954.
HAMILTON, WARREN, 1969, Mesozoic California and the underflow of Pacific mantle: *ibid.*, v. 80, p. 2409–2429.
HOTZ, P. E., 1971, Plutonic rocks of the Klamath Mountains, California and Oregon: U.S. Geol. Survey Prof. Paper 684B, p. 1–20.
HSÜ, K. J., 1971, Franciscan mélanges as a model for eugeosynclinal sedimentation and underthrusting tectonics: Jour. Geophys. Research, v. 76, p. 1162–1170.
HYNDMAN, R. D., 1972, Plate motions relative to the deep mantle and the development of subduction zones: Nature, v. 238, p. 263–265.
IRWIN, W. P., 1957, Franciscan Group in Coast Ranges and its equivalents in Sacramento Valley, California: Am. Assoc. Petroleum Geologists Bull., v. 41, p. 2284–2297.
———, 1964, Late Mesozoic orogenies in the ultramafic belts of northwestern California and southwestern Oregon, *in* Geological Survey Research 1964: U.S. Geol. Survey Prof. Paper 501-C, p. 1–9.
———, 1966, Geology of the Klamath Mountains province: California Div. Mines and Geology Bull. 190, p. 19–38.
———, AND OTHERS, in press, Geologic Map of the Pickett Peak Quadrangle, Trinity County, California: U.S. Geol. Survey Map GQ-1111.
JONES, D. L., BAILEY, E. H., AND IMLAY, R. W., 1969, Structural and stratigraphic significance of the *Buchia* zones in the Colyear Spring-Paskenta area, California: *ibid.*, Prof. Paper 647-A, p. 1–24.
———, AND BLAKE, M. C., JR., in preparation, Fossil localities in Franciscan Assemblage, northern California.
———, AND IRWIN, W. P., 1971, Structural implications of an offset early Cretaceous shoreline in northern California: Geol. Soc. America Bull., v. 82, p. 815–822.
KARIG, D. E., 1972, Remnant arcs: *ibid.*, v. 83, p. 1057–1068.
LANPHERE, M. A., 1971, Age of the Mesozoic oceanic crust in the California Coast Ranges: *ibid.*, v. 82, p. 3209–3212.
LARSON, R. L., AND CHASE, C. G., 1972, Late Mesozoic evolution of the western Pacific Ocean: *ibid.*, v. 83, p. 3627–3644.
———, AND PITMAN, W. C., III, 1972, World-wide correlation of Mesozoic magnetic anomalies, and its implications: *ibid.*, p. 3645–3662.
LAWTON, J. E., 1956, Geology of the north half of the Morgan Valley Quadrangle and the south half of the Wilbur Springs Quadrangle, California (Ph.D. thesis): Stanford, California, Stanford Univ.
LEE, D. E., AND OTHERS, 1964, Isotopic ages of glaucophane schists from the area of Cazadero, California: U.S. Geol. Survey Prof. Paper 475-D, p. 105–107.
MOISEYEV, A. N., 1966, The geology and geochemistry of the Wilbur Springs quicksilver district, Colusa and Lake Counties, California (Ph.D. thesis): Stanford, California, Stanford Univ., 214 p.
O'DAY, MICHAEL, AND KRAMER, J. C., 1972, Geologic guide to the northern Coast Ranges—Lake, Sonoma, and Mendocino Counties, California: Sacramento, California, Sacramento Geol. Soc. Ann. Field Trip 1972, Guidebook, p. 51–56.
OJAKANGAS, R. W., 1968, Cretaceous sedimentation, Sacramento Valley, California: Geol. Soc. America Bull., v. 79, p. 976–1008.
RICH, E. I., 1971, Geologic map of the Wilbur Springs Quadrangle, Colusa and Lake Counties, California: U.S. Geol. Survey Misc. Geol. Inv. Map I-538, scale 1:48,000.

Ross, D. C., Wentworth, C. M., and McKee, E. H., 1973, Cretaceous mafic conglomerate near Gualala offset 350 miles by San Andreas Fault from oceanic crustal source near Eagle Rest Peak, California: U.S. Geol. Survey Jour. Research, v. 1, p. 45–52.

Sims, J. D., and others, 1973, Preliminary geologic map of Solano County and parts of Napa, Contra Costa, Marin and Yolo Counties, California: *ibid.*, Map MF-484, scale 1:62,500.

Soliman, S. M., 1965, Geology of the east half of the Mount Hamilton Quadrangle, California: California Div. Mines and Geology Bull. 185, 32 p.

Suppe, John, in press, Geology of the Leech Lake Mountain-Ball Mountain region, California: a cross section of the northeastern Franciscan belt: California Univ. Publs. Geol. Sci.

———, and Armstrong, R. L., 1972, Potassium-argon dating of Franciscan metamorphic rocks: Am. Jour. Sci., v. 272, p. 217–233.

Swe, Win, and Dickinson, W. R., 1970, Sedimentation and thrusting of late Mesozoic rocks in the Coast Ranges near Clear Lake, California: Geol. Soc. America Bull., v. 81, p. 165–188.

Taliaferro, N. L., 1943, Franciscan-Knoxville problem: Am. Assoc. Petroleum Geologists Bull., v. 27, p. 109–219.

Weaver, C. E., 1943, Point Arena-Fort Ross region: California Div. Mines Bull. 118, pt. 3, p. 628–632.

25

Reprinted from *Geol. Soc. America Bull.* **85:**1623-1634 (1974)

Deformation and Metamorphism of the Franciscan Subduction Zone Complex Northwest of Pacheco Pass, California

DARREL S. COWAN* *Shell Oil Company, P.O. Box 527, Houston, Texas 77001*

[*Editor's Note:* Figure 2 has been omitted.]

ABSTRACT

The Franciscan Complex northwest of Pacheco Pass, California, includes three fault-bounded units, each characterized by a different deformational style and suite of metamorphic mineral assemblages. Structurally highest is jadeitic pyroxene-bearing metagraywacke semischist. The areally extensive Garzas tectonic mélange separates the semischist from the structurally lowest pumpellyite-bearing Orestimba metagraywacke. The Garzas mélange is representative of Franciscan mélanges in general. These mappable bodies have an internal fabric dominated by penetrative, mesoscopic shear fractures and contain tectonic inclusions of all sizes immersed in a pervasively sheared, generally fine-grained matrix. The shear fractures record brittle deformation of consolidated rock bodies. Many mélanges contain exotic inclusions, clearly not derived from adjacent units, and metamorphic mineral assemblages and textures indicate that they were metamorphosed before being tectonically mixed with more voluminous, generally lower grade, metagraywacke inclusions.

The structural units are grossly sheetlike in external form and are separated from one another by gently to steeply dipping major faults. Unlike low-angle thrusts in imbricated Cordilleran terranes, the faults do not systematically repeat or offset a normal stratigraphic sequence but rather juxtapose rock units that bear no apparent stratigraphic, deformational, or metamorphic relation to one another. The structural units were separately deformed and metamorphosed under a variety of conditions prior to their tectonic juxtaposition during late Mesozoic continental margin subduction. Field and petrographic evidence permit, but do not prove, the hypothesis that both the semischist and exotic, high-grade mélange inclusions once were more deeply buried and have been emplaced upward into their present anomalously shallow structural positions.
Key words: structural geology, Franciscan, blueschist metamorphism, tectonic mélange, subduction, California.

INTRODUCTION

The Franciscan Complex, a heterogeneous assemblage of metagraywacke, chert, mafic volcanic rocks, and minor isolated blocks and slabs of high-grade blueschist- and amphibolite-facies metamorphic rocks, is widely exposed in the California Coast Ranges and parts of southwestern Oregon and Baja California. The complex is currently interpreted as having accumulated in a late Mesozoic–early Tertiary subduction zone at the western margin of the North American plate (for example, Hamilton, 1969; Bailey and Blake, 1969; Ernst, 1970; and many others). Franciscan "trench" materials were carried against and beneath the coeval Great Valley sequence, a thick succession of clastic rocks of Tithonian to Maestrichtian age that contrasts strikingly with the Franciscan in its lesser deformation and metamorphism, along a major fault system of regional extent that is an important crustal expression of lithospheric consumption at depth.

Miyashiro (1961) suggested that the Franciscan and grossly parallel Sierra Nevada batholith and related high-temperature, low-pressure metamorphic rocks are part of a series of circum-Pacific paired metamorphic belts. As such, they are in essence late Mesozoic analogs of rock assemblages forming in active subduction zones and associated magmatic arcs, respectively (Takeuchi and Uyeda, 1965). When viewed in the context of plate tectonic theory and paired metamorphic belts, the Franciscan is hardly a unique assemblage, nor is it a particularly anomalous one. It offers an excellent opportunity to learn about the metamorphic and deformational history of subduction zone assemblages and to further our understanding of how plate interactions are expressed in the geologic record.

The purpose of this paper is to summarize structural and petrologic data from a representative Franciscan terrane in the Diablo Range northwest of Pacheco Pass, California, and to emphasize certain aspects of the tectonic framework of the study area that may characterize the Franciscan, and perhaps some other subduction zone complexes, in general. A more detailed presentation of lithologic and petrographic data is in Cowan (1972), and the reader is referred to Bailey and others (1964) for a comprehensive summary of Franciscan geology.

* Present address: Department of Geological Sciences, University of Washington, Seattle, Washington 98195.

FRANCISCAN COMPLEX NORTHWEST OF PACHECO PASS

The core of the northern part of the Diablo Range (Fig. 1) consists of the Franciscan Complex, flanked by Upper Jurassic to Upper Cretaceous sedimentary and volcanic rocks of the Great Valley sequence that are lithologically similar to Great Valley strata elsewhere in the Coast Ranges (Dickinson and Rich, 1972). Vickery (1924), Huey (1948), Leith (1949), Briggs (1953), Schilling (1962), Maddock (1964), and Raymond (1973a) mapped parts of the northern Diablo Range and together established that the entire boundary of the Franciscan mass is a series of steeply dipping faults now collectively termed the "Tesla-Ortigalita fault" (Page, 1966; Raymond, 1973b).

The area mapped during this study is approximately 20 km northwest of Pacheco Pass (Fig. 1). The northern part of the area slightly overlaps the southern edge of the Mount Boardman quad mapped by Maddock (1964). Raymond (1973a, 1973b) mapped the Mount Oso area to the north, and Cotton (1972) differentiated Franciscan bedrock units in the western half of the range north of California Route 152. The Franciscan Complex at Pacheco Pass has been exhaustively studied by McKee (1962a, 1962b), Ernst and Seki (1967), and Ernst and others (1970). Ernst (1971b) later made a petrographic reconnaissance of Franciscan metagraywacke from the core of the northern Diablo Range and determined the approximate areal distribution of metaclastic rocks containing jadeitic pyroxene. Investigations to date indicate the Franciscan in the Diablo Range is representative, both petrologically and structurally, of the complex in general.

The Franciscan in the study area includes three major fault-bounded structural units, each characterized by a distinctive deformational style and suite of metamorphic mineral assemblages. In descending structural order, the three semitabular units include (1) metagraywacke semischist, (2) "Garzas" tectonic mélange, and (3) gently folded "Orestimba" metagraywacke. The deformational styles represented here are not unique to the area but probably characterize the complex throughout the Coast Ranges. The Tesla-Ortigalita fault that separates the Franciscan from Great Valley rocks today is probably a steeply dipping to vertical modification of the regional, presumably low-angle, "Coast Range thrust" along which little-deformed strata of the Great Valley sequence were originally juxtaposed over the complexly deformed Franciscan (Bailey and others, 1970; Raymond, 1973b). Tertiary and Quaternary uplift of the Franciscan core of the Diablo Range was accommodated at least in part by displacements along this fault.

Northwest of the Garzas Creek Narrows, the Tesla-Ortigalita fault is sharp and nearly vertical (Fig. 2). Southeast of the narrows, an unconsolidated, locally powdery, bluish-gray matrix of thoroughly comminuted rock debris containing isolated blocks of graywacke, semischist, greenstone, thinly bedded red and green chert, and glaucophane-lawsonite schist occupies the fault zone and expands to a width of 400 m at Richard Creek. The contact of the shear zone debris with fractured Great Valley strata is approximately vertical or dips steeply southwest. The sandstone, conglomerate, and shale of the Great Valley sequence were not examined in detail during this study, but the age of the strata immediately adjacent to the Franciscan is probably Turonian. Both Maddock (1964) and Schilling (1962) found fossils of this age in stratigraphically equivalent rocks to the northwest and southeast.

METAGRAYWACKE SEMISCHIST

The structurally highest unit within the Franciscan locally is a semischistose metagraywacke that ranges in thickness from approximately 200 m to 600 m. This unit occurs immediately adjacent to the sharp trace of the Tesla-Ortigalita fault, or accompanying shear zone debris, and is separated from structurally underlying tectonic mélange by a fault that dips moderately to steeply eastward. No fossils were found in the unit, but whole-rock K-Ar dates (Table 1) indicate it is at least as old as Albian age. The dates also show the unit was metamorphosed before the deposition of presently adjacent Great Valley strata.

Bedding in the semischist is generally well preserved and becomes only locally disrupted near the Tesla-Ortigalita fault. Medium- to coarse-grained graywacke predominates in beds that average several centimeters, but range up to 2 m, in thickness. A single, discontinuous layer of thin-bedded, white to pale-green chert several meters thick crops out in Richard Creek. Transverse, concordant, and ptygmatic veins of quartz and albite are abundant in metagraywacke.

The most striking and distinctive structural feature of the metagraywacke is a foliation, with an accompanying incipient cleavage, that is exclusively parallel to bedding. Although the foliation ranges from moderate to strong, no systematic zonal variations in textural reconstitution related to the faults bounding the unit were observed. Mesoscopically, original detrital grains and larger clasts in conglomerates, especially lithic fragments, are noticeably flattened parallel to the foliation. Although the clastic nature of the rocks is still obvious, grain boundaries are diffuse, and the graywackes have a semischistose fabric. Even though bedding and foliation strike northwest and dip consistently northeast, both *s* surfaces locally have been folded on a small scale. Interbedded graywacke and phyllite display groups of kinklike folds with sharp hinges and amplitudes of a few centimeters; whereas individual graywacke beds commonly contain isolated shear bands, transverse to the foliation, that define the axial planes of small asymmetric angular folds with amplitudes that range from a few millimeters to several centimeters. Apparently, no secondary axial plane foliation was developed in either case. The rather limited, nonrepresentative exposures of this unit prevented a systematic analysis of the folds, but their existence does not detract from the remarkably consistent attitude of bedding and foliation, which grossly parallel the lower tectonic contact of the unit and seem to preclude the presence of any large folds within it.

Microscopically, the foliation that characterizes the entire unit is defined by the planar preferred orientation of the boundaries of mineral grains and grain aggregates (Fig. 3A). It is clear that the rocks are tectonites with planar fabrics acquired during recrystallization and flow under stress. There is very little evidence for simple cataclasis or mechanical granulation unaccompanied by concurrent recrystallization. Metamorphic white mica, chlorite, and lawsonite have grown parallel and subparallel to the foliation. Recrystallized volcanic and metamorphic lithic fragments and chert grains in metagraywacke and conglomerate are markedly flattened and attenuated in this plane. Deformed clastic quartz grains, the most abundant detrital

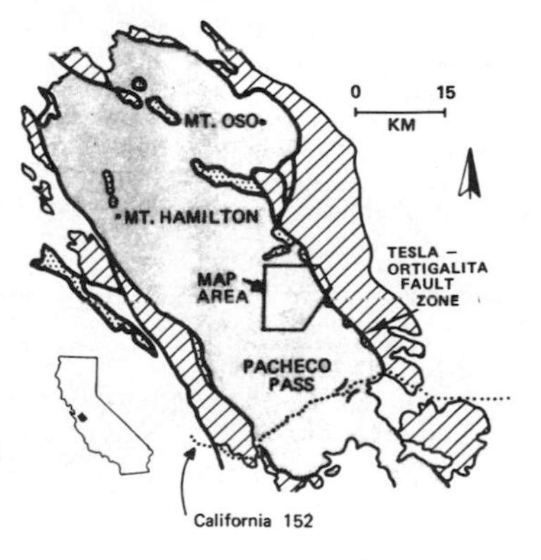

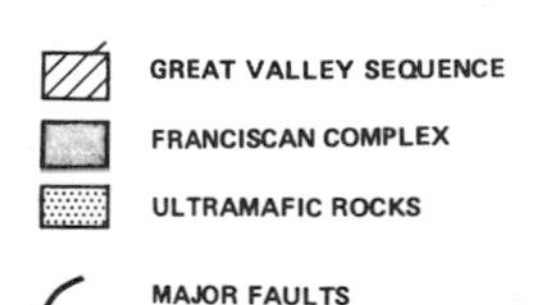

Figure 1. Generalized geologic map of northern Diablo Range, California. Simplified from San Jose Sheet, Geologic Map of California (Rogers, 1966).

constituent in the rocks, are the most important elements of the tectonite fabric. They display strong undulatory extinction and deformation lamellae, and the ends of tabular grains have polygonized into fine-grained granoblastic aggregates that in part define the foliation (Fig. 3A). The aggregates, which often contain oriented plates of white mica and chlorite, are well developed in most metagraywacke of the unit, and quartz in even the least recrystallized rocks is incipiently polygonized.

The metaclastic rocks in the unit contain the metamorphic mineral assemblages quartz + albite + lawsonite ± jadeitic pyroxene ± sodic amphibole ± aragonite + white mica + chlorite + stilpnomelane + sphene. Albite and lawsonite are present in each sample studied petrographically, and incipient to well-developed jadeitic pyroxene occurs in most, but not all metagraywacke samples. The pyroxene always coexists with albite; samples with the highest modal percentages of lawsonite and pyroxene contain some albite grains that completely lack these metamorphic minerals. Pyroxene prisms have grown within plagioclase grains and apparently do not have any preferred orientation parallel to the foliation, nor have they been granulated by postmetamorphic cataclasis. The distribution of the assemblages is shown in Figure 2. Bloxam (1960), McKee (1962a), Suppe (1969), and Ernst (1965; 1971a, 1971b; Ernst and others, 1970) have described in detail mineralogically similar jadeitic pyroxene-bearing metagraywacke from elsewhere in the Franciscan.

GARZAS TECTONIC MÉLANGE

Structurally beneath the metagraywacke semischist is an areally extensive unit informally termed the "Garzas tectonic mélange." The unit displays a distinctive deformational style characterized by small angular phacoids and larger geometrically and dimensionally variable blocks of resistant rock types enclosed in a pervasively sheared, fine-grained matrix. The blocks include both a wide variety of metamorphically exotic high-grade amphibolite- and blueschist-facies rocks, as well as lower grade greenstone and chert that typify an ocean-floor assemblage. Metagraywacke blocks are most abundant by far. Although the rock types and degree of shearing vary within the unit, the essential components of its chaotic structural geometry are penetrative mesoscopic shear surfaces that have a tectonic origin (Fig. 4). No further genetic interpretations are implied for the moment in the term tectoniç mélange. A more complete discussion of mélanges is deferred to a later section.

Mapping Procedure

Because of the extreme range in size, shape, and lithology of the resistant fragments and blocks encountered in the Garzas mélange, it is important to carefully select only the most meaningful data for inclusion in the geologic map. During this investigation, the mélange was treated as a single structural unit, and with the exception of two large, stratigraphically coherent blocks of metagraywacke with dimensions of several kilometers, tectonic inclusions were not differentiated from the fine-grained sheared matrix that encloses them. On this scale, it was impossible to accurately map the boundaries of all but the largest inclusions. Instead, I made several reconnaissance traverses through the mélange in an effort to establish gross patterns in the distribution of rock types. Some of the more resistant inclusions, including foliated glaucophane-lawsonite schist and albite-chlorite greenstone, form the familiar craggy outcrops, averaging several meters in diameter, that protrude from grassy slopes and are commonly referred to as "knockers" (Coleman and Lanphere, 1971, p. 2401; Photo 65 in Bailey and others, 1964). The strongly sheared fine-grained matrix in the mélange is semicontinuously exposed only along major intermittent-stream canyons.

Because the original stratal continuity of metagraywacke in the tectonic mélange has been destroyed, bedding is rarely observed. Bedding is preserved in the interiors of some large inclusions of metagraywacke, but these attitudes are not reliable guides to structural relations, because the inclusions are but fragments of larger blocks or layers and are not necessarily related to surrounding inclusions. The most prominent measurable feature in the mélange is the marked foliation defined by abundant subparallel surfaces of shear and dislocation in the matrix. This foliation replaces bedding as the most important mesoscopic fabric element. Systematic measurement of foliation attitudes throughout the entire mélange in the area was precluded by the limited, nonrepresentative exposures. Often, the attitudes vary within a single outcrop, and the shear foliation wraps around larger inclusions. Attitudes plotted on the geologic map (Fig. 2) are visual averages. Strikes vary from northeast to northwest, but the foliations measured in Garzas Creek dip consistently eastward, grossly parallel to the eastward-dipping fault contacts of the mélange with adjacent units. In the field, I placed these contacts at the first appearance of tectonic inclusions (such as greenstone, chert, and high-grade schist) that were clearly not derived from adjacent metagraywacke terranes. The characteristic sheared mélange matrix is only rarely exposed, and stratal continuity in adjacent metagraywacke becomes locally disrupted by similar shearing within a few meters of the contact with mélange.

Matrix and Inclusions

The matrix of the Garzas mélange is predominantly sheared siltstone and mudstone and is similar to the pelitic matrix of other mélanges described by Reed (1957, p. 31–32), Dickinson (1966, p. 462), and Hsü (1968). Weathering commonly emphasizes discrete shear planes. The matrix is cohesive and contrasts significantly with the powdery, completely milled gougelike debris that occupies the Tesla-Ortigalita shear zone and other high-angle fault zones in the Franciscan. On close inspection, small elongate lenses of fine-grained graywacke are interleaved with the sheared pelitic matrix, and a small part of the dominantly argillaceous matrix may have been derived from cataclasis and comminution of larger graywacke inclusions.

The tectonic inclusions in the mélange include metagraywacke with well-preserved clastic textures and a great variety of metaigneous and schistose metaclastic rocks and chert. Their heterogeneous dis-

TABLE 1. POTASSIUM-ARGON DATES

Sample no.	Date (m.y.)	Type date
Metagraywacke Semischist		
185	110 ± 2	Whole rock
202B	111 ± 2	do.
203A	76 ± 2	do.
Inclusions from Garzas Tectonic Mélange		
55	106 ± 2	Whole rock
291	116 ± 2	do.
64 (coarse)	130 ± 2.6	White mica concentrate; K_2O%: 5.46, 5.42; 29.15 x $10^{-6}cm^3$ radiogenic ^{40}Ar; 16% air correction
64 (60-150 mesh)	131 ± 2.6	White mica concentrate; K_2O%: 6.84, 6.88; 37.22 x $10^{-6}cm^3$ radiogenic ^{40}Ar; 10% air correction
Orestimba Metagraywacke		
7	93 ± 2	Whole rock
8	90 ± 2	do.
79	90 ± 3	do.
249	88 ± 2	do.

Note: Dates determined by R. L. Armstrong. See Suppe and Armstrong (1972) for brief petrographic descriptions and complete analytical data for all samples except number 64. Locations of dated samples indicated on Figure 2 of this report.

tribution is immediately apparent in Figure 2, which displays only the locations of samples collected on traverses. Hsü (1968, p. 1065), followed by Berkland and others (1972), classified inclusions in mélanges as *native* and *exotic*: his native inclusions are fragments of rocks originally interbedded with fine-grained pelitic material that now forms mélange matrix, and exotic inclusions are foreign to this now-disrupted terrane. In practice, however, it is often difficult to decide which rock types now present in a tectonic mélange were originally interbedded in a stratigraphically coherent rock-stratigraphic precursor; chert and greenstone are particularly troublesome examples. In this paper, I will restrict the term "exotic" to blocks of any protolith that were metamorphosed prior to their association with the sandstones and mudstones that form the bulk of the mélange.

Metagraywacke Inclusions

The most abundant inclusions in the Garzas mélange are metagraywacke. Although recrystallized, they have well-preserved, essentially isotropic, clastic fabrics. The metagraywackes are quite variable petrologically; detrital modes of several samples, selected to illustrate the variability, are shown in Figure 5. No two samples are from a single continuous inclusion. Only one inclusion was found that contained interbedded conglomerate. The inclusions range in size from a few millimeters to approximately 1 km in their longest dimension; the largest inclusion, the Mustang Peak metagraywacke slab, is at least 4 km long and is discussed separately below. Fragments averaging a few centimeters or less in diameter are irregularly rounded to elongate and are scattered through a strongly sheared fine-grained matrix. Inclusions several meters or less in length are most commonly elongate angular phacoids with their longest dimensions approximately parallel to the shear-plane foliation; this particular geometry is most characteristic of tectonic mélanges (Fig. 4; figures in Hsü, 1968, 1969; Dickinson, 1971b). Most of the largest inclusions have indeterminate shapes but are probably elongate with long axes subparallel to the regional strike of the shear-plane foliation. These large inclusions, which are clearly surrounded by sheared mélange, have sheared margins and are fractured throughout, but bedding is often preserved intact in their interiors. Apparently, inclusions of interbedded graywacke and siltstone must be attenuated to a certain critical size by fracturing and fragmentation before stratal continuity becomes completely disrupted by penetrative mesoscopic shearing. No fossils have been found in the inclusions.

The metagraywacke inclusions contain a variety of metamorphic mineral assemblages with markedly heterogeneous distribution (Fig. 2):

albite albite + pumpellyite albite + pumpellyite + lawsonite albite + lawsonite	} + quartz + chlorite + white mica ± stilpnomelane

Sodic amphibole was not identified in any inclusions. Metamorphic calcium carbonate is present in several samples; aragonite was positively identified only in assemblages containing lawsonite ± pumpellyite and was often partly recrystallized to calcite. Because of the rapid postmetamorphic inversion of aragonite to calcite (Brown and others, 1962), it is probably impossible to determine petrographically whether calcite coexisted stably with any or all of these as-

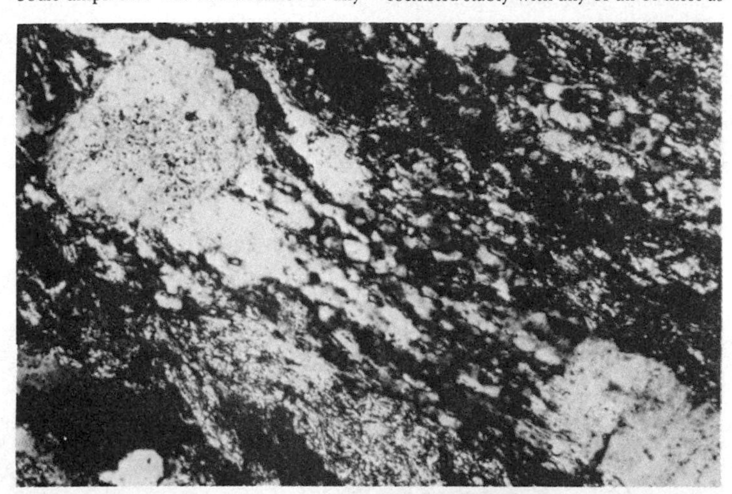

B.
0.5 MM

Figure 3. A. Deformed clastic quartz grains in metagraywacke semischist. Crossed nicols. Large relict grains are undulose and have grossly lensoid shapes. Ends of grains have polygonized into fine-grained granoblastic aggregates and contain plates of newly grown white mica and chlorite. Tabular aggregates, relict grains, and phyllosilicates define metamorphic foliation. B. Undeformed isotropic clastic fabric in Orestimba metagraywacke. Crossed nicols. Nearly all grains visible are quartz and detrital plagioclase, now altered to albite. Sample contains abundant metamorphic pumpellyite, white mica, and chlorite replacing framework grains.

semblages. Veins of quartz and carbonate are abundant.

Mustang Peak and Zimba Metagraywacke Slabs

The Mustang Peak and Zimba metagraywacke slabs, informally named after the topographic highs they respectively uphold, are the largest inclusions in the tectonic mélange (Fig. 2). Together they illustrate, on a large scale, the geometric relations of metagraywacke inclusions to surrounding pervasively sheared mélange matrix and associated smaller inclusions.

The Mustang Peak inclusion is at least 4 km long and consists entirely of well-bedded fine- to medium-grained metagraywacke and siltstone. Some sedimentary structures, including small-scale cross-laminations, sole markings, and convolute bedding, are present. The metagraywacke units have detrital modes distinct from both the Orestimba unit and most other metagraywacke inclusions in the mélange (Fig. 5); they contain more volcanic lithic fragments and detrital biotite grains, partly to completely replaced by chlorite stilpnomelane, than Orestimba metagraywackes. Both the Mustang Peak and Zimba inclusions contain metamorphic mineral assemblages that characterize other mélange metagraywacke units (Fig. 2); two samples from the Zimba inclusion contain minor glaucophane in addition to pumpellyite and lawsonite.

Bedding is gently folded and generally intact in the interior of the Mustang Peak inclusion, but within a few meters of its margins, stratal continuity becomes progressively disrupted by penetrative shearing geometrically identical to that displayed by the surrounding mélange. The disruption is best exposed on the ridge that culminates in Mustang Peak, and other unmappable zones of sheared metagraywacke and siltstone less than 10 m wide were encountered sporadically on traverses through the interior of the inclusion. These zones may represent the incipient fragmentation of the large inclusion into smaller tectonic blocks. Contacts of the Mustang Peak metagraywacke with surrounding mélange have been, in most places, only approximately located. The eastern tectonic contact dips gently eastward but has been modified by a nearly vertical fault south of Mustang Canyon. The western contact dips steeply to the southeast. The Mustang Peak inclusion thins progressively to the northeast and can best be visualized as an eastward-dipping, largely coherent, slab of metagraywacke embedded in sheared mélange.

The Zimba metagraywacke is separated from underlying mélange by a tectonic contact that dips gently northeastward. The metagraywacke has a weak bedding-plane foliation produced by compaction of soft lithic fragments and rotation of tabular detrital grains of quartz and plagioclase. Bed-

A.

B.

Figure 4. A. Tectonic mélange consisting predominantly of metagraywacke with small amounts of sheared siltstone. Bedding and stratal continuity have been completely disrupted by penetrative brittle-shear fractures that define foliation oriented approximately parallel to hammer handle. B. Tectonic mélange exposed along Garzas Creek. Lens-shaped inclusions of relatively resistant metagraywacke embedded in cohesive, strongly sheared fine-grained matrix.

ding becomes disrupted by penetrative shearing within 1 to 2 m of the contact with mélange; smaller inclusions similar to the Zimba metagraywacke — and probably derived from it — are locally abundant below this contact. In contrast to the Mustang Peak inclusion, the Zimba metagraywacke is essentially a structurally and topographically high klippe. There is no field evidence that it was originally overlain by tectonic mélange, and the present structural position suggests that the Zimba unit may once have been a part of the metagraywacke semischist, even though their mineral assemblages are different and they are now separated by tectonic mélange. The strong fabric of the semischist precludes point counting for comparison with Zimba detrital modes.

Other Inclusions

In addition to the metagraywacke blocks and slabs, the mélange contains inclusions of metavolcanic rocks, semischistose and schistose metaclastic rocks, and metacherts. A variety of protoliths and metamorphic mineral assemblages are represented. They range in size from approximately 10 cm to perhaps 50 m in their longest dimension; smaller inclusions are often demonstrably elongate with long axes oriented subparallel to the shear foliation in the surrounding fine-grained matrix. The larger inclusions project above grassy hillsides as rugged outcrops that are irregularly rounded in plan; their actual dimensions can only be estimated.

A detailed petrographic description of the inclusions sampled in the field is beyond the scope of this report; their distribution is indicated in Figure 2, and Table 2[1] lists their mineral assemblages and protolith where determinable.

Metavolcanic rocks are most abundant. Similar inclusions are widespread in the Franciscan, and excellent descriptions of their petrology and distribution are in Coleman and Lanphere (1971), Coleman and Lee (1963), Ernst (1965), and Ernst and others (1970). In the study area, metavolcanic rocks with the lowest grade mineral assemblages typically contain albite + chlorite + pumpellyite + sphene. Relict pilotaxitic and porphyritic textures are commonly preserved in these rocks, but unequivocal pillow structures were observed at only one locality.

Some of the inclusions in the area are fine-grained foliated schists that contain sodic amphibole, lawsonite, chlorite, and sphene; quartz and phengite are sometimes present. These are the familiar "blueschists" that occur as small tectonic blocks and larger structural units throughout the Franciscan. Switzer (1951), Brothers (1954), Borg (1956), McKee (1962a), Coleman and Lee (1963), Bailey and others (1964), Ernst (1965; Ernst and others, 1970), and Suppe (1969) have described the petrology and mineralogy of these and higher grade mafic blueschist rocks in detail. Even though igneous textures were generally obliterated by recrystallization, the composition of a tectonic block from locality 64 in the area (Table 3, see footnote 1) is similar to mafic blueschist from other areas with predominantly spilitic compositions. Some samples have relict fragmental, agglomeratic, and diabasic textures. Two samples contain pale-green omphacite, chlorite, and incipient glaucophane; similar omphacitic metabasalt occurs at Pacheco Pass (Ernst and others, 1970).

The highest grade inclusions in the mélange are generally coarse-grained schist that contains the characteristic assemblage sodic amphibole + garnet + phengite and, less commonly, sodic amphibole + epidote and omphacite + garnet + glaucophane. Lawsonite is also present in some assemblages, and chlorite commonly replaces garnet. These minerals and their textures indicate retrogression, and all originally eclogitic samples containing both omphacite and garnet also contain glaucophane and lawsonite. One coarse-grained amphibolite contains green hornblende + epidote + quartz; the hornblende is fringed with blue amphibole. These assemblages also characterize the highest grade blueschist- and amphibolite-facies metamorphic rocks found elsewhere in the Franciscan; they occur in other parts of the Franciscan only as isolated, clearly tectonic blocks often associated with serpentinite and lower grade metabasalt, metagraywacke, and chert in tectonic mélanges (Coleman and Lee, 1963; Coleman and Lanphere, 1971).

Semischistose and schistose metaclastic inclusions in the mélange contain quartz + lawsonite + white mica ± albite ± chlorite ± glaucophane. Chert and metachert inclusions are uncommon. Some are thin-bedded, white, light green, tan, or red in color, and they contain ovoid recrystallized relict radiolarian tests. These inclusions are identical to the essentially unmetamorphosed well-bedded chert that is widespread in the Franciscan Complex (Bailey and others, 1964, p. 55–68). In addition, a few strongly foliated, lineated, and completely recrystallized metachert samples contain small porphyroblasts of garnet, lepidoblastic sodic amphibole, and locally pale-green phengite plates in a mosaic of medium-grained granoblastic quartz. Similar metachert has been described by McKee (1962a), Coleman and Lee (1963), and Ernst (1965; Ernst and others, 1970).

Only two small bodies of sheared serpentinite, neither of which exceeds 100 m in its longest dimension, were located in the field, and each lies on the projected extension of a high-angle fault that terminates the metagraywacke tectonite unit east of Days Pass (Fig. 2).

Discussion of Exotic Blocks

Even though rocks of similar petrology and mineralogy to the inclusions described here are discussed in detail in the literature cited above, their structural relations to adjacent Franciscan rocks in other areas are locally ambiguous. Some are clearly tectonic inclusions in serpentinite (Brothers, 1954), whereas others occur as isolated tectonic blocks in linear "shear zones" of uncertain origin and structural geometry (McKee, 1962a; Borg, 1956). Some larger masses are essentially mappable, discrete structural units (Coleman and Lee, 1963; Suppe, 1969). It is important, therefore, to emphasize again the tectonic environment of these inclusions in the study area.

Exposures along Garzas Creek show conclusively that the inclusions, like the metagraywacke described above, are completely enclosed in a cohesive, pervasively sheared fine-grained matrix. The matrix, and associated metagraywacke, displays the generally well-developed shear-plane folia-

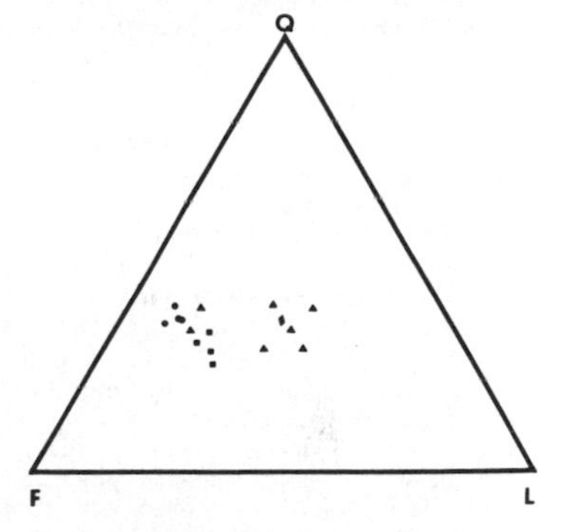

Figure 5. Volumetric Q-F-L percentages in metagraywacke from Orestimba unit, tectonic mélange, and Zimba slab. See Dickinson (1970a) for general discussion of point-counting procedures and philosophy and interpretations of detrital sandstone modes. Q = total quartzose grains; F = feldspar (plagioclase) grains; L = total unstable lithic grains of volcanic, metamorphic, and sedimentary origins. 400 framework points (Q+F+L) counted in each sample.

[1]See NAPS document 02420 for four pages of supplementary material. Order from ASIS/NAPS c/o Microfiche Publications, 305 E. 46th Street, New York, New York 10017. Remit in advance for each NAPS accession number $1.50 for microfiche or $7.25 for photocopies. Make checks payable to Microfiche Publications.

tion that characterizes tectonic mélanges. Because the sheared matrix weathers easily, it is only rarely observed as polished remnants plastered around the margins of inclusions exposed on hillsides. It is apparent from Figure 2 and Table 2 (see footnote 1) that inclusions of markedly different metamorphic grades and lithologies have been intimately mixed together. Many inclusions have well-developed, and often complex, internal metamorphic structures; these foliations and lineations record an earlier metamorphic history, and, as pointed out by Coleman and Lee (1963), they have no systematic structural or genetic relation to the deformational geometry of the surrounding mélange or adjacent inclusions. Finally, it is clear from field evidence in the area that exotic inclusions, which are in metamorphic disequilibrium with surrounding metagraywacke, are not restricted to narrow, high-angle fault zones.

ORESTIMBA METAGRAYWACKE

The structurally lowest unit in the Franciscan Complex consists predominantly of massive to thin-bedded metagraywacke with minor amounts of interbedded siltstone. This unit is informally termed "Orestimba metagraywacke"; excellent exposures occur along Orestimba Creek and its major intermittent tributaries. The Orestimba metagraywacke lies structurally beneath the Garzas tectonic mélange and is separated from it by a fault that dips gently to moderately eastward. Bedding is folded but not dislocated, and metamorphic foliation is generally absent. Although I found no fossils in the study area, whole-rock K-Ar dates on four samples indicate it must be at least as old as Cenomanian (Table 1). Cotton (1972) found two fossils of probably Alban or younger age in a lithologically similar unit 10 km from the southwesternmost exposures of the Orestimba metagraywacke. The units, which may be equivalent, are presently separated by faults and mélange.

Massively bedded, monotonously medium-grained metagraywacke predominates but is interbedded with a thin-bedded, fine-grained metagraywacke and phyllitic siltstone facies east of Orestimba Creek. Although faint laminations parallel to bedding and size grading are extremely rare in the former facies, sedimentary structures, including wavy current laminations, fine-scale cross-laminations, and graded bedding, are abundant in the latter. Conglomerate and pebbly mudstone apparently are absent from the entire unit.

A few thin discontinuous lenses of thin-bedded, red-brown radiolarian chert are interbedded with graywacke southwest of Gill Ranch. The largest of these is 175 m thick and holds up the Rooster Comb (Fig. 2). This lens, and bedding within it, dip moderately north, but constituent chert beds are folded, fractured, and brecciated. Thin, discontinuous pods of fine-grained, pumpellyite-bearing green diabase occur along its upper surface.

The metagraywacke in the unit is distinctly quartzofeldspathic and contains a very small amount of lithic detritus (Fig. 6, see footnote 1). It contrasts significantly in this respect with metagraywacke inclusions in the tectonic mélange and with the metagraywacke semischist, although determination of detrital modes is difficult because of metamorphic recrystallization. K-feldspar is completely absent from these rocks, even though the high quartz-feldspar content and a heavy mineral suite that includes allanite and zircon signal a plutonic provenance.

In marked contrast to the structural style of the Garzas mélange, bedding in the Orestimba unit is intact and gently folded. Stratal continuity becomes disrupted only locally within a few meters of the contact with structurally overlying mélange. The discontinuous and massively bedded nature of most outcrops makes accurate determinations of attitudes impossible; they commonly vary in dip and strike over distances of as little as 30 m. The impression is one of gently undulating fold surfaces with amplitudes of this order and larger. Thin-bedded metagraywacke and phyllite locally display cylindrical parallel folds with wave lengths and amplitudes that average 0.5 m. Even smaller folds occur on the limbs of these larger mesoscopic features. Apparently, fold wave lengths are a function of bed thicknesses in the unit. Microscopically, most sandstone in the unit has generally well-preserved isotropic clastic fabrics (Fig. 3B), in contrast to the metagraywacke semischist. A planar metamorphic fabric defined by the preferred orientation of lensoid quartz grains and newly formed phyllosilicates is incipiently developed in Orestimba graywacke only in the northeastern part of the unit.

The Orestimba metagraywacke contains the following metamorphic mineral assemblages arranged in a crudely zonal pattern from southwest to northeast (Fig. 2): albite + pumpellyite, albite + pumpellyite + lawsonite ± glaucophone, albite + lawsonite, albite + lawsonite + jadeitic pyroxene ± glaucophane. All assemblages contain quartz + chlorite + white mica ± calcium carbonate + sphene ± stilpnomelane. Aragonite was positively identified only in samples containing lawsonite. Two samples, shown on Figure 2, contain pumpellyite, lawsonite, glaucophane, and very rare incipient needles of jadeitic pyroxene in albite. Pumpellyite is well developed in the central and southwestern parts of the unit. To the northeast, pyroxene always occurs sparingly as radiating sprays of fine prisms in albite that locally are intergrown with pale, yellow-green chlorite or with golden brown stilpnomelane. Glaucophane locally coexists with chlorite.

METAMORPHISM

Metamorphic Mineral Assemblages

Franciscan metaclastic rocks in the study area contain the following metamorphic mineral assemblages: (1) albite, (2) albite + pumpellyite, (3) albite + pumpellyite + lawsonite, (4) albite + lawsonite, and (5) albite + lawsonite + jadeitic pyroxene. All assemblages contain quartz + chlorite + white phengitic mica. Assemblages (1), (2), and (3) may contain calcite; (3) through (5) may contain aragonite. Sodic amphibole may occur locally in all but (1). Experimental studies strongly suggest that assemblages (1) through (5) form in response to increasing pressures at nearly constant, relatively low, temperatures (see Ernst 1971a, for a comprehensive review). Metastable recrystallization and metasomatic changes in bulk composition are of only minor and local importance in Franciscan metamorphism.

Each structural unit contains a characteristic suite of metamorphic mineral assemblages. The metagraywacke semischist is dominated by assemblage (5), which apparently is stable at pressures of 7 to 9 kb at temperatures of 200° to 300°C. Individual metagraywacke inclusions from the Garzas mélange contain (1), (2), (3), or (4). These assemblages, lacking jadeitic pyroxene, are probably stable at correspondingly lower pressures of 3 to 7 kb and temperatures of 150° to 250°C. Pumpellyite-bearing assemblages (2) and (3) predominate in the Orestimba metagraywacke, but jadeitic pyroxene is developed locally at its northeastern margin (Fig. 2). I found no evidence that major structures within the latter unit control the distribution of each assemblage, and chemical analyses (Table 3, Fig. 6, see footnote 1) indicate that metamorphism was isochemical. The assemblages might have formed contemporaneously in response to a simple lithostatic pressure gradient; the close spatial proximity of pumpellyite- and pyroxene-bearing assemblages would then suggest that pumpellyite can be stable at pressures somewhat higher than previously assumed. Conversely, pumpellyite-bearing assemblages might have been converted to higher pressure assemblages during a later, separate metamorphic event. The metastable coexistence of pumpellyite with jadeitic pyroxene might then be due to sluggish reaction rates.

Tectonic inclusions of metaigneous rocks in the Garzas mélange include low-grade greenstone with relict igneous textures and containing albite + chlorite + pumpellyite; foliated and lineated blueschists containing sodic amphibole + lawsonite; and high-grade, completely recrystallized rocks containing sodic amphibole, garnet, epidote, hornblende, and omphacite in various combinations. These assemblages record progressively higher temperatures of metamorphism (Ernst and others, 1970; Taylor and Coleman, 1968), possibly as high as 400° to 535°C in the latter case. On the basis of metamorphic mineral assemblages alone, it is probable that all metaigneous inclusions but those containing albite + chlorite + pumpellyite are exotic and did not recrystallize in metamorphic equilibrium with associated albite-bearing and pumpellyite + lawsonite-bearing metagraywacke. However, mineral foliations and lineations that characterize many of the exotic inclusions contrast strikingly with the essentially unmodified relict clastic textures of the graywacke. This fundamental textural contrast proves that the exotic inclusions could not have been in their present structural positions during at least part, if not all, of their metamorphic history.

Not a single grain of K-feldspar was found in graywacke or other rocks in the study area. It is difficult to imagine a source terrane that could have provided hundreds of cubic kilometers of sediment completely devoid of K-feldspar. K_2O contents of quartzofeldspathic Orestimba metagraywacke (Table 3, see footnote 1) are about the same as "Coastal Belt" Franciscan graywacke samples that contain abundant modal K-feldspar (Bailey and others, 1964, p. 33). Although original K-feldspar content of Franciscan graywacke may *in part* be a function of its stratigraphic age (Bailey and others, 1964, p. 140–141), a more reasonable explanation for the total absence of this phase in rocks from the study area and at least some other Franciscan metagraywacke is that K-feldspar is unstable in rocks of a wide range of bulk compositions under the high-pressure–low-temperature metamorphic conditions accompanying recrystallization to blueschist facies and prehnite-pumpellyite facies mineral assemblages.

Metamorphic Ages

Ten K-Ar isotopic dates from the area are presented in Table 1. With the exception of locality 64, all dates were obtained from whole-rock samples. Suppe and Armstrong (1972) suggested that whole-rock dates apparently reflect the time span during which metamorphic minerals, over an entire region consisting of several structural units, became closed systems with respect to argon diffusion. White mica is more argon-retentive than whole rocks. Although whole-rock ages from the region range from 116 to 76 m.y., the four dates from the Orestimba metagraywacke are clustered about a mean of 90 ± 2 m.y. These dates overlap with dates from the metagraywacke semischist, but, as a group, they are significantly younger than the two oldest dates from the semischist. This age difference permits, but does not prove, the hypothesis that the semischist was initially recrystallized earlier than the Orestimba metagraywacke, even though rocks in at least the former unit remained open systems until Late Cretaceous time.

Dates from exotic inclusions in the mélange range from 131 to 106 m.y. Coarse and fine fractions of white mica separated from a foliated glaucophane-lawsonite inclusion in Garzas Creek (locality 64) yielded dates of 130 m.y. and 131 m.y., respectively. This striking concordance suggests that argon loss ceased rather abruptly in the rock and was not a function of grain size. Although the two whole-rock dates obtained from mélange inclusions are comparable to dates from the semischist, white mica from sample 64 is significantly older than all other samples dated. Similar high-grade inclusions from mélanges elsewhere in the Franciscan generally yield older dates for white mica than adjacent lower grade units (Coleman and Lanphere, 1971; Suppe and Armstrong, 1972).

MAJOR FAULTS

The Franciscan in the study area can best be visualized as a series of discrete, grossly sheetlike structural units separated from one another by gently to steeply dipping major faults. The preceding descriptions of each unit clearly show that rocks with different deformational and metamorphic histories and different ages have been juxtaposed. The faults separating the Garzas tectonic mélange from the metagraywacke semischist and the Orestimba metagraywacke are each required by three mutually independent criteria: a striking contrast in deformational styles, different metamorphic mineral assemblages in metaclastic rocks, and juxtaposition of graywackes of different detrital modes. Suppe (1969, 1973) mapped similar faults in the Franciscan in the northern Coast Ranges. These enigmatic structures, also noted by Brown (1964a, 1964b) and Ghent (1965), probably characterize the entire complex.

Unfortunately, the ages and displacement histories of these faults may be as obscure as their existence is obvious. Unlike low-angle Cordilleran thrusts and conventional high-angle normal, reverse, and strike-slip faults, they do not systematically repeat or offset a dated stratigraphic succession but rather juxtapose rock units that bear no apparent stratigraphic, deformational, or metamorphic relation to one another. However, a comparison of metamorphic mineral assemblages across a given fault may help to decipher the final relative displacements of adjacent structural units.

Faulting and Metamorphism

In the study area, the jadeitic pyroxene-bearing metagraywacke semischist structurally overlies the Garzas mélange and Orestimba metagraywacke, both of which consist predominantly of pumpellyite- and lawsonite-bearing metaclastic rocks (Fig. 2). This structural arrangement, noted elsewhere in the Franciscan (Blake, 1965; Suppe, 1973), is the reverse of what would be expected had the rocks recrystallized in response to simple burial, because the highest pressure assemblages occur in rocks above and at shallower structural depths than rocks containing presumably lower pressure assemblages.

At least two hypotheses could explain the observed anomaly. The semischist either was once more deeply buried and has been displaced several kilometers upward along major faults since its metamorphism, or was metamorphosed in its present structural position at depths less than those required for production of observed high-pressure assemblages. The second hypothesis would in turn require that lithostatic pressures be supplemented by tectonic overpressures, perhaps generated beneath major low-angle faults (Blake and others, 1967, 1969). Theoretically, this process would produce an inverted sequence of mineral assemblages within a structural unit. However, the upper fault boundary of the semischist, to which assemblages within the unit could be referred, is not preserved. Although the Great Valley sequence may have originally overlain the semischist along the low-angle "Coast Range thrust" (Bailey and others, 1970; Raymond, 1973b), they are at present juxtaposed along the later, high-angle Tesla-Ortigalita fault.

The most important evidence in support of the first hypothesis is the major faults themselves. In view of the striking contrast in mineral assemblages and metamorphic ages across the faults throughout the study area, it seems unlikely that the tectonic units could have been metamorphosed while they occupied their present relative structural positions. Postmetamorphic displacements along the fault separating the semischist from underlying mélange could have accommodated the emplacement of the semischist into its present anomalously shallow level. Suppe (1970) first proposed such a hypothesis to explain the juxtaposition of low- and high-grade Franciscan rocks.

Ages of Faulting

The absolute age of the final displacements along major faults in the study area is impossible to specify because the stratigraphic age of displaced units is unknown. Substantial displacements probably did not occur after argon diffusion from metamorphic minerals ceased approximately 76 m.y. ago. Similar dates are the youngest reported from the Diablo Range as a whole (Suppe and Armstrong, 1972).

Several high-angle faults have a weak erosional topographic expression and are visible on air photos (Fig. 2). These faults are clearly younger than the major, generally lower angle, faults they locally offset. Faults in the mélange have pulverized mélange matrix to a bluish-gray gouge containing small inclusions of vein quartz and other rock types. The gouge resembles the finely comminuted debris in the Tesla-Ortigalita fault zone. Raymond (1973a) mapped similar faults within the Franciscan in the Mount Oso area. I interpret these faults as broadly contemporaneous with the Tesla-Ortigalita fault system. Most of them probably formed during late Cenozoic diapiric uplift of the Franciscan core of the Diablo Range.

NATURE AND ORIGIN OF TECTONIC MÉLANGES

Definitions

The structural units juxtaposed along major faults in the study area display contrasting deformational styles. Conventional foliated metamorphic tectonites and simply folded sequences of metagraywacke with relict clastic textures and intact bedding are common throughout the Franciscan and are represented by the metagraywacke semischist and Orestimba metagraywacke, respectively. However, it is obvious from field evidence alone that vast chaotic terranes cannot be described by conventional labels such as semischist and bedded sequence. These mappable bodies, which display a third fundamental style of deformation, have a characteristic internal fabric dominated by penetrative mesoscopic shear fractures and contain tectonic inclusions of all sizes immersed in a pervasively sheared, generally fine-grained matrix. The essence of the shear-fracture fabric is that it records a deformation. Tectonic inclusions comprise a range of rock types, and the deformational style in question is not limited to terranes containing demonstrably exotic blocks.

Given that such deformed rock bodies exist, we must decide what to call them. I have chosen the term "tectonic mélange," even though "mélange" itself is in danger of becoming a useless, wastebasket term used to describe all manner of chaotic rocks and even the Franciscan Complex as a whole. Clearly, Hsü (1966; 1968, p. 1065) recognized the unique fabric of the chaotic terranes when he carefully defined Franciscan mélanges as "mappable bodies of deformed rocks characterized by the inclusion of tectonically mixed fragments or blocks. . . in a pervasively sheared, fine-grained, and commonly pelitic matrix." Greenly (1919, p. 193–195, Pl. VIII) had first used the term "autoclastic mélange" to describe structurally analogous rocks in Anglesey consisting of lenticular fragments of all sizes embedded in a foliated matrix. Both Hsü and Greenly believed that tectonic processes were responsible for the distinctive fabric of mélanges, but Hsü restricted the term to pervasively sheared bodies containing certain types of inclusions. In effect, I have broadened Hsü's definition of mélanges to include *all* similarly deformed rock bodies, regardless of the types of inclusions or the earlier history of the affected rocks. I have also added the modifier "tectonic" to emphasize that, by definition, mélanges record a deformation. No other genetic implications hinder use of the term tectonic mélange to describe rocks displaying a particular, easily recognizable, deformational style.

The internal features of tectonic mélanges are characteristic and provide the most important criteria for recognition and differentiation of chaotic terranes in the field. Pervasive, mesoscopic shear fractures are the essential elements of their internal fabric. The shear surfaces occur as subparallel anastomosing fractures in brittle graywacke and more closely spaced fractures in less competent argillaceous rock (Fig. 4; Hsü, 1968, Pl. 1; Hsü, 1969, Photos 2 and 8; Dickinson, 1971b, Fig. 4). The latter commonly have a scaly appearance, and individual lenticular chips are polished and slickensided. The fracturing resulted in disruption of original stratal continuity in bedded sequences of sandstone and mudstone. In incipiently deformed sequences, individual sandstone beds are simply displaced along fractures that initially form at low angles to bedding; disruption progresses until only isolated, phacoidal fragments of beds, completely surrounded by a sheared argillite matrix, remain. Because the shearing is progressive in nature, there is no established minimum degree of deformation that distinguishes a tectonic mélange from an undeformed precursor. The division is rather arbitrary and depends on the judgment of the observer.

By accepting an expanded, more general definition of tectonic mélange, we can emphasize the structural kinship of similarly deformed bodies and thus avoid a whole spectrum of specialized terms. For example, some mélanges consist predominantly of graywacke and may contain minor inclusions of chert and greenstone. Other mélanges, such as the Garzas mélange, contain *exotic* blocks that have clearly had different metamorphic histories than the more voluminous graywacke inclusions with which they are now intimately associated. Hsü (1968, p. 1065–1066) termed the former "broken formations" and the latter true "mélanges" and characterized "broken formations" as disrupted rock-stratigraphic units. Berkland and others (1972) accepted Hsü's definition of broken formation. I propose that this term be abandoned because the style of deformation in both broken formations and exotic-bearing mélanges is identical. Moreover, it is very difficult in practice to establish whether certain inclusions in a mélange are fragments of rocks formerly interbedded in a discrete rock-stratigraphic unit. If necessary, modifiers such as "graywacke-chert-greenstone" or "exotic" can precede "tectonic mélange" to indicate rock types present as inclusions.

Some needless confusion has also arisen regarding the distinction between mélanges and olistostromes. Olistostromes are sedimentary accumulations of generally rounded, resistant inclusions embedded in massive pelitic, sandy, or marly matrices (Abbate and others, 1970). They are probably deposited from submarine slides and debris flows. Because the deformation that produces tectonic mélanges can in theory affect rock bodies of any kind, olistostromes, as well as nonchaotic bedded sequences of sandstone and mudstone, can be deformed after deposition by surfaces of brittle shear and thus become tectonic mélanges. As discussed below, it may in practice be difficult to establish whether a given mélange was originally an olistostrome or a bedded sequence.

Origin of Tectonic Mélanges

Unfortunately, our knowledge of tectonic mélanges is too scanty to allow a comprehensive and convincing hypothesis for their origin. Instead, I will focus on three important questions (and discuss some possible solutions): (1) What was the nature of a given tectonic mélange prior to its deformation? (2) How are exotic inclusions introduced into mélanges? and (3) What do observed structures imply about the deformational environment of mélanges? This entire discussion of mélanges draws heavily on my experience in the study area and my reconnaissance observations elsewhere in the Franciscan and on Hsü's (1968, 1969) description of mélanges in the Morro Bay area, California.

In some areas, it is obvious that a tectonic mélange was derived from either a bedded sequence or an olistostrome. The Mustang Peak slab, for example, has an interior of folded, but unsheared, interbedded graywacke and mudstone; bedding becomes progressively disrupted by penetrative shearing near its margins. This mélange is *structurally* indistinguishable from the Garzas mélange enclosing the slab, and the two can be separated only on the basis of exotic inclusions in the Garzas. In contrast, most of the "Williams chaos facies," an olistostrome in the northern Coast Ranges (Suppe, 1969), is a penetratively sheared mélange, but soft-sediment deformational features and primary chaotic structures in some parts are unaffected by shearing. However, it may be difficult or impossible to determine the parent of a mélange that cannot be traced into an undeformed precursor. Shearing may have completely obliterated diagnostic sedimentary features. The Garzas mélange typifies the dilemma. Abundant inclusions of Mustang Peak and less common Zimba-type metagraywacke are concentrated near each slab, suggesting that at least part of the mélange formed as a result of systematic fragmentation and disruption of bedded sandstone-mudstone sequences. However, the mélange apparently contains few inclusions derived from the adjacent metagraywacke semischist and Orestimba unit. In fact, the range in graywacke modes (Fig. 5) and the variety of nongraywacke inclusions (Table 2, see footnote 1) suggest that parts of several formerly undeformed, but presently unexposed rock-stratigraphic units may have been incorporated into the mélange. Although I did not observe any soft-sediment deformational structures in the matrix or inclusions of the mélange, it is possible that parts of the deformed body were initially deposited as olistostromes. Thus, even though the geometry and relative structural positions of inclusions are at present demonstrably or inferentially relatable to surfaces of shear, the distinctive inclusion-in-matrix fabric, and the exotic inclusions themselves, may have been inherited from chaotic submarine slumps. It seems that similar reasoning must be applied to similar mélanges if positive evidence of a parent is lacking.

The origin of exotic inclusions and the processes responsible for their emplacement in Franciscan mélanges are outstanding problems. Most of the controversy centers on the familiar isolated blocks of foliated high-grade "blueschists" that are widespread, but volumetrically insignificant, in the Franciscan (Bailey and others, 1964, p. 89–111). These blocks were clearly not derived from units adjacent to the enclosing mélange and have no exposed source. Metamorphic mineral assemblages, textures, and, in some cases, ages together prove that the inclusions were metamorphosed before they were mixed with associated mélange metagraywacke. The Garzas mélange contains a representative suite of these inclusions. Coarse-grained blueschist and amphibolite with the highest grade mineral assemblages are rare. Coleman and Lanphere (1971) measured K-Ar ages of 155 ± 8 m.y. on actinolite and 149 ± 5 m.y. on hornblende from a coarse-grained garnet amphibolite of this general type collected from the Garzas mélange just southeast of the mapped area. Mineralogically similar high-grade blocks elsewhere in the Franciscan consistently give K-Ar mineral ages of about 150 m.y. More common are foliated glaucophane–lawsonite–white mica metavolcanic rocks; white mica from a typical example on Garzas Creek (locality 64) yielded K-Ar ages of 130 and 131 ± 2.6 m.y. (Table 1). Blueschist samples with similar mineral assemblages from other Franciscan terranes have yielded nearly identical K-Ar dates (Lee and others, 1964; Suppe and Armstrong, 1972). For the Franciscan in general, the demonstrably exotic inclusions are older and of distinctly higher metamorphic grade than associated metagraywacke. Although some rare metachert is included in the exotic group, most inclusions have "basaltic" compositions and are inferred to have had igneous protoliths.

If all inclusions of this type were simply large blocks in olistostromes that developed into tectonic mélanges after deposition and consolidation, their origin and emplacement would be rather straightforward. Rocks that once were deeply buried and subjected to moderate temperatures and high pressures were elevated tectonically above the ocean floor. The uplifted terranes provided detritus to submarine chaotic debris-flow deposits interbedded with stratified sandstone and mudstone. The "olistostrome" hypothesis is unsatisfactory for several reasons. In the Garzas mélange, for example, there is no positive evidence that any part of the deformed body was initially an olistostrome. Significantly, the highest grade inclusions seemingly always "occupy disturbed zones (mélange or shear dislocation)" in the Franciscan as a whole (Coleman and Lanphere, 1971, p. 2398). Unless we accept the premise that all tectonic mélanges containing exotic inclusions with no presently exposed source were originally olistostromes, we must conclude, as did Coleman and Lanphere, that most exotic inclusions are fragments of relatively old metamorphic terranes that have been tectonically mixed with younger Franciscan clastic rocks.

Although the exact nature of the mixing process is unknown, the mineralogy of exotic inclusions requires that relatively dense blocks from deeply buried terranes have been tectonically transported upward from their sites of metamorphism to their present structural positions among less dense, lower grade metagraywacke. Some blocks in other Franciscan mélanges have actinolite ± talc rinds that indicate they were once immersed in serpentinite (Coleman and Lanphere, 1971). It is conceivable that exotic inclusions could have been carried upward some distance in pods of serpentinite that were rising through relatively more dense metasediments, but mélange inclusions in the study area are enclosed in strongly sheared metagraywacke and argillite, and serpentinite hosts, if ever present, have become comminuted or have migrated away and disappeared. A more appealing suggestion is that the blocks have been emplaced upward along systems of innumerable shear planes in tectonic mélanges that intersect high-grade parent terranes at depth (Suppe, 1972). This process is kinematically analogous to the upward emplacement of much larger, fault-bounded slabs of high-grade rocks such as the metagraywacke semischist and the Taliaferro metamorphic complex in the northern Coast Ranges (Suppe, 1969, 1970).

Finally, it is interesting to speculate on the nature of the deformational environment that produced Franciscan tectonic mélanges. Several observations are pertinent. Clearly, the internal surfaces of shear and dislocation in mélanges represent the postdepositional deformation of consolidated, lithified rock bodies. Sandstone beds, and, less obviously, associated mudstone, record large-scale brittle deformation. Brittle behavior is not included in the catalog of structures usually ascribed to submarine slumping of "soft," unlithified sediments. Sandstone beds were initially fragmented by shear fractures that formed subparallel to bedding (Fig. 4); this compressional strain contrasts with extensional strain recorded by fractures perpendicular to bedding. On a larger scale, mélanges typically, but not necessarily, separate discrete tectonic units, such as the semischist and Orestimba unit, with ages and metamorphic mineral assemblages that preclude a normal stratigraphic, depositional relation.

In view of these observations, I believe tectonic mélanges are structurally equivalent to faults. Mélanges can best be characterized as sheetlike zones of distributed shear formed during brittle deformation of consolidated rock under a considerable, but as yet unspecified, overburden. Displacements have occurred along innumerable subparallel shear fractures, and the locus of tectonic dislocation has in effect expanded from a plane (=fault) to a zone several meters to kilometers in width (=tectonic mélange). Remnants of rock-stratigraphic units that became progressively attenuated as their margins contributed inclusions to a mélange are sometimes preserved within or adjacent to it; the Mustang Peak and Zimba slabs are examples. However, a mélange need not be directly related to adjacent units. The Orestimba metagraywacke and the semischist apparently contributed little, if any, material to the Garzas mélange; they were juxtaposed with it by displacements on the major faults that presently separate the units from the mélange.

SUMMARY

The Franciscan Complex is now interpreted as having accumulated in a late Mesozoic–early Tertiary subduction zone at the North American continental margin. This interpretation, with minor modifications, has been reiterated so often it has become hoary with age (for example, Hamilton, 1969; Bailey and Blake, 1969; Ernst, 1970, 1971a, 1971b; Ernst and others, 1970; Page, 1970; Dickinson, 1970b, 1971a), but it offers an elegant and readily acceptable explanation for both the metamorphic and tectonic features of the complex and its relation to adjacent coeval rocks. Ideally, we could directly compare the Franciscan with its most likely modern analogs, the structurally complex inner walls of oceanic trenches, but available marine geophysical data offer only tantalizing hints of major arcward-dipping faults within highly deformed, but otherwise irresolvable, acoustic basement (for example, Beck and Lehner, 1974).

Although field work in other ancient subduction zone assemblages will doubtless reveal additional facets of the deformation accompanying subduction, several fundamental features of the study area may characterize both the Franciscan elsewhere and, to the extent that the Franciscan is a representative example, subduction zone assemblages in general: (1) the complex consists of discrete structural units, grossly sheetlike in geometry and separated from one another by gently to steeply dipping major faults; (2) the structural units contain rocks of different ages, bear no apparent stratigraphic relation to one another, and have each experienced a different deformational and metamorphic history; (3) tectonic mélanges are major zones of distributed shear that record large-scale tectonic transport and deformation of consolidated rock bodies; and (4) rocks that were once more deeply buried may have been displaced upward from structural positions occupied during metamorphism.

The essence of subduction zone deformation become apparent when one compares Franciscan structures to those of more conventional Cordilleran fold-and-thrust belts where low-angle thrusts systematically repeat imbricate slices of strata that can generally be related to an autochthonous succession. In the Franciscan, adjacent structural units need never have been parts of an orderly stratigraphic sequence. Instead, the units include rocks of different ages derived from different sources; terrigenous sediments have been mixed with oceanic basalts and pelagic cherts on both a large and small scale. Rocks were deformed and metamorphosed under a variety of conditions and then juxtaposed along major faults. There is as yet no satisfactory dynamic interpretation to account for the broadly distributed shear recorded in mélanges or the apparent upward displacement of rocks that were once more deeply buried. Perhaps, as Ernst (1971c) proposed, such deformation occurred during buoyant upward rise of subducted materials in response to an over-all isostatic necessity as new materials were added incrementally, or "underplated," to the western margin of the subduction zone.

In the broadest sense, Cordilleran structures record crustal shortening accomplished by imbrication and repetition along low-angle faults. Tectonic mélanges and major faults that have juxtaposed unrelated rocks record continental margin accretion accompanying subduction. These features may prove as useful as blueschist-facies minerals in identifying ancient subduction zone assemblages in the geologic record.

ACKNOWLEDGMENTS

I am indebted to W. R. Dickinson for his stimulating advice, criticism, and unflagging interest in this research. I also thank M. C. Blake, Jr., R. G. Coleman, K. Crawford, W. G. Ernst, D. L. Jones, C. F. Mansfield, B. M. Page, L. A. Raymond, R. A. Schweickert, and John Suppe, each of whom freely discussed with me many of the ideas and observations presented here. Blake, Ernst, and Suppe critically read earlier versions of the manuscript. R. L. Armstrong dated a suite of samples from the area, and the University of Mexico provided chemical analyses under a cooperative agreement with the Department of Geology at Stanford University.

I am grateful to Mrs. Ernest Gill and Kaiser-Aetna for permission to enter the Gill Ranch.

This research was supported by a National Science Foundation Graduate Fellowship and research grants from the Geological Society of America and the Shell Companies Foundation.

REFERENCES CITED

Abbate, E., Bortolotti, V., and Passerini, P., 1970, Olistostromes and olistoliths: Sed. Geology, v. 4, p. 521–557.

Bailey, E. H., and Blake, M. C., Jr., 1969, Late Mesozoic tectonic development of western California: Geotectonics, no. 3, p. 148–154.

Bailey, E. H., Irwin, W. P., and Jones, D. L., 1964, Franciscan and related rocks and their significance in the geology of western California: California Div. Mines and Geology Bull. 183, 177 p.

Bailey, E. H., Blake, M. C., Jr., and Jones, D. L., 1970, On-land Mesozoic oceanic crust in California Coast Ranges: U.S. Geol. Survey Prof. Paper 700–C, p. C70–C81.

Beck, R. H., and Lehner, P., 1974, Oceans, new frontier in exploration: Am. Assoc. Petroleum Geologists Bull., v. 58, p. 376–395.

Berkland, J. O., Raymond, L. A., Kramer, J. C., Moores, E. M., and O'Day, M., 1972, What is Franciscan?: Am. Assoc. Petroleum Geologists Bull., v. 56, p. 2295–2302.

Blake, M. C., Jr., 1965, Structure and petrology of low-grade metamorphic rocks, blueschist-facies, Yolla Bolly area, northern California [Ph.D. thesis]: Stanford, Calif., Stanford Univ., 91 p.

Blake, M. C., Jr., Irwin, W. P., and Coleman, R. G., 1967, Upside-down metamorphic zonation, blueschist facies, along a regional thrust in California and Oregon: U.S. Geol. Survey Prof. Paper 575–C, p. C1–C9.

——1969, Blueschist-facies metamorphism related to regional thrust faulting: Tectonophysics, v. 8, p. 237–246.

Bloxam, T. W., 1960, Jadeite-rocks and glaucophane-schists from Angel Island, San Francisco Bay, California: Am. Jour. Sci., v. 258, p. 555–573.

Borg, I. Y., 1956, Glaucophane schists and eclogites near Healdsburg, California: Geol. Soc. America Bull., v. 67, p. 1563–1584.

Briggs, L. I., Jr., 1953, Geology of the Ortigalita Peak quadrangle, California: California Div. Mines Bull. 167, 61 p.

Brothers, R. N., 1954, Glaucophane schists from the North Berkeley Hills, California: Am. Jour. Sci., v. 252, p. 614–626.

Brown, R. D., Jr., 1964a, Geologic map of the Stonyford quadrangle, Glenn, Colusa, and Lake Counties, California: U.S. Geol. Survey Mineral Inv. Map MF–279.

——1964b, Thrust-fault relations in the northern Coast Ranges, California: U.S. Geol. Survey Prof. Paper 475–D, p. D7–D13.

Brown, W. H., Fyfe, W. S., and Turner, F. J., 1962, Aragonite in California glaucophane schists, and the kinetics of the aragonite-calcite transformation: Jour. Petrology, v. 3, p. 566–582.

Coleman, R. G., and Lanphere, M. A., 1971, Distribution and age of high-grade blueschists, associated eclogites, and amphibolites from Oregon and California: Geol. Soc. America Bull., v. 82, p. 2397–2412.

Coleman, R. G., and Lee, D. E., 1963, Glaucophane-bearing metamorphic rock types of the Cazadero area, California: Jour. Petrology, v. 4, p. 260–301.

Cotton, W. R., 1972, Preliminary geologic map of the Franciscan rocks in the central part of the Diablo Range, Santa Clara and Alameda Counties, California: U.S. Geol. Survey Misc. Field Studies Map MF–343.

Cowan, D. S., 1972, Petrology and structure of the Franciscan assemblage northwest of Pacheco Pass, California [Ph.D. thesis]: Stanford, Calif., Stanford Univ., 74 p.

Dickinson, W. R., 1966, Table Mountain serpentinite extrusion in California Coast Ranges:

Geol. Soc. America Bull., v. 77, p. 451–472.
Dickinson, W. R., 1970a, Interpreting detrital modes of graywacke and arkose: Jour. Sed. Petrology, v. 40, p. 695–707.
——1970b, Relations of andesites, granites, and derivative sandstones to arc-trench tectonics: Rev. Geophysics and Space Physics, v. 8, p. 813–860.
——1971a, Clastic sedimentary sequences deposited in shelf, slope, and trough settings between magmatic arcs and associated trenches: Pacific Geology, v. 3, p. 15–30.
——1971b, Detrital modes of New Zealand graywackes: Sed. Geology, v. 5, p. 37–56.
Dickinson, W. R., and Rich, E. I., 1972, Petrologic intervals and petrofacies in the Great Valley sequence, Sacramento Valley, California: Geol. Soc. America Bull., v. 83, p. 3007–3024.
Ernst, W. G., 1965, Mineral parageneses of Franciscan metamorphic rocks, Panoche Pass, California: Geol. Soc. America Bull., v. 76, p. 879–914.
——1970, Tectonic contact between the Franciscan mélange and the Great Valley sequence, crustal expression of a late Mesozoic Benioff zone: Jour. Geophys. Research, v. 75, p. 886–902.
——1971a, Do mineral parageneses reflect unusually high-pressure conditions of Franciscan metamorphism?: Am. Jour. Sci., v. 270, p. 81–108.
——1971b, Petrologic reconnaissance of Franciscan metagraywackes from the Diablo Range, central California Coast Ranges: Jour. Petrology, v. 12, p. 413–437.
——1971c, Metamorphic zonations on presumably subducted lithospheric plates from Japan, California, and the Alps: Contr. Mineralogy and Petrology, v. 34, p. 43–59.
Ernst, W. G., and Seki, Y., 1967, Petrologic comparison of the Franciscan and Sanbagawa metamorphic terranes: Tectonophysics, v. 4, p. 463–478.
Ernst, W. G., Seki, Y., Onuki, H., and Gilbert, M. C., 1970, Comparative study of low-grade metamorphism in the California Coast Ranges and the Outer Metamorphic Belt of Japan: Geol. Soc. America Mem. 124, 276 p.
Ghent, E. D., 1965, Glaucophane-schist facies metamorphism in the Black Butte area, northern Coast Ranges, California: Am. Jour. Sci., v. 263, p. 385–400.
Greenly, E., 1919, The geology of Anglesey: Great Britain Geol. Survey Mem., 980 p.
Hamilton, W., 1969, Mesozoic California and the underflow of Pacific mantle: Geol. Soc. America Bull., v. 80, p. 2409–2430.
Hsü, K. J., 1966, Mélange concept and its application to an interpretation of the California Coast Range geology [abs.]: Geol. Soc. America Spec. Paper 101, p. 99–100.
——1968, Principles of mélanges and their bearing on the Franciscan-Knoxville paradox: Geol. Soc. America Bull., v. 79, p. 1063–1074.
——1969, Preliminary report and geologic guide to Franciscan mélanges of the Morro Bay–San Simeon area, California: California Div. Mines and Geology Spec. Pub. 35, 46 p.
Huey, A. S., 1948, Geology of the Tesla quadrangle, California: California Div. Mines Bull. 140, 75 p.
Lee, D. E., Thomas, H. H., Marvin, R. F., and Coleman, R. G., 1964, Isotope ages of glaucophane schists from the area of Cazadero, California: U.S. Geol. Survey Prof. Paper 475-D, p. D105–D107.
Leith, C. J., 1949, Geology of the Quien Sabe quadrangle, California: California Div. Mines Bull. 147, 60 p.
Maddock, M. E., 1964, Geology of the Mt. Boardman quadrangle, Santa Clara and Stanislaus Counties, California: California Div. Mines and Geology, Map Sheet 3.
McKee, B., 1962a, Widespread occurrence of jadeite, lawsonite, and glaucophane in central California: Am. Jour. Sci., v. 260, p. 596–610.
——1962b, Aragonite in the Franciscan rocks of the Pacheco Pass area, California: Am. Mineralogist, v. 47, p. 379–387.
Miyashiro, A., 1961, Evolution of metamorphic belts: Jour. Petrology, v. 2, p. 277–311.
Page, B. M., 1966, Geology of the Coast Ranges of California, *in* Bailey, E. H., ed., Geology of northern California: California Div. Mines and Geology Bull. 190, p. 255–276.
——1970, Sur-Nacimiento fault zone of California: Continental margin tectonics: Geol. Soc. America Bull., v. 81, p. 667–690.
Raymond, L. A., 1973a, Franciscan geology of the Mount Oso area, central California [Ph.D. thesis]: Davis, Calif., Univ. California, Davis, 185 p.
——1973b, Tesla-Ortigalita fault, Coast Range thrust fault, and Franciscan metamorphism, northeastern Diablo Range, California: Geol. Soc. America Bull., v. 84, p. 3547–3562.
Reed, J. J., 1957, Petrology of the lower Mesozoic rocks of the Wellington district: New Zealand Geol. Survey Bull. (New Series) 57, 60 p.
Rogers, Thomas H., compiler, 1966, Geologic map of California: California Div. Mines and Geology, San Jose sheet, scale 1:250,000.
Schilling, F. A., Jr., 1962, The Upper Cretaceous stratigraphy of the Pacheco Pass quadrangle, California [Ph.D. thesis]: Stanford, Calif., Stanford Univ., 253 p.
Suppe, J., 1969, Franciscan geology of the Leech Lake Mountain–Anthony Peak region, northeastern Coast Ranges, California [Ph.D. thesis]: New Haven, Conn., Yale Univ., 99 p.
——1970, Relationship between low-angle faulting and metamorphism in Franciscan tectonics, California: Geol. Soc. America, Abs. with Programs, (Ann. Mtg.), v. 2, no. 7, p. 756–757.
——1972, Interrelationships of high-pressure metamorphism, deformation and sedimentation in Franciscan tectonics, U.S.A.: Internat. Geol. Cong., 24th, Montreal 1972, Comptes Rendus, sec. 3., p. 552–559.
——1973, Geology of the Leech Lake Mountain–Ball Mountain region, California: California Univ. Pubs. Geol. Sci., v. 107, 82 p.
Suppe, J., and Armstrong, R. L., 1972, Potassium-argon dating of Franciscan metamorphic rocks: Am. Jour. Sci., v. 272, p. 217–233.
Switzer, G., 1951, Mineralogy of the California glaucophane schists: California Div. Mines Bull. 161, p. 51–70.
Takeuchi, H., and Uyeda, S., 1965, A possibility of present-day regional metamorphism: Tectonophysics, v. 2, p. 59–68.
Taylor, H. P., Jr., and Coleman, R. G., 1968, O^{18}/O^{16} ratios of coexisting minerals in glaucophane-bearing metamorphic rocks: Geol. Soc. America Bull., v. 79, p. 1727–1756.
Vickery, F. P., 1924, Structural dynamics of the Livermore region: Jour. Geology, v. 33, p. 608–628.

Manuscript Received by the Society June 18, 1973
Revised Manuscript Received March 28, 1974

Editor's Comments on Papers 26, 27, and 28

The three papers presented in this section describe mélanges of the Pacific region outside California. Connelly, in a joint study with Hill and Gill (1976) and Moore and Wheeler (1978), investigated a chaotic assemblage of chert (including radiolarite), argillite, wacke, greenstone, gabbro, and ultramafic rocks of the Kodiak Islands—the Uyak Complex, which includes typical deep sea sequences (Paper 26). The deformation is consistent with a shear couple related to underthrusting toward the northwest. It is suggested that the prehnite-pumpellyite facies metamorphism is related to a pressure-temperature (P-T) regime that prevailed in the subduction zone. The complex is supposed to represent one phase in a prolonged history of seaward accretion and subduction. Moore and Karig (1976) defined this process. Gabbro and ultramafic rocks account for 6% of the mélange and other greenstone lithologies for 20%. They are interpreted as fragments of dismembered oceanic crust. The greenstone fragments are of diverse ages (Hill and Gill, 1976). Radiolarian cherts account for 2% of the mélange and have a close spatial relationship to pillow lavas, and some unsheared contacts with them are preserved. The radiolarian microfaunas are of ?Paleozoic, late Jurassic-Cretaceous, and late Valanginian-Aptian age. Sedimentary structures—parallel lamination, sole marks, and graded and cross bedding—are preserved in some of the chert beds, also palagonite partings. The radiolarites were deposited directly on the basalts but are believed to include both primary pelagic sediments and secondarily redeposited turbidites. The favored model requires a source in an elevated mid-ocean rise for both volcanics and radiolarites; the lack of clastic sediments and air-fall tuffs is consistent with such a model. The very old fossil ages are explained by supposing that they represent the base of the oceanic sediment column. Anomalous radiolarite ages are also recorded from the ophiolite-dominated mélange of the Makran, Iran, (the Coloured Mélange, Paper 21), a somewhat different type of mélange to the sediment-dominated mélange of Alaska.

Limestone of varied ages, including shelf types, accounts for 1% of the blocks in this mélange. There is an anomalous occurrence of fusulinid limestone of Tethyan affinities. Wacke and thinly bedded argillite make up 20% of the mélange, but no primary sedimentary structures are preserved in these fragments. Gray chert and argillite form the matrix (45%), in which the phacoids of the other lithologies float. The more competent chert layers in the matrix have become folded, boudined, thoroughly shattered, and suspended in the argillite fraction. Emplacement during late Cretaceous subduction is favored, though early to middle Cretaceous (post-Valanginian) emplacement cannot be ruled out.

Moore and Wheeler (1978) described the deformation of this mélange in detail. They considered that mélanges are a product of nongenetic deformation and are defined by a peculiar structural style. In their view a mélange need not contain exotic components. This is a questionable standpoint (see introduction to this volume). Moore (1973) previously described Cretaceous turbidite sequences from this region, probably trench deposits as broken formations, entertaining an origin in either gravity sliding or underthrusting at the inner trench wall. He remarked on the homoclinal dips in thick sections, all toward the continent, tilting being superimposed independently to the folding and probably related to uplift along major high angle faults parallel to the continental margin. In the later paper, Moore and Wheeler suggested that the lack of knocker topography, produced by blocks and slabs exposed in strong relief, is either due to silicification of the matrix (rendering it also resistant to erosion) or to the horizontal nature of the slab/matrix contacts. There are no large-scale folds and the mélange consists essentially of isolated slabs, separated into lenses by brittle fracture and set in a chert-argillite matrix. All the pillow facings are toward the northwest, but there is a lack of facing indication in the massive graywackes. Large-scale deformation by thrusting may well have occurred, but if large fold structures were ever present, they have been sheared out. The mélange zones are fault bounded and there are probably smaller-scale imbrications within them. Foliation is parallel to most small-scale, first generation folds, which are closed, moderately disharmonic, and tend to parallel geometry. There is some similarity to Appalachian styles of folding, related to shear zones in major anticlines. There is a lack of evidence of soft sediment deformation, and this, combined with the presence of foliation formed under significant containing pressure, suggests tectonic origin for the first generation folds. A model involving transfer from the underthrust to the overthrust plate, offscraping and passage through a master shear zone separating two plates, was suggested—emplacement being rapid and prior to the metamorphic peak (thermal equilibration). Such a chronology would be consistent with underthrusting in a subduction zone, the first deformation being related to emplacement within the hanging wall of the subduction zone. Later steepening is attributed to landward tilting, possibly consequent on imbrication beneath the seaward margin of the mélange. This steepening produced the second generation structures that are not directly related to mélange formation. These authors mention other possible models, involving dynamic uplift combined with landward tilting (Silling and Cowan, 1977), compression of the accreted wedge (Seely, 1977), and gravitational flow (Elliott, 1976; Hamilton, 1977).

Suzuki and Hada (Paper 27) describe mélange from the Cretaceous of Shikoku, Japan— a tectonic, chaotically sheared assemblage of mudstone and slabs and blocks of sandstone, greenstone, and chert of deep sea environment. This mélange has thrust-fault boundaries with a second turbidite sequence of weakly deformed sandstone and mudstone, the two comprising an imbrication assemblage with polarity to the south. The greenstones include pillowed and unpillowed basic lavas, pillow breccia, hyaloclastite, and dolerite (possibly dike feeders to the volcanics). The basalt is alkalic and sea-mount provenance is suggested. Sandstone and chert are interbedded in the mélange. The entire mélange assemblage is related to an uplifted subduction complex. The feldspathic and lithic arenite of the second, turbidite sequence is believed to be quite distinct from the clastic sediments fraction of the mélange, which is well sorted and consists of better rounded grains. Polycrystalline quartz and rock fragments make up 86–95% of the detrital grains, basaltic rock makes up 5–14%, and metamorphic and granite clastic grains are absent. In contrast, the sandstones of the second, turbidite sequence include lithic grains of intermediate volcanics, metamorphic rocks, granite, and pelitic rocks.

The model favored for the mélange is imbrication of an accretion complex in the trench slope region during subduction. The imbrics of turbidite are related to the mid-slope terrace or slope basin in an arc-trench gap situation. Chaotization of the mélange is believed to have occurred during subduction, phacoids of various sizes and shapes being suspended in a less competent matrix of mudstone. There was subsequent imbrication with the turbidite sequence and rotation.

A feature of this region is the contrast in metamorphic state between the greenstone and turbidite imbrics. The former display prehnite-pumpellyite facies metamorphism, which may either relate to metamorphism on the ocean floor or later, during subduction and underthrusting. The interleaved turbidite imbrics display no metamorphism.

Radiolarian chert in the mélange commonly displays local preservation of sedimentary contacts with pillow lavas. The radiolarian microfaunas are Albian-Coniacian.

The model presented is similar to that of Moore and Wheeler (1978) for the Kodiak mélange. Moore and Karig (1976) also described the sedimentology, structural geology and tectonics of the Shikoku subduction zone, their paper being concerned with the modern circum-Pacific subduction. Shallow deformation in the zone has been found by deep sea drilling to be characterized by ordered folding with some possible minor thrusting but neither pervasive, irregular minor folds nor widespread structurally chaotic zones. There is, however, some evidence of olistostromes, which may be underthrust beneath the overturned fold immediately after derivation of the fold scarp. These give rise to stratigraphically discontinuous sheared units that would be indistinguishable from tectonically broken formations. These authors conclude that if broken formations or mélanges are being produced in this subduction zone, then they must be developing in regions of deeper underthrusting or larger finite strain than those sampled by deep sea core 298. This paper, perhaps, emphasizes the dangers inherent in extrapolating models for subduction processes from one geological time to another. Such processes clearly display wide variety in their character, both in different parts of a single ocean at any one time and from one geological time to another.

Iwasaki (1979) described a gabbroic breccia from Mikabu, Shikoku, which he interpreted as a gabbroic olistostrome.

A most important paper relating to mélange formation in the circum-Pacific region is that of Scholl and Marlow (1974). It cannot be presented here in facsimile because of space limitations, but the salient points relevant to the theme of this volume are briefly summarized here. Most important is the definition of (1) *subducted deposits*—strata and rocks deeply injected below the crust and only able to return to the surface as igneous or highly metamorphosed rock—and (2) *offscraped deposits*—strata and rocks skimmed off the upper part of the oceanic lithosphere and folded against, and partly thrust beneath, the base of the continental crust. These offscraped deposits retain much of their original formational characteristics but may have an imprint of high P-T mineralization due to partial subduction. In their Figure 5, these authors present hypothetical models of offscraping involving tectonic mélange formation. They stated that mixing of ocean floor pelagic sediments with terrigenous and pyroclastic sediments shed from island-arc or continental regions is a dramatic consequence of sea floor spreading. This realization has led to the concept of offscraping from the descending plate for deformed Mesozoic rocks of the circum-Pacific fold belt. This paper is mainly concerned with the anomalous dominance of continental and arc-derived turbidites, rather than ocean floor sequences, in the circum-Pacific eugeosynclines, and the authors concluded that deposits of modern eugeosynclines of this region did not interact with those of the oceanic plate in the bowels of the trench, but in an upslope, more landward situation, possible inner-arc, reverse-arc basin, mid-slope terrace, or continental/insular slope. The peculiar mélange terrains and high pressure metamorphism characteristic of many circum-Pacific eugeosynclines are believed to be due to subsequent or contemporaneous deformation, partial subduction, and offscraping of these basinal deposits, not deposits of the actual trench. This conclusion emphasizes the incorrectness of referring all tectonic mélanges to the actual trench zone and accounts for the fact that many subduction-related tectonic and olistostrome mélanges are not dominantly ophiolitic.

The third paper in this section, by Ballance and Spörli (Paper 28), has been selected to represent mélange developments in New Zealand. There is an excellent description of a purely sedimentary olistostrome mélange (Kear and Waterhouse, 1967), which has been briefly mentioned in Part III of this book (editorial comments), and Ballance and Spörli describe another chaotic assemblage, below rather than above the Lower Miocene Waitemata Group (quite distinct from the post- or syn-Waitemata Onerahi chaos breccia).

Whereas the Onerahi chaos-breccia contains crushed and broken Upper Cretaceous and Lower Tertiary sedimentary rocks and blocks of serpentinite, some very large, the allochthon consists of noncalcareous and calcareous mudstone, argillaceous micrite, muddy and micaceous siltstone, and fine micaceous sandstone, much of it internally sheared and brecciated—mixed lithologies and abnormal superposition are characteristic. There is, however, no overturning of large blocks. The age range is Upper Cretaceous to Upper Oligocene. Basic igneous rocks form large, rootless bodies up to tens of kilometers long. These are especially numerous at the northern end of the outcrop. Serpentinite is certainly not abundant, and it has not been identified with certainty. The thickness of the allochthon is believed to be up to 5 km, the body being wedge shaped and thickening from zero in the east to at least 2.86 km in the west. It has a length of 370 km from north to south, and an east-west extent of 60 km. It may have once been more extensive. Emplacement took place between Oligocene and Miocene. The mechanism was not the sort of tectonics that forms great recumbent folds. Jostling in a near-surface situation is suggested by lack of induration and metamorphism, but persistent high water pressures in depth could have inhibited induration and metamorphism. The authors believe that these features may not, therefore, provide a reliable guide to near-surface chaotization. They favor final emplacement by gravity sliding, but apply neither the term *olistostrome* nor the term *mélange*. They do not favor genetic distinction on the basis of internal structure such as the presence or absence of shearing (Hsü, Paper 2) because, in large bodies of rock disrupted at low confining pressures, there is almost complete convergence between features produced by gravity sliding and more compressive tectonics. Clearly, however, the allochthon is a product of tectonic causes, not random slope failure.

Spörli and Bell (1976) described another mélange occurrence from west of Napier in the Ruahine Range of North Island. This consists of lenses of sandstone, green and red sheared volcanics, unsheared red amygdaloidal volcanics, red and green argillite, and red bedded cherts, all embedded in a matrix of dark argillite and siltstone. There are minor developments of white crystalline limestone in contact with the volcanics, forming lenses up to 50 m long, with bedded cherts, and tens of meters thick. There are only rare occurrences of thickly bedded coarse sandstones, most of the sedimentary rocks in this mélange being fine grained and argillaceous. The outcrop zone is 2 km wide. All traces of bedding have been virtually obliterated, the more competent rocks being disrupted into lenses, lozenges, and subrounded ellipsoids and discs, with complex fracture patterns. There are swarms of lenses of both similar dimensions and identical lithology. That formation of the mélange occurred prior to metamorphism is suggested by the equal grade of matrix and fragments. A mudflow (gravity slide) origin is considered possible.

Bradshaw (1973) described a mélange from South Island, the Esk Head Mélange, for which he invoked a tectonic origin that he believed was consistent with its great thickness. Spörli and Bell, however, doubted whether thickness alone can be invoked to exclude submarine gravity sliding. Even so, they believed that the content and structural framework of this mélange speak for an origin closely related to tectonic deformation. They also discussed the superimposed rotation of this mélange to its present steep attitude.

Grant-Mackie (1971), Spörli and Barter (1973), Hill (1975), and

Spörli (1978) have all referred to mélanges in New Zealand. In general, these do not appear to include ophiolite-dominated mélanges comparable with the colored mélanges of Iran (Paper 21) or the serpentinite mélanges of the USSR (Papers 29 and 30).

REFERENCES

Bradshaw, J. D., 1973, Allochthonous Mesozoic Fossil Localities in Mélange within Torlesse Rocks of North Canterbury, *Royal Soc. New Zealand Jour.* **3:**161-167.

Elliott, D., 1976, The Motion of Thrust Sheets, *Jour. Geophys. Research* **81:**849-963.

Grant-Mackie, J. A., 1971, The Probably Autochthonous Nature of Norian and Newly Found Carnian Rocks of the Oroua Valley, Western Ruahine Range (abstract), in *Symposium on Torlesse Supergroup, Wellington, 1971,* p. 271.

Hamilton, W., 1977, Subduction in the Indonesian Region, in *Island Arcs, Deep Sea Trenches and Back-arc Basins,* M. Talwani and W. C. Pitman, eds., American Geophysical Union Maurice Ewing Series, vol. 1, pp. 15-31.

Hill, M. D., and J. B. Gill, 1976, Mesozoic Greenstones of Diverse Ages from Kodiak Islands, Alaska (abstract), *EOS (Am. Geophys. Union Trans.)* **57:**21.

Hill, P. H., 1975, *Taitai Series Rocks at Te Kaha,* M.Sc. thesis, University of Auckland, New Zealand, 149p.

Iwasaki, M., 1979, Gabbroic Breccia (Olistostrome) in the Mikabu Greenstone Belt of the Eastern Shikoku, *Geol. Soc. Japan Jour.* **85:**481-487.

Kear, D., and B. C. Waterhouse, 1967, Onerahi Chaos-Breccia of Northland, *New Zealand Jour. Geology and Geophysics* **10:**629-646.

Moore, G. F., and D. E. Karig, 1976, Development of Sedimentary Basins on the Lower Trench Slope, *Geology* **4:**693-697.

Moore, J. C., 1973, Complex Deformation of Cretaceous Trench Deposits, Southwestern Alaska, *Geol. Soc. America Bull.* **84:**2005:2020.

Moore, J. C., and R. L. Wheeler, 1978, Structural Fabric of a Mélange, Kodiak Islands, Alaska, *Am. Jour. Sci.* **278:**739-765.

Scholl, D. W., and M. S. Marlow, 1974, Sedimentary Sequence in Modern Pacific Trenches and the Deformed Circum-Pacific Eugeosynclines, in *Modern and Ancient Geosynclinal Sedimentation,* R. H. Dott, Jr. and R. H. Shaver, eds., Soc. Econ. Paleontologists and Mineralogists Spec. Pub. No. 19, pp. 193-211.

Seely, D. R., 1977, The Significance of Landward Vergence and Oblique Structural Trends on Trench Inner Slopes, in *Island Arcs, Deep Sea Trenches and Back-arc Basins,* M. Talwani and W. C. Pitman, eds., American Geophysical Union Maurice Ewing Series, vol. 1, pp. 187-198.

Spörli, K. B., 1978, Mesozoic Tectonics, North Island, New Zealand, *Geol. Soc. America Bull.* **89:**415-425.

Spörli, K. B., and T. P. Barter, 1973, Geological Reconnaissance in the Torlesse Supergroup in the Kaimanawa Ranges along the Lower Reaches of the Waipakihi River, North Island, New Zealand, *Royal Soc. New Zealand Jour.* **3:**363-380.

Spörli, K. B., and A. B. Bell, 1976, Torlesse Mélange and Coherent Sequences, Eastern Ruahine Range, North Island, New Zealand, *New Zealand Jour. Geology and Geophysics* **19:**427-447.

26

Reprinted from pages 755-761, 763-764, and 766-769 of *Geol. Soc. America Bull.* **89:**755-769 (1978)

Uyak Complex, Kodiak Islands, Alaska: A Cretaceous subduction complex

WILLIAM CONNELLY* *Earth Sciences Board, University of California, Santa Cruz, Santa Cruz, California 95064*

[*Editor's Note:* Table 1, analytical data for some Kodiak Island rocks, has been omitted.]

ABSTRACT

The Uyak Complex is a chaotic assemblage of gray chert and argillite, wacke, greenstone, radiolarian chert, and gabbroic and ultramafic rocks. The simplest interpretation of these rocks is that the gabbroic and ultramafic rocks and greenstone represent basal oceanic crust upon which the radiolarian chert, gray chert and argillite, and wacke were deposited, respectively, at a mid-ocean rise, on the abyssal ocean floor, and in an oceanic trench. Fossils are scarce in the complex and range in age from mid-Permian to mid–Early Cretaceous. The Uyak Complex was emplaced by underthrusting to the northwest beneath lower Mesozoic metamorphic, igneous, and sedimentary rocks. During underthrusting, brittle rock types were broken into phacoids of all sizes and suspended in the less competent matrix of gray chert and argillite with their longest dimensions aligned subparallel to cataclastic foliation. Prehnite and pumpellyite developed extensively in lithologies of suitable composition.

The Uyak Complex correlates to the northeast with a similar assemblage of deep-sea rocks on the Barren Islands and with the McHugh Complex on the Kenai Peninsula and near Anchorage. The Uyak-McHugh belt defines a probable subduction complex trending northeast for at least 600 km along the margin of southwestern Alaska. The time of emplacement of this mélange is uncertain, but fossils present indicate that it occurred after mid–Early Cretaceous time. To the southeast, the Uyak is underthrust by deformed turbidites of the Kodiak Formation, which are interpreted to have been deposited in an oceanic trench and accreted to the Alaskan margin in Late Cretaceous time. The relationship between the Uyak Complex and Kodiak Formation is uncertain, but they may represent two phases, or two facies, of Late Cretaceous accretion.

Two large bodies of schist, referred to as the Kodiak Islands schist terrane, occur along the northwest border of the Uyak. The Kodiak Islands schist terrane is a blueschist-bearing metamorphic belt that has yielded mainly Early Jurassic K-Ar mineral ages. Similarities in K-Ar ages, metamorphism, and tectonic setting support a correlation between the Kodiak Islands schists and the Seldovia schist terrane on southern Kenai Peninsula. The Upper Triassic Shuyak Formation structurally overlies the Kodiak Islands schist terrane and Uyak Complex but is separated from them by a long, narrow pluton emplaced in Early Jurassic time. The Shuyak is a little-deformed formation of volcanic and volcaniclastic rocks. It correlates with similar rocks on the Alaska and Kenai Peninsulas, which together outline a lower Mesozoic forearc basin; K-Ar ages show that much of the Alaska-Aleutian Range batholith was intruded coeval with deposition in this forearc basin. A likely interpretation of these rocks is that the Kodiak-Seldovia schists are the only vestige of a subduction complex emplaced along the margin of southwestern Alaska during the prominent early Mesozoic volcanoplutonic activity recorded on the Alaska and Kenai Peninsulas and the Kodiak Islands.

INTRODUCTION

Plate-tectonic theory is now on firm ground and is providing the genetic concepts necessary to unravel the complex geology of ancient continental margins. Geophysical and geological studies of modern plate margins indicate that oceanic trenches are the site of large-scale underthrusting of oceanic lithosphere (Isacks and Molnar, 1971) and that deep-sea sediments are scraped from the downgoing plate and accreted to the overlying plate in the process (Kulm and others, 1973). Seismic profiles across these convergent margins suggest that the thick wedges of sediment making up the trench inner wall are structured by compressional folds and landward-dipping thrust faults (Seely and others, 1974; Silver, 1971). Although these sediments are complexly deformed, they often are compositionally identical or similar to those at the trench axis (Kulm and others, 1973; von Huene, 1972; Fisher and Engle, 1969). However, because acoustic methods cannot resolve the deep internal structures of subduction complexes and because drilling and dredge hauls are capable of sampling only the uppermost part of these belts, it is informative to study ancient uplifted subduction complexes.

The margin of southwestern Alaska has been the site of accretion of several belts of deep-sea deposits since early Mesozoic time. These accretionary belts are well exposed along the fiord-indented coastlines of the Kodiak Islands (Fig. 1) and apparently have not been affected by subsequent strike-slip faulting. To better understand the nature of subduction complexes in general and to clarify the geology of this little-studied region, Casey Moore, Malcolm Hill, Betsy Hill, James Gill, and myself have conducted an integrated study of the Uyak Complex. This paper deals primarily with the sedimentary geology of the Uyak and a sedimentation model to explain its origin.

The Uyak is interpreted here as a subduction complex because of rock types present, style of deformation, grade of metamorphism, and tectonic setting. Rock types included in the Uyak are similar to lithologies that occur in typical deep-sea sequences. Deformation of these rocks is intense, but structural analyses indicate that deformation occurred in a shear-couple consistent with emplacement by underthrusting to the northwest beneath an existing continental margin (Moore and Wheeler, 1975). Prehnite-pumpellyite–facies

* Present address: Amoco Production Company, Security Life Bldg., Denver, Colorado 80202.

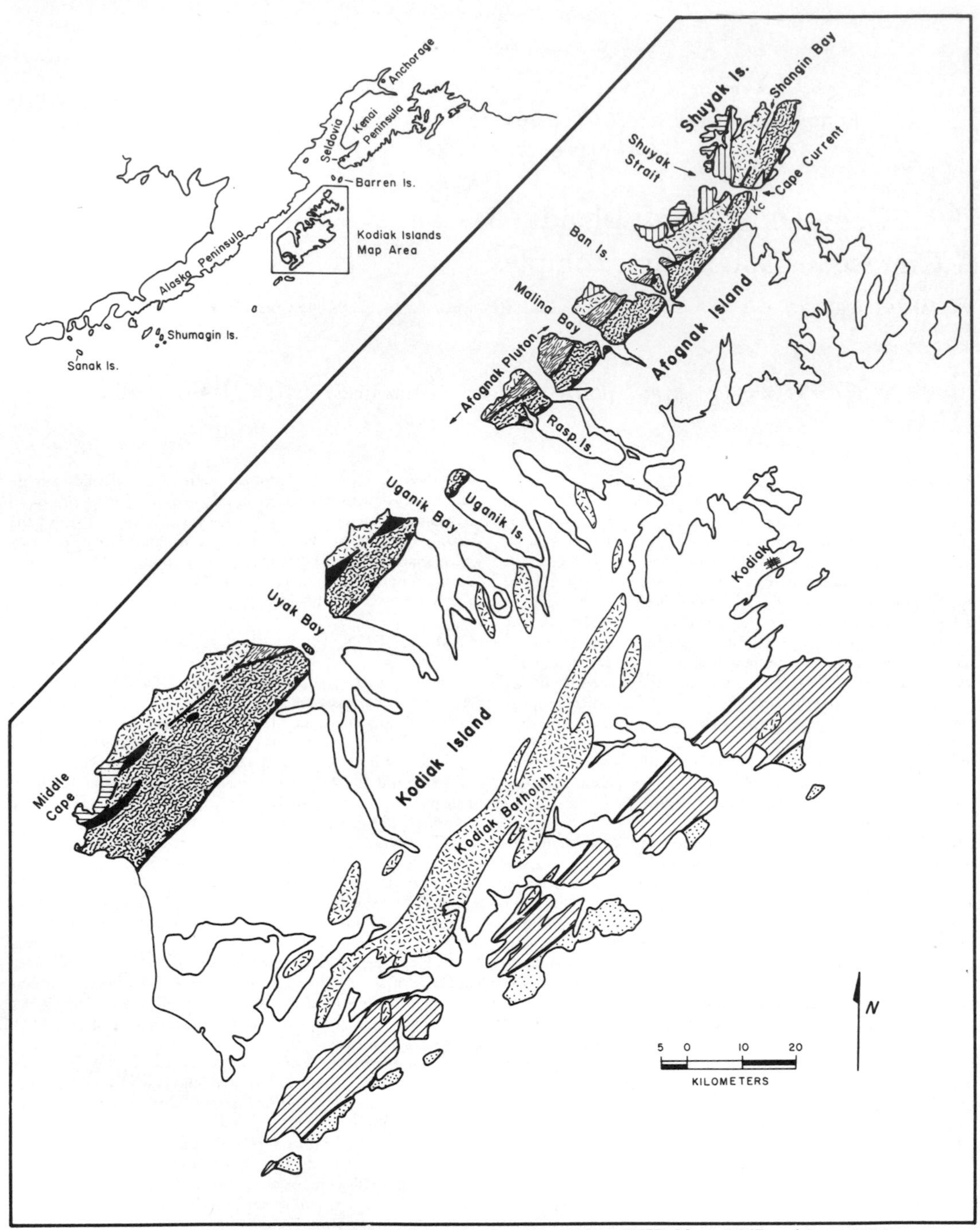

Figure 1. Generalized geologic map of Kodiak Islands; after Connelly and Moore (1977), Moore (1967), and Capps (1937).

metamorphism of the Uyak could have occurred in a pressure-temperature regime thought to exist in subduction zones. Finally, the Uyak Complex and correlative McHugh Complex define a long, narrow outcrop belt along the margin of southwestern Alaska that is sandwiched between a younger belt of deep-sea rocks seaward and an older belt landward, suggesting a long history of seaward accretion by subduction.

GEOLOGIC SETTING

The oldest rock unit on the Kodiak Islands is a sequence of vesicular pillow lava and openly folded volcaniclastic turbidites at least 7 km thick (Fig. 1). This previously unmapped volcanogenic section is here named the Shuyak Formation after the type section in western Shuyak Strait (see Connelly and Moore, 1977). The Shuyak Formation is divided into a lower volcanic member and an upper sedimentary member (discussed below); the Late Triassic pelecypod *Halobia* cf. *H. halorica* (N. J. Silberling, 1976, written commun.) has been collected from several localities in the sedimentary member. It is likely that a lithologically similar unit of rocks in the Middle Cape area on westernmost Kodiak Island is an isolated remnant of the Shuyak Formation, but no fossils were found there. The Shuyak Formation correlates to the northeast with lithologically and biostratigraphically equivalent unnamed rocks on the Barren Islands (Cowan and Boss, 1978) and near Seldovia on southern Kenai Peninsula (Martin, 1915), and to the northwest with similar rocks on the Alaska Peninsula at Puale Bay and in the Lake Iliamna–Kamishak Bay area (Burk, 1965; Detterman and Hartsock, 1966).

To the southeast, the Shuyak Formation is structurally underlain by the Kodiak Islands schist terrane along the inferred Shuyak fault (Fig. 1). The Kodiak Islands schist terrane consists of thinly layered and intricately folded quartz-mica schist, greenschist, blueschist, and epidote amphibolite, and it has yielded mainly Early Jurassic K-Ar mineral ages. Similarities in K-Ar ages, metamorphism, and tectonic setting support a correlation between the Kodiak Islands schists and the Seldovia schist terrane on southern Kenai Peninsula (Carden and others, 1977). The inferred Shuyak fault is now intruded by the dioritic Early Jurassic Afognak pluton.

The Uyak Complex structurally underlies the Kodiak Islands schist terrane and Afognak pluton along the Raspberry fault (Fig. 1). The Uyak Formation of G. W. Moore (1969) is here modified to the Uyak Complex (see Berkland and others, 1972). The term "complex" is used in recognition of the highly complicated internal structural relations of the Uyak rocks and is in accordance with the American Commission on Stratigraphic Nomenclature (1970, art. 6, p. 6–7). Similarities in lithology, style of deformation, degree of metamorphism, and tectonic setting suggest a correlation of the Uyak Complex with an unnamed unit on the Barren Islands and with rocks on the southern Kenai Peninsula presumed to be an extension of the McHugh Complex of the Anchorage area (Clark, 1973; Moore and Connelly, 1976; Magoon and others, 1976; Cowan and Boss, 1978).

The highly deformed turbidites of the Kodiak Formation underthrust the southeast side of the Uyak along the Uganik thrust (Capps, 1937; Moore, 1967, 1969). The Kodiak Formation is part of an extensive turbidite sequence that is interpreted to have been deposited in an oceanic trench along the margin of southwestern Alaska in Late Cretaceous time (Burk, 1965; Moore, 1972, 1973a, 1973b; Jones and Clark, 1973; Budnik, 1974).

The Paleocene(?) to Eocene Ghost Rocks Formation is faulted against the seaward side of the Kodiak Formation (Fig. 1). The Ghost Rocks Formation consists primarily of tightly folded thin- to medium-bedded argillite and wacke with occasional occurrences of nonvesicular pillowed greenstone and agglomerate (Moore, 1969); intense localized shearing has produced many zones of tectonic mélange. The only fossils recovered from the Ghost Rocks Formation are Eocene planktonic foraminifera (J. L. Thompson, 1975, written commun.) collected by me from a pelagic limestone at Ghost Rocks, Sitkalidak Strait, Kodiak Island. The metamorphic grade of the Ghost Rocks Formation locally reaches the prehnite-pumpellyite facies. The oceanic lithologies and style of deformation

EXPLANATION

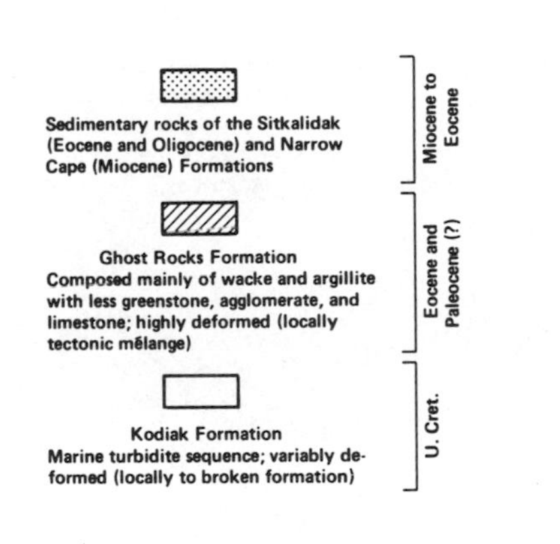

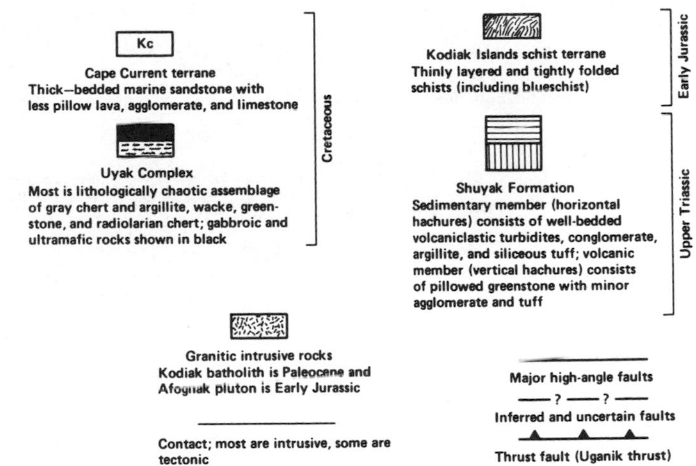

Figure 1. *(Continued).*

of the Ghost Rocks suggest that it is a subduction complex. Younger accreted deep-sea sequences occur seaward of the Kodiak Islands (von Huene, 1972) and indicate that this region has been the site of continual accretion up to Holocene time.

In the Alaska Peninsula are exposed volcanic and plutonic rocks with ages that apparently coincide with phases of accretion on the Kodiak Islands (Burk, 1965; Detterman and Hartsock, 1966; Reed and Lanphere, 1973). In addition, in the Kodiak Islands, two anomalously seaward plutonic belts are exposed: (1) the Early Jurassic Afognak pluton, which intrudes between the Shuyak Formation and Kodiak Islands schist terrane, and (2) the Paleocene Kodiak batholith, which intrudes the Kodiak Formation in central Kodiak Island (Burk, 1965; Moore, 1969; Hill and Morris, 1977).

LITHOLOGY OF UYAK COMPLEX

Intense shearing in the Uyak has destroyed most stratal continuity and tectonically juxtaposed different lithologies. Nevertheless, isolated clues and modern analogs suggest a sedimentation model based on plate tectonics to explain the relationships between the diverse deep-sea lithologies included in this mélange (Fig. 2).

Ultramafic, Gabbroic, and Basaltic Rocks

Gabbroic and ultramafic rocks occur as kilometre-sized slabs in the northwesternmost exposures of the Uyak (Fig. 1) and constitute about 6% of the mélange. These slabs contain layered gabbro, clinopyroxenite, dunite, and plagioclase peridotite; no harzburgite or sheeted-dike complexes have been observed (Hill, 1975). The bodies are always fault bounded, and serpentinization is pronounced near their margins. On the basis of mode of occurrence and mineralogy, these rocks tentatively are interpreted as fragments of dismembered oceanic crust (Hill and Brannon, 1976).

Greenstones generally occur as fault-bounded blocks tens to hundreds of metres in size and constitute about 20% of the complex. Nonvesicular pillowed greenstone with a thin chloritic interpillow matrix is most common, but massive greenstone occurs locally. The greenstones are partially altered to chlorite, albite, pumpellyite, and calcite, but relict phenocrysts of plagioclase and clinopyroxene locally are recognizable (M. Hill, 1976, personal commun.). Quench textures, variolitic structures, and nonvesicularity suggest extrusion in relatively deep water (Jones, 1969; Moore, 1970; Furnes, 1973). Major- and trace-element analyses suggest that the greenstones are ocean-floor basalts (Hill and Gill, 1976). Thus, the greenstones also appear to be fragments of dismembered oceanic crust.

Radiolarian Chert

Small tectonic blocks of rhythmically bedded radiolarian chert are scattered throughout the Uyak and account for about 2% of the mélange. Unsheared sedimentary contacts between radiolarian chert and underlying pillow basalt occur only at a few localities, but blocks of these two rock types are commonly in close proximity. The chert typically is red (although sometimes green or mottled) and occurs as homogeneous contorted sequences of 2- to 7-cm-thick beds separated by thin shaly partings. Individual chert beds locally contain small-scale laminations (Fig. 3), and X-ray diffraction of the shaly partings between beds indicates the presence of hematite and probably palagonite (see Nisbet and Price, 1974). Petrographic examination of this chert reveals abundant quartz-filled radiolaria in various stages of preservation set in a very fine-grained matrix of microcyrstalline quartz.

E. A. Pessagno (1976, written commun.) has examined radiolaria extracted from Uyak chert collected at five localities by myself and by G. W. Moore, and has identified (1) cryptoceptic Nassellariina forms of probable Paleozoic age; (2) *Archaeodictyomitra* sp. of probable Late Jurassic or Early Cretaceous age; (3) *Parvicingula boesi, Thanarla conica,* and *Archaeodictyomitra* cf. *A. vulgaris,* of late Valanginian age; (4) *Thanarla conica, Parvicingula*?, and *Pseudodictyomitra*?, of late Valanginian to Aptian age; and (5) *Thanarla conica* and *Archaeodictyomitra* sp. of late Valanginian to late Aptian age.

Sedimentary structures such as graded bedding, cross-bedding, parallel laminations, and sole marks have been described in radiolarian chert beds at many localities (Nisbet and Price, 1974; Garrison, 1974; Imoto and Saito, 1973). One interpretation of these structures is that they formed by redeposition of the radiolaria and their matrix by turbidity currents flowing down the tectonically active slopes of mid-ocean rises (Fig. 2). These currents may have flowed directly over exposed oceanic basalt and finally ponded as lenticular deposits between fault blocks along the rise. The model therefore explains the characteristic but enigmatic thin bedding and lenticularity of radiolarian chert deposits the world over. The presence of palagonite in the shaly partings between chert beds that is quite similar to palagonite of weathered submarine basalts (Matthews, 1971; Nisbet and Price, 1974) supports this interpretation.

I have adopted this general model for the Uyak radiolarian chert and suggest that it was deposited directly on pillow basalt of oceanic crust along the flanks of a mid-ocean spreading center, both as redeposited radiolarian turbidites and as primary pelagic sediments. Deposition on an elevated mid-ocean rise rather than

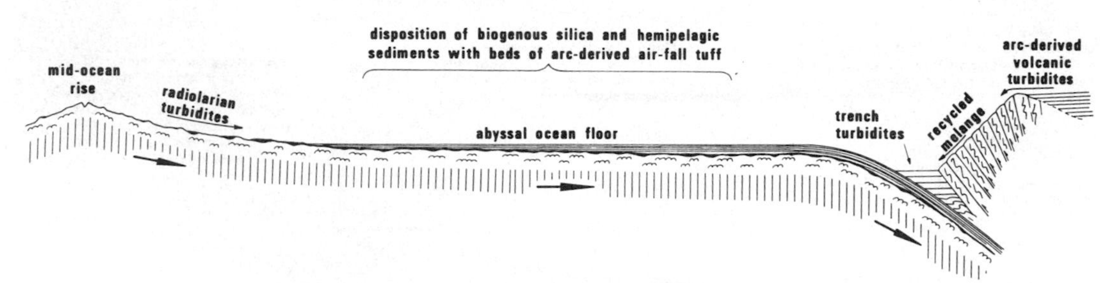

Figure 2. Sedimentation model for Uyak Complex, showing primary sedimentary regimes for oceanic and continental-margin environments. Model portrays sedimentary environments preceding Jurassic and Cretaceous proliferations of pelagic calcareous micro-organisms.

the abyssal ocean floor is supported by the lack of clastic sediments and air-fall tuff in the chert deposits. Furthermore, deposition directly on the pillow basalt would place the chert at the base of the ocean sediment column and therefore explain why some of the included radiolaria are among the oldest fossils from the complex.

Gray Chert and Argillite

Rocks included in our mapping unit "gray chert and argillite" account for about 45% of the Uyak Complex. This is an intensely deformed, thinly layered unit typically containing 55% to 65% gray chert, 25% to 35% black argillite, and 0% to 10% green tuff or limestone; a similar lithology containing more than 40% tuff accounts for about 6% of the mélange. Deformation has caused the more competent chert layers to become folded, boudinaged, or thoroughly shattered and suspended in the argillaceous matrix (Fig. 4). The green tuff occurs in these deformed sections as discontinuous layers generally less than 2 cm thick, but locally greater than 10 cm thick. At outcrop and map scale, the gray chert and argillite unit comprises the mélange matrix in which phacoids of more competent rock types are enclosed (Fig. 5).

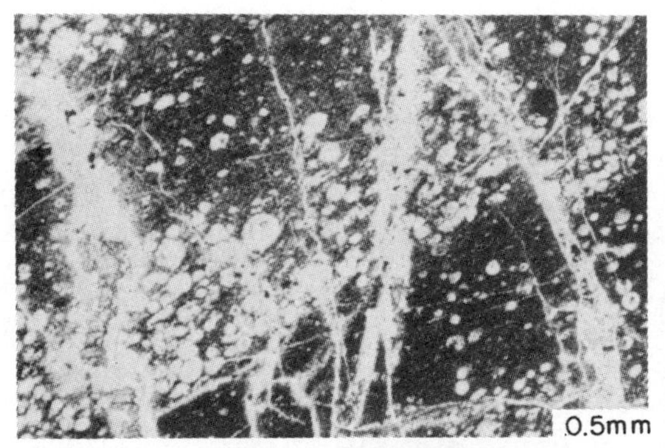

Figure 3. Photomicrograph of sample from radiolarian chert bed, collected from east Uganik Bay. Note how radiolaria define small-scale laminations.

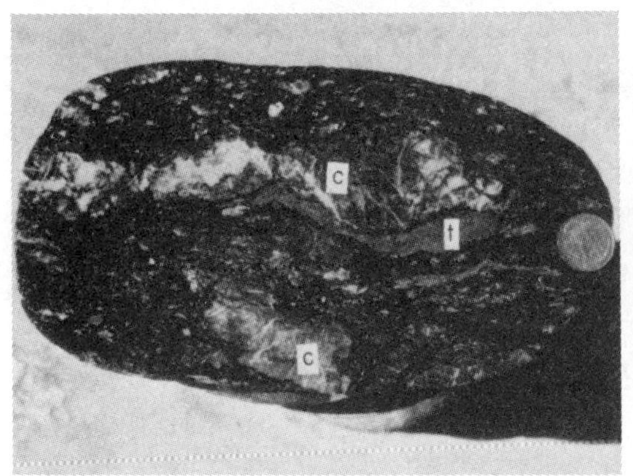

Figure 4. Cut slab of highly deformed "gray chert and argillite" from Malina Bay, showing how individual chert layers (c) are shattered and suspended in argillaceous matrix; note thin layer of green tuff (t).

The gray chert is highly recrystallized and cut by abundant quartz-filled fracture veinlets. No traces of volcanic debris such as shards, clasts, or phenocrysts have been observed in chert layers, and ghosts of radiolaria occur only rarely. X-ray diffraction analyses indicate that this chert is composed entirely of quartz. The argillite and tuff in this unit are highly altered to pumpellyite, prehnite, and chlorite. The green tuff is petrographically distinct from the argillite because of its green color and greater abundance of plagioclase and volcanic fragments.

The Deep Sea Drilling Project has found that "bedded cherts," especially of Eocene age, are a volumetrically important rock type in abyssal sequences of most deep-ocean basins (see Davies and Supko, 1973). These cherts generally are devoid of biogenous remains and are thinly interbedded with pelagic sediment types such as abyssal clay and air-fall tuff (Heath and Moberly, 1971; Davies and Supko, 1973). These abyssal "bedded cherts" are lithologically similar to the Uyak gray chert and represent a probable Eocene analog. However, calcareous sediments often are important constituents in Tertiary abyssal sequences but are lacking in the ancient counterpart, perhaps because pelagic calcareous organisms did not proliferate and diversify until late Mesozoic and early Cenozoic

Figure 5. "Gray chert and argillite" at Uyak Bay; note phacoid of wacke (w). Scale bar = 15 cm.

time, whereas deposition of most of the Uyak sediments may have occurred before this time.

In the North Pacific, layers of air-fall tuff as much as 6 cm thick often occur interbedded with normal abyssal sediments more than 1,000 km seaward of the nearest volcanic arc, and the percentage of tuff increases toward the arc (Ninkovich and Robertson, 1975; Scheidegger and Kulm, 1975). This relation suggests that the green tuff interlayered with the Uyak gray chert and argillite was blown seaward from a volcanic arc and deposited with other deep-sea sediments. The absence of tuff interbedded with the radiolarian chert implies that the tuff did not originate as an aquagene tuff at a mid-ocean spreading center.

The distinctly different characteristics of radiolarian chert and gray chert in the Uyak Complex raise interesting questions in regard to their origin and diagenesis. The abundant radiolaria preserved in the radiolarian chert clearly attest to its biogenic origin. However, the scarcity of biogenous remains in the Uyak gray chert and the "bedded cherts" of modern ocean basins leaves their origin less clear. Wise and Weaver (1974) reviewed the many hypotheses proposed for the origin of silica in deep-sea chert deposits and concluded that most silica was originally deposited as the opaline skeletal material of microplankton (Berger and von Rad, 1972; von Rad and Rosch, 1974; Garrison, 1974). If both forms of chert are biogenic, why then are radiolaria preserved in one, while in the other occasional ghosts and molds of radiolaria are the only remaining biogenous traces? The answer is not clear, but it may reflect significant differences in diagenetic processes on a mid-ocean rise and on the abyssal ocean floor. Perhaps the high heat flow at a mid-ocean rise favors preservation of the forms of siliceous microfossils by increasing the rate of diagenetic reactions as the biogenic opaline remains are converted to opal-CT and subsequently to quartz (see Ernst and Calvert, 1969; Heath and Moberly, 1971). Preservation may also be enhanced by the addition of secondary silica to pore water from the diagenetic and hydrothermal alteration of underlying ocean–floor basalt (Garrison, 1974; Hart, 1973; Fein, 1973). Alternatively, it may be that certain sediments interbedded with the "bedded cherts" and the Uyak gray chert contain unstable silicate minerals that have low silicon/aluminum ratios and readily react with silica released from microfossils. Johnson (1976) reported that regions in the deep ocean with anomalously poorly preserved siliceous microfossils approximately correspond to regions with a high input of hemipelagic sediments; this suggests that the sediments are a potential "silica sink" that lowers interstitial silica concentrations, thereby favoring the solution of siliceous remains.

Limestone

Limestone is uncommon in the Uyak and generally occurs as small recrystallized phacoids in gray chert and argillite; mappable bodies were encountered at only three localities (see Connelly and Moore, 1977) and constitute about 1% of the mélange. The limestones exposed on southeastern Ban Island (a body 0.2 by 2.4 km) and on the northeastern tip of Shuyak Island (0.3 km thick) consist of contorted sequences of thin-bedded white limestone separated by thin layers of black argillite. This limestone is recrystallized and lacks preserved fossils.

The third mappable body is exposed on the northeast shore of Uyak Bay. This tectonic block contains a sequence of thick-bedded, variably recrystallized grainstone with occasional thin layers of tuff and tuffaceous limestone. Fusulinids collected from this block were identified by G. Wilde (1975, written commun.) as *Neoschwagerina* sp.; crinoid stems are also present. In a second sample of fossiliferous limestone (micritic-skeletal wackestone), which was collected from a small phacoid in gray chert and argillite on the northeast shore of Uganik Bay, G. Wilde identified the fusulinids *Neoschwagerina* sp., *Cancellina*? sp., and *Codonofusiella* sp. and the foraminifers *Colaniella, Pachyphloria,* and *Nodosinella,* as well as Dasyclad alga?. These fusulinids are mid-Permian in age (pre-middle Guadalupian) and of the Verbeckinid association; this association indicates a Tethyan affinity (Ross, 1967).

G. W. Moore (1969) located an additional phacoid of fossiliferous limestone on the west shore of Uyak Bay that yielded fragments of gastropods, pelecypods, echinoderms, a solitary coral, and the hydrozoan *Spongiomorpha* (N. J. Silberling, 1966, 1976, written communs.). Most fossil material here is of indeterminate age, but the *Spongiomorpha* is a form strikingly similar to a Late Triassic (Norian) hydrozoan on the Alaska Peninsula.

Perhaps limestone is rare in the Uyak Complex because most of the sediments were deposited in late Paleozoic and early Mesozoic time before the appearance of pelagic calcareous organisms: the proliferation of Coccolithophorids occurred in Jurassic time, and the first appearance of planktonic Foraminifera was not until the Cretaceous (Garrison and Fisher, 1969; Bosellini and Winterer, 1975). Alternatively, Uyak sediments may have been deposited after the proliferation of pelagic calcareous organisms, but below the carbonate compensation depth; thus they could be deep-water upper Mesozoic rocks. The thin-bedded limestone and argillite sequences described on Ban and Shuyak Islands are probable examples of Uyak sediments deposited after the appearance of pelagic calcareous organisms. The significance of the Tethyan affinity of the fusulinid limestone is not clear, but it may indicate that the limestone was conveyed to the Uyak plate margin from warmer latitudes, or perhaps that it formed as carbonate banks or reefs along the Alaskan margin in fairly shallow water (see Monger and Ross, 1971). The *Spongiomorpha* limestone may have originated on the Alaska Peninsula and then been introduced subsequently into the Uyak trench as a slide block.

Wacke

Massive wacke and contorted thinly interbedded wacke and argillite make up about 20% of the Uyak Complex and occur as variously sized tectonic blocks. Primary sedimentary structures are no longer recognizable in either variety. Wispy shear planes and white shatter veinlets of prehnite and/or calcite commonly are visible in hand specimens.

The two varieties of wacke are petrographically quite similar. They typically are medium grained, poorly sorted, and have diffuse grain boundaries surrounded by fine-grained matrices of phyllosilicates (mostly chlorite), prehnite, and/or pumpellyite. In thin section, fabrics of unsheared domains are isotropic to slightly flattened but locally are semischistose. These relatively well-preserved domains are separated by narrow anastomosing shear zones, which generally are 1 or 2 mm wide but which may make up an entire thin section or hand specimen. Movement along these shear fractures has cataclastically reduced grain size and produced a high percentage of granulated matrix (now recrystallized); rounded to augen-shaped relict grains and rock fragments are suspended in the mylonitic matrix. Point counting of 20 samples of Uyak wacke indicates that fine-grained matrix averages 28%, but this ranges from 16% to 41%. Secondary prehnite and pumpellyite are abundant and occur in veinlets and as replacement of matrix and plagioclase.

The determination of the modal mineralogy by point counting is hampered by the low-grade metamorphism and internal shearing. An average of 600 counts each was made on 20 samples and plotted on a standard quartz-feldspar-lithic diagram (Fig. 6), following the techniques of Dickinson (1970). Samples group into a more abundant arkosic wacke and a less abundant chert-clast wacke. Potassium feldspar accounts for less than 1.0% of all samples, so the ratio of plagioclase to total feldspar (P/F) is always near unity. The presence or absence of potassium feldspar is independent of the occurrence of secondary prehnite or pumpellyite, so its deficiency cannot be explained by its loss during low-grade metamorphic reactions. Unstable rock fragments consist almost entirely of andesitic clasts in the arkosic wacke and of both andesitic and basaltic clasts in the chert-clast wacke, giving rise to a ratio of volcanic rock fragments to total unstable lithic fragments (V/L) of about 1.0 in both. The ratio of chert to total chert plus quartz (C/Q) is low in the arkosic wacke and high in the chert-clast wacke. Quartz grains in all samples generally are monocrystalline, have nonundulatory extinction, and are cut by simple fractures; these characteristics suggest a volcanic parent rock (Blatt and others, 1972, p. 270–278). Clasts of radiolarian chert and variolitic greenstone are sparsely present in the arkosic wacke, whereas in the chert-clast wacke they are the main constituents. In addition to an almost complete lack of metamorphic and sedimentary rock fragments, secondary indicators of metamorphic and sedimentary provenance such as primary mica, gneissose or undulatory quartz grains, and well-rounded multicycled grains also are lacking. It is clear, then, that these clastic rocks were derived predominantly from an andesitic terrane. I interpret the Uyak wacke as having been deposited in an oceanic trench adjacent to an active volcanic arc.

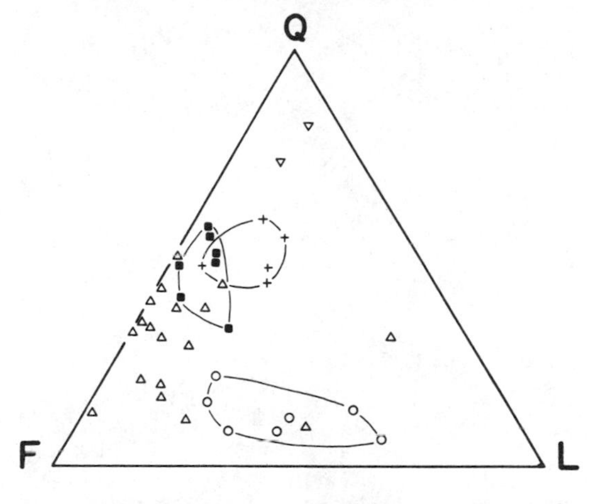

	C/Q	P/F	V/L
△ Uyak Arkosic Wacke	0.3	1.0	1.0
▽ Uyak Chert-Clast Wacke	0.9	1.0	1.0
○ Cape Current Sandstone	0.2	1.0	1.0
+ Kodiak Fm. Wacke	0.3	1.0	–
■ Sliver Wacke	0.2	1.0	–

Figure 6. Modal mineralogies of various clastic detrital rocks from northwest Kodiak Islands plotted on Q-F-L diagram; Q includes quartz and chert, F includes plagioclase and potassium feldspar, and L includes unstable lithic fragments.

The occurrence of radiolarian chert and greenstone fragments in the wacke may indicate some recycling of Uyak rocks whereby the deformed and uplifted "sea-floor scrapings" forming the trench inner wall were syntectonically eroded and redeposited with normal trench turbidites (Fig. 2). This mechanism of recycling was invoked by Cowan and Page (1975) to explain the occurrence of blueschist cobbles in a conglomerate of the Franciscan Complex which has only a prehnite-pumpellyite–facies background metamorphism. They believed that the most likely source for these cobbles was the syntectonic uplift and subaqueous erosion of landward wedges of the subduction complex. Cobbles such as radiolarian chert and greenstone are also present in this Franciscan conglomerate but are not as clearly exotic as the blueschist. Nothing as dramatic as this Franciscan example has been recognized in the Uyak Complex, but a similar origin is suggested for the chert-clast wacke and for the sparse clasts of greenstone and radiolarian chert in the more typical arkosic wacke.

Table 1 includes chemical analyses of ten samples of arkosic wacke and one sample of chert-clast wacke collected from widely spaced localities in the Uyak. Crook (1974) and Schwab (1975) postulated a correlation between the composition of wacke (framework components and volatile-free chemistry) and the geotectonic setting of the continental margin at which the wacke accumulates. There are two primary source areas for continental-margin wackes: quartz-rich sediments from a craton and quartz-poor sediments from a volcanic arc. Thus, on the basis of framework quartz, total SiO_2, and K_2O/Na_2O ratio, wacke should group into categories corresponding to Atlantic-type rifted margins, Andean-type consumptive margins (adjacent to a craton), and western Pacific-type consumptive margins (adjacent to an island arc). Atlantic-type margins are characterized by >65% quartz, average 70% SiO_2, and K_2O/Na_2O greater than 1.0; Andean-type margins have 15% to 65% quartz, average 68% to 74% SiO_2, and K_2O/Na_2O less than 1.0; and west Pacific–type margins have <15% quartz, average 58% SiO_2, and K_2O/Na_2O much less than 1.0 (Crook, 1974; Schwab, 1975). Uyak wacke has an average of 17% framework quartz, 66% SiO_2, and K_2O/Na_2O of 0.25, and therefore it appears to be an Andean-type wacke. The composition of Uyak wacke verges on the characteristics of a west Pacific–type wacke, possibly reflecting the primitive nature of the southwestern Alaskan craton in Uyak time. Note that the composition of the Upper Cretaceous Kodiak Formation (average 36% quartz, 73% SiO_2, K_2O/Na_2O = 0.26) reflects a greater contribution of cratonic sediment.

METAMORPHISM OF UYAK COMPLEX

Wacke in the Uyak Complex characteristically contains the metamorphic assemblage quartz-albite-chlorite-prehnite-pumpellyite-sphene ± calcite, pyrite, and celadonite. Relict detrital minerals in these rocks include hornblende, clinopyroxene, epidote, potassium feldspar, biotite, zircon, and apatite. Greenstone is characterized by the assemblage albite-pumpellyite-chlorite-sphene ± calcite, prehnite, quartz, pyrite, and epidote, commonly with relict phenocrysts of clinopyroxene and plagioclase. Many of these metamorphic minerals are fine grained and petrographically evasive, so X-ray diffraction analyses of heavy-mineral separates were conducted on 24 samples to confirm identifications.

Prehnite is nearly ubiquitous in wacke but is absent in most greenstone. It occurs as fanning or intergrown idioblastic laths in veinlets with quartz or albite, as spongy amoeboid porphyroblasts replacing the sheared matrix, or as intergrown laths replacing plagioclase grains (Fig. 7). Single crystals commonly exceed 0.5 mm in length, and clusters of crystals may exceed 2.0 cm. Prehnite is often a major component in wacke and may constitute more than 13% of a sample.

Pumpellyite occurs in most samples of wacke and greenstone. It usually is colorless and occurs as very fine-grained acicular gray aggregates in sheared matrix and in plagioclase. In addition, greenstone sometimes contains small needles of green pleochroic pumpellyite in veinlets with albite.

Sphene also occurs as fine-grained mats and is often distinguishable from pumpellyite only by X-ray diffraction. Chlorite typically occurs as tiny plates in the matrix of wacke or as an alteration product of detrital biotite; pseudomorphs of chlorite after clinopyroxene are common in greenstone. Plagioclase has been albitized and has a characteristic cloudy appearance; secondary albite commonly fills shatter veinlets. Secondary epidote is quite rare in wacke and greenstone, but rounded detrital clasts are not uncommon in wacke.

The metamorphic mineral assemblages in these rocks are characteristic of the prehnite-pumpellyite facies; conditions did not exceed the prehnite-out isograd that marks the lower limit of the pumpellyite-actinolite facies (Coombs and others, 1970; Liou, 1971; Bishop, 1972; Surdam, 1973). The pressure and temperature conditions for the prehnite-pumpellyite facies are not well known and are influenced by variations in activities of components such as H_2O, CO_2, SiO_2, Ca^{++}, and H^+ (Coombs and others, 1970; Bishop, 1972), but estimates range from 200 to 350 °C and 3 to 5 kb.

The pervasively sheared nature of these rocks makes it difficult to establish a relationship between metamorphism and deformation. Shatter veinlets containing prehnite commonly cut foliation and are therefore posttectonic. However some veinlets cutting foliation are in turn truncated by later brittle fractures, indicating some postmetamorphic deformation. Prehnite growing in the sheared matrix in wacke generally is undeformed and therefore posttectonic, but the local occurrence of deformed prehnite in shear zones suggests that metamorphism was syntectonic as well. In a subduction

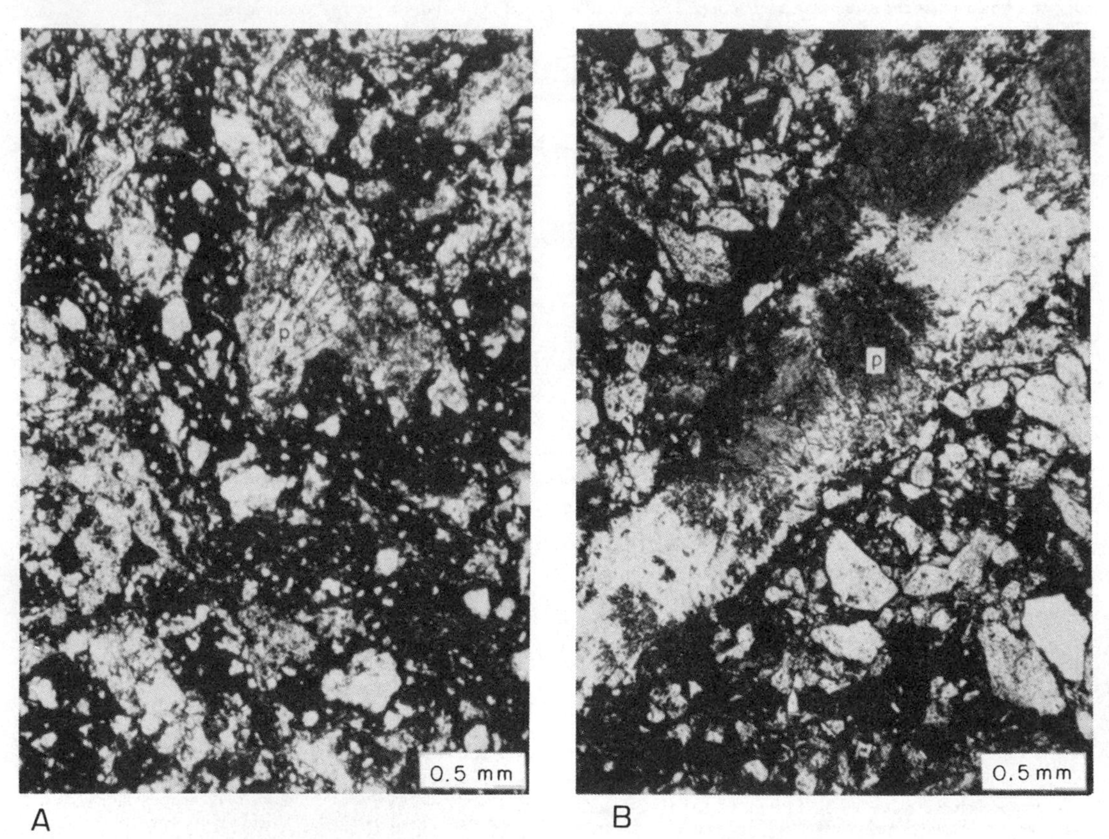

Figure 7. Photomicrographs of well-developed prehnite (p) in Uyak wacke collected from Gurney Bay area, (A) replacing matrix with fanning laths, and (B) filling shatter veinlet with quartz.

model, this may indicate that metamorphism occurred during active underthrusting (syntectonic metamorphism), continued after accretion of the subduction wedge onto the overthrust plate (posttectonic metamorphism), and was followed by late fracturing and cataclasis during uplift of the subduction complex (postmetamorphic deformation). A similar progression of deformation was suggested by Glassley and Cowan (1975) for Franciscan rocks.

[*Editor's Note:* Text on the lithology and metamorphism of related rocks has been omitted.]

STRUCTURAL GEOLOGY

Uyak Complex

The Uyak Complex is a chaotically dislocated and pervasively sheared assemblage of deep-sea lithologies. Mesoscopic shear fractures and innumerable faults of unknown magnitude have disrupted stratal continuity and juxtaposed contrasting sedimentary and igneous rocks. Shear surfaces contain slickensides and occur both as subparallel anastomosing fractures in competent lithologies such as wacke and greenstone and as more closely spaced fractures in the less competent argillaceous rocks (Fig. 9). The more brittle rock types are broken into angular phacoids and either juxtaposed against other tectonic blocks or enclosed in the highly deformed gray chert and argillite matrix. These tectonic inclusions range from millimetres to kilometres in size and are aligned with their longest dimensions subparallel to foliation. The structural style of this terrane is best described as *tectonic mélange* (Cowan, 1974; Hsu, 1974). Although lithologically chaotic, the fabric of this mélange is surprisingly orderly (Moore and Wheeler, 1975).

The orientation of fracture cleavages and faults in the complex define a foliation that trends consistently northeast and dips steeply northwest (Connelly and Moore, 1977). This foliation dips much more steeply than the inclination of modern Benioff zones at shallow depths. This apparent anomaly may be explained by Seely and others' (1974) model for accretion at convergent plate margins whereby thin wedge-shaped slices of deep-sea sediments are progressively inserted along landward-dipping thrust surfaces at the foot of the trench inner wall, causing the progressive landward rotation of the subduction complex (Fig. 2).

A detailed structural analysis of the Uyak indicates that the mean direction of slip during deformation was N38° ± 11°W and that structural transport was southeast under northwest (Moore and Wheeler, 1975). Thus, it appears that the Uyak was thrust beneath the early Mesozoic rocks on the northwest side of the Kodiak Islands; this underthrusting was toward the presumably coeval volcanoplutonic arc to the northwest on the Alaska Peninsula.

Major Faults

The contrasting characteristics of the Kodiak Islands schist terrane and the Shuyak Formation indicate that a major fault must separate them, but this fault is now obscured by the Afognak pluton that occurs between the two units (Fig. 1). I believe that this fault is a major early Mesozoic thrust; I shall call it the Shuyak fault. The Shuyak fault correlates with the Port Graham fault on southern Kenai Peninsula which juxtaposes Upper Triassic and Lower Jurassic volcanogenic rocks on the northwest against blueschist-bearing metamorphic rocks of the Seldovia schist terrane (Carden and others, 1977; Magoon and others, 1976; Forbes and Lanphere, 1973). The combined Shuyak–Port Graham fault may represent a trace of the early Mesozoic plate boundary along the margin of southwestern Alaska.

The Uyak Complex is bounded on the northwest by a fault juxtaposing it with either the Kodiak Islands schist terrane or the Afognak pluton; this is named the Raspberry fault. The Raspberry fault marks the southeastern extent of early Mesozoic rocks along this margin and may therefore be the trace of the continental margin at the time of Uyak accretion.

The Uyak Complex is underthrust by the Kodiak Formation along a steeply northwest-dipping thrust fault, here named the Uganik thrust (Fig. 1; Moore and Connelly, 1977). The Uyak and Kodiak rocks responded differently to the faulting that juxtaposed them. The underthrust turbidite sequence was transformed into a 1-km-thick broken formation adjacent to the thrust contact, whereas deformation in the more rigid Uyak rocks was concentrated in a zone 1 to 500 m wide and is expressed by an overprinted fracture cleavage but no remobilization of layering. This broken formation is identical in structural style to those occurring throughout the Kodiak Formation, and it apparently developed during underthrusting. Tectonic mixing in the Uganik thrust zone inserted kilometre-sized "slivers" of Kodiak Formation lithology into the Uyak as far as 3 km structurally above the thrust contact, and mixed smaller blocks of Uyak lithology as far as 0.5 km into the Kodiak broken formation. The local occurrence of well-bedded Kodiak turbidites separated from the Uyak by a simple fault trace suggests some recent reactivation of this fault surface.

The Uganik thrust correlates with the Chugach Bay fault on southern Kenai Peninsula (Cowan and Boss, 1978) and the Eagle

Figure 9. Foliated, matrix-rich fabric characteristic of most Uyak mélange.

River thrust in the Anchorage area (Clark, 1972), where it juxtaposes the McHugh Complex against the Valdez Group. Because of uncertainties as to the age of emplacement of the Uyak-McHugh belt (discussed below), the significance of the Uganik–Eagle River thrust is unclear. If the Uyak-McHugh belt was emplaced in late Early Cretaceous time, then these rocks are older than the Upper Cretaceous Kodiak-Valdez trench deposits, and the thrust must mark a trace of the Late Cretaceous plate boundary. However if the Uyak-McHugh belt was emplaced in Late Cretaceous time and is genetically related to the Kodiak-Valdez (preferred model), then the thrust juxtaposes the more deeply subducted and more highly metamorphosed Uyak-McHugh rocks against the Kodiak-Valdez trench deposits. The occurrence of more highly metamorphosed rocks along the arcward side of subduction complexes has been noted at other convergent margins (see, for example, Suppe, 1972; Cowan, 1974) and may be attributed to differential uplift of the subduction complex from continued underplating of deep-sea sediments at the foot of the trench inner wall (Ernst, 1975; Seely and others, 1974).

AGE OF EMPLACEMENT

Limits on the age of emplacement of a subduction complex may be determined from its incorporated fossils. However these include pelagic fossils such as radiolaria, coccoliths, and foraminifera, which are deposited with abyssal sediments on oceanic crust before being offscraped and accreted at a consuming plate margin, and fossils interbedded with trench turbidites that accumulate during subduction. The former provide only a lower limit on the age of emplacement, whereas the latter closely approximates that age. Either fossil type may occur in structurally coherent offscraped packages that young seaward, or in extraformational slivers that are inserted into previously accreted packages. More definitive evidence for the time of active subduction may be obtained from (1) radiometric ages of high-pressure subduction-zone metamorphic rocks, (2) biostratigraphic ages from the associated forearc-basin deposits, and (3) radiometric ages from the associated magmatic arc.

Mainly Early Jurassic K-Ar ages from blueschists of the Kodiak-Seldovia schist terrane apparently provide a reliable measure for its age of emplacement (Fig. 10). This estimate approximately coincides with the prominent Late Triassic to Middle Jurassic andesitic volcanism recorded in the rocks of the arc and forearc basin on the Alaska and Kenai Peninsulas (Burk, 1965; Detterman and Hartsock, 1966; Martin, 1915) and on the Kodiak and Barren Islands (Cowan and Boss, 1978; Moore and Connelly, 1976); and with the 176 to 154 m.y. B.P. phase of plutonism in the Alaska-Aleutian Range batholith (Reed and Lanphere, 1973).

The age of emplacement of the Uyak-McHugh complex is problematical. Radiolaria from Uyak cherts range in age from Paleozoic to mid–Early Cretaceous (Valanginian to Aptian) and provide a lower limit for the age of accretion (Fig. 10). The Permian fusulinid limestone and Upper Triassic *Spongiomorpha* limestone also provide lower limits. The youngest fossils included in the complex are *Inoceramus* prisms from an extraformational "sliver" of probable Kodiak Formation. Thus fossil evidence alone can limit the age of emplacement only to post-Valanginian. It is therefore useful to identify the associated magmatic arc and attempt to indirectly determine the age of emplacement.

The early Mesozoic phase of magamtism on the Alaska Peninsula was followed by the uplift and erosion of pre-existing plutons and deposition of Upper Jurassic (Oxfordian) to Lower Cretaceous (Valanginian) nonvolcanic plutoniclastic beds (Burk, 1965; Detterman and Hartsock, 1966). These rocks are truncated upsection by a strong unconformity that locally is succeeded by uppermost Cretaceous (Campanian and Maestrichtian) volcanic-rich strata. Similarly, early Mesozoic plutonism on the Alaska Peninsula was succeeded by a hiatus, which in turn was succeeded by a Late Cretaceous to Paleocene event yielding K-Ar ages of 83 to 58 m.y. (Reed and Lanphere, 1973).

I suggest that the Uyak-McHugh complex was emplaced during Late Cretaceous subduction coincident with magmatism on the Alaska Peninsula. The only likely alternative is Early to middle Cretaceous (post-Valanginian) emplacement, but if such an event occurred, it produced insignificant magnatism.

The Kodiak Formation and Valdez Group are interpreted as

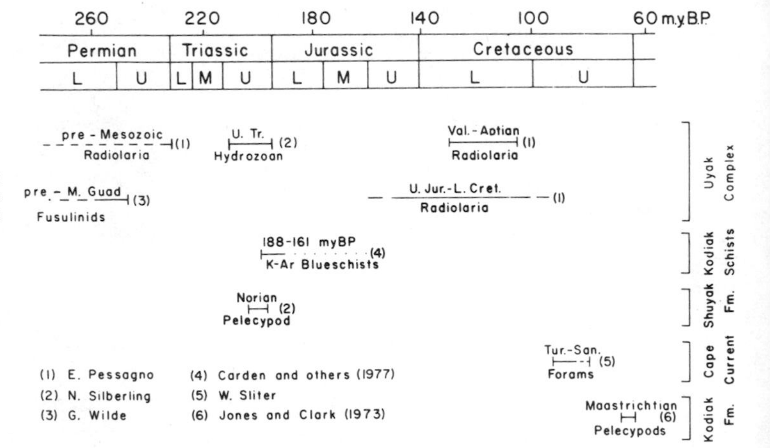

Figure 10. Summary of age data from Uyak Complex, Kodiak Islands schist terrane, Shuyak Formation, Cape Current terrane, and Kodiak Formation.

trench turbidites, so the Upper Cretaceous (Maestrichtian) pelecypods in these rocks presumably provide a close estimate for the time of active subduction and accretion. This estimate coincides with the Late Cretaceous magmatism on the Alaska Peninsula. Therefore, the relationship between the Uyak-McHugh and Kodiak-Valdez rocks is unclear. Perhaps they represent two phases of Late Cretaceous accretion, or perhaps the Uyak-McHugh is a more deeply subducted abyssal facies and the Kodiak-Valdez a late-stage trench-fill facies of the same Late Cretaceous accretionary event.

ACKNOWLEDGMENTS

Special thanks go to my fellow Kodiak workers Casey Moore, Mal Hill, Betsy Hill, and Jim Gill for their many contributions to this manuscript. Casey Moore, Bob Garrison, and Eli Silver served on my thesis committee and provided sound advice during preparation of this paper. Gene Gonzales provided expert technical assistance. Paleontological work was kindly done by Emile Pessagno, Norm Silberling, Garner Wilde, Bill Sliter, and Jon Thompson. Finally, I thank the many fishermen and trappers of the Kodiak Islands who provided food, shelter, and colorful conversation during our field work. This work was supported by National Science Foundation Grant GA-43266, the Alaskan Branch of the U.S. Geological Survey, Exxon Production Research, and Sigma Xi.

REFERENCES CITED

American Commission on Stratigraphic Nomenclature, 1970, Code of stratigraphic nomenclature: Tulsa, Okla., Am. Assoc. Petroleum Geologists, 22 p.

Berger, W. H., and von Rad, U., 1972, Cretaceous and Cenozoic sediments from the Atlantic Ocean, *in* Hayes, D. E., Pimm, A. C., and others, Initial reports of the Deep Sea Drilling Project, Vol. 14: Washington, D.C., U.S. Govt. Printing Office, p. 787–954.

Berkland, J. O., Raymond, L.A., Kraemer, J. C., Moores, E. M., and O'Day, M., 1972, What is Franciscan?: Am. Assoc. Petroleum Geologists Bull., v. 56, p. 2295–2302.

Bishop, D. G., 1972, Progressive metamorphism from prehnite-pumpellyite to greenschist facies in the Dansey Pass area, Otago, New Zealand: Geol. Soc. America Bull., v. 83, p. 3177–3198.

Blatt, H., Middleton, G., and Murray, R., 1972, Origin of sedimentary rocks: Englewood Cliffs, N.J., Prentice-Hall, 634 p.

Bosellini, A., and Winterer, E. L., 1975, Pelagic limestone and radiolarite of the Tethyan Mesozoic: A genetic model: Geology, v. 3, p. 279–272.

Budnik, R., 1974, The geologic history of the Valdez Group, Kenai Peninsula, Alaska: Deposition and deformation at a Late Cretaceous consumptive plate margin [Ph.D. thesis]: Los Angeles, Univ. California, Los Angeles, 139 p.

Burk, C. A., 1965, Geology of the Alaska Peninsula — Island arc and continental margin: Geol. Soc. America Mem. 99, 250 p.

Capps, S. R., 1937, Kodiak and adjacent Islands, Alaska: U.S. Geol. Survey Bull. 880-C, p. C111–C184.

Carden, J. R., 1977, The geology and comparative petrology and geochemistry of the Seldovia/Kodiak and Brooks Range blueschist terranes, Alaska [Ph.D. thesis]: Fairbanks, Univ. Alaska.

Carden, J. R., Connelly, W., Forbes, R. B., and Turner, D. L., 1977, Blueschists of the Kodiak Islands, Alaska: An extension of the Seldovia schist terrane: Geology, v. 5, p. 529–533.

Clark, S.H.B., 1972, Reconnaissance bedrock geologic map of the Chugach Mountains near Anchorage, Alaska: U.S. Geol. Survey Misc. Geol. Inv. Map MF-350.

——1973, The McHugh Complex of south-central Alaska: U.S. Geol. Survey Bull. 1372-D, P. D1–D11.

Connelly, W., 1976, Mesozoic geology of the Kodiak Islands and its bearing on the tectonics of southern Alaska [Ph.D. thesis]: Santa Cruz, Univ. California, Santa Cruz, 197 p.

Connelly, W., and Moore, J. C., 1977, Geologic map of the northwest side of the Kodiak Islands, Alaska: U.S. Geol. Survey Open-File Map 77-382.

Coombs, D. S., Horodyski, R. J., and Naylor, R. S., 1970, Occurrences of prehnite-pumpellyite facies metamorphism in northern Maine: Am. Jour. Sci., v. 268, p. 142–156.

Cowan, D. S., 1974, Deformation and metamorphism of the Franciscan subduction zone complex northwest of Pacheco Pass, California: Geol. Soc. America Bull., v. 85, p. 1623–1634.

Cowan, D. S., and Boss, R. F., 1978, Tectonic framework of the southern Kenai Peninsula, Alaska: Geol. Soc. America Bull., v. 89, p. 155–158.

Cowan, D. S., and Page, B. M., 1975, Recycled Franciscan material in Franciscan mélange west of Paso Robles, California: Geol. Soc. America Bull., v. 86, p. 1089–1095.

Crook, K. A., 1974, Lithogenesis and geotectonics: The significance of compositional variation in flysch arenites (graywackes), *in* Dott, R. H., Jr., and Shaver, R. H., eds., Modern and ancient geosynclinal sedimentation: Soc. Econ. Paleontologists and Mineralogists Spec. Pub. 19, p. 304–310.

Davies, T. A., and Supko, P. R., 1973, Oceanic sediments and their diagenesis: Some examples from deep-sea drilling: Jour. Sed. Petrology, v. 43, p. 381–390.

Detterman, R. L., and Hartsock, J. K., 1966, Geology of the Iniskin-Tuxedni region, Alaska: U.S. Geol. Survey Prof. Paper 512, 78 p.

Dickinson, W. R., 1970, Interpreting detrital modes of graywacke and arkose: Jour. Sed. Petrology, v. 40, p. 695–707.

Ernst, W. G., 1973, Blueschist metamorphism and P-T regimes in active subduction zones: Tectonophysics, v. 17, p. 255–272.

——1975, Systematics of large-scale tectonics and age progressions in Alpine and circum-Pacific blueschist belts: Tectonophysics, v. 26, p. 229–246.

Ernst, W. G., and Calvert, S. E., 1969, An experimental study of the recrystallization of porcelanite and its bearing on the origin of some bedded cherts: Am. Jour. Sci., v. 267-A, p. 114–133.

Fein, C. D., 1973, Chemical composition and submarine weathering of some oceanic tholeiites from the East Pacific Rise: Geol. Soc. America Abs. with Programs, v. 5, p. 41.

Fisher, R. L., and Engel, C. G., 1969, Ultramafic and basaltic rocks dredged from the nearshore flank of the Tonga Trench: Geol. Soc. America Bull., v. 80, p. 1373–1378.

Forbes, R. B., and Lanphere, M. A., 1973, Tectonic significance of mineral ages of blueschists near Seldovia, Alaska: Jour. Geophys. Research, v. 78, p. 1383–1386.

Furnes, H., 1973, Variolitic structure in Ordovician pillow lava and its possible significance as an environmental indicator: Geology, v. 1, p. 27–30.

Garrison, R. E., 1974, Radiolarian cherts, pelagic limestones, and igneous rocks in eugeosynclinal assemblages, *in* Hsu, K. J., and Jenkyns, H. C., eds., Pelagic sedimentation: On land and under the sea: Internat. Assoc. Sedimentologists Spec. Pub. 1, p. 367–399.

Garrison, R. E., and Fisher, A. G., 1969, Deep-water limestones and radiolarities of the Alpine Jurassic, *in* Friedman, G. M., ed., Depositional environments in carbonate rocks: Soc. Econ. Paleontologists and Mineralogists Spec. Pub. 14, p. 20–56.

Glassley, W. F., and Cowan, D. S., 1975, Metamorphic history of Franciscan greenstone, blueschist, and eclogite [abs.]: EOS (Am. Geophys. Union Trans.), v. 56, p. 1081.

Hart, R., 1973, Geochemical and geophysical implications of the reaction between sea water and the oceanic crust: Nature, v. 243, p. 76–78.

Heath, G. R., and Moberly, R., 1971, Cherts from the western Pacific, Leg 7, Deep Sea Drilling Project, *in* Winterer, E. L., Riedel, W. R., and others, Initial reports of the Deep Sea Drilling Project, Vol. 7: Washington, D.C., U.S. Govt. Printing Office, p. 991–1007.

Hill, B. B., 1975, Layered gabbroic and ultramafic rocks in a tectonic mélange, Kodiak Islands, Alaska: Geol. Soc. America Abs. with Programs, v. 7, p. 1116.

Hill, B. B., and Brannon, J., 1976, Layered basic and ultrabasic rocks, Kodiak Island, Alaska: The lower portion of a dismembered ophiolite? [abs.]: EOS (Am. Geophys. Union Trans.), v. 57, p. 1027.

Hill, M. D., and Gill, J. B., 1976, Mesozoic greenstones of diverse ages from the Kodiak Islands, Alaska [abs.]: EOS (Am. Geophys. Union Trans.), v. 57, p. 1021.

Hill, M. D., and Morris, J. D., 1977, Near-trench plutonism in southwestern Alaska: Geol. Soc. America Abs. with Programs, v. 9, p. 436–437.

Hsü, K. J., 1974, Mélanges and their distinction from olistostromes, *in*

Dott, R. H., Jr., and Shaver, R. H., eds., Modern and ancient geosynclinal sedimentation: Soc. Econ. Paleontologists and Mineralogists Spec. Pub. 19, p. 321–333.
Imoto, N., and Saito, Y., 1973, Scanning electron microscopy of chert: [Tokyo] Nat. Sci. Mus. Bull., v. 16, p. 397–402.
Isacks, B., and Molnar, P., 1971, Distribution of stresses in the descending lithosphere from a global survey of focal-mechanism solutions of mantle earthquakes: Rev. Geophysics and Space Physics, v. 9, p. 103–125.
Jakeš, P., and Gill, J., 1970, Rare earth elements and the island arc series: Earth and Planetary Sci. Letters, v. 9, p. 17–28.
Johnson, T. C., 1976, Controls on the preservation of biogenic opal in sediments of the eastern tropical Pacific: Science, v. 192, p. 887–890.
Jones, D. L., and Clark, S.H.B., 1973, Upper Cretaceous (Maestrichtian) fossils from the Kenai-Chugach Mountains, Kodiak and Shumagin Islands, southern Alaska: U.S. Geol. Survey Jour. Research, v. 1, p. 125–136.
Jones, J. G., 1969, Pillow lavas as depth indicators: Am. Jour. Sci., v. 267, p. 181–195.
Kulm, L. D., von Huene, R., and others, 1973, Initial reports of the Deep Sea Drilling Project, Vol. 18: Washington D.C., U.S. Govt. Printing Office, 1,077 p.
Liou, J. G., 1971, Synthesis and stability relations of prehnite, $Ca_2Al_2Si_3O_{10}(OH)_2$: Am. Mineralogist, v. 56, p. 507–531.
MacKevett, E. M., Jr., 1970, Geology of the McCarthy B-4 quadrangle, Alaska: U.S. Geol. Survey Bull. 1333, 31 p.
Magoon, L. B., Adkison, W. L., and Egbert, R. M., 1976, Map showing geology, wildcat wells, Tertiary plant fossil localities, K-Ar age dates, and petroleum operations, Cook Inlet area, Alaska: U.S. Geol. Survey Misc. Geol. Inv. Map I-1019.
Maresch, W. V., 1973, New data on the synthesis and stability relations of glaucophane: Earth and Planetary Sci. Letters, v. 20, p. 385–390.
Martin, G. C., 1913, The mineral deposits of Kodiak and the neighboring Islands: U.S. Geol. Survey Bull. 542, p. 125–136.
——1915, The western part of the Kenai Peninsula: U.S. Geol. Survey Bull. 587, p. 41–112.
Matthews, D. H., 1971, Altered basalts from Shallow Bank, an abyssal hill in the NE Atlantic and from a nearby seamount: Royal Soc. London Philos. Trans., v. 268, p. 551–572.
Monger, J.W.H., and Ross, C. A., 1971, Distribution of fusulinaceans in the western Canadian Cordillera: Canadian Jour. Earth Sci., v. 8, p. 259–278.
Moore, G. W., 1967, Preliminary geologic map of Kodiak Island and vicinity, Alaska: U.S. Geol. Survey Open-File Rept. 271.
——1969, New formations on Kodiak and adjacent Islands, Alaska: U.S. Geol. Survey Bull. 1274-A, p. A27–A35.
Moore, J. C., 1972, Uplifted trench sediments: Southwestern Alaska–Bering Shelf edge: Science, v. 175, p. 1103–1105.
——1973a, Cretaceous continental margin sedimentation, southwestern Alaska: Geol. Soc. America Bull., v. 84, p. 595–614.
——1973b, Complex deformation of Cretaceous trench deposits, southwestern Alaska: Geol. Soc. America Bull., v. 84, p. 2005–2020.
Moore, J. C., and Connelly, W., 1976, Subduction, arc volcanism, and forearc sedimentation during the early Mesozoic, S.W. Alaska: Geol. Soc. America Abs. with Programs, v. 8, p. 397–398.
——1977, Mesozoic tectonics of the southern Alaska margin, *in* Talwani, M., and Pitman, W. C., III, eds., Island arcs, deep sea trenches and back-arc basins: Jour. Geophys. Research Maurice Ewing ser. vol. 1, p. 71–82.
Moore, J. C., and Wheeler, R. L., 1975, Orientation of slip during early Mesozoic subduction, Kodiak Islands, Alaska: Geol. Soc. America Abs. with Programs, v. 7, p. 1203–1204.
Moore, J. G., 1970, Water content of basalt erupted on the ocean floor: Contr. Mineralogy and Petrology, v. 28, p. 272–292.
Ninkovich, D., and Robertson, J. H., 1975, Volcanogenic effects on the rates of deposition of sediments in the northwest Pacific Ocean: Earth and Planetary Sci. Letters, v. 27, p. 127–136.
Nisbet, E. G., and Price, I., 1974, Siliceous turbidites: Bedded cherts as redeposited, ocean ridge–derived sediments, *in* Hsu, K. J., and Jenkyns, H. C., eds., Pelagic sedimentation: On land and under the sea: Internat. Assoc. Sedimentologists Spec. Pub. 1, p. 351–366.
Plafker, G., 1972, Alaskan earthquake of 1964 and Chilean earthquake of 1960: Implications for arc tectonics: Jour. Geophys. Research, v. 77, p. 901–925.
Reed, B. L., and Lanphere, M. A., 1973, Alaska-Aleutian Range batholith: Geochronology, chemistry, and relation to circum-Pacific plutonism: Geol. Soc. America Bull., v. 84, p. 2583–2610.
Ross, C. A., 1967, Development of fusulinid (Foraminiferid) faunal realms: Jour. Paleontology, v. 41, p. 1341–1354.
Scheidegger, K. F., and Kulm, L. D., 1975, Late Cenozoic volcanism in the Aleutian arc: Information from ash layers in the northeastern Gulf of Alaska: Geol. Soc. America Bull., v. 86, p. 1407–1412.
Schwab, F. L., 1975, Framework mineralogy and chemical composition of continental margin–type sandstone: Geology, v. 3, p. 487–490.
Seely, D. R., Vail, P. R., and Walton, G. G., 1974, Trench slope model, *in* Burk, C. A., and Drake, C. L., eds., The geology of continental margins: New York, Springer-Verlag, p. 249–260.
Silver, E. A., 1971, Transitional tectonics and late Cenozoic structure of the continental margin off northernmost California: Geol. Soc. America Bull., v. 82, p. 1–22.
Suppe, J., 1972, Interrelationships of high-pressure metamorphism, deformation and sedimentation in Franciscan tectonics, U.S.A.: Internat. Geol. Cong., 24th, Montreal 1972, Comptes Rendus, sec. 3, p. 552–559.
Surdam, R. C., 1973, Low-grade metamorphism of tuffaceous rocks in the Karmutsen Group, Vancouver Island, British Columbia: Geol. Soc. America Bull., v. 84, p. 1911–1922.
von Huene, R., 1972, Structure of the continental margin and tectonism at the eastern Aleutian trench: Geol. Soc. America Bull., v. 83, p. 3613–3626.
von Rad, U., and Rosch, H., 1974, Petrography and diagenesis of deep-sea cherts from the central Atlantic, *in* Hsu, K. J., and Jenkyns, H. C., eds., Pelagic sedimentation: On land and under the sea: Internat. Assoc. Sedimentologists Spec. Pub. 1, p. 327–347.
Wise, S. W., Jr., and Weaver, F. M., 1974, Chertification of oceanic sediments, *in* Hsu, K. J., and Jenkyns, H. C., eds., Pelagic sedimentation: On land and under the sea: Internat. Assoc. Sedimentologists Spec. Pub. 1, p. 301–326.

Manuscript Received by the Society November 17, 1976
Revised Manuscript Received June 27, 1977
Manuscript Accepted August 1, 1977

27

Reprinted from *Geol. Soc. Japan Jour.* **85**:467–478 (1979)

CRETACEOUS TECTONIC MÉLANGE OF THE SHIMANTO BELT IN SHIKOKU, JAPAN

TAKASHI SUZUKI* and SHIGEKI HADA*

Abstract The Cretaceous Susaki Formation in Shikoku, Japan, consists of two main assemblages which have a contrasting lithology, style of deformation, and degree of metamorphism with each other. One assemblage is a chaotically and pervasively sheared mudstone which includes various sizes of slabs and blocks composed of sandstones, greenstones and cherts. This is best described as a tectonic mélange and is similar to the lithologies that occur in deep sea environment. The other assemblage is a turbidite sequence of weakly deformed sandstone and mudstone alternating in various thickness. Both assemblages are generally bounded with each other by the conspicuous northerly dipping thrust faults which gave rise to the imbrication with a polarity proceeding to the south.

The greenstones of restricted distribution to the tectonic mélanges contain massive basaltic lava, almost nonvesicular pillowed basalt, pillow breccia and hyaloclastite. Besides, there are dykes of dolerite which are probably feeder channels of surrounding pillow lavas. The result of a study of chemical composition of clinopyroxenes from basaltic rocks shows the tendency of magma type towards the alkali basalt type in volcanism. This is interested in correspondence to that of volcanic islands and seamounts.

Sandstone inclusions in the tectonic mélanges have contrasting characteristics of compositions and degree of sorting and roundness as compared with the sandstones of coherently bedded layers in the turbidite sequences. The sandstones of the tectonic mélange include unusual chert-rich lithic wackes and uplifted subduction complexes may form sources for these sandstones. Those of the turbidite sequences are feldspathic arenite and lithic arenite. Thus, the source areas of the sandstones of two assemblages are considered to be different from each other.

Sedimentary and structural characteristics of the two assemblages of the Susaki Formation indicate that they might represent the products in the tectonic framework of the Pacific-type continental margins. Namely, the tectonic mélanges correspond to an accretionary complex imbricated into the trench slope during the process of subduction and the turbidite sequences to the sediments of the midslope terrace or slope basin in an arc-trench gap environment. During subduction, brittle rocks such as sandstone, greenstones and chert were broken into phacoids of various sizes and suspended in the less competent matrix of mudstone. The highly sheared tectonic mélanges have been progressively rotated as new materials were underplated at the trench, and an enlarging body of turbidite sequence was ponded on the "basement" of the accretionary complex with the relationship of a tectonic unconformity. Some of major thrust faults that originally bounded accreted packets remained active even as they rotated and the tectonic mélanges were tectonically imbricated with the turbidite sequences.

Received May 23, 1979
* Department of Geology, Faculty of Science, Kochi University, Kochi, 780 Japan.

Introduction

One of the key subjects concerning the tectonic pattern of Southwest Japan is to appropriately explain the tectogenesis of the complex geology of the Shimanto Belt, that is considered to have developed along an ancient continental margin during Late Cretaceous to mid-Tertiary time.

The Shimanto Belt of Southwest Japan is occupied by the poorly fossiliferous strata composed mainly of sandstone and mudstone with minor amounts of submarine basic volcanic rocks. Although the sedimentological characteristics of a part of the Shimanto Belt have been fairly well known (KISHU SHIMANTO RESEARCH GROUP, 1968, 1975; MIYAMOTO, 1976; TERAOKA, 1977), very little information is available concerning the petrological and geochemical features of greenstones and its related rocks, and the geological structures throughout the Shimanto Belt.

The tectonic interpretations of this belt hinge on two models which stand on completely different viewpoints; geosynclinal concept (KATTO, 1961; IMAI *et al.*, 1971; KISHU SHIMANTO RESEARCH GROUP, 1968, 1975; HARATA & TOKUOKA, 1978) or plate tectonic explanation (KANMERA, 1976a, b; SUZUKI *et al.*, 1978, 1979; SAKAI, 1978). The former explains that the Shimanto Belt was the site of thick geosynclinal sedimentation attaining more than 10,000 m in thickness in each of the Cretaceous and Paleogene period, sometimes with submarine basic volcanic episodes, and that a rising cordillera must have existed on the southern (oceanward) side and the Shimanto geosyncline was a shallow trough-like basin not including an abyssal or deeper bathyal environment (HARATA & TOKUOKA, 1978). On the other hand, the latter explains that the Shimanto Belt consists of two formations of comparatively thin deposits. The two formations are considerably different from each other in lithologic facies and geologic structure, and were formed under very different environments from each other. These include typical deep sea sequences. Moreover, the fundamental geologic structures of these formations are characterized by thrust faulting which gave rise to the imbrication dipping to the north and remarkable shortening of the sedimentary bodies (KANMERA & SAKAI, 1975; SAKAI, 1978).

Two fundamental questions regarding the genesis of complex geology of the Shimanto Belt are (1) to what part of the Recent ocean bottom does the site of sedimentation of the Shimantogawa Group correspond?, as discussed by KANMERA & SAKAI (1975), and (2) whether the accreted materials as explained by KARIG (1974) and MOORE, G. F. & KARIG (1976) are present in the Shimanto Belt or not?. Seismic and geologic profiles across the modern convergent margins suggest that the thick wedges of sediments making up the trench inner wall are structured by the compressional folds and landward-dipping thrust faults (SEELY *et al.*, 1974; MOORE, J. C. & KARIG, 1976).

The purpose of this paper is to summarize structural and petrologic data from the Susaki Formation of the Shimanto Belt in Shikoku, and to emphasize certain aspects concerning the tectonic framework of the studied areas that are characterized by the tectonic mélanges and the turbidite sequence.

Geologic and tectonic setting

The Shimanto Belt in the outer side of Southwest Japan is separated from the Chichibu Belt to the north by the Butsuzo Tectonic Line and faces to the Pacific Ocean to the south. The thick sequences of "geosynclinal" sediments which has been generally called the "Shimantogawa Group" are extensively distributed in the belt. The northern part of the belt is occupied generously by the Cretaceous sediments and the southern part by the Paleogene (including the Lower Miocene), although their detailed successions are not yet established in many places because of the complicated geologic structure and pausity of fossils.

The Shimantogawa Group of the studied areas belongs to the Susaki Formation of approximately Gyliakian to Urakawan i.e. Cenomanian to Santonian in age (KATTO, 1969). The Susaki Formation structurally underlies the Lower Cretaceous Hayama Formation on its landward margin and is succeeded on its seaward margin by the Eocene Oyamamisaki Formation in the eastern part of Shikoku. The Hayama Formation consists primarily of medium- to thick-bedded sandstone and mudstone which were deformed into tight large-scale folds, with occasional occurrences of red shale and conglomerate. The sedimentary features of the formation show many of the characteristics of turbidite. The Oyamamisaki Formation consists of sandstone and mudstone alternating in various thickness with occasional interbeds of conglomerate. KATTO & TAIRA (1978) recently regarded the formation as the deposits of the submarine fan and channel.

The Susaki Formation has been interpreted previously by KATTO (1961, 1969 and others) as a thick sequence which consists of sandstone and mudstone of various thickness and proportions accompanying occasional interbeds or lentils of red shale, chert and greenstones. However, recent study revealed that the strata of the formation are assigned to two main assemblages which have a contrasting lithology, style of deformation and degree of metamorphism with each other. One assemblage consists mainly of intensively sheared and steeply dipping strata of mudstone, the other of weakly deformed sandstone and mudstone alternating in various thickness. Although both assemblages have hitherto been regarded to represent a basically conformable sequence of extensive "eugeosynclinal" sediments of the "Shimanto geosyncline", the striking differences in stratigraphic and structural characteristics between both assemblages indicate that their relationship is not so simple as previously considered.

The former assemblage is a chaotically and pervasively sheared assemblage of diverse lithology. It includes mappable rock units composed of greenstones and cherts in relatively isolated lenticular bodies as compared to the more abundant argillite matrix (Figs. 1 and 2). Slivers of coherently bedded alternation of sandstone and mudstone having tectonic contact also occur within the argillite matrix. Cliff-scale observations indicate that the argillite matrix also include various sizes of slabs and blocks composed of dismembered sandstones, greenstones and cherts. The penetrative foliations are developed in the argillite matrix. The more brittle lithologies of lensoid-shape or boudin-shape are aligned with their longest dimension subparallel to foliation of more ductile argillite matrix. Judging from these structural style, this assemblage is best described as "tectonic mélange" (COWAN, 1974; HSÜ, 1974; CONNELLY, 1978). MOORE & WHEELER (1978) have been interpreted that mélange designates a mappable rock unit composed of variously deformed blocks of brittle materials dispersed in a foliated, ductile matrix. The foliation trends consistently east-west and dips steeply north in general. No large-scale fold is

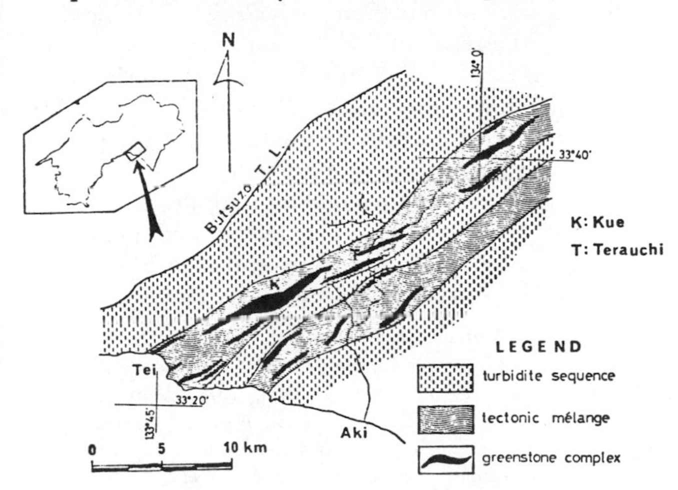

Fig. 1. Map showing the distribution of the turbidite sequences, tectonic mélanges and greenstone complex in the Cretaceous of the Shimantogawa Group (eastern part).

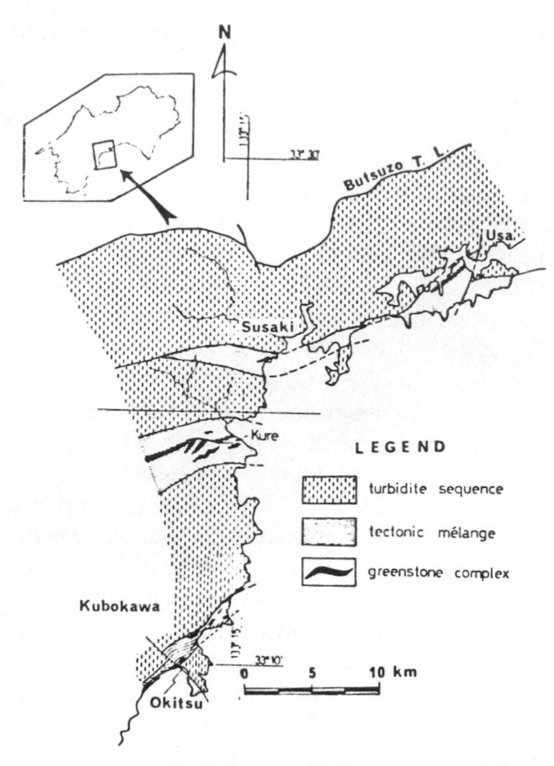

Fig. 2. Map showing the distribution of the turbidite sequences, tectonic mélanges and greenstone complex in the Cretaceous of the Shimantogawa Group (western part).

observed and all top directions determined from pillow lavas face consistently to the north, although the sandstone units are usually massive and lacking facing indicators. On the contrary, the latter assemblage consists of monotonous lithology and its degree of deformation is very weak. It keeps sufficient continuity of layers except in the part of slumping masses, and deformed into open to close large-scale folds. Sedimentary structures of such as graded bedding and sole marks are sometimes recognizable and the assemblage here designated "turbidite sequence".

Both assemblages of contrasting lithology and tectonic style are generally bounded each other by conspicuous strike faults at many places and form a sandwitch structure i.e. the highly sheared tectonic mélanges are tectonically imbricated with the turbidite sequences (SUZUKI *et al.*, 1978). Faults between the tectonic mélange situating the landward side of the turbidite sequence and the turbidite sequence distributing the oceanward side of the tectonic mélange are especially remarkable and these strike east-west and dip steeply to the north. However, when the pattern of the distribution is inverse order mentioned above i.e. the turbidite sequence apparently overlies the tectonic mélange dipping steeply to the north, faults are obscure or may not exist, although the structural obliqueness has still been observed. Considering from the characteristics in structural style and the relationship between both assemblages, it is reasonable to consider that the tectonic mélange is a stratigraphically lower member than the turbidite sequence and originally the latter unconformably overlaid the former. We suppose that this relationship may comparable to the tectonic unconformity proposed by KANMERA (1976 a, b) as a characteristic tectonic feature in the arc-trench gap environment.

Lithology of greenstones and its related rocks

Greenstones

The mappable rock units composed of greenstones and cherts as relatively isolated lenticular bodies, which are 20—300 m thick*, are scattered in the mélange zone of the Susaki Formation (Figs. 1 and 2). The greenstones in the Susaki Formation contain massive basaltic lava, almost nonvesicular pillowed basalt, pillow breccia, hyaloclastite and reworked sediments which is considered to be the result of the resedimentation of volcanic products. Variolitic structure consisting of divergent plagioclase fibers and quench texture are generally characteristic of the pillows. The greenstones are partially altered to chlorite, pumpellyite, prehnite, albite, calcite and quartz, but relic plagioclase and clinopyroxene are generally recognizable as phenocrysts and groundmasses. Many types of pillows,

* The greenstones of the Sakihama body in the Muroto Formation of the Shimanto Belt, eastern Shikoku, are more than 1,000 m in thickness, including gabbroic, picritic and serpentinized ultramafic rocks, and show a kind of "ophiolitic" igneous sequence (SUZUKI *et al.*, 1978).

such as cylinder, trapdoor, bulbous, flattened, elongated and hollow pillows (BALLAND & MOORE, 1977), are recognized. From the pillow forms (especially draping and drooping of pillow), the greenstones in the Susaki Formation show facing consistently to the north.

Nonvesicularity, variolitic structure and quench texture in the basaltic rocks suggest extrusion in relatively deep water environment, such as ocean floor (JONES, 1969; MOORE, 1970; CONNELLY, 1978). This consideration is not inconsistent with the conclusion drawn from the bulk chemical compositions (SUGISAKI *et al.*, 1979), as described later.

Many dykes of dolerites, which were probably feeder channels of surrounding pillow lavas, are found along the Aki River in the Terauchi area, Aki City. They intruded into the sheared mudstones parallel or oblique to their foliations with partly chilled margin

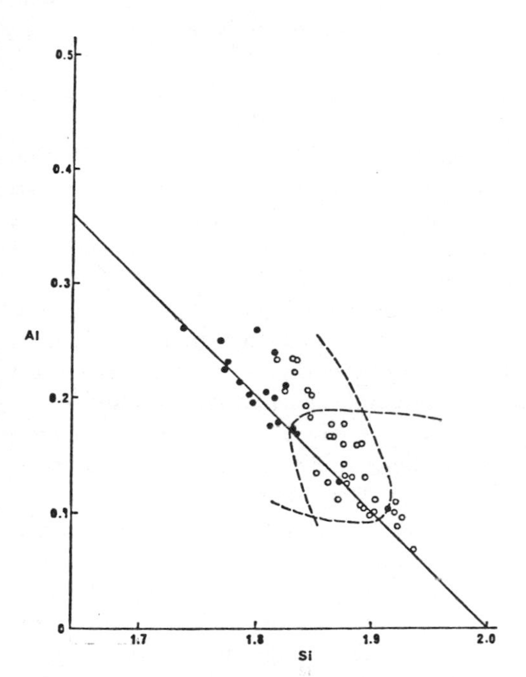

Fig. 3. Relation between the Al and Si contents of the clinopyroxenes from the basaltic rock complex in the Susaki Formation at Kue. Open circle: clinopyroxene from the lower and middle horizons of the basaltic rocks. Solid circle: clinopyroxene from the uppermost horizon of the basaltic rocks and the dolerite sheet.

(SUZUKI *et al.*, 1978). These are more or less deformed blocks and the rock has a cataclastic texture. It is inferred that the greenstones and sheared mudstones in the tectonic mélange of the Susaki Formation were fundamentally the products in the same depositional environment and then were suffered the same tectonic history.

SUGISAKI *et al.* (1979) examined the bulk chemical compositions of volcanic rocks from the Shimanto Belt formed during the Cretaceous to the early Tertiary time in Southwest Japan. According to them, the volcanics from the Cretaceous system apparently resemble the ocean floor basalts in terms of major compositions, and they might have been formed under a tension tectonics (SCHOLTZ *et al.*, 1971; SUGISAKI, 1976).

A study of the distribution of aluminium in soda-poor pyroxenes occurring in igneous rocks reveals characteristics which differ in the various magma types (KUSHIRO, 1960; LABAS, 1962). Ca-rich relic clinopyroxenes in eleven basaltic rock samples preferred from the lower to the upper horizon at Kue, Geisei-mura, are chemically analysed by means of EPMA method*. The Al-Si and Ti-Al relationships in Ca-rich relic clinopyroxenes were examined using the diagrams given by MARUYAMA (1976). As a result, clinopyroxenes from the lower and middle horizons of the basaltic rock complex (about 200 m thick) are plotted in the wide composition area, especially much occupied by the transitional area between the alkali basalt and tholeiitic magma types (Figs. 3 and 4). So it is difficult to estimate the magma type only by compositions of clinopyroxenes. However, clinopyroxenes from pillow lavas of the uppermost horizon and a doleritic sheet (about 10 m thick) which is distributed near the basaltic rock complex in question are distinctly plotted in the composition area for the alkali basalt magma type (Figs. 3 and 4).

The change of magma type from tholeiitic to alkaline during the volcanism is common,

* Space does not permit us to insert the data as to chemical compositions of the analysed clinopyroxenes. These data will be published SUZUKI & HIKOSAKA (in preparation).

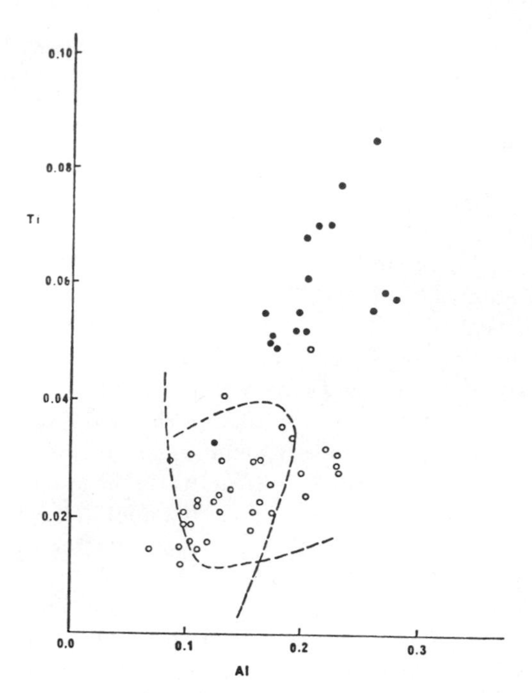

Fig. 4. Relation between the Ti and Al contents of the clinopyroxenes from the rocks as same as Fig. 3. Symbols are same as Fig. 3.

as shown in the volcanism of volcanic islands such as Hawaii (MACDONALD, 1968; JACKSON & WRIGHT, 1970) and seamounts such as the Emperor seamounts (KONO, 1979). The tendency of magma type toward the alkal· basalt type in volcanism at Kue is interested in correspondence to that of volcanic islands and seamounts, although more detailed chemical information is necessary.

Metamorphic zonation of the northeastern part of the Shimanto Belt in Kyushu were studied by IMAI *et al.* (1971) and they recognized the zonation from the prehnite–pumpellyite zone to the actinolite zone. In the tectonic mélange zones of the Susaki Formation in Shikoku, a low–grade regional metamorphism of prehnite–pumpellyite facies mineral assemblage is also recognized not only in the greenstones but also in its related rocks as described later. We consider that this metamorphism is comparable to ocean–floor metamorphism and/or subduction one during active underthrusting which may have continued after accretion of the subduction wedge onto the overthrust plate, as shown by CONNELLY (1978). On the contrary, no such metamorphic minerals is recognized in the matrix of sandstone and mudstone in the turbidite sequences which are tectonically imbricated with the tectonic mélanges, although it is not clear whether these rocks were suffered regional metamorphism or not. Therefore, it is possible to consider that the degree of metamorphism of the both assemblages changes discontinuously.

Related rocks

These greenstones are generally accompanied with red chert, with or without radiolaria, which are often thinly interbedded with reddish shales. Conformable contacts between red chert and underlying pillow basalt were observed in many localities. Petrographic examination of radiolarian chert shows abundant quartz–filled radiolaria in various stages of preservation set in a very fine–grained matrix of hematite and microcrystalline quartz. Recently, well–preserved remains of radiolarians are collected from reddish impure chert in the tectonic mélange at Kure, Nakatosa–cho, Kochi Prefecture. The following species are identified by Dr. A. YAO of Osaka City University; *Alievium superbum* (SQUINABOL), *Cryptamphorella* sp., *Dictyomitra macrocephala* SQUINABOL, *D. torquata* FOREMAN, *Amphipyndax stocki* (CAMPBELL & CLARK), *Stichomitra asymbatos* FOREMAN. These radiolarians indicate Albian to Coniacian ages.

The greenstone complex including red chert and reddish shale is always surrounded by the intensively sheared mudstone which contains inclusions of dismembered fine–grained sandstone and grey or green chert. Sheared mudstone often includes thinly and discontinuously layered unit of greenish tuff also. This sheared rock is tentatively termed "pebbly shale" (SUZUKI *et al.*, 1978). This mudstone is characterized by fracture cleavages, crenuration and microfold. Deformation has caused the more competent rock units to become boudinaged or shattered in the argillite matrix. In thin section, veinlets of pumpellyite, chlorite and calcite are commonly visible in sheared mudstone.

As the sandstones in the tectonic mélanges

are characterized by those of fine-grained and are distinguishable from the "greywacke" type sandstones in the turbidite sequence with the naked eye, we examined the composition of framework grains of sandstones in each zone. The sandstones in the mélanges are fairly well sorted with comparatively better rounded grains. On the other hand, those in the turbidite sequences have mostly sub-angular grains and sorting is not always good (SUZUKI *et al.*, 1978). Rock-fragments in the sandstones of the mélanges (Susaki Formation) at Terauchi, Aki City, consist of polycrystalline quartz (86—95%) and basaltic rocks (5—14%) alone. There is no rock-fragment derived from granitic or metamorphic rocks. On the contrary, variation of rock-fragments in the sandstones of the turbidite sequence at Terauchi is small amounts of chert (15—37%) and characterized by considerable amounts of volcanic, granitic, metamorphic and pelitic rock-fragments (SUZUKI *et al.*, 1978).

We show another new example on the sandstone analysis from the Susaki Formation. Sandstones from the tectonic mélanges occur as variously sized tectonic blocks in intensively sheared mudstone. These are medium- to fine-grained, well sorted and have narrow shear zones, which generally are 1 or 2 mm wide but which may make an entire thin section. Movement along these shear fractures has cataclastically reduced grain size. Calcite is characteristic in matrix of many sandstones and a considerable amounts of sandstones have diffuse grain boundaries surrounded by fine-grained matrices of recrystallized quartz. Secondary prehnite occurs in veinlets. On the other hand, sandstones of the turbidite sequences are collected from medium- to coarse-grained massive parts of well-bedded sandstone layers. These are poorly sorted, lack penetrative deformation and characteristically contain mud crusts.

The modal mineralogy was determined by automatic point counter and an average 300 counts each was made on 50 samples, and plotted on a standard quartz-feldspar-rock fragments diagrams following the techniques of FUJII (1955), OKADA (1968) and TERAOKA (1977) (Figs. 5, 6 and 7). Sandstones of the tectonic mélanges indicate that the fine-grained matrix is there about 30%, but this ranges from 18 to 42%. These contain extremely low contents of potassium feldspar; the average amounts of potassium feldspar and the average ratio of potassium feldspar to total feldspar (K/F) are 2.58% and 0.17 each, whereas those of the turbidite sequences are 10.24% and 0.41 each. These contain

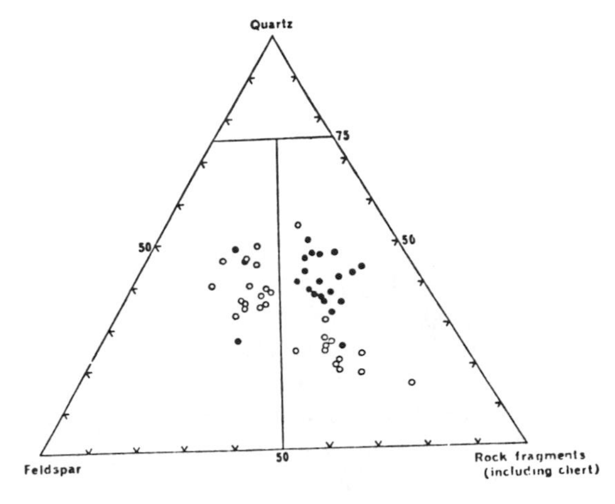

Fig. 5. Q-F-R diagram by OKADA (1968) for the sandstones of the Susaki Formation. Open circle: sandstone of the turbidite sequence. Solid circle: sandstone of the tectonic mélange.

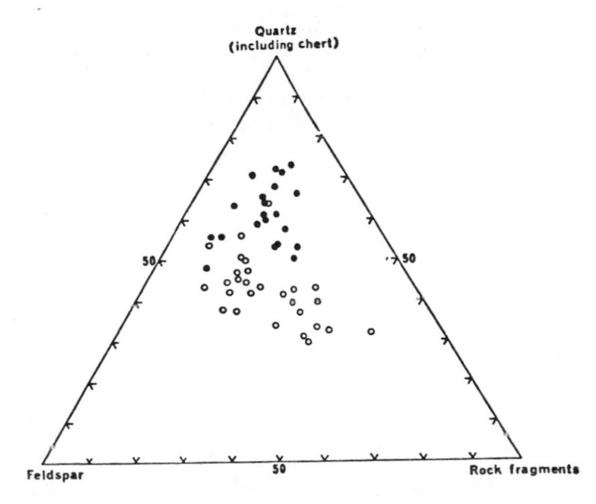

Fig. 6. Q-F-R diagram by FUJII (1955) for the sandstones of the Susaki Formation. Symbols are identical with those of Fig. 5.

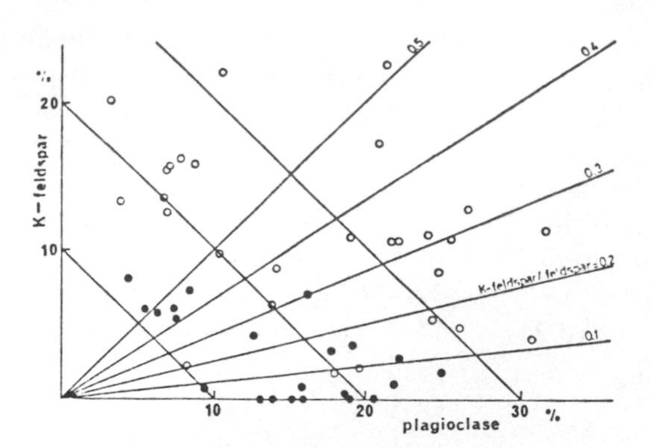

Fig. 7. Relation between potassium feldspar and plagioclase contents in the sandstones of the Susaki Formation. Symbols are identical with those of Fig. 5.

very high ratio of chert clast (10—21%). Consequently, almost all sandstones of the tectonic mélanges are grouped into lithic wacke according to the OKADA's definition. Unusual chert-rich lithic sandstones were reported from the mélange zone of Alaska (CONNELLY, 1978). DICKINSON (1978) and CONNELLY (1978) considered that recycling of uplifted subduction complexes form the important sources for these sandstones. On the contrary, samples from the turbidite sequences are grouped into more abundant arenite-type sandstones according to the OKADA's definition (matrix: 13—27%). These sandstones belong in two variants, the one lithic sandstones with low content of quartz and the other feldspathic sandstones. We suppose that the source rock of the sandstones in the turbidite sequences of the Susaki Formation were igneous rocks, mainly acid plutonic and volcanic rocks, associated with minor sedimentary and metamorphic rocks. These have strong resemblance to the Lower Cretaceous sandstones of the Shimanto Terrane in Southwest Japan as shown by TERAOKA (1977). From the modal analysis of more than 900 samples of sandstones from Kyushu and West Shikoku, TERAOKA (1977) compared with the Cretaceous sandstones between the Shimanto Terrane and the Median Zone of Southwest Japan. He collected and analyzed the samples from comparatively coarse-grained and massive parts of sandstone layers over several 10 cm in thickness in the turbidite sequences. The data on petrography and paleocurrents of these sandstones indicate that at least during Cretaceous time the main source area for the Shimanto geosynclinal sediments may have been in the Inner Zone. This conclusion is opposed to the prevailing opinion (KISHU SHIMANTO RESEARCH GROUP, 1968, 1975) that the great quantities of detritus were supplied to the geosyncline not only from the northern inner side but also from the southern outer side. We support the TERAOKA's opinion as far as the main source area for the turbidite sequence in the Cretaceous Shimanto sediments is concerned, because of the precise resemblance of the sandstone composition. Considering from the nature of the tectonic mélange as will be discussed later, it is difficult to hypothesize the southern source area (Kuroshio paleoland).

Judging from these characteristics of the composition and the degree of sorting and roundness, there is a possibility that the source areas of the sandstones of the tectonic mélanges and those of the turbidite sequences have been different from each other and it is also suggested that the sandstones of the tectonic mélanges have natures of more distal facies as compared with those of the turbidite sequences, because the former is characterized by the higher content of stable grains.

Cretaceous tectogenesis in the Shimanto Belt

In recent years, the continental margins have been the subject of increasing attention because of their fundamentally great geological significance. Modern continental margins can be basically classified as Atlantic-type or Pacific-type (DEWEY & BIRD, 1970; and others) depending on whether they have experienced a relatively long period of stability (Atlantic-type) or whether they have undergone active tectonism during later geological times (Pacific-type) (HEEZEN, 1974).

The geology of the Shimanto Belt has long been interpreted by the tectonic framework of "geosyncline". It explains that the Shimanto geosyncline was brought to birth probably in the Late Jurassic and was developed through Cretaceous. However, a wide divergence of views still exists concerning to an actualistic model of the Shimanto geosyncline, although it is generally agreed that this is a linear belt of thick sediments deposited on a subsiding crust. The tectonic environment of the Shimanto Belt, for example, has been considered as follows; between continents (geosynclinal model) (KISHU SHIMANTO RESEARCH GROUP, 1975; TATEISHI, 1974; HARATA & TOKUOKA, 1978), between continent and island arc (marginal sea model) (ICHIKAWA *et al.*, 1972) and between island arc or continent and ocean (MATSUMOTO, 1967; KIMURA & TOKUYAMA, 1971; KIMURA, 1973). Recently, KANMERA & SAKAI (1975) and KANMERA (1976 a, b) discussed the site of the sedimentation of the Shimantogawa Group and proposed a totally new model for the Shimanto Belt. They revealed that the development of the Shimanto Belt includes very different process of sedimentation and tectonism from that of geosynclines and corresponded to the site of the tectonic environment of modern arc–trench gap. Recent geophysical and geological studies on modern continental margins indicate that the oceanic trenches are the loci of plate convergence. Our conclusion concerning the Cretaceous tectogenesis in the Shimanto Belt in Shikoku is also comparable with their model.

The Susaki Formation consists of two assemblages of the turbidite sequence and the tectonic mélange. Both were experienced the completely different metamorphic and deformational history. We suppose that the geosynclinal model for the Shimanto Belt was based mainly on the study of the turbidite sequence. However, the tectonic mélange is similar to lithologies that occur in the deep sea environment and probably indicates a subduction complex. Namely, mélanges have been reported from many orogenic belt throughout the world. DEWEY (1969) suggested that the Dunnage Mélange of Cambrian age in Newfoundland Appalachians represents a deposit which is characteristic of an oceanic trench and a subduction zone. Mélanges in the Franciscan Complex, California, and Uyak Complex, southeastern Alaska, have also been interpreted as remnants of ancient subduction zones (ERNST, 1970; COWAN, 1974, 1978; CONNELLY, 1978; MOORE & WHEELER, 1978). Therefore we consider that it is difficult to explain the Cretaceous tectogenesis in the Shimanto Belt by the tectonic framework of geosyncline. Preferably, the structural and sedimentary characteristics of these assemblages mentioned above indicate that they may represent the products in the tectonic framework of the Pacific–type continental margin.

KARIG (1974) and KARIG & SHARMAN (1975) have proposed a model for the evolution of trench margin. Their model is that pelagic sediments, trench fill sediments and sometimes parts of underlying oceanic crust are striped off the downgoing oceanic plate and accreted to inner wall in arcward dipping wedge–shaped, thrust–bound packets characterized internally by folding and intensive shearing. With continued subduction, older, previously accreted materials is passively uplifted and eventually may form a linear ridge (trench slope break), and older thrusts are rotated arcward by the addition of successive wedge or slices of younger, newly accreted sediments at the base of the inner wall. A progressively enlarging body of turbidite sequence is commonly ponded in the midslope terrace behind this ridge. SEELY *et al.* (1974) emphasized a hypothetical systematic repetition of imbricated trench and pelagic facies within the wedge based on the seismic profiles. KANMERA (1976a, b) summarized these earlier works and emphasized the characteristics of the tectogenesis in the frontal part of the arc–trench gap which is very different from the process for the metamorphism and deformation basing on the concept of "geosyncline".

As the concrete example of these models, MOORE, J. C. & KARIG (1976) have proposed that this type of sedimentary assemblage is associated with the Shikoku subduction zone of Southwest Japan, along the Nankai Trough where the Phillippine Sea plate is underthrusted beneath the Asian plate. The accretionary complex is a presumably fault–

bound wedge of an overturned anticline involving consolidated, cleaved and dewatered sediments, despite Quaternary age. MOORE, G. F. & KARIG (1976) studied a mid–Tertiary subduction complex, constituting part of the trench slope break of the active Sunda Trench, which was exposed on Nias Island, southwest Sumatra. Highly sheared tectonic mélanges, comprising sediments accreted at the base of the inner wall, are tectonically imbricated with strata deposited in slope basins that developed on the lower trench slope. In southwestern Alaska, greenstones the Uyak Mélange Complex represent basal oceanic crust upon which the radiolarian chert, gray chert and argillite, and wacke were deposited, respectively, at a mid–ocean rise, on the abyssal ocean floor, and in an oceanic trench (CONNELLY, 1978). To the southeast, the Uyak Complex is under–thrusted by turbidite sequences of the Kodiak deposits in an oceanic trench and accreted to the Alaskan margin in Late Cretaceous time (MOORE, 1973; CONNELLY, 1978).

Considering from these studies and the characteristics of the Susaki Formation, the two assemblages of the formation may correspond respectively to an accretionary complex imbricated to the trench slope and sediments of the midslope terrace or slope basin on the "basement" of the accretionary complex in an arc–trench gap environment. It is interpreted that the site of volcanism and sedimentation of the tectonic mélanges was originally pelagic and comparatively deeper water environments, such as ocean floor, nearby seamounts or a trench. Results of Leg 58 of Deep Sea Drilling Project (GEOTIMES, 1978) showed that the occurrence of basalts intercalated with claystones, ash, siltstone in Early Eocene time at 446 site (at the Daito Ridge and Basin Province) and mud interposed between pillow and massive basalts in Early Miocene time at 442 site (south of the Nankai Trough). These sequences might be similar to the original assemblage of the tectonic mélange before subduction. Deep–sea sequences including trench fill sediments were deformed to tectonic mélange under the process of tectonic compression such as subduction or accretion. During underthrusting, brittle rock types were broken into phacoids of all sizes and suspended in the less competent matrix of mudstones.

On the other hand, it is conceivable that the turbidite sequence of the Susaki Formation was originally developed on the midslope terrace behind the trench slope break or slope basins on the lower trench slope. We are uncertain whether the turbidite sequence includes the sediments which have been deposited in an oceanic trench directly as in Alaska (CONNELLY, 1978) or not. KANMERA (1976 a,b) revealed that these sediments on midslope terrace or slope basins were deposited on the "basement" of the tectonic mélange with the relationship of tectonic unconformity.

The highly sheared tectonic mélanges are tectonically imbricated with the folded turbidite sequences in the Susaki Formation and thrust faults bounding them are conspicuous northerly and steeply dipping faults. This may indicate that thrust faults originally bounded accreted packets have progressively been rotated as new materials was underplated at the trench and some of major faults remained active even as they rotated.

Acknowledgements

The authors wish to express their sincere gratitude to Prof. K. KANMERA of Kyushu University for critical reading of the manuscript. We are indebted to Dr. R. SUGISAKI of Nagoya University for chemical analyses of the greenstones, and to Dr. A. YAO of Osaka City University for paleontological work of radioralia. We also thank Mr. S. YOSHIKURA of Kochi University, who freely discussed with us many of observations presented here. Thanks also due to Messrs. M. HIKOSAKA, H. MATSUMOTO and E. TAKAMATSU of Kochi University for their helps in the field work. This work was supported by Scientific Research from the Ministry of Education of Japan.

References

BALLARD, R. D. and MOORE, J. G., 1977 : *Photographic atlas of the Mid-Atlantic ridge rift valley.* 114p., Springer-Verlag, New York.

CONNELLY, W., 1978 : Uyak Complex, Kodiak Islands, Alaska: A Cretaceous subduction complex. *Geol. Soc. Amer., Bull.*, **89**, 755—769.

COWAN, D. S., 1974 : Deformation and metamorphism of the Franciscan subduction zone complex northwest of Pacheco Pass, California. *Ibid.*, **85**, 1623—1634.

————, 1978 : Origin of blueschist-bearing chaotic rocks in the Franciscan Complex, San Simeon, California. *Ibid.*, **89**, 1415—1423.

DEWEY, J. F., 1969 : Evolution of the Appalachian-Caledonian orogen. *Nature*, **222**, 124—129.

———— and BIRD, J. M., 1970 : Mountain belts and the new global tectonics. *Jour. Geophys. Res.*, **75**, 2625—2647.

DICKINSON, W. R., 1978 : Plate tectonic influences on sandstone compositions. *Abstracts with Programs, Geol. Soc. Canada, Min. Assoc. Canada and Geol. Soc. Amer.*, **3**, 389.

ERNST, W. G., 1970 : Tectonic contact between the Franciscan mélange and the Great Valley sequence, crustal expression of a late Mesozoic Benioff zone. *Jour. Geophys. Res.*, **75**, 886—902.

FUJII, K., 1955 : Some problems on the studies of sandstone. *Chikyu kagaku (Earth Sciences)*, **20**, 9—18 (in Japanese with English abstract).

GEOTIMES, 1978 : On Leg 58, Philippine Sea drilled. 23—25.

HARATA, T. and TOKUOKA, T., 1978 : A consideration on the Paleogene paleogeography in Southwest Japan. *In* HUZITA, K. *et al.*, eds.: *Cenozoic Geology of Japan* (Prof. N. IKEBE memorial volume), 1—12 (in Japanese).

HEEZEN, B. C., 1974 : Atlantic-type continental margins. *In* BURKE, C. A. and DRAKE, C. L., eds.: *The Geology of Continental Margins*, 13—24. Springer-Verlag, New York.

HSÜ, K. J., 1974 : Mélanges and their distinction from olistrostromes. *In* DOTT, R. H. Jr. and SHAVER, R. H., eds.: *Modern and Ancient Geosynclinal Sedimentation*, **19**, 321—333. Soc. Econ. Paleontologists and Mineralogists Spec. Pub.

ICHIKAWA, K., MATSUMOTO, T. and IWASAKI, M., 1972 : Geologic history of Japanese Islands. *Kagaku (Science)*, **42**, 181—191 (in Japanese).

IMAI, I., TERAOKA, Y. and OKUMURA, K., 1971 : Geologic structure and metamorphic zonation of the northeastern part of the Shimanto terrane in Kyushu, Japan. *Jour. Geol. Soc. Japan*, **77**, 207—220 (in Japanese with English abstract).

JACKSON, E. D. and WRIGHT, T. L., 1970 : Xenoliths in the Honolulu volcanic series, Hawaii. *Jour. Petrol.*, **11**, 405—430.

JONES, J. G., 1969 : Pillow lavas as depth indicators. *Amer. Jour. Sci.*, **267**, 181—196.

KANMERA, K., 1976a : Comparison between past and present geosynclinal sedimentary bodies I. *Kagaku (Science)*, **46**, 284—291 (in Japanese).

————, 1976 b: Comparison between past and present geosynclinal sedimentary bodies II. *Ibid.*, **46**, 371—378 (in Japanese).

———— and SAKAI, T., 1975 : To what part of the recent ocean bottom does correspond to the site of the sedimentation of the Shimantogawa Group. *Report of GDP in Japan, II-1-(7) Structural Geology*, no. 3, 55—64 (in Japanese).

KARIG, D. E., 1974 : Evolution of arc systems in the western Pacific. *Annu. Rev. Earth Planet. Sci.*, **2**, 51—75.

———— and SHARMAN, G. F., 1975 : Accretion and subduction in trenches. *Geol. Soc. Amer., Bull.*, **86**, 377—389.

KATTO, J., 1961 : Shimanto Belt. *In* KATTO, J. *et al.*, eds.: *Explanation text of the geology and mineral resources of Kochi Prefecture*, 56—90. Kochi Pref. Govern (in Japanese).

————, 1969 : *Geology of Kochi Prefecture*, 316 p., Kochi City (in Japanese).

———— and TAIRA, A., 1978 : Lithofacies and sedimentary environments of the Murotohanto Group. *Geologic News*, no. 287, 21—31 (in Japanese).

KIMURA, T., 1973 : The old 'inner' arc and its deformation in Japan. *In* COLEMAN, P. J., ed.: *The Western Pacific-Island Arcs, marginal seas geochemistry-*, 255—273. University of Western Australia Press.

———— and TOKUYAMA, A., 1971: Geosynclinal prisms and tectonics in Japan. *Mem. Geol. Soc. Japan*, no. 6, 9—20.

KISHU SHIMANTO RESEARCH GROUP, 1968 : The study of the Shimanto Terrain in the Kii Peninsula, Southwest Japan (part 2), The present status of the research and the southern formerland in the Pacific Ocean. *Chikyu kagaku (Earth Science)*, **22**, 224—231 (in Japanese with English abstract).

————, 1975 : The Development of the Shimanto Geosyncline. *Monograph Assoc. Geol.*

Collab. Japan, no. 19, 143—156 (in Japanese with English abstract).

KONO, M., 1979 : Verification of hot spot theory-Ocean-floor drilling by the 55th navigation of DSDP. *Kagaku (Science)*, **49**, 180—181 (in Japanese).

KUSHIRO, I., 1960 : Si-Al relation in clinopyroxenes from igneous rocks. *Amer. Jour. Sci.*, **258**, 548—554.

LEBAS, M. J., 1962 : The role of aluminium in igneous clinopyroxenes with relation to their parentage. *Ibid.*, **260**, 267—288.

MACDONALD, G. A., 1968 : Composition and origin of Hawaii lavas. *Geol. Soc. Amer., Mem.*, **116**, 477—522.

MARUYAMA, S., 1976 : Chemical natures of the Sawadani greenstone complex in the Chichibu Belt, eastern Shikoku. *Jour. Geol. Soc. Japan*, **82**, 183—197 (in Japanese with English abstract).

MATSUMOTO, T., 1967 : Fundamental problems in the circum-Pacific orogenesis. *Tectonophysics*, **4**, 595—613.

MIYAMOTO, T., 1976 : Comparison of the Cretaceous sandstones from the Chichibu and Shimanto Terrains in the Odochi area, Kochi Prefecture, Shikoku. *Jour. Geol. Soc. Japan*, **82**, 449—462 (in Japanese with English abstract).

MOORE, G. F. and KARIG, D. E., 1976 : Development of sedimentary basins on the lower trench slope. *Geology*, **4**, 693—697.

MOORE, J. C., 1973 : Complex deformation of Cretaceous trench deposits, southwestern Alaska. *Geol. Soc. Amer., Bull.*, **84**, 2005—2020.

——— and KARIG, D. E., 1976 : Sedimentology, structural geology, and tectonics of the Shikoku subduction zone, southwestern Japan. *Ibid.*, **87**, 1259—1268.

——— and WHEELER, R. L., 1978 : Structural fabric of a mélange Kodiak Islands, Alaska. *Amer. Jour. Sci.*, **278**, 739—765.

MOORE, J. G., 1970 : Water content of basalt erupted on the ocean floor. *Contr. Mineral. Petrol.*, **28**, 272—292.

OKADA, H., 1968 : Classification and nomenclature of sandstones. *Jour. Geol. Soc. Japan*, **74**, 371—384 (in Japanese with English abstract).

SAKAI, T., 1978: Geologic structure and stratigraphy of the Shimantogawa Group in the middle reaches of the Gokase River, Miyazaki Prefecture. *Sci. Repts., Dept. Geol., Kyushu Univ.*, **13**, 23—38 (in Japanese with English abstract).

SCHOLTZ, C. H., BARAZANGI, M. and SBAR, L., 1971: Late Cenozoic evolution of the Great Basin, Western United States, as an ensialic interarc basin. *Geol. Soc. Amer., Bull.*, **82**, 2979—2990.

SEELY, D. R., VAIL, P. R. and WALTON, G. G., 1974 : Trench-slope model. *In* BURK, C. A. and DRAKE, C. L., eds.: *The Geology of Continental Margins*, 249—260, Springer-Verlag, New York.

SUGISAKI, R., 1976 : Chemical characteristics of volcanic rocks-relation to plate movements. *Lithos.*, **9**, 17—30.

———, SUZUKI, T., KANMERA, K., SAKAI, T. and SANO, H., 1979 : Chemical compositions of green rocks in the Shimanto Belt, southwest Japan. *Jour. Geol. Soc. Japan*, **85**, 455—466.

SUZUKI, T., HADA, S., UMEMURA, H., KADO, K., SAKAMOTO, Y. and NAKAGAWA, H., 1978: Genetic consideration of the green rocks in the Shimanto Belt-with special reference to mode of occurrence-. *Chikyu kagaku (Earth Science)*, **32**, 321—330 (in Japanese with English abstract).

———, ——— and YOSHIKURA, S., 1979 : Tectonic environment of the outer side of Southwest Japan. *Chikyu (The Earth Monthly)*, **1**, 57—62 (in Japanese).

——— and HIKOSAKA, M.: Petrological study of greenstones and surrounding rocks of the Shimanto Belt in Geisei-mura, Aki-gun, Kochi Prefecture (in preparation).

TATEISHI, M., 1974 : Restoration of sedimentary basin of the Muro Group and the Kuroshio paleoland problem. *Papers at the "Symposium on the Shimanto geosyncline"*, 84—91 (in Japanese).

TERAOKA, Y., 1977 : Comparison of the Cretaceous sandstones between the Shimanto Terrane and the Median Zone of Southwest Japan, with reference to the provenance of the Shimanto geosynclinal sediments. *Jour. Geol. Soc. Japan*, **83**, 795—810 (in Japanese with English abstract).

28

Reprinted from pages 259-267 and 272-275 of *Royal Soc. New Zealand Jour.*
9:259-275 (1979)

Northland Allochthon

P. F. Ballance and K. B. Spörli
Department of Geology, University of Auckland

Abstract

The name Northland Allochthon is proposed for the thick and extensive assemblage of Upper Cretaceous to Upper Oligocene sediments, plus Tangihua volcanic masses, occurring on the western side of Northland, shown by drillholes at Ngawha, Waimamaku and Batley to be allochthonous and to contain masses of more-or-less coherent sedimentary rock possibly up to 2.5 km thick. The Allochthon is unconformably overlain by the Waitemata and correlative Groups of Waitakian to Southland age (uppermost Oligocene to mid-Miocene) and was itself emplaced by gravity sliding during the Waitakian. It is thus quite distinct from the post-or-syn-Waitemata Onerahi Chaos-breccia of Kear and Waterhouse. The Allochthon now extends from the Waikato Fault northwards probably to North Cape, a distance of more than 350 km, and westwards beneath the present continental shelf, a width approaching 100 km. It may originally have been much wider. It contains a stratigraphic sequence with an original thickness of several kilometres, now present as olistoliths of sedimentary and volcanic rocks of all sizes up to hundreds of cubic kilometres, irregularly juxtaposed in terms of stratigraphic age, and separated by zones of strongly transposed clay-rich sediment. It is inferred to have been emplaced from the present north, from an everted marine basin with oceanic volcanic basement, which lay to the north of the present Raukumara Peninsula on a pre-Miocene reconstruction of the North Island. Parts of the Allochthon were transported at least 350 km; the maximum estimate of speed is 350 mm per year, following uplift of the depositional basin of perhaps as much as 14 km. Uplift and perhaps compression of the depositional basin, and downsinking of the receiving basin by 3 km or more, were two of the events marking the first appearance through the North Island of the India/Pacific subducting plate boundary.

A probable extension of the Allochthon is present at East Cape (Matakaoa Volcanics), and it can possibly be traced along the east coast of the North Island and into the Kaikoura area of the South Island.

Introduction

The thick Upper Cretaceous and Lower Tertiary strata of western Northland, New Zealand, have long presented difficulties in geological mapping. They are in marked contrast to the autochthonous thin lower Tertiary sequences on the uplifted basement greywacke blocks of eastern Northland. Kear and Waterhouse (1967) and Katz (1968) suggested, and recent drillholes at Ngawha (Bowen and Skinner 1972), Waimamaku (Hornibrook *et al.* 1976) and Batley (unpublished Geological Survey information) (Fig. 1) have established, that substantial thicknesses of these western rocks have reversed and often disordered stratigraphic sequences and are allochthonous. This paper discusses the nature, extent and significance of this allochthonous unit, the major part of which we name the Northland Allochthon.

Relationship to Onerahi Chaos-breccia

Kear and Waterhouse (1967) established the existence in Northland of a widespread, thin and fragmentary sheet of olistostrome, which overlies rocks

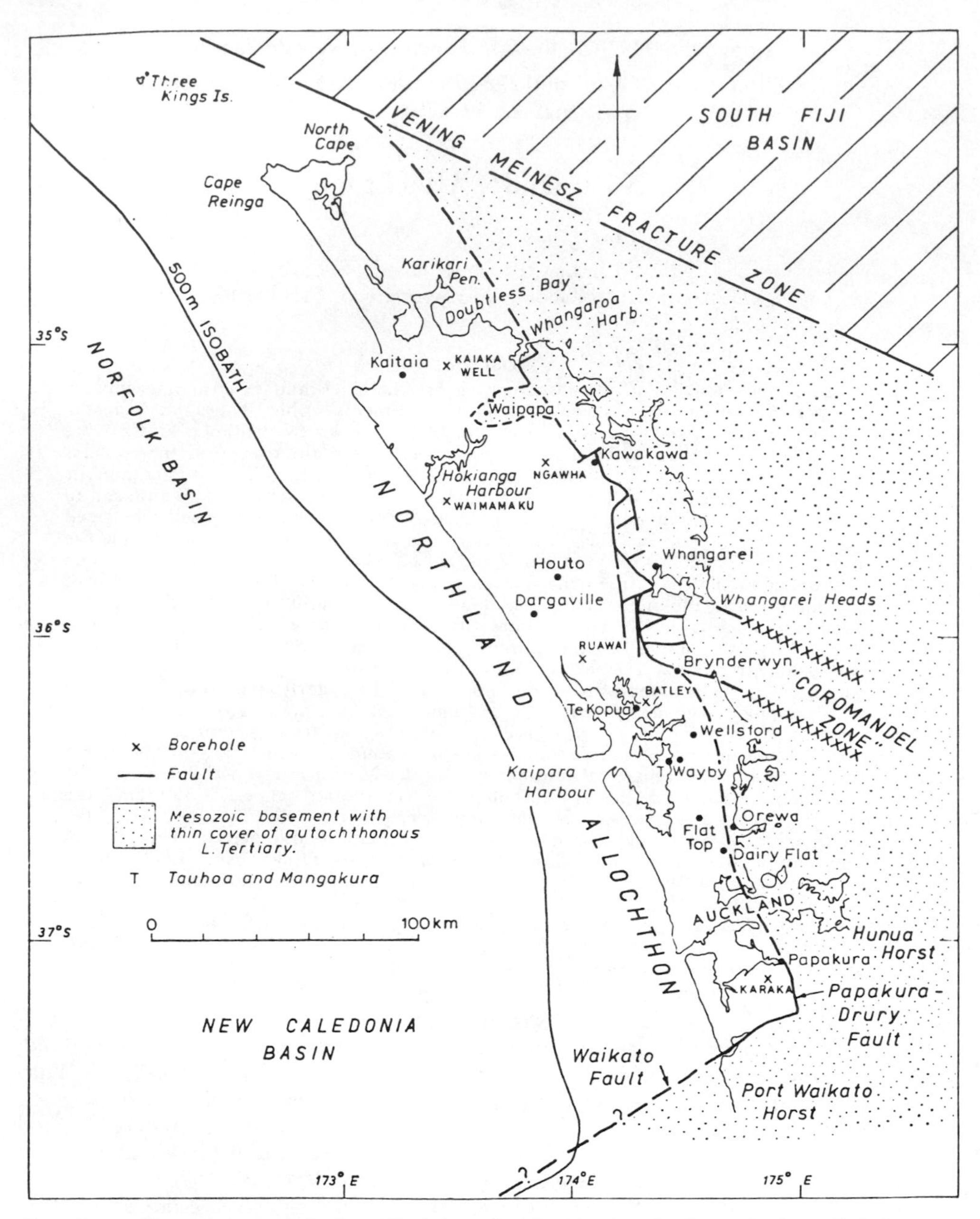

FIG. 1. — Present extent of the Allochthon in Northland. Outliers of the Allochthon at Kawakawa and Whangarei not shown. "Coromandel Zone" is a belt of Upper Cenozoic volcanics and intrusions which is apparently offset from the north end of Coromandel Peninsula to Northland; it may influence the change in direction of the eastern boundary of the Allochthon at Whangarei.

of the lower Miocene Waitemata Group, and underlies the Pleistocene Kerikeri Volcanic Group. Later (Kear and Waterhouse 1977) they extended the Onerahi Chaos-breccia to include all the known chaotic Cretaceous to Lower Tertiary sedimentary rocks of Northland.

However, we consider that a very important distinction exists between those allochthonous rocks which underlie the Lower Miocene Waitemata and Otaua Groups (Ballance *et al.* 1977) (our Northland Allochthon), and those which overlie them. We propose that the name Onerahi Chaos-breccia be restricted to the latter. This distinction was foreshadowed by Spörli and Issaac (1973) and Ballance and McCarthy (1975).

Noteworthy characteristics of the Onerahi Chaos-breccia are that it consists of crushed and broken Upper Cretaceous and Lower Tertiary sedimentary rocks and contains blocks of serpentinite, some of which are very large. Distinction between the two allochthons is difficult where Waitemata and correlative strata are absent, as noted by Kear and Waterhouse (1977).

Relationship of the Two Allochthons to the Autochthonous Waitemata and Correlative Groups

There is abundant evidence that Onerahi Chaos-breccia overlies the Waitemata Group (Waterhouse 1966). Kear and Waterhouse (1967) noted that the contact is often near the base of the Waitemata Group, and later (1977) suggested that Onerahi Chaos-breccia is actually interbedded with near-basal Waitemata Group.

The evidence that the Northland Allochthon underlies the Waitemata and correlative Groups has not been adequately discussed in previous publications. The proven allochthonous rocks at Waimamaku (Hornibrook *et al.* 1976) stratigraphically underlie the Lower Miocene Otaua Group. In spite of the complication caused by a patch of chaotic rocks at a higher elevation than the Otaua Group at the eastern end of the Waimamaku Valley, mapping by Thompson (1961) and Wakefield (1976) shows that over most of the valley the allochthonous rocks clearly underlie the Otaua Group. The same situation holds with respect to the Waitemata Group throughout the northern Kaipara Harbour (Arlidge 1955); the allochthonous nature of the underlying rocks was shown by the Batley drillhole (Thompson, in Katz 1968).

The basal contact of the Waitemata Group on the Allochthon is exposed on either limb of the synclinal outcrop of Puriri Formation (Ballance 1964; 1976b) at Te Kopua Point, northern Kaipara Harbour. Puriri Formation rests on brecciated, non-calcareous claystone of the Upper Cretaceous Mangakahia Group, in part with an undisturbed sedimentary contact and a basal pebble bed of Mangakahia Group claystone. On the northern side of the syncline, irregular dykes and pods of brecciated Mangakahia Group claystone, up to a metre in width, intrude the Puriri Formation; one dyke of brecciated Puriri Formation penetrates downwards into Mangakahia Group.

THE ALLOCHTHON IN DRILLHOLES

Sedimentary Rocks

The definitely allochthonous sedimentary rocks in drillholes consist of non-calcareous and calcareous mudstones, argillaceous micrites, micrites, muddy siltstones, micaceous siltstones, and micaceous fine sandstones. Cores from the Waimamaku and Ngawha wells (Cores 1 to 3, Hornibrook *et al.* 1976; Bowen and Skinner 1972) show mixed lithologies and abnormal superposition. Many of the lithologies are internally sheared or brecciated. Large coherent blocks appear to be separated by more strongly deformed shear zones. Ages range from Haumurian (Upper Cretaceous) to Duntroonian (mid to Upper Oligocene).

Igneous Rocks

The only igneous rock penetrated by a drillhole is the dolerite body 29 m thick, near the base of Waimamaku No. 2. It is interpreted as a sill by Hornibrook *et al.* (1976) and dated at 31.1 ± 0.4 m.y. by Adams (1975, p. 446). It is not clear whether the dolerite and its containing strata are in or out of place.

THE ALLOCHTHON AT OUTCROP

Sedimentary Rocks

Sedimentary rocks of Upper Cretaceous to Lower Tertiary age outcrop widely in northwestern Northland, from Kaipara Harbour northwards. They have been mapped by many authors (see Hay 1960; Kear and Hay 1961; Thompson 1961; Katz 1968); the strata occur in coherent units large enough to be mapped on scales of up to 1:1,000,000. Contacts between adjacent formations are rarely exposed; formations sometimes follow one another in stratigraphic sequence (e.g. Hay 1960).

Elsewhere, however, sequences are not stratigraphic. For example, Upper Cretaceous rocks are juxtaposed against Eocene, and Upper Cretaceous against Oligocene, in the northern Kaipara Harbour (Arlidge 1955; Thompson 1961; Carter 1967, 1969; Spörli and Isaac 1973). Such rock bodies are often separated by areas of strongly sheared and mixed rock (Ballance and McCarthy 1975). None of the large slabs are known to be overturned. Thus the surface geology seems to correspond closely with that revealed in the Waimamaku drillhole.

Igneous Rocks

We also consider that the Tangihua and related basic igneous bodies are blocks within the Allochthon (Brothers 1974; Brothers and Delaloye 1976).

Most Tangihua bodies have a negligible expression on gravity maps (Reilly 1965a, b). Quennell and Hay (1964) considered them to be rootless, and there seems little doubt that this interpretation is correct. The largest of these rootless blocks are some tens of kilometres in length. They seem to be more common in the north and at the surface of the Allochthon.

Large serpentinite blocks occur in the younger Onerahi Chaos-breccia (Kear and Waterhouse 1967; O'Brien and Rodgers 1973). Serpentinite pebbles are also known in the Albany Conglomerate of the Waitemata Group (Bunting 1970; Ballance 1974), which also contains debris from Tangihua igneous masses. Serpentinite has not been recorded to date in the Northland Allochthon. However, it is possible that the serpentinite bodies at Wellsford, Tauhoa and Mangakura (Thompson 1961) belong to the Allochthon. Whatever their origin, it appears that serpentinites are much less abundant in the Northland Allochthon than they are in the Onerahi Chaos-breccia.

THICKNESS AND EXTENT

The Allochthon is known to be at least 2.86 km thick in Waimamaku No. 2 well (Hornibrook *et al.* 1976). Geophysical data suggest a thickness of as much as 5 km to basement at nearby Hokianga (Woodward 1970). Hornibrook *et al.* (1976) interpreted as autochthonous the lowermost 500 m of the Waimamaku No. 2 well, consisting of a condensed and incomplete section of Cretaceous and Eocene sediments. The nature of the unexplored 1700 m of section down to geophysical basement, which we infer to consist of Triassic and Jurassic rocks, is not known. It may be autochthonous Cretaceous. Alternatively, it may actually be allochthonous.

There is little other evidenc of maximum thickness of the Allochthon. The present thinning towards the east is largely due to pre-Waitemata erosion.

Southern Extent

The Allochthon is unconformably overlain by the Waitemata Group around the southern Kaipara Harbour. Rocks underlying the Waitemata Group are next seen at Silverdale and Dairy Flat. Here the sheared Mahurangi Limestone, of Whaingaroan to Duntroonian age (Lower to mid-Oligocene) is more than 300 m thick (Waterhouse 1966), and we interpret it as a large block in the Allochthon. Patches mapped as Onerahi Chaos-breccia west of Silverdale, which outcrop on low ground adjacent to escarpments of Waitemata Group rocks, may actually be inliers of Northland Allochthon.

Lower Tertiary rocks underlying the Waitemata Group are next exposed to the south on the basement horsts of Hunua and Port Waikato. They consist of thin, autochthonous shelf sandstones and carbonates of Oligocene age overlying basal coal measures (Purser 1961; Schofield 1967). We believe, however, that the Allochthon extends south beneath the Waitemata Group as far as the Waikato Fault (Fig. 1), on the basis of the following facts.

(a) The basal conglomerates of the Waitemata Group at Hays Creek and Winstone's Quarry, east of Papakura, contain blocks up to one metre across of unindurated, coral-bearing Eocene greensand, argillaceous limestone and chocolate shale. These lithologies are not present in the local autochthonous sequences, and a nearby source in the pre-Waitemata Allochthon, now buried beneath the adjacent Manukau structural basin, is inferred.

(b) The Karaka Bore (Kear 1954), south of Manukau Harbour, penetrated 600^+ m of Waitemata Group (Ballance 1974), before intersecting Dannevirke (Lower Eocene) rocks at the bottom of the hole (Kear 1959) that have been correlated with Onerahi Chaos-breccia by Kear and Waterhouse (1977). These are considered to be Northland Allochthon beneath the Waitemata Group.

(c) Blocks of argillaceous limestone identical to Mahurangi Limestone occur occasionally as xenoliths in the Pleistocene lavas of Auckland and the Manukau basin (Searle 1964); a source in the Allochthon at depth is inferred.

(d) The Waikato Fault was a major palaeogeographic influence during deposition of the succeeding Waitemata Group (Ballance 1974).

Eastern Extent

The present eastern boundary of the Allochthon is mostly faulted (Fig. 1). It is concealed for 120 km by the overlying Waitemata Group, between Brynderwyn and Papakura, where drillhole information shows that it passes between Dairy Flat and Waiwera.

East of the bounding faults remnants of the Allochthon are very few. Some is preserved at Whangarei Heads, 30 km east of the faults, where it is intruded by dykes feeding the overlying Lower Miocene Wairakau Andesites (Brothers 1974).

Thus at the time of emplacement the Allochthon probably covered all of Northland.

The relatively undisturbed basal contact of the Allochthon therefore only outcrops at Whangarei Heads, south of Whangaroa Harbour (Kear and Hay 1961; Maehl 1970), south of Kawakawa (Hay's (1960) Kawakawa overthrust) and in the vicinity of Whangarei.

Western and Northern Extent

The Allochthon passes beyond the present west coast and underlies the continental shelf (Katz 1974).

The Cretaceous rocks near North Cape (Leitch 1970) and the Wangakea Volcanics (Farnell 1973), are here included in the Allochthon and are overlain by autochthonous Upper Oligocene to Lower Miocene rocks (Leitch 1970). The Allochthon is thus thought to pass beyond northern New Zealand.

The Older Cretaceous Rocks

The position of the older to mid-Cretaceous (Clarence to Lower Raukumara) rocks north of Whangaroa Harbour, with respect to the Allochthon, needs discussion. They are the indurated sandstones, mudstones and conglomerates of Hay's (1975) Tokerau, Whatuwhiwhi, Tupou and Awapoko Formations and Houhora Volcanic Group. These rocks occur along the present margin of the Allochthon, they are older than any lithologies definitely included in the Allochthon, they contain keratophyres which are not known in the main body of the Allochthon, and they are locally schistose (Le Couteur 1967), while in the Allochthon there is no true schistosity developed apart from a shear foliation. Thus the age, and the paleogeographic and tectonic environments of these rocks, appear to be different from those of the Allochthon. It is uncertain whether they are allochthonous. In either case they are different from all the other adjacent rocks and have considerable tectonic significance, as yet unexplored.

Summary

The Northland Allochthon is thus a wedge-shaped body thickening from zero in the east to at least 2.86 km in the west. It extends more than 370 km from north to south and more than 60 km from east to west. It may originally have been much more extensive.

THE ALLOCHTHON IN EASTERN NORTH ISLAND

The former eastern extension of the allochthon is probably present in the Matakaoa volcanics of East Cape, as suggested by Brothers (1974), and in the nappe structures in the Cretaceous-Tertiary south of East Cape, as described by Stoneley (1968), Ridd (1964, 1968) and Speden (1976). Moreover, the question arises whether most of the Cretaceous-lower Tertiary rocks along the East Coast of the North Island, possibly extending even to Kaikoura — which are usually much more complexly deformed and imbricated than the overlying younger Tertiary and Pleistocene sequences (Ridd 1964, 1968; Kingma 1971; Johnston 1975) — were not also once part of the allochthonous sheet (but with few Tangihua and Matakaoa-type volcanics) and have subsequently been strung out into slices elongated in the northeast direction by dextral shear on strands of the Alpine Fault system (Fig. 2E, F).

TIME OF EMPLACEMENT

The autochthonous rocks underlying the Northland Allochthon are thought to be late Eocene in Waimamaku No. 2 (Hornibrook *et al.* 1976), Upper Eocene at Ngawha (Bowen and Skinner 1972) and mid-Oligocene at Kawakawa (Hay 1960).

The strata in the Allochthon range from late Cretaceous to mid-Oligocene in Waimamaku No. 2 (Hornibrook *et al.* 1976), and from late Cretaceous to Upper Eocene in the Ngawha hole (Bowen and Skinner 1972). At outcrop the strata range from Ngaterian (mid-Cretaceous) to Waitakian (late Oligocene to Lower Miocene, Hornibrook 1974) (Arlidge 1955; Milligan 1959; Hay 1960; Kear 1964). Emplacement in Waitakian times or later is therefore indicated. We know of no evidence for more than one period of emplacement.

The stratigraphically overlying, autochthonous Waitemata Group includes units dated as Waitakian (Siltstone unit (a) of Kear 1974; Wainui Siltstone of Waterhouse 1966) and Waitakian to Otaian (Timber Bay Formation of Carter 1971; Puriri Formation, Ballance 1976b). That restricts the time of emplacement to the Waitakian, approximately at the Oligocene-Miocene boundary.

MECHANISM OF EMPLACEMENT

The lack of overturning of large blocks, and the disordered juxtaposition of blocks of differing age, rule out tectonics involving great recumbent folds. The

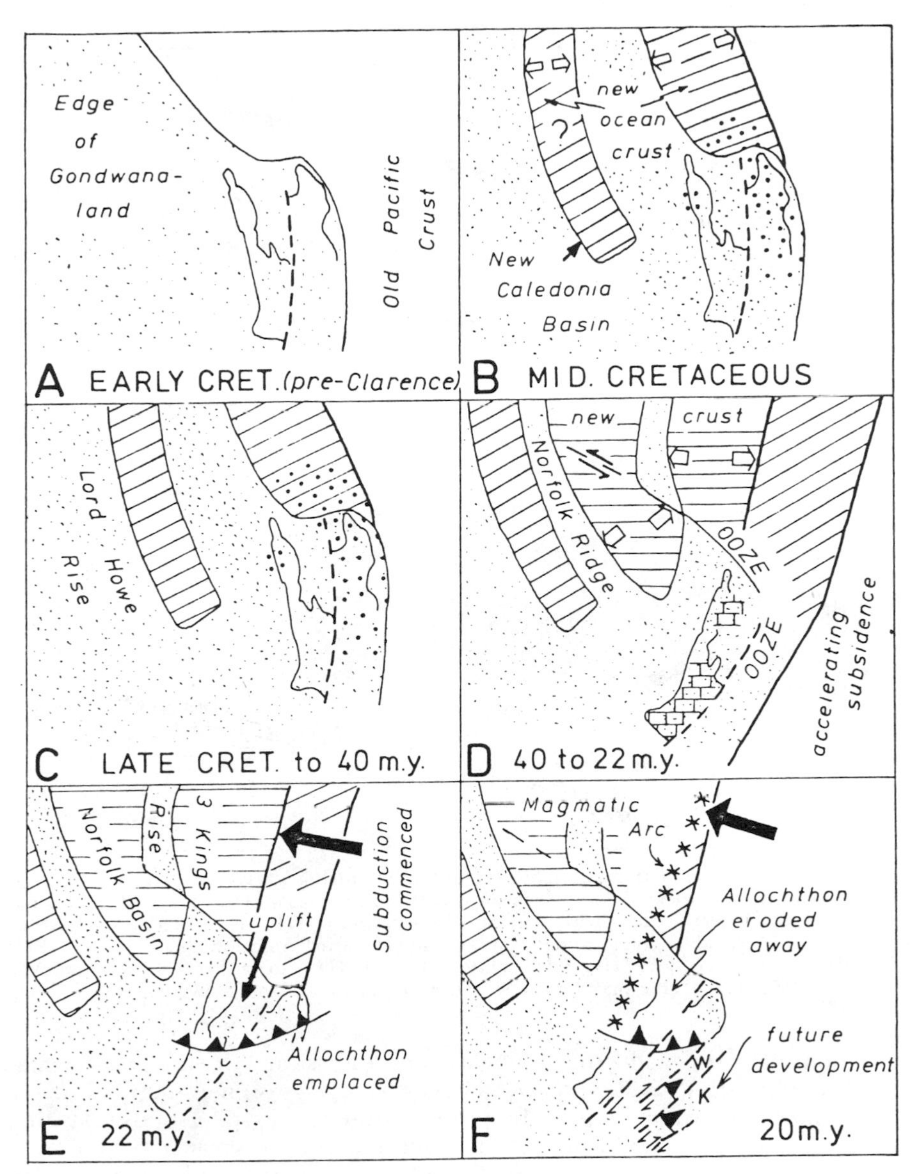

FIG. 2. — Deposition and emplacement of the Northland Allochthon, shown schematically, using the pre-Miocene reconstruction of Ballance (1976a). Light stipple indicates continental crust, heavy stipple clastic sedimentation, brick pattern shelf carbonate sedimentation. Parallel lines show newly created ocean crust; open arrows indicate inferred seafloor spreading, but do not necessarily represent the vectors of relative movement. Teeth indicate southern limit of Allochthon. Map F shows the possible transposition of the Allochthon along the Alpine Fault Zone during the Upper Cenozoic. Sites of future Wairarapa and Kaikoura areas indicated by letters W and K respectively. Scale very approximate.

widespread internal shearing and brecciation, and general lack of induration and metamorphism, suggest jostling in a near-surface situation, although these features may have been influenced by persistent high pore water pressures at depth and may not be a reliable guide to a near-surface situation.

These facts favour final emplacement by gravity sliding. The overall

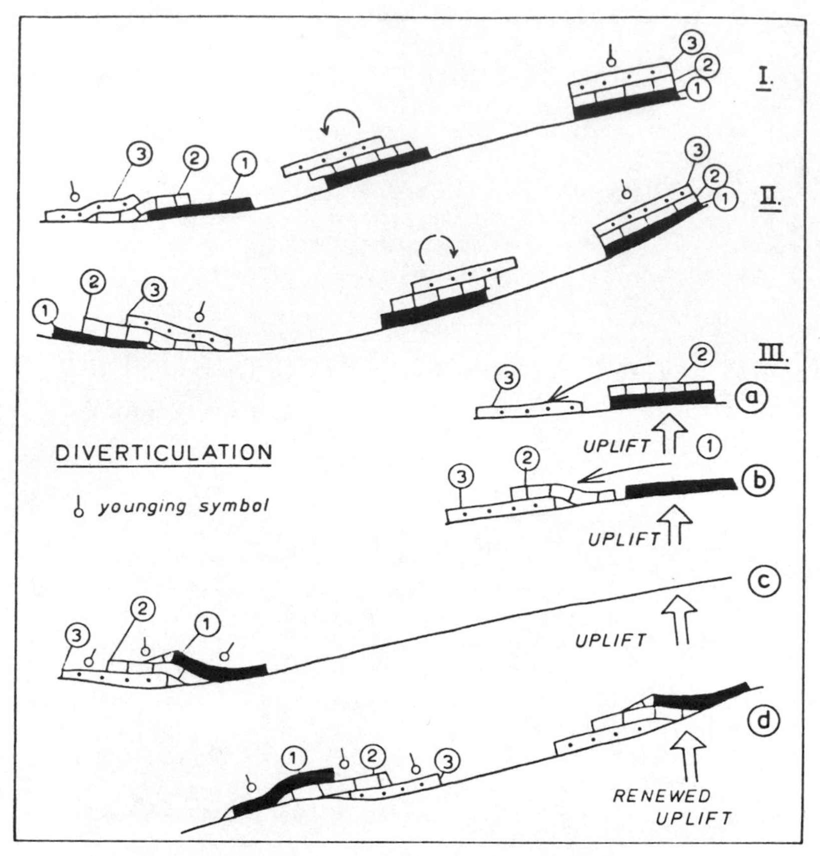

FIG. 3. — Schematic models of decollement processes during gravity sliding. 1, 2, 3 = age sequence in stratified rocks. I: "deck of cards" mechanism with forward rotation. II: "deck of cards" mechanism with backwards rotation. III: diverticulation. Note that diverticulation may also come about by extreme forward rotation in model I. IIId is a possible continuation of IIIc. In models I and IIIa to c the youngest rocks slide furthest. IIIc may be the situation of the Northland Allochthon.

distribution of ages, with the younger constituents apparently concentrated in the south, suggests derivation from the north. This assumes that all the rocks in the Allochthon were derived from one stratigraphic pile, and that the top sediment in the pile slid farthest, either in the simple shear mode of a tilted deck of cards or in a more complicated diverticulation mechanism (Lugeon 1943) which tends to invert the sequence of ages but not the direction of younging (Fig. 3).

NAME AND STATUS OF THE UNIT

Olistostrome or Mélange or Neither?

We do not think that a distinction can be made between mélange and olistostrome on the basis of internal structure, such as presence or absence of pervasive shearing, or of content, such as the presence or absence of ophiolite slabs (e.g. Hsu 1974), because in large bodies of disrupted rocks at low confining pressures convergence between features produced by gravity sliding and by more compressive tectonics can be practically complete. Furthermore, it is quite possible that the early movements which led to the formation of the Allochthon were indeed compressive, because considerable uplift had to take place before the sediments, and specially the Tangihua slab, were able to slide to their present location. Finally, it is not admissible to equate exclusively the term "olistostrome" with a "non-tectonic" and term "mélange" with a "tectonic" origin. Clearly the Northland Allochthon is a product of tectonic causes and is not just a random slope failure.

We therefore chose the name *Northland Allochthon,* in preference to the alternatives — olistostrome, mélange, argille scagliose, wild flysch, scaly clay, megabreccia, chaos-breccia — because the deposit is not really chaotic, and because the name is non-genetic and merely records the allochthonous nature of the deposits. This name had already been used by Griffiths (1975), following Temple (1972), but under considerable misconception as to the content and to the stratigraphic relationships of the rocks involved. A preliminary name, West Northland Chaos-breccia, was used by Ballance and McCarthy (1975) for what is here called Northland Allochthon. It was also called West Northland Allochthon by Ballance *et al.* (1977).

RELATION OF TANGIHUA IGNEOUS MASSES TO THE SEDIMENTS

The Tangihua blocks were interpreted as seamounts by Quennell and Hay (1964) and Farquhar (1969), but with the recognition of their allochthonous nature it is apparent that they may originally have formed a continuous sheet. In composition and layered structure they correspond to oceanic crust (Brothers 1974; Briggs and Searle 1975; Brothers and Delaloye 1976). Fossils from intercalated sediments at Cape Reinga and Doubtless Bay are Cretaceous and at Houto Jurassic (Farnell 1973; Brothers 1974; Hay 1975; Hornibrook and Hay 1978). On the fossil evidence alone, a basin model equating the Tangihuas with oceanic crust beneath the sedimentary sequence was adequate, but to accommodate younger K/Ar dates Brothers and Delaloye (1976) postulate a zone or suture of oceanic crustal generation lying adjacent to the sedimentary basin. Alternatively, the K/Ar dates may reflect a mid-Tertiary metamorphic event related to the uplift and emplacement of the Allochthon (R. N. Brothers, pers. comm. 1976). It is difficult to perceive how the Tangihua lavas could have accumulated during the Tertiary without acquiring interbedded fossiliferous Tertiary sediments. The lower Miocene volcanic grits reported by Hughes (1966) are regarded (by Brothers 1974) as part of the Waitakere Group (Ballance *et al.* 1977) lying unconformably on the Tangihua rocks.

It makes little difference to the discussion of provenance of the Allochthon which of the two models is accepted. It does, however, affect the interpretation of the tectonic status of the basin of origin.

NATURE OF THE SOURCE ROCKS

If we assume that the rocks in the Allochthon were derived from a single depositional basin, a stratigraphic compilation of that basin can be made, which is presented in Table 1. The essential features of it are: (a) at least some Tangihua basic igneous masses with interbedded sediments are the oldest rocks present, (b) the next youngest rocks, of early to late Cretaceous age, are thick, terrigenous sediments, (c) in the lowest Tertiary, terrigenous sediments are very fine grained and include probable abyssal red clays, (d) Eocene glauconitic sediments indicated renewed influx of terrigneous sediment but are interbedded with micrite and are probably redeposited (the full geotectonic significance of this change is not yet understood), (e) Oligocene sediments are predominantly argillaceous nannofossil oozes. Maximum recorded thicknesses of the sediments total several kilometres.

[*Editor's Note:* Material has been omitted at this point.]

ACKNOWLEDGMENTS

We acknowledge many helpful discussions with Auckland colleagues, and the generous provision of unpublished data by R. N. Brothers. N. de B. Hornibrook kindly kept us informed of his work on the Waimamaku wells. D. N. B. Skinner generously shared his local knowledge of Northland geology and reviewed the manuscript.

REFERENCES

ADAMS, C. J. D. 1975. New Zealand potassium-argon list — 2. *N.Z. Journal of Geology and Geophysics* 18: 443-467.

ARLIDGE, E. Z. 1955. The geology of Hukatere Peninsula, North Kaipara Harbour. Unpublished MSc thesis, University of Auckland.

BALLANCE, P. F. 1964. Streaked-out mud ripples below Miocene turbidites, Puriri Formation, New Zealand. *Journal of Sedimentary Petrology* 34: 91-101.

——— 1974. An interarc flysch basin in Northern New Zealand: Waitemata Group (Upper Oligocene — Lower Miocene). *Journal of Geology* 82: 439-471.

——— 1976a. Evolution of the Upper Cenozoic magmatic arc and plate boundary in Northern New Zealand. *Earth and Planetary Science Letters* 28: 356-370.

——— 1976b. Stratigraphy and bibliography of the Waitemata Group of Auckland, New Zealand. *N.Z. Journal of Geology and Geophysics* 19: 897-932.

BALLANCE, P. F., and MCCARTHY, J. A. 1975. Geology of Okahukura Peninsula, Kaipara Harbour, New Zealand. *N.Z. Journal of Geology and Geophysics* 18: 721-43.

BALLANCE, P. F.; HAYWARD, B. W.; WAKEFIELD, L. L. 1977. Group Nomenclature of Late Oligocene and early Miocene rocks in Auckland and Northland, New Zealand; and an Akarana Supergroup. *N.Z. Journal of Geology and Geophysics* 20: 673-686.

BERGGREN, W. A. 1972. A Cenozoic time-scale — some implications for regional geology and paleobiography. *Lethaia* 5: 195-215.

BOWEN, F. E., and SKINNER, D. N. B. 1972. Geological interpretation of Ngawha deep drillhole, Kaikohe, Northland (N15), New Zealand. *N.Z. Journal of Geology and Geophysics* 15: 129-39.

BRIGGS, R. M., and SEARLE, E. J. 1975. Tangihua volcanics in the Opouteke-Pakotai area, Northland, New Zealand. *N.Z. Journal of Geology and Geophysics* 18: 327-341.

BROTHERS, R. N. 1974. Kaikoura Orogeny in Northland, New Zealand. *N.Z. Journal of Geology and Geophysics* 17: 1-18.

BROTHERS, R. N., and DELALOYE, M. 1976. On-land ophiolites, North Auckland, New Zealand. *25th International Geological Congress Abstracts* 1 (2): 45-46.

BUNTING, F. J. L. 1970. An account of the Albany Conglomerate in the Kaukapakapa-Wainui district. Unpublished BSc hons. thesis, University of Auckland.
BURNS, R. E.; ANDREWS, J. E., *et al.* 1973. *Initial reports of the Deep Sea Drilling Project.* Washington (US Government Printing Office) 21.
CARTER, L. 1967. Geology of Puketotara Peninsula, Kaipara, Northland. Unpublished MSc thesis, University of Auckland.
——— 1969. The Mahurangi Limestone from Puketotara Peninsula, Northland, New Zealand. *N.Z. Journal of Geology and Geophysics* 12: 104-18.
——— 1971. Stratigraphy and sedimentology of the Waitemata Group, Puketotara Peninsula, Northland. *N.Z. Journal of Geology and Geophysics* 14: 169-91.
DE JONG, K. A., and SCHOLTEN, R. (Eds.). 1973. *Gravity and Tectonics.* Wiley-Interscience.
FARNELL, E. J. 1973. Geology of the Cape Reinga area, Northland. Unpublished MSc Thesis, University of Auckland.
FARQUHAR, O. C. 1969. Former Seamounts in New Zealand and the volcanoes of modern oceans. *Oceanography and Marine Biology, an Annual Review.* 7: 101-72.
GRIFFITHS, J. R. 1975. New Zealand and the Southwest Pacific margin of Gondwanaland *in* K. S. W. Campbell (Editor). *"Gondwana Geology" Third Gondwana Symposium, Canberra, Australia 1973.* Australian National University Press. 619-637.
HATHERTON, T., and SIBSON, R. H. 1970. Junction Magnetic Anomaly north of Waikato River. *N.Z. Journal of Geology and Geophysics* 13: 655-62.
HAY, R. F. 1960. The geology of Mangakahia Subdivision. *N.Z. Geological Survey Bulletin n.s. 61.*
——— 1975. Sheet N7 Doubtless Bay (1st Edition). *Geological Map of New Zealand 1: 63360. Map (1 sheet) and Notes* (24 pp.). Department of Scientific and Industrial Research, Wellington.
HAYES, D. C., and RINGIS, J. 1973. Seafloor spreading in the Tasman Sea. *Nature* 243: 454-58.
HEALY, J. 1949. The age and distribution of the coal measures in North Auckland. *Transactions of the Royal Society of N.Z.* 77: 281-8.
HILL, P. H. 1974. Taitai Series rocks at Te Kaha. Unpublished MSc thesis, Auckland University.
HORNIBROOK, N. DE B. 1974. *Globigerinoides* (Foraminifera) in the Waitakian Stage (mid Tertiary) Raglan Harbour, New Zealand. *N.Z. Journal of Geology and Geophysics* 17: 500-508.
HORNIBROOK, N. DE B.; EDWARDS, A. R.; MILDENHALL, D. C.; WEBB, P. N.; WILSON, G. J. 1976. Major displacements in Northland, New Zealand; micropaleontology and stratigraphy of Waimamaku 1 and 2 wells. *N.Z. Journal of Geology and Geophysics* 19: 233-63.
HORNIBROOK, N. DE B., and HAY, R. F. 1978. Late Cretaceous agglutinated foraminifera from sediments interbedded with the Tangihua Volcanics, Northland, New Zealand. *In* Belford, D. J., Scheibnerova, V. (Eds.). "The Crespin Volume: Essays in Honour of Irene Crespin" B.M.R. Bulletin 192: 67-71.
HSU, K. J. 1969. A preliminary analysis of the statics and kinetics of the Glarus overthrust. *Eclogae Geologicae Helveticae* 62: 143-154.
——— 1974. Melanges and their distinction from olistostromes. *In* DOTT, R. H., and SHAVER, R. H. (Eds.). *"Modern and Ancient Geosynclinal Sedimentation."* Society of Economic Paleontologists and Mineralogists, Special Publication 19: 321-333.
HUGHES, W. S. 1966. Igneous rocks from the northern Wairoa district. Unpublished MSc thesis, University of Auckland.
JOHNSTON, M. R. 1975. Sheet N159 and part Sheet N158 Tinui-Awatoitoi (1st Edition) *"Geological Map of New Zealand 1:63360". Map (1 sheet) and Notes (16 pp.).* Department of Scientific and Industrial Research, Wellington.
KARIG, D. E. 1970. Ridges and basins of the Tonga-Kermadec Island arc system. *Journal of Geophysical Research* 75: 239-254.
KATZ, H. R. 1968. Potential oil formations in New Zealand and their stratigraphic position as related to basin evolution. *N.Z. Journal of Geology and Geophysics* 11: 1077-1133.
——— 1974. Recent exploration for oil and gas. Chapter 24 *in* WILLIAMS, G. J. (Ed.). *Economic Geology of New Zealand* (2nd ed., T. J. MCKEE Memorial Volume). Australasian Institute of Mining and Metallurgy Monograph 4: 463-80.
KEAR, D. 1954. Karaka drillhole: Unpublished report New Zealand Geological Survey.
——— 1959. Geology of the Kamo Mine area. *N.Z. Journal of Geology and Geophysics* 2: 541-568.
——— 1964. Notes from the New Zealand Geological Survey — 2. Kaiaka drillhole near Kaitaia, Northland. *N.Z. Journal of Geology and Geophysics* 7: 903-10.

——— 1974. Waitemata Group near Whangarei. *Journal of the Royal Society of N.Z.* 4: 277-82.

KEAR, D., and HAY, R. F. 1961. Sheet 1 North Cape. *"Geological Map of New Zealand, 1:250,000"*. N.Z. Department of Scientific and Industrial Research, Wellington.

KEAR, D., and SCHOFIELD, J. C. 1959. Te Kuiti Group. *N.Z. Journal of Geology and Geophysics* 2: 685-717.

KEAR, D., and WATERHOUSE, B. C. 1967. Onerahi Chaos-breccia of Northland. *N.Z. Journal of Geology and Geophysics* 10: 629-46.

——— 1977. Onerahi Chaos-breccia: further thoughts. *N.Z. Journal of Geology and Geophysics* 20: 205-9.

KEHLE, R. O. 1970. Analysis of gravity sliding and orogenic translation. *Bulletin of the Geological Society of America* 81: 1641-1664.

KINGMA J. T. 1965. Sheet 6, East Cape (1st Edition) *"Geological Map of New Zealand 1:250,000."* Department of Scientific and Industrial Research, Wellington.

——— 1971. Geology of the Te Aute subdivision. *New Zealand Geological Survey Bulletin n.s. 70.*

LE COUTEUR, P. C. 1967. The geology of a region north-west of Whangaroa Harbour, Northland. Unpublished MSc thesis, University of Auckland.

LEITCH, E. C. 1970. Contributions to the geology of northernmost New Zealand: II — The stratigraphy of the North Cape district. *Transactions of the Royal Society of N.Z. (Earth Sciences)* 8: 45-68.

LEMOINE, M. 1973. About gravity gliding tectonics in the Western Alps *in* DE JONG, K. A., SCHOLTEN, R. (eds.). *Gravity and Tectonics.* John Wiley, New York.

LUGEON, M. 1943. Une nouvelle hypothèse tectonique: La diverticulation. *Bulletin Société Vaudoise des Sciences Naturelles* 62: 260-261.

MAEHL, H. W. R. 1970. Geology of the Whangaroa-Marble Bay-Te Ngaere district, Northland. Unpublished MSc thesis, University of Auckland.

MILLIGAN, E. N. 1959. The geology of North Hokianga district. Unpublished MSc thesis, University of Auckland.

MOLNAR, P.: ATWATER, T.; MAMMERICKX, J.; SMITH, S. M. 1975. Magnetic anomalies, bathymetry and tectonic evolution of the South Pacific since the Late Cretaceous. *Royal Astronomical Society Geophysical Journal* 60: 383-420.

MOORE, P. R. 1974. Cretaceous stratigraphy and structure of western Koranga Valley, Raukumara Peninsula. Unpublished MSc thesis, University of Auckland.

O'BRIEN, J. P., and RODGERS, K. A. 1973. Alpine type serpentinites from Auckland Province — II. North Auckland Serpentinites. *Journal of the Royal Society of N.Z.* 3: 263-280.

PACKHAM, G. 1975. Aspects of the geological history of the New Hebrides and South Fiji basins. *Bulletin of the Australian Society of Exploration Geophysicists* 6: 50-51

PACKHAM, G., and TERRILL, A. 1975. Submarine geology of the South Fiji Basin. *In* ANDREWS, J. G., PACKHAM, G. *et al. Initial Reports of the Deep Sea Drilling Project.* Washington (U.S. Government Printing Office) 30.

PURSER, B. H. 1961. Geology of the Port Waikato region (Onewhero sheet N51). *New Zealand Geological Survey Bulletin n.s. 69.*

QUENNELL, A. M., and HAY, R. F. 1964. Origin of the Tangihua Group of North Auckland. *N.Z. Journal of Geology and Geophysics* 7: 638-9.

REILLY, W. I. 1965, a. Sheet 1, North Cape (1st Edition). *"Gravity Map of New Zealand 1:250,000, Bouguer and Isostatic Anomalies"*. Department of Scientific and Industrial Research, Wellington.

——— 1965, b. Sheet 2a Whangarei (1st Edition). *"Gravity Map of New Zealand 1:250,000, Bouguer and Isostatic Anomalies"*. Department of Scientific and Industrial Research, Wellington.

RIDD, M. F. 1964. Succession and structural interpretation of the Whangara-Waimate Area. *N.Z. Journal of Geology and Geophysics* 7: 279-98.

——— 1968. Gravity gliding on the Raukumara Peninsula (letter). *N.Z. Journal of Geology and Geophysics* 11: 547-8.

SCHOFIELD, J. C. 1967. Sheet 3 Auckland (1st edition). *"Geological Map of New Zealand 1:250,000"*. Department of Scientific and Industrial Research, Wellington.

SEARLE, E. J. 1964. City of Volcanoes: a geology of Auckland. Paul's Book Arcade, Auckland.

SPEDEN, I. G. 1973. Distribution, stratigraphy, and stratigraphic relationships of Cretaceous Sediments, Western Raukumara Peninsula, New Zealand. *N.Z. Journal of Geology and Geophysics* 16: 243-68.

——— 1976. Geology of Mt Taitai, Tapuaeroa Valley, Raukumara Peninsula. *N.Z. Journal of Geology and Geophysics* 19: 71-119.

SPÖRLI, K. B., and ISAAC, M. 1973. Road log for the field trip Auckland-Warkworth-Batley/Bull Point (Kaipara Harbour) — Helensville-Auckland. *Conference on the New Zealand Cretaceous Abstracts.* University of Auckland, July 1973: 4-24.
SQUIRES, D. F. 1960. Relative durations of the Tertiary Series and Stages in New Zealand. *N.Z. Journal of Geology and Geophysics* 3: 137-40.
STONELEY, R. 1962. Marl diapirism near Gisborne, New Zealand. *N.Z. Journal of Geology and Geophysics* 5: 630-41.
———— 1968. A lower Tertiary decollement on the east coast, North Island, New Zealand. *N.Z. Journal of Geology and Geophysics* 11: 128-56.
TEMPLE, P. G. 1972. Structure of island arcs bounding the Australian Continental Plate. *The Australain Petroleum Exploration Association Journal* 2: 74-80.
THOMPSON, B. N. 1961. Sheet 2A Whangarei. " *Geological Map of New Zealand 1:250,000* ". Department of Scientific and Industrial Research, Wellington.
WAKEFIELD, L. L. 1976. Lower Miocene paleogeography and molluscan taxonomy of Northland, New Zealand. Unpublished PhD thesis, University of Auckland.
WATERHOUSE, B. C. 1966. Mid Tertiary stratigraphy of the Silverdale district. *N.Z. Journal of Geology and Geophysics* 9: 153-72.
———— 1968. Upper Jurassic "sandwacke" at Waiwera, Auckland, and regional stratigraphic implications. *N.Z. Journal of Geology and Geophysics* 11: 248-52.
WOODWARD, D. J. 1970. Interpretation of gravity and magnetic Anomalies at Hokianga, Northland. *N.Z. Journal of Geology and Geophysics* 13: 364-69.

DR P. F. BALLANCE
Dr K. B. SPÖRLI
Department of Geology
University of Auckland
Private Bag
Auckland
New Zealand

Part VII

MÉLANGES OF THE USSR

Editor's Comments on Papers 29 and 30

29 KNIPPER
Constitution and Age of Serpentinite Mélange in the Lesser Caucasus

30 KNIPPER
Development of Serpentinite Mélange in the Lesser Caucasus

Mélanges are recorded from several widely separated fold mountain ranges in the USSR—for example, the Caucasus, Tien-Shan, Polar Urals, and Pacific Ranges. The two papers by Knipper (Papers 29 and 30) have been selected because they encapsulate the essentials of the dominant Russian thinking and methodology (in conjunction with Paper 7 by Leonov). Knipper, in Paper 29, described a serpentinite mélange from the Sevan-Ankera zone in the lesser Caucasus, a mélange with "a cement of serpentinite schists and breccias." Blocks within it are up to 2 km in maximum dimension and range down to microscopic dimensions. They are ellipsoidal, near spherical, brick shaped, tabular of flattened, and characteristically display smoothed over, sheared surfaces. All have thin serpentinite jackets, with slickensides intersecting on the polished surfaces. Inclusions of other rock types stand out in relief and are lenticular in plan, wedge shaped in cross section. They include ophiolitic gabbro and volcano-sedimentary lithologies, as well as clastic sediments of Senonian age—sandstones, limestones, and conglomerates. There are also much older (?Paleozoic) bituminous limestones, organic limestones (?Jurassic), and pink, sandy Upper Triassic limestones, all otherwise unknown from the area. Minor metamorphic rock enclaves include marble, schists, amphibolite, and rodingite.

Knipper entertained the possibility that the tectonic churning up of the ophiolite sequence of volcano-sedimentary rocks and related intrusions may not be as intense as the observed chaotization suggests. The intrusive rock components display ordered igneous differentiation trends in the layering over substantial sections of the larger enclaves. Knipper also described argillite-matrixed olistostrome members associated with the serpentinite mélange.

The metamorphic rock components are believed to be exotic crystalline inclusions, tectonically introduced. The serpentinite matrix is ubiquitous, but nowhere does it display thermal contacts. The essential originating process is believed to have been crushing and tectonic mixing of ophiolitic rocks with sedimentary and metamorphic rocks. The gigantic tectonic breccia or mélange was a product of strong Neocomian (Austrian phase) movements, but there were later protrusions of mélange during the late Cretaceous and early Eocene tectonic events, further complicating it. The peculiar protrusion-forming capability of mélange is attributed by Knipper to the unusual behavior of the serpentinite *mélange cement* under strong compression.

Knipper (Paper 30) described in detail the olistostrome intervals that differ from the serpentinite mélange in having a terrigenous, not a serpentinite, matrix and also in not being intensely deformed but displaying graded stratification and evidence of cosedimentational flowage of material. The serpentinite mélange contains Upper Senonian and Paleogene blocks, whereas the olistostrome complex contains only pre–Upper Senonian blocks.

Knipper related the Sevan-Ankera zone to a continent/ocean interface to the north of Paleo-Tethys and erected a model involving northward overthrusting of the oceanic crust. A concluding discussion of the term *mélange* favors restriction to tectonically originated chaotic assemblages. Knipper rejects any possibility of the serpentinite mélange being of high-level, superficial, submarine gravity slide origin (i.e., an olistostrome).

Gasanov (1974) described a similar serpentinite mélange from the Shakh-Dag Range, lesser Caucasus, advancing arguments based on radiometric age determinations on cross-cutting plagioclase granites for a pre-late Jurassic age for the serpentinite matrix—that is, older than the blocks enclosed by it, which are late Jurassic–Neocomian volcanics and cherts. He regarded the mélange as of tectonic origin, a product of disintegration of an ophiolite complex and associated cherty sediments, and distinguished what he called *monomict* and *polymict* mélange, though his terminology is puzzling. Monomict mélange, the earlier formed, has less included material as it preceded the effusive-radiolarian chert emergence. Polymict mélange contains much more block material because it postdated further tectonic movements and the emergence of volcanics and radiolarian cherts.

Sokolov (1973) described a colored mélange of tectonic origin from the Amasiya region, lesser Caucasus. He observed that the ultrabasics are not intrusive bodies that cut and have metamorphosed the enclosed rocks but rather tectonically disrupted blocks and massifs that intruded and mixed with the enclosed rocks in the cold state. Garnet eclogite metamorphic rocks are recorded, but there is controversy whether they are older (?Precambrian) exotics or products of Alpine metamorphism of gabbro. The mélange was produced by a series of tectonic reactivations over a long period of time, and the main pre-Senonian ophiolitic mélange component suffered an erosional phase during the Senonian followed by repeated tectonic activation and capture of Upper Cretaceous and Paleogene rock components in the late Eocene movements.

Sabdyushev and Usmanov (1971) described a belt of mélange of tectonic origin up to 5 km wide and 2 km thick from Tamdytau in the Southern Tien-Shan, Western Uzbekistan. It consists of chert, dolomite, limestone, plagioclase granite, albitophyre, gabbro, and sandstone (possibly also serpentinized ultrabasics), set in a matrix of thinly bedded sandstone and shale. This mélange, of Carboniferous age, is attributed to late Carboniferous compression at an ancient ocean/continent suture. Another mélange, also of tectonic origin, is described from nearby; however, this is really a broken formation consisting of an irregularly bedded shale matrix enclosing disintegrated sandstone beds. Mélange of this type is produced from both Silurian and middle to Upper Carboniferous sedimentary sequences and is not obviously a mélange because it lacks conspicuous relief contrast.

Kurenkov (1978) described both serpentinite mélange and olistostrome complexes, also of Carboniferous age, from the foothills of the Alai Range, Southern Tien-Shan. The Kan serpentinite mélange forms a thick northward-overthurst tectonic plate, and an olistostrome sequence has been overthrust onto this in turn. The olistostromes are well stratified bodies up to 55 m in individual thickness and quite distinct from the mélange, though they also contain abundant ophiolitic fragments. Another mélange occurrence is recorded 150 km away from the Kan mélange—the East Karachatyr mélange—and this extends for 50 km in a belt up to 10 km wide. It consists of large blocks separated by narrow, wedging out zones of intensely tectonized serpentinite. Blocks are mostly equant, polygonal, and up to several kilometers in maximum dimension. They consist of pillowed and variolitic spilite diabase, quartzose or tuffaceous sandstone, gray or pink limestone, green and gray chert, andesitic porphyry, and siliceous schists. From another part of the same range, olistostromes with a sand-clay matrix are recorded, and this mélange includes blocks of such olistostrome material, being of subsequent formation. The olistostromes are related to an earlier phase of regional constriction, and the mélange is related to a later phase when tectonic compressional effects were dominant.

Dergunov et al. (1975) described a serpentinite mélange from the Ray-Iz massif in the Polar Urals, an unusual mélange of much smaller block scale than other serpentinite mélanges of the USSR. The range is from several centimeters to several meters in maximum dimension. Boulders and blocks of various rock types are enclosed in a kneaded serpentinite matrix. This mélange occupies the contact zone between the dominantly ultramafic complex forming the greater part of the range and Upper Devonian to Carboniferous cherty, carbonaceous, and phyllitic shales, siltstones, sandstones, and limestone lenses. The zone is up to 350 m thick, and the enclosed angular to rounded blocks include harzburgite, dunite, chromitite, altered basalt, basaltic tuff, diabase, carbonaceous shale, quartzite, jadeitite, and other metamorphic rocks (assemblages of chlorite, amphibole, pyroxene, olivine, and garnet, also rodingite *sensu lato*). Blueschists carrying glaucophane are in contact with the mélange zone. Both these and the mélange occupy the base of a great allochthonous sheet of ultramafic rocks. The formation of the mélange must at least have partly postdated the blueschist metamorphism, for the blueschists occur as enclaves in the mélange. The authors attribute this mélange to tectonic processes at considerable crustal depth, beneath a major thrust sheet along a paleo-ocean margin suture.

Aleksandrov (1973) described a serpentinite mélange zone up to 14 km wide and 50 km long from the Koryak Highlands (Chirynay River) in the Pacific Ranges of the USSR. The matrix is intensely sheared serpentinite. Blocks in it include pyroxenite and interbanded pyroxenite and dunite, unaffected by serpentinization and up to 150 m diameter. The shape is elongated, ellipsoidal, or flattened, with sharp, sheared, and slickensided contacts with the serpentinite. A few very large ultrabasic blocks have areas up to 25 square km. Also present are blocks of jasper, chert, graywacke, basic-to-acid lava, tuff, gabbro, granodiorite, marble, sandstone, and conglomerate—also lawsonite-glaucophane schists (blueschists) and mercury-ore-bearing listvenites. The age of the ophiolite complex is unknown, but suggested ranges are from middle Paleozoic to late Cretaceous. Tectonic mixing and squeezing of rocks of diverse ages in a deep seated overthrust zone is suggested, the main stage of mélange development being pre-Jurassic, though further complications were later imposed on the structure. Aleksandrov mentions four other mélange developments in this region.

An extraordinary feature of Aleksandrov's paper is the mention of Bogdanov's (1969) desire to rename this the Franciscan series.

Contrary to all tenets of international stratigraphy, such a course would only plunge the mélange problem into an even deeper morass of confusion. Surprisingly, Aleksandrov favored such a change.

Raznitsin (1978) described mélange from East Sakhalin, forming a zone bordering the Sea of Okhotsk in the Pacific Margin region of the USSR. Two narrow zones extend for 40 km blocks, commonly sheathed by slickensided serpentinite jackets, and include spilite, diabase, quartzite, and brecciated basalt. They are up to 5 X 20 m dimension. Lherzolite, websterite, enstatite, wehrlite, troctolite, and chromitite are recorded, also massive serpentinite, banded ultrabasics and gabbro, and granite. The blocks mostly come from a characteristic ophiolite suite. However, metamorphic rock inclusions also occur—for example, amphibolite, prehnite schist, listvenite, and rodingite. Locally, this mélange belt is up to 800 m wide. Cataclasis is not marked in the mélange zones, though in the Berezovka zone there is intense mylonization at the base of the zone and rather more cataclasis in the zone overall. The ophiolites of the mélange in the Berezovka zone have been radiometrically dated as late Jurassic-late Cretaceous, and the mélange is believed to be the product of late Cretaceous tectonic movements.

Olistostromes associated with these mélange zones include fragments of diabase, green and red radiolarite, limestone, and sandstone, set in a matrix of sheared and churned up silty argillite. The olistoliths are up to 2 m in diameter, rounded or ellipsoidal. Numerous faults and slickensides pervade the entire olistostrome mass. Amphibolite and serpentinite blocks are possible inclusions also, but field relationships are not clear. There are other types of olistostrome, one with sedimentary blocks dominant—of jasper—and another containing marble, jasper, sandstone, and pillow lava blocks. Olistostromes are recorded up to 2,000 m thick from this area, much thicker than the classic Apennine olistostromes (see Paper 5). The matrix is dated on fossil evidence as late Cretaceous-Paleocene (Danian). The olistostrome sequence is related to the frontal part of a tectonic thrust sheet, moving forward during the late Cretaceous event that produced the serpentinite mélange, considered to be of purely tectonic origin.

REFERENCES

Aleksandrov, A. A., 1973, Serpentinite Mélange in the Upper Course of the Chirynay River, Koryak Highlands, *Geotectonics* **4:**232-236.

Bogdanov, N. A., 1969, Thalassogeosynclines in the Circum-Pacific Belt, *Geotectonics* **3.**

Dergunov, A. B., A. P. Kazak, and Yu. Ye. Moldvantsev, 1975, Serpentinite Mélange and the Structural Position of the Ray-Iz Ultrabasic Massif, *Geotectonics* **1:**15–18.
Gasanov, T. A., 1974, Mélange in the Shakh-Dag Range, Lesser Caucasus, *Geotectonics* **5:**306–309.
Kurenkov, S. A., 1978, The Serpentinite Mélange and Olistostrome Complexes of the Alai Range, Southern Tien-Shan, *Geotectonics* **5:**376–383.
Raznitsin, Yu. S., 1978, Serpentine Mélange and Olistostrome in the Southeastern Part of the East Sakhalin Mountains, *Geotectonics* **5:**333–342.
Sabdyushev, Sh. Sh., and R. R. Usmanov, 1971, Tectonic Sheets, Mélange, and the Ancient Ocean Crust in Tamdytau (Western Uzbekistan), *Geotectonics* **5:**283–287.
Sokolov, S. D., 1973, Tectonic Mélange in the Amasiya Region, Lesser Caucasus, *Geotectonics* **1:**35–38.

29

Reprinted from *Geotectonics* **5**:275-282 (1971)

CONSTITUTION AND AGE OF SERPENTINITE MÉLANGE IN THE LESSER CAUCASUS

A. L. KNIPPER

Serpentinite mélange of the Sevan-Akera zone of the Lesser Caucasus is the result of tectonic crushing and mixing of non-contemporaneous rocks. The mélange is cemented by serpentinites containing blocks of a pre-Albian ophiolite complex (assorted gabbro rocks, gabbro-amphibolites, basic volcanics, radiolarian cherts), including the Precambrian (or Lower Paleozoic) crystalline schists and also tectonic blocks of various sedimentary rocks: Paleozoic (?), Triassic, Jurassic, Cretaceous, and Paleogene. Although the mélange was formed at the end of the Neocomian (Austrian tectonic phase), its structure and constitution was further complicated in connection with the subsequent forward movements of the Lower Senonian and post-Upper Eocene.

It was quite a surprise for Soviet geologists, participants in the most recent International Colloquium on the Alpine Belt, 1967, to discover after a field trip to Iran and Turkey that the unusual tectonic mélange identified as the "Ankara mélange" by Bailey and McCallien (1956) in the Turkish sector of the Alpine belt of Eurasia is extensively developed in the Soviet Union, at least in the Lesser Caucasus. Regardless of the concept of its origin, the unusual character of this formation calls for a thorough reexamination of many conventional views of the stratigraphy, evolution, and tectonics of the Lesser Caucasus — a task accomplished in part by V. Ye. Khain (1968, 1969) and A. V. Peyve (1969).

The peculiar array of rocks, their lateral and vertical relationship, and of conditions of occurrence — all mark the mélange as a unique geologic formation warranting a most careful study.

This article is an attempt to describe the constitution of the ophiolite complex* of Armenia and Azerbaydzhan, and to date it, on the basis of field data we have gathered during our 1963-1969 work in the Lesser Caucasus. This point is of primary importance in clearing up the meaning of the term, "mélange", and in understanding the history of the origin and development of this formation. Some of our previous conclusions are set forth in our earlier articles and in that of Peyve (1969).

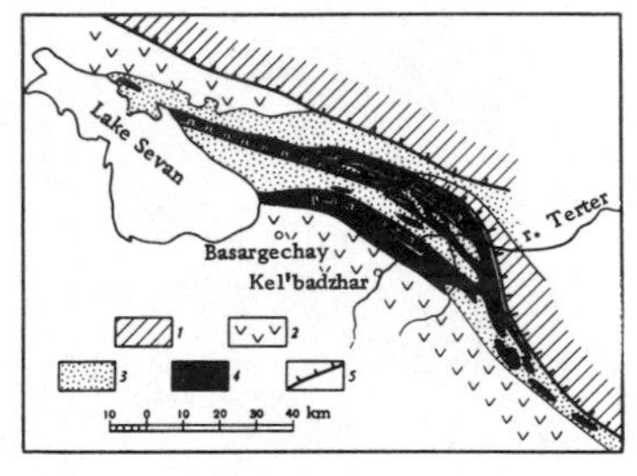

Fig. 1. Index map of ophiolite outcrops in the Sevan-Akera zone.

1) Jurassic basic volcanic rocks of the Somkhet-Karabakh zone. Sevan-Akera zone: 2) Paleogene volcanic rocks; 3) Cretaceous-Paleogene sedimentary rocks; 4) ophiolites; 5) thrusts.

CONSTITUTION OF THE OPHIOLITE COMPLEX OF THE SEVAN-AKERA ZONE

Outcrops of the ophiolite series have long since been known in the Lesser Caucasus. Their petrography, stratigraphy, and tectonics are adequately described by S. B. Abovyan, Sh. A. Azizbekov, A. A. Cabrielan, V. L. Yegoyan, M. A. Kashkay, L. N. Leont'yev, K. N. Paffengol'ts, V. Ye. Khain, E. Sh. Shikhalibeyli, and many others. All geologists group this formation in two large belts. The northern, Sevan-Akera, belt (Shikhalibeyli, 1964) extends southeast of the northeast shore of Lake Sevan, as far as the middle course of the Akera River (Fig. 1). Its western continuation is represented by ophiolite outcrops in the Leninakan area. The southern belt, known as the Vedinsk (Yegoyan, 1953), is considerably less extensive (in the modern structure); it is located on the left bank of the Araks, southwest and southeast of Yerevan.

All the students of these two belts describe the conventional geosynclinal ophiolite complexes of basic effusives and radiolarian cherts cut by basic and ultrabasic intrusions. The age of this series and its components is still controversial, interpreted differently by different groups of investigators.

All mentions of the presence of ophiolites in the Sevan-Akera and Vedinsk belts of the Lesser Caucasus are very recent (Khain, 1968, 1969; Peyve, 1969; Lomize, 1970). We now consider the distinctive features of Lesser Caucasian ophiolite complexes, with emphasis on the Sevan-Akera zone. The mélange of the Vedinsk belt was recently described by Lomize (1970).

All outcrops of the ophiolite series of the Sevan-Akera zone lie within the so-called Sarybaba synclinorium (Shikhalibeyli, 1964) and its northwestern extension in Armenia. The ophiolite outcrops appear in the south, in the area of the Lysogorsk Pass,

---

*In our opinion, the term, "formation",** is in no way applicable to the group of rocks usually included by geologists in the ophiolite formation. Our views on this subject have been adequately stated (Knipper, 1968, 1970). Therefore, in this exposition, we speak of the "ophiolite complex", "ophiolite series", or "ophiolite association".

**T. N.: The Russian term, "formation", as defined by N. S. Shatskiy, means "a geologic body represented by a complex of genetic depositional types, closely related paragenetically and formed in the same tectonic and climatic environment".

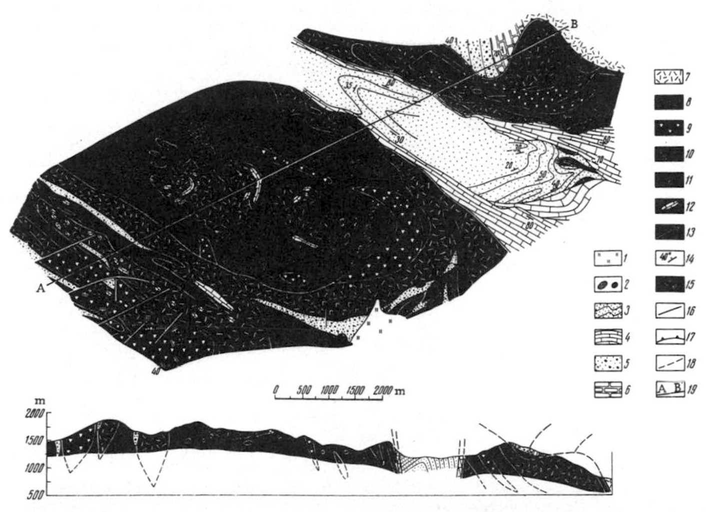

Fig. 2. Geologic map of the Terter-Levchay watershed between the villages of Cheykend and Kamyshly:

1) Quaternary deposits, undifferentiated; 2) Tertiary andesite necks; 3) Paleogene flysch; 4) Upper Senonian limestones; 5) Albian-Cenomanian terrigenous section; 6) Tithonian to Valanginian limestones and cherty shales; 7) Upper Jurassic volcanic section, basic in composition. Ophiolite complex: 8) volcanic-sedimentary section; 9) gabbro-amphibolite complex; 10) same, in fine blocks amidst the serpentinites; 11) serpentinites; 12) limestone lenses in the volcanic-sedimentary rocks, and limestone blocks in the serpentinites; 13) listwänites; 14) dip of the beds; 15) dip of the gneissoid banding within the gabbro-amphibolite complex; 16) normal and reverse faults; 17) thrusts; 18) probable boundaries; 19) line of the profile.

and are traceable in a single zone about as far as the Terter River valley, where they appear to split up into two parallel branches separated by the area of Upper Senonian limestones and Paleogene flysch (Fig. 1). The northern branch passes across the Lev village area toward the northern foothills of the Shakhdag range. Its northwestern extension is the Dzhil-Satanakhach ultrabasic massif (Abovyan, 1966) and the Shordzha and Tokhludzha serpentinites. The southern branch is traceable across the Geydara area and Zodsk Pass, as far as the village of Shishkay where the ophiolite outcrops dip under the waters of Lake Sevan.

Turning now to the constitution of ophiolites northwest of the Terter River, we see that it contains here, as in many places of the Alpine zone of Eurasia, a great variety of rocks different in composition and origin, with an ultrabasic cement.

Predominant among these rocks are serpentinites formed mainly from peridotites. Considerably less common are serpentinized dunites and slightly altered pyroxenites (Azizbekov, et al., 1950; Abovyan, 1966). These rocks, bottle- to salad-green, locally almost black, are quasi-conglomeratic: assorted blocks of relatively slightly serpentinized ultrabasic rocks in a "cement" of serpentinite schists and breccias. The blocks vary in size from microscopic to 1-2 km across. Their form is just as variable: ellipsoidal, brick-shaped, tabular, and shaped like a pressing iron. Their faces always are smoothed over and slick. Nearly spherical inclusions are fairly common. They all are enclosed in thin serpentinite jackets whose polished surfaces exhibit numerous slickensides intersecting at different angles.

The relationship between the massive slightly serpentinized ultrabasic and serpentinite schists and mylonites varies greatly from place to place. Massive harzburgites are predominant in some localities (Geydara; northwestern part of the Dzhil-Satanakhach massif), with the serpentinite schists and mylonites decidedly subordinate, forming a sort of fringe about the huge blocks; they separate the blocks or else mark the zone of numerous fractures within the massive peridotites.

This serpentinite mass encloses rocks of most diverse composition and origin. These rocks, usually harder than the serpentinites, stand out in relief, in isolated knolls and pointed knobs. Rounded to ellipsoidal, they locally form sizable massifs, 1-2 km wide and extending for 10-15 km.

In all instances, these large massifs consist of the ophiolitic rocks: gabbros and volcanic-sedimentary formations. They usually are lenticular in plan and wedge-shaped in cross section (Fig. 2).

The less common forms are saucer-shaped; their contacts with the serpentinites are very intricate in plan. The minor blocks, measuring one to 300-400 m^3 show a great variety of boundaries, in plan: circular, ellipsoidal, tabular, etc. Their form is in no way different from that of the slightly serpentinized ultrabasic blocks described before. Their surface is either lined with a highly polished serpentinite sheath or marked by numerous striae, every which way. These inclusions are cut by numerous fractures, with some of them showing evidence of serpentinite mylonite being injected into the block.

The most common components of the blocks are various gabbro rocks, diabases, basic effusives and their tuffs, and cherty sedimentary rocks, including red jaspers. They all are randomly scattered in the serpentinites, so that any member of the ophiolite group may be associated with others in a variety of combinations

(Fig. 2). The red jaspers and cherty tuffs of the tectonic blocks form a complex system of folds threaded through with numerous quartz and carbonate veinlets which never continue in serpentinites outside the blocks. In some of the blocks, the porphyrites and gabbro rocks are exceedingly strongly brecciated, locally appearing to consist of sharp fragments closely ground against each other.

At the same time, the volcanic-sedimentary and gabbro rocks may occur in large massifs or lenses, 10-15 km long by 1-2 km wide. Such are the Levchay and Chaykend gabbro-amphibolite massifs and the tectonic lenses of porphyrites in the upper reaches of the Levchay and Shamkirchay. These gabbro rocks, described by Azizbekov (1950), are identical with those from the Dzhil-Satanakhach massif (Abovyan, 1966). They often show conspicuous banding, particularly so in several massifs in Azerbaydzhan and Armenia. The marked change in the trend of banding within the massif locally suggests that the relatively large massifs of these rocks consists of blocks closely ground into each other (Fig. 2). At the same time, gabbros in some of the segments and zones present a continuous body of tectonic breccia dissected by feldspar veinlets of various thicknesses. These veinlets have not been observed either in the serpentinites or in the volcanic-sedimentary rocks. They always end at the faces of the gabbro block. The interior of the large volcanic-sedimentary blocks is, on the whole, not as intensively reworked as that of the gabbro massifs. Their individual sedimentary horizons (tuffs, lenses of reef limestones) are often discernible, thus allowing an interpretation of the constitution of these tectonic wedges.

The ophiolite rocks, especially the serpentinites, contain inclusions of most diverse sedimentary rocks unrelated to the ophiolite series. These are Lower Senonian conglomerates and sandstones (left bank of the Agkayachay), Upper Senonian limestones (Kamyshla village, right bank of the Shympyrtdar). All these tectonic inclusions usually occur in marginal parts of the ophiolite areas, near the Upper Cretaceous and Paleogene outcrops. Their age is readily determined from the fauna or from the perfect lithological similarity with the adjacent sections, faunally dated.

At the same time, the serpentinites also contain blocks of sedimentary rocks unknown from the Sevan-Amera zone. These are, first of all, gray to black bituminous limestones reminiscent externally of Paleozoic formations in the Nakhichevan ASSR, also assorted organic limestones, possibly Upper Jurassic (Khain, et al., 1949), and Upper Triassic pink sandy limestones (Solovkin, 1950). There are few of these inclusions.

Metamorphic rocks, although present everywhere in the ophiolite complex of the Lesser Caucasus, also are subordinate. These are sugar-white and lenticularly stratified marbles. The second variety often contains intercalations of green quartz-chlorite-carbonate schists. Also very common are tectonic xenoliths of dark-green, banded amphibolites, easily mapped amidst serpentinites on the right bank of the Agmayachay River. Blocks of assorted crystalline schists are present everywhere, with quartz-sericite- and chlorite-sericite green schists conspicuous among them.

A place of their own among these metamorphic formations belongs to rocks usually identified as the gabbro-amphibolite complex (Azizbekov, 1950). These are metamorphosed basic effusives most closely associated with banded gabbro complex — to such an extent that they can not always be told apart, even in detailed mapping. These rocks are exposed north of the village of Kylchaly and in the Dzhanakhmed village area. Most often these are tough dark amphibolites (locally with garnet) gradually changing to the typical banded gabbro by way of gabbro-amphibolites. The relatively thick members of amphibolite rocks contain lenses of crystalline schists which in turn include marble lenses.

Another special group of metamorphic rocks comprises the numerous bodies of white rodingites, the products of metasomatism of the diabase and gabbro-diabase dikes in the course of serpentinization of the ultrabasic rocks (Coleman, 1963; Dal Piaz, 1969). These dikes never continue beyond the serpentinite bodies. They usually show a strong tectonic effect: breaks, boudinage, and intricate folding.

The material presented leaves no doubt about a tectonic reworking of all rocks of the ophiolite complex of the Sevan-Akera zone. The only question is the extent to which these movements have masked the original relationship between the several members of the ophiolite association. Perhaps all that tectonic churning was not as strong as it appears, what with the described picture perhaps brought about by consecutive intrusions of ultrabasic and gabbro rocks into an effusive-sedimentary complex, and with all the inclusions in the serpentinite being simply xenoliths. The answer to this question hinges on the age of the ophiolite series as a whole, and of the serpentinites and gabbro rocks in particular, as will be presently explained. For the time being, we consider another point, also quite important for an understanding of the original relationship between the members of the ophiolite complex.

The point is that the present relationship between rocks of the ophiolite complex of Armenia and Azerbaydzhan sets it apart from the ophiolite section of the Ligurian coast in Italy, and the Vurinos massif. These two sections may be regarded as standard, following the work of Passerini (1965) and Brunn (1960); their ophiolites are known to exhibit a definite sequence. These sections recently were referred to by Peyve (1969) and this author (1970), both of whom visited the Ligurian exposures and were convinced of the validity of Passerini's views.

The ophiolites of the Ligurian coast form a stratified section with ultrabasic rocks at the base. The middle part is made up of various gabbro rocks, with predominantly basic assorted volcanics higher up. The section is capped by siliceous rocks, including radiolarian ones.

In any interpretation of the origin of this section — be it the result of magmatic differentiation (Brunn, 1960) or a block of the oceanic basement with its volcanic-sedimentary cover (Peyve, 1969; Knipper, 1970) — the ultrabasics are the oldest rocks, and the radiolarian rocks the youngest. This picture, conveyed by the most complete sections, is exactly opposite to that inferred for the Sevan-Akera zone on the assumption of an intrusive origin for the ultrabasics and gabbro. In this interpretation, the rock sequence in the ophiolite series of the Lesser Caucasus should be as follows (from older to younger): basic effusives → radiolarian rocks → ultrabasic intrusions → gabbro intrusions. Quite naturally, this suggests some special conditions of formation of the ophiolites in Armenia and Azerbaydzhan. That, however, is not the case.

The fact is that some of the large blocks of ophiolite rocks exhibit a fairly distinct sequence of their components. That was first established by Abovyan (1961) for ultrabasic and gabbro rocks on the northeast shores of Lake Sevan, where their section is quite complete: assorted ultrabasic rocks at the base gradually changing upward to various gabbro rocks, by way of an alternation of troctolites and anorthosites. These are followed by diorites and even plagioclase granites.

Likewise, Shikhalibeyli (1964) noted some time ago the presence of a definitive sequence for the effusive-sedimentary section in the upper reaches of the Terter. Here, the lower part of the section is made up mainly of effusive rocks, while the upper part is represented by shales and siliceous rocks.

During our work in the Sevan-Akera zone, we have not seen any single complete section comparable to those in Italy and Greece. The contacts of the ultrabasic and gabbro complexes with the effusive-sedimentary series are always tectonic. On purely structural considerations (frequent proximity of the gabbro-amphibolite complex and the effusive-sedimentary series), and judging from the startling similarity of this ophiolite complex to the rocks of the Ligurian coast and those of the Vurinos massif, and other sections (Knipper, 1970), we may assume that the ophiolite series of the Sevan-Akera zone formerly occurred in the same sequence as in Italy and Greece.

If this is true, the presently observed relationship of rocks within the ophiolite complex could have originated only as the result of very strong tectonic activity.

In concluding our description of the ophiolite series from the Sevan-Amera zone, Lesser Caucasus, we note that the material presented above suggests that the present-day ophiolite outcrops within the Sevan-Akera zone consititute a sort of tectonic hodgepodge of rocks of most diverse ages and origins. These formations are quite identical with the mélange described from many regions of the Alpine zone, and should be regarded as such. Any other name for them (for example, "ophiolite formation") would only render ambiguous its constitution and origin.

THE AGE OF COMPONENT ROCKS OF THE MÉLANGE

A prerequisite for a correct understanding of the origin and development of a mélange is the dating of its components. We have already touched upon the age of some of the sedimentary components of the mélange in question. The most important, and most difficult, problem is the age of the ophiolite complex, these rocks being the principal and essential components of the mélange.

Author	Age							
	Upper Jurassic	Neocomian	Albian	Cenomanian	Turonian	Lower Senonian	Upper Senonian	Upper Eocene
K.N. Paffengol'ts (1941)					----			——
V.P. Rentgarten (1959)	----					----		——
L.N. Leont'yev and V.Ye. Khain (1949)						——		——
E.Sh. Shikalibeyli et al. (1964)						----		
S.B. Abovyan (1961)								——
S.A. Palandzhyan (1965)	——							
L.S. Melikyan et al. (1967)	----							
G.S. Arutyunyan (1967)					—— ----			
A.L. Knipper ($1965_{1,2}$)	——	——	——					

Note: Age of rocks of the ophiolite association in the Sevan-Akera zone, according to different authors. Solid line — serpentinites and gabbros; dashed line -- volcanic-sedimentary series.

In the Table are listed the views of a number of geologists who have done field work in the Sevan-Akera zone and whose dating of the several rock components of the ophiolite formation constitute the basis of many geological concepts. The advocates of the Upper Eocene age for the ultrabasic and gabbro rocks cite evidence of ultrabasic intrusions in the faunally dated Upper Eocene deposits in the area of villages Shorzha and Dzhil. The followers of Leont'yev, Khain, and Shikhalibeyli emphasize the transgressive position of basal Campanian horizons on the serpentinites and volcanic rocks in many sections of the Sevan-Akera zone. The volcanic series itself is dated from its occurrence above the faunally dated Cenomanian deposits. A group of Armenian geologists (Palandzhyan, 1965; Melikyan, et al., 1967) base their conclusions on the presence of an immense number of ophiolite fragments in the Lower Senonian conglomerates which overlie an eroded gabbro-amphibolite surface. My own opinion is that the serpentinites and gabbro-amphibolites are pre-Cenomanian, because their fragments are present in the Cenomanian conglomerates (Knipper, $1965_{1,2}$). The same conglomerates are known to contain fragments of red jaspers and porphyrites.

Geologists try several ways to get out of this maze of contradictory data. Some of them recognize three phases of ultrabasic and gabbro magma intrusions (Aslanyan, 1958). Others postulate a single phase of ultrabasic emplacement (pre-Cenomanian or Upper Jurassic) and invoke the protrusive hypothesis in explaining the intrusions into the younger deposits (Knipper and Kostanyan, 1964; Knipper, 1965_1). Still others concede the existence of another ophilite complex (Upper Jurassic, according to Rentgarten, 1959; Precambrian, according to Shikhalibeyli, 1967) in addition to the Lower Senonian. It is the products of erosion of that other complex that fill up the described clastic intervals (Shikhalibeyli, 1967).

We now consider the geologic facts which indicate unambiguously a pre-Cenomanian (Albian-Cenomanian?) age for the serpentinites, gabbro rocks, and the volcanic-sedimentary series. The following is a description of sections where, according to Shikhalibeyli (1964), volcanic-sedimentary rocks unquestionably overlie the Cenomanian deposits.*

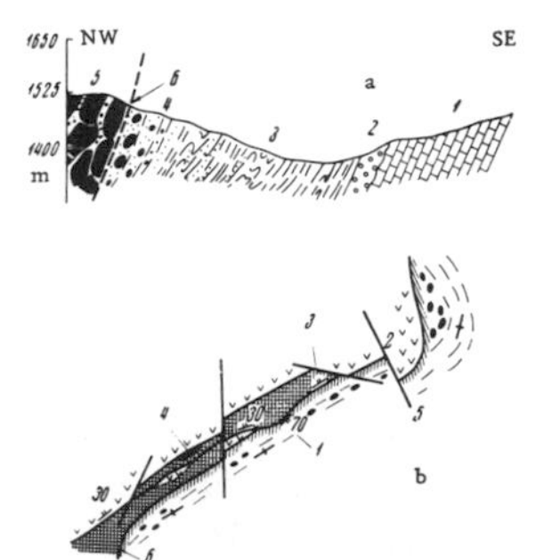

Fig. 3-a. Exposed section in the left bank of the Levchay River, Chaplya village area.

1) Upper Jurassic to Valanginian limestones; 2) conglomerates; 3) shales, sandstones, marls; 4) olistostromous interval; 5) breccia of ophiolite rocks; 6) position of Fig. 3-b.

Fig. 3-b. Contact of the ophiolite complex with the Cenomanian olistostromous interval, right flank of the Chaplya ravine.

1) Olistostromous interval; 2) diabase porphyrites; 3) gabbros and gabbro-amphibolites; 4) serpentinites; 5) normal fault; 6) mylonitized zone.

One such section is situated on the Levchay River, in the right flank of the Chaplya ravine (Fig. 3). Here, the Albian-Cenomanian section is clearly subdivided into three members, the lower one, conglomerates and sandstones, rests with an erosional and structural break on thin-tabular cherty limestones readily

*In our opinion, it has not been rigorously demonstrated that all of these deposits are Cenomanian (see below). Faunal determinations by Khalivov, cited in Shikhalibeyli's primary work (1964) suggest their Albian-Cenomanian age. Our own field data show that it is impossible to draw a sharp boundary between the Albian and the Cenomanian intervals as designated by Shikhalibeyli: Albian and Cenomanian faunas often occur in one and the same member. I cannot dwell on this point for lack of space in this short article, but I hope to take it up in a forthcoming article. For the time being, I refer to the Cenomanian complex of Shakhalibeyli (1964) as the Albian-Cenomanian.

correlative with the Lower Valanginian limestone of the Sosuzluk ridge (Aliyev and Aliyulla, 1963). It contains numerous fragments of Jurassic limestones, porphyrites, and quartz porphyries.

The middle member is made up mostly of argillites and siltstones, with marly argillites in the upper part. Sandstones, rhythmically alternating with the finer-grained terrigenous rocks, appear higher up, along with isolated beds of fine-pebble conglomerates and gravelites with abundant fragments of porphyrites, red jaspers, and limestones. The argillaceous rocks of this interval contain floating pebbles and boulders of Lower Valanginian cherty limestones, gray crystalline limestones, pink siliceous tuffs, red jaspers, red and green amygduloidal porphyrites, gabbros, and gabbro-amphibolites. All these fragments, in an argillaceous silty cement, become more numerous and increase in size, going up the section where they measure up to 6-10 m. This is a typical olistostromous interval with numerous olistoliths. A significant point is that the olistoliths exhibit fractures which do not continue in the cement. This is particularly noticeable in the red jaspers where the calcite and quartz veinlets are abruptly cut off at the olistolith faces. This suggests that tectonic deformations preceded the period of erosion.

Exposed higher up in the right side of the Chaplya ravine are blocks of the ophiolite complex, closely ground together. Present here are all the formations present in the olistoliths of the olistostromous member. The ophiolite blocks contact the olistostromous member along the steep normal fault, with the argillites and siltstones standing on edge and crushed to a fine rubble, in most instances (Fig. 3-b).

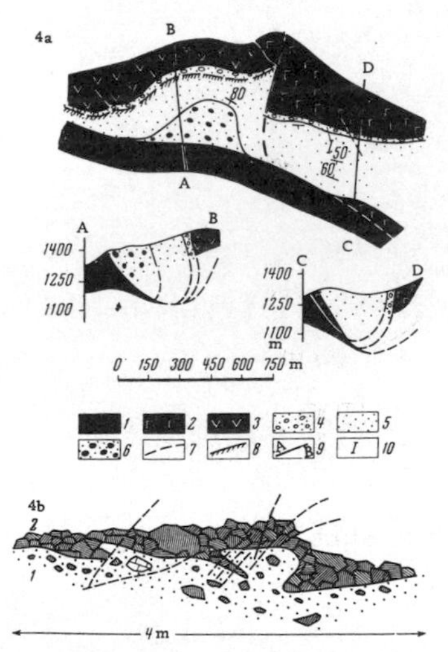

Fig. 4-a. Geologic map of the Kylychla village area:

1) Serpentinites; 2) gabbro-amphibolites; 3) diabase-porphyrites. Albian-Cenomanian: 4) basal conglomerates; 5) siltstones, argillites, sandstones; 6) olistostromous member; 7) fault lines (in map); 8) zone of mylonites; 9) profile lines; 10) position of Fig. 4-b.

Fig. 4-b. Contact of the Albian-Cenomanian basal conglomerate-breccia (1) with brecciated gabbro-amphibolites (2), (surface).

We do not believe that this section is indicative in any way of a Lower Senonian age for the ophiolite complex, because the conglomeratic member identified here by Shikhalibeyli (1964) obviously occurs in the upper interval of rocks assigned by him to the Cenomanian. Besides, it contains fragments of ophiolite rocks, which is indicative of its pre-Cenomanian age.

We have studied an even more illustrative section, northwest of the village of Kylychla where — according to Shikhalibeyli (1964) — the Lower Senonian ophiolite series has basal conglomerates and overlies the eroded Cenomanian deposits.

It is clear from the geologic map of the Kylychla area (Fig. 4) that the basal conglomerates occur here not at the base of the volcanic rocks but rather at the base of its overlying sedimentary section. The conglomerates overlie a rough, pitted surface of rocks of variable composition (spheroidal lavas, diabases, gabbro-amphibolites) and consist wholly of the products of their disintegration (Fig. 4-b).

The Albian-Cenomanian section of the Kylychla area is similar, on the whole, to that of the Chaplya ravine. The basal conglomerates are present here, too; the middle part of the section consists of argillites, siltstones, and marly shales; while the upper part is coarse clastic. Only the large olistoliths of the ophiolite complex are missing from the uppermost part of the section. Instead, the areno-argillaceous cement contains large (up to 1 m), well-rounded boulders of porphyrites, red jaspers, gabbro-amphibolites, amphibolites, and serpentinites — i.e., all those rocks assigned to the Lower Senonian ophiolite complex by Shikhalibeyli (1964).

Another point is that bodies of the gabbro-diabase and gabbro-amphibolite complex within this section often are intensively crushed and contain lenses of crystalline schists and marbles. The conglomerates broken up by occasional fractures are clearly seen to overlie a tectonic breccia. Considering in addition that the fragments in these basal conglomerates and in the olistostromous member also show evidence of intense cleavage, in no way related to the fractures observed in the silty argillaceous cement, the conclusion is inescapable: the formation of the Albian-Cenomanian basal conglomerates was preceded by strong tectonic deformation.

While the basal conglomerates of the described section overlie the eroded surface of a gabbro-amphibolite complex, in the Kylycha area, N.A. Musayev and myself have observed them resting directly on serpentinites, farther south (left bank of the Shalva River, in the area of the villages of Vagazin and Ardashevi). All that unambiguously indicates a pre-Alvian-Cenomanian age for the Sevan-Akera ophiolite complex. And this does not mean any Precambrian ophiolite complex hypothesized by Shikhalibeyli (1967) but the very same volcanic-sedimentary series assigned, along with the ultrabasic and gabbro rocks now assigned to the Lower Senonian by all students of the geology of Azerbaydzhan. Thus, we believe that Melikyan and his coauthors (Melikyan, et al., 1967) are absolutely right in asserting that all rocks of the Lesser Caucasian ophiolite complex are pre-Lower Senonian.

The material presented here affirms a pre-Albian-Cenomanian age for the ophiolite series of the Sevan-Akera zone. In that respect, the development of this region is very close to the development of the entire Alpine zone of Eurasia where a pre-Albian and even pre-Neocomian and pre-Tithonian age for the ophiolite complex has been confirmed everywhere (Kaz'min, 1966; Peyve, 1969).

However, the question of the lower age limit for this ophiolite series remains unanswered because the scanty fauna. Dating of the volcanic-sedimentary series of the ophiolite complex, comes from the limestones and radiolarian rocks, i.e., from the uppermost part of the section. The fossils from these rocks are Upper Triassic (Syria, Cyprus) and Upper Jurassic to Lower Cretaceous

(Oman, Iran).* It is therefore quite possible that the volcanic interval of this section, below the radiolarian rocks, is even older (Lower to Middle Triassic or even Paleozoic).

The age of metamorphic inclusions in the mélange presents a special problem. The current view, popular among the Azerbaydzhanian geologists is that these rocks are young, Lower Senonian. In the opinion of Shikalibeyli (1956, 1964), their metamorphism is caused by Lower Senonian ultrabasic intrusions which altered the sedimentary and volcanic formations of the same age. Evidence cited in this article for the pre-Albian age of the ultrabasic rocks is hardly compatible with that hypothesis.

Our view is supported also by the fact that all the varities of crystalline schists which occur in the serpentinites are also abundant in the fragments of the Lower Senonian conglomerates (Rentgarten, 1959; Palandzhyan, 1965). The most illustrative in that respect is a section between the villages of Geysy and Shishkay (northeast shore of Lake Sevan).

Such being the case, we can speak only of a pre-Albian-Cenomanian phase of metamorphism, and the question is whether this metamorphism was the effect of a high-temperature ultrabasic magma, or had the crystalline schists been metamorphosed prior to their entry as fragments into the ultrabasic rocks.

For an answer to this question we turn to the Adzhiras area where "... Lower Senonian rocks (mainly argillites) in contact with the serpentinite intrusions are altered to epidote-, chlorite-, talc-muscovite-, and muscovite-albite schists reminiscent of the pre-Paleozoic crystalline schists for which they have been and are now mistaken" (Shikhalibeyli, 1964).

The geologic map of Adzharis (Fig. 5) clearly shows a gentle brachyanticlinal fold traceable on the base of the Campanian limestones, and with a typical mélange exposed in the core. The southern part of the core is made up of strongly brecciated spilites; the northern part — of serpentinites which enclose numerous blocks and lenses of metamorphic formations and rocks of the ophiolite series. The latter form a large lens, mostly of pink and green spilites, in the extreme northern part of the core. In the southern part of this lens the volcanic rocks alternate with members of radiolarian rocks, gray cherts, and pinkish-gray cherty pelitomorphous limestones with well-preserved radiolaria tests.

In direct contact with these rocks, a great variety of metamorphic rocks are enclosed in the serpentinite.** They can be assigned to two principal groups. The first group comprises chlorite- and graphite-sericite schists that originated in a metamorphism of fine argillaceous sediments. These schists often also contain glaucophane and silicified and carbonatized sills. The second group consists of porphyroids, often muscovitized and containing garnet, formed from the acidic volcanic rocks. The porphyroids often are strongly cataclastic.

Two obvious conclusions can be made. First, the close proximity of metamorphosed and unmetamorphosed rocks suggests that this metamorphism is in no way related to the ultrabasic rocks. Second, the marked difference between the petrographic composition of the metamorphic rocks and that of the ophiolite series indicates that originally different rocks coexist in close proximity in the Adzharis area.

None of these points lends support to the concept of the schists having been formed from rocks of the ophiolite series. Apart from the described rocks, the mélange of the Sevan-Akera zone contains chlorite-, epidote-chlorite, and chlorite-muscovite-quartz schists, and also amphibolites some of which bear glaucophane.

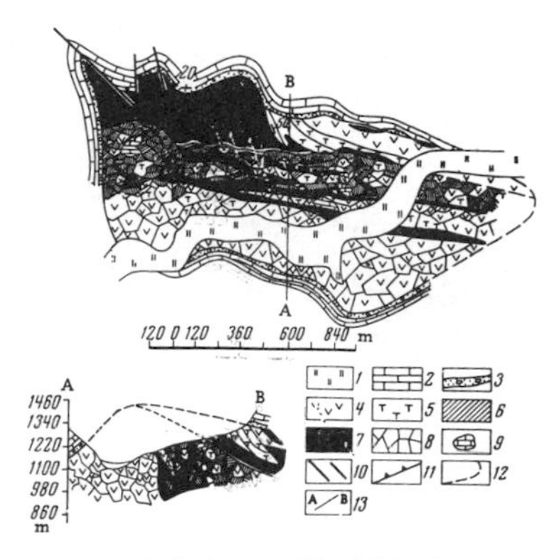

Fig. 5. Geologic map of the Adzharis area (Azerbaydzhan):

1) Quaternary deposits, undifferentiated; 2) Campanian limestones and marls; 3) Campanian basal conglomerates and sandy limestones; 4) spilites; 5) red jaspers, cherty limestones with spilite stringers; 6) crystalline schists; 7) serpentinites; 8) brecciated rocks; 9) marbles; 10) normal faults; 11) reverse faults; 12) probable strike of rocks; 13) line of profile.

All the schists are extremely carbonatized, silicified in addition to being mylonitized. It is often possible to establish that carbonatization and silification were superimposed on the already mylonitized schists.

We see then that the metamorphic rocks of the Sevan-Akera mélange do not show any evidence of high-temperature metamorphism. This substantially refutes the hypothesis of emplacement of an ultrabasic melt, as we have repeatedly stated (Knipper, 1968, 1969). On the contrary, mylonitization of the metamorphic rocks lends support to the protrusive origin for the ultrabasic rocks in the present structure. This view is supported also by the presence of secondary glaucophane — a high-pressure and low-temperature mineral (Markov, 1970) — in the crystalline schists.

Therefore the material presented here suggests that the blocks of crystalline rocks in the mélange are tectonic inclusion brought to the near-surface parts of the crust by the ultrabasic protrusions. The transfer of these blocks under the conditions of strong tectonic compression was accompanied by their mylonitization and glaucophanization. The subsequent silicification and carbonization of the metamorphic rocks may have been associated with one of the phases of serpentinization of ultrabasic rocks in the Sevan-Akera zone.

In concluding this analysis, we reiterate that all inclusions in the mélange are cemented solely by the schistose and mylonitized serpentinites. These blocks have never been seen anywhere to occur amidst the massive slightly-serpentinized ultrabasic rocks. Nor have we ever observed any fragments of sedimentary, volcanic, and metamorphic rocks welded into the ultrabasics. Consequently, and considering the total absence of high-temperature contacts with any of the afore-mentioned rocks, a magmatic origin for the Sevan-Akera ultrabasics, at any stage of geologic development for this zone is simply ruled out.

---

*R. Kh. Lipman, who has identified the radiolaria from the Rentgarten collection (1959) believes that radiolarian rocks from the ophiolite complex of the Sevan-Akera zone are most probably Upper Jurassic to Lower Cretaceous.

**I. I. Vishnevskaya and A. M. Leytes kindly looked over the thin sections of metamorphic rocks from my collection.

Consequently, we concur with K. P. Paffengol'ts (1929) and Sh. A. Azizbekov, N.V. Pashala, and E.Sh. Shikhalibeyli (1950) with regard to a Precambrian or Lower Paleozoic age for these metamorphic rocks.

A special problem is that of the age of the banded gabbros and gabbro-amphibolites. G.S. Arutyunyan (1968) and V.F. Morkovkina and S.I. Gavrilova (1969) ascribe their origin to metasomatic processes which altered the ultrabasic rocks and a certain ancient volcanic complex. According to Morkovkina (1962) and A.A. Yefimov and L.P. Yefimova (1967), who studied a similar gabbroid complex in the Urals, these rocks could have originated only under conditions of the granulite facies of metamorphism. If this is true, the complex of banded gabbros (or gabbro with a gneissoid structure) and gabbro-amphibolites could have originated as far back as the ancient Precambrian (for more information see Knipper, 1970).

CONCLUSIONS

We can assert on the basis of this material that the Sevan-Akera zone of the Lesser Caucasus is a typical mélange originated in a crushing of ophiolite rocks and in their tectonic mixing with various metamorphic and sedimentary formations, the latter including the Upper Cretaceous and Paleogene. The field data cited constitute convincing evidence for a pre-Albian age for ophiolite rocks from the Sevan-Akera zone, which indicates a development history similar to that of other, so-called eugeosynclinal, troughs of the Alpine belt of Eurasia where the ophiolites are of the same age, or nearly so (Kaz'min, 1966; Peyve, 1969). The main stage of the formation of the Sevan-Akera mélange should be assigned the same pre-Albian age. The strong terminal Neocomian tectonic movements (Austrian phase) radically modified the original relationship between rocks of the ophiolite series and formed a gigantic tectonic breccia — mélange. These tectonic movements evidently produced a surface heterogeneous in the composition of its rocks, with the individual blocks forming a kind of gigantic tectonic mosaic. We are led to this conclusion by the geologic map of the Kylychla village area where the pre-Albian structure is preserved underneath the basal conglomerates of the Albian-Cenomanian terrigenous section. Other supporting evidence is the composition of basal conglomerates overlying the ophiolite series: solely the products of erosion of the volcanic-sedimentary series (Kylychla area) in some places; only gabbro-amphibolite complex rocks in others (Kylychla and Chaykend areas; and serpentinites alone in still others (villages of Vaguaz and Ardashevi). Predominant in Albian-Cenomanian terrigenous deposits in the upper reaches of Gorchu River is arkosic material originated in the erosion of granite-metamorphic rocks.

For all these reasons, we are justified in asserting that surface relief at the beginning of the Albian-Cenomanian was made up of tectonically reworked ophiolite formations with isolated massifs and blocks of metamorphic rocks. As we see it, the mélange and its constitution had already been completed to a considerable extent by the beginning of the Albian. During the subsequent ages there was a complication of its constitution. As the effect of tectonic forces in the Upper Cretaceous and Lower Eocene, mélange protrusions twice cut the overlying deposits and incorporated the blocks of various Upper Cretaceous and Paleogene rocks. The protrusion-forming capability of mélange is determined by the unusual behavior of "mélange cement" — serpentinite under conditions of strong compression.

REFERENCES

ABOVYAN, S.B. Anorthosites of the Shorzha and Dzhil-Satanokhach gabbro-peridotite massifs in the Armenian SSR. Izv. Akad. nauk ArmSSR. Ser. geol. i geogr. nauk, 14, No. 5, 1961.

ABONYAN, S.B. Ultrabasic and basic rocks of the ophiolite formation. In: Geologiya ArmSSR (Geology of the Armenian SSR), Vol. 3. Izd. Akad. nauk ArmSSR, 1966.

AZIZBEKOV, SH.A. Gabbro-amphibolite intrusions of the ophiolite formation in the southeast of the Lesser Caucasus. Dokl. Akad. nauk ArmSSR, 6, No. 6, 1950.

AZIZBEKOV, SH.A., N.V. PASHALY and E.SH SHIKHALIBEYLI. Peridotite intrusions of the ophiolite formation in the southeast of the Lesser Caucasus (Azerbaydzhan). Izv. Akad. nauk AzerbSSR, No. 4, 1950.

ALIYEV, O.B. and KH. ALIYULLA. Materials bearing on a Lower Valanginian age of the Suzlug Mountain deposits (Lesser Caucasus). Dokl. Akad. nauk AzerbSSR, 19, No. 11, 1963.

ARUTYUNYAN, G.S. Classification of intrusions in the northwest part of the Sevan range by their age. Izv. Akad. nauk ArmSSR. Ser. Nauki o Zemle, 20, No. 1-2, 1967.

ARUTYUNYAN, G.S. Origin of gabbro rocks associated with ultrabasic intrusions (in the northwestern part of the Sevan range). Izv. Akad. nauk ArmSSR. Ser. Nauki o Zemle, No. 5, 1968.

ASLANYAN, A.T. Regional'naya geologiya Armenii. Yerevan (Regional geology of Armenia. Yerevan). Izd. Aypetrat, 1958.

YEGOYAN, V.L. Upper Cretaceous volcanism and ultrabasic intrusions in the east of the Lesser Caucasus. Izv. Akad. nauk AzerbSSR, No. 6, 1953.

YEFIMOV, A.A. and L.P. YEFIMOVA. The Kytlym platinum-bearing massif. Materialy po geol. i polezn. iskop. Urala, No. 13. Izd. Nedra, 1967.

KAZ'MIN, V.G. The position of ophiolite formations in the tectonic development of the western sector of the Alpine-Himalayan system. Geotectonika, No. 3, 1966.

KNIPPER, A.L. The age of gabbroids in the Cheykend massif (Sevan-Akera zone, Lesser Caucasus). Dokl. Akad. nauk SSSR, 162, No. 2, 1965_1.

KNIPPER, A.L. Distinctive features of formation of anticlines with serpentinite cores (Sevan-Akera zone, Lesser Caucasus). Byull. MOIP, Otd. geol., 50, 1965_2.

KNIPPER, A.L. Tectonic position of ultrabasic rocks in geosynclinal regions and some problems of the initial magmatism. In: Problemy svyazi tektoniki i magmatizma (Problems of the relationship between tectonics and magmatism). Izd. Nauka, 1968.

KNIPPER, A.L. Mantle rocks at the surface. Priroda, No. 7, 1969.

KNIPPER, A.L. Gabbroids of the ophiolite "formation" in the section of the oceanic crust. Geotektonika, No. 2, 1970.

KNIPPER, A.L. and YU.L. KOSTANYAN. The age of ultrabasic rocks on the northeast shores of Lake Sevan. Izv. Akad. nauk SSSR. Ser. geol., No. 10, 1964.

LEONT'YEV, L.N. and V.YE. KHAIN. Upper Cretaceous ultrabasic rocks and the ophiolite formation in the Lesser Caucasus. Dokl. Akad. nauk SSSR, 65, No. 1, 1949.

LOMIZE, M.G. The Vedinsk ophiolite tectonic zone (Lesser Caucasus). Byull. MOIP. Otd. geol., 45, No. 6, 1970.

MARKOV, M.S. Metamorphic complexes and their position in the development of island arcs. Geotektonika, No. 2, 1970.

MELIKYAN, L.S., S.A. PALANDZHMAN, Z.O. CHIBUKHCHYAN and ZH.S. VARTAZAR'YAN. Concerning the geologic position and age of the ophiolite series in the Shirak-Sevan-Akera zone of the Lesser Caucasus. Izv. Akad. nauk Arm SSR. Ser. Nauki o Zemle, 20, No. 1-2, 1967.

MORKOVKINA, V.F. Metasomatic alterations of ultrabasic rocks in the Polar Urals. Trudy Inst. geol. rudn. mestorozhd., petrogr., mineralog. i geokhimii. Izd. Akad. nauk SSSR, No. 77, 1962.

MORKOVKINA, V.F. and S.I. GAVRILOVA. Origin of rocks from the ophilite belt of the Lesser Caucasus. Avtoreferaty rabot sotrudnikov IGEM, 1968. Trudy Inst. geol. rudn. mestorozhd., petrogr., mineralog. i geokhimii. Izd. Akad. nauk SSSR, 1969.

PALANDZHYAN, S.A. Contribution to the geology of ultrabasic and basic intrusive rocks on the northeast shores of Lake Sevan. Izv. Akad. nauk. ArmSSR. Ser. geol. geogr., No. 1, 1965.

PAFFENGOL'TS, K.N. Main features of the geology and tectonics of the Gandzhi region, Azerbaydzhan. Izv. geol. komit., 48, No. 3, 1929.

PEYVE, A.V. Oceanic crust of the geologic past. Geotektonika, No. 4, 1969.

RENTGARTEN, V.P. Cretaceous stratigraphy of the Lesser Caucasus. Regional'naya stratigrafiya SSSR, Vol. 6 (Regional stratigraphy of the USSR, Vol. 6). Izd. Akad. nauk SSSR, 1959.

SOLOVKIN, A.N. The Triassic of the upper reaches of the Akera (Azerbaydzhanian SSR). Dokl. Akad. nauk AzerbSSR, 6, No. 9, 1950.

KHAIN, V.YE., R.N. ABDULLAYEV and E.SH. SHIKHALIBEYLI. Exotic cliffs of the sedimentary klippe type in the Lesser Caucasus. Dokl. Akad. nauk SSSR, 17, No. 2, 1949.

KHAIN, V.YE. Main structural features of the Alpine belt of Eurasia in the Near and Middle East. Art. 1. Vestn. Mosk. univ. Ser. IV, Geologiya, No. 6, 1968. Articles 2-3: Ibid., Nos. 1-2, 1969.

SHIKHALIBEYLI, E.SH. Tectonics of the Sevan-Akera synclinorium in Azerbaydzhan. Trudy Soveshch. po tektonike Al'piyskoy geosyncl. oblasti yuga SSSR (Proceeding of the Conference on the tectonics of the Alpine geosynclinal region in the south of the USSR). Izd. Akad. nauk AzerbSSR, Baku, 1956.

SHIKHALIBEYLI, E.SH. Geologicheskoye stroyeniye i istoriya tektonni-cheskogo razvitiya vostochnoy chasty Malogo Kavkaza, 1 (Geology and tectonic development of the eastern part of the Lesser Caucasus, Vol. 1). Izd. Akad. nauk AzerbSSR, Baku, 1964. Vol. 3, 1967.

BAILEY, E.B. and W.J. McCALLIEN. J. Serpentine lavas, the Ankara mélange and the Anatolian thrust. Trans. Roy. Soc. Edinburgh, B. 42, 1956.

BRUNN, J.H. Mise en place et differenciation de l'association pluto-volcanique du cortege ophiolitique. Rev. de Géographie Physique (2), v. III, fasc. 3, Paris, 1960.

COLEMAN, R.G. Serpentinites, rodingites and tectonic inclusions in Alpine-type mountain chains. Geol. Soc. Amer. Sp. Papers, 73, 1963.

DAL PIAZ, G.V. Filoni rodingitici e zone di reazione a bassa temperatura al contatto tettonico fra serpentine e rocce incassanti nelle Alpi occidentali italiane. Rendiconti della Soc. Italiana di Minoralogie e Petrologia, Vol. 25, 1969.

PASSERINI, P. Rapporti fra le ofioliti e le formazioni sedimentarie fra Placenza e il Mare Tirreno. Boll. Soc. geol. ital., Vol. 84, 1965.

Received November 5, 1970

Geological Institute,
USSR Academy of Sciences

30

Reprinted from *Geotectonics* **6**:384-390 (1970)

DEVELOPMENT OF SERPENTINITE MÉLANGE IN THE LESSER CAUCASUS

A. L. KNIPPER

The history of development of serpentinite mélange was associated with the period of tectonic compression, with the principal phase dating back to the end of the Neocomian. In the Upper Cretaceous (sub-Hercynian phase), the progress of intrusions and tectonic cover was accompanied by the formation of an olistostrome complex. The post-Upper Eocene (Pyreneean) tectonic phase reactivated the protrusions of mélange. An attempt is made to clarify the mélange.

As demonstrated in our earlier article (Knipper, 1971), mélange in the Sevan-Akera zone of the Lesser Caucasus had been formed to a considerable extent by the beginning of the Alpine stage, and the origin of this giant tectonic breccia should be related, as recently stated by A. V. Peyve (1969), to the Austrian tectonic phase.

The present article is an attempt to make a detailed reconstruction of the history of the Sevan-Akera mélange zone from the time of its inception to the present day, with arguments adduced for our view of its tectonic character.

Before beginning our presentation, we will describe an extremely unusual sedimentary section of the Sevan-Akera zone, similar in many respects to mélange but differing from it in origin. Without understanding the particulars of the origin of this complex it is impossible to understand the history of the development of mélange in Upper Cretaceous time.

OLISTOSTROME SERIES OF THE SEVAN-AKERA ZONE

We now consider a thin but very illustrative section of the south slope of the Malaya Kalaboynu Mountain, Azerbaydzhan (Fig. 1). A sequence exposed at the foot of the slope is represented by an alternation of light-gray argillites and siltstones with a visible thickness of 100-120 m. The fossils identified here by A. G. Khalilov (1954) date these deposits as Albian. Present higher up in this section is a lens of large boulder conglomerates with fragments and chunks of weathered serpentinites, red jaspers, porphyrites, and gray stratified sandy limestones. The conglomerates are in fault contact with a narrow linear body of serpentinites which appear to girdle the entire south slope of the Malaya Kalaboynu Mountain. It is about 3 km long, with a maximum thickness of 50 m. These are common dark-green serpentinites in a boudinage reminiscent of conglomerates.

These rocks are overlain by serpentinite conglomerate-breccias cemented with a serpentinite sand in a carbonate groundmass. The bulk of the conglomerate-breccias is represented by fragments and chunks of serpentinites (95% of the volume of the rock). In addition, these formations contain small fragments of dark-gray cherty marles and limestones. Going up the section, the conglomerate-breccias change to serpeninite gravelites and serpentinite sandstones. On the whole, these rocks constitute a large rhythm, 15 m thick.

The transition from the sedimentary to igneous rocks appears to be so gradual as to make it difficult to draw a boundary between the serpentinite conglomerate-breccias and serpentinite boudinage. This boundary is identified out in the field by the appearance of limestone and marl fragments among the well-rounded serpentinite fragments (externally very similar to the serpentine boudins) and by the presence of incursions and break-throughs of carbonate-cemented serpentinite sandstones which appear to envelop the boudins.

The material presented suggests that the serpentinite body in the south slope of the Malaya Kalaboynu Mountain is a large submarine slump plate, and that this part of the section is on the whole a typical olistostrome with a serpentinite olistolith.

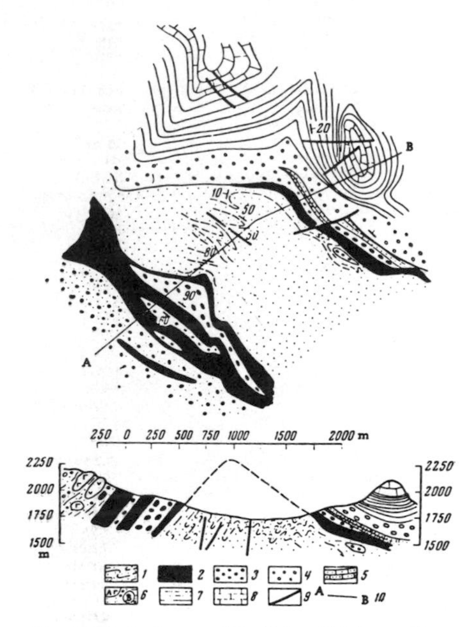

Fig. 1. Geologic map of the Malaya Kalaboynu Mountain area, Azerbaydzhan:

1) Albian-Cenomanian siltstones and argillites; 2) serpentinite olistoliths. Cenomanian to Lower Senonian: 3) sedimentary boulder breccias and conglomerates; 4) conglomerates and coarse-grained sandstones; 5) calcareous sandstones and sandy limestones; 6) olistostrome complex; 7) marly claystones (Upper Santonian?); 8) Campanian marls and limestones; 9) faults; 10) line of profile.

The overlying section is as follows, reading upward:

1. Calcareous sandstones with abundant serpentinite fragments and containing lenses of sedimentary breccias full of fragments of limestones and green siliceous rocks. The sandstone

intervals contain beds of organoclastic limestones. This sequence is 20 m thick.

2. Gray to pink sandy organoclastic limestones with small fragments of serpentinites and red jaspers. The limestones are lumpy to conglomeratic, becoming thin-bedded higher up where they contain numerous fragments of red jaspers and serpentinites. Thickness, 10 m.

3. A 60 m thick interval with shales at the base; these alternate with light-green sandstones in the middle part. Sandstones are predominant in the upper part. Imbedded in these rocks are boulders and blocks (1-5 m^3) of sandstones full of serpentinite fragments.

4. Sandy limestones, 2 m thick.

5. Poorly stratified medium-grained sandstones, 80 m thick.

6. Well-stratified marly shales and marls, 100 m thick.

7. Red jaspers resting with a sharp contact on the underlying marly shales and changing upward to typical Campanian marls.

The olistostrome interval of this section (serpentinite conglomerates and members 1-5) show very rapid lateral changes in lithology and thickness. For example, conglomerate-breccias overlying the olistolith wedge out westward over a distance of 300 m, while member 2 thickens eastward to 50 m, over a distance of 200 m.

A very similar but very thicker olistostrome member is exposed in the upper reaches of the Meydanchay River, within the so-called Touragaychay synclinorium (Shikhalibeyli, 1964). Here a quite similar section of Albian-Cenomanian deposits can be observed from the Zindzhirli River in the west and to the Buzulukh area in the east (Fig. 2). Its base locally shows 5-10 m of conglomerates resting with an erosional break on the Tithonian-Valanginian limestones or on Neocomian to Albian pyroclastics (north slope of the Sosuzluk ridge). Higher up the section, the conglomerates are replaced by argillites and siltstones with marl beds at the base, and beds and lenses of green sandstones and fine-pebble conglomerates. This member is up to 500 m thick. These deposits bear a rich fauna, Middle to Upper Albian according to A.G. Khalilov (cited from Shikhaliveyli, 1964). They are overlain by a thick terrigenous olistostrome interval similar to that in the Malaya Kalaboyny Mountain section.

This interval contains gigantic olistoliths, up to 3-4 km long by 300-400 m thick, of nearly all the mélange rocks. Some of these slabs are uniform in composition (porphyrites, red jaspers, gabbro-amphibolites, serpentinites). Others consist of numerous fragments and chunks of all the ophiolite rocks of the complex, closely ground into each other. The lower contacts of such slabs are tectonic, while they are encased on all other sides of chaotically mixed boulder breccia of all the rocks of these slabs. These phenomena are very conspicuous northwest of the village of Ayerk, along the trail to Lake Karatel. Where the olistoliths are large, the olistostrome interval can be mistaken for the ophiolite complex in situ; in other places it is a typical layered conglomerate of the ophiolite fragments. Evidently this was the reason for designating the conglomerate facies of the Lower Senonian volcanics, with the individual olistoliths mistaken for layers of volcanic rocks (Rengarten, 1959; Shikhalibeyli, 1964).

The contact of the olistostrome interval with the underlying shales varies from place to place. It may be gradual, as in the upper reaches of the Meydanchay, with beds of fine-pebble conglomerates and sedimentary breccias of all the products of erosion of the ophiolite association, and pebbles of metamorphic rocks and associated limestones appearing in the upper part of the shale intervals. The conglomerate and breccia fragments become gradually coarser going up the section; isolated large olistoliths appear; the cement loses its layering; and we finally see a chaotically constituted olistostrome complex. In other places, such as the north end of the Sosuzluk ridge, small blocks of mélange rocks, unaccompanied by terrigenous rocks, appear among the siltstones and argillites. Locally, the transition from the very fine sediments of the middle part of the section to the olistostrome complex is very sharp. In that event, the shales contact the piled-up large blocks of most diverse rocks (marbles, serpentinites, radiolarian earths, etc.).

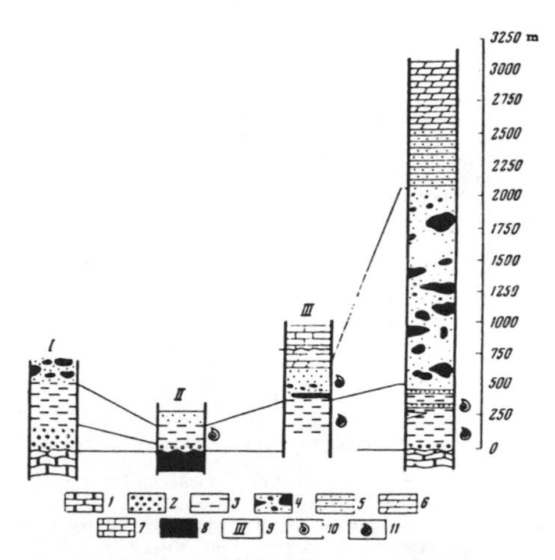

Fig. 2. Correlation of some of the Cretaceous sections of the Sevan-Akera zone:

I) Village of Chapli area; II) Village of Kylchaly area; III) Malaya Kalaboynu Mountain; IV) upper reaches of the Meydanchay River. 1) Tithonian-Valanginian limesones. Albian-Cenomanian deposits: 2) basal conglomerates; 3) argillites and siltstones; 4) Cenomanian to Lower Senonian olistostrome interval. Lower Senonian: 5) sandstones; 6) marls; 7) Campanian limestones; 8) mélange; 9) section Nos.; 10) fauna indicative of a Cenomanian age for the rocks; 11) fauna indicative of an Albian age.

The olistostrome complex of the Touragaychay zone is identifiable southeast as far as Naldbandkhan where it is replaced, over a distance of 2-3 km, by sandstones and fine-pebble conglomerates.

The olistostrome complex is extremely common within the Sarybaba zone. It outcrops best in the middle and lower courses of the Tutkhum River where it occupies the same stratigraphic position as in the Touragaychay zone. The huge olistoliths of mélange disappear going northwest, and the olistostrome complex is replaced by a thick section of well-stratified conglomerates with numerous fragments of rocks from the ophiolite association, crystalline schists, etc.*

As to the age of olistostrome deposits in the Sevan-Akera zone, we do not believe it can be determined, as yet, with an adequate certainty — because the extant fauna determinations are ambiguous. Turning to Fig. 2, a correlation of some of the Cretaceous sections of the Touragaychay and Sarybaba zones, we see that

*This refers to sections in the west and east slopes of the Shakhdag range.

these sections are readily correlated, lithologically from the shale and siltstone intervals with sandstone beds, common to all of them. This interval is underlain by thin basal conglomerates, and overlain by the olistostrome part of the section. According to Shikhalibeyli (1964), the interval common to all the sections is Albian in the Malaya Kalaboynu Mountain and within the Touragaychay synclinorium, and Cenomanian in the Chapli-Kalychla area (the two least-named sections were described in detail in the preceding publication; Knipper, 1971). Furthermore, we collected pelecypods from the middle part of the shale interval north of the Socyzluk Mountain. They were identified by O. B. Aliyev as Amphidonta columba plicatula Lam., Haustator cf. obscuricostatus Pčel, Neithea quinquecostata Sow., Avellala, sp., and Cardium, sp. This fauna is indicative of a Cenomanian age for the enclosing rocks. It is therefore best, pending additional paleontological study, to speak of an Albian-Cenomanian age for this part of the section, so much so because its division into two independent intervals is hardly feasible. Shikhalibeyli (1964) spearates in several sections Albial argillaceous rocks from the Cenomanian areno-argillaceous. The latter overlie, in his opinion, the Lower Cretaceous rocks. We have not observed such an unconformity anywhere. This point evidently was not too clear to Shikhalibeyli himself. In discussing the relationship between the Albian and Cenomanian deposits north of the Sosuzluk range, he speaks of an unconformable occurrence of Cenomanian rocks (p. 155 of his publication), while stating on page 142 that they rest without a visible angular unconformity on the Albian areno-argillaceous deposits. The same is true with reference to the Kalaboynu Mountain section. Shikhalibeyly states on page 126 that the position of the Cenomanian conglomerates is unconformable; he reverses himself on page 149 when he says that the Albian-Cenomanian boundary, here, is difficult to define. Consequently, the question of the lower age boundary of the olistostrome interval remains moot. As we see it, this part of the section was formed during the time from the uppermost Cenomanian through the Lower Senonian. This view is based on findings of a Cenomanian fauna in coarse boulder conglomerates in the area of the Malaya Kalaboynu Mountain (O.D. Gamzayev cited by Shikhalibeyli, 1964); on the fact that the olistostrome complex is overlain everywhere by the Campanian marls and limestones; and on correlation of these sections with conglomerates along the northeast shores of Lake Sevan whose Lower Senonian age has been known since V. P. Rengarten's publication (1959). Both Rengarten and Shikhalibeyli assign to the Lower Senonian the same deposits within the Touragaychay zone (conglomerate facies of the Lower Senonian volcanics, according to Shikhalibeyli). On the basis of what has been said, we assign the development of the entire terrigenous complex of the Sevan-Akera zone to a period extending from the Albian through the Lower Senonian.* Additional paleontological studies are needed to draw definite stratigraphic boundaries within this section. This is because of the possible presence here of lithologic boundaries that are time transgressive, as is the case in an identical section of Upper Cretaceous sediments along the Ligurian coast of Italy (Passerini, 1965).

In concluding the description of the olistostrome complex, we reiterate that the olistostrome formation in the zones with a great unloading of a large number of ophiolite blocks often is quite reminiscent of a fragmented volcanic-sedimentary series. It is therefore not surprising that all the earlier investigators, ourselves included, mistook it for an independent Upper Cretaceous volcanic formation. However, a closer study clearly shows a difference between it and the mélange of the Sevan-Akera zone. The mélange is cemented with serpentinite; the olistostrome complex has a terrigenous cement. The first is intensely deformed; the second always shows a graded stratification and evidence of a cosedimentation flow of material. Finally, the mélange contains exotic Upper Senonian and Paleogene blocks (Knipper, 1971), while only the pre-Upper Senonian rocks are present in the olistostrome complex.

*We stress again that there are no breaks and unconformities within this section. It is therefore hardly reasonable to speak of a regional Turonian unconformity, as Shikhalibeyli does (Shikhalibeyli, 1964, 1967). Nor does the absence of Turonian fauna — the argument used by Shikhalibeyly — necessarily mean the absence of the corresponding deposits.

DEVELOPMENT OF MÉLANGE IN THE SEVAN-AKERA ZONE

Field data cited in my earlier article (Knipper, 1971) show that the accumulation of the volcanic-sedimentary association was interrupted at the end of the Neocomian by strong tectonic movements which turned these rocks into a gigantic mash — the mélange. As we shall presently see, during this time there was a major remodeling of the structure of the Sevan-Akera zone, so that the history of its development can be divided into two major stages: pre-Albian and post-Albian.

An understanding of the pre-Albian history of the Sevan-Akera zone is based, to a considerable extent, on an understanding of the history of movements of its segment of crystalline basement, the presence of which is now inferred for this part of the Alpine belt, mainly from the geophysical data. A majority of the geologists believe that this sialic body is part of the African-Arabian platform.

However, opinion is sharply divided as to the pre-Albian position of this basement. The overwhelming majority of investigators (Stecklin, 1966; Belov, 1966; Muratov, 1969; and others) believe that the Early Mesozoic position of this basement was close to the present one, and that the basement was subjected to an intense fragmentation which led to the formation of deep troughs within its confines. These troughs were the sites of the ophiolite association. According to the other view (Peyve, 1969), the site of the present Alpine belt was occupied in the Early Mesozoic by a vast oceanic trough with a typically oceanic crust. At the end of the Jurassic and in the Early Cretaceous this area was thrust over by an immense sheet of rocks of the African-Arabian platform, which formed a gigantic underthrust at the junction with the Hercynian folded structures of the Caucasus. In his article Peyve cites numerous arguments in support of a drift for the African-Arabian platform. Without repeating these arguments, I only note that the Mesozoic paleogeography of the modern Lesser Caucasus is amazingly reminiscent of the marginal zones of the continents on their transition to the ocean. Indeed, the differentiated Jurassic series of spilite-keratophyres and andesite volcanics in the Somkhet-Karabakh zone, where they overlie a Hercynian basement and are cut by bodies of Jurassic and Cretaceous granites, are very similar to the rock series in island arcs of the eastern Pacific margin, recently described by M. S. Markov (1970).

This similarity is enhanced by the presence of an M-surface high beneath the Kura trough — a phenomenon known to be typical of interior seas which separate the island arcs from the continent (the continent represented, in this instance, by the Hercycian folded structures of the Greater Caucasus). The assumption of an oceanic trough having existed at the site of the present Alpine folded zone of Eurasia is supported also by the presence of an extended Upper Paleozoic volcanic belt similar to that along the east coast of the Pacific. According to A. A. Mossakovskiy (1970), this belt was situated along the north coast of the pre-Tethys oceanic trough.

These considerations, apart from Peyve's arguments, suggest that, prior to the Neocomian, the area south of the Touragaychay zone was the northern part of the paleo-Tethys oceanic trough with a typical oceanic floor section. The base of this section was represented by the ultrabasic rocks (mantle); the middle part, by gabbro-amphibolites (layer 3); and the upper part, by basic effusives and siliceous-carbonate rocks (analogue of layers 2 and 1, respectively).

The history of the origin of mélange in the Alpine zone of Eurasia can be visualized as follows, in conformity with the concepts of Peyve (1969) and Kazmin (1971): a drift of thick sialic rocks of the African-Arabian platform, and their thrust over the oceanic floor section which caused a fragmentation of the latter.

The breaking-up, attendant to the movement of the sialic plate, led to a collapse of blocks of the crystalline basement and its sedimentary mantle in front of this gigantic overthrust. The creeping massifs, blocks, and slabs were crushed and mixed with fragmentation products of the ophiolite complex during their burial underneath the oncoming tabular sialic body. At the same time, mélange protrusions — originated because of the high mobility of serpentinites in the tangential compression — penetrated the upper reaches of the crust by way of isolated fractures in the allochthonous body. During the course of this movement, the mélange was enriched in new exotic blocks torn off the crystalline basement and its sedimentary mantle.

At the end of the Neocomian, sialic masses of the African-Arabian platform in the area corresponding to the present Sevan-Akera zone were thrust under the crystalline basement of the Hercynian margin along the north Tethys coast (under the Touragaychay and Somkhet-Karabakh zones). This underthrust was accompanied by a squeezing-out of the serpentinite mélange along its surface. While the movement of the crystalline blocks of the African-Arabian platform was to the north, the mélange flowed in the opposite direction. The squeezing-out of the mélange may ve visualized as having taken place along the contact between two crystalline masses (Fig. 3). In this setup, the underthrust plate can be named, in Aubouin's terminology (1967), the relative autochthon, with the tectonic sheets of mélange in the Somkhet-Karabakh zone being the allochthonous bodies.

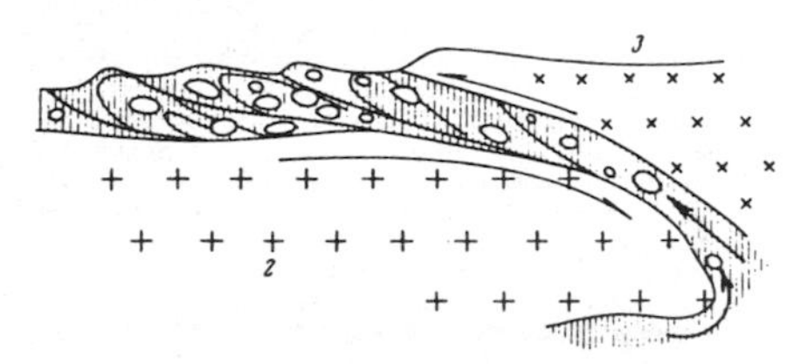

Fig. 3. Diagram of the development of mélange formation (1) in the Sevan-Akera zone at the end of the Neocomian, as the result of crystalline masses of the African-Arabian platform (2) being thrust under the Hercynian folded structures of the Somkhet-Karabakh zone (3).

The geophysical data available for this part of the Lesser Caucasus do not contradict these conclusions. The gravimetric maps of the pre-Alpine basement surface show that this surface is subsided 2-3 km, in the Sevan-Akera zone (Shikhalibeyli, 1967). Consequently, this is the thickness of the allochthonous mélange. According to the gravimetric and magnetometric surveys, "the belt of ultrabasic intrusions is not marked by higher magnetic or gravity fields, possibly because of a small thickness of the ultrabasic rocks" (Khain, 1968). The same data indicate that "the deep fault, which provided a channel for the ultrabasic intrusions, evidently is a high-angle thrust with its surface dipping north-northeast. ... The excessively steep magnetic and gravity gradients reflect large intrusions in somewhat deeper reaches of the crust rather than the present outcrops of ultrabasic intrusions torn from the main body of ultrabasics and brought to the surface along the faults" (Gabriyelyan and Tatevosyan, 1966).

Except for the assumption of magmatic roots for the ultrabasic rocks, Gabriyelyan's and Tatevosyan's views are very close to those presented in this article.

Thus, we assume that a compound imbricate structure had been formed by the beginning of the Albian in that segment of the crust now corresponding to the Sarybaba zone, with an allochthonous mélange plate overlying a relative autochthon of sialic masses of the African-Arabian platform. An early Albian subsidence and marine transgression resulted in a burial of the mélange surface with its Hauterivian complex (Knipper, 1971) underneath the Albian basal conglomerates (Fig. 4).

The continued subsidence was accompanied in the Middle and Upper Albian by deposition of a relatively uniform section of claystones and siltstones. The underthrusting of the relative autochthon was resumed in the Cenomanian and went on uninterrupted until the end of the Lower Senonian. These movements squeezed new batches of mélange into the Sarybaba zone. With the uninterrupted flow of mélange from the deeper reaches of the underthrust, it was directed upward in the zone with a sharp drop in static pressure associated with the great weight of the Samkhet-Karabakh allochthon, and intruded the Albian to Cenomanian sediments accumulated by that time. An extended linear uplift originated at the junction of the Sarybaba and Karabakh zones, with mélange in the core of the uplift reaching the surface of the sea floor.

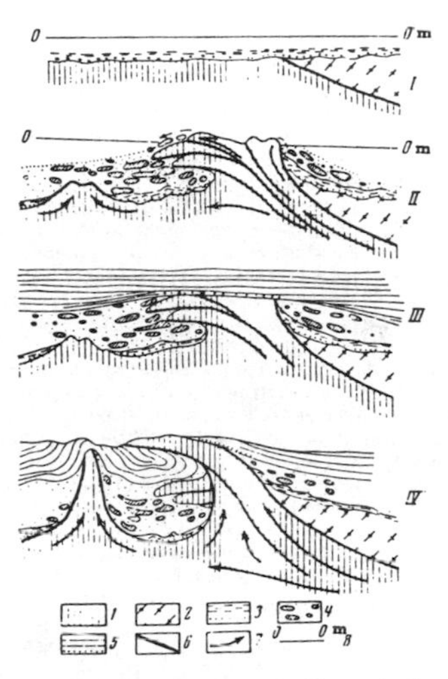

Fig. 4. Development of mélange in the Sevan-Akera zone, beginning with the Albian:

Paleogeographic profiles: I) by the Middle Cenomanian; II) by the end of the Lower Senonian; III) by the end of the Upper Eocene; IV) by the end of the Oligocene. 1) Melange; 2) Samkhet-Karabakh complex; 3) Albian to Cenomanian terrigenous rocks; 4) olistostrome complex; 5) undifferentiated Upper Senonian to Upper Eocene deposits; 6) faults; 7) direction of mass movement; 8) sea level.

As the underthrust of the relative autochthon continued, and the inflow of new batches of mélange into the axial part of the uplift continued with it, a series of gravity sheets originated and crept down the slopes of the uplift, over the sea floor. The movement of these sheets was accompanied by their disintegration and erosion, submarine slides of the sediments, and the emergence of rocks with a graded stratification. As the result of the intermittent movement of the sheets, some of the larger slabs of mélange rocks, and blocks originated in their disintegration, were buried in sandstones, and fine-pebble conglomerates deposited over the entire trough. This is the process that led to the development of the olistostrome complex described in the beginning of this article. The influx of a huge volume of mélange into the area of erosion led to an overloading of the Lower Senonian basin with clastic material, so that by the Middle Santonian the marginal part of the Sarybaba zone was a shallow sea with an archipelago of numerous islands of mélange.

The underthrust of the relative autochthon ceased by the end of the Senonian, and evolution of the olistostrome complex ceased with it. The vast Upper Senonian transgression eroded the nontronite type of weathering crust which had formed in the meantime over the peneplaned surface of the islands. Carbonate deposits were laid down everywhere in the Upper Senonian, under open sea conditions. These deposits were followed in the Paleogene by carbonate-terrigenous and volcanic sequences. Movement over the deep underthrust surface was resumed during the Pyreneean phase of folding (the end of the Upper Eocene and Early Oligocene), and again mélange protrusions with their captured blocks of Upper Senonian and Paleogene rocks broke through the upper horizons of the Sevan-Akera sedimentary and volcanic section. That was the final event in the history of the folded structure in this part of the Lesser Caucasus.

Judging from this exposition, there were at least three stages of mélange formation in the Sevan-Akera zone, each stage was associated with a large tectonic phase. These phases are, from older to younger, Austrian (Neocomian), sub-Hercynian (Lower Senonian), and Pyreneean (terminal Upper Eocene). The Austrian phase evidently was the most effective: a typical mélange with its unusual structure and array of rocks was already in existence in the Sevan-Akera zone by the end of the Neocomian. The subsequent phases merely complicated the composition of the mélange and structure due to the effect of compressive stresses.

A significant point is that the mélange that cuts the Upper Eocene deposits or else is thrust over them (for example, in the Shorzha and Dzhil-Satanakhach massifs; Knipper, 1965, 1966) consists almost wholly of serpentinites with only a very few exotic inclusions. It appears that such a composition of mélange in these massifs is the result of the separation of serpentinite material, in compression, from large bodies of mélange, now buried. This "squeezing-out" and "out-flow" of serpentinite material is related to its great mobility at high pressures — much greater than for the other components of the mélange. This implies the existence somewhere of a body of residual serpentinite-free mélange (possibly with a small amount of serpentinite cement) — a sort of "dry residue" of serpentinite squeezings.

CONCERNING THE TERM "MÉLANGE"

The term "mélange", was introduced in the geological literature by E. Greenly (Greenly, 1919) who gave the name "autoclastic mélange" to a complex Precambrian tectonic breccia in Anglesey, Wales, consisting of blocks of spilites, diabases, quartzites, limestones, jaspers, and coarse-grained sandstones cemented by an argillaceous material. This mélange, formed in the Precambrian, was tectonically mixed with Ordovician shales, in Caledonian time. The result was, in Greenly's terminology, a "polykinematic mélange", i.e., containing the fragments of an older mélange. According to Greenly's concept, a mélange originates in the tectonic fragmentation and mixing of rocks in large overthrusts.

In the following years, the term mélange was completely forgotten, and was not revived until 1952 when Bailey and McCallien described as "Ankara mélange", a tectonically fragmented heterogeneous complex in the vicinity of Ankara, Turkey. This complex consisted of assorted blocks of ultrabasic and gabbro rocks, basic volcanics, radiolarite-earths, limestones, flysch, and crystalline schists (Bailey and McCallien, 1956). These two authors believed after Greenly that mélange is a purely tectonic formation that originated in the movement of large allochthonous plates of crystalline rocks.

A majority of geologists conceded a tectonic character for mélange in the outcrops (Gansser, 1960; Khain, 1968; Peyve, 1969; Kazmin, 1971), so much so because these outcrops always occur in the zones of strongest tectonic deformations. However, conditions of formation of the original sequence, subsequently turned into a mélange, are subject to discussion.

According to one school of thought (Hsü, 1968; Peyve, 1969; Knipper, 1970; Kazmin, 1971), mélange is a purely tectonic formation, as believed by Greenly, Bailey, and McCallien. It originates in a fragmentation and mixing of diverse stratigraphic complexes; hence its variable composition. According to this view, the diversity of mélange rocks is a secondary phenomenon fully dependent on tectonic movements.

In another view, the heterogeneous petrographic composition of mélange is a primary phenomenon, the result of a slump of large olistoliths down the steep slopes of a trough and to its floor. The olistostrome hypothesis was advanced by A. Gansser, in 1959 (Gansser, 1960), and was recently supported by Khain (1969). This view is based on firm field evidence: a tectonically deformed olistostrome is difficult, often impossible, to tell apart from a typical mélange. At the same time, it is quite necessary to distinguish between these formations of different origins, because their identification influences the interpretation of tectonic history of the region under study.

K. Hsü (1968) recently discussed this subject along with several criteria for defining the term, "mélange". Some of his conclusions are cited below.

As defined by Hsü on the basis of his study of sedimentary mélange in California, "mélange is a body of deformed rocks, mappable and clearly outlined, characterized by the presence of inclusions of tectonically mixed fragments and blocks, up to several miles long, and enclosed everywhere in a fine-grained, usually pelitic, sheared cement. The mélange component blocks are both indigenous and exotic, in a cement. The indigenous blocks are disintegrated brittle layers interbedded with a deformed pliable cement. The exotic blocks are tectonic inclusions severed from rocks enclosing the melange body", (Hsü, 1968, p. 1065).

We stress again that Hsü describes mélange cemented with purely sedimentary rocks: argillites and siltstones. Naturally, in zones of strong tectonic deformations, such as mélange is difficult to tell apart from the original olistostrome. Nonetheless, Hsü believes this can be done after a careful study of the cement of the deformed sequence and of its component exotic blocks. The salient points of Hsü's presentation are as follows: mélange cement always constitutes evidence of deformation in the already consolidated rocks; olistostrome cement, on the other hand, always shows evidence of flowage of the unconsolidated sediments, submarine-slope deformation, and graded stratification. As to the exotic blocks, those in the mélange may be younger than the cement, which can never be the case in the olistostrome complexes. This last point, stressed by Hsü, is so important that it should be included in the definition of mélange.

Hsü believes that the mélange complexes should be mapped in conformity with five rules, three of which I believe to be the most important: 1) geological mapping of mélange zones cannot be done on the assumption of a great extent of the beds; 2) the sequence of the mélange section cannot be inferred from a normal stratification; and 3) determination of the time of deposition of mélange rocks on the basis of the oldest and youngest fauna identified from the mélange is wrong.

These three rules of Hsü's should be supplemented by another one, noted by Khain (1968): the lower contact of the mélange with any underlying formations is always tectonic.

This presentation adequately illustrates, in my opinion, the difference between mélange and olistostrome. Thus, according to Hsü, mélange is a tectonically fragmented and mixed rock consisting of a deformed cement (sedimentary in the instance described by Hsü) and exotic blocks, some of them younger than the cement. The standard laws of sedimentation are not applicable to mélange complexes.

The question arises in connection with what has been said: is it correct to designate as mélange those tectonic formations whose cement is serpentinite — the very same complex we have described as mélange within the Sevan-Akera zone.

As we have seen, the component rocks of this tectonic formation are fragmented and tectonically mixed; its cement shows evidence of a strong tectonic deformation and contains exotic blocks younger than the cement.* This tectonic breccia makes up a large body within the area of the Lesser Caucasus under study, and all its features are those of a typical mélange.

However, applying this term to such formations as the mélange of California and that of the Lesser Caucasus, different as they are in their original and present aspects, is hardly conducive to a clarification of the meaning of the term. Bailey and McCallien, who named the Ankara mélange, probably were conscious of that, as was Gansser who designated as "variegated mélange" the variety with a serpentinite cement; and the very same Hsü who proposed the term "Franciscan mélange".

In my view, it is wrong to affix a proper name to the term "mélange", mainly because a geographic name tells the reader nothing of the petrography of the tectonic formation in question, and consequently nothing of the mode of its origin. Indeed, it is hardly proper to designate as Franciscan mélange that one described by I.I. Belostotskiy (1967) or the "argille scagliose" of the Apennines; nor is it proper to describe the Lesser Caucasus mélange as the Ankara mélange. Just as meaningless is Gansser's term, "variegated mélange", because any mélange is variegated in the array and color of the rocks.

I believe that mélange should be classified on more substantial criteria reflecting the character of its site and formation conditions. The composition of exotic blocks is rather irrelevant in that respect, being accidental to a considerable extent. The best classification criterion seems to be the composition of the mélange cement, never an accidental component of this giant tectonic breccia.

In this classification, the mélange of the Sevan-Akera zone of the Lesser Caucasus should be called serpentinite mélange. This term takes in the Ankara mélange of Bailey and McCallien, the variegated mélange of Gansser, the "mixed tectonic facies" of Blumental (Pamir, 1960), and Erol's series of "boulder beds" (Erol, 1956). All these formations are designated by the symbol "Mof" (Mesozoic ophiolite series) on the 1:500,000 Tectonic Map of Turkey.

Serpentinite mélange consists, to a considerable extent, of the products of fragmentation of the ophiolite association — the fact which should substantially influence our paleotectonic reconstructions.

The clear implication of our exposition is that serpentinite mélange (and any mélange) is a purely tectonic formation. Thus, we concur with the author of this term, E. Greenly, and with the large group of investigators sharing this view (Bailey, McCallien, Hsü, Peyve, and Kazmin).

The olistostrome hypothesis of serpentinite mélange origin, proposed by Gansser (1960) is hardly plausible, based as it is on the assumption, difficult to substantiate, that thick basic volcanic series can be formed.

*Even if we reject the concept of a markedly different age for rock of the ophiolite association (Knipper, 1970; Kazmin, 1971), the presence of Upper Cretaceous and Paleogene sedimentary rocks in the serpentinite blocks indicates a marked difference in the age of the cement and some of the exotic inclusions.

Thus the term, serpentinite mélange, should be attached to bodies of deformed rocks with mappable boundaries, in tectonic contact with the underlying formations, and marked by the presence of a brecciated serpentinite cement which is a mixture of blocks (including the exotic) of various sizes, with some of the blocks (if not all of them) younger than the serpentinite cement.

An important feature of serpentinite mélange, setting it apart from the sedimentary series (including the olistostrome complexes), is that its formation is not restricted to a single period of geologic time. Serpentinite mélange is formed by forces associated with the phases of tectonic compression, which reactivate the mélange, complicate its structure, and modify its composition through the addition of tectonic xenoliths. Consequently, such prime features of any rocks as structure and composition are inconsistent in a mélange, in the course of time. It is conceivable that the geologic "life" of serpentinite mélange from the Sevan-Akera zone has not ended, and that its protrusions will be reactivated by future tectonic compressions.

REFERENCES

BELOV, A.A. Contribution to the history of the development of the northern margin of the Iranian epi-Baikalian subplatform in the Lesser Caucasus. Izv. Akad. nauk SSSR. Ser. geol., No. 10, 1968.

BELOSTOTSKIY, I.I. Tectonic sheets in the Devola River basin, Dinaric Alps. Geotektonika, No. 6, 1967.

GABRIYELYAN, A.A. and L.K. TATEVOSYAN. Geological and geophysical subdivision of the Armenian SSR and the adjacent parts of the Anti-Caucasus. Izv. Akad. nauk ArmSSR. Nauki o Zemle, No. 12, 1966.

KAZMIN, V.G. The problem of the Alpine mélange. Geotektonika, No. 2, 1971.

KNIPPER, A.L. Distinctive features of the development of anticlines with serpentinite cores (Sevan-Akera zone of the Lesser Caucasus). Byull. MOIP, Otd. geol., No. 2, 1965.

KNIPPER, A.L. Thrusts and tectonic sheets along the northeastern shores of Lake Sevan. Geotektonika, No. 3, 1966.

KNIPPER, A.L. Gabbro rocks of the ophiolite formation in the oceanic crust. Geotektonika, No. 2, 1970.

KNIPPER, A.L. Serpentinite mélange of the Lesser Caucasus (constitution and age). Geotektonika, No. 5, 1971.

MARKOV, M.S. Metamorphic complexes and their role in the development of the island arcs. Geotektonika, No. 2, 1970.

MASSAKOVSKIY, A.A. Upper Paleozoic volcanic belt of Europe and Asia. Geotektonika, No. 4, 1970.

MURATOV, M.V. Structure of the folded basement in the Mediterranean belt of Europe and western Asia, and the principal stages of its development. Geotektonika, No. 2, 1969.

AUBOUIN, J. Geosinklinali (Geosynclines). Izd. Mir., 1967.

PEYVE, A.V. Oceanic crust of the geologic past. Geotektonika, No. 4, 1969.

RENGARTEN, V.P. Cretaceous stratigraphy of the Lesser Caucasus. In: Regional'naya stratigrafiya SSSR (Regional stratigraphy of the USSR). Izd. Akad. nauk SSSR, 1959.

KHAIN, V.YE. Main structural features of the Alpine belt of Europe in the Near and Middle East. Article 1. Vestn. Mosk. univ., Ser. IV (Geologiya), No. 6, 1968.

KHALILOV, A.G. Fauna and stratigraphy of Albian deposits in the Tutgun River basin (Lesser Caucasus). Dokl. Akad. nauk AzerbSSR, 10, No. 3, 1954.

KHESIN, B.E. Geophysical properties of tectono-magmatic zones in Azerbaydzhan. Geotektonika, No. 6, 1968.

SHIKHALIBEYLI, E.SH. Geologicheskoye stroyeniye i istoriya tektonicheskogo razvitiya vostochnoy chasti Malogo Kavkaza (Geology and history of tectonic development in the east of the Lesser Caucasus), Vol. 1, Izd. Akad. nauk AzerbSSR, 1964.

SHIKHALIBEYLI, E.SH. Geologicheskoye stroyeniye i istoriya tektonicheskogo razvitiya vostochnoy chasti Malogo Kavkaza (Geology history of tectonic development in the east of the Lesser Caucasus), Vol. 3. Izd. Akad. nauk AzerbSSR, 1967.

STECKLIN, D. Tectonics of Iran. Geotektonika, No. 1, 1956.

BAILEY, E.B. and W.I. McCALLIEN. Serpentine lavas, the Ankara mêlange and the Anatolian thrust. Trans. R. Soc., Edinburgh, B. 42, 1956.

EROL, O. A study of the geology and the geomorphology of the region southeast of Ankara in Elman Daği and its surrounding area. MTA yayinil, ser. D, No. 9, 1956.

GANSSER, A. Ausseralpine Ophiolithprobleme, Eclogae. geol. helv., 52, No. 2, 1960.

GREENLY, E. The geology of Angelsy. Great Britian geol. Surv., mem. 1919.

HSÜ, K.J. Principles of mélanges and their bearing at the Franciscan-Knoxville Paradox. Bull. Geol. Soc. America, 79, No. 8, 1968.

PAMIR, H.N. Lexique stratigraphique international. No. 3, Asie, fasc. 9 c, Turquie CNRS, 1960.

PASSERINI, P. Raporti fra le ofioliti e le formazioni sedimentarie fra Piacenza e il Mare Tirreno. Bolletino della societa Geologica Italiana, 84, fasc. 5, 1965.

Geological Institute
USSR Academy of Sciences

Received November 5, 1970

AUTHOR CITATION INDEX

SUBJECT INDEX

About the Editor

GERALD JOSEPH HOME McCALL is presently engaged as a full-time consulting geologist with Malartic Hygrade Gold Mines (Canada) Ltd. and has been their senior representative at a major gold mining operation near Val d'Or, Quebec for the past two and one-half years. From 1976 to 1980, he was senior consulting geologist for Paragon-Contech, Teheran, where he engaged in a major regional mapping program for the Geological Survey of Iran. In addition to having served as an independent consultant to various other organizations, he was director of exploration at Research and Exploration Management Pty Ltd. in Melbourne from 1970 to 1972 and spent ten years prior to this as Reader in Geology at the University of West Australia and ten as geologist in the Kenya Geological Survey.

Dr. McCall earned the Bachelor of Science degree in geology from London University in 1949 and also holds the Ph.D (1951) and D.Sc. (1968) degrees in geology from that institution. The author of over 90 scientific papers, of *Meteorites and Their Origins* (David & Charles Ltd.), of *The Archean: Search for the Beginning* (Dowden, Hutchinson & Ross, 1977), of *Meteorite Craters* (Dowden, Hutchinson & Ross, 1977), and *Astroblemes–Cryptoexplosion Structures* (Dowden, Hutchinson & Ross, 1979), he is an honorary Fellow of Liverpool University, where the library research necessary to produce the present volume was carried out.

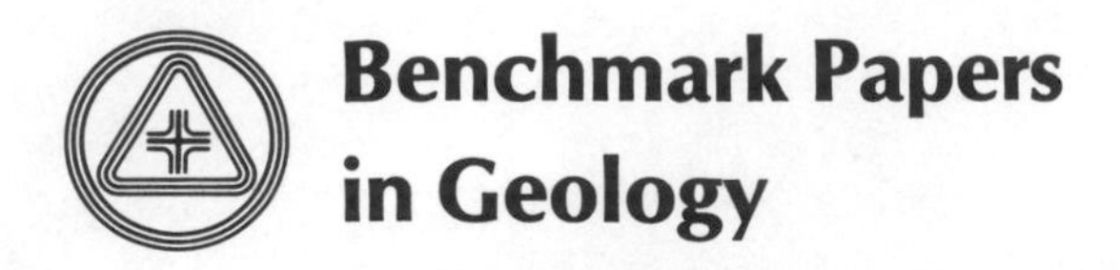

Benchmark Papers in Geology

Series Editor: Rhodes W. Fairbridge
Columbia University

Volume